The Minerals, Metals & Materials Series

Jiann-Yang Hwang · Tao Jiang
Mark William Kennedy · Onuralp Yücel
P. Chris Pistorius · Varadarajan Seshadri
Baojun Zhao · Dean Gregurek
Ender Keskinkilic
Editors

8th International Symposium on High-Temperature Metallurgical Processing

TMS

Springer

Editors
Jiann-Yang Hwang
Michigan Technological University
Houghton, MI
USA

Varadarajan Seshadri
Universidade Federal de Minas Gerais
Belo Horizonte
Brazil

Tao Jiang
Central South University
Changsha, Hunan
China

Baojun Zhao
The University of Queensland
Brisbane
Australia

Mark William Kennedy
Proval Partners SA
Lausanne
Switzerland

Dean Gregurek
RHI AG
Leoben
Austria

Onuralp Yücel
Istanbul Technical University
Istanbul
Turkey

Ender Keskinkilic
Atilim University
Ankara
Turkey

P. Chris Pistorius
Carnegie Mellon University
Pittsburgh, PA
USA

ISSN 2367-1181 ISSN 2367-1696 (electronic)
The Minerals, Metals & Materials Series
ISBN 978-3-319-51339-3 ISBN 978-3-319-51340-9 (eBook)
DOI 10.1007/978-3-319-51340-9

Library of Congress Control Number: 2016961669

Printed on acid-free paper

This Springer imprint is published by Springer Nature
The registered company is Springer International Publishing AG
The registered company address is: Gewerbestrasse 11, 6330 Cham, Switzerland

Preface

This book collects selected papers presented at the 8th International Symposium on High-Temperature Metallurgical Processing organized in conjunction with the TMS 2017 Annual Meeting & Exhibition in San Diego, California, USA. More than 142 abstracts were submitted. Among them, 72 were selected for oral presentation and 60 were provided with poster presentation opportunity. After reviewing the 98 submitted manuscripts, 77 of them were accepted for publication on this book.

As the title of symposium suggests, the interest of the symposium is on thermal processing of minerals, metals, and materials that intends to promote physical and chemical transformations of materials to enable the extraction and production of valuable materials such as metals, alloys, ceramics, and compounds.

The symposium was open to participants from both industry and academia and focused on innovative high-temperature technologies including those based on nontraditional heating methods as well as their environmental aspects such as handling and treatment of emission gases and by-products. Because high-temperature processes require high energy input to sustain the temperature at which the processes take place, the symposium intends to address the needs for sustainable technologies with reduced energy consumption and reduced emission of pollutants. The symposium also welcomed contributions on thermodynamics and kinetics of chemical reactions and phase transformations that take place at elevated temperatures.

This is the seventh book exclusively dedicated to this important and burgeoning topic. We hope the book will serve as a reference for both new and current metallurgists, particularly those who are actively engaged in exploring innovative technologies and routes that lead to more energy efficient and environmental sustainable solutions.

There could not be this book without contributions from the authors of included papers, time and effort that reviewers dedicated to the manuscripts, and help from the publisher. We thank them all! We also want to thank Mrs. Feng Chen for her assistance in collating the submitted abstracts and manuscripts.

Jiann-Yang Hwang
Tao Jiang
Mark William Kennedy
Onuralp Yücel
P. Chris Pistorius
Varadarajan Seshadri
Baojun Zhao
Dean Gregurek
Ender Keskinkilic

Contents

Part II Simulation of High Temperature Process

Part III Fundamental Research of Metallurgical Process

Part V Extraction and Recovery of Metals

Part VI Ironmaking and Steelmaking

About the Editors

Jiann-Yang Hwang is Professor in the Department of Materials Science and Engineering at Michigan Technological University. He is also the Chief Energy and Environment Advisor at the Wuhan Iron and Steel Group Company, a Fortune Global 500 company. He has been the Editor-in-Chief of the *Journal of Minerals and Materials Characterization and Engineering* since 2002. Dr. Hwang has founded several enterprises in areas including water desalination and treatment equipment, microwave steel production, chemicals, flyash processing, antimicrobial materials, and plating wastes treatment. Several universities have honored him as a Guest Professor, including the Central South University, University of Science and Technology Beijing, Chongqing University, Kunming University of Science and Technology, Hebei United University, etc.

Dr. Hwang received his B.S. degree from National Cheng Kung University 1974, M.S. in 1980 and Ph.D. in 1982, both from Purdue University. He joined Michigan Technological University in 1984 and has served as its Director of the Institute of Materials Processing from 1992 to 2011 and the Chair of Mining Engineering Department in 1995. He has been a TMS member since 1985. His research interests include the characterization and processing of materials and their applications. He has been actively involved in the areas of separation technologies, pyrometallurgy, microwaves, hydrogen storages, ceramics, recycling, water treatment, environmental protection, biomaterials, and energy and fuels. He has more than 28 patents and has published more than 200 papers. He has chaired the

Materials Characterization Committee and the Pyrometallurgy Committee in TMS and has organized several symposiums. He is the recipient of the TMS Technology Award and the Michigan Tech Bhata Rath Research Award.

Tao Jiang received his M.S. in 1986 and Ph.D. in 1990, both from Central South University of Technology. He then joined the university and served as Assistant Professor (1990–1992) and Full Professor (1992–2000). From 2000 to 2003, he was a Visiting Scientist to the Department of Metallurgical Engineering, the University of Utah. Since 2003, Dr. Jiang has been a Professor in the School of Minerals Processing & Bioengineering at Central South University. He has been Specially-Appointed Professor of Chang Jiang Scholar Program of China since 2008 and dean of the school since 2010.

Dr. Jiang's research interests include agglomeration and direct reduction of iron ores, and extraction of refractory gold ores. He has undertaken more than 50 projects from government and industry, including the National Science Fund for Distinguished Young Scholars Program. He and co-workers invented the direct reduction process of composite binder pellets and three plants were set up in China based on the invention. He proposed the innovative composite agglomeration process of iron ore fines, which was put into production in Baotou Steel Company, China. He is actively involved in the areas of utilization of nontraditional ferrous resources such as complex ores and various solid wastes. Dr. Jiang has published 340 technical papers, 6 books including *Direct Reduction of Composite Binder Pellets and Use of DRI*, *Principle & Technology of Agglomeration of Iron Ores*, *Chemistry of Extractive Metallurgy of Gold*, *Electrochemistry and Technology of Catalytical Leaching of Gold*. He holds 39 patents and has more than 35 conference presentations.

Mark William Kennedy is Chief Technology Officer at Proval Partners SA of Lausanne Switzerland and a part time Associate Professor at the Norwegian University of Science and Technology (NTNU), Department of Materials Science and Engineering (as of January 2017).

Onuralp Yücel completed his technical education with a Ph.D. in Metallurgical Engineering from Istanbul Technical University (ITU) where he has held the post of Professor since 2002. He was a Visiting Scientist in Berlin Technical University between 1987 and 1988. He carried out postdoctoral studies at New Mexico Institute of Mining and Technology, Socorro, USA between 1993 and 1994. Professor Yücel has as many as 250 publications/presentations to his credit, which include topics such as technological developments in the production of wide range of metals, ferroalloys, advanced ceramic powders and application of carbothermic and metallothermic processes among others. He was the vice chairman of ITU, Metallurgical and Materials Engineering Department between 2004 and 2007. He was director of ITU, Applied Research Center of Material Science & Production Technologies between 2006 and 2012.

Dr. Yücel's areas of interest include: *Pyrometallurgy:* Pretreatment of concentrates (production of WO_3, Sb_2O_3, As_2O_3, MoO_3), smelting and reduction of slags, production ferroalloys, alloys and metals carbothermic and metallothermic processes in EAF or in ladle (copper, cobalt, vanadium, chromium, ferroboron, cobalt boron, nickel boron, ferromolybdenum, ferromanganese, silicomanganese, ferrovanadium, ferrotungsten, ferrochromium, nickel-chromium-molybdenum-iron and aluminum-titanium-boron alloys)

Ceramic Powder Production and Processing: Production of carbide, nitride, boride powders and their processing by explosive consolidation or sintering techniques (B_4C, TiB_2, ZrB_2, SiC, CrB_2).

Beneficiation of Industrial Wastes; Production of metals and compounds from galvanizing ash, brass production wastes and vanadium sludges produced aluminum production. Grit production from aluminum, copper and steel slags.

P. Chris Pistorius is POSCO Professor and Co-Director of the Center for Iron and Steelmaking Research in the Department of Materials Science and Engineering at Carnegie Mellon University, Pittsburgh, Pennsylvania. His research focuses on pyrometallurgy and electrochemistry. He was Associate Professor (1991–1996) and professor (1997–2008) in the Department of Materials Science and Metallurgical Engineering, University of Pretoria, South Africa, and served as the chair of that department from May 2002 to June 2008. Chris has a Master's degree in Metallurgical Engineering from the University of Pretoria, and a Ph.D. from the University of Cambridge, United Kingdom.

Varadarajan Seshadri had his early education in Madras University, India, did postgraduation in metallurgy from Indian Institute of Science Bangalore in India and he received his doctorate from Technical University of Freiberg in Germany . He served in the Department of Metallurgical Engineering, Indian Institute of Technology, Kanpur and became Professor and Head of the Department of Metallurgical Engineering in Maharaja Sayajirao University of Baroda (1971). He was invited visiting professor in 1978 by the Department of Metallurgical Engineering and Materials, Universidade Federal de Minas Gerais, Belo Horizonte, Brazil, where he has continued ever since. He was also senior Humboldt Fellow in Max-Planck-Institut für Eisenforschung, Düsseldorf, Germany (1971–1973), Visiting Professor/Scientist in Technical University, Aachen,Germany, University of Alberta, Canada and University of Missouri, Rolla, USA, and has also served as UNDP specialist in the National Metllurgical Laboratory India. He has won many awards for his publications in the field of iron- and steelmaking, transport phenomena, physical and mathematical modeling. In these areas he has guided many masters and Ph.D. students and has authored more than 250 papers in addition to 3 books. He received the "Distinguished Professor" award from the Universidade Federal de Minas Gerais, Brazil in 2006 for his contributions to undergraduate and postgraduate education as well as industry oriented research in the field of metallurgical engineering. He has served as an organizing/advisory committte member for many international conferences and reviewer for many journals. At present he is Professor Emeritus in the Department of Metallurgical Engineering, Universidade Federal de Minas Gerais, Belo Horizonte, Brazil.

Baojun Zhao is Codelco-Fangyuan Professor in the School of Chemical Engineering at The University of Queensland, Brisbane, Australia. His primary fields of research are experimental investigation of physio-chemical properties of high temperature melts; i.e., phase equilibria and viscosity of oxide systems. He has developed a number of novel research techniques to enable the phase equilibria and viscosity to be accurately determined for complex systems. Systematic fundamental studies have been carried out to directly support industrial operations of copper, zinc, lead, ferroalloys, iron and steel.

Zhao has received the Spriggs Phase Equilibria Award from The American Ceramic Society (2014), the Billiton Gold Medal from Institute of Materials, Minerals and Mining, UK (2009) and the Non-Ferrous Pyrometallurgy Best Paper Award from Metallurgical Society of CIM, Canada (2009). He has long-term close collaborations with international metallurgical companies including Baosteel, Codelco, Dongying Fangyuan Nonferrous, Rio Tinto Iron Ore, Shougang and Luzhong Refractories.

Dean Gregurek is a Senior Mineralogist in the RHI AG Technology Center Leoben, Austria since 2001. Dr. Gregurek received his Master of Science degree at the University of Graz in 1995 and his doctorate degree in Applied Mineralogy from the University of Leoben in 1999. Prior to RHI he worked two years for Luzenac Europe in the talc business. His current research interests and technical expertise are focused on chemical and mineralogical studies related to interactions between refractories, molten metals and slags from pyrometallurgical furnaces. Dr. Gregurek has been a TMS member since 2012, is currently a *JOM* advisor, vice-chair for the Pyrometallurgy Committee and a co-organizer for the 7th and 8th

International Symposiums on High-Temperature Metallurgical Processing (TMS Annual Meetings 2016 and 2017).

Ender Keskinkilic earned his undergraduate degree from the Department of Metallurgical and Materials Engineering of Middle East Technical University (METU), Ankara—the capital city of Turkey—in 1999. He continued M.S. and Ph.D. studies in the same department. He worked as a research assistant in METU between 1999 and 2003. After receiving the master's degree in 2001, he progressed further in the field of extractive metallurgy. During the Ph.D. period, he moved to Eregli-Zonguldak in 2003 and he worked in Quality Metallurgy & RD Department of Eregli Iron and Steel Works Co. (ERDEMIR), which is the leading steel company in Turkey regarding the qualities produced and the production capacity. After earning the Ph.D. degree in 2007, Dr. Keskinkilic returned to university and he started to work in the Department of Metallurgical and Materials Engineering of Atilim University, Ankara, in 2008. He has been working as a faculty in Atilim since then. He was Assistant Professor between 2009 and 2014. He has been working as an associate professor since 2014. Dr. Keskinkilic's primary field of interest is extractive metallurgy and more specifically pyrometallurgical processes such as iron- and steelmaking, ladle metallurgy, ferroalloy production and nonferrous extractive metallurgy.

Part I
Energy Efficient Clean Metallurgical Technology

Flash Ironmaking from Magnetite Concentrate in a Laboratory Reactor: Experimental and CFD Work

M. Elzohiery, D.Q. Fan, Y. Mohassab and H.Y. Sohn

Abstract A flash ironmaking process is being developed at the University of Utah in which iron is produced directly from magnetite concentrate. The kinetics of magnetite reduction by a mixture of H_2 and CO gases was determined in the temperature range 1150–1600 °C. Over 90% reduction degree was achieved at a temperature as low as 1250 °C within 4–6 s which is the typical residence time available in a flash reactor. The kinetics results were applied to experimental data obtained in a laboratory flash reactor in which H_2 and CO gas mixtures were produced from in situ partial oxidation of natural gas with oxygen. CFD was used to simulate the laboratory reactor, and the results were in good agreement with the experimental data.

Keywords Reduction kinetics · Magnetite concentrate · Hydrogen reduction · Ironmaking · Flash reduction

Introduction

As a part of developing the ironmaking industry, a lot of efforts has been dedicated to reduce the energy consumption as well as the greenhouse gas CO_2 production. The blast furnace (BF) is most dominant ironmaking process which produces over 90% of the world iron from iron ore. The novel flash ironmaking process being developed at the University of Utah [1–8] aims at producing iron directly from iron oxide concentrates by a gas-solid flash reaction utilizing natural gas as the reductant and fuel in temperature range 1150–1600 °C. The direct use of iron ore concentrate (<100 μm) in the flash process will bypass pelletizing and sintering steps in blast

M. Elzohiery · D.Q. Fan · Y. Mohassab · H.Y. Sohn (✉)
Department of Metallurgical Engineering, University of Utah,
Salt Lake City, UT 84112, USA
e-mail: h.y.sohn@utah.edu

J.-Y. Hwang et al. (eds.), *8th International Symposium on High-Temperature Metallurgical Processing*, The Minerals, Metals & Materials Series,
DOI 10.1007/978-3-319-51340-9_1

furnace lowering the energy consumption by 30% [9]. Using natural gas will eliminate the cokemaking step in the blast furnace reducing the CO_2 gas emission by 39–51% [6]. In order to develop and design the industrial flash ironmaking reactor, the reduction kinetics of magnetite concentrates by hydrogen and carbon monoxide gas mixture in temperature range 1150–1600 °C was studied. Using the developed single rate equations, a rate equation for the reduction kinetics by the H_2 and CO gas mixture was established. CFD simulation was used to verify the results of the developed rate equations on the results of a laboratory-scale flash reactor which was built and used for producing iron directly from iron oxide concentrate using natural gas partially combusted with oxygen. The results showed good agreement between the developed rate equations and the experimental results.

Experimental Work

The kinetics of magnetite reduction by H_2 and CO gas mixture was investigated in the temperature range 1150–1600 °C which is the proposed temperature range for the flash ironmaking technology. Most of the previous kinetics studies were either at lower temperature ranges or using hematite ore. In order to get the rate equation for the gas mixture reduction kinetics, it is important to study the reduction kinetics by the single gases.

The magnetite concentrate particles used in this study are of irregular shape and contain oxide gangues. This gangue content results in the particle experiencing fusion at temperatures higher than 1350 °C as was found by experiments. Therefore, it was important to study the kinetics at temperature lower than 1350 °C separately from that at higher temperatures.

Experiments were designed and rate equations were formulated for the reduction of magnetite by H_2, CO and H_2 + CO gas mixture in the temperature range 1150–1350 °C in which the particles remain as solids, and at 1350–1600 °C in which the particles melt. Different size fractions of concentrate particles were used in this work in order to study the effect of particle size. High temperature laminar-flow reactor (LFR) was used to determine the reduction rate of the individual particles. The reactor had a dilute particle-gas conveyed system in which the gas composition was maintained essentially uniform throughout the reactor by using a large excess amount of reducing gas. Water cooled feeding line was inserted into a ceramic tube to avoid any soot formation due to the catalytic effect of the copper tubing that enhances the reduction of CO by H_2 gas and that results in negligible amount of soot formed. A schematic diagram is shown in Fig. 1.

An ICP method, which was developed in our laboratory, was used to determine the iron content in the produced samples. This method has high accuracy in determining iron content in the samples containing high amounts of oxide gangues as in the concentrate used in this work [10].

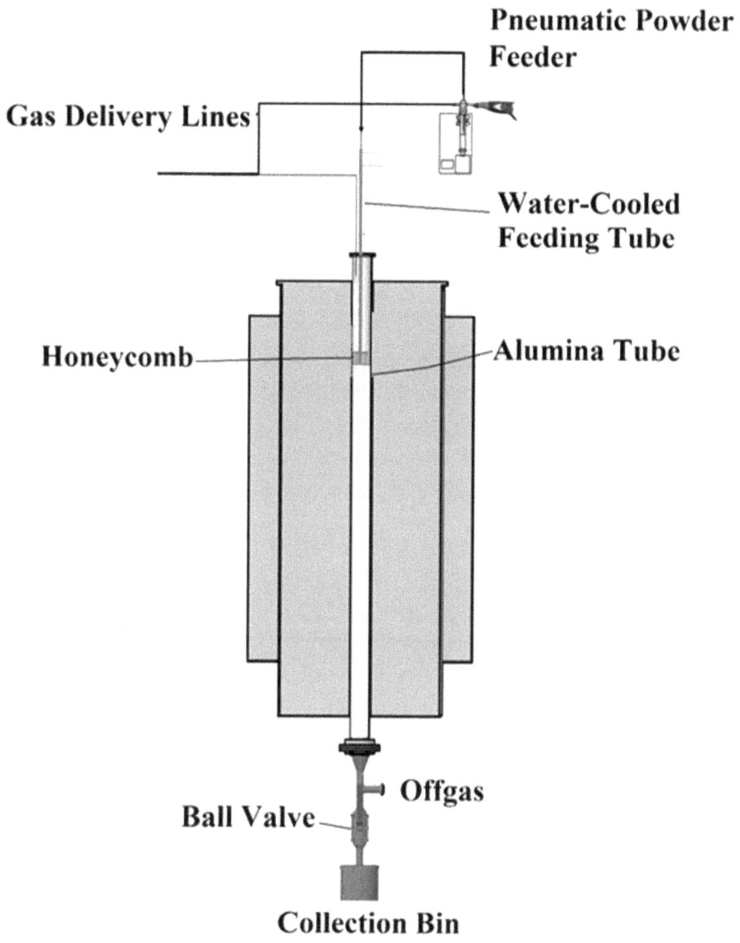

Fig. 1 Schematic diagram for the laminar flow reactor (LFR)

Different experimental conditions were designed in a way to mimic the most used industrial ratios of the syngas such as: H_2:CO = 2:1, 1:1, and 0.5:1 (molar ratio). Similar detailed experimental work, relevant equations for calculating the nominal particles residence time, % excess hydrogen, excess driving force (EDF), reduction degree (RD%), and the length of the reaction zone are described in previous work [8, 11]. The average temperature was calculated up to a certain position where the difference between the average and the actual temperatures was greater than 30 °C. The reaction zone was defined as the zone with temperature $T_a \pm 30$ °C where T_a was the average temperature.

The developed single rate equations were used to formulate a combined rate equation describing the reduction kinetics by the gas mixture. CFD simulation was used to verify the results of these rate equations with the experimental results obtained from a laboratory scale flash reactor. In this reactor, oxygen/methane form a non-premixed flame where the heat produced is used to heat up the reactor up to 1175 °C. Due to the high volume to surface ratio, heat loss was compensated by electric heating. The partially oxidized methane produced a hydrogen and carbon monoxide mixture that performed the reduction reaction. Different experimental conditions were investigated as the feeding ports and the flame configuration.

Results and Discussion

The aim of this work is to formulate rate equations for magnetite reduction by H_2 and CO gas mixture at the temperature ranges 1150–1350 and 1350–1600 °C. The rate equations for single gas reduction were determined and used for formulating the mixture rate equation. In all rate equations, the effect of reducing gas partial pressure, temperature, particle size, and residence time were studied. These factors represent the main controlling factors in the flash ironmaking technology. The reduction degree of each experiment was calculated based on the amount of oxygen removed after reduction. The reproducibility of the experiments were found to be within ±5%.

Reduction Kinetics Determination

The rate equation for magnetite reduction by hydrogen was found to follow the nucleation and growth kinetics equation with the Avrami parameter $n = 1$. The rate equation had a first order dependence the H_2 gas partial pressure with no particle size effect. The complete rate equation in the temperature range 1150–1350 °C was formulated and the activation energy was 193 kJ as shown in Eq. (1):

$$[-Ln(1 - X)] = 8.65 \times 10^6 \times e^{\frac{-193{,}000}{RT}} \times \left[p_{H_2} - \frac{p_{H_2O}}{K_{eq}}\right]_{lm} \times t \quad (1)$$

where R is 8.314 J/mol K, T is in K, p is in atm, and t is in s.

The kinetics of magnetite reduction by CO was found to be much slower than that with H_2. The nucleation and growth equation was used to describe the reduction kinetics with the Avrami parameter $n = 0.4$. Although the Avrami parameter is typically greater than 1, there are some examples of reactions in which it is less than 1. This phenomena occurs when the nuclei growth rate decreases with growth as described by Kim et al. [12] and Hedmark et al. [13]. Wunderlich [14] has analyzed a large amount of data on the rates that follow the Avrami equation, and reported that the Avrami parameter can range from lower than 1 to greater than 6.

The rate equation for magnetite reduction by CO had a first order dependence on CO partial pressure and the activation energy was 479 kJ/mol. There was no effect also for the particle size. The complete rate equation obtained for the reduction of magnetite with carbon monoxide gas in the temperature range 1150–1350 °C was

$$[-Ln(1-X)]^{1/0.4} = 3.41 \times 10^{14} \times e^{\frac{-479000}{RT}} \times \left[p_{CO} - \frac{p_{CO_2}}{K_{eq}}\right]_{lm} \times t \qquad (2)$$

where R is 8.314 J/mol K, T is in K, p is in atm, and t is in s.

The absence of size effect at that temperature range was explained as the particles experience thermal stress and develop cracks of similar dimensions when rapidly heated as they are fed into the reactor at a high temperature. The reaction rate thus depends on the dimension of the solid between the cracks not on the original size. Figure 2 shows that good agreement was obtained between the experimental results and the results obtained by the rate equations.

These rate equations were used to develop a rate equation for the reduction kinetics by CO + H_2 gas mixtures. The presence of CO increased the reduction rate to be higher than the sum of the rates by individual gases. In order to account for the enhanced effects, the reduction contribution due to H_2 was amplified by a constant for the case of reduction by H_2 + CO mixtures. The rate expression for the CO + H_2 gas mixtures in the temperature range 1150–1350 °C was obtained as

$$\frac{dX}{dt} = 1.3 \cdot \left.\frac{dX}{dt}\right|_{H_2} + \left.\frac{dX}{dt}\right|_{CO} \qquad (3)$$

The comparison of the calculated and experimental results was shown in Fig. 3.

The 1st order model in term of the fraction of solid remaining unreacted describes the reduction kinetics of magnetite concentrate by H_2 in the temperature range 1350–1600 °C. The reaction kinetics had a first-order dependence on the

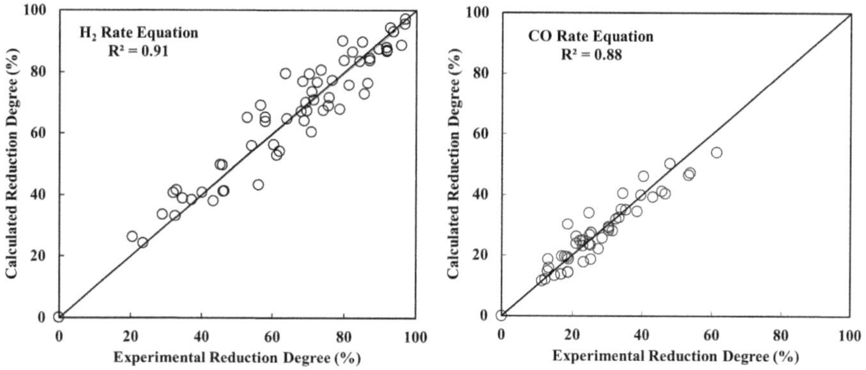

Fig. 2 The agreement between the experimental results and the obtained results by the developed rate equations

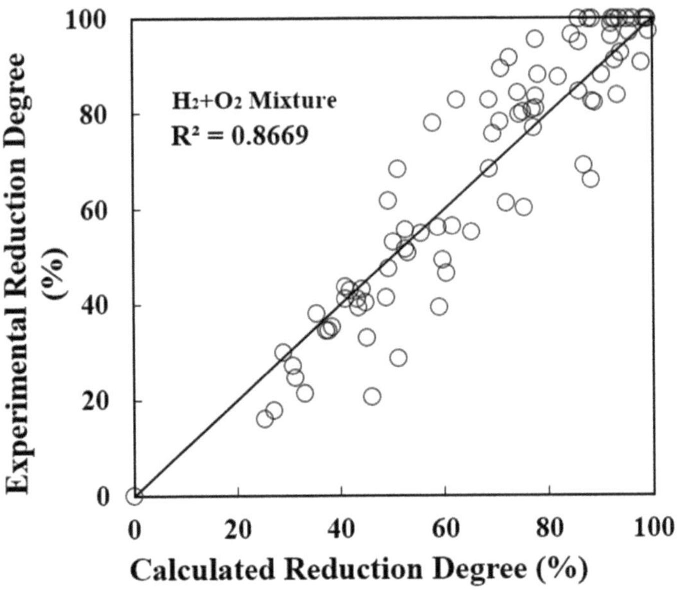

Fig. 3 Comparisons between the calculated reduction degrees versus experimental results with H_2 + CO mixtures as the reducing gas

hydrogen partial pressure with a significant particle size effect on the reduction due to the fusion status of the particles. The activation energy was 177 kJ/mol and the complete rate equation obtained for the reduction of magnetite with hydrogen gas in the temperature range 1350–1600 °C is

$$[-Ln(1-X)] = 4.2 \times 10^7 \times e^{\frac{-177{,}000}{RT}} \times (d_p)^{-1}\left[p_{H_2} - \frac{p_{H_2O}}{K_{eq}}\right]_{lm} \times t \quad (4)$$

where R is 8.314 J/mol K, T is in K, d_p is in μm, p is in atm, and t is in s.

The nucleation and growth model described the reduction kinetics of magnetite by CO gas and it was similar to the lower temperature range with an Avrami parameter less than 1 ($n = 0.5$). The dependence on CO partial pressure was found to be of the first order and the activation energy was 49 kJ/mol. The kinetics of reduction by CO had the same dependence on particle size as in the case of reduction by hydrogen. The complete rate equation obtained for the reduction of magnetite with carbon monoxide gas in the temperature range 1350–1600 °C is

$$[-Ln(1-X)]^{1/0.5} = 4.2 \times 10^2 \times e^{\frac{-49{,}000}{RT}} \times (d_p)^{-1}\left[p_{CO} - \frac{p_{CO_2}}{K_{eq}}\right]_{lm} \times t \quad (5)$$

where R is 8.314 J/mol K, T is in K, d_p is in μm, p is in atm, and t is in s.

Figure 4 shows good agreement between the experimental and the calculated results.

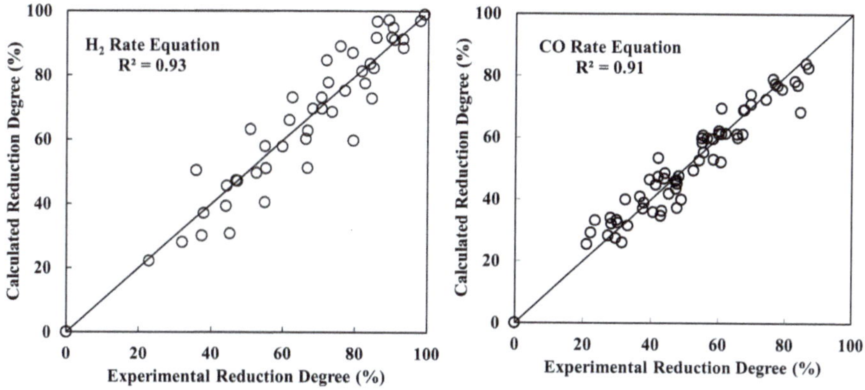

Fig. 4 Agreement between the experimental and calculated reduction degree by the developed rate equations in the temperature range 1350–1600 °C

Table 1 Experimental and calculated reduction degree (%)

CH_4 flow rate ($L\ h^{-1}$)	O_2 flow rate ($L\ h^{-1}$)	Experimental (pct)	Calculated (pct)	Residence time (s)
300	270	83 ± 5	90	7.2
600	570	75 ± 5	80	4.9

CFD Simulation of Laboratory Scale Flash Reactor

A laboratory scale flash reactor was used to verify the rate expressions under an actual flash reaction condition. Two typical experimental conditions were simulated and the calculated reduction degrees were compared with the experimental values, which is shown in Table 1.

Acknowledgments The authors acknowledge the financial support from the U.S. Department of Energy under Award Number DEEE0005751 with cost share by the American Iron and Steel Institute (AISI) and the University of Utah.

References

1. H.Y. Sohn, Suspension ironmaking technology with greatly reduced energy requirement and CO_2 emissions. Steel Times Int. **31**, 68–72 (2007)
2. H.Y. Sohn et al., Suspension reduction technology for ironmaking with low CO_2 emission and energy requirement. Iron Steel Technol (AIST Trans.) **6**, 158–165 (2009)
3. M.E. Choi, H.Y. Sohn, Development of green suspension ironmaking technology based on hydrogen reduction of iron oxide concentrate: rate measurements. Ironmaking Steelmaking **37**, 81–88 (2010)
4. H.K. Pinegar, M.S. Moats, H.Y. Sohn, Flowsheet development, process simulation and economic feasibility analysis for novel suspension ironmaking technology based on natural gas: part 1—flowsheet and simulation for ironmaking with reformerless natural gas. Ironmaking Steelmaking **39**, 398–408 (2012)
5. M.Y.M. Ahmed, Phase equilibria between iron and slag in $Co/CO_2/H_2/H_2O$ atmospheres relevant to a novel flash ironmaking technology (Phd Dissertation), (University of Utah, 2013)
6. H.K. Pinegar, M.S. Moats, H.Y. Sohn, Flowsheet development, process simulation and economic feasibility analysis for novel suspension ironmaking technology based on natural gas: part 3—economic feasibility analysis. Ironmaking Steelmaking **40**, 44–49 (2013)
7. Y. Mohassab, H.Y. Sohn, Effect of water vapor on sulfur distribution between liquid Fe and Mgo-saturated slag relevant to a flash ironmaking technology. Steel Res. Int. (2014)
8. F. Chen et al., Hydrogen reduction kinetics of hematite concentrate particles relevant to a novel flash ironmaking process. Metall. Mater. Trans. B **46**, 1133–1145 (2015)
9. H.Y. Sohn, M. Olivas-Martinez, Methods for calculating energy requirements for processes in which a reactant is also a fuel: need for standardization. JOM **66**, 1557–1564 (2014)
10. Y. Mohassab, et al., *Determination of total iron content in iron ore and Dri: titrimetric method versus ICP-OES analysis* (Wiley, London, 2016)
11. F. Chen et al., Kinetics of the reduction of hematite concentrate particles by carbon monoxide relevant to a novel flash ironmaking process. Metall. Mater. Trans. B (2015)
12. T.K. Kim, S.O. Kim, I.J. Chung, Crystallization kinetics of thermotropic liquid crystalline poly(ester-imide)s. Eur. Polymer J. **33**, 1613–1626 (1997)
13. P.G. Hedmark et al., Cold-crystallization in liquid crystalline poly(p-hydroxybenzoic acid-co-ethylene terephthalate). Polymer **30**, 2068–2073 (1989)
14. B. Wunderlich, *Chapter V—the Nucleation Step* (Academic Press, London, 1976)

Synthesis of Chromite for Subsequent Carburization by Methane-Hydrogen Gas Mixture

Vincent Canaguier, Ingeborg-Helene Svenum and Leiv Kolbeinsen

Abstract The syntheses of stoichiometric iron chromite and chromites with chosen impurity content have been developed using the induction skull melting technique. The aim for this synthetic material is to obtain pure samples to study the carburization by methane-hydrogen gas mixtures. For each synthesis, the parameters influencing the skull melting experiment have been considered. The product phases have been analyzed using X-Ray Diffraction (XRD), Scanning Electron Microscopy (SEM), Energy-Dispersive X-ray Spectroscopy (EDS) and X-ray Photoelectron Spectroscopy (XPS). The synthetic material was subsequently pelletized and sintered in argon. The use of excess iron was found to improve the quality of the final product as well as to help the melting. The difference in magnesium content between first and last solids formed seem to indicate a segregation phenomenon, while no evidence could support a similar trend for aluminum substitution in chromite.

Keywords Synthesis · Chromite · Skull melting

Introduction

Mineral ores are the starting point for most metal production processes. These ores often contain a wide range of impurities in addition to the valuable oxides. It is known that the composition of these ores vary from deposit to deposit, and also within a single deposit. For chromite ores, which are the focus of this article, significant variations can be observed. The contents of diverse oxides, such as: Cr_2O_3, FeO, Fe_2O_3, Al_2O_3, MgO, SiO_2 and CaO in natural chromites are subject to

V. Canaguier (✉) · L. Kolbeinsen
NTNU, Norwegian University of Science and Technology, Alfred Getz vei 2, Trondheim 7491, Norway
e-mail: vincent.y.canaguier@ntnu.no

I.-H. Svenum
SINTEF Materials and Chemistry, Høgskoleringen 5, Trondheim 7034, Norway

J.-Y. Hwang et al. (eds.), *8th International Symposium on High-Temperature Metallurgical Processing*, The Minerals, Metals & Materials Series,
DOI 10.1007/978-3-319-51340-9_2

important fluctuations [1]. These make the investigation of their individual effect, e.g. on reduction rates, particularly complicated as it is not possible to vary one oxide content at a time using natural ores.

Chromites of various compositions have been investigated in several former studies using synthetic materials. Other studies focused on the synthesis of these oxides in a pure form. Solid-state synthesis, using various protocols, was used to produce synthetic chromite ($FeCr_2O_4$) [2–4] and synthetic magnesiochromite ($MgCr_2O_4$) [2]. In addition, Qayyum et al. synthetized various compositions of the (Fe, Mg)(Cr, Al)$_2$O$_4$ solid solution in solid-state [3], while (Fe, Mg)(Fe, Cr, Al)$_2$O$_4$ solid solution synthesis was studied by Wu et al. [5].

The work presented in this article aimed at developing a liquid state synthesis route allowing the production of large quantities of material at once and in a short time. A high production capacity is required to allow pelletization and sintering of the material prior to reduction and carburization experiments. The quality and variability of the composition of the synthesized oxides are used to evaluate this synthesis route.

Experimental Details

The Skull Melting technique, using a cold crucible induction furnace, is utilized for the synthesis of the $(Fe_{1-x}Mg_x)(Cr_{1-y}Al_y)_2O_4$ solid solution samples.

Apparatus

The equipment used for the synthesis is a cold crucible induction furnace. It consists in two parts: a high frequency induction coil supplied by Farfield Electronics PTY Ltd (750 kHz, 75 kVA) for the heating and a water-cooled copper crucible supplied by ANSTO. The control of the furnace is performed by changing the current input in the induction coil. A schematic of the equipment is given in Fig. 1.

The copper crucible is cylindrically shaped, and its wall is built from "fingers" separated from each other by an interstitial space. The inside and the top surfaces of the crucible are coated before use with a boron-nitride coating applied either as a paste or with a spray. A top view schematic of a loaded crucible is given in Fig. 2. The mass load capacity is then determined by the shape and density of the raw materials and varies between 200 and 300 g.

The inside of the furnace, delineated by the external glass tube, is evacuated by a rotary pump. The atmosphere can be replaced by an inert gas. The temperature measurements are performed by a spectropyrometer type FMP 2 from FAR Associates.

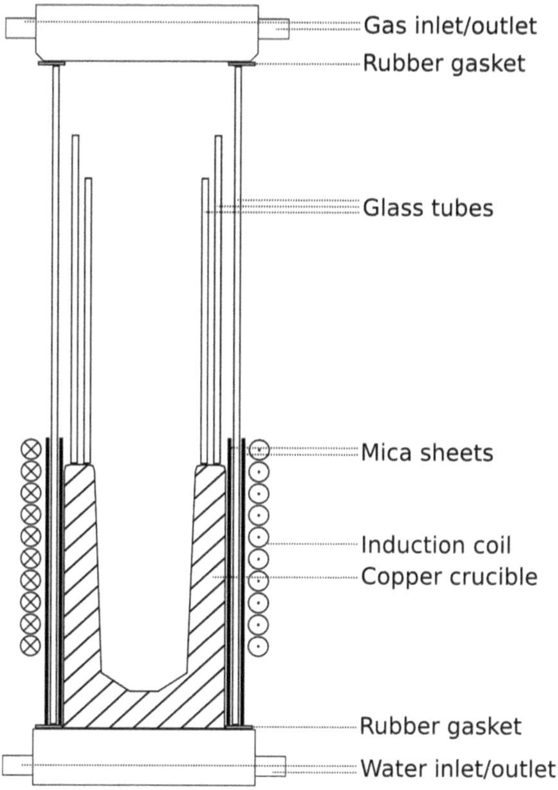

Fig. 1 Schematics of the cold crucible induction furnace with a cut view

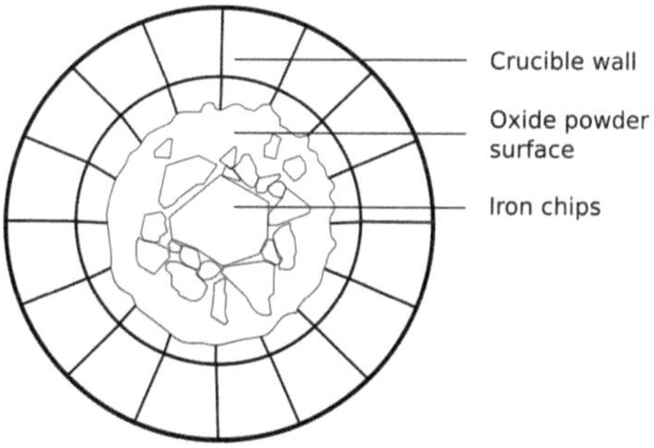

Fig. 2 Schematics of the crucible and loading with a top view

Chromite Solid Solution Synthesis

The reactants used for the synthesis are Cr_2O_3 powder (purity 98%), Fe_2O_3 powder (size $\leq$ 5 μm, purity $\geq$ 99%), pure Fe pieces (chips, purity 99.98%), and Al_2O_3 powder (size $\leq$ 10 μm, purity $\geq$ 99.5%). The quantities are based on a stoichiometric reaction synthesis plus excess iron if relevant. The gas atmosphere used is argon, blown in at 1.19 NL/min.

Table 1 Synthesis parameters for the $(Fe_{1-x}Mg_x)(Cr_{1-y}Al_y)_2O_4$ samples

Ref.	Sample	Power [kW]	Fe [g]	Fe^{a}_{xs} [g]	Fe_2O_3 [g]	Cr_2O_3 [g]	Al_2O_3 [g]	MgO [g]
1	x = 0, y = 0	67.50[b]	16.74	0	47.56	135.87	0	0
2	x = 0, y = 0	52.50	13.30	0	38.03	108.63	0	0
3	x = 0, y = 0	63.75	16.63	0	47.48	135.85	0	0
4	x = 0, y = 0	63.75	16.63	4.08	51.76	135.81	0	0
5	x = 0, y = 0	63.75	16.61	2.00	47.57	135.81	0	0
6	x = 0, y = 0.25	63.75	17.65	0	50.38	107.97	24.12	0
7	x = 0.25, y = 0	63.75	12.92	0	36.96	140.72	0	9.34
8	x = 0.25, y = 0	63.75	12.93	7.35	36.99	140.80	0	9.50

[a] Fe_{xs} is the iron excess added with regard to the stoichiometric composition
[b] Reduced to 52.50 kW during the experiment due to splashing

The oxide powders were mixed in a plastic bag and were placed and packed in the crucible. The iron chips were placed on the top of the oxides so that the biggest chips were in the center of the crucible and a maximum of the surface was covered. For several stoichiometric chromite samples (x = 0, y = 0) and one chromite-magnesiochromite solid solution sample (x = 0.25, y = 0), excess iron was added to ensure good melting. A sketch of the iron placement is given in Fig. 2. The synthesis parameters for the $(Fe_{1-x}Mg_x)(Cr_{1-y}Al_y)_2O_4$ solid solution samples are provided in Table 1.

Analyses

The composition of the synthesized phases was analyzed by X-ray diffraction (XRD) using a Bruker D8 DaVinci. Scanning ranges of $2\theta = 5–60°$ and $2\theta = 15–75°$ were used. The radiation was obtained from a copper anode. The peaks obtained were compared to the Powder Diffraction File (PDF) database.

Parts of the sample representative of the section of the synthesis product were embedded in epoxy and polished using SiC paper and diamond suspensions. These samples mounted in epoxy were cleaned with ethanol, partly wrapped in aluminum and degassed overnight. Secondary electron (SE), backscattered electron (BSE) and energy-dispersive x-ray spectroscopy (EDS) images were obtained using a field emission scanning electron microscope (FESEM Zeiss Supra or FESEM Zeiss Ultra). X-ray photoelectron spectroscopy (XPS) analysis was performed using a Kratos Axis Ultra spectrometer with a monochromatic Al Kα source. Individual core levels were measured with a pass energy of 20 eV. XPS analysis was carried out on powder samples pressed onto carbon tape.

Pelletizing and Sintering

Once the material has been synthesized and verified, it was crushed and pelletized without binding additives. The pellets were sized between 1 and 1.6 mm of diameter. After drying the pellets were sintered in argon at 870 °C for 2 h.

Results and Discussion

Product Appearance

The synthesis product was first observed with the naked eye to evaluate the success of the melting. A first criterion for a successful melting is the absence of crust on the top of the crucible. When the iron chips placed on the top of the charge melt, surrounding oxides will start melting too, and ultimately a molten bath is formed in the bottom of the crucible. If oxides are left on the top they might fall down in the melt and dissolve into it. However, if these oxides heated by the radiating iron chips sinter before the formation of the melt, a part of the oxides is left unaffected and hence does not fall into the melt. This will damage the product's composition as the stoichiometry of the reaction cannot be fulfilled. Pictures of a sample after synthesis and of a sintered oxide crust are given in Fig. 3.

To solve the crust formation problem, excess iron is used. Iron, when melting, pierces the oxide charge. With extra iron, and a good spread of the iron chips on the oxide charge surface, the crust is avoided. The downside is that the stoichiometry is once more at risk. But from a practical point of view, it is easier to remove extra iron that will form a distinct product phase than to remove M_2O_3 phases within the spinel solid solution. In addition to facilitate the melting, having excess iron is useful to compensate for the losses due to splashing.

All meltings except one were successful. The chromite synthesis (x = 0, y = 0) carried out at 70% power failed and only a small fraction of the reactants melted. This sample evidenced the minimum requirements in terms of power input.

Fig. 3 Pictures of samples after synthesis: **a** Chromite sample (x = 0, y = 0). **b** Unreacted crust in a chromite-magnesiochromite synthesis experiment (x = 0.25, y = 0)

X-Ray Diffraction Analysis

The crystal structure was analyzed for each sample using XRD. The detected major and minor phases are reported in Table 2. The phases mentioned in the table are not necessarily present in the whole sample, as different part of the product were crushed down and analyzed independently. However, it shows the deviations from the expected material. The sample n°2 was discarded due to limited melting.

The sample n°1 shows a lot of unreacted chromium oxide which can be explained by two phenomena: the losses of pure iron due to splashing, and the subsequent modification of the power input. The sample n°3 showed the same trend, to a lower extent, even though the power input was reduced from the start and kept constant. The undesired phases were observed in small quantities as seen in samples n°4 and 5 only when iron is added. Concerning sample n°6, some unreacted M_2O_3 phase was observed, in addition to a dominant spinel phase. The main peaks fit to the pattern of $Fe(Cr_{1.4}Al_{0.6})O_4$ whose composition is close to the target. The minor phases left are a solid solution of aluminum and chromium oxide. For the sample n°7 large amounts of crust were observed which explains why some iron and chromium oxide were left unreacted. The spinel composition was found to be close to $(Fe_{0.67}Mg_{0.33})Cr_2O_4$, which makes sense given the unreacted iron in the sample. For the sample n°8, the patterns showed peaks matching the compositions of $(Fe_{0.91}Mg_{0.09})Cr_2O_4$ (inner sample) and $(Fe_{0.76}Mg_{0.24})Cr_2O_4$ (border sample). Traces of iron were observed by XRD analysis, but large amounts were remove before the crushing.

These observations have to be considered carefully as the different compositions of the spinel solid-solution studied are not easily distinguishable by XRD analysis.

Table 2 X-ray diffraction results for the $(Fe_{1-x}Mg_x)(Cr_{1-y}Al_y)_2O_4$ samples

Ref.	Sample	Power [kW]	Crust (Y/N)	Major phases	Minor phases
1	x = 0, y = 0	67.50[a]	N	$FeCr_2O_4$; Cr_2O_3	FeO
3	x = 0, y = 0	63.75	N	$FeCr_2O_4$; Cr_2O_3	–
4	x = 0, y = 0	63.75	N	$FeCr_2O_4$	Fe
5	x = 0, y = 0	63.75	N	$FeCr_2O_4$	Cr_2O_3
6	x = 0, y = 0.25	63.75	Y	$Fe(Cr_{1.4}Al_{0.6})O_4$	$(Cr_{1.25}Al_{0.72})O_3$
7	x = 0.25, y = 0	63.75	Y	$(Fe_{0.67}Mg_{0.33})Cr_2O_4$; Cr_2O_3;Fe	–
8	x = 0.25, y = 0	63.75	N	$(Fe_{0.91}Mg_{0.09})Cr_2O_4$; $(Fe_{0.67}Mg_{0.33})Cr_2O_4$; Fe[b]	–

[a]Reduced to 52.50 kW during the experiment due to splashing
[b]Only traces were detected by XRD but large amounts were removed before the analysis

However, the addition of pure iron seemed generally beneficial to the synthesis. For the samples n°4 and 5, where the iron addition was moderate, the product is very close to the desired composition. In the sample n°8 where a large amount of excess iron was necessary, a good part can be recovered before further processing of the ores. Also, within sample n°8, whose composition was analyzed close to the edge and close to the center, different patterns fitted the observed peaks. This suggests a different content in magnesium across the sample and is further investigated by elemental analysis.

SEM Imaging, EDS and XPS Analysis

The samples n°5, 6 and 8, which are considered the most successful based on their composition (x = 0, y = 0), (x = 0.25, y = 0) and (x = 0, y = 0.25) respectively are studied by electron microscopy. XPS analysis was also performed on a nominal $FeCr_2O_4$ sample. The obtained XPS spectra is consistent with a Fe/Cr ratio of 0.48 confirming the bulk composition. A picture of sample n°5 obtained using a backscattered electron detector is shown in Fig. 4. Further EDS analysis confirm that the bulk is a chromite spinel phase, but some impurities are also present. In sample n°5 and 8, a metallic Fe–Cr alloy phase and a FeO phase have been evidenced in addition to the spinel. In sample n°6, a $(Cr, Al)_2O_3$ phase was evidenced in addition to FeO and a Fe–Cr alloy.

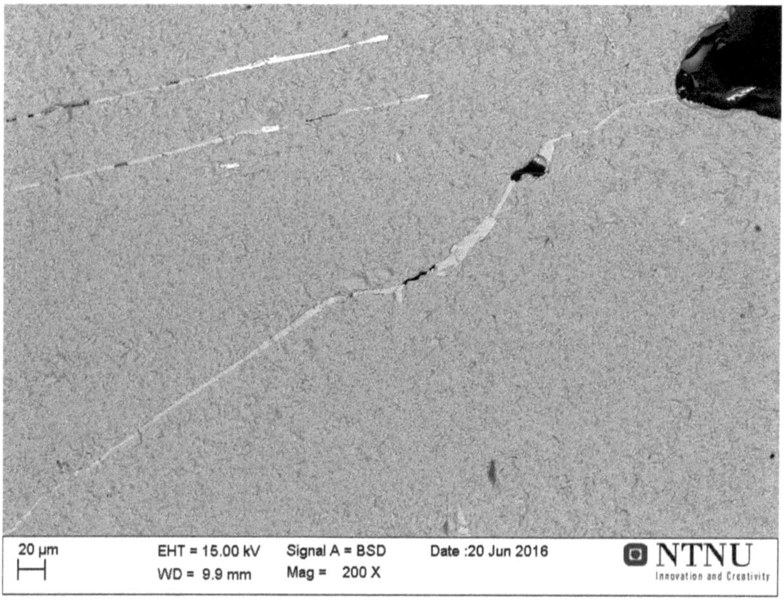

Fig. 4 Picture of a chromite sample (x = 0, y = 0)

Table 3 Metallic ratios from EDS measurements and targeted values for the $(Fe_{1-x}Mg_x)(Cr_{1-y}Al_y)_2O_4$ samples

Ref.	Sample	(Fe + Mg)/(Cr + Al) [0.500]	Al/Cr [0.333]	Mg/Fe [0.333]
5	x = 0, y = 0	0.427–0.457	–	–
6	x = 0, y = 0.25	0.456–0.456	0.313–0.305	–
8	x = 0.25, y = 0	0.452–0.426	–	0.420–0.254

Point analysis were carried out on the samples to verify the metal ratios within the spinel phase. Two sets of point, near the center of the sample and near the edge, have been measured. The concept was to investigate whether a possible compositional gradient in aluminum or magnesium content exist between the first and last solid phases formed. The average metallic ratios are given in Table 3. The first value represents the average ratio for the edge sample (first solid), while the second value is the average for the center sample (last solid). The value in square brackets is the targeted metallic ratio. From these measurements, only the magnesium to iron ratio is distinct enough and may show a segregation phenomenon. The difference between the Fe/Cr ratio given by XPS and EDS for sample n°5 could be explained by the relatively low accuracy of EDS measurements.

Conclusions

Synthetic chromite, chromite-hercynite and chromite-magnesiochromite have been synthesized successfully by induction skull melting of pure reagents. The formation of a crust of sintered materials was found to be the foremost cause of failure during the synthesis, but addition of iron in excess facilitates the melting.

The XRD studies showed that the product phase is, to a large extent, the expected spinel phase. Though, the electron microscopy showed variations in the magnesium content within the spinel solid solution that may be attributed to a segregation phenomenon. Impurities such as FeO and a metallic phase are also detected by electron microscopy.

Acknowledgements This work was financed by the Norwegian Research Council (NRC-project 233825) through the CORALSEA project. The authors would also like to thank Edith Thomassen and Ingeborg Solheim for the pellet preparation.

References

1. J.F. Papp, B.R. Lipin, "Chromium" (Open-File Report 01-381, U.S. Geological Survey, 2001)
2. S. Klemme, H.St.C. O'Neill, W. Schnelle, E. Gmelin, The heat capacity of $MgCr_2O_4$, $FeCr_2O_4$, and Cr_2O_3 at low temperatures and derived thermodynamic properties. Am. Mineral. **85**, 1686–1693 (2000)

3. M.A. Qayyum, D.A. Reeve, Reduction of chromites to sponge ferrochromium in methane-hydrogen mixtures. Can. Metall. Q. **15**(3), 193–200 (1976)
4. N.S. Sundar Murti, V. Seshadri, Kinetics of reduction of synthetic chromite with carbon. Trans. Iron Steel Inst. Jpn. **22**(12), 925–933 (1982)
5. X.R. Wu, C. Chen, H.H. Lü, L.S. Li, Synthesis, structure characterization and magnetic property of $(Mg_{1-y}Fe_y)(Al_{0.4}Cr_xFe_{1.6-x})O_4$ spinel solid solution. *Material Science and Environmental Engineering*, ed. by Ping Chen (Wuhan, China: International Conference on Material Science and Environmental Engineering, 2015), pp. 109–113

Effects of Hydrogen-Enriched Reduction on Metallurgical Properties of Iron-Bearing Burdens Under BF Operation with COG Injection

Hongtao Wang, Mansheng Chu, Chuanguang Bi, Zhenggen Liu and Wenbin Dai

Abstract Coke oven gas (COG) injection is considered as one of feasible measures to reduce carbon emission of BF. The effects of COG injection on the reduction of iron-bearing burdens were numerically simulated by means of multi-fluid BF model, which revealed that with increasing COG injection rate, solid temperature gradually decrease, in-furnace H_2 concentration increases, the reduction of magnetite and wustite by H_2 is evidently accelerated, the reaction rate of wustite by H_2 reduction is enhanced and the reduction degree of mixed burdens is strengthened. Under this hydrogen-enriched condition, metallurgical properties of iron-bearing burdens were experimentally investigated. The GB/T13242-91 test results showed that with increasing COG injection rate, the reduction swelling index of oxidized pellets decreases while the reduction ratio of pellets increases. The softening-melting tests suggested that softening interval of mixed burdens increases from 163.6 to 228.3 °C while melting interval decreases from 147.3 to 92.4 °C. The cohesive zone moves downwards and becomes narrow. The permeability of mixed burdens is improved. In terms of metallurgical properties of iron-bearing burdens, COG injection benefits to improve BF performance.

Keywords COG injection · Multi-fluid BF model · Hydrogen-enriched reduction · Reduction swelling behaviors · Softening-melting properties · Carbon emission reduction

H. Wang · M. Chu (✉) · Z. Liu · W. Dai
School of Metallurgy, Northeastern University, Shenyang 110819, China
e-mail: chums@smm.neu.edu.cn

Z. Liu
e-mail: liuzg@smm.neu.edu.cn

C. Bi
Shanghai Meishan Iron and Steel Corporation Ltd, Nanjing 210039, China

J.-Y. Hwang et al. (eds.), *8th International Symposium on High-Temperature Metallurgical Processing*, The Minerals, Metals & Materials Series,
DOI 10.1007/978-3-319-51340-9_3

Introduction

Nowadays, in order to mitigate global greenhouse effect, various countries are necessary to reduce CO_2 emission based on The Paris Agreement produced from Paris Climate Conference in 2015. As one of the most important pillar industries, steel industry should be responsible for reducing CO_2 emission since steel plants use huge fossil fuel [1]. For steel production, blast furnace-converter process will be dominant for a long time in China. In 2015, 93.9% crude steel is produced by this process in China [2]. Additionally, in this process, BF is the main source of CO_2 emission and the main consumer of energy [3]. Therefore, BF ironmaking is the crucial step for saving energy and reducing CO_2 emission in steel works.

After past decades developments, it is quite difficult for BF ironmaking to realize low carbon and ultrahigh efficiency just through traditional technologies (ore beneficiation, burden distribution, high top pressure, pulverized coal injection, etc.) since the operational efficiency of BF ironmaking has been close to its physical limitation [4]. Standing upon this background, some innovative technologies [5–8] has been proposed or applied in BF ironmaking. Among these technologies, coke oven gas (COG) injection is considered as one of feasible measures to reduce carbon emission of BF. For instance, Japan began a serious of low carbon emission innovative technologies COURSE50 Project in 2008 to reduce CO_2 emission by further developing the technique of injecting COG or reformed COG into the BF shaft [9]. It comprises technologies for reducing CO_2 emission from BF and technologies for separating and recovering CO_2, which will aim to contribute to 30% reduction in CO_2 emission. Additionally, the metallurgical properties of iron-bearing burdens under COG injection are greatly important since the content of in-furnace hydrogen is accelerated after COG injection and some researches on the reduction behaviors of iron-bearing burdens under COG injection have been investigated. Kemppainen et al. [10] studied the effects of H_2/H_2O on the reduction behaviors of pellets in a CO/CO_2 atmosphere with equal reducing potentials of H_2 and CO, and found that pellets have a higher reduction rate at lower temperature under H_2–H_2O–CO–CO_2 atmosphere but it has no significant differences in two atmospheres at high temperature. Yang et al. [11] studied the influence of hydrogen concentration of bosh gas on the softening-melting properties of iron-bearing burdens under condition of keeping CO content constant at 30% and revealed that softening interval first widens then narrows, and melting interval narrows and burden permeability is improved.

However, most of the researches only aimed at the reduction behaviors of iron-bearing burdens in simple hydrogen-enriched atmosphere, but the metallurgical properties of iron-bearing burdens in reasonable atmosphere corresponding to COG injection have not been systematically investigated so far. In this paper, the effects of COG injection on the reduction of mixed burdens were numerically simulated by means of multi-fluid BF model [12, 13]. According to the simulation results, the effects of hydrogen-enriched reduction on the reduction swelling behaviors of pellets and the softening-melting properties of mixed burdens under

COG injection were experimentally investigated. This research work could provide theoretical fundament and technological support for practical application of low carbon BF ironmaking through COG injection.

Simulated and Experimental Schemes

Effects of COG Injection on Reduction of Mixed Burdens

Effects of COG injection on reduction of mixed burdens were numerically simulated by means of multi-fluid BF model. The simulated object is an actual running BF (diameter 11.5 m, height 29.4 m, inner volume 2580 m^3). The reference case is the operation without COG injection. On basis of it, three cases with varied COG injection rates of 58.89, 110.08 and 152.34 Nm^3/tHM were evaluated. The above four cases are indexed as Case 1, 2, 3 and 4, respectively. For comparing all 4 cases under consistent conditions, raceway temperature and bosh gas flow rate are maintained unchanged through reasonably adjusting blast flow rate from 1111.47 to 638.51 Nm^3/tHM and oxygen enrichment from 3.00 to 19.22%.

Reduction Swelling Behaviors of Pellets Under COG Injection

The effects of hydrogen-enriched reduction on the swelling behaviors of pellets in actual atmosphere corresponding to COG injection were experimentally investigated according to the Chinese National Standard GB/T13242-91 [14]. The chemical composition of pellet is listed in Table 1. The gas compositions in tests are one of the above simulation results. After tests, the reduction swelling index (*RSI*) and the reduction ratio of pellets were calculated.

Softening-Melting Properties of Mixed Burdens Under COG Injection

The effects of hydrogen-enriched reduction on the softening-melting properties of mixed burdens in actual atmosphere corresponding to COG injection were experimentally studied. Table 1 shows the chemical composition of sinter and pellet used

Table 1 Chemical composition of sinter and pellet used in the tests (mass fraction)

Composition	TFe/%	FeO/%	SiO_2/%	CaO/%	MgO/%	Al_2O_3/%	R/−
Sinter	56.21	11.41	5.50	12.00	2.52	0.95	2.18
Pellet	65.48	0.48	5.81	0.30	0.05	0.36	0.05

in experiments. In tests, the weight of sinter is 300 g and the weight of pellet is 200 g. The major experimental conditions and some indexes refer to the references [15] and the gas compositions in tests at different temperature are one of the simulation results.

Results and Discussion

Effects of COG Injection on Reduction of Mixed Burdens

Distribution of Solid Temperature and in-Furnace Hydrogen Concentration

Figure 1 shows the distribution of solid temperature and in-furnace hydrogen concentration for all 4 evaluation cases according to simulation results. With increasing COG injection rate, solid temperature level gradually decreases, which is mainly due to the increase of burdens declining and charging rate. In-furnace H_2 concentration increases remarkably with the increase of COG injection rate as large amount of H_2, CH_4 and other hydrocarbons contained in COG decompose to generate hydrogen. Additionally, there are little changes for in-furnace CO content.

Ratios of H_2 Reduction Involving in the Whole Indirect Reduction

The ratios of H_2 reduction to the entire indirect reduction of iron oxides for four cases are shown in Fig. 2. For the operation with COG injection, the ratios of

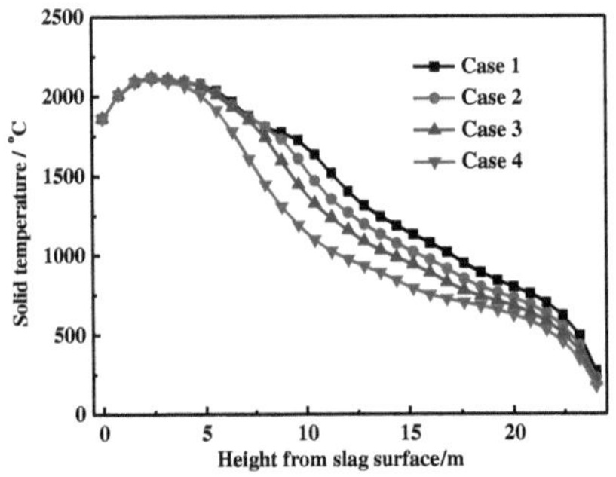

(a) Distribution of solid temperature

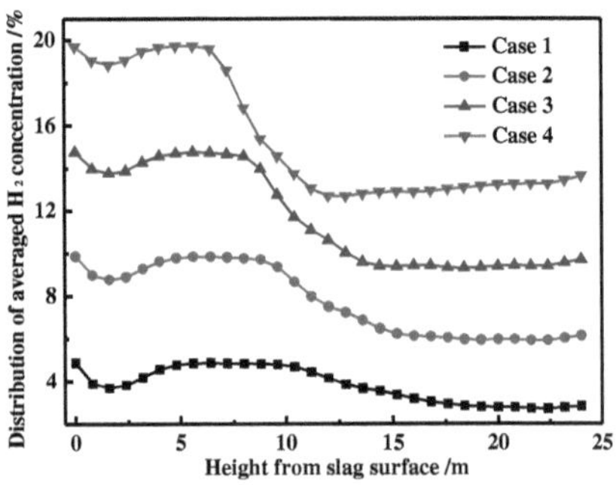

(b) Distribution of in-furnace H_2 concentration

Fig. 1 Distribution of solid temperature and in-furnace hydrogen concentration hydrogen reduction increases, which is much more remarkable for the involvement ratios of hydrogen in the reductions of magnetite and wustite in mixed burdens. Therefore, with increasing COG injection, the hydrogen-enriched reduction for iron oxides is accelerated and the reduction rate is increased before melting.

Effects of COG Injection on Reaction Rate of Wustite by H_2 Reduction

Figure 3 shows the effects of COG injection on reaction rate of wustite by H_2 reduction. With increasing COG injection, the reaction rate of wustite by H_2 reduction is enhanced. The main reason is that in-furnace hydrogen content is evidently increased under COG injection.

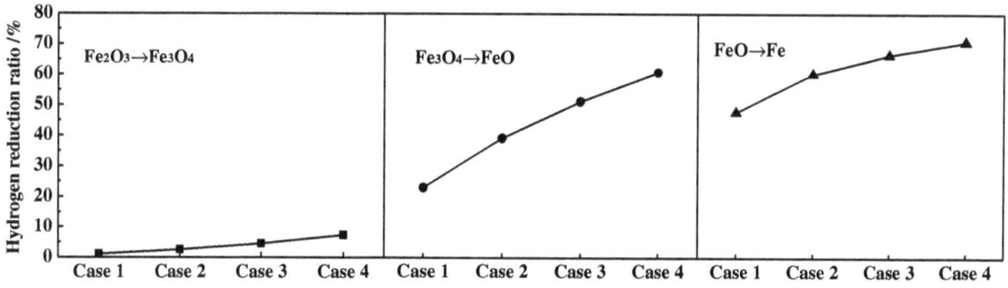

Fig. 2 Ratios of H_2 reduction to the entire indirect reduction

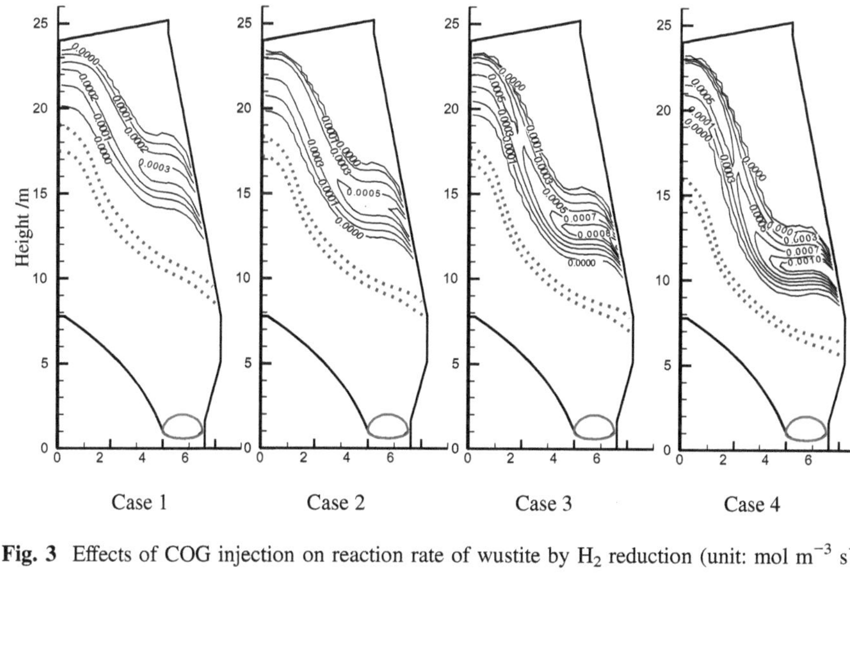

Fig. 3 Effects of COG injection on reaction rate of wustite by H_2 reduction (unit: mol m^{-3} s^{-1})

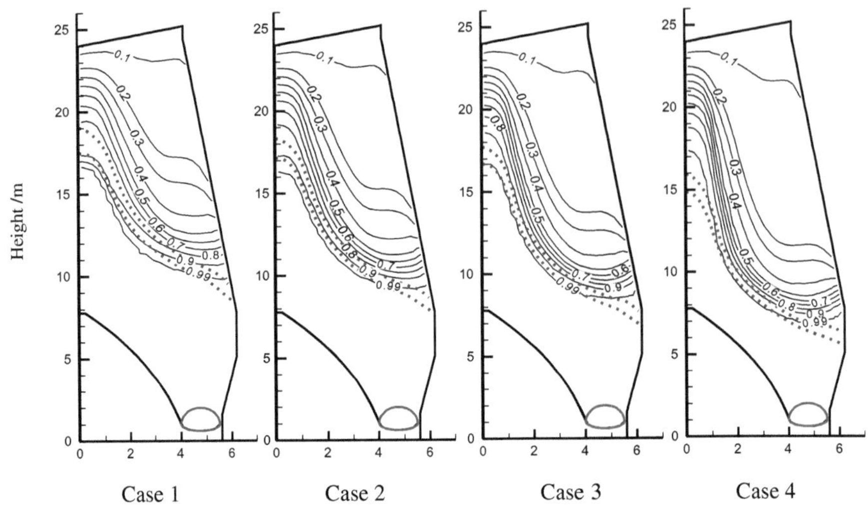

Fig. 4 Effects of COG injection on reduction degree of mixed burdens (unit: –)

Effects of COG Injection on Reduction Degree of Mixed Burdens

Effects of COG injection on reduction degree of mixed burdens is shown in Fig. 4. With the increase of COG injection rate, the reducing rate of mixed burdens is strengthened. In Case 1, the reduction degree of mixed burdens is about 75% before entering the cohesive zone while that is more than 90%, nearly 99% in local region for Case 4. The main reason is that the reducing atmosphere is enhanced under COG injection. The acceleration of the reduction rate of wustite (seen in Fig. 3) further enhances the reduction degree of mixed burdens.

Reduction Swelling Behaviors of Pellets Under COG Injection

Figure 5 shows the effects of hydrogen-enriched reduction on the swelling behaviors of pellets in an actual atmosphere corresponding to COG injection. With COG injection rate increasing, *RSI* of pellets decreases slightly while the reduction ratio of pellets increases. So the reduction swelling behaviors of pellets is improved under COG injection.

Scanning electron microscopy (SEM) photos of pellets after reduction is shown in Fig. 6. In the figure, slag-io represents slag phase containing little iron oxide. The center structures of pellets after reduction have little change, and the some variations converge on surface. With COG injection rate increasing, the amounts of platy structure in pellets surface after reduction reduce while flocculent structure increases gradually as well as Fe particle. The interaction force of metal iron crystal

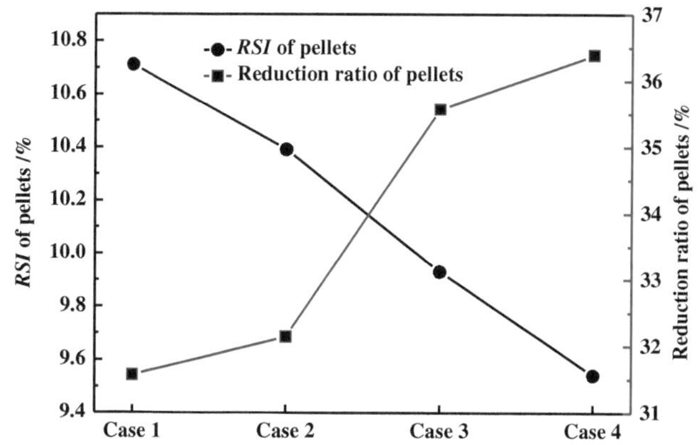

Fig. 5 Reduction swelling behaviors of pellets under COG injection

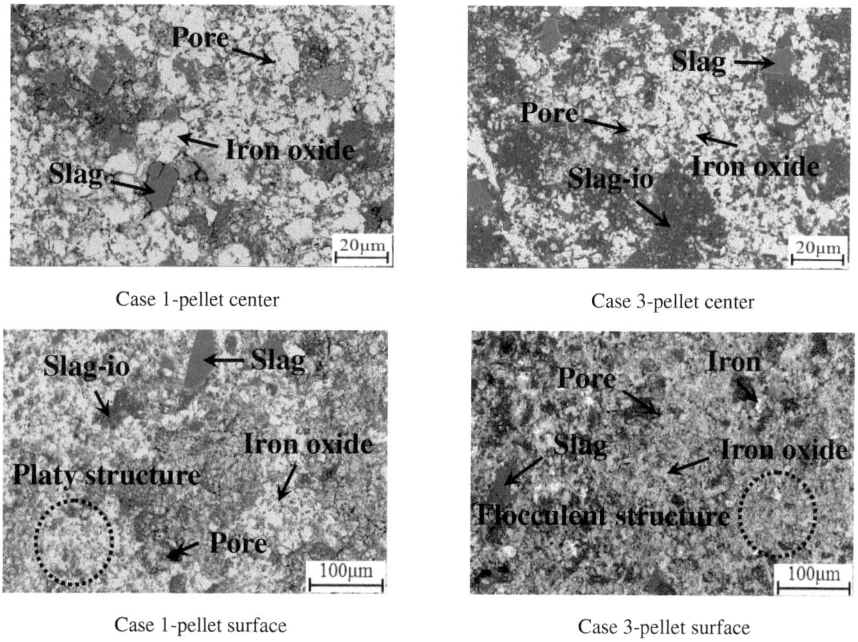

Fig. 6 SEM photos of pellets after reduction with different COG injection rate

is enhanced with Fe particle generated in pellets, which result in the decrease of pellets swelling. Additionally, the H_2 content in reducing gas is increased with COG injection rate increasing, which contributes to the increase of pellets reduction ratio and the decrease of pellets swelling.

Softening-Melting Properties of Mixed Burdens Under COG Injection

Softening Property of Mixed Burdens

Figure 7 shows the softening property of mixed burdens in an atmosphere corresponding to COG injection. With COG injection rate increasing, T_4 (softening start temperature) of mixed burdens decreases from 1134.8 to 1104.4 °C mainly due to the increase of reduction rate of mixed burdens under COG injection. However, T_{40} (softening end temperature) increases from 1298.4 to 1332.7 °C since the melting-point of slag phase increases due to the decrease of FeO content. In addition, T_{40}–T_4 (softening interval) widens from 163.6 to 228.3 °C.

Melting Property of Mixed Burdens

The melting property of mixed burdens in an atmosphere corresponding to COG injection is shown in Fig. 8. T_S (melting start temperature) of mixed burdens increases from 1272.7 to 1333.6 °C since FeO is gradually reduced to Fe around T_S temperature, and T_D (dripping temperature) increases slightly from 1420.0 to 1426.0 °C mainly due to the carburization amount in primary iron phase decreasing and the slag phase melting-point increasing under COG injection. Additionally, $T_D - T_S$ (melting interval, cohesive zone) tends to narrow obviously from 147.3 to 92.4 °C. So the cohesive zone moves downwards and becomes narrower.

Gas Permeability of Mixed Burdens

The softening-dripping properties characteristic value (indexed as *S-value*) can be introduced to evaluate the gas permeability of mixed burdens [15]. Figure 9 shows

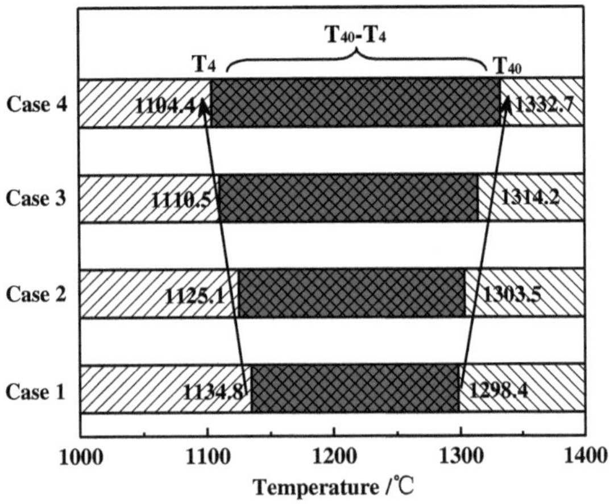

Fig. 7 Softening property of mixed burdens under COG injection

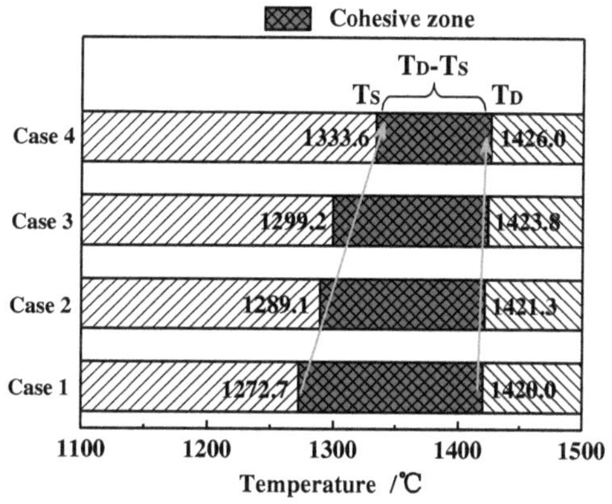

Fig. 8 Melting property of mixed burdens under COG injection

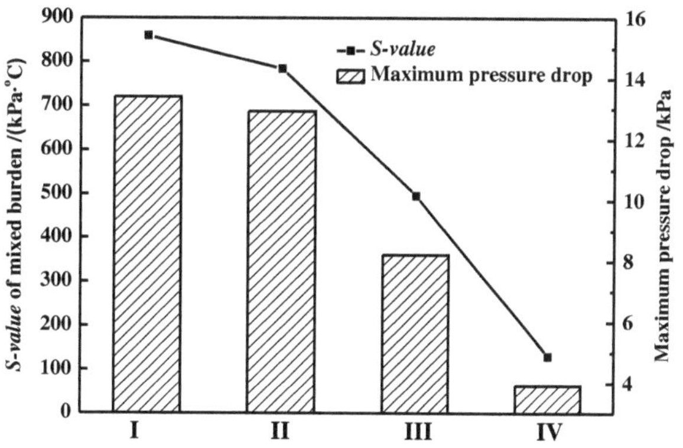

Fig. 9 *S-value* and maximum pressure drop of mixed burdens under COG injection

the *S-value* and the maximum pressure drop of mixed burdens in an atmosphere corresponding to COG injection. With the increase of COG injection rate, *S-value* of mixed burdens decreases as well as maximum pressure drop, which suggest that the permeability of mixed burdens is improved under COG injection. The main reason is that the reduction rate of mixed burdens accelerates as well as the diffusivity of gas phase due to lower viscosity coefficient of H_2. Additionally, narrower cohesive zone under COG injection is another reason.

Conclusions

(1) With COG injection rate increasing, in-furnace H_2 concentration increases evidently and the solid temperature decreases. The ratios of H_2 reduction to the entire indirect reduction increase, much more evidently for the reduction of magnetite and wustite. Additionally, the reaction rate of wustite by H_2 reduction is accelerated and the reduction degree of mixed burdens is strengthened.

(2) The GB/T13242-91 test results show that with COG injection rate increasing, *RSI* of pellets decreases while the reduction ratio of pellets increases. The reduction swelling properties of pellets is improved under COG injection.
(3) With the increase of COG injection, softening interval widens from 163.6 to 228.3 °C while melting interval tends to narrow from 147.3 to 92.4 °C. The cohesive zone moves downwards and becomes narrower. The permeability of mixed burden is improved.
(4) The results of simulation and experiments reveal that BF operation performance could be improved through COG injection in terms of metallurgical properties of iron-bearing burdens.

Acknowledgements The authors wish to acknowledge gratefully the financial support provided by the National Science Foundation of China (50804008) and Shanghai Meishan Iron and Steel Corporation Ltd.

References

1. Y. Tanaka et al., Reaction behavior of coal rich composite iron ore hot briquettes under load at high temperatures until 1400 °C. ISIJ Int. **51**(8), 1240–1246 (2011)
2. World Steel Association, "World Steel in Figures 2016" (Report, 2016, 10–11)
3. Q. Zhang et al., Carbon flow analysis and CO_2 emission reduction strategies of iron-making system in steel enterprise. J. Northeastern Univ. (Natural Sci) **34**(3), 392–394 (2013)
4. Z. Liu et al., Numerical simulation on novel blast furnace operation of combining coke oven gas injection with hot burden charging. Ironmaking Steelmaking **43**(1), 64–73 (2016)
5. G.D. Wang, Technology innovation and development direction of iron and steel industry. Iron Steel **50**(9), 1–10 (2015)
6. H. Wang et al., Fundamental research on iron coke hot briquette—A new type burden used in blast furnace. Ironmaking Steelmaking **43**(8), 571–580 (2016)
7. H.T. Wang et al., Mathematical simulation on blast furnace operation of coke oven gas injection in combination with top gas recycling. Steel Res. Int. **87**(5), 539–549 (2016)
8. E.A. Mousa, A. Babich, D. Senk, Enhancement of iron ore sinter reducibility through coke oven gas injection into the modern blast furnace. ISIJ Int. **53**(8), 1372–1380 (2013)
9. M. Naito, K. Takeda, Y. Matsui, Ironmaking technology for the last 100 years: development to advanced technologies from introduction of technological know-how, and evolution to next-generation process. ISIJ Int. **55**(1), 7–35 (2015)
10. A. Kemppainen et al., Effect of H_2–H_2O on the reduction of olivine pellets in CO–CO_2 gas. ISIJ Int. **52**(11), 1973–1978 (2012)
11. G.Q. Yang et al., Influence of hydrogen content of bosh gas on softening-melting property of iron-bearing burden material. Iron Steel **47**(9), 14–18 (2012)
12. P.R. Austin, H. Nogami, J.I. Yagi, Analysis of actual blast furnace operations and evaluation of static liquid holdup effects by the four fluid model. ISIJ Int. **38**(3), 246–255 (1998)
13. M.S. Chu et al., Research progress on mathematical modeling of blast furnace ironmaking process. Iron and Steel **49**(11), 1–8 (2014)
14. Chinese National Standard. Iron ore pellets—determination of relative free—swelling index (Standard, GB/T13240-91, 1992)
15. Z.G. Liu et al., Effect of MgO content in sinter on the softening-melting behavior of mixed burden made from chromium-bearing vanadium-titanium magnetite. Int J Miner Metall Mater. **23**(1), 25–32 (2016)

Microwave-Intensified Reduction of Biochar-Containing Briquettes

Zhiwei Peng, Xiaolong Lin, Tiancheng Nie, Zhizhong Li, Yuanbo Zhang, Guanghui Li and Tao Jiang

Abstract The coal-based direct reduction process has gradually become the focus of research concerning direct reduced iron production due to its advantages, including high reduction rate, stable chemical composition, and good metallurgical properties of products. Based on coal-based reduction theory and briquetting technique, this paper reports the use of microwave energy for intensifying the reduction of carbon-containing briquettes mainly consisting of magnetite concentrate and biochar. The heating behaviors of the raw materials show that under the conditions of microwave power of 1250 W, N_2 flow rate of 4 L min^{-1} and briquetting pressure of 400 N, the briquettes present the highest heating rate. The average heating rate reaches 35.16 °C min^{-1}. When the briquettes are irradiated by microwaves at 1050 °C for 45 min, the metallization degree of the reduced briquettes reaches 93.58%.

Keywords Direct reduction · Microwave irradiation · Biochar · Briquette · Metallization degree

Introduction

Iron ores are one of the most important resources for development of human society. In the past decades, much work has been made concerning reduction of iron ores. With the continuous development of technology of ferrous metallurgy, there exist three processes, namely blast furnace process, direct reduction method and smelting reduction process. Currently, the blast furnace process dominates current iron production [1]. However, it has high dependence on coke and usually causes severe environmental pollution which limits its further development [2]. Unlike the blast furnace process, the direct reduction technology uses gas, liquid

---

Z. Peng (✉) · X. Lin · T. Nie · Z. Li · Y. Zhang · G. Li · T. Jiang
School of Minerals Processing and Bioengineering, Central South University, Changsha, Hunan 410083, China
e-mail: zwpeng@csu.edu.cn

J.-Y. Hwang et al. (eds.), *8th International Symposium on High-Temperature Metallurgical Processing*, The Minerals, Metals & Materials Series,
DOI 10.1007/978-3-319-51340-9_4

fuel or coal as reducing agent and fuel sources for reduction of iron oxides. It produces good-quality products (e.g., direct reduced iron) with much less pollution [3]. By now, use of coal in direct reduction process has evolved into an important iron making process next to blast furnace process due to high reduction degree, stable chemical composition and good metallurgical properties of products. However, its production scale is still limited because of insufficient heating efficiency [4, 5]. In order to solve this problem enormous attention has been devoted to seeking clean and sustainable energy sources for the reduction. Biomass-derived carbonaceous materials, such as biochar, were found to be a viable alternative to coal in view of its high carbon content, low impurities and good reactivity [6]. Its performance in iron ore reduction can be further improved in a microwave field despite lack of in-depth exploration [6, 7]. This paper focuses on the reduction of biochar-containing briquette under microwave irradiation. It will offer new thoughts to direct ironmaking technology and contribute to sustainable development of iron and steel industry.

Experimental

Materials

The materials included magnetite concentrate (from Iran), biochar (charcoal) and binder. The chemical composition, particle size composition, specific surface area and particle morphology of the concentrate were, respectively, determined by chemical titration, sieve screening, Brunauer Emmett Teller (BET) method and scanning electron microscopy. The main chemical composition of magnetite concentrate is shown in Table 1. The magnetite concentrate contained 67.61 wt% TFe and 20.06 wt% FeO. Few gangue minerals were found in the concentrate. The contents of SiO_2, Al_2O_3, CaO and MgO were 2.00 wt%, 0.65 wt%, 0.80 wt% and 0.44 wt%, respectively. Evidently, it was a high-grade iron ore with low impurities.

Table 2 shows the particle size distribution and specific surface area of magnetite concentrate. The concentrate was very fine as the fraction of grain size of less than 0.074 mm reached 92.29% and that below 0.045 mm approached 83.28%.

Figure 1 shows the surface morphology of the concentrate. It is seen that the particles were irregular and the surface of the particles was pretty smooth.

Biochar was purchased from a commercial company in Shanghai, China. More than 95% of the biochar particles were less than 0.074 mm after ball milling. Table 3 shows the industrial analysis of the biochar. It had over 80 wt% fixed carbon and 4.48 wt% ash, satisfying the requirements of iron ore reduction.

Table 1 Main chemical composition of magnetite concentrate/wt%

TFe	FeO	SiO_2	Al_2O_3	CaO	MgO	K_2O	Na_2O	Cu	TiO_2	P	S	LOI
67.61	20.06	2.00	0.65	0.80	0.44	0.028	0.18	0.0034	0.72	0.067	0.01	0.51

Table 2 Size distribution and specific surface area of magnetite concentrate/wt%

–0.045 mm	+0.045–0.074 mm	+0.074–0.15 mm	+0.15 mm	Specific surface area/$cm^2\ g^{-1}$
83.28	9.01	7.20	0.51	1716

Fig. 1 Surface morphology of iron ore particles

Table 3 Industrial analysis of biochar/wt%

w (FC_{ad})	w (A_{ad})	w (V_{ad})	S
80.45	4.48	15.07	0.085

Table 4 Technical indices of sodium carboxymethyl cellulose

As/wt%	pH	Chlorides/wt %	Viscosity/mPa s	Fe/wt %	Drying loss/wt %	Na/wt %	Heavy metals/wt %
≦0.0002	6.0–8.5	≦3.0	800–1200	≦0.03	≦10	6.5–8.5	≦0.002

The binder used in this experiment was sodium carboxymethyl cellulose (AR), referred to as CMC-Na. It was one of cellulose derivatives with 100–2000 degrees of polymerization of glucose. The relative molecular mass of the binder was 242.16. Table 4 shows its main technical indices.

Experimental Methods

Magnetite concentrate powder, binder and biochar were homogeneously mixed and briquetted. The briquettes (Ø 1 cm × H 1.5 cm) were sent into a ZG1516 microwave tube furnace (2.45 GHz, 1250 W) after drying for a given period of time. The temperature rising characteristics of briquettes (represented by plot of temperature against heating time) were examined until the sample temperature exceeded 1000 °C in 4 L min^{-1} nitrogen as protective gas. The briquettes reduced under specified conditions (see below) were then taken out and cooled in liquid nitrogen rapidly to prevent reoxidation of metallic iron formed. By chemical titration, the metallization degrees of the biochar-bearing briquettes were determined.

Results and Discussion

Temperature Rising Characteristics of Briquettes

Figure 2 shows the temperature rising curve of biochar-containing briquettes. The temperature of briquettes gradually increases with time. The maximum heating rate occurs in the time interval from 5 to 25 min. The temperature of briquettes reaches the highest temperature of 1054.7 °C in 40 min and the average temperature rising rate is 35.16 °C min^{-1}. This observation shows the strong microwave absorption in both magnetite concentrate and biochar, which agrees well with the findings in previous reports [7, 8].

Microwave Reduction of Briquettes

Microwave heating usually accelerate the process of reduction because of its volumetric and selective heating effects [6]. As both the concentrate and the reducing agent are good microwave absorbers, the reduction can be completed in a few minutes. To assess the process, the effects of different parameters on the metallization of iron ore have been performed. Figure 3 shows the effect of reduction temperature on the metallization degree of briquettes after reduction for 15 min. When the reduction temperature is increased from 900 to 1050 °C, the metallization degree increases from 47.98 to 89.65%. In the low temperature range up to

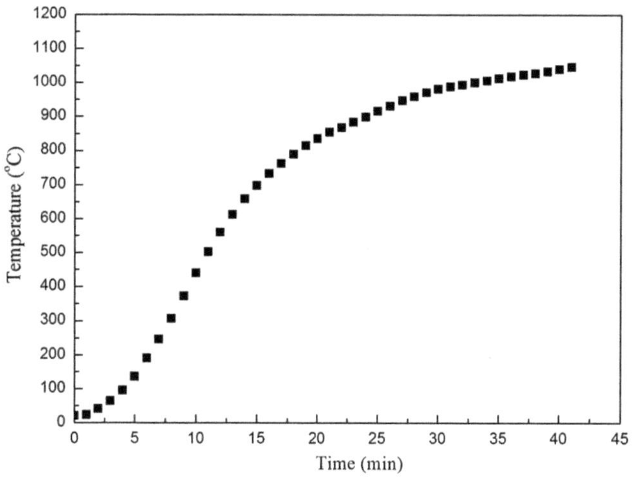

Fig. 2 Temperature rising curve of briquettes

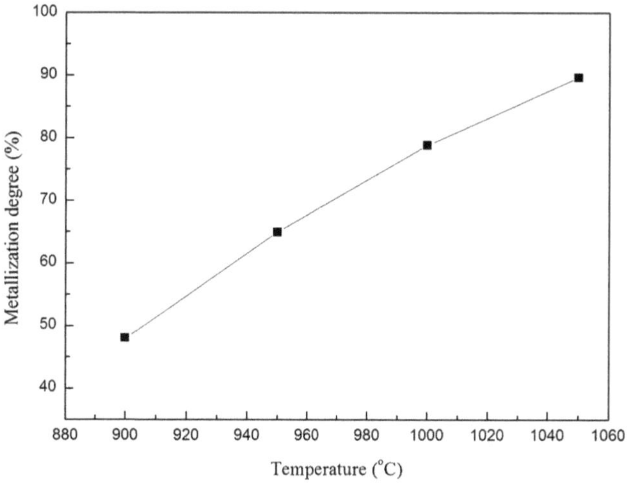

Fig. 3 Effect of reduction temperature on metallization degree of briquettes after reduction for 15 min

900 °C, the solid-solid reduction plays a major role in the reduction in which wüstite is formed due to reduction of the briquettes and only little metallic iron is formed. When the temperature is elevated to above 900 °C, the gas-solid reduction dominates the whole process. The wüstite formed can be reduced rapidly to metallic iron. As the whole process is a heat absorption process, elevating temperature is beneficial to the metallization.

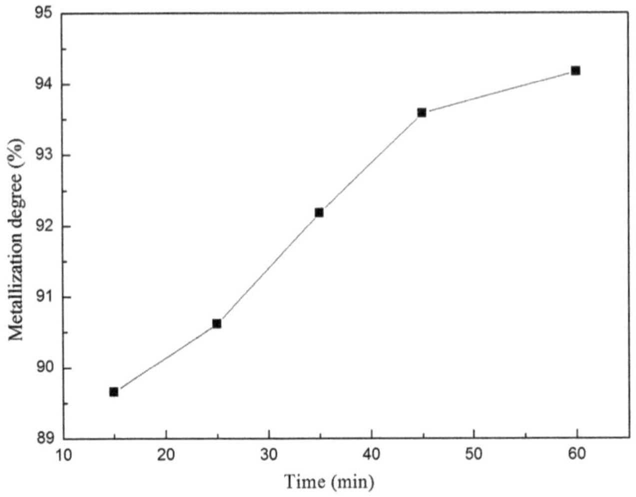

Fig. 4 Effect of reduction time on metallization degree of briquettes at 1050 °C

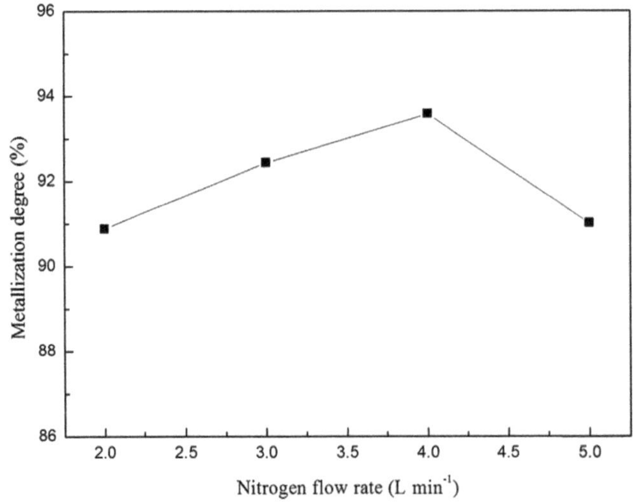

Fig. 5 Effect of nitrogen flow rate on metallization degree of briquettes after reduction at 1050 °C for 45 min

Figure 4 shows the effect of reduction time on the metallization degree of briquettes at 1050 °C. When the reduction time is increased from 15 to 60 min the metallization degree increases from 89.65 to 94.17%. The degree increases relatively rapidly in 45 min, while its growth rate decreases when the time is extended to 60 min. The decline in the growth rate is mainly ascribed to the decreasing content of biochar that is consumed continuously during reduction. Another reason why differences occur is due to possible microwave reflection of enlarged metallic grains formed within the longer time period.

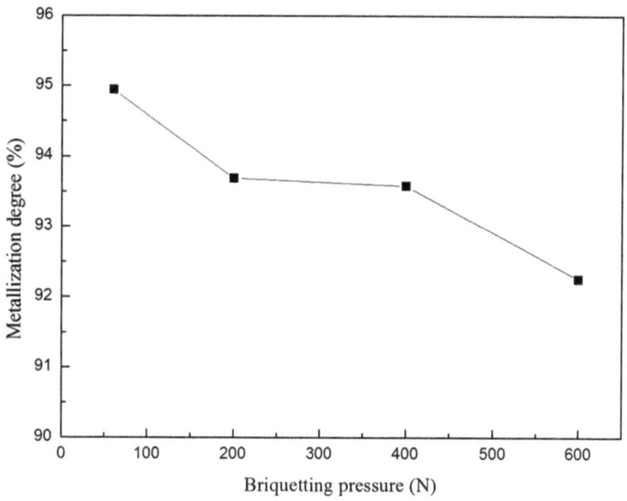

Fig. 6 Effect of briquetting pressure on metallization degree of brequettes after reduction at 1050 °C for 45 min

Figure 5 shows the effect of flow rate of nitrogen gas on the metallization degree of briquettes after reduction at 1050 °C for 45 min. When the flow rate is increased from 2 to 5 L min^{-1}, the degree initially increases and then decreases, varying between 90.88 and 93.58%. The maximum degree occurs at 4 L min^{-1} with the value of 93.58%. The change is mainly associated with the balance between the protective gas and the gases produced by the reduction reactions. Suitable flow rate will speed up the reduction process by bringing away CO_2 produced during reduction. However, with too high flow rate, heat transfer does not fit the process, lowering the metallization degree.

Figure 6 shows the effect of briquetting pressure on the metallization degree of briquettes after reduction at 1050 °C for 45 min. The degree decreases continuously when the pressure increases. It decreases from 94.94 to 92.25% when the pressure is elevated from 60 to 600 N. This is, as expected, because of the compacted structure which inhibits the diffusion of CO within the briquettes when the pressure is increased.

Conclusion

The effects of various parameters on the metallization degree of the biochar-containing briquettes were investigated in this study. The metallization degree rises from 47.98 to 89.65% in 15 min when the temperature increases from 900 to 1050 °C. At low temperatures, iron oxides are reduced to wüstite. At high temperatures, wüstite is further reduced and the metallization degree increases with reduction time. When the reduction time is 45 min, the metallization degree reaches 93.58% at 1050 °C. Further increase in time causes excessive consumption of carbon. The metallic grains formed are expected to undergo re-oxidation. When the flow rate is 4 L min^{-1}, the metallization degree remains at the maximum value and a larger flow rate adversely affects the heat transfer which reduces the degree of metallization. As the metallization degree declines with increasing briquetting pressure, it is indicated that the compacted structure will restrain the diffusion of CO that is unfavorable to the reduction. A compromise should be obtained between the requirements of reduction efficiency and product transportation.

Acknowledgements This work was supported by the National Natural Science Foundation of China under Grant 51504297, the Shenghua Lieying Program of Central South University under Grant 502035001, and the Fundamental Research Funds for the Central Universities of Central South University under Grant 2016zzts464.

References

1. M. Smith, Blast furnace ironmaking: view on future developments. Ironmaking Steelmaking **42**, 734–742 (2015)
2. F.M. Zhang, C.Z. Cao, X.L. Meng, L. Li, Technological development orientation on ironmaking of contemporary blast furnace. Adv. Mater. Res. **875–877**, 1138–1142 (2014)
3. D. Guo, M. Hu, C. Pu, B. Xiao, Z. Hu, Kinetics and mechanisms of direct reduction of iron ore-biomass composite pellets with hydrogen gas. Int. J. Hydrogen Energy **40**, 4733–4740 (2005)
4. J.Y. Shi, E. Donskoi, D.L.S. Mcelwain, L.J. Wibberley, Modelling novel coal based direct reduction process. Ironmaking Steelmaking **35**, 3–13 (2013)
5. H. Michishita, H. Tanaka, Prospects for coal-based direct reduction process. Kobelco Technol Rev **29**, 69–76 (2010)
6. Z. Peng, J.Y. Hwang, Microwave-assisted metallurgy. Int. Mater. Rev. **60**, 30–63 (2015)
7. Z. Peng, J.Y. Hwang, J. Mouris, R. Hutcheon, X. Huang, Microwave penetration depth in materials with non-zero magnetic susceptibility. ISIJ Int. **50**, 1590–1596 (2010)
8. J.A. Menéndez, A. Arenillas, B. Fidalgo, Y. Fernández, L. Zubizarreta, E.G. Calvo, J.M. Bermúdez, Microwave heating processes involving carbon materials. Fuel Process. Technol. **91**, 1–8 (2010)

Refractory Challenges in Lead Recycling Furnaces

D. Gregurek, K. Reinharter, V. Reiter, C. Wenzl and A. Spanring

Abstract The refractory linings in lead furnaces are exposed to several stresses rather complex in their interaction. In the present study the magnesia-chromite brick out of a lead recycling furnace suffered from a high chemical attack by the process slag. The high CaO, BaO and sulfur bearing silicate slag, as well as a high Na_2O supply from soda resulted both in a deep-reaching infiltration of the brick microstructure and a severe corrosion of the brick components. Both the sintered magnesia and chromite were attacked chemically. The FactSage calculations showed the formation of high amounts of liquid phase in the infiltrated microstructure. Further phenomena affecting the refractory performance such as iron oxide attack, redox effects, and boron attack are also discussed in the paper. The obtained information and insights serve as a basis for improving refractory materials (i.e., choice of refractories for individual process and new developments) and consequently furnace operations.

Keywords Refractories · Post mortem study · Lead recycling furnace · Magnesia-chromite brick

Introduction

Lead is one of the metals with the highest recycling rate worldwide [1]—therefore, the importance of secondary lead production, namely recycling, has increased over the last years and decades and the amount of secondary production has already surpassed the primary production. Various furnace types can be used for lead recycling, for example the Short Rotary Furnace, Reverberatory Furnace, TSL (Top Submerged Lance) reactor and TBRC. Despite the different furnace types, the

D. Gregurek (✉) · K. Reinharter · V. Reiter
RHI AG, TC Leoben, Magnesitstrasse 2, Leoben 8700, Austria
e-mail: dean.gregurek@rhi-ag.com

C. Wenzl · A. Spanring
RHI AG, Wienerbergstrasse 9, Vienna 1100, Austria

J.-Y. Hwang et al. (eds.), *8th International Symposium on High-Temperature Metallurgical Processing*, The Minerals, Metals & Materials Series,
DOI 10.1007/978-3-319-51340-9_5

metallurgical process of lead recycling is basically the same. The most important recycling material and hence input material for secondary lead smelters are lead batteries. For recovering the lead, the used batteries undergo separation and pre-treatment steps to divide the battery components into the different material fractions. Lead recycling therefore has to deal not only with metallic lead (battery grids), but also with lead components (i.e., sulfates and oxides) that require suitable metallurgical processing. Additionally, other lead-containing residues (e.g., slags from other metal production processes) may also be used in the recycling process, leading to varying input compositions that may require process adaptations. The slags used in secondary lead production are generally based on the system FeOx–SiO_2–CaO, also the use of soda as slag additive is common practice. Many smelters also use a combination of different slag types, namely FeO_x–SiO_2-slag and soda slag in separate process stages [1–3].

The refractory lining of the different lead recycling furnaces is exposed to several stresses and wear mechanisms, including chemical, thermal, and mechanical factors [4]. As the lead processes are carried out at temperatures far higher than the lead melting temperature (327 °C), namely around 1000 °C and higher to achieve a liquid slag, the metallic lead is highly overheated and due to its low viscosity causes deep-reaching infiltration into the refractory lining. This causes changes in thermal properties, which are of special importance in batch processes that comprise repeated heating and cooling cycles of the furnace. Additionally, the slags cause chemical attack on the basic magnesia-chromite bricks, especially in the case of high SiO_2 loads, namely more acidic slags. Even though soda slags are very efficient from a metallurgical point of view, they lead to refractory wear by both alkali and sulphur. Furthermore, the mechanical wear factors play an important role in refractory life, especially in rotating furnace types, but also due to liquid flows, for example during tapping operations.

A general overview of wear phenomena in the lead melting furnaces is given in several papers [5–7]. Particularly in the lead industry, many investigations were done regarding refractory corrosion testing in different pilot-scale and industrial furnaces. For instance, Prestes et al. analyzed wear phenomena on magnesia-chromite bricks from a lead short rotary furnace [5]. Additionally, some experimental work using crucible corrosion testing was also carried out. Wei reviewed available literature concerning corrosion of refractories in lead-smelting reactors such as the KIVCET furnace and the TBRC (Top Blowing Rotary Converter), and evaluated in the laboratory the corrosion behavior of various refractory materials in contact with industrial slags [6]. Finally, Scheunis et al. [7] investigated the effect of phase formation during chemical corrosion of magnesia-chromite refractories in contact with PbO–SiO_2 based slag from non-ferrous metal production.

The present work gives an example of a detailed refractory wear analysis of a used magnesia-chromite brick from a lead recycling furnace. To support mineralogical findings and chemical analyses, thermochemical calculations by FactSage software package were also performed.

Analytical Procedure

Generally, every single post mortem study starts with the visual inspection carried out on the brick cross section followed by selection of samples for chemical analysis and mineralogical investigation. The chemical analysis was carried out by using X-ray fluorescence analysis (Bruker S8 TIGER). The mineralogical investigation was performed on polished sections using a reflected light microscope, X-ray diffraction (Bruker D8 ADVANCE), and a scanning electron microscope (SEM) (JEOL JSM-6460) combined with an energy-dispersive and wave-dispersive X-ray analyser.

Results and Discussion

Visual Inspection

The investigated refractory product represents a silicate bonded magnesia-chromite brick of type MCr60 (ISO 10081-2), which is based on sintered magnesia and chromite. The residual brick thickness of the received sample after six months of operation is 200 mm (see Fig. 1). The original brick thickness was 390 mm. The immediate brick hot face is covered with a thin slag coating. On the cross section, the macroscopically visible infiltration depth (pore infiltration—mainly slag, but also metal) reaches up to the brick cold face. In different distances from the hot face, cracks running parallel to the hot face can be observed, partly filled with metallic lead and slag.

Chemical Analysis

Generally, in lead recycling furnaces the low-melting, partly alkali-rich silicate slag is able to penetrate deeply into the brick microstructure [8]. The silica-rich slag and the reaction products formed by the corrosion of the brick-inherent components can be traced up to the cold face. This is a big difference compared to slag attack and infiltration in copper furnaces, where the infiltration depth is just a few mm from the hot face.

Table 1 shows the chemical composition of a deeply infiltrated magnesia-chromite brick. According to the chemical analysis, in the first 110 mm from the hot face a high sodium, sulfur and silica content were determined. Nevertheless, the highest concentrations of SiO_2, Fe_2O_3, BaO, and CaO, were confirmed to be located in the first 25 mm from the hot face. Some minor phosphorous and lead content could also be identified.

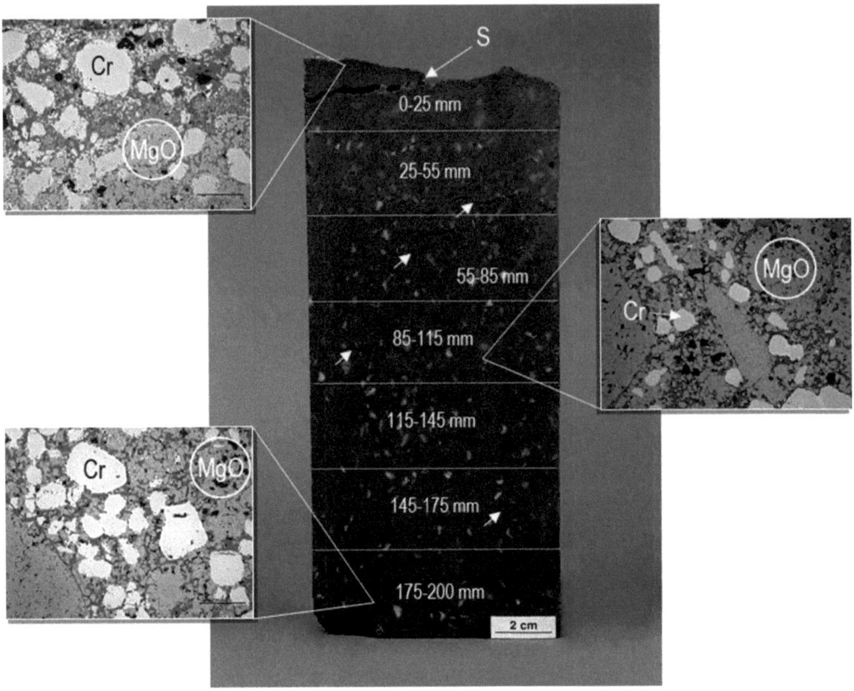

Fig. 1 Cross sectional view. Magnesia-chromite brick from a lead secondary recycling furnace. The residual brick thickness is 200 mm. The immediate brick hot face is covered with a thin slag coating (S). Crack formation (*arrows*). Polished sections for mineralogical investigation of segments over the brick thickness. Microstructural overview of degenerated and corroded brick microstructure (enlarged images). Corroded sintered magnesia (MgO). Chromite (Cr)

Mineralogical Investigation

For mineralogical investigation several polished sections were taken (see also the marked areas in Fig. 1). On a microscopic scale, several microstructural changes were detected and can be summarized as follows:

The immediate brick hot face shows a 2 mm thin reaction zone consisting of phosphorous- and lead-rich Na–Ca–Mg–Fe–Ba-silicate, phosphorous-rich Na–Ca–Mg–Fe–Ba–Al-silicate and Na–Ca–Mg–Fe–Al-silicate. The magnesia component is completely dissolved leaving relics of primary and secondary chromite precipitations (Fig. 2a). The chromite relics are enriched with Fe- and Sn-oxide.

Below the reaction zone there is a pore-filling infiltration and corrosion of the brick microstructure (Fig. 2b). Generally, slag infiltration depends on several parameters like viscosity, pore size distribution and wetting angle [8]. It is assumed that these kinetic considerations can be neglected here as the ratio of the wear rate to the infiltration velocity is much smaller than one. As a consequence, microscopic investigations of used bricks always show a total infiltration of the brick pores up to

Table 1 Chemical analysis of the magnesia-chromite brick out of a lead recycling furnace

wt%	Na_2O	MgO	Al_2O_3	SiO_2	P_2O_5	SO_3	CaO	TiO_2	Cr_2O_3	Fe_2O_3[a]	BaO	PbO
Hot face (0–25 mm)	9	47	5	10	0.8	2	2	0.3	11	9	2	0.8
Middle part (90–110 mm)	9	55	4	3	0.5	7	1	0.2	12	8	0.4	
Cold end (180–200 mm)	0.9	55	6	2	0.1	0.3	1	0.3	21	13	0.3	
Unused		60.4	6.3	1.8			1.1		19.0	12.0		

[a]Total iron calculated as Fe_2O_3

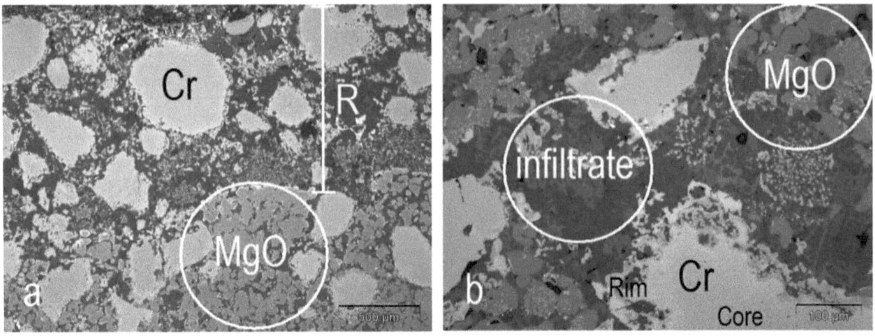

Fig. 2 Mineralogical investigation by reflected light microscopy and scanning electron microscope. Magnesia-chromite brick out of lead recycling furnace. 2 mm thin reaction zone (R) at the immediate brick hot face (**a**). Corroded sintered magnesia (MgO) and chromite (Cr). Approx. 30 mm from hot face (**b**). Infiltrated and corroded brick microstructure. Severely corroded sintered magnesia. Initial corrosion of chromite rims

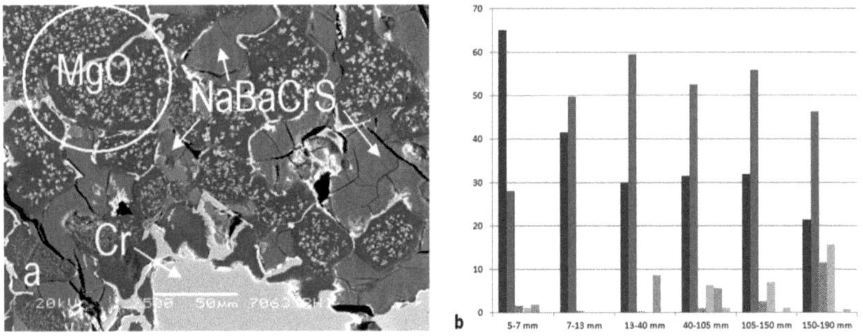

Fig. 3 Mineralogical investigation by reflected light microscopy and scanning electron microscope. Approx. 150 mm from hot face (**a**). Infiltrated brick microstructure. Corroded sintered magnesia (MgO). Pores filled with barium- and phosphorous-enriched Na–Ca–Cr-sulfate (NaBaCrS). Chromite (Cr). Histogram showing differences in the chemical composition of Na–Ca–Cr-sulfate from the hot to cold face (**b**). This phase contains also some barium

the cold end. This moreover means that the invariant point of the respective mineral phase assemblages is below the cold face temperature.

Within the infiltrated brick microstructure, the sintered magnesia shows severe intra-granular corrosion particularly in the area 2–5 mm from the hot face. That means that the slag also penetrated along the crystal boundaries of the sintered magnesia and corroded the interstitial phases of the magnesia.

Due to the corrosion of the magnesia, the main reaction products are Mg-silicate of type forsterite (Mg_2SiO_4) slightly enriched with CaO, Ca–Mg-silicate of type monticellite ($CaMgSiO_4$) and Na–Ca–Mg–Fe–Al-silicate (Fig. 2 right). By help of the scanning electron microscope, phosphorous and lead rich Na–Mg–Ca–Ba-silicates could also be identified (Fig. 3a). Corrosion of sintered magnesia, but also of the interstitial phases between the MgO grains, can be detected up to a depth of approximately 90–100 mm from the hot face. Additionally, primary and secondary chromite precipitations within the sintered magnesia, which are usually typical for this brick brand (original and unused condition), could not be observed any more.

In the infiltrated brick microstructure (e.g., 30–200 mm from the hot face) the chromite rims are strongly depleted in Cr_2O_3 and Al_2O_3, as well as highly enriched with Fe_2O_3. The corrosion of chromite was obviously caused by high sodium supply, namely a surplus of sodium not saturated by sulfur. Based on the SEM-EDX analysis the highest Cr_2O_3 content in Na–Ca-sulfate (i.e., >10 wt%) could be observed at the brick cold face (e.g. 150–190 mm from hot face, see Fig. 3b).

As already macroscopically observed, at different distances from the hot face several cracks partly filled with metal and slag can be observed.

FactSage Calculations

The phase composition of the unused brick, the change of the phase composition and newly formed phases due to slag infiltration, and the amount of liquid phase, were described thermodynamically for the slag-penetrated brick. The software package FactSage 6.3 was used for the thermodynamic calculations [9]. The focus was on the description of the phase composition of both the immediate hot face and the temperature dependent infiltrated of the brick microstructure in the range of 700–1600 °C. The calculations were performed with the average chemical analysis. For slags containing bivalent oxides, such as those of iron and titanium, the atmospheric conditions (i.e., oxygen partial pressure) can represent an important factor as the phase constitution strongly depends on the valence of the elements. The calculations were performed for both oxidizing and reducing conditions by varying the oxygen partial pressure from $pO_2 = 0.21$ atm to $pO_2 = 10^{-8}$ atm. As the differences of the results of the calculations under oxidising and reducing conditions are marginal, only the results of the reducing atmosphere are presented.

The databases FactPS, FToxide and FTmisc were used for the calculations. The database offers two different liquid phases for the chemical form and the solubility of sulfur, denominated as slag A and slag B. Slag A involves the solubility of sulfide whereas slag B comprises the sulfate solubility (SO_3). Slag A is recommended for high Na_2O-levels and was used for the calculations, as slag B has the additional disadvantage that no chromium is included in the optimized dataset for the liquid phase. The following oxides were incorporated in the calculations: Na_2O, MgO, SiO_2, Al_2O_3, SO_3, CaO, Cr_2O_3 and Fe_2O_3. Phosphorus and barium oxide were neglected.

The calculated main phases of the unused brick are monoxide (Mg, Ca, Fe)O and spinel (Mg, Fe,)(Fe, Al, Cr)$_2$O$_4$, additionally olivine ((Mg, Ca, Fe)$_2$ SiO$_4$) appears in minor amounts. The solidus temperature (i.e., the temperature of first formation

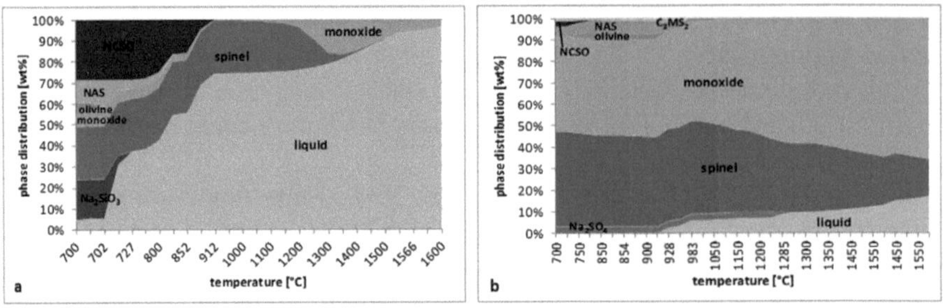

Fig. 4 Temperature dependent phase distribution calculated with FactSage for reducing conditions. Hot face (**a**). Brick infiltration depth 0–25 mm. Cold side (**b**). Brick infiltration depth 200 mm. Meaning of abbreviations: [(Na_2, Ca)$Na_2Ca_3SiO_9$] NCSO. [$NaAlO_2$–$NaAlSiO_4$] NAS. [$Ca_3MgSi_2O_8$] C_3MS_2. [(Mg, Fe, Ca)O] monoxide. [(Mg, Ca, Fe)$_2SiO_4$] olivine. [(Mg, Fe)(Al, Fe, Cr)$_2O_4$] spinel

of liquid phases), is 1416 °C. The phase composition at the hot face and cold side are shown in Fig. 4. Besides the original brick phases, the thermodynamically stable phases are a sodium-calcium silicate named NCSO [(Na_2, Ca) Na_2Ca_3 SiO_9], a phase named NAS (solid solution $NaAlO_2$–$NaAlSiO_4$) and Na_2SiO_3. As expected, at the hot face a high amount of liquid phase is stable at 1000 °C, 70% of the considered system is molten. With increasing infiltration depth, the amount of liquid phase decreases and the solidus temperature shifts to higher temperatures. At 1000 °C, the amount of liquid phase decreases to 6 wt%, which is at an infiltration depth of 200 mm (cold side).

The thermodynamic calculations show a good agreement with the practical investigations, as also determined in the mineralogical investigations, the stable phases at the hot face are: olivine and different types of Na-silicates (Na–Al-silicate, Na-silicate, Na–Ca-silicates). At the cold side additionally sodium sulfate is stable. The dissolution of chromium in Na-sulfate that was found in the mineralogical investigations cannot be described, due to the lack of thermodynamic data.

Conclusions

The main wear mechanism of the silicate bonded magnesia-chromite brick from the lead recycling furnace is chemical attack due to the CaO, BaO and sulfur bearing silicate slag, as well as high Na_2O supply from soda. This resulted in a deep reaching infiltration and corrosion of the brick microstructure—not only of the sintered magnesia, but also of the chromite. Additionally, the primary and secondary chromite precipitations within the sintered magnesia, which are usually typical for this brick brand, could not be observed anymore. This can be explained by a reduction of the chromite precipitations, possibly caused by the reducing atmosphere during the furnace operation.

Obviously, a high soda supply and a surplus of sodium (i.e., not saturated by sulfur) resulted additionally in corrosion of the chromite and the interstitial phase within and between, respectively, the sintered magnesia. Consequently, such deeply infiltrated and corroded brick microstructure is highly susceptible to crack formation and spalling, especially in case of discontinuous (i.e., batch-wise) furnace operation.

The thermochemical calculations are in good agreement with the mineralogical investigations and also show the formation of different types of Na-silicates and Na-sulfate due to slag infiltration, as well as high amounts of liquid phase in the infiltrated microstructure corresponding to deep-reaching infiltration.

For all metallurgical processes and vessels, adequate refractory linings have to be chosen in order to guarantee safe furnace operation and satisfactory furnace lifetime with short downtimes for repairs. In order to consider all these aspects, holistic refractory solutions are required: starting with proper material selection, functional lining concepts for the individual process and refractory engineering, and also taking into account a time-saving and safe installation methods. Furthermore, to ensure future improvements to the refractory products and hence production processes, post-mortem investigations of used refractory materials from production vessels help to understand the wear mechanisms and process challenges.

References

1. Recycling Rates of Metals—A Status Report. Available online: http://www.unep.org/resourcepanel/Portals/24102/PDFs/Metals_Recycling_Rates_110412-1.pdf (May 2011)
2. R.D. Prengaman, Secondray lead production—a survey of smelters and refineries, in *Lead & Zinc '05. Proceedings on the International Symposium on Lead and Zinc Processing* (Kyoto, Japan, 2005), pp. 73–87
3. A. Spanring, Refractory application in the in the secondary lead industry. RHI Bull **1**(56), 56–62 (2010)
4. G. Routschka, *Refractory Materials*, 4th edn. (Vulkan-Verlag, Essen, 2011), p. 505
5. E. Prestes, A.S.A. Chinelatto, W.S. Resende, Post mortem analysis of burned magnesia-chromite brick used in short rotary furnace of secondary lead smelting. Ceramica **55**(61), 61–66 (2009)
6. L. Wei, *Corrosion of Refractories in lead smelting reactors.* University of British Columbia, MSc thesis, 2000
7. L. Scheunis, A.F. Mehrjardi, M. Campforts, P.T. Jones, B. Blanpain, E. Jak, The effect of phase formation during use on the chemical corrosion of magnesia-chromite refractories in contact with a nonferrous $PbSiO_2$ based slag. J. Eur. Ceram. Soc. **34**(1599), 1599–1610 (2014)
8. D. Gregurek, K. Reinharter, C. Majcenovic, C. Wenzl, A. Spanring, Overview of wear phenomena in lead processing furnaces. J. Eur. Ceram. Soc. **35**(1684), 1693–1698 (2015)
9. C.W. Bale et al., FactSage thermochemical software and databases. Calphad **26**(2), 189–228 (2002)

Synthesis of Carbide Ceramics via Reduction of Adsorbed Anions on an Activated Carbon Matrix

Grant C. Wallace, Jerome P. Downey, Jannette Chorney, Alaina Mallard and David Hutchins

Abstract Current commercial processes for producing ceramic carbides require temperatures of 1400–2000 °C and often use extensive milling operations to produce a powder product. A process that could reduce the energy requirements of commercial carbide production could allow for these materials to be implemented in a greater number of applications. In this study, tungstate (WO_4^{2-}) and silicate (SiO_3^{2-}) anions were adsorbed onto activated carbon and converted into silicon carbide (SiC) whiskers and a mixture of tungsten and tungsten carbide (W/WC) crystals via carbothermal reduction using inert and reducing gas atmospheres at temperatures much lower than what is required by current commercial processes (950 °C for W/WC/W_2C and 1200 °C for SiC). The adsorption process was statistically-optimized via a central composite response surface analysis using DesignExpert 9. Inductive coupled plasma optical emission spectroscopy (ICP-OES) was used to measure, and optimize, adsorption efficiency while the carburization products were characterized using X-ray diffraction and scanning electron microscopy.

Keywords Tungsten carbide · Silicon carbide · Carburization · Adsorption optimization

Introduction

Ceramic carbides have been valued as industrial materials due to their high hardness values and their resistance to degradation under extreme process conditions. Tungsten carbide and silicon carbide are both examples of these extremely durable ceramics and are commonly used in a number of industrial applications such as

G.C. Wallace (✉) · J.P. Downey · J. Chorney · A. Mallard · D. Hutchins
Montana Tech of the University of Montana, 1300 W. Park St., Butte, MT 59701, USA
e-mail: gwallace@mtech.edu

J.P. Downey
e-mail: jdowney@mtech.edu

J.-Y. Hwang et al. (eds.), *8th International Symposium on High-Temperature Metallurgical Processing*, The Minerals, Metals & Materials Series,
DOI 10.1007/978-3-319-51340-9_6

cutting tools, nozzles, and drill bits [1]. In addition to these applications, the electrical properties of silicon carbide have made SiC a promising material in the production of high-temperature electronic devices [2] while nanocrystalline tungsten carbide is a possible alternative to platinum based catalysts in many chemical processes.

Current commercial processes for producing tungsten carbide, silicon carbide, and other ceramic carbides use temperatures ranging between 1400 and 2000 °C [3] and, in the case of tungsten carbide, require extensive milling in order to produce particles approaching nano-scale [1]. Alternative ceramic carbide synthesis methods, such as the spray conversion process, have been implemented to improve upon the conventional methods of producing tungsten carbide [4].

Tungsten, in the form of tungstate anions, has been found to adsorb well to activated carbon from acidic solutions in the presence of sodium chloride [5, 6]. However, these studies provide little information on the optimal conditions for adsorbing tungsten anions for use as a carbide precursor.

The use of aqueous tungsten solutions to form tungsten carbide has been reported in the production of substrate-mounted tungsten carbide [7] and WC-Co alloys [8]. Carbon substrates, in the form of pine wood [9] and carbon nanotube sheets [10], were used to synthesize silicon carbide nanostructures via adsorption of silicate anions and reduction under an Ar (1400 °C) or an Ar/H_2 (1300–1500 °C) atmosphere.

The Metallurgical and Materials Engineering Department at Montana Tech of the University of Montana is currently investigating a method for synthesizing nanocrystalline ceramic carbides through the adsorption of anions on an activated carbon matrix and subsequent carburization under inert and reducing gas atmospheres. To date, this research has focused on the synthesis of tungsten carbide and silicon carbide in an attempt to reduce the energy requirements of current commercial operations, and to statistically optimize the parameters for preparing ceramic carbide precursor materials.

Experimental Procedure

Anion Adsorption

Tungstate anions were adsorbed onto activated carbon (Sigma-Aldrich −100/+400 Mesh) by dissolving sodium tungstate dihydrate (Na_2WO_4 * $2H_2O$) in 100 mL of deionized water. Sodium chloride was also added so that the solution possessed a NaCl concentration of approximately 0.2 M. Hydrochloric acid (HCl) was then used to bring the solution pH to 2.0. Once the pH of the solution was established, a 2.5 g sample of activated carbon was added to the solution and the solution(s) was then placed on an orbital shaker table and allowed to mix at 480 rpm. The duration of mixing time varied between 1 and 24 h with each adsorption experiment. Following mixing, the activated carbon was separated from the solution by vacuum filtration. Samples of the remaining solution were collected for ICP-OES/ICP-MS analysis and the activated carbon was placed in a drying oven to remove any moisture in preparation for carburization. A series of six scoping experiments was completed in this manner to establish upper and lower boundary conditions for tungstate concentrations for an experimental design matrix that was used to optimize tungstate anion adsorption.

A similar procedure was developed for the synthesis of a silicon carbide precursor material. Sodium metasilicate (Na_2SiO_3) was dissolved in 100 mL of deionized water to produce silicate solutions of varying concentrations: 0.1, 1, 2, 5, 10, 20, and 50 g/L. Sodium chloride was added to half of the planned scoping experiments to determine its effect on silicate adsorption. A series of 18 scoping experiments was performed to assess the effect of pH on silicate adsorption onto activated carbon. Solutions were prepared at pH 2.0, 6.0, and the ambient silicate solution pH (11.6–11.9). A 2.5 g sample of activated carbon was added to each solution and the solutions were agitated at 480 rpm on an orbital shaker table and allowed to mix for a series of reaction times (1, 12, or 24 h). In a manner similar to the tungstate adsorption experiments, the carbon samples were separated from the remaining solution by vacuum filtration. HCl was used to lower the pH of the silicate solutions, but experiments at higher concentrations were inhibited by the formation of silica gel as the HCl additions induced a rapid precipitation of the silicate ions in sample solutions that contained more than 0.1 g/L sodium metasilicate. A series of 12 additional scoping experiments was performed at the ambient pH of the silicate solutions to determine the effect of increasing silicate concentrations on adsorption. Half of these experiments were carried out for 12 h and the remaining six experiments were carried out an adsorption time of 24 h. For all of the silicate adsorption experiments, the solutions were analyzed via ICP-OES.

Adsorption Optimization and Scale-up Experimentation

The adsorption of tungstate anions onto activated carbon was optimized using a response surface experimental design matrix consisting of 30 individual experiments, six of which were established as mid-points. This design matrix, prepared using the statistical analysis software, DesignExpert9, analyzed the effects of temperature (°C), pH, adsorption time (h), and initial tungsten concentration (mg/L) on tungsten adsorption which was measured in grams of adsorbed tungsten per gram of activated carbon (gW/gC). For this design matrix, Temperature was held at 20, 40, and 60 °C, pH was held at approximately 2, 4, and 6, time was held at 1, 12.5, and 24 h, and initial tungsten concentration was held at approximately 1000, 10,500, and 20,000 mg/L. The amount of tungsten adsorption was determined using ICP-OES.

Conditions for optimizing tungstate anion adsorption were determined using the statistical model produced by DesignExpert 9 (StatEase Inc.). Solutions containing 18 g/L W were prepared in 2 and 4 L batches for loading 50 and 100 g charges of activated carbon respectively. The carbon charges were mixed into the solution and allowed to interact with the tungstate solution for 2 h before being filtered and dried. These optimized larger-scale batches of activated carbon were then used as tungsten carbide precursor material in the tungsten carbide synthesis experiments performed under reducing gas mixtures.

Ceramic Carbide Synthesis

Initial carburization experiments were conducted using an inert argon atmosphere in a MTI GSL 1100X tube furnace. Samples (1.25 g) of tungsten-loaded activated carbon were placed in ceramic boats and carburized at 1000 °C under an inert argon atmosphere (250 mL/min) for a reaction time of 20 h. Further inert atmosphere carburization experiments were carried out at 1400 °C for 20 h using a MTI GSL 1500X tube furnace.

Carburization experiments were also conducted using reducing gas atmospheres consisting of a mixture of H_2, CH_4, and CO. These carburization experiments were conducted using an ATS rotary furnace with a quartz kiln body, and the gas mixtures were controlled with Omega flowmeters. Initial reducing gas carburization experiments consisted of four individual experiments (RGC-1 to RGC-4). The composition of the gas mixture used in these experiments was set at approximately 60% H_2, 25% CO, and 15% CH_4 by volume at a flow rate of 0.5 L/min. Experiments were carried out at this gas composition at both 850 and 950 °C for 6 and 8 h. The tungstate-loaded activated carbon charges (approx. 50 g) were heated to the specified temperature under a 0.5 L/min flow of argon purge gas.

The composition of the reducing gas mixture was altered in an attempt to improve the conversion of tungstate to tungsten carbide (RGC-9). A sample of precursor material weighing 27.22 g was added to the rotary kiln and heated to 950 °C at a rate of 6 °C/min under argon gas. Once the inside of the kiln reached the desired temperature, a reducing gas mixture consisting of 40% H_2, 35% CO, and 25% CH_4 by volume was introduced to the furnace and the precursor was heated at this temperature for 8 h.

Carburization of the silicon carbide precursor was conducted in a MTI GSL 1500X tube furnace using both argon and H_2 as inert and reducing gas atmospheres. Samples of silicate-loaded activated carbon weighing approx. 1.25 g were and heated under atmospheres of argon or H_2 at flow rates of 0.5 L/min. Inert gas carburization experiments were conducted at 1400 °C for 1, 4, 12, and 24 h. Reducing gas carburization experiments were carried out over a range of temperatures between 1100 and 1400°C for 1, 4, and 6 h. The H_2 gas atmosphere was introduced to the tube furnace at a flowrate of approx. 0.5 L/min.

Following carburization, all samples were ground to −325 Mesh and packed onto a powder XRD slide. A Rigaku Ultima IV X-ray diffractometer using Cu-Kα radiation at 40 kV and 40 mA was used to analyze the samples. In addition to XRD, the samples were also analyzed using an FEI tabletop SEM to obtain images of the carburized products.

Results and Discussion

Adsorption

The results from the silicate adsorption scoping experiments are provided in Figs. 1 and 2. From Fig. 1, it is evident that the addition of NaCl has little to no effect on the adsorption of silicate anions from solution. This finding differs from the behavior observed in tungstate anions [6]. Silicate adsorption also decreased as the adsorption time was increased beyond 12 h and increased significantly with increasing pH.

From Fig. 2, silicate adsorption was found to increase with increasing initial silicate concentration in solution. Again, silicate adsorption was found to decrease when adsorption was allowed to continue beyond 12 h of adsorption time. This behavior could be due to silicate anions desorbing as the activated carbon is allowed to interact with the solution for longer than 12 h.

A three-dimensional response surface model of tungsten adsorption was developed using the experimental design matrix prepared using DesignExpert9. Tungsten adsorption was modelled using an inverse square root transform and a modified quadratic relationship. The model for tungstate adsorption can be expressed in the form of an equation which incorporates all of the statistically significant variables (Eq. 1):

$$\begin{aligned}\frac{1}{\sqrt{W}} &= 0.034320 - 3.09986 \times 10^{-6}A + 9.72219 \times 10^{-6}B \\ &\quad + 6.50037 \times 10^{-8}AB + 8.4568 \times 10^{-11}A^2\end{aligned} \tag{1}$$

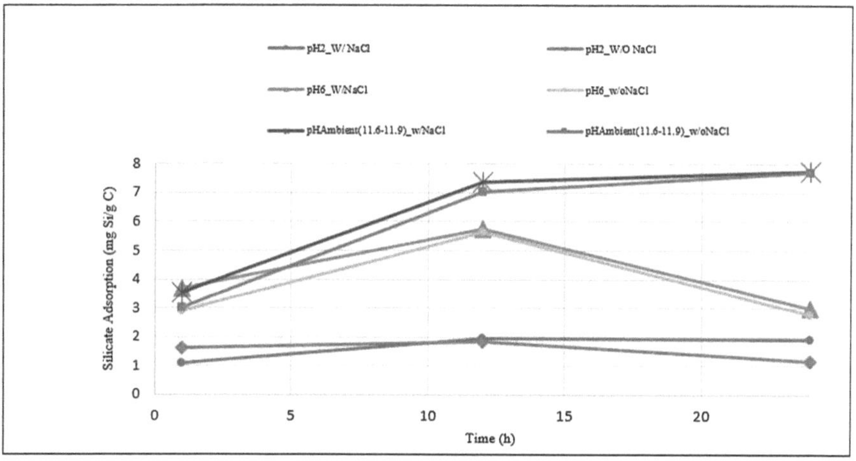

Fig. 1 Silicate adsorption scoping experiments measuring the effect of time and pH

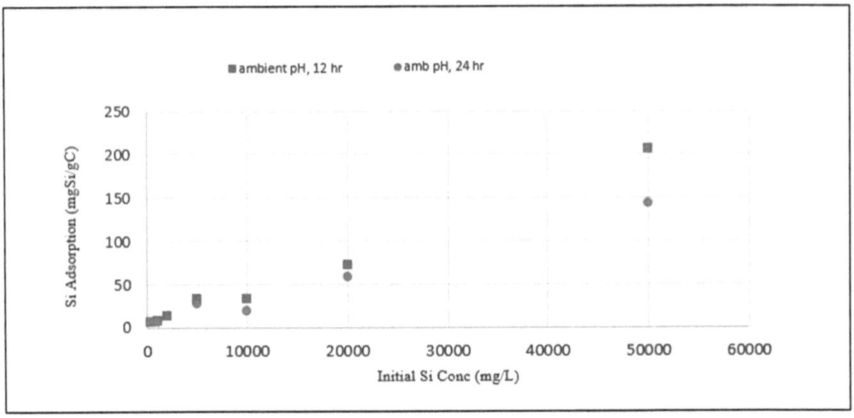

Fig. 2 Silicate adsorption Scoping experiments measuring the effect of initial Si concentration on adsorption

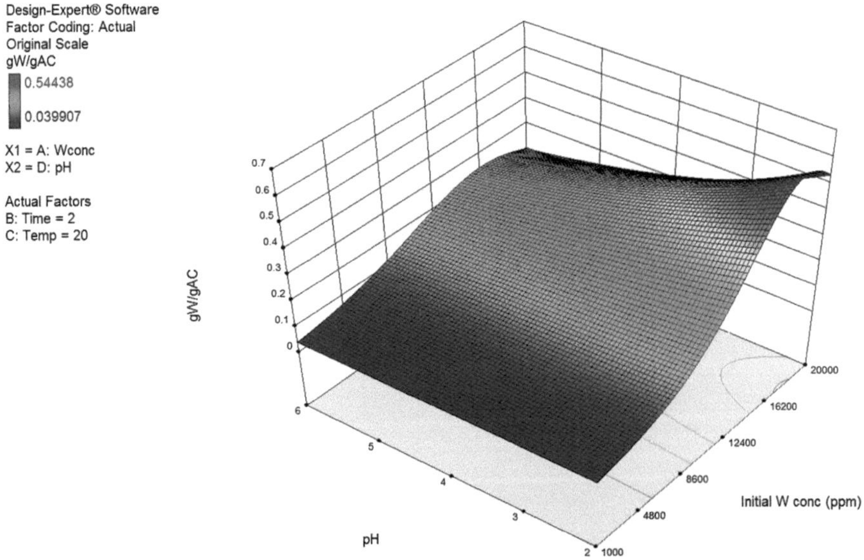

Fig. 3 Response surface model of tungsten adsorption

where, *W* represents the amount of tungsten loaded onto activated carbon (g W/g C), *A* represents initial tungsten concentration (mg/L), and *B* represents the initial solution pH. The experimental data, revealed that adsorption time and solution temperature were not statistically significant variables.

Figure 3 shows the adsorption behavior of tungsten in response to changes in pH and initial tungsten concentration. Adsorption time was held constant at 1 h and temperature was fixed at 20 °C.

From the design matrix data, tungstate anion adsorption improves as pH decreases when the initial tungsten solution concentration remains between 17,000 and 19,000 mg/L. In order to maximize tungstate adsorption, optimal parameters were determined to be a solution pH of 2, an initial tungsten concentration of 18,000 mg/L, an adsorption time of 2 h, and a solution temperature of 20 °C. These conditions were used for the scaled-up adsorption experiments to produce 50 and 100 g batches of precursor material for the carburization experiments. ICP-OES results from these scaled-up adsorption experiments determined that 0.35 gW/gC was achievable under these conditions.

Carbide Synthesis

Examples of XRD patterns for carburization products from inert and reducing gas carburization experiments are provided in Fig. 4. Figure 4a shows experiment WC-12 which involved the carburization of the precursor material under argon gas at 1400 °C for 20 h. Figure 4b shows experiment RGC-9, in which the precursor material was carburized for 6 h under a reducing gas mixture consisting of 40% H_2, 35% CO, and 25% CH_4.

A comparison of the XRD results from inert and reducing gas carburization experiments revealed that the reducing gas carburization experiments produced greater amounts of WC in the carburization product. Carburization was also achieved at lower temperatures and a lower reaction time.

Silicon carbide was detected in silicate-loaded carbon samples carburized at 1400 °C for as little as 1 h under argon gas. Under H_2 atmospheres, silicon carbide

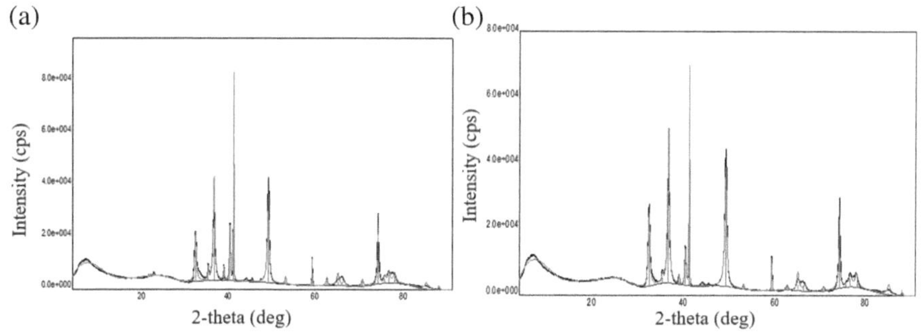

WPPF Quantitative Analysis (Rel. Wt %)			
Phase	RGC8-1	RGC8-2	RGC8-3
WC	70.7	67.5	63.3
W	17.0	20.7	23.0
W_2C	12.3	11.8	13.7

WPPF Quantitative Analysis (Rel. Wt %)			
Phase	RGC9-1	RGC9-2	RGC9-3
WC	71.1	75.6	74.1
W	19.6	7.2	17.6
W_2C	9.3	17.2	8.3

Fig. 4 XRD results for inert gas carburization (**a**) and reducing gas carburization (**b**) of tungstate-loaded carbon precursor

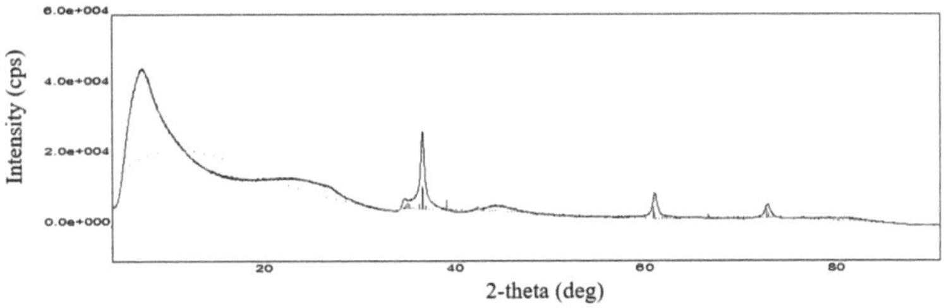

Fig. 5 XRD pattern for SiC synthesized on the activated carbon matrix (1400 °C, 20 h, Ar)

Fig. 6 SEM images of mixed WC/W_2C/W crystals (*left*) and silicon carbide whiskers (*right*) formed on an activated carbon matrix

was detected in precursor samples carburized at 1200 °C for 6 h. An example of the XRD pattern produced by the silicate carburization product is provided in Fig. 5.

An example of the analysis of the carburized silicate-loaded precursor material by SEM is shown in Fig. 6.

Figure 6 shows the formation of silicon carbide whiskers on the activated carbon matrix. These silicon carbide whiskers were observed in all samples of carburized silicate-loaded precursor material.

Conclusions

A mixture of WC, W and W_2C crystals was synthesized by adsorption of WO_4^{2-} anions onto activated carbon and carburization under argon gas at 1400 °C for 20 h. Mixtures of WC, W_2C and W crystals were formed using a reducing gas atmosphere (40% H_2, 35% CO, and 25% CH_4) at temperatures as low as 850 °C, but with improved WC conversion at 950 °C. The use of the reducing gas mixture significantly lowered the energy requirement for the tungsten carburization reaction in this system and also reduced the production of metastable W_2C. Silicate anion adsorption was found to improve with increasing pH and initial silicate concentration, and silicon carbide whiskers were formed under H_2 at temperatures as low as 1200 °C and a reaction time of 6 h. Optimized conditions for tungstate anion adsorption were established and implemented in scaled-up adsorption experiments which yielded tungsten loadings at approximately 0.35 g W/g C. Although a significant amount of tungsten was loaded onto the activated carbon charges, this value is lower than what was predicted by the design matrix for the optimal conditions. Scale-up of the adsorption process may have induced elements into the system that may have slowed adsorption kinetics. Future work will be directed toward maximizing WC production efficiency and the development of methods to separate carburized products from the remaining carbon matrix.

Acknowledgements Research was sponsored by the Army Research Laboratory and was accomplished under Cooperative Agreement Number **W911NF-15-2-0020**. The views and conclusions contained in this document are those of the authors and should not be interpreted as representing the official policies, either expressed or implied, of the Army Research Laboratory or the U.S. Government. The U.S. Government is authorized to reproduce and distribute reprints for Government purposes notwithstanding any copyright notation herein.

References

1. M.F. Zawrah, Synthesis and characterization of WC-Co nanocomposites by novel chemical Method. Ceram. Int. **33**, 155–161 (2007)
2. J.W. Palmour et al., 6H-silicon carbide devices and applications. Physica B **185**(1–4), 461–465 (1993)
3. G.S. Upadhyaya, *Cemented Tungsten Carbides: Production, Properties and Testing* (Noyes Publications, Westwood, New Jersey, 1998), pp. 77–79
4. P. Seegopaul, L.E. McCandlish, F.M. Shinneman, Production capability and powder processing methods for nanostructured WC-Co powder. Int. J. Refract Metal Hard Mater. **15**, 133–138 (1997)
5. J.J. Cruywagen, A.T. Pienaar, The adsorption of tungsten(VI) on activated carbon from 1.0 M Na(H)Cl solution. Polyhedron **8**(1), 71–76 (1989)
6. K. Ohashi, K. Murakami, K. Yamamoto, Adsorption behavior of tungsten (VI) onto activated carbon. Jpn. Soc. Anal. Chem. **32**, 313–319 (1983)
7. S. Xiaoliang, Y. Hua, S. Peng, S. Gangqin, D. Xinlong, Z. Xiong, Synthesis of multi-walled carbon nanotube-tungsten carbide composites by the reduction and carbonization process. Carbon **45**, 1735–1742 (2007)
8. T. Danny Xiao, T. Xinglong, Y. Maozhong, P. Shigao, P. Fangcai, Y. Jiangao, D. Yu, Synthesis of commercial-scale tungsten carbide-cobalt (WC/Co) nanocomposite using aquesous solutions of tungsten (W), cobalt (Co), and carbon (C) precursors. J. Mater. Sci. Chem. Eng. **2**(7), 1–15 (2014)
9. K. Vyshnyakova, G. Yushin, L. Pereselentseva, Y. Gogotsi, Formation of porous SiC ceramics by pyrolysis of wood impregnated with silica. Int. J. Appl. Ceram. Technol. **3**(6), 485–490 (2006)
10. M. Ding, A. Star, Synthesis of one-dimensional SiC nanostructures from a glassy buckypaper. ACS Appl. Mater. Interfaces **5**(6) (2013)

Part II
Simulation of High Temperature Process

A CFD Based Algorithm for Kinetics Analysis of the Reduction of Hematite Concentrate by H_2 + CO Mixtures in a Drop Tube Reactor

De-Qiu Fan, M. Elzohiery, Y. Mohassab and H.Y. Sohn

Abstract A novel flash ironmaking process is under development at the University of Utah. A computational fluid dynamics (CFD) approach, coupled with experimental results, was developed for the analysis of the kinetics of reduction of hematite concentrate by H_2 + CO mixtures at 1150–1350 °C in a drop tube reactor. Rate expressions of hematite concentrate reduction by single gases (H_2 or CO) were used in the analysis of the experimental data under H_2 + CO mixtures. Synergistic effects, which were found in the H_2 + CO mixture experiments compared with simple summation of contributions of component gases, was considered in this paper. The gas phase was treated as a continuum in the Eulerian frame of reference, and the particles are tracked using a Lagrangian approach in the evaluation of the particle residence time and temperature profile inside the reactor. The calculated reduction degrees based on this approach are in good agreement with the experimental values.

Keywords Flash reaction · Concentrate particle · Reduction kinetics · Ironmaking · CFD

Introduction

The blast furnace process has produced more than 90 pct of primary iron until 2014 according the World Steel Association, with the balance by alternate processes such as direct-reduced iron (DRI) and smelting reduction. Although the blast furnace has a high production rate and other advantages, it suffers from high energy consumption and CO_2 emissions. A novel ironmaking technology is under development at the University of Utah [1–7]. In this process, iron is produced from iron oxide concentrate in a flash reactor, utilizing hydrogen, natural gas, or coal gas as

D.-Q. Fan · M. Elzohiery · Y. Mohassab · H.Y. Sohn (✉)
Department of Metallurgical Engineering, University of Utah, Salt Lake City, UT 84112, USA
e-mail: h.y.sohn@utah.edu

J.-Y. Hwang et al. (eds.), *8th International Symposium on High-Temperature Metallurgical Processing*, The Minerals, Metals & Materials Series,
DOI 10.1007/978-3-319-51340-9_7

the reducing agent as well as a fuel. This new Flash Ironmaking Technology (FIT) eliminates the cokemaking and pelletization/sintering processes associated with the traditional blast furnace process, significantly reducing the CO_2 emissions and energy consumption compared with the blast furnace process.

A thermogravimetric analysis (TGA) system and a drop tube reactor (DTR) are the two often used experimental apparatuses for studying the kinetics of gas-solid reactions. TGA is usually used when the time range of the gas-solid reaction is in the order of minutes or longer. For fast gas-solid reactions that are completed in only a few seconds, however, the DTR is typically used. In this work, the reduction of hematite concentrate particles by H_2 + CO mixtures was investigated. This reaction was completed within a few seconds in the temperature range 1423–1623 K. In order to analyze the overall kinetics, the reduction of by pure H_2 or CO gas was investigated separately in previous work [5, 7]. Rate expressions of hematite reduction by single reducing gases were obtained using a CFD based algorithm in this work. These rate expressions were then used to analyze the experimental data obtained with H_2 + CO mixtures.

Temperature measurements in the reactor during the experiments in this work as well as CFD simulations found that there was a short, low-temperature region near the tip of the water-cooled tube through which the carrier gas and particles were injected at room temperature. The presence of this region affects the temperature and flow of the particles. Direct measurement of particle temperature and real-time tracking of particle inside the DTR would require sophisticated and intrusive experimental techniques, e.g. optical access to the DTR interior and particle image velocimetry (PIV). Moreover the use of H_2 and/or CO gas at high temperatures adds an extra complexity. The use of CFD to incorporate such details allows realistic calculations of particle temperature and residence time as functions of position and improves the accuracy of the kinetics analysis. The purpose of this work was to develop a reliable rate expression of the gaseous reduction of hematite concentrate particles that would be useful in the design of an industrial flash ironmaking reactor.

Experimental Work

A schematic representation of the DTR is shown in Fig. 1. The reactor consisted of a vertical split tubular furnace with a maximum working temperature of 1813 K and a cylindrical alumina tube with an inner diameter of 5.6 and 193 cm in length. The isothermal zone for all the experiments was maintained at a temperature between 1423 and 1623 K by bar-type SiC heating elements. In general, the typical temperature profile exhibited a long isothermal zone with short lower temperature zones at the two ends of the reactor. A cylindrical alumina honeycomb was installed at the top part of the reactor. The particles together with N_2 as the carrier gas were fed from the water-cooled powder feeding tube inserted through the center of the alumina honeycomb. Reducing gases like H_2, CO or a H_2 + CO mixture at room temperature were fed from through the alumina honeycomb to enter the reaction

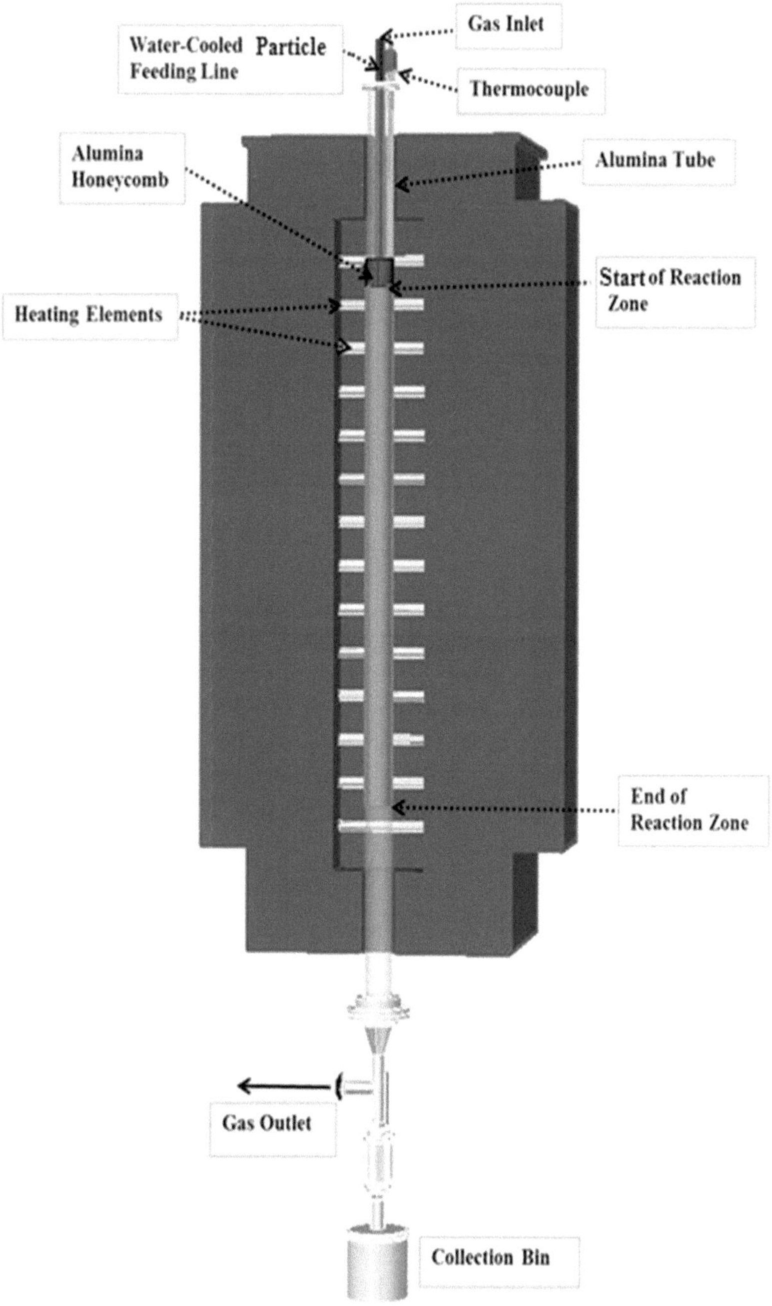

Fig. 1 Schematic representation of the laminar flow reactor (DTR)

zone. The main purpose of the honeycomb was to straighten gas flow and improve the heating of the reducing gases before they entered the reaction zone. Wide ranges of experimental conditions of hematite reduction by H_2, CO and H_2 + CO were tested. Dried hematite concentrate particles were delivered continuously with a carrier gas (N_2) at a constant flow rate. The average particle size of the concentrate particles was 21.3 μm. In the flash ironmaking process, the particle residence time in the reactor is expected to be between 2 and 10 s. The gas flow rates were adjusted in this work to provide these residence times while allowing desired reduction degrees depending on other experimental conditions such as temperature and gas composition. During the experiment, the particle feeding rate was controlled in the range of 30–100 mg/min. The mass ratio of particle to gas was usually kept at <0.05. The reacted particles were collected in a collector at the bottom of the reactor.

Mathematical Model

Governing Equation

The Euler-Lagrange approach was used to model the two-phase flow, in which the gas phase was treated as continuum in the Eulerian frame of reference while the solid phase was tracked in the Lagrangian frame of reference. For the gas phase, the realizable *k*-ε model [8] was chosen for simulating the spread of the carrier gas jet. Radiation from the wall was taken into account using the discrete ordinate (DO) model [9]. The general steady state transport equation is given by

$$\nabla \cdot \rho \vec{u} \phi - \nabla \cdot \Gamma_\phi \nabla \phi = S^\phi \tag{1}$$

where ϕ represents the generic dependent variable, Γ_ϕ is the effective transport coefficient, and S^ϕ is the source term.

As the volume fraction of the solid phase in the reactor is of the order of 10^{-6}, the effects of particle collisions were neglected. The particle addition at this low level does not significantly affect the gas phase velocity [10]. The particles were mostly concentrated in a narrow region near the reactor centerline as observed by a PIV system. Therefore, the centerline velocity and temperature of the gas phase from the CFD results were used to track the particle. In this sense, only a one-way coupling approach was needed in tracking the concentrate particles. The solid phase governing equations are listed in Table 1. The reduction degree in Eq. (4) is in terms of oxygen removal due to the fact that different phases of iron oxides coexist in a particle during the reduction of hematite. For all the experiments, the feed rates of the reducing gases (H_2, CO or H_2 + CO mixture) were in more than 500% excess of the minimum required amounts of reducing gases to ensure that the gaseous reactant concentrations remain essentially constant over the entire reactor length for

Table 1 Solid phase governing equations

Particle movement	$\frac{du_{p,i}}{dt} = F_D\left(u_i - u_{p,i}\right) + \frac{g_i\left(\rho_p - \rho\right)}{\rho_p}$ (2)
Energy	$m_p c_{p,d} \frac{dT_p}{dt} = hA_p\left(T - T_p\right) - \frac{dm_p}{dt} \Delta H_{reac} + \varepsilon_p A_p \sigma\left(T^4 - T_p^4\right)$ (3)
Oxygen removal	$\frac{dX}{dt} = k_{oj} \exp\left(-\frac{E_j}{RT}\right) \cdot \left(p_j - p_{jO}/K_{ej}\right) \cdot (1 - X)$; $j = H_2$ or CO; Eq. (5) for $H_2 + CO$ (4)

accurate determination of the kinetics, and the amount of water vapor and CO_2 generated by reaction only accounted for less than 2%.

Algorithm

The algorithm used in this work for the kinetics analysis of reduction by single reducing gases is shown in Fig. 2, which has been detailed in Ref. [6].

After obtaining the kinetic parameters of the gaseous reduction of hematite particles by a single component gas, the rate expressions were used for obtaining the rate expression of the reduction by H_2 + CO mixtures. A simple additive relationship between the reduction rates by the component gases was initially assumed, but under H_2 + CO mixtures apparent synergistic effects of the two gases were found. In order to account for the enhanced effects and considering that the reduction is dominated by H_2 at the prevailing conditions, the contribution of H_2 in the rate equation was amplified by the presence of CO in reduction by H_2 + CO mixtures. The contribution of CO was kept the same as in the case of pure CO. This is because at the temperature range investigated most of the reduction work is done by H_2. The coefficients were found to be dependent on the partial pressure of CO and reaction temperature. Therefore, the following rate equation was assumed for reduction by an H_2 + CO mixture:

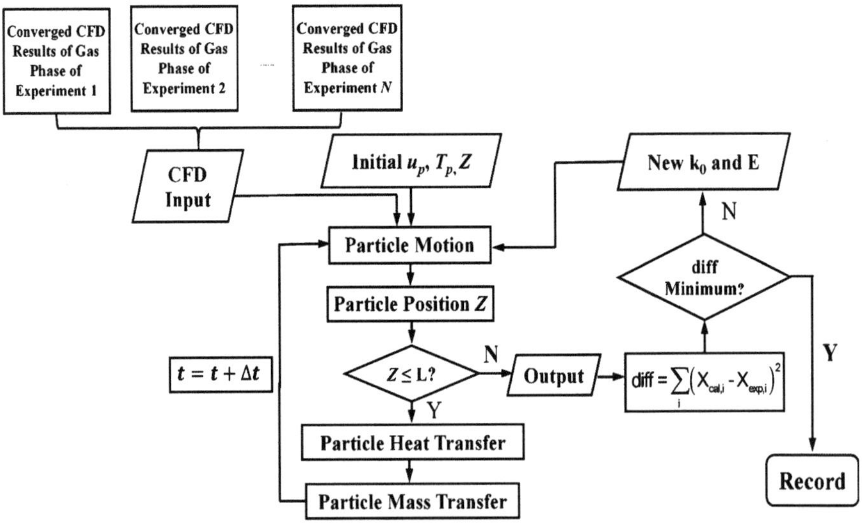

Fig. 2 A block diagram of the algorithm for kinetics parameter optimization

$$\frac{dX}{dt} = [1 + m(T) \cdot \frac{P_{CO}}{P_{CO} + P_{H_2}}] \cdot \left.\frac{dX}{dt}\right|_{H_2} + \left.\frac{dX}{dt}\right|_{CO} \quad (5)$$

where $m(T) = aT + b$.

A series of a and b values were tested against the experimental data to determine the best-fit values.

Results and Discussion

The optimum activation energy and pre-exponential factor for the reduction of hematite concentrate particles singly by H_2 or CO were obtained following the algorithm as shown in Fig. 2. The optimum values are presented in Table 2.

The comparisons between the calculated reduction degrees using the optimum rate parameters and the experimental results are shown in Figs. 3 and 4 for the reduction of hematite concentrate particle by single reducing gas H_2 or CO, respectively. It is noted that the optimum kinetic parameters obtained predict

Table 2 Searching range and optimized values of the kinetic parameters

	Reduction by H_2	Reduction by CO
E range (kJ/mol)	107–321	115–347
k_o range(atm^{-1} s^{-1}) (1 atm = 101.325 kPa)	4410–2.93 × 10^{11}	1910–9.66 × 10^{11}
Optimized values	E = 218, k_o = 8.47 × 10^7	E = 241, k_o = 5.18 × 10^7

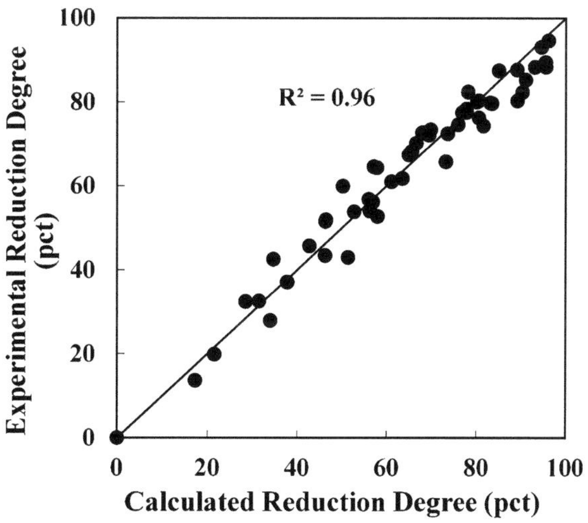

Fig. 3 Comparisons between the calculated and experimental reduction degrees with H_2 as the single reducing gas

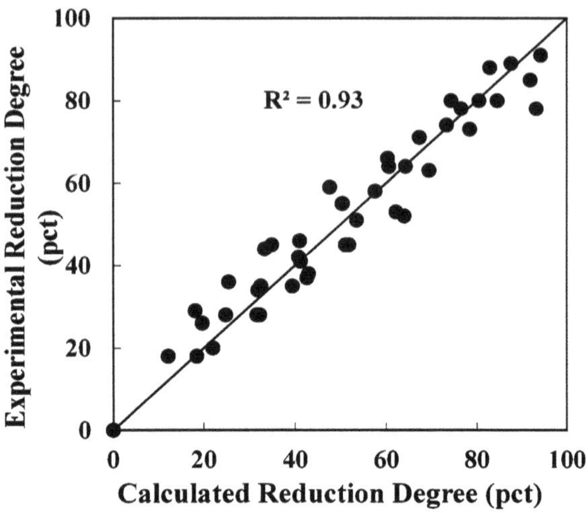

Fig. 4 Comparisons between the calculated and experimental reduction degrees with CO as the single reducing gas

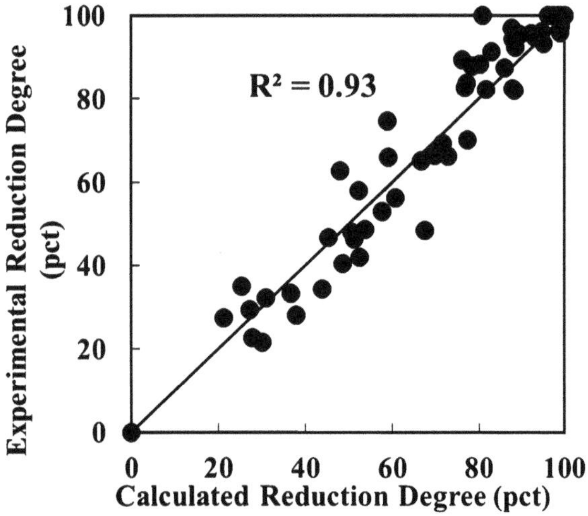

Fig. 5 Comparisons between the calculated and experimental reduction degrees with H_2 + CO as the reducing gas

the reduction degrees very well as the correlation coefficients between the calculated values and corresponding experimental results are higher than 0.93 in both cases.

The values of *a* and *b* that give the best fit to experimental reduction degrees were—0.004 and 7.004 for the H_2 + CO mixture cases. The comparisons between the calculated reduction degrees and the experimental values are shown in Fig. 5.

It is seen from Fig. 5 that the reduction rate by H_2 + CO mixtures can be adequately represented by the following expression:

$$\frac{dX}{dt} = [1 + (-0.004T + 7.004) \cdot \frac{p_{CO}}{p_{CO} + p_{H_2}}] \cdot \left.\frac{dX}{dt}\right|_{H_2} + \left.\frac{dX}{dt}\right|_{CO}; \quad T \text{ in K and } t \text{ in s} \tag{6}$$

with $\left.\frac{dX}{dt}\right|_{H_2}$ and $\left.\frac{dX}{dt}\right|_{CO}$ given by the corresponding expressions of Eq. (4) and rate parameters given in Table 2.

Conclusions

An algorithm for kinetics analysis of the gaseous reduction of hematite particles in a DTR that corrects for the velocity and temperature changes was proposed in this paper. The rate parameters obtained through global search provided an activation energy value of 218 kJ/mol and 241 kJ/mol for the reduction of hematite particle by hydrogen and carbon monoxide, respectively. The synergistic effect of CO on the reduction rate in the case H_2 + CO mixtures was determined to be dependent on CO partial pressure and reaction temperature.

Acknowledgements The authors acknowledge the financial support from American Iron and Steel Institute (AISI) through a Research Service Agreement with the University of Utah under AISI's CO_2 Breakthrough Program. This material also contains results of work supported by the U.S. Department of Energy under Award Number DE-EE0005751.

References

1. H.Y. Sohn, Suspension ironmaking technology with greatly reduced energy requirement and CO_2 emissions. Steel Times Int. **31**, 68–72 (2007)
2. H.Y. Sohn et al., Suspension reduction technology for ironmaking with low CO_2 emission and energy requirement. Iron Steel Technol. (AIST Trans.) **6**, 158–165 (2009)
3. M.E. Choi, H.Y. Sohn, Development of green suspension ironmaking technology based on hydrogen reduction of iron oxide concentrate: rate measurements. Ironmaking Steelmaking **37**, 81–88 (2010)
4. H.K. Pinegar, M.S. Moats, H.Y. Sohn, Flowsheet development, process simulation and economic feasibility analysis for novel suspension ironmaking technology based on natural gas: part 1—flowsheet and simulation for ironmaking with reformerless natural gas. Ironmaking Steelmaking **39**, 398–408 (2012)
5. F. Chen et al., Kinetics of the reduction of hematite concentrate particles by carbon monoxide relevant to a novel flash ironmaking process. Metall. Mater. Trans. B **46**, 1133–1145 (2015)
6. D. Fan, Y. Mohassab, M. Elzohiery, H.Y. Sohn, Analysis of the hydrogen reduction rate of magnetite concentrate particles in a drop tube reactor through CFD modeling. Metall. Mater. Trans. B **47**, 1669–1680 (2016)

7. F. Chen, Y. Mohassab, S. Zhang, H.Y. Sohn, Kinetics of the reduction of hematite concentrate particles by carbon monoxide relevant to a novel flash ironmaking process. Metall. Mater. Trans. B **46**, 1716–1728 (2015)
8. T.-H. Shih et al., A new k-ϵ eddy viscosity model for high reynolds number turbulent flows. Comput. Fluids **24**, 227–238 (1995)
9. E.H. Chui, G.D. Raithby, Computation of radiant heat transfer on a nonorthogonal mesh using the finite volume method. Numer. Heat Trans. B-Fund. **23**, 269–288 (1993)
10. A.Y. Varaksin, *Turbulent Particle-Laden Gas Flows* (Springer, New York, 2007), pp. 22–23

A Continuous Dynamic Process Model to Design a Carbon Profile Toward Yield Improvement

Mohammed A. Tayeb, Narottam Behera and Raja Mathu

Abstract With the struggling steel market nowadays, steel producers are under intense pressure to reduce their operational cost. Therefore, optimization in the melt shops becomes not only preferable, but mandatory. A continuous dynamic process model for scrap-DRI EAF process was developed. The model was then used to propose a new carbon profile, implemented successfully to decrease (%FeO) in the slag and improve the yield. While this work is continued at HADEED, this article reviews the basics of the model and presents a summary of the preliminary plant findings.

Keywords Carbon · EAF · FeO · Yield · Model

Introduction and Background

During the operation of an Electric Arc Furnace (EAF), liquid slag is desirable in order to improve the process and the electrical efficiency. The formation of a liquid slag reduces the effect of arc radiation, slag splashing onto the furnace lining and the water cooled panels, and decreases electrode consumption. By forming a foamy slag, the positive effect of a liquid slag may be further enhanced leading to a higher productivity and lower tap-to-tap time. Slag foaming is normally described as a process with two basic demands, powerful gas generation in the slag combined with a suitable slag viscosity. In the EAF production of steel, oxygen is injected to create a well foaming slag [1]. The iron oxide formed by the oxygen injection will react with injected carbon, reducing the FeO and forming CO gas. It has been reported [2] that roughly 40–50% of FeO losses during production occur in the EAF. This oxidation will not only decrease the Fe recovery, but it will also negatively affect

M.A. Tayeb (✉) · N. Behera
Metals Technology, SABIC, Jubail, Saudi Arabia
e-mail: Tayebmab@sabic.com

R. Mathu
SPDR Division, HADEED, Jubail, Saudi Arabia

J.-Y. Hwang et al. (eds.), *8th International Symposium on High-Temperature Metallurgical Processing*, The Minerals, Metals & Materials Series,
DOI 10.1007/978-3-319-51340-9_8

the overall performance of the melting process. It is therefore important to limit the Fe oxidation from the scrap and DRI. High FeO content in the slag leads to a very fluid slag, which prevents slag foaming and consequently harms the overall performance of the melting process in the EAF. The main purpose of this project was to investigate the causes of high FeO level in the EAF slag and optimize the process to reduce the FeO concentration and improve the metallic yield.

EAF slag practices throughout the industry vary considerably. The slag binary basicity (CaO/SiO_2) in the EAF process usually varies from 1 to 3 and FeO levels ranges between 20 and 50%, while still achieving good slag foaming behavior. Furnace slag viscosity is determined by the ratio of slag components and how those components react to make the slag thinner or thicker. It is the balance of these components that creates the correct slag viscosity to hold the CO bubbles longer. The FeO in the slag is proportional to the oxygen dissolved in the steel. In addition, there is also a relationship between the carbon and oxygen contents in the steel. At low carbon levels, oxygen reacts with iron to form FeO in the slag, which makes the slag thinner, and considerably impacts yield losses [1].

Carbon is injected to the EAF to control the slag foaming and reduce slag FeO content. The reaction between carbon and slag includes four separate steps:

(i) Iron cations transfer from the bulk slag phase to the slag-gas interface,
(ii) CO reacts with oxygen in the slag, $CO + O^{2-} = CO_2 + 2e^-$,
(iii) CO_2 transfer through the gas phase to the surface of carbon, and
(iv) CO_2 reaction with carbon to form CO.

These processes involve gas-liquid, gas-solid and even solid-liquid reactions, as well as mass transfer in the gas and slag phases [3, 4].

The addition of charge carbon with the scrap to achieve the target carbon level in the bath has been widely used as a cost effective way to obtain the required carbon level. Unfortunately, the recovery of carbon from charged carbon is not consistent and varies from heat-to-heat. The amount of charged carbon depends on several factors, including: the carbon content of scrap, the carbon content of DRI, the amount of oxygen injection, the desired tap carbon content, and the economics of iron yield versus carbon cost [1, 4].

In this work, the carbon lump addition practice was reviewed including its benefits to carbon recovery and slag FeO concentration.

Thermodynamic Model

The DRI-EAF process involves continuous feeding of DRI, fluxes, oxygen and carbon, as well as semi-continuous flushing of the slag. We have developed a continuous dynamic process model for the DRI-EAF operation based on a continuous stirred tank reactor (CSTR) model for the slag phase and accumulative batch reactor model for the metal phase [5]. It uses a dynamic mass balance and first

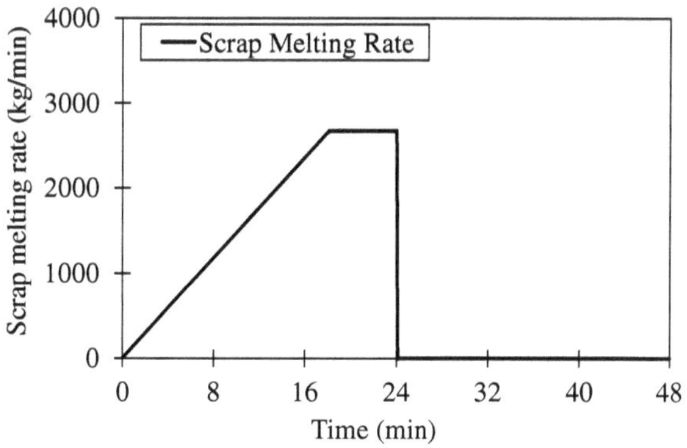

Fig. 1 Input scrap melting profile

order rate equations to compute the slag composition, with the ultimate goal of developing a model that would assist EAF operators. The model is supported by an input plant data and back engineered to calculate the required fitting parameters.

The assumptions involved in the model are:

- Local thermodynamic equilibrium is established at the slag/steel interface,
- The transport of species occurs by mass transfer to the interface,
- The bulk liquid steel and slag remain homogeneous throughout the reaction, and
- Iron oxide activity at the interface is mainly determined by the carbon reaction.

DRI comes in relatively small size and therefore its melting was assumed to be instantaneous. On the other hand, the melting pattern for scrap was assumed considering the energy balance, as shown in Fig. 1. The temperature is an input parameter and the profile is adjusted to match the one observed in industry. The carbon addition/injection, as well as the oxygen and natural gas injections, were adjusted to reflect the observations.

Dynamic Mass Balance

The slag composition is computed using a dynamic mass balance for CaO, SiO_2, MgO, Al_2O_3, MnO and FeO. The mass balance includes the content in the DRI gangue, plus that coming from flux addition, and subtracted from those leaving the furnace when slag is flushed. Scrap and pig iron contain Si, Mn, and Al, which undergo complete or partial oxidation to enter the slag phase. The FeO mass balance is based on the assumption that all oxygen not used for C, Si, Mn, and Al oxidation reacts with iron. For example, the calcium oxide composition is calculated as:

$$m_{SL}^{\text{CaO}}(t+\Delta t) = m_{SL}^{\text{CaO}}(t) + FR^{Lime}(t+\Delta t)\Delta t\frac{(\%\text{CaO})^{Lime}}{100} + FR^{D.Lime}(t+\Delta t)\Delta t\frac{(\%\text{CaO})^{D.Lime}}{100} - (SFR)\Delta t\frac{(\%\text{CaO}_{Slag})^{t}}{100} + FR^{DRI}(t+\Delta t)\Delta t\frac{(\%\text{CaO})^{DRI}}{100} \quad (1)$$

where FR denotes the feeding rate, D.Lime is the dolime and SFR represents the slag flushing rate.

Fe–C–O Reactions

Five reactions govern iron–carbon–oxygen interactions [6–9]. The first reaction is pre-reduction of iron oxide by carbon in the DRI; 1% of carbon in DRI will reduce around 6% of iron oxide. Lancing oxygen in the bath reacts with carbon and iron, according to Eqs. (2) and (3). The carbon reaction is very fast and favorable at high carbon content, while iron oxidation is promoted and predominates, due to its abundance at lower carbon levels. Carbon powder is injected simultaneously into the slag phase, reducing iron oxide, as shown in Eq. (4). Likewise, carbon in the liquid metal diffuses to the slag/metal interface reducing FeO, according to Eq. (5).

$$\underline{\text{C}} + 0.5\text{O}_2 = \text{CO}_{(g)} \quad (2)$$

$$\underline{\text{Fe}} + 0.5\text{O}_2 = (\text{FeO}) \quad (3)$$

$$(\text{FeO}) + \underline{\text{C}}_{dissolved} = \underline{\text{Fe}} + \text{CO}_{(g)} \quad (4)$$

$$(\text{FeO}) + \text{C}_{injected} = \underline{\text{Fe}} + \text{CO}_{(g)} \quad (5)$$

The weight of iron oxide can be calculated by a complex set of kinetic equations for the reduction and oxidation rates of FeO, Fe, and C; this would rely on kinetic parameters, which are not fully known. Instead, simple mass balance—assuming that all oxygen not used for carbon oxidation reacts with iron—was used to predict the iron oxide content of the slag. The carbon feed rate and carbon content of the steel bath are both model inputs.

Model Results

The scrap-DRI EAF model was applied to a one-scrap bucket charge operation furnace with continuous DRI feeding. The furnace has a 180 ton capacity with

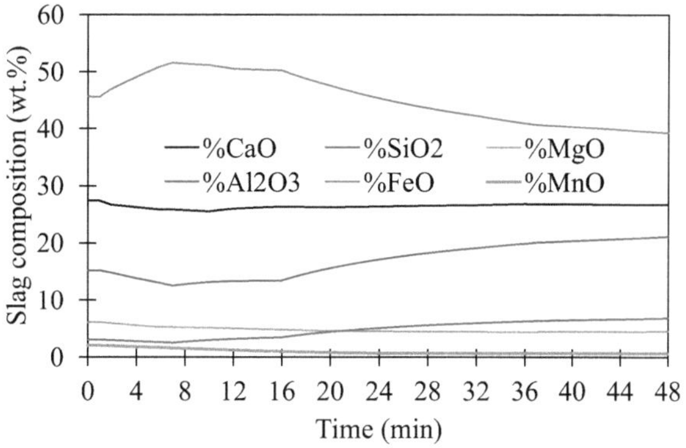

Fig. 2 Model predictions for slag composition evolution over time

around 150 ton tap weight. The pig iron, if present, is charged with the scrap bucket. Fluxes (dolime and lime) are injected continuously, while coke is injected almost throughout the heat for foaming the slag. Natural gas is mostly injected early in the heat to provide energy. Oxygen injection is as well continuous. The slag is semi-continuously discharged from the furnace through a slag door. The model is executed for a heat typical of the current operation at the HADEED SPLPII plant, EAF#06 [10]. Figure 2 shows an example of the result, indicating the slag components compositional evolution over time.

Results and Discussion

Current Operational Practice at HADEED SPLPII

The HADEED steel plant Long product II composes: one electric arc furnace, one ladle furnace and one billet caster, designed to produce 1 million tons of billets per annum. A batch of 150 ton of steel is produced around every 60 min by melting scrap and DRI. During scrap melting, some amount of lump carbon is initially charged to maintain the carbon level of the batch, while a similar amount of carbon is injected to obtain the foamy slag with the aid of oxygen injection during the melting process.

Proposed Changes

The developed model was used to perform a sensitivity analysis. Based on the model predictions, it was recommended to reduce the initial lump carbon by 75%. Additionally, it was recommended to increase the overall carbon injection profile. Comparing the new proposed and current practice, the FeO concentration of the slag should be reduced by around 11%, as per the model predictions.

Trials Results and Discussion

Following the above proposed new carbon profile, several heats were conducted with well controlled parameters. Slag samples were collected and analyzed by X-ray Fluorescence Testing at SABIC Technology Center-Jubail laboratories. The results indicate an overall reduction of FeO concentration of the slag by around 18%. This corresponds to about 1.5 tons of Fe and accordingly the yield should be similarly improved. The reduction trend in the slag FeO concentration agrees well with the model predictions.

Based on the above promising results, an extended trial was proposed and executed for several months. For evaluation, 202 randomly selected heats data were collected after the implementation of the new carbon profile. The followings were observed/noted:

- Although the input DRI quality varies, DRI report shows that all the charged DRI during the analysis period is coming from Module E with relatively similar quality.
- Slag data with unreasonable values, such as very low iron oxide, were removed from the analysis. Similarly, any repeated data in the slag report were removed. For the evaluation, any reported yield above 97% was not considered to be sound and therefore ignored in the comparison.
- Reported slag data might not be representative, because 2–3 slag samples are collected daily and not for each heat. Therefore, one slag sample per heat was collected from several heats before the trial and during the trial for comparison.
- The results show that around one-third of the heats (73 heats) follow the new injection carbon profile and only one-tenth of the heats are consistent with the restricted initial carbon input. The yield for the heats that follow the new carbon profile is improved. The results imply that the new profile is indeed working and the yield is improved (i.e. 2% yield improvement).

Conclusion

This project is intended to reduce the high reported FeO concentration in the slag by proposing a new carbon feeding profile. Preliminary and extended trials both showed very good improvement in the yield and reduced slag FeO concentration by applying the new carbon profile, which considers a lower initial amount of lump carbon and higher injection profile. The proposed changes are based on theoretical and practical considerations by utilizing the developed dynamic process model for EAF#06 at HADEED. The calculated benefits range between $1.3 and $3.4 million per year, which is significant considering that there was only a 50% implementation rate for the new carbon feeding profile. It is therefore recommended to continue utilizing the new carbon profile with a special direction to melt shop operators to follow the new carbon profile with special attention to the initial lump input to be restricted to a maximum of around 30% of the current level. This work is a continuous improvement work at HADEED to optimize the process and reduce the overall operational cost.

Acknowledgements The authors are grateful to the support from HADEED as well as SABIC Metals SBU management.

References

1. R.J. Fruehan, *The Making, Shaping and Treating of Steel* (The AISE Steel Foundation, Pittsburgh, PA, 1998)
2. Private Communication. HADEED
3. Electric Arc Furnace Simulation User Guide (2006). steeluniversity.org
4. E. Pretorius, H. Oltmann, EAF fundamentals, in *LWB Refractories*
5. M.A. Tayeb, Phosphorus Control in DRI-EAF Steelmaking: Thermodynamics, Effect of Alumina and Process Modeling. Ph.D. thesis, Carnegie Mellon University, 2015
6. D.J. Min, J.W. Han, W.S. Chung, A study of the reduction rate of FeO in slag by solid carbon. Metall. Mater. Trans. B **30**(2), 215–221 (1999)
7. B. Sarma, A.W. Cramb, R.J. Fruehan, Reduction of FeO in smelting slags by solid carbon: experimental results. Metall. Mater. Trans. B **27**(5), 717–730 (1996)
8. D.-J. Min, R.J. Fruehan, Rate of reduction of FeO in slag by Fe-C drops. Metall. Mater. Trans. B **23**(1), 29–37 (1992)
9. P. Wei et al., Kinetics of carbon oxidation reaction between molten iron of high carbon concentration and iron oxide containing slag. ISIJ Int. **31**(4), 358–365 (1991)
10. Saudi Iron & Steel Company HADEED (1979–2017)

Alloy Yield Prediction Model Based on the Data Analysis in EAF Steelmaking Process

Ling-zhi Yang, Yu-feng Guo, Feng Chen, Ze-Qiang Xie and Bai-hui Wang

Abstract In this paper, based on the big data in steelmaking process, equation about the relationship between steel composition (especially the oxygen content) and the element yield is established by using mathematical analysis method to analyze the factors influencing yield. Meanwhile, combining the alloy yield database that were established based on self-learning of historical data and computer technology in steelmaking process, a method of yield prediction of alloy elements is studied. To provide support for this method, alloying elements yield prediction software is developed to realize the rapid and intuitive site operation guidance, and to realize the precise control of dosage of alloy material.

Keywords Electric arc furnace · Alloy elements · Alloy yield · Oxygen content

Introduction

In EAF steelmaking process [1] for special steel, molten steel in refining process need to add a variety of alloy material to adjust the composition of molten steel to steel technical standard requirements. Metallurgical workers have made a lot of work in research and development of alloy charging model [2–7].

Alloy material usage is calculated in accordance with the requirements of steel components, steel grade and yield of alloy elements. Yield is calculated based on a long period of time of alloy material record in the actual process of iron and steel enterprise production in usual. There is not considering the factors of oxygen content to affect yield in different furnaces, which reduce the precision of the dosage of alloy material.

L. Yang (✉) · Y. Guo · F. Chen · Z.-Q. Xie · B. Wang
School of Minerals Processing and Bioengineering, Central South University, Changsha, Hunan, People's Republic of China
e-mail: yanglingzhi@163.com

J.-Y. Hwang et al. (eds.), *8th International Symposium on High-Temperature Metallurgical Processing*, The Minerals, Metals & Materials Series,
DOI 10.1007/978-3-319-51340-9_9

The influence of the steel oxygen content on the alloy yield was studied, which is analyzed after smelting. Meanwhile, as a result of the fast rhythm of steelmaking process, refining steel liquid oxygen content is usually lack of test. At last, there is no software system as a support to achieve the actual quick reference guide.

In order to solve the technical problems, the paper provides a kind of alloy yield correction method to improve precise control of dosage of alloy material in the refining process, which is based on the analysis of the data of alloy charging. The paper is including: ladle furnace steel oxygen content prediction model, ladle furnace yield prediction model and database of alloying elements yield [8].

The Research of LF Oxygen Content Prediction Model

In order to study each element to impact oxygen content in the high temperature liquid steel, the thermodynamic data are analyzed in the process of steelmaking. The affinity of each element to oxygen is studied in theory. It is concluded that C, Si, Al are the biggest influence on oxygen content. Related thermodynamic equation is shown as Eqs. (1)–(3). ΔG^{θ} is the standard gibbs free energy, and the unit is J/mol.

$$[C] + [O] = CO_{(g)} \Delta G^{\theta} = -2340 - 39.20T \tag{1}$$

$$[Si] + 2[O] = SiO_{2(l)} \Delta G^{\theta} = -570580 + 215.82T \tag{2}$$

$$2[Al] + 3[O] = Al_2O_{3(s)} \Delta G^{\theta} = -1202070 + 386.28T \tag{3}$$

Theoretical Analysis of the Influential Factors of Oxygen Content

The oxygen content in steel is not affected by a single equilibrium reaction. According to the principle of thermodynamics and the Gibbs free energy of related reaction, the corresponding temperature equilibrium constant is found out. According to the C–O equilibrium constant K_{C-O}, Si–O equilibrium constant K_{Si-O} and Al–O equilibrium constant K_{Al-O}, oxygen content was calculated respectively as follows:

1. According to the C–O balance and carbon content of steel, oxygen content is:

$$\omega_O(C) = K_{C-O}/\omega_C \tag{4}$$

2. According to the Si–O balance and silicon content of steel, oxygen content is:

$$\omega_{\mathrm{O}}(\mathrm{Si}) = \sqrt{K_{\mathrm{Si-O}}/\omega_{\mathrm{Si}}} \tag{5}$$

3. According to the Al–O balance and aluminum content of steel, oxygen content is:

$$\omega_{\mathrm{O}}(\mathrm{Al}) = \sqrt[3]{K_{\mathrm{Al-O}}/(\omega_{\mathrm{Al}})^2} \tag{6}$$

Oxygen Content Prediction

In order to quantitatively determine the different elements (C, Si, Al) for effect weight, the oxygen content after ladle furnace process is analyzed. Combined with the actual production process a lot of package composition data (including oxygen content), the relationship between actual content of C, Si, Al $(\omega_{\mathrm{C}}, \omega_{\mathrm{Si}}, \omega_{\mathrm{Al}})$ and the actual oxygen content ω_{O} were linear coupling respectively. It is concluded that the linear relation is shown as follow:

$$\omega_{\mathrm{O}} = K_{\mathrm{C}} - \mathrm{f}_{\mathrm{C}}\omega_{\mathrm{C}} \tag{7}$$

$$\omega_{\mathrm{O}} = K_{\mathrm{Si}} - \mathrm{f}_{\mathrm{Si}}\omega_{\mathrm{Si}} \tag{8}$$

$$\omega_{\mathrm{O}} = K_{\mathrm{Al}} - \mathrm{f}_{\mathrm{Al}}\omega_{\mathrm{Al}} \tag{9}$$

$\mathrm{f}_{\mathrm{C}}, \mathrm{f}_{\mathrm{Si}}, \mathrm{f}_{\mathrm{Al}}$ is the corresponding influence factor.

According to the different conditions of equilibrium calculation of oxygen content (Eqs. 7–9) and impact factor, it is concluded that the prediction equation of oxygen content as follows:

$$\omega_{\mathrm{O}}^{F} = \frac{\mathrm{f}_{\mathrm{C}}\omega_{\mathrm{O}}(\mathrm{C}) + \mathrm{f}_{\mathrm{Si}}\omega_{\mathrm{O}}(\mathrm{Si}) + \mathrm{f}_{\mathrm{Al}}\omega_{\mathrm{O}}(\mathrm{Al})}{\mathrm{f}_{\mathrm{C}} + \mathrm{f}_{\mathrm{Si}} + \mathrm{f}_{\mathrm{Al}}} \tag{10}$$

Development of Alloy Yield Prediction Model in LF Process

In EAF steelmaking process for special steel, molten steel in refining process need to add a variety of alloy material to adjust the composition of liquid steel to steel technical standard requirements. Meanwhile, the alloy yield is not fixed, which is influenced by the composition of molten steel in furnace (mainly is influenced by oxygen). In order to get accurate yield to accurate guidance alloy charging operation, relation between oxygen content in steel and alloy yield is established.

$$Y^{\mathrm{F}}_{\mathrm{A}_n} = F_{A_n}(\omega^{\mathrm{F}}_{\mathrm{O}}) \tag{11}$$

When the oxygen content in molten steel increased, the yield of alloy reduces. For quantitative research on influence on yield by oxygen content in molten steel in refining process, a large number of actual production data have been counted to obtain the specific parameters of the equation $\left(Y^{\mathrm{F}}_{\mathrm{A}_n} = F_{A_n}(\omega^{\mathrm{F}}_{\mathrm{O}})\right)$.

The alloy elements are including: C, Si, Mn, Cr, Mo, V, Ti, Ni, etc., the specific production enterprises is set according to the actual production need to adjust types of alloy.

The relation between the steel liquid oxygen content and yield was obtained by data analysis. In the process of practical production, oxygen content $(\omega^{\mathrm{F}}_{\mathrm{O}})$ is calculated by means of ladle furnace oxygen content prediction model.

Database of Alloy Yield

In order to get accurate yield of alloy elements, steel grade and the historical data also need to be considered except for oxygen content. In this paper, database of alloying elements yield is establishing with different steel grade. Database of alloying elements yield is established by data collection for charging and element in history to calculate the actual yield for each furnace.

When steel grade X in smelting furnace refining process is finished, model respectively calculated total actual charging amount of alloy (A1, A2...Am) for this steel grade X based on the alloy charging data. Then model respectively calculated total actual amount of alloy (A1, A2...Am) that enter molten steel for this steel grade X based on the change of the composition of molten steel. At last, alloy yield (A1, A2...Am) of steel grade X is obtained. And the database alloy yield is the average or weighted average value of history actual alloy yield, which is classified by steel grade.

When smelting, the model can obtain the corresponding yield value of each element as a reference from the database by steel grade.

In order to consider the effect on alloy yield by the oxygen content and steel grade, $Y^{\mathrm{F}}_{\mathrm{A}_n}$ calculated by oxygen, $Y^{D}_{\mathrm{A}_n}$ in the database and the weight parameter m are jointly decided to prediction for alloy yield $Y^{\mathrm{P}}_{\mathrm{A}_n}$. The equation is shown as follow:

$$Y^{P}_{\mathrm{A}_n} = (1 - m)Y^{D}_{\mathrm{A}_n} + mY^{\mathrm{F}}_{\mathrm{A}_n} \tag{12}$$

Application of Alloy Yield Prediction Model

According to introduced above in this paper, the flow chart of the model can be summarized as table Fig. 1.

Alloy yield prediction model includes 5 steps.

Step 1: Related data about alloy yield is collected, which include the alloy composition, molten steel composition and alloy charging; Molten steel composition includes composition before ladle furnace process and composition before ladle furnace process.

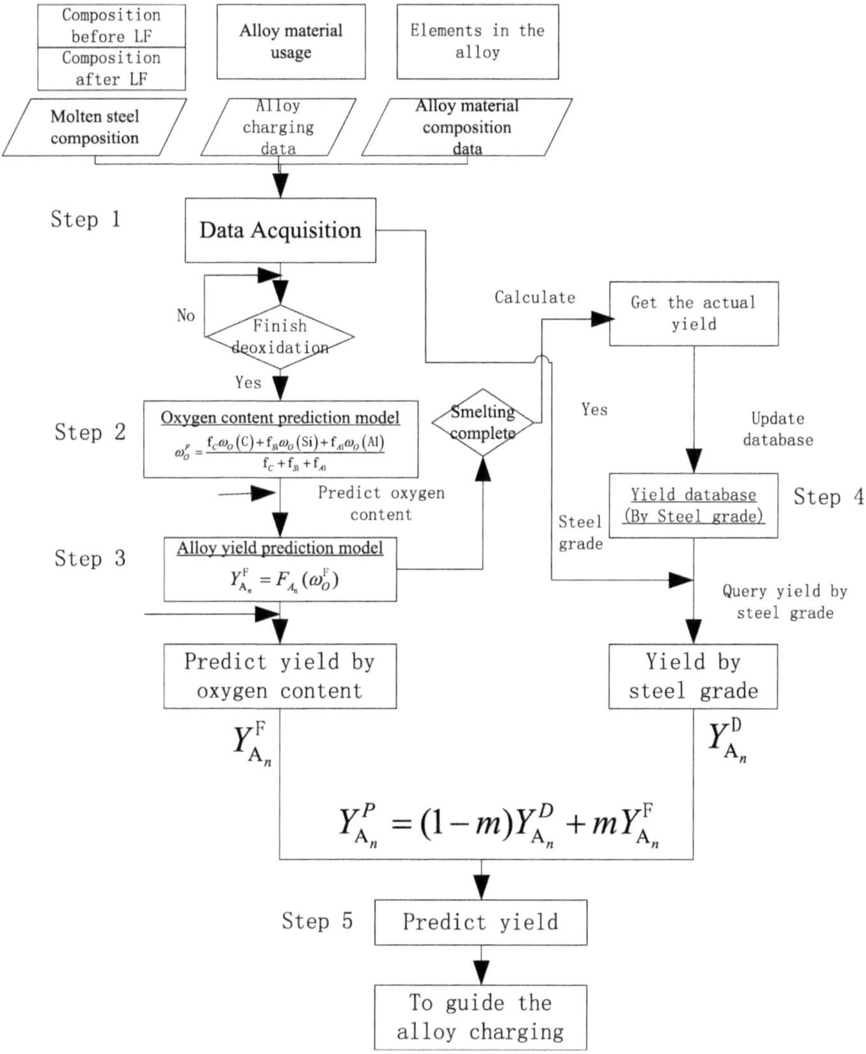

Fig. 1 The flow chart of the alloy yield prediction model

Step 2: When deoxidization is finish, prediction of oxygen content in molten steel can be calculated according to the Eq. (10) and composition data in ladle furnace.

The relationship between actual content of C, Si, Al ($\omega_{C}, \omega_{Si}, \omega_{Al}$) and the actual oxygen content ω_{O} were linear coupling as shown in Fig. 2. In Fig. 2, it is concluded that Al has bigger influence on the oxygen content than C and Si.

According to Section "Theoretical Analysis of the Influential Factors of Oxygen Content" in this paper, the historical data of the smelting furnace is studied to get the parameters in Eq. (10), which is shown as Eq. (13).

$$\omega_{O}^{F} = \frac{20.8\omega_{O}(C) + 75.8\omega_{O}(Si) + 2236.1\omega_{O}(Al)}{20.8 + 75.8 + 2236.1} \tag{13}$$

Step 3: According to the prediction of oxygen content ω_{O}^{F}, the alloy yield $Y_{A_n}^{F}$ is predicted.

The relationship between actual alloy yield and the actual oxygen content ω_{O} were linear coupling as shown in Fig. 3.

According to Section "Development of Alloy Yield Prediction Model in LF Process" in this paper, the historical data of the smelting furnace is studied to get the parameters in Eq. (11), which is shown as Eqs. (14)–(18).

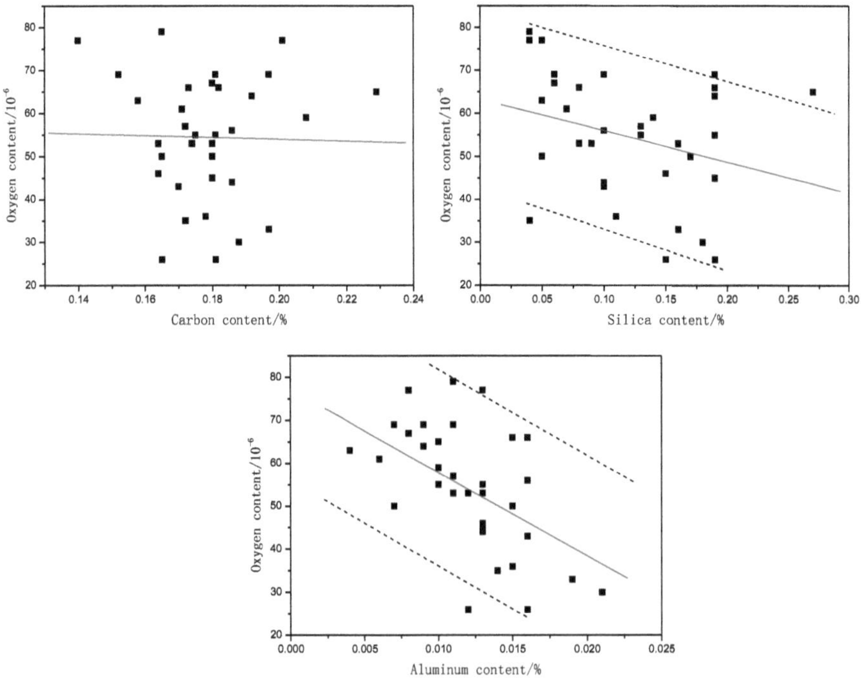

Fig. 2 The relationship between actual content of C, Si, Al and the actual oxygen content

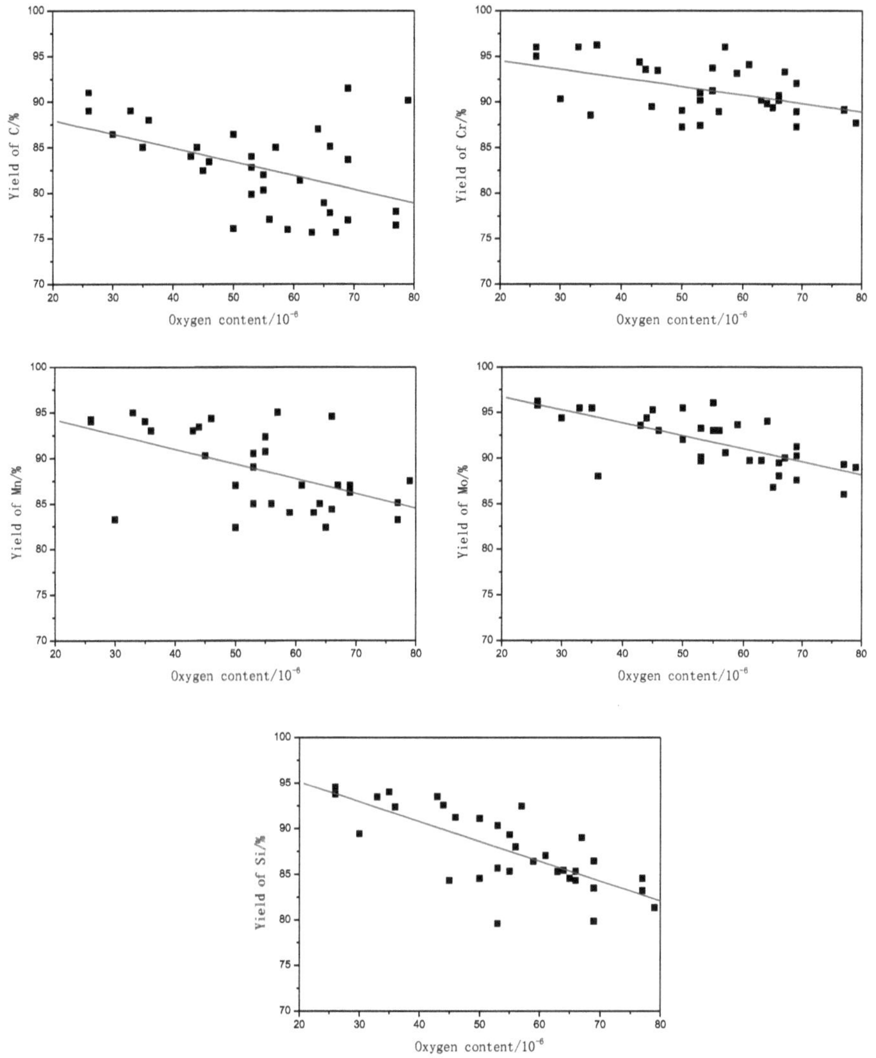

Fig. 3 The relationship between actual alloy yield and the actual oxygen content

$$Y_{\mathrm{C}}^{F} = -0.150\omega_{\mathrm{O}}^{\mathrm{F}} + 90.95 \tag{14}$$

$$Y_{\mathrm{Si}}^{F} = -0.217\omega_{\mathrm{O}}^{\mathrm{F}} + 99.43 \tag{15}$$

$$Y_{\mathrm{Mn}} = -0.161\omega_{\mathrm{O}}^{\mathrm{F}} + 97.41 \tag{16}$$

$$Y^{F}_{\mathrm{Cr}} = -0.094\omega^{\mathrm{F}}_{\mathrm{O}} + 96.43 \tag{17}$$

$$Y^{\mathrm{F}}_{\mathrm{Mo}} = -0.142\omega^{\mathrm{F}}_{\mathrm{O}} + 99.54 \tag{18}$$

Step 4: When smelting, the model can obtain the yield value $Y^{\mathrm{D}}_{\mathrm{A}_n}$ from the database by steel grade. Database of alloy yield is established with getting the actual yield for each furnace by data collection for charging and element in history. And the actual yield is classified by steel grade to get the average or weighted average value of alloy yield. Part of alloy yield classified by steel grade is shown in Table 1.

The interface of alloy yield database and model calculation is shown in Figs. 4 and 5.

Table 1 Part of alloy yield classified by steel grade in the database of alloy yield

Steel grade	C	Si	Mn	Cr	Mo	V
10MnVNb/2	80.3	89.2	94.9	\	\	\
10MnVNbMo/2	79.2	90.0	95.0	\	92.7	\
12Cr1MoVG	80.0	88.5	94.8	91.3	93.6	98.7
12MnNbVs/1	79.4	91.6	95.0	\	\	99.0
16Mn4/1	80.1	90.6	95.9	\	\	\
24CrMo42	81.0	90.0	94.9	90.9	91.2	\
26CrMo4s/2	78.6	91.2	94.3	92.0	93.0	\
26CrMo4V/2	80.0	91.3	96.6	91.7	92.2	98.6
27CrMo27s/1	80.4	91.5	94.6	91.8	93.0	99.8
39Mn2V/1	79.4	90.0	95.8	\	\	99.1
39Mn2V	81.0	89.8	95.0	\	\	99.3

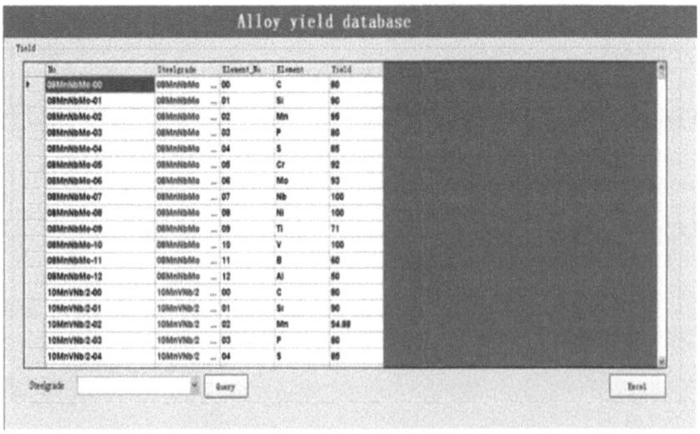

Fig. 4 The interface of alloy yield database

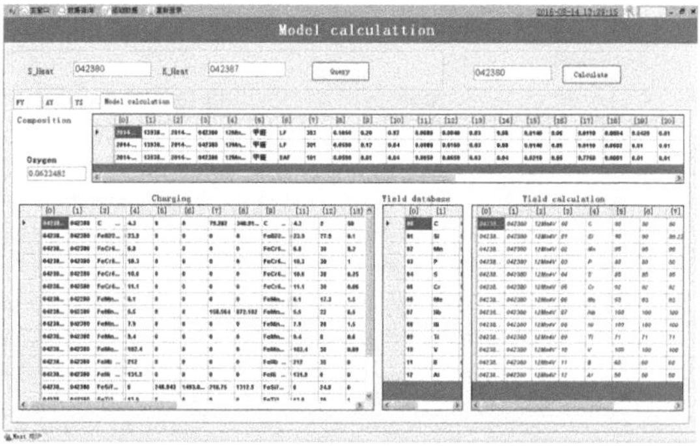

Fig. 5 The interface of model calculation

Step 5: Considering the effect on alloy yield by the oxygen content and steel grade, $Y^{\mathrm{F}}_{\mathrm{A}_n}$ and $Y^{D}_{\mathrm{A}_n}$ are comprehensive analyzed (set the parameters m = 0.5) for predicting the comprehensive yield $Y^{\mathrm{P}}_{\mathrm{A}_n}$. So Eq. (12) is shown as follow:

$$Y^{P}_{\mathrm{A}_n} = 0.5Y^{D}_{\mathrm{A}_n} + 0.5Y^{\mathrm{F}}_{\mathrm{A}_n} \tag{19}$$

Conclusions

1. Based on the big data in steelmaking process, the paper gets the equation about the relationship between steel composition (especially the oxygen content) and the element yield is established by using mathematical analysis method to analyze the factors influencing yield. Meanwhile, alloy yield database were established based on self-learning of historical data and computer technology, through which accurate real-time alloy yield was obtained.
2. Alloy yield prediction model is developed to realize the rapid and intuitive site operation guidance to provide support for this method, which improved the accuracy of dosage of the alloy. When smelting, the model calculate the yield value $Y^{F}_{\mathrm{A}_n}$ by oxygen content. Then, the model obtains the yield value $Y^{\mathrm{D}}_{\mathrm{A}_n}$ from the database by steel grade. At last, considering the effect on alloy yield by the oxygen content and steel grade, the comprehensive yield $Y^{\mathrm{P}}_{\mathrm{A}_n}$ is calculated.

References

1. L.Z. Yang et al., Study of optimizing combined-blowing in EAF based on K-medoids clustering algorithm. Adv. Mater. Res. **881–883**, 1540–1544 (2014)
2. K.W. Huang et al., LD converter alloy minimum cost control mode. Metall. Ind. Autom. **2**, 11–13 (2003)

3. G.B. Li et al., Development of LF refining composition prediction model. Angang Technol. **4**, 26 (2009)
4. W. Gong et al., Component controlling model in BOF steelmaking process. J. Northeast. Univ. **23**(12), 1155 (2002)
5. R.H. Luo et al., A study on prediction models for alloying charging and end analysis during ladle furnace refining. Spec. Steel **32**(2), 8–10 (2011)
6. A.N. Conejo et al., Optimization of aluminum deoxidation practice in the fimiace. Mater. Manuf. Processes **21**(8), 796–803 (2006)
7. I. Ekmekci et al., Mass balance modeling for electric arc furnace and ladle furnace system in steelmaking facility in turkey. J. Iron. Steel Res. Int. **14**(5), 1–55 (2007)
8. L.Z. Yang et al., Alloy charging optimization model based on the yield dynamic libraries. J. Univ. Sci. Technol. Beijing **36**(s1), 104–109 (2014)

Analysis of Jet Behavior and Surface Fluctuations in the Meniscus of Fluid in a Physical Model of a Beam Blank Mold and CFD Modelling

Johne Jesus Mol Peixoto, Weslei Viana Gabriel, Ciro Azevedo Silva, Leticia Queiroz Ribeiro, Carlos Antonio da Silva, Itavahn Alves da Silva and Varadarajan Seshadri

Abstract In this study, the influence of the immersion depth, casting speed on the flow characteristics in a beam blank mold with two tubular nozzles have been investigated using a physical model as well as CFD simulations. The results of CFD simulations were found to be consistent with the results of experiments in the physical model. The liquid jet spreads faster the flange region compared to the web region. This can lead to a non-uniform shell thickness along the cross section resulting in uneven solidification of the skin. The jet penetration depth ranged between 66 and 77 cm. The average intensity of meniscus fluctuation was 0.22 mm, and the maximum fluctuation was found to be 0.85 mm. Based on these results, the suggested optimum operating parameters for immersion depth and casting speed were 75 mm and maximum casting speed of 1 m/min.

Keywords Beam blank · Flow · PIV · PLIF

Introduction

Yoon [1] considers the efficiency of the continuous casting process as a result of many technological advances in recent years. One such development is get the product to the nearest geometry for the final form of application thus reducing operating costs and with increased productivity.

J.J.M. Peixoto · W.V. Gabriel · C.A. Silva · L.Q. Ribeiro · C.A. da Silva · I.A. da Silva
Department of Metallurgical Engineering and Materials,
Federal University of Ouro Preto, Ouro Preto, MG 35400-000, Brazil

V. Seshadri (✉)
Department of Metallurgical Engineering and Materials, Federal University of Minas Gerais, 6627, Av. Antonio Carlos, Belo Horizonte 31270-901, Brazil
e-mail: seshadri@demet.ufmg.br

J.-Y. Hwang et al. (eds.), *8th International Symposium on High-Temperature Metallurgical Processing*, The Minerals, Metals & Materials Series,
DOI 10.1007/978-3-319-51340-9_10

Onishi et al. [2] point out that due to the difference in shape of the product, the continuous casting of beam blank differs from the casting of slabs and billets and also from the point of view of the internal tensions and solidification mechanisms. Chen et al. [3] and Lee et al. [4], among others, point out that the transfer of heat between the nozzle and the walls of the mold, profile of the velocity field (particularly in the meniscus region), oscillation pattern of the mold and segregation phenomena during the solidification regime have a great influence on the internal and surface quality of the product. Kim et al. [5], Lee et al. [4] and Hibeller et al. [6] detected the formation of an air gap at the tip of the flange during the initial stage of solidification. This impairs the heat extraction process, decreasing the solidification rate. Similarly, the impact of the liquid steel jet retards the growth of the skin in the fillet region. In the secondary cooling zone of beam blank continuous casting process, Zhao et al. [7] found that the longitudinal cracks in the fillet region are generated due to the large temperature difference between the fillet and other regions on the surface of the shaft. These researchers have proposed an improvement in this refrigeration system, in order to equalize the surface temperature and optimize the consumption of water.

The geometry of the SEN (Submerged Entry Nozzle) influences the velocity field and heat dissipation in the mold region. Xu and Zhu [8] compared the use of straight tubular nozzles and those with nozzles fitted with three lateral holes on the output. In the latter, two holes point towards the flange and one towards central region of the mold. The results showed that with the use of tubular nozzles, the temperature profile enables the solidified skin to be thinner in the fillet region and thicker in the web; while in the case of nozzles with three side holes, there is greater fluctuation of the meniscus. The meeting of the jets in the central region of the web creates a high temperature zone, which leads to a thinner skin and hence is prone to breakages. In this work, physical and mathematical modeling were applied to show the effects of casting speed and SEN immersion depth on the flow behavior in the mold of continuous casting of beam blank.

Materials and Methods

Physical Modeling

The physical modeling experiments were performed in an acrylic replica of a beam blank continuous casting mold, having a scale factor of 1:1, and with the dimensions of $499 \times 415 \times 125\ mm^3$ and height of 1.5 m. The submerged nozzles were tubular with an internal diameter of 34.6 mm. The mold supply flows were 100, 125 and 150 l/min, which correspond to the casting speed of 0.78, 0.98 and 1.2 m/min in the prototype respectively. These parameters were determined applying the usual geometrical and dynamic similarity principles. The SEN immersion depths were 100, 75 and 50 mm. The flow pattern was analyzed over the

AA and BB planes, and also in the web center and flange region (Fig. 1b, c). The fluctuations of the meniscus were captured with the help of an ultrasonic distance measurement sensor (Sick UM30-211118) positioned at points P1 and P2 in the web center and the fillet region, as shown in Fig. 1c. Two tests were conducted for each flow rate and for different immersion depths of the submerged nozzles. The position of the meniscus was measured as a function of time over 60 s, and the data were used to calculate the mean and maximum intensity of fluctuation for each test.

The dye injection technique was used to characterize the behavior of the jet in the model (simulating the steel jet discharged into the mold). This method consists in injecting the tracer in the water inlet pipe and following its trajectory inside the mold until its total dispersion. This enables visualization of the flow and the depth of jet penetration. The technique PLIF (Planar Laser-Induced Fluorescence) was used (Dynamics software; PIV assembly by DANTEC®) for a qualitative analysis of the flow inside the mold. This technique makes use of the fluorescence property of Rhodamine 6G as tracer to evaluate the development and course of the flow. The technique consists in illuminating a particular region of the mold (Fig. 2) with a pulsed laser plane (wavelength 532 nm), which acts on the tracer molecules (Rhodamine 6G) exciting and causing the same to emit fluorescent light (wave length between 540 and 640 nm). Images are captured by a CCD camera equipped

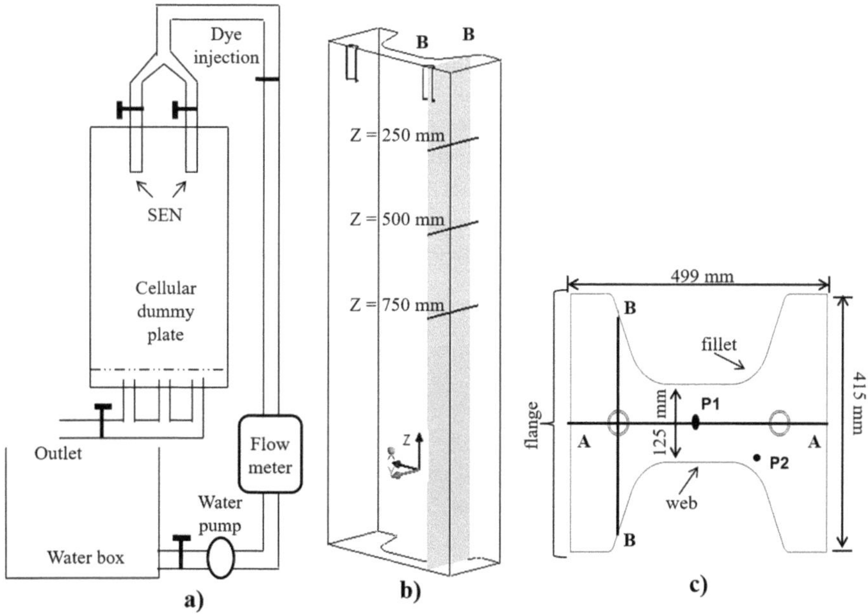

Fig. 1 **a** Schematic representation of physical model; **b** view of beam blank mold at the AA plane of symmetry; **c** dimension of beam blank mold and cross sections of beam blank mold in different regions and AA and BB planes sections with measuring points of the meniscus fluctuation, P1 and P2

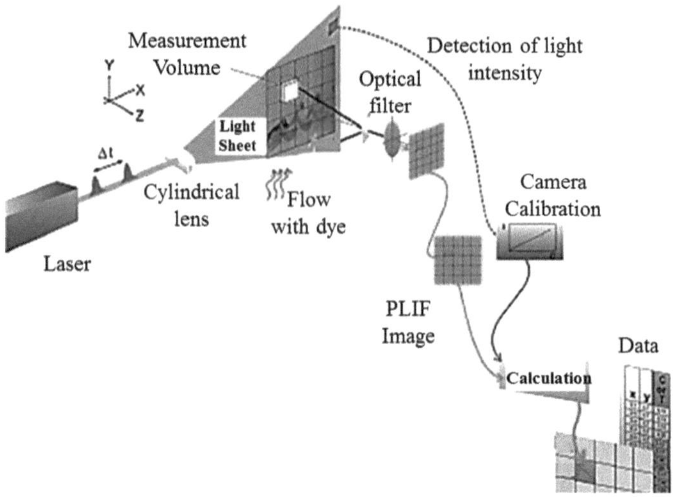

Fig. 2 Schematic representation of Planar LIF apparatus according to DANTEC®. *Source* http://www.dantecdynamics.com/measurement-principles-of-planar-lif

with a light filter permeable to wavelength (570 nm), enabling mapping of the flow through the correlation between the intensity of the emitted radiation and concentration of the tracer, previously established by a calibration curve.

To quantitatively analyze the flow through physical simulations, PIV technique (Particle Image Velocimetry) was used. This method consists in the dispersion of particle with density equal to the water and incidence of a laser beam in pulses, which allows mapping the path within the fluid through an interpolation process. A CCD camera captures a sequence of images that are processed by a specific software (Dynamics DANTEC®), for evaluating the velocity values and vorticity fluctuations, amongst other flow parameters.

Mathematical Modeling

Steady state computer simulations of the flow field were performed with the software CFX Ansys® 15.0, considering water as incompressible Newtonian fluid, at 25 °C. The continuity equation, the Navier-Stokes and the equations of (k-ε) turbulence model for evaluation of effective viscosity were solved. More details regard these and boundary conditions can be found elsewhere [9]. The mass flow rate of fluid at the entrance amounted to 0.833, 1.042 and 1.25 kg/s for casting speeds equivalent to 0.78, 0.98 and 1.2 m/min, respectively.

To evaluate the tracer behavior, a mathematical simulation of transient state for a total time of 15 s was carried out. The result of this mathematical simulation in respect of reaching steady state under the same conditions of flow rate and SEN immersion depth was used as an initial condition to ensure that the tracer was injected into the mold with fully developed flow.

Results and Discussion

Analysis of Velocity Profile

The velocity profile was determined in three parts of BB plane by PIV technique, covering a total depth of 800 mm. In Fig. 3a, b, the results of these tests and those computational simulation are compared. There seems to be good consistency in respect of the structure of the jet and location of the vortices. Similarly, Fig. 4 shows the vertical velocity component along a horizontal line that bisects the BB plane, at distances of 250, 500 and 750 mm from the meniscus. Comparison of the results obtained from physical simulations (PIV) with those obtained by computer simulation (CFD) (Fig. 4b), show reasonable consistency.

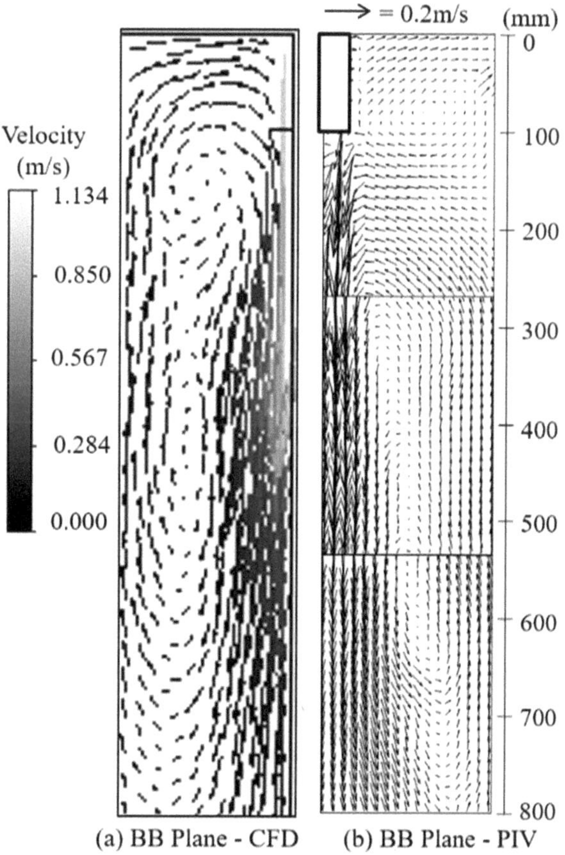

Fig. 3 Velocity profile for flow rate of 100 l/min and 100 mm of SEN immersion: **a** plane BB—CFD—results, **b** plane BB—PIV—results

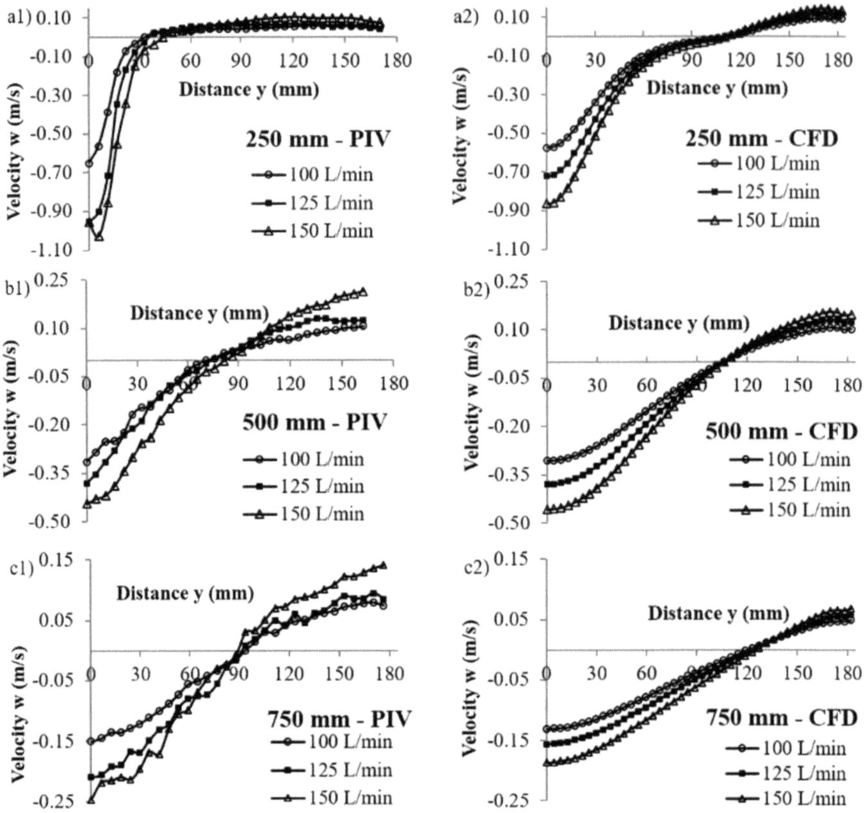

Fig. 4 Chart speed depending on the distance y from the center of the SEN until the tip of the flange obtained: by PIV technique at a distance from the meniscus **a1** 250 mm, **b1** 500 mm and **c1** 750 mm; CFD numerical simulation in a distance from the meniscus **a2** 250 mm, **b2** 500 mm and **c2** 750 mm

Figure 5 shows that the change in SEN immersion depth from 50 to 75 mm does not alter significantly the flow pattern. The position of vortices and velocity distribution in the vector field seem to be virtually the same under both the conditions. For these immersion depths, there are six vortices in total, two at the center and four in the flanges even for an immersion of 100 mm [9]. This flow pattern also is in agreement with the results of Lee et al. [4] and Chen et al. [10].

Characterization of Jet

Figure 6 shows that, under such flow conditions, the depth of the jets discharged into the mold amounted to approximately 0.8 m, for a time interval of 2 s. Until this moment apparently, the action of the drag force from the jet, leaving the nozzles,

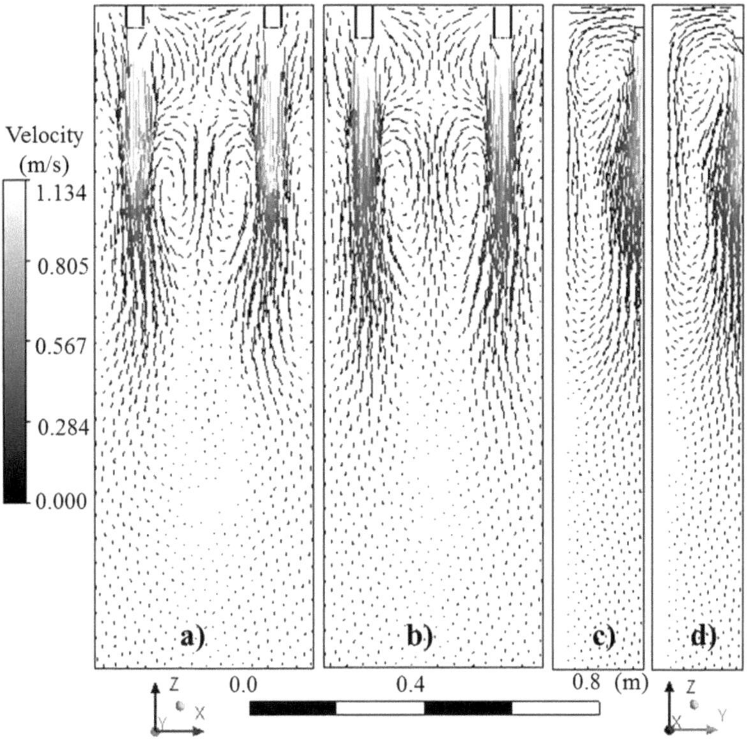

Fig. 5 Velocity profile obtained by mathematical simulation for flow rate of 125 l/min. AA Plane, immersion depth: **a** 50 mm; **b** 75 mm; BB plane, immersion depth: **c** 50 mm; **d** 75 mm

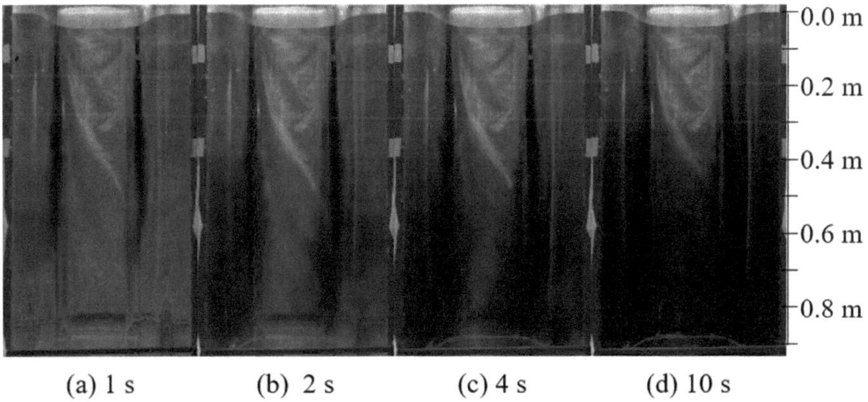

Fig. 6 Scattering of tracer (ink) inserted in the mold feed pipe—**a** t = 1 s; **b** t = 2 s; **c** t = 4 s and **d** t = 10 s. Flow rate of 125 l/min, SEN immersion 100 mm

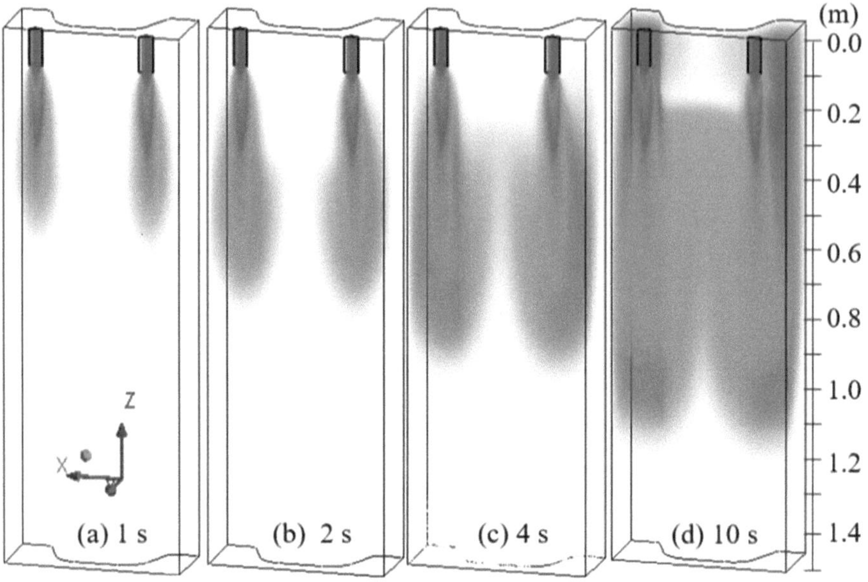

Fig. 7 Tracer scattering as obtained by mathematical simulation in transient state—**a** t = 1 s; **b** t = 2 s; **c** t = 4 s and **d** t = 10 s. Flow 125 l/min with 100 mm depth of immersion

on the tracer (ink) prevails. At longer times, there is an increasing lateral spreading of the ink and no further penetration. Computer simulation considering transient tracer dispersion under the same conditions of immersion and flow, shows a penetration depth of the jet coming out of the SEN of around 0.75 m for a time interval of 2 s (Fig. 7c). Both the results of the physical and mathematical modeling showed that the tracer spreads more quickly to the flanges and rather slowly at the center region. It can be noted that at the time interval of 10 s, the tracer does not yet cover the web center region at the top of the meniscus, showing that replacement of the fluid in this region is slow. In the prototype, this implies a lower heat input in this region leading to a thicker skin in the central region. The impact of steel jet on the skin slows the skin growth in the regions of the flange, center and fillet. This effect on the flange and central region was also observed by Xu and Zhu [8], who called it a "washing effect".

The values of the jet penetration depths obtained by mathematical simulation are shown in Table 1. The increased flow rate and SEN immersion resulted in increase in the depth of the jet. Chen et al. [10] found the depth of impact of more than 1 m for immersions depths of SEN ranging from 25 to 125 mm; Zhang et al. [11] obtained in their physical modeling experiments values around 50 and 60 cm, for immersion depths of SEN between 50 and 110 mm.

Figure 8 shows the comparison of the tracer scattering (Rhodamine 6G) for immersion depths of 75 and 100 mm. One can note the transient character of the jet, which for t = 1 s shows small deviations from its vertical symmetry axis.

Table 1 Jet penetration depth in relation to the flow and immersion of SEN

SEN immersion depth	50 mm			75 mm			100 mm*		
Flow rate (l/min)	100	125	150	100	125	150	100	125	150
Jet penetration depth (cm)	66.3	70.8	75.3	69.7	75.3	77.5	74.0	75.5	77.3

* Peixoto et al. [9]

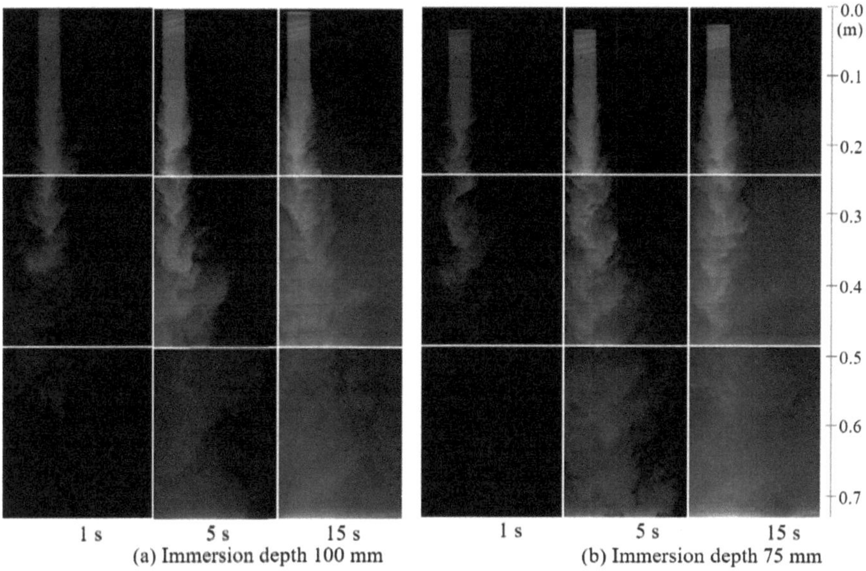

Fig. 8 Scattering of tracer (Rhodamine 6G) obtained via PLIF technique in BB plane for flow rate 125 l/min and SEN immersion depth: **a** 100 mm and **b** 75 mm

After spreading of the jet, the inertial force, responsible for the flow in the vertical direction, is no longer predominant. The spreading is due to convection and the fluid returns toward the meniscus, following the movement of the flange region of vortices. It can be noted that the reduction of the immersion depth of the SEN results in greater spreading of the jet. At t = 15 s the upper region of the tip of the flange is filled with tracer for the SEN immersion depth of 75 mm (Fig. 8b) and this is not the case for the SEN immersion depth of 100 mm (Fig. 8a).

Evaluation of the Fluctuation of the Meniscus

Figure 9 shows the results of analysis of fluctuations in the meniscus. The average fluctuation intensity was of the order of 0.22 mm and the maximum intensity was slightly higher in the central region (0.85 mm) than in the fillet region (0.75 mm).

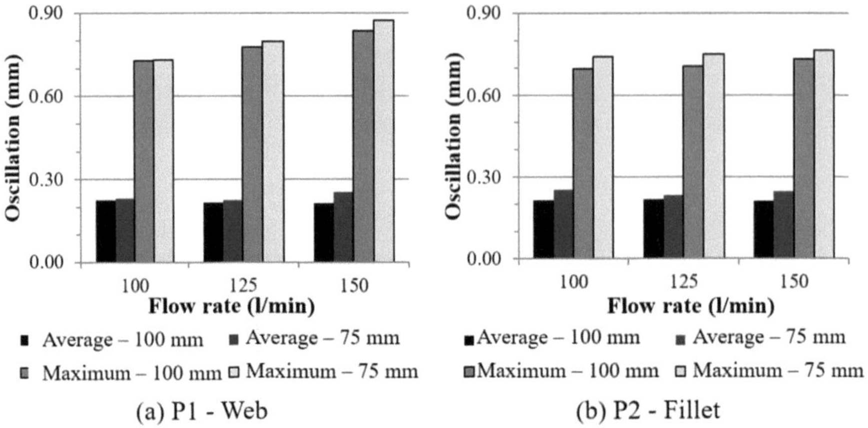

Fig. 9 Meniscus oscillation amplitude for different flow rates and immersions: **a** Web-P1; **b** Fillet-P2

The reduction of the SEN immersion depth did not result in significant variation in the fluctuation intensity of meniscus. These data suggested that using the straight tubular nozzle, the position of the free surface of the meniscus can't be promptly controlled by adjusting the nozzle immersion, which would be detrimental to the process of fusion of flux powder and flotation of inclusions.

There was substantial consistency between the results obtained with those available in the literature with regard to the fluctuation of the meniscus. By mathematical simulation, Xu and Zhu [8] observed that for tubular nozzles, the oscillation level of the meniscus was lower than 1 mm; while Zhang et al. [11] via physical modeling obtained an average fluctuation of around 0.2 mm for the case of SEN of 40 mm of inner diameter and 70 mm immersion depth. On the other side, Chen et al. [10] obtained by mathematical simulation with SEN immersion depth of 75 mm, an oscillation fluctuation on the order of 0.8 and 2.2 mm in the region of web and flange respectively. From the results of the present work, it is suggested that an immersion depth for the SEN's should be around 75 mm and a maximum casting speed of 1 m/min.

Conclusions

The results of the mathematical simulations are consistent with the results of the physical modeling using two tubular nozzles with internal diameter of 34.6 mm. It was observed that:

- The tracer spreads quickly to the edge flanges and slowly at the center region;
- The jet penetration depth varied between 66 and 77 cm;

- The average intensity meniscus fluctuation was of the order of 0.22 mm and the maximum intensity fluctuation was 0.85 mm. This parameter was not significantly affected by the variation of the casting speed and the depth of immersion of SEN.

Acknowledgements The authors wish to thank the National Council of Scientific and Technological Development (CNPq), the Ministry of Higher Education (CAPES), Gorceix Foundation and FAPEMIG of Brazil for their support.

References

1. J.K. Yoon, Applications of numerical simulation to continuous casting technology. ISIJ Int. **48**(7), 879–884 (2008)
2. M. Onishi et al., Continuous casting of beam blanks. Kawasaki Steel Tech. Rep. **3**, 13–25 (1981)
3. W. Chen et al., Numerical simulation of the thermo-mechanical process for beam blank continuous casting. ACTA Metall. Sinica (Engl. Lett.) **20**(4), 241–250 (2007)
4. J.E. Lee et al., Prediction of cracks in continuously cast steel beam blank through fully coupled analysis of fluid flow, heat transfer, and deformation behavior of a solidifying shell. Metall. Mater. Trans. A, **31**(A), 225–237 (2000)
5. K. Kim et al., Analysis of surface and internal cracks in continuously cast beam blank. Ironmaking Steelmaking **24**(3), 249–256 (1997)
6. L.C. Hibbeler et al., Thermomechanical modeling of beam blank casting. Iron Steel Technol. **6**(7), 60–73 (2009)
7. Y. Zhao et al., Two-dimensional heat transfer model for secondary cooling of continuously cast beam blanks. Ironmaking Steelmaking **41**(5), 377–386 (2014)
8. M. Xu, M. Zhu, Transport phenomena in a beam-blank continuous casting mold with two types of submerged entry nozzle. ISIJ Int. **55**(4), 791–798 (2015)
9. J.J.M. Peixoto et al., Computational and physical simulation of fluid flow inside a beam blank continuous casting mold. J. Mater. Process. Technol. **233**, 89–99 (2016)
10. W. Chen et al., Three-dimensional FEM study of fluid flow in mould for beam blank continuous casting: influence of straight through conduit type SEN. Ironmaking Steelmaking **39**(8), 551–559 (2012)
11. L. Zhang et al., Hydraulic simulations of fluid flow in beam blank casting mold with double nozzles, in *Proceedings EPD Congress* (2014), pp. 375–384

CFD Study of Gas-Liquid Phase Interaction Inside a Submerged Lance Smelting Furnace for Copper Smelting

Guangwu Tang, Armin K. Silaen, Hongjie Yan, Zhixiang Cui, Zhi Wang, Haibin Wang, Kaile Tang, Ping Zhou and Chenn Q. Zhou

Abstract Bath smelting technology, such as the submerged lance smelting furnace, is used in the modern copper making industry. The first submerged lance smelting furnace developed by Dongying Fangyuan Nonferrous Metals Co., Ltd has shown potential for high productivity and energy savings. In this study, computational fluid dynamics (CFD) was applied to investigate the current design of the furnace. A three-dimensional multiphase CFD model was developed to study the interaction between injected gas and the liquid bath. The multiphase Eulerian model was used for simulating gas/liquid two-phase flow. The flow pattern in the submerged lance smelting furnace indicated rapid flow development and strong turbulent interactions between the gas and liquid phases. The model was validated based on a water model experiment of the mixing process. Mixing times from the simulation results show good agreement with experimental data. Additionally, based on this model, the gas residence time and liquid copper matte splashing phenomena under varying gas flow rates were investigated.

Keywords Submerged lance smelting furnace · CFD · Multiphase flow · Mixing time

---

G. Tang · A.K. Silaen · C.Q. Zhou (✉)
Center for Innovation through Visualization and Simulation,
Purdue University Northwest, 2200 169th Street, Hammond, IN 46323, USA
e-mail: czhou@pnw.edu

H. Yan · K. Tang · P. Zhou
School of Energy Science and Engineering, Central South University,
Changsha 410083, Hunan, China

Z. Cui · Z. Wang · H. Wang
Dongying Fangyuan Nonferrous Metals,
386 Yunhe Rd, Dongying 257091, Shandong, China

J.-Y. Hwang et al. (eds.), *8th International Symposium on High-Temperature Metallurgical Processing*, The Minerals, Metals & Materials Series,
DOI 10.1007/978-3-319-51340-9_11

Introduction

Bath smelting is a crucial technology in the copper making industry due to its innate high smelting efficiency, low capital cost, low energy consumption, and reduced dust production. From the perspective of gas blown regime, developed bath smelting practices can be categorized into three general types: (1) top-submerged blown, including Ausmelt and Isasmelt; (2) top-suspended blown, such as Mitsubishi smelting; and (3) submerged-side blown, such as Noranda and Teniente smelting.

The submerged lance smelting furnace utilizes the bottom blown oxygen copper smelting process developed in Dongying, China by Dongying Fangyuan Nonferrous Metals Co., Ltd, in a commercial scale, as shown in Fig. 1. This is the first modern copper smelting technology [1] developed in China with the advantages of high oxygen enrichment, low temperature, and auto-thermal operation. During the copper smelting process, the mixing gases, including oxygen and air, are injected from lances located at the bottom of the bath. This injection induces a significant stirring effect, serving to support chemical reactions and heat, mass, and momentum transfer. Guo et al. [2] constructed the mechanism model of copper oxygen-enriched bottom blowing smelting, and provided a better understanding of bowing patterns in the new bottom blown furnace for the copper industry. Bottom blowing is affected by the amount of gas injected, the number and configuration of tuyeres, and several other parameters. Several researchers have studied these parameters to improve bath mixing effects. Lang et al. [3] used a water model to study mixing phenomena in the copper smelting furnace. Gonzalez-Trejo et al. [4] combined top blowing of air and nitrogen bottom stirring to obtain an adequate mixing of the copper bath in a Peirce-Smith converter. Zhang [5] analyzed the gas-liquid flow using numerical simulation and optimized the lance structure parameters. But the flow pattern in the submerged lance smelting furnace has not been fully investigated.

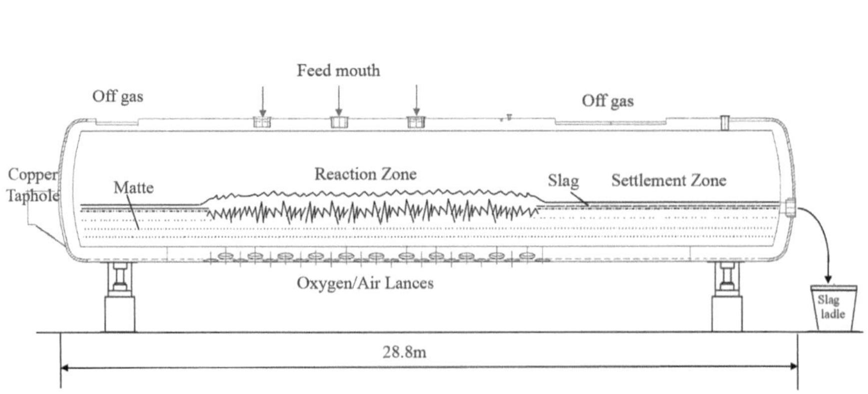

Fig. 1 The submerged lance smelting furnace at Dongying Fangyuan Nonferrous Metals

A three-dimensional multiphase CFD model was developed to study the interaction between injected gas and liquid bath and verified using a water model experiment. The multiphase Eulerian model was adopted in this study, and the gas residence time and liquid matte splashing were investigated based on the numerical simulation.

CFD Model

In this study, the Eulerian multiphase model was employed to simulate gas and copper phase interactions. Under the Eulerian model, the multi-fluid VOF sub-model was applied for providing a framework to couple the VOF model and describe the interphase behavior. This approach will overcome some limitations of the VOF model that arise due to the shared velocity and temperature formulation. The Realizable k-ε model was employed to model the effects of turbulence. In this gas-liquid two phase flow, two sets of governing equations are solved with specific drag model to capture the sharp interface between the gas and liquid phases. The governing equations used are as follows:

Conservation of Mass:
Gas-phase continuity equation:

$$\frac{\partial}{\partial t}\left(\alpha_g \rho_g\right) + \nabla \cdot \left(\alpha_g \rho_g \vec{v}_g\right) = 0 \tag{1}$$

Liquid-phase continuity equation:

$$\frac{\partial}{\partial t}\left(\alpha_q \rho_q\right) + \nabla \cdot \left(\alpha_q \rho_q \vec{v}_q\right) = 0 \tag{2}$$

where α_g is the gas phase volume fraction, ρ_g is gas phase density, $\vec{v}_g$ is the gas phase velocity, the subscript q represents the liquid phase. The first term is the time rate of change of either the gas phase or liquid phase, and the second term is the advection of the gas phase or liquid phase. Each computational cell is shared by gas and liquid phase with the total of liquid and gas volume fractions summing to 1.

$$\alpha_g + \alpha_q = 1 \tag{3}$$

Conservation of Momentum:
Gas-phase momentum equation:

$$\frac{\partial}{\partial t}\left(\alpha_g \rho_g \vec{v_g}\right) + \nabla \cdot \left(\alpha_g \rho_g \vec{v}_g \vec{v}_g\right) = -\alpha_g \nabla p + \nabla \cdot \bar{\bar{\tau}}_g + \alpha_g \rho_g \vec{g} + \sum_{p=1}^{n} \vec{R}_{pg} + \vec{F} \tag{4}$$

where $\bar{\bar{\tau}}_g$ is the gas phase stress-strain tensor, defined as:

$$\bar{\bar{\tau}}_g = \alpha_g \mu_g \left(\nabla \vec{v}_g + \nabla \vec{v}_g^T \right) + \alpha_g (\lambda_g - \frac{2}{3} \mu_g) \nabla \cdot \vec{v}_g \bar{\bar{I}} \quad (5)$$

Here μ_g and λ_g are the shear and bulk viscosity of phase g, $\vec{F}$ is a turbulent dispersion force. $\vec{R}_{pg}$ is an interaction force between phases, and p is the pressure shared by all phases. $\bar{\bar{I}}$ is the unit stress tensor.

Liquid-phase momentum equation:

$$\frac{\partial}{\partial t} (\alpha_q \rho_q \vec{v}_q) + \nabla \cdot (\alpha_q \rho_q \vec{v}_q \vec{v}_q) = -\alpha_q \nabla p + \nabla \cdot \bar{\bar{\tau}}_q + \alpha_q \rho_q \vec{g} + \sum_{p=1}^{n} \vec{R}_{qp} + \vec{F} \quad (6)$$

$$\sum_{p=1}^{n} \vec{R}_{pq} = \sum_{p=1}^{n} K_{pq} (\vec{v}_p - \vec{v}_q) \quad (7)$$

where K_{pq} is the interphase momentum exchange coefficient. The symmetric drag law, which is used in the sharp interface regime, is isotropic and tends to approach the behavior of the VOF model for high drag coefficients. In the multi-fluid VOF model, the effective drag coefficient in the principle direction p is described as:

$$K_p = K_{symmetric} \lambda_p \quad (8)$$

where λ_p is the friction factor vector in the principal direction. $K_{symmetric}$ is the isotropic drag coefficient obtained from the symmetric drag law.

In the symmetric drag law, the density and the viscosity are calculated from volume averaged properties:

$$\rho_{pq} = \alpha_p \rho_p + \alpha_q \rho_q \quad (9)$$

$$\mu_{pq} = \alpha_p \mu_p + \alpha_q \mu_q \quad (10)$$

and the diameter is defined as:

$$d_{pq} = \frac{1}{2} (d_p + d_q) \quad (11)$$

in turn:

$$K_{pq} = \frac{\rho_{pq} f}{6 \tau_{pq}} d_p A_i \quad (12)$$

where τ_{pq} and f are defined as:

$$\tau_{pq} = \frac{\rho_{pq}(d_{pq})^2}{18\mu_{pq}} \tag{13}$$

$$f = \frac{C_D Re}{24} \tag{14}$$

where Re are defined as:

$$Re = \frac{\rho_{pq}|v_p - v_q|d_{pq}}{\mu_{pq}} \tag{15}$$

where K_{pq} is the fluid-fluid exchange coefficient, A_i is the interfacial area, f is the drag function, C_D is the drag coefficient that based on the relative Reynolds number Re. τ_{pq} is the "particulate relaxation time." d_p is the diameter of the bubbles or droplets of phase p.

Species transport
To evaluate the mixing characteristics for the model validation with water model and gas residence time, a species transport equation is solved. For each phase g, predict the local mass fraction of each species, Y_i^g, through the solution of a convection-diffusion equation for i species, when applied to multiphase mixture can be expressed in the following form:

$$\frac{\partial}{\partial_t}(\rho_g \alpha_g Y_i^g) + \nabla \cdot (\rho_g \alpha_g Y_i^g \vec{v}_g) = -\nabla \cdot \alpha_g \vec{J_i}^g \tag{16}$$

where, $\vec{J_i}^g$ is the diffusion flux of species i, which arises due to the gradients of concentration.

$$\vec{J_i}^g = -(\rho_g D_{i,m} + \frac{\mu_t}{Sc_t})\nabla Y_i^g \tag{17}$$

where, $D_{i,m}$ is the diffusion coefficient for species i in the mixture, Sc_t is the turbulent Schmidt number, which is 0.7. μ_t is turbulent diffusivity.

The numerical prediction of the mixing process was achieved through a two-step procedure. First, boundary conditions were defined to calculate the flow field. Then tracer material was patched to recording mixing time based on the current flow field. The recovery of the tracer, defined as chemistry changes from the initial value, was used for mixing analysis. This value was normalized to the total chemistry change, using the equation:

$$tracer\ recovery = \frac{C - C_0}{C_\infty - C_0} * 100\% \tag{18}$$

where C is the current mass value of tracer at the sampling position, C_0 is the initial mass value of tracer in the melt, C_∞ is the mass value of tracer at equilibrium point set by the last time point.

Geometrical Model and Conditions

Figure 2 shows the three-dimensional model based on the drawings provided by Dongying Fangyuan Nonferrous Metals Co., Ltd. The furnace is a cylindrical vessel with 23 oxygen lances on the bottom. The oxygen lances are placed in two rows at angles of 7° and 22° respectively, corresponding to the center point of the cylindrical vessel. There are three feed mouths at the top of the furnace above the oxygen lances. Two exhaust outlets are located near each end of the furnace. During operation, raw materials are charged into the furnace from the feed mouth at the top. The raw materials fall into the reaction zone and are quickly decomposed and melted to react with the oxygen injected through the lances. The slag and copper matte is discharged from the left and right ends of the furnace.

This simulation was completed using the commercial software package ANSYS Fluent 16.2. The Finite Volume method (FVM) was used to solve the partial differential equations by discretizing the equations with an upwind differencing scheme over the finite volumes. The Phase Coupled SIMPLE algorithm was used to adjust the pressure and velocities after each iteration when solving the gas-liquid momentum equations. Each simulation was performed for 10 s of flow time to reach a quasi-steady state on the High Performance Computing (HPC) cluster at Purdue University Northwest. The no-slip boundary condition and standard wall functions are applied at both the bottom wall and the side-wall for fluid flow. The time step is 0.0001 s, and the convergence criteria for all of the residuals are set to 10^{-3}.

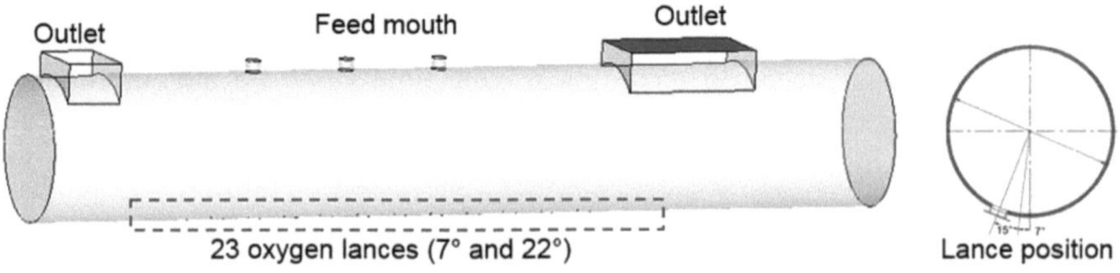

Fig. 2 Computational geometry of the submerged lance smelting furnace

Results and Discussions

Model Validation

Due to the difficulty of conducting measurements on the fluid flow inside the submerged lance furnace, the numerical model was firstly compared with a water model experiment carried out by Lang et al. [3]. A scaled-down 1:12 model was established to represent the flow behavior and assist in the understanding of mixing phenomena in a submerged lance smelting furnace. In the experiment, only a single lance was used, as shown in Fig. 3a. In order to measure the mixing time under various conditions, a potassium chloride aqueous solution was used as a tracer and electric resistance was measured at different locations. More details can be found in Lang et al. [3]. In order to validate the developed numerical model with the water model, a geometry identical to the experimental setup (scaled-down 1:12) was developed, as shown in Fig. 3b. A single lance was used in the numerical model and a tracer with same fluid properties was introduced at the same location as the experiment. The tracer concentration was measured at four different points at uniform time intervals throughout the simulation. A quasi-steady state flow field was first achieved before introducing the tracer. The mixing time was measured after the introduction of tracer based on the tracer concentrations measured with respect to time, as shown in Fig. 4.

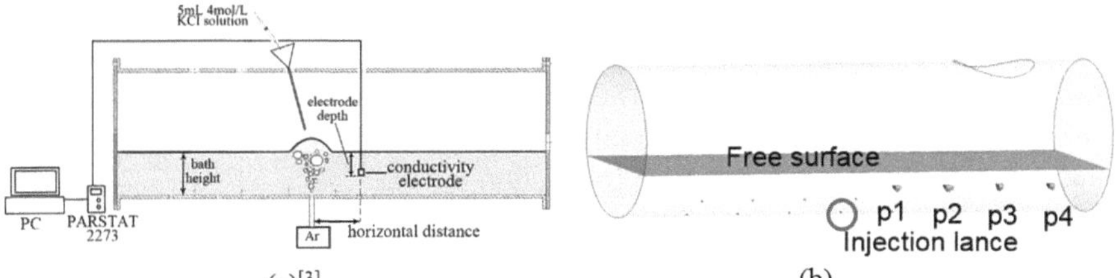

(a)[3] (b)

Fig. 3 Experimental set up (**a**), and numerical set up (**b**)

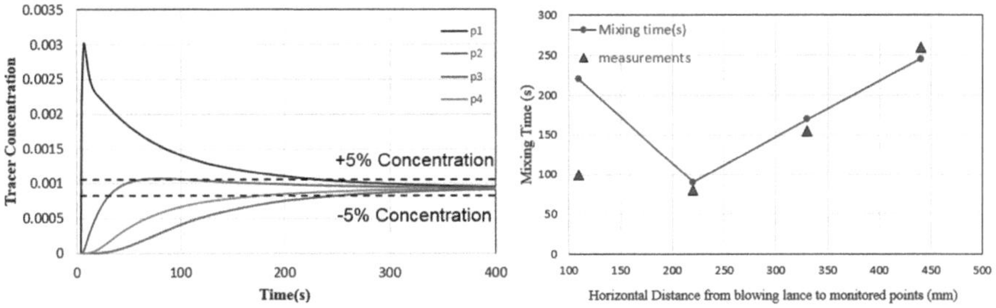

(a) (b)

Fig. 4 Tracer concentration with respect to time (**a**), comparison with measurements (**b**)

Figure 4a indicates that the tracer concentration profile with respect to time varies at different measurement points. The closest measurement point $p1$ experienced the highest tracer concentration after the tracer was introduced, before gradually reducing to the final homogenized concentration. Measurement point $p2$ also shows the tracer concentration first increasing to values larger than the homogenized concentration, before falling off. The tracer concentration profile with respect to time measured at points $p3$ and $p4$ indicate a similar mixing process. Due to these tracking locations being further away from the injection lance, the tracer concentration gradually increased to the homogenized concentration. By comparing the 95% mixing time predicted by the CFD model with measurements, it can be seen that the numerical model prediction agrees with the measurements for point $p2$, $p3$ and $p4$ as shown in Fig. 4b. The discrepancies observed at point $p1$ occur due to the numerical patch process, where the closest point to the introduced tracer will experience high concentration. Similar profiles can be find in other published research [6].

Flow Characteristics

By using the multi-fluid VOF model coupled with the Eulerian multiphase model, the transient flow field was obtained. The boundary condition of each lance mass flow rate was set to 0.48 kg/s, and the average copper phase velocity under the free surface at three different levels are shown in Fig. 5.

The average velocity at the three different levels below the free surface shows a similar profile according to Fig. 5. The flow field reached a quasi-steady state after approximately 4 s. The average velocity is approximately 0.3 m/s, indicating rapid flow development and strong phase interaction effects.

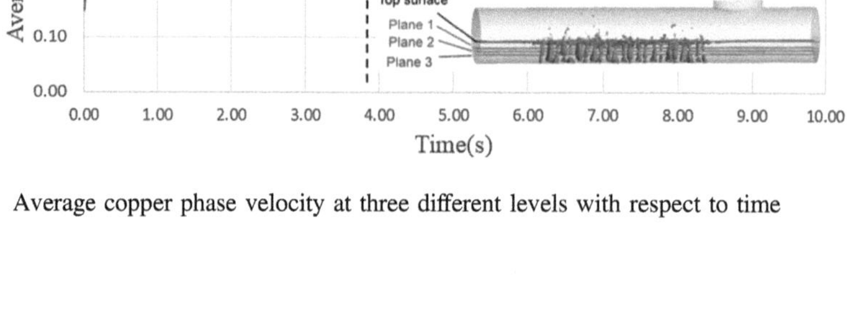

Fig. 5 Average copper phase velocity at three different levels with respect to time

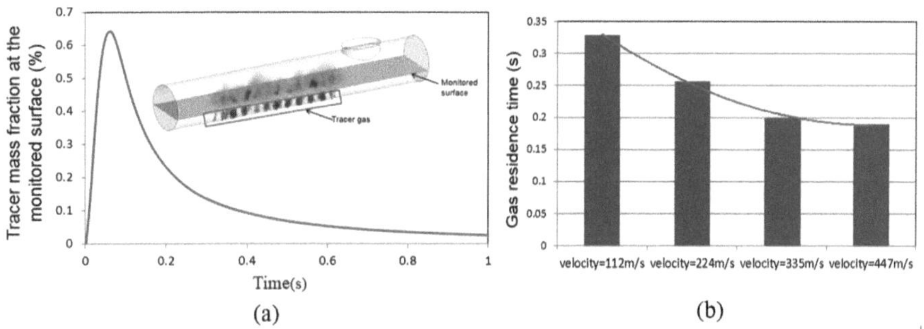

Fig. 6 Tracer (gas) mass fraction at the monitored surface with respect to time (**a**), tracer (gas) residence time under different gas velocity (**b**)

Gas Residence Time

In the bath smelting process, the residence time of dissolved oxygen enriched air in the bath will directly affect the efficiency of the smelting process and the energy efficiency. In general, the longer the reactants stay in the reactor, the more complete the reaction. Therefore, it is necessary to analyze the residence time of oxygen enriched air in the bath.

After the flow field reached a quasi-steady state, a numerical approach was used to measure the gas residence time. A tracer gas with the same properties as the oxygen-enriched air was introduced in the flow field at the lance inlets. By monitoring the tracer gas concentration with respect to time as shown in Fig. 6a, the gas residence time can be calculated based on Eq. 19, giving a value of 0.247 s.

$$\bar{t} = \frac{\sum t_i m_i}{\sum m_i} \tag{19}$$

The calculated residence time of the gas shows good agreement with times observed in previous literature [5, 7]. Since the gas flow rate under actual operating conditions is different at each oxygen lance, ranging from 87.65 to 560 m/s range, a study of the effects of different gas flow rates on the gas residence time was conducted. This study utilized four different lance velocities: 112, 224, 335, and 447 m/s. The results in Fig. 6b show that as the lance gas velocity (flow rate) increases, the gas residence time is reduced. When the lance flow exceeds 335 m/s, there is no longer a significant reduction in gas residence time.

Effect of Gas Velocity on Splashing

During the bath smelting process, the air bubbles rise through the high-temperature melt. The gas volume is affected by the process of thermal expansion and

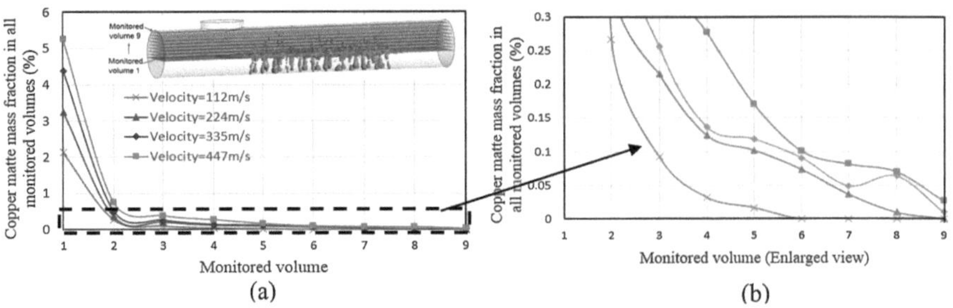

Fig. 7 Copper matte splashing under different gas velocities (**a**), enlarged view (**b**)

contraction. The air bubbles break up when they reach the free surface. Because the internal pressure of the air mass is greater than the upper portion of the flue gas, the bubbles break instantly and melt splashing occurs. The melt splashing effect is directly related to the gas phase kinetic energy. When splashing occurs, it will reduce copper content and also increase heat loss. In addition, copper splashes to the feed mouth could potentially block it, thus negatively impacting equipment safety and smelting efficiency. Figure 7 shows the splashing under various inlet gas velocities.

In order to evaluate the degree of splashing, the liquid level above the area was divided into nine volumes with equal height. When the flow field reached quasi-steady state, the matte weight was detected in each volume. Since under actual operating condition the gas flow rates are different from each oxygen lance, gas flow rates are varied in the range of 87.65–560 m/s. For the study of gas velocity impacts on splashing, four lances speeds (112, 224, 335, and 447 m/s) were utilized. The simulation results in Fig. 7 show that with an increase in lance gas velocity (flow rate), the amount of liquid splashing also increased. When the gas velocity exceeds 224 m/s, splashing began to reach top positions.

Conclusions

In this study, a three dimensional multiphase flow model was developed for a submerged lance smelting furnace developed by Dongying Fangyuan Nonferrous Metals Co., Ltd. The multi-fluid VOF model was coupled with Eulerian multiphase model to describe the interphase behavior. The mixing time predicted by the developed CFD model shows good agreement with water model experiments. The simulation results indicate that the submerged lance smelting furnace has rapid flow development characteristics. The residence time of the injected oxygen enriched air under different injection velocities indicate that when the lance flow exceeds 335 m/s, the gas residence time reduces to less than 0.2 s. Beyond this velocity, the residence time is no longer reduced significantly with increasing gas velocity.

The splashing study has shown that when the gas velocity exceeds 224 m/s, copper begins to reach the top part of the submerged lance smelting furnace.

Acknowledgements The authors would like to thank Dongying Fangyuan Nonferrous Metals Co., Ltd. for offering this research opportunity, providing funding and guidance during the course of this work.

References

1. Z. Cui, D. Shen, Z. Wang, New process of copper smelting with oxygen enriched bottom blowing technology. Youse Jinshu (3), 17–20 (2010)
2. X.Y. Guo, Q.M. Wang, L. Liao, Q. Tian, Y.Z. Zhang, Mechanism and multiphase interface behavior of copper sulfide concentrate smelting in oxygen-enriched bottom blowing furnace. Nonferrous Met. Sci. Eng. **5**, 28–34 (2014)
3. S. Lang, Z.X. Cui, X.D. Ma, M. Rhamdhani, A. Nguyen, B.J. Zhao, Mixing phenomena in a bottom blown copper smelter: a water model study. Metall. Mater. Trans. **B46**, 1218–1225 (2015)
4. J. Gonzalez-Trejo et al., CFD simulation gas-liquid flow in a copper converter with bottom air injection. Int. J. Chem. Reactor Eng. **6**(1), 1–20 (2008)
5. Z.Y. Zhang, Numerical simulation and optimization of multiphase flows behavior in Bottom-blown furnace. Masters Thesis, Central South University, 2012
6. H. Liu, Z. Qi, M. Xu, Numerical simulation of fluid flow and interfacial behavior in three-phase argon-stirred ladles with one plug and dual plugs. Steel Res. Int. **82**(4), 440–458 (2011)
7. F.K. Liu, Numerical simulation and optimization of multiphase flow in bottom blowing lead smelting furnace. Masters Thesis, Central South University, 2013, pp. 24–25

Debottlenecking High Temperature Metallurgical Plants Through Modeling and Simulation

K. Adham and T. Buchholz

Abstract Debottlenecking of high temperature processes within an existing metallurgical plant can significantly increase the throughput to meet new production demands. Modeling and simulation can help to fulfil that aim through incremental changes and operational improvements, rather than the more expensive option of building new plants and facing new environmental and infrastructure challenges. This paper describes a methodology of how modeling tools can be used to simulate the processing steps, how to estimate their existing and ultimate capacities, and how to determine the minimum cost points for a plant capacity expansion. This technique is presented through a generic description that can be applied to different metallurgical processing studies. We start with heat-and-mass balances, draw a bottleneck diagram, brainstorm capacity increase options, and prepare a cost-capacity curve. Hatch has successfully applied this technique to a number of plant capacity studies, with great benefits to our industrial clients (Adham et al. in A Success Path from Process R&D to Commercial Plants, Montreal, Canada, [1]).

Keywords Debottlenecking · Metallurgical plants · Ultimate capacity · Process modeling · Optimized investment · Bottleneck diagram · Cost-capacity curve

Introduction

Often it is economically more viable to improve and debottleneck the existing facilities, rather than build new ones. However, spending new money on old plants requires careful considerations. Process modeling, hand in hand with plant measurements, can identify the best scenarios to balance the three aspects of plant capacity (nameplate, utilization and yield), both at the global and the individual

K. Adham (✉) · T. Buchholz
Hatch Ltd., Mississauga, Canada
e-mail: kadham@hatch.ca

J.-Y. Hwang et al. (eds.), *8th International Symposium on High-Temperature Metallurgical Processing*, The Minerals, Metals & Materials Series,
DOI 10.1007/978-3-319-51340-9_12

system and subsystem levels [2]. Hatch has performed a number of debottlenecking projects for various minerals and metals plants, applying process modeling to provide an interactive framework that can examine and estimate a range of options for debottlenecking [3].

Defining the Debottlenecking Requirements

A successful debottlenecking project finds alignment of the solutions to a plant's capacity increase with the expectations and limitations of the type of plant within particular industry segments. For example, a Client for a Plant may mandate a production increase of at least 15% in order to justify a performance-driven debottlenecking project. A larger project scope for a Client-producer with a global distribution system and production spread over multiple plants may allow for a somewhat lower target capacity increase for the individual plant. A plant's capacity is constrained by its weakest links in the production chain. In order to eventually identify and strengthen these links, the plant's process has to be broken down into discrete unit operations. These can be visualized in Block Flow Diagrams (BFD) or Process Flow Diagrams (PFD), and can be grouped in logical combined-systems, based on their interactions.

Estimating Component Capacities

For a simple representation of our methodology, the plant capacity is considered as a function of up-time (UT) and yield. UT represents the utilization of the equipment, and is the product of on-stream-efficiency (OSE) and rate-factor (RF). The OSE reflects the amount of time that the equipment is actually producing (hours per year) while the RF reflects the ratio of actual capacity versus the maximum (nameplate) capacity. Yield in this case is the conversion factor (product/feed) between what the equipment or subsystem under consideration is processing, versus the product at its battery limit. It should be noted that, in the context of this debottlenecking methodology, yield and nameplate are not necessarily following their conventional definitions. Improving on any of these two factors (up-time and yield) is an act of capacity debottlenecking, as long as the impact of modifying one does not deteriorate the other. For example, consider a hypothetical copper roaster plant (Fig. 1 example showing roaster, off-gas and cooler). That roaster plant is conditioning feed to an electro-refinery with a yield of 15% (cathode/feed mass ratio) while running at an OSE of 85% (7446 h/year out of 8760). The 85% on-stream may appear acceptable. A debottlenecking team, through plant investigation and process modeling, determines that the maximum (nameplate) capacity of the system is equivalent to 25 tph cathodes, however, its actual average production capacity has only corresponded to 20 tph. Therefore, this plant area has a rate-factor of 80% (20 over 25) and an uptime (OSE × RF) of only 69.7%. The focus for debottlenecking such a plant would be to increase a combination of the three variables; increasing the 15% yield (less loses), raise the OSE above 85% (more hours/year) and the 80% RF (up to nameplate capacity).

Other than process simulation, estimating the nameplate capacity of the process equipment would typically require some degree of plant data analysis, investigation of operator experience, in-plant test campaigns, and review with the equipment designers and suppliers. For example, the rate factor can be determined based on analysis of production data pulled from the plant historian (data gathering and storage system, e.g. PI ProcessBook® or Wonderware®).

The average throughput while online is extracted straight off the historian, while an estimate of the nameplate capacity can be determined from identification of the best demonstrated periods (BDP) of highest throughput. The BDP, though an excellent practical representation of the bottom line Nameplate, should however be qualified by considering whether or not the equipment under study is limited by downstream or upstream interactions. It is therefore critical that the engineer performing the assessment keeps a holistic view of the process in mind.

OSE, as well as some factors affecting the RF, are determined from analysis of the historian, maintenance logs and failure mode analysis (Fig. 2).

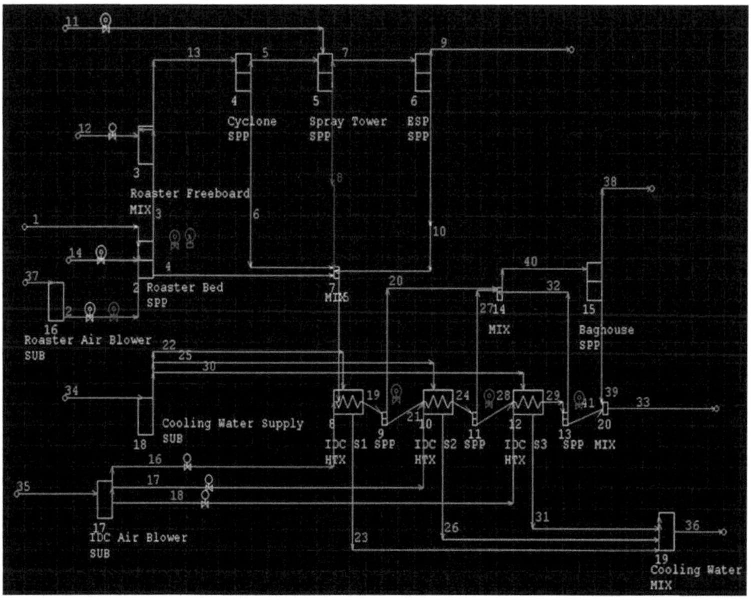

Fig. 1 Metsim process model example of a roasting plant [1]

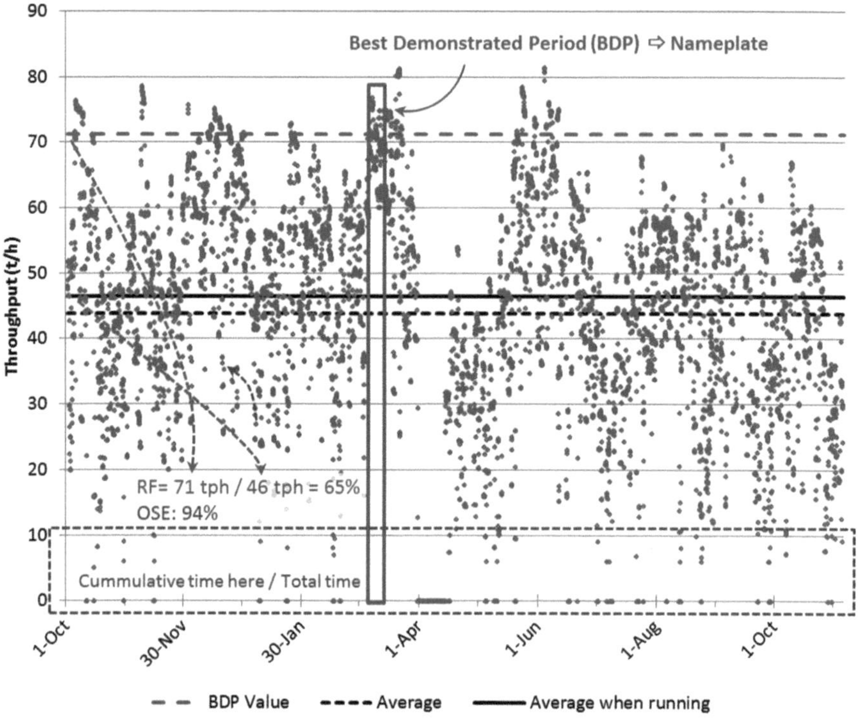

Fig. 2 Illustration of BDP, OSE and RF estimation from operating data

Identifying Options and Innovations

The output of the process modelling, data analysis and engineering (described above) is an equipment (or sub-system) capacity table, which is eventually graphically represented as a bottleneck ranking diagram (BRD), which lists all items within the battery limits by their capacity (Fig. 3). This BRD represents the existing production facility's baseline.

Identifying potential improvements and innovations can best be realized through sharing the results of process modeling and equipment capacity determinations (BRD), with the plant's operating personnel through open and frank idea-generation sessions (ideation). Participants should be made at ease and feel free to express their opinions and ideas, without any limit or consideration; because ideation can best work in a free-flowing format, without judgment and curtailment (Fig. 4). A tool, developed by Hatch and acting as a catalyst for creativity in these ideation sessions, is the UTC Diagram (Utilization–Throughput–Capacity), which maps out the impact of the proposed changes on the nameplate, OSE, RF (hence UT) and yield—and all together on capacity (Fig. 5).

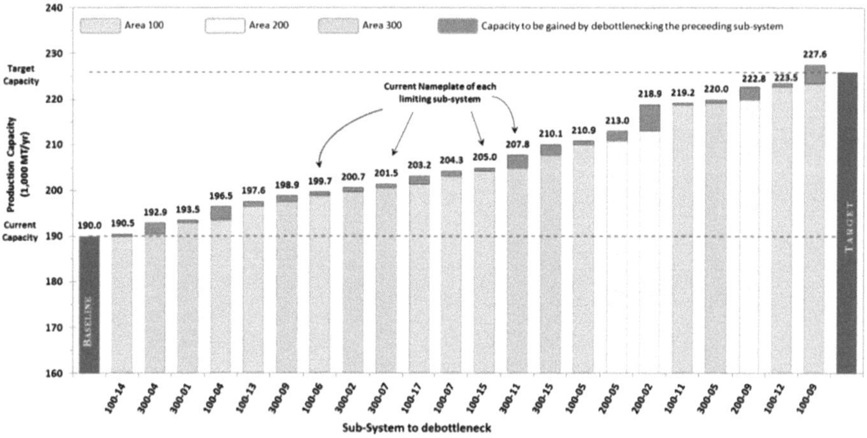

Fig. 3 BRD example for equipment with capacity higher than baseline but less than target in each subsystem of a plant

Fig. 4 Scope convergence pyramid, from brainstorming to estimation

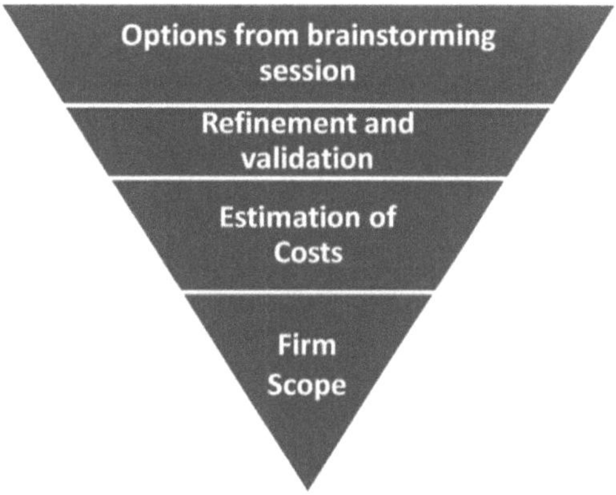

Firstly a range of ideas are generated, for each plant processing area, and major equipment components. This same team should then be tasked with reviewing and ranking their own ideas. At this point, critiquing and debating the merits and difficulties associated with each option and idea is not only allowed, but actually required. An option evaluation matrix may be used for, qualifying, and roughly quantifying (if possible), the impact of all proposed ideas in terms of the key operating parameters, (e.g. capacity, quality, capital and operating costs, safety, and risks).

The UTC diagrams (Fig. 5) can be updated to show results of the analysis. The final product is a ranked shortlist of the promising options and innovations, which will then be studied and estimated, by the engineering team.

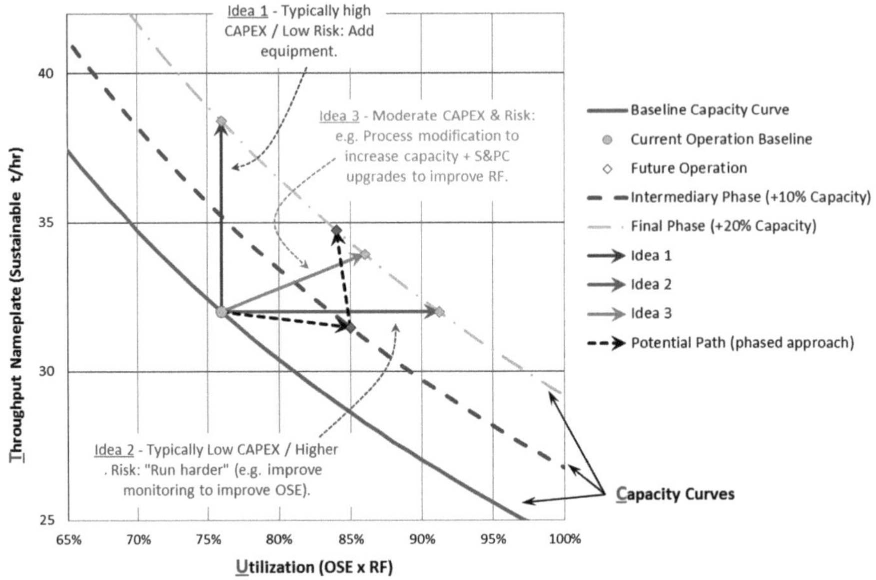

Fig. 5 UTC chart (utilization–throughput–capacity)

Estimating Cost Versus Benefit

Mass-and-energy balances should be prepared for each of the study options, in order to estimate the increase in Variable Costs associated with the process changes (e.g. higher reagent consumption, increased power consumption, increased fuel requirements and etc.). The same balances are used to estimate the benefits that can be realized, in terms of production capacity and product quality.

The process simulator is then run under the anticipated case for each option (or a grouping of options), in order to establish the interaction among the major equipment versus the upper limit on their capacities. The impact of new and innovative ideas on the equipment performance should be inputted to the simulator (ideally after prototype or demonstrative testing), in order to determine their estimated benefits.

Once a debottlenecking solution is selected for each sub-system, the BRD is updated to include the new capacities, as well as the cumulative cost of implementation. This chart along with a cost/capacity chart (Fig. 6) plotting the cost per capacity gained versus the future plant capacity, are used for determining the business case sweet-spot (highest gain for lowest price). A quantifiable road map is generated between the current plant's operation and achieving the intended capacity by way of debottlenecking. Depending on the project-specific peculiarities, a phased approach over time may then be advisable.

With visual clarity of the examples of BRD and cost/capacity curves in mind, upgrade to each capacity-limiting sub-system is kept as single-step debottlenecking

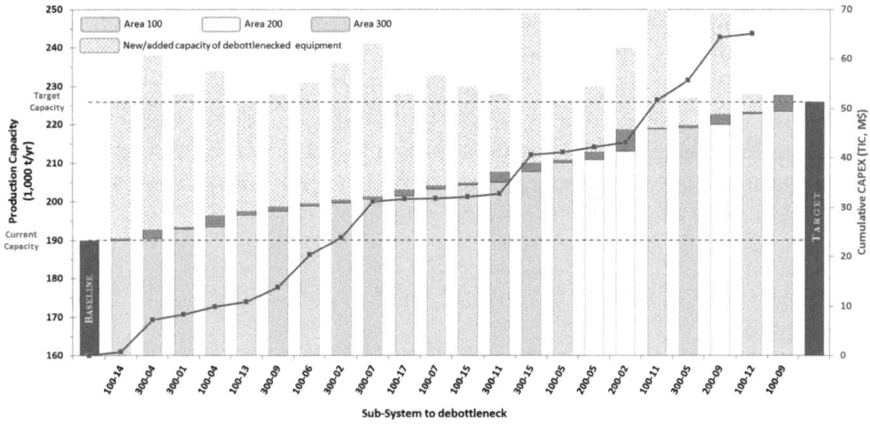

Fig. 6 BRD after capacity increase versus capital cost evaluation

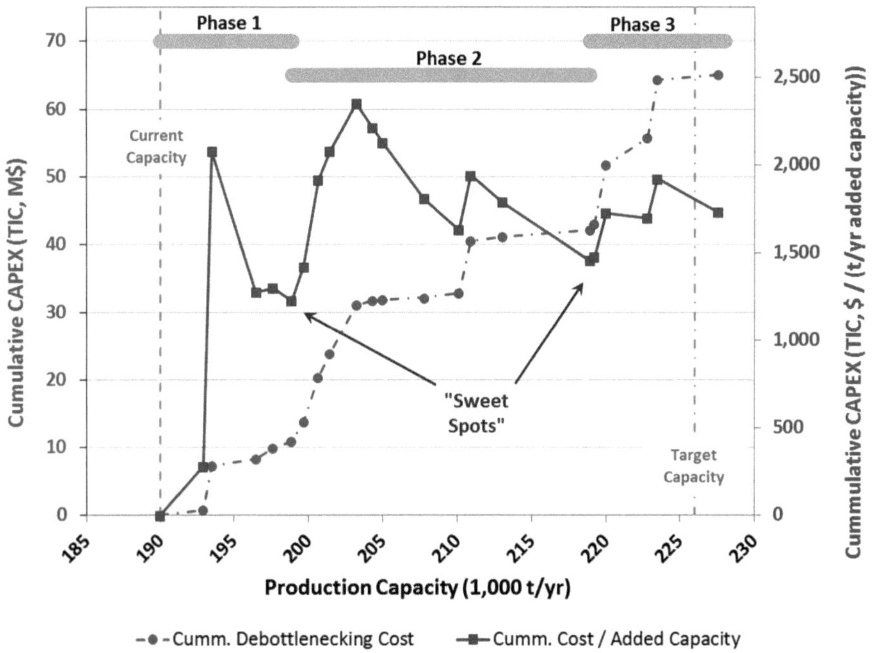

Fig. 7 Determining sweet-spots on a cost versus capacity curve

to a capacity greater than the target capacity. In practical terms however, a phased approach of debottlenecking may be applicable to each sub-system (Fig. 7). For example, a roaster may first see its capacity increased by way of additional air-injection capacity; then followed by a later debottlenecking phase, an upgrade of its cyclone-catch recirculation system.

All of the work associated with a debottlenecking project will undergo rigorous quality controls. The scope of these projects will typically cover a wide variety of types of equipment, and the lead engineers overseeing the study should be robust generalists. They should not hesitate to implicate specific subject matter experts for additional input and review. Proposed costs and performances should be benchmarked against past experiences and industry standards, with frequent reviews set out as clear milestones along the project schedule.

Conclusion

A method has been described for debottlenecking high-temperature metallurgical processes, through identifying incremental operational improvements. Our methodology can be used to estimate the existing and ultimate plant capacities, and determine the minimum cost points for expansion. This technique starts with heat-and-mass balances, draws a bottleneck diagram, identifies capacity increase options, and finally leads to a cost-capacity curve.

References

1. K. Adham, L. May, A. Tohn, A success path from process R&D to commercial plants. Presented at the Conference of Metallurgists (COM 2011), Montreal, Canada (2011)
2. T. De Mull, S. Saich, K. Sobel, Development of a mineral processing flowsheet—case history, Batu Hijau, in *Mineral Processing Plant Design, Practice and Control*, ed. by A.L. Mular, D. N. Halbe, D.J. Barratt (SME Publishing, Vancouver, 2002), pp. 2211–2237
3. M. Dionne, P. Lavoie, An approach for disc filter debottlenecking and flotation plant improvements. Presented at the AICHE 34th International Phosphate Fertilizer and Sulfuric Acid Technology Conference, Clearwater Beach, FL, USA (2010)

Assessment of Slag Entrainment in a RH Degasser Through Physical Modelling Using Circulating Fluids of Different Densities/Oil Systems for Simulating Steel Melt/Slag

Johne Jesus Mol Peixoto, Natalia Barros Barony, Heric Henrique Souza e Silva, Weslei Viana Gabriel, Carlos Antonio da Silva, Itavahn Alves da Silva and Varadarajan Seshadri

Abstract In this work the circulation rates have been determined in a 1:7.5 scale physical model of RH reactor (Ruhrstal Heraeus) and using oil/water and oil/zinc chloride solution to simulate steel melt/slag in the prototype with the objective of evaluating the influence of density and viscosity on the entrainment of oil simulating slag. A technique based on strain gage bridge and particle image velocimetry was developed to assess the circulation rate, as the usual tracer method is not feasible due to the presence of chloride ions in $ZnCl_2$. The critical velocity and critical diameter of oil drops for slag (oil) entrainment in water/oil and $ZnCl_2$/oil systems were estimated using equations from the literature. Due to the difference in density, the $ZnCl_2$/oil system, showed a critical velocity that is 28% higher than the silicone oil/water system while the maximum size of the entrained droplets was about 65% lower than that for water/oil system.

Keywords RH degassing · Model · Circulation rate · Slag entrainment

J.J.M. Peixoto · N.B. Barony · H.H.S.eSilva · W.V. Gabriel · C.A. da Silva · I.A. da Silva
Department of Metallurgical Engineering and Materials, Federal University of Ouro Preto, 35400-000 Ouro Preto, MG, Brazil

V. Seshadri (✉)
Department of Metallurgical Engineering and Materials, Federal University of Minas Gerais, 6627, Av. Antonio Carlos, 31.270-901 Belo Horizonte, Brazil
e-mail: seshadri@demet.ufmg.br

J.-Y. Hwang et al. (eds.), *8th International Symposium on High-Temperature Metallurgical Processing*, The Minerals, Metals & Materials Series,
DOI 10.1007/978-3-319-51340-9_13

Introduction

Yang et al. [1] have simulated the addition of desulfurizing slag at a RH vacuum chamber, using a 1:2.8 scale model. Analyzing the evolution of the oil layer at the top of the ladle, they were able to evaluate the residence time and the amount of oil dispersed within the liquid due to the gas flow. Costa et al. [2] studied the desulfurization process in RH by adding a supernatant layer of slag in the vacuum chamber after the rise of the liquid steel level due to the effect of vacuum. A mathematical model was used under industrial conditions to analyze the desulfurization of steel and evaluate the influence of parameters such as chemical composition and volume of the slag. It should be noted that these and other works on desulphurization of steel in RH do not take in consideration the conditions for the dragging of slag added to the vacuum chamber through the down leg. The formation of a slag-metal emulsion is of considerable practical importance, as the increase in the interfacial area caused by dispersion of one liquid phase into the other is responsible for the high reaction rates, found in processes such as BOF, the ladle furnace etc.

In the physical modeling, for studying the interactions between metal and slag, it is common practice to use water/oil system. This has a limitation in respect of density difference between the two phases. In order to modify this parameter, aqueous solutions of zinc chloride [3] or sodium chloride [4] can be used instead of water. If $ZnCl_2$ solution is used for the simulation of liquid steel it is necessary to develop a method to determine the circulation rate since the usual method of addition of tracers like KCl etc. together with monitoring the conductivity values over a period of time cannot be used as the solution already contain ions. Hence in this study, an indirect measurement system based on strain gage bridge and PIV technique have been used to determine the circulation rate of zinc chloride ($ZnCl_2$) in the physical model for investigating the effect of gas flow rate and other parameters on the circulation rate.

Materials and Methods

Laboratory simulation tests were conducted in a physical model of the RH degassing unit simulating a prototype (RH degassing unit of a steel plant). It was built in acrylic with a scale factor $\lambda = 1{:}7.5$. Dimensions and parameters of the prototype and model are given in Table 1.

The parameters of the model presented in the last column of Table 1 were determined using the scale ratio and appropriate similarity criteria, resulting in dimensionless numbers with the application of Buckingham π theorem to the variables relevant to RH process as in earlier publications [5]. As is seen the injection nozzles do not obey the scale ratio. This feature requires adapting these criteria with respect to modified Froude number. The dimensionless parameters

Table 1 Parameters of RH reactor and those of physical model as determined through similarity criteria

Parameters		Prototype	Physical model
Fluid data	Fluid	Steel	Water
	Density of fluid (kg/m^3)	7000	1000
Reactor dimensions	Internal diameter (mm)	2245	299.33
	Height (mm)	7600	65[a]
Snorkels	Length (mm)	1820	242.67
	Internal diameter (mm)	750	100
	Distance between center lines (mm)	1540	205.33
	Depth of immersion (mm)	450	60
System of gas injection	Gas	Argon	Compressed air
	Flow rate (NL/min)	2500	120
	Density of gas (kg/m^3)	0.13	1.123
	Gas injection points (Upleg)	16 nozzles, distributed in 2 levels of injection	Same as prototype
	Distance between gas injection points	175 mm	23.3 mm
	Diameter of injection nozzle (mm)	8	2.4

[a]Water column height compatible with prototype similarity

namely the Froud number (Fr), Flow Number (N_{Va}) and modified Froud number for nozzle (Fr_m) for calculating the parameters in Table 1, are given below:

$$Fr = \frac{V^2}{gD};\ N_{Va} = \frac{Q}{D^2 V};\ Fr_m = \frac{\rho_g\, Q(Q/\varphi^2)}{\rho_l D^3\, g} \tag{1}$$

where: ρ_l and ρ_g are the densities of liquid and gas respectively; D is the snorkel diameter; V is the liquid velocity; Q is the gas flow rate; φ is the nozzle diameter and g is the gravity acceleration.

In the physical model water and $ZnCl_2$ solution were used for simulating the steel melt and oil for the slag. Gas (compressed air) flow rates employed varied from 30 to 120 l/min (increases of 10 l/min). Immersion depths of the legs were 60, 100 and 140 mm and the height of liquid in the vacuum chamber was fixed at 70 mm. In the physical model the ladle was replaced by a rectangular box (lower vessel), which enables elimination of curvature effects from the ladle walls. A planar surface allows the utilization of the PIV technique without distortion of the laser beam. A schematic diagram of the physical model is given in Fig. 1a.

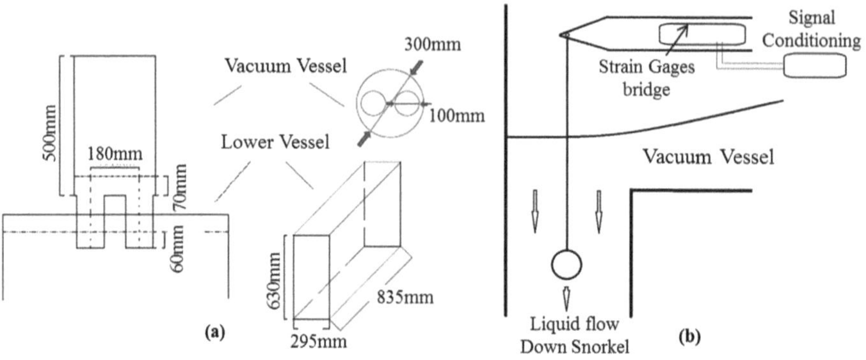

Fig. 1 **a** Schematic representation of the geometric parameters of the physical model; **b** Schematic diagram of set up for determination of circulation rate with strain gage bridge method

Determination of Circulation Rate of Fluid by Strain Gage Bridge Method (SGBM)

Determination of circulation rate in RH physical model can be done using an indirect measuring system based on strain gage bridge. The experimental set up is given in Fig. 1b. It consists of a sphere with known mass (m), volume (v) and radius (R) that hangs from the end of a flexible blade. The sphere is positioned in the down snorkel and thus subjected to the interaction with the fluid that comes from the vacuum chamber to the ladle. The force acting on the blade should be such that it is in the elastic range of flexion. Two Strain Gages are glued on top of the blade face and two on the lower face. They are interconnected in the form of a Wheatstone bridge. The primary signal from the strain gage bridge is processed through a signal conditioner. The resulting e.m.f (Volts) output signal is proportional to the blade flexion. This can be fed into a A/D acquisition board. The electrical signal of the Wheatstone bridge is proportional to the deformation and the latter can related to the sphere liquid interaction forces. The sphere positioned inside the down leg is subject to forces corresponding to the weight, buoyancy and drag force due to the flowing fluid. Greater the rate of fluid flow, greater is the drag force, which in turn causes increased deformation of the blade. This gives rise to higher intensity of signal generated by the bridge. The rate of variation of the electrical signal (Volts) can be used to estimate the flow rate (circulation rate).

Three series of readings are necessary for processing the data. The first one is initial reading when there is no deformation (Zero point), measuring only the effect of the own weight of the blade without coupling to the sphere. The second reading is taken from the signal of the bridge after the coupling with the sphere in the air without any flow of fluid. The difference between the second reading and the first one corresponds to the deformation caused by the weight of the sphere. Therefore these two first signals are meant for calibration of the apparatus. The third reading is taken with the fluid flow and this reading accounts for the consequent frictional effect on the sphere depending on the flow rate. The velocity of the fluid in the downleg and the circulation rate can be evaluated from the scheme of equations given below

$$(WS - B) + f\frac{1}{2}\rho_l V^2 \pi R^2 = (3rd\,reading - 1st\,reading) \times \frac{weight\,of\,sphere(WS)}{(2nd\,reading - 1st\,reading)} \tag{2}$$

where B (buoyancy—4/3 $\pi R^3 \rho_l$) and WS (weight of sphere) are in Newton's; 1st, 2nd and 3rd readings are strain gage bridge signals given in volts; μ *and* ρ_l *are* viscosity and density of the liquid respectively.

For values of Reynolds number $Re = \rho_l VR/\mu$ above 1000 the friction factor (f) value is almost constant [6], namely $f = 0.44$ for Reynolds > 1000 The liquid velocity (V) in the downleg with diameter D, can be evaluated from Eq. (2) and the flow rate or circulation rate is given by:

$$[kg/s] = \rho_l (V\frac{\pi D^2}{4}) \tag{3}$$

Two points need to be highlighted on the application of this methodology. Beyond gas flow rates of 90 l/min there is significant presence of gas bubbles in the down snorkel, i.e. the flow ceases to be monophasic. Furthermore, the flow in the down snorkel is shown extremely turbulent with strong recirculating currents (swirls); thus it is difficult to apply the unidirectional and symmetrical flow conditions around the spherical body. Thus, it is necessary to use a second method for determining the flow rate in order to calibrate and validate the results of the SGBM.

PIV (Particle Image Velocimetry)

PIV technique can be used to assess the liquid velocity profile in the down leg. It is a straightforward procedure, if the liquid is transparent and if no bubbles are present. The velocity profiles were determined along the downward leg for preselected heights. For each combination of flow rate and immersion three series of 100 images were acquired at 10 Hz frequency. Polyamide particles were used as seeding and a PIV device provided by Dantec was used [7]. Each velocity profile depicts the downward velocity component along the diameter of the down snorkel. The average velocity of the liquid in the down snorkel can be evaluated as:

$$\overline{V} = \int_{o}^{R} \frac{2\,\pi\, r\, V(r) dr}{\pi\, R^2} \tag{4}$$

where R is the radius of down leg; V is the liquid velocity at the point r as determined by PIV technique; and r is radial the distance from the center of the down leg towards the wall.

The flow values (flow rate) obtained by the two techniques were compared, for the calibration of SGBM. As already pointed out this additional calibration is required because the flow in the down leg is turbulent with recirculating currents; which is contrary to assumptions of unidirectional and symmetrical flow conditions around the sphere. It should be noted, when the gas flow rate increases, a fraction of gas may be drawn into the down leg and thus hindering measurements with PIV which is sensitive to presence of gas bubbles.

Oil Entrainment

Tests were carried out using water/silicon oil and $ZnCl_2$ aqueous solution/silicon oil in order to evaluate the density influence on the slag drag conditions. Gas flow rates from 30 to 70 l/min (increases of 10 l/min) were employed for immersion depth of the snorkels 100 mm and the height of liquid 70 mm in the vacuum chamber. A 10 mm oil layer was added at the vacuum chamber on top of the circulation fluid (water or $ZnCl_2$ solution), before starting the gas injection. A sequence of images of the vacuum chamber as well as of the down snorkel just below its tip has been captured by high speed camera (300 fps) in order to highlight the oil dragging process. These videos were recorded from the beginning of gas injection. Relevant fluid properties are as follows: $\rho_{water} = 1.0$ g/cm^3 and $\mu_{water} = 1.0$ mPa s, $\rho_{ZnCl2} = 1.22$ g/cm^3 and $\mu_{ZnCl2} = 15.33$ mPa s, $\rho_{oil} = 0.95$ g/cm^3, $\mu_{oil} = 526$ mPa s, $\sigma_{water/oil} = 35.7$ mN/m, $\sigma_{ZnCl2/oil} = 26.1$ mN/m.

Results and Discussion

Circulation Rate

The velocity profiles obtained via PIV velocimetry for different gas flow rates are shown in Fig. 2a, c. From these, three horizontal lines coinciding with the diameter of the leg were drawn at 15, 30 and 45 mm above the downward leg lower end (Fig. 2b, d). It can be seen that the velocity values do not vary significantly as the measuring line moves away from the down leg tip (15–45 mm). It should also be noted that for the flow rate of 110 l/min, the velocity distribution is completely asymmetric due to the presence of large amounts of bubbles. From these values, flow rate of water can be determined in the down leg (circulation rate) for each gas flow rate, see Eq. 4. Figure 3a shows a comparison between circulation rate values obtained using the strain gages bridge method and PIV technique for water.

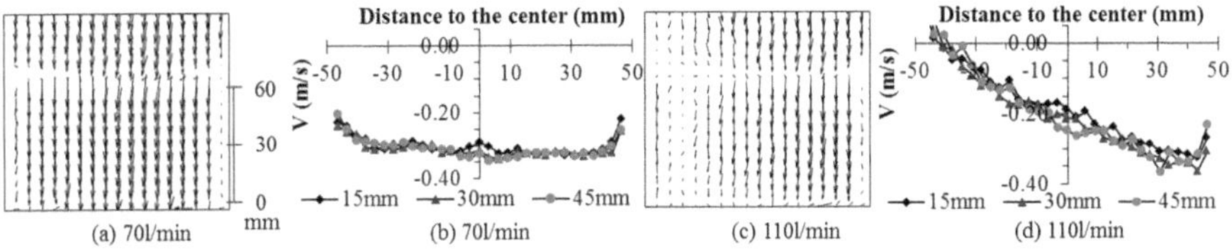

Fig. 2 Profile of velocities in the down leg obtained through PIV technique using leg immersion equal to 60 mm and for gas flow rates: **a**, **b** 70 l/min; **c**, **d** 110 l/min

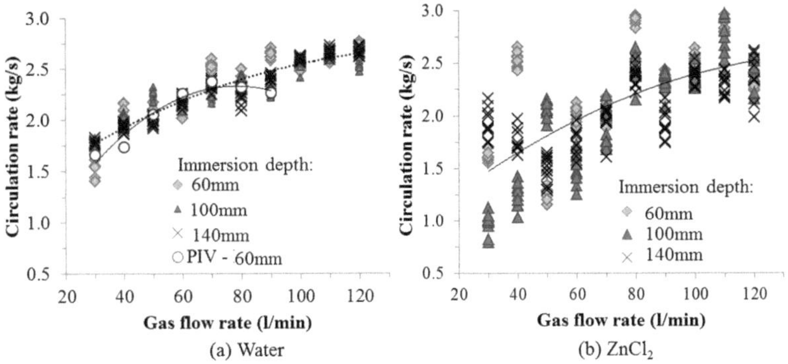

Fig. 3 **a** Circulation rate values from strain gage bridge and PIV methods after correction of the former with a calibration factor $\alpha = 0.6$, in the case of water; **b** circulation rates for aqueous solution of $ZnCl_2$ with a calibration factor $\alpha = 0.6$

The values obtained by strain gage bridge method were higher than those obtained by PIV, requiring the use of a calibration factor for conciliating the measurements. The calibration factor was found to be 0.6, and this value was thereafter applied to $ZnCl_2$ solution, as shown in Fig. 3.b.

In Fig. 3 it is noticed that the flow rate increases with increasing gas flow rates as expected and according to earlier works [8–10]. It was not possible to identify the influence of the depth of immersion leg on the circulation rate. Comparing the results of experiments using water with the values that resulted using the $ZnCl_2$ solution it is observed that the circulation rate was slightly higher in the case of water as circulating fluid. However it should be noted that there is a tendency for the circulation rate values to stabilize at for higher gas flow rates. Existence of a saturation value for the gas flow, from which the circulation rate tends to decrease because excess gas in the upleg reduces upward pushing of the liquid contained therein [10].

Estimation of Slag Entrainment

Table 2 presents some of the expressions from literature to compute minimum critical velocity beyond which slag entrainment can be caused by the shearing force due to turbulence on the surface as well as critical diameter (maximum) of the entrained droplets [11]. Although the geometry and flow conditions are not exactly the same as in the RH system these expressions highlight the influence of physical properties on the entrainment phenomena. Physical properties of silicon oil and circulating fluids can be used to estimate the slag entrainment. The values of critical velocities calculated for the present case as well as the critical diameter are given in the last columns of Table 2. It is observed, that due to the difference in density, the $ZnCl_2$/oil system shows a critical velocity that is 28% higher than the silicone oil/water system while the maximum size of the entrained droplets is about 65% lower than that of water/oil system.

In the experiments using water/oil, the oil drag occurred in less than 10 s for all gas flow rates from 30 to 70 l/min; oil droplets promptly returned from the lower vessel to the vacuum chamber as it can be seen on the last images of Fig. 4a. When $ZnCl_2$ solution/oil system is used to simulate steel/slag, the total time length required for totally dragging the oil from the vacuum chamber increases, (Fig. 4b)

Table 2 Critical entrainment velocities and droplet sizes for $ZnCl_2$/silicone oil and water/silicone oil

Equations for critical velocity (Ref. [11])	Remarks	V_{ZnCl2} (m/s)	V_{water} (m/s)	V_{ZnCl2}/V_{water} (%)
$V_{crit} = \sqrt[4]{4g(\rho_l - \rho_u)\sigma_{ul}\frac{(\mu_l+\mu_u)^4}{(\rho_u\mu_l^2+\rho_l\mu_u^2)^2}}$	Viscosity effect according to Kelvin-Helmholtz theory	0.120	0.092	131
$V_{crit} = \sqrt{8}\cdot\sqrt[4]{\frac{g\sigma_{ul}}{\rho_l}(1-r)}$	–	0.233	0.183	128
$V_{crit} = \sqrt[4]{48}\cdot\sqrt[4]{\frac{g\sigma_{ul}}{\rho_l}(\frac{1}{r^2}-\frac{1}{r})}$	Energy balance for droplet dragging	0.246	0.175	141
$V_{crit} = \sqrt{12.3}\cdot\sqrt[4]{\frac{g\sigma_{ul}}{\rho_l}(1-r)}$	Ladle with gas injection	0.328	0.233	141

Equations for critical diameter (Ref. [11])	Remarks	d_{ZnCl2} (mm)	d_{water} (mm)	d_{ZnCl2}/d_{water} (%)
$d = \frac{0.534.\sigma_{ul}^{0.693}}{g^{0.564}(\rho_l-\rho_u)^{0.130}\rho_u^{0.306}}\frac{\mu_u^{0.114}}{\mu_l^{0.372}}$	Water/oil interface	3.1	13.1	23
$d = 2\cdot\sqrt{\frac{\sigma_{ul}}{g(\rho_l-\rho_u)}}$	–	6.3	17.1	37
$d = \sqrt{12}\cdot\sqrt{\frac{\sigma_{ul}}{g(\rho_l-\rho_u)}}$	Energy balance	10.9	29.6	37

Where: g gravity acceleration, ρ_l denser phase density, ρ_u density of supernatant phase, μ_l viscosity of denser phase, μ_u viscosity of supernatant phase, σ_{ul} interfacial tension; $r = \frac{\rho_u}{\rho_l}$; R = radius

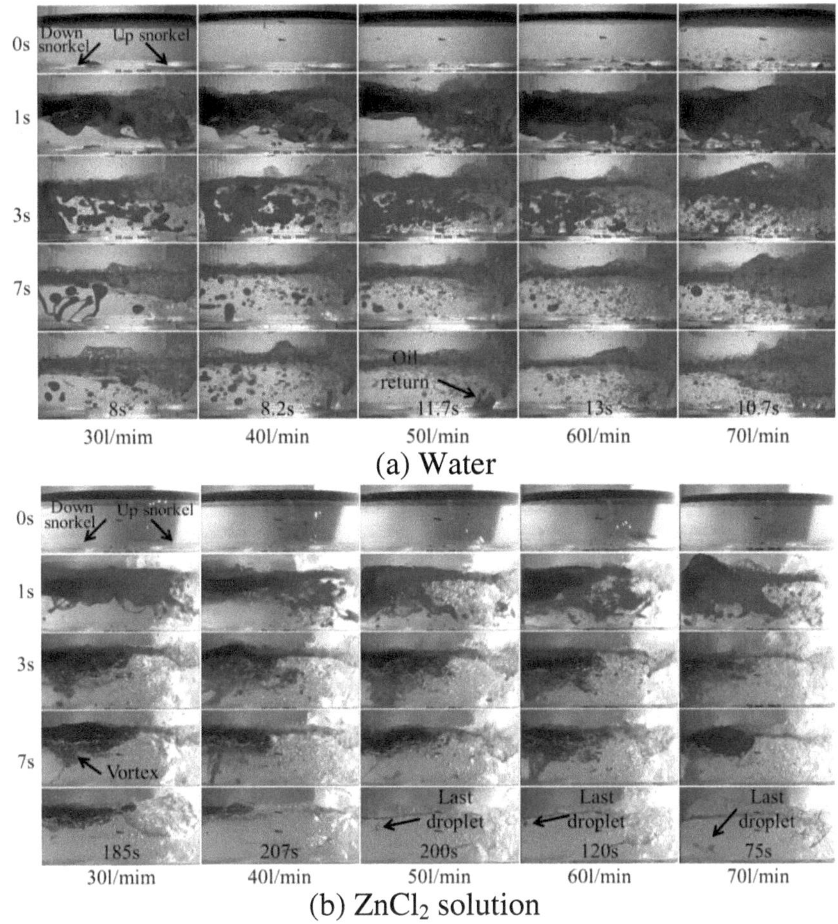

Fig. 4 Comparison of drag oil at the vacuum chamber as function of time: using **a** Water; **b** aqueous solution of $ZnCl_2$

A significant oil layer still remains in the chamber even after 200 s for gas flow rate of 30 and 40 l/min. The oil is intermittently dragged by the vortexes that occur above the down snorkel and no significant recirculation of oil droplets to the chamber has been noticed. Increasing the gas flow rate results in more oil entrained since the vortex frequency of occurrence also increases. Thus, for gas flow rate of 70 l/min there is no oil at the vacuum chamber after 75 s. These results highlight that the influence of density difference of the phases on entrainment conditions.

Looking at the oil being dragged in the down snorkel a remarkable difference can be noticed in the cases of water and $ZnCl_2$ solution as circulating fluids (Fig. 5). When water is used the oil layer is almost completely sucked as a large chunk,

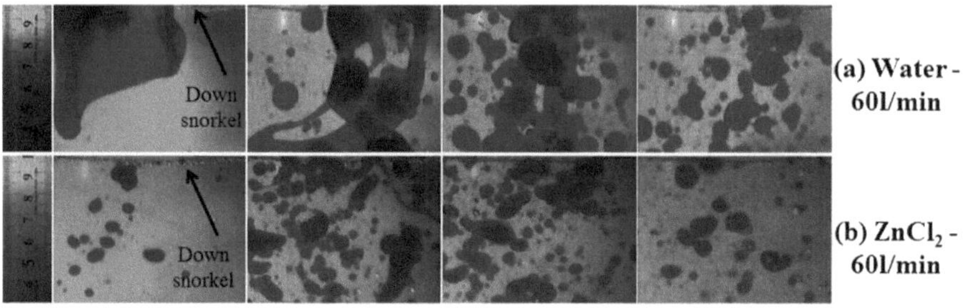

Fig. 5 Comparison of oil drops at the lower vessel below the tip of down snorkel and gas flow rate—60 l/min: **a** Water; **b** $ZnCl_2$

(Fig. 5a). Afterwards droplets as large as 25 mm show up. When using $ZnCl_2$ solution, (Fig. 5.b), the oil is dragged as droplets with a maximum size of 13 mm, which is in reasonable agreement with the calculations presented in Table 2.

Conclusions

The circulation rate in a RH degasser model was determined using the method of strain gage bridge (SGBM) and the PIV technique. Data from the latter were used to calibrate and validate results from SGBM. The calibration factor was determined as 0.6. Increasing gas flow from 30 to 120 l/min seems to increase of circulation rate from 1.5 to 2.75 kg/s in the water system and from 1.5 to 2.5 kg/s for $ZnCl_2$ solution system. Calculations using equations in the literature for the extent of slag entrainment show that there is remarkable difference in terms of entrainment of slag between the systems oil/water and oil/$ZnCl_2$. Oil drag (slag drag) occurs at all gas flow rates for the oil/water system while the $ZnCl_2$/oil system this occurs only in the case of higher gas flow rates (beyond 70 l/min) and the maximum diameter of the entrained droplet was about 65% lower. Thus is important to take in consideration the slag/metal density difference in order to evaluate the likelihood of dispersing slag formed in the vacuum chamber into the liquid steel in the ladle.

Acknowledgements The authors wish to thank CNPq, CAPES, the Gorceix Foundation and FAPEMIG of Brazil for the support.

References

1. H. Yang et al., A mathematical model to characterize RH desulfurization process. J. Iron Steel Res. Int. **21**(11), 995–1001 (2014)
2. L.T. Costa, R.P. Tavares, Dessulfuração do aço em desgaseificador RH por meio de modelo matemático, *Paper presented at the 46° Seminário de Aciaria - Internacional*. (Rio de janeiro, RJ, 2015), pp. 552–561

3. Z. Lin, R.I.L. Guthrie, Modeling of metallurgical emulsions. Metall. Mater. Trans. B **25**(B), 855–864 (1994)
4. J. Savolanein, T. Fabritius, O. Mattila, Effect of fluid physical properties on the emulsification. ISIJ Int. **49**(1), 29–36 (2009)
5. V. Seshadri, S.L.S. Costa, Cold model studies of RH degassing process. Trans. ISIJ **26**, 133–138 (1986)
6. R.B. Bird, W.E. Stewart, E.N. Lightfoot, *Transport phenomena*. (Wiley, New York, NY 2002), p. 187
7. http://www.dantecdynamics.com/measurement-principles-of-piv
8. V. Seshadri et al., Decarburisation rates in RH–KTB degasser of CST steel plant through physical modelling study. Ironmaking Steelmaking **33**(1), 34–38 (2006)
9. A.T.P. Almeida et al., Physical modeling of vacuum decarburization in an RH degasser. *Paper presented at the AISTech*. (Cleveland, Ohio, 1–4 May 2006), pp. 761–770
10. Y.G. Park, K.W. Yi, S.B. Ahn, The numerical calculation of circulation flow rate in the degassing Rheinstahl-Heraeus process. ISIJ Int. **40**(8), 749–755 (2000)
11. L.C. Hibbeler, B.G. Thomas, Mold slag entrainment mechanisms in continuous casting molds, in *Proceedings of AISTech* (2013), pp. 1215–1230

Part III
Fundamental Research of Metallurgical Process

Investigate on the Phase Composition of Vanadium Slag with High CaO Content and Influence of P_2O_5 on Crystallization Kinetics of Spinels

Wang Zhou, Bing Xie, Zhao-Qun Ke, Jiang Diao, Wen-Feng Tan, Yu-Hao Liu, Hong-Yi Li and Zhang Tao

Abstract An effective way to decrease the dephosphorization burden in the current vanadium extraction-steelmaking process is slagging with CaO in vanadium extraction step. The phase composition of vanadium slag with different basicity and the regularities distribution of elements have been investigated by X-ray diffraction analysis and scanning electron microscopy combining with energy disperse X-ray spectrometry, respectively. The results indicate that V mainly concentrates in the spinel phases; Ca and P are rich in the silicate phases. Fe_2SiO_4 transform to $CaFeSiO_4$, while the basicity of vanadium slag is 1.3. With further increase of basicity, $2CaO{\cdot}SiO_2$ could form in the silicate phases. Moreover, the phosphorus-containing phase will transform from $Ca_3(PO_4)_2$ to $Ca_3(PO_4)_2$-$n$$Ca_2SiO_4$. With the increase of P_2O_5 content, the crystallization rate of spinels decrease obviously. Furthermore, the peak temperature for crystallization of spinels decrease 5 K with the P_2O_5 content increasing 1 wt%.

Keywords Vanadium slag with high CaO content · Phase · Spinels · Crystallization

Introduction

Vanadium plays an extraordinarily important role in steelmaking, chemical industry and aviation fields [1]. However, there is no single recoverable vanadium mineral, it is mainly associated to patronite, vanadinite, carnotite and vanadium-titanium magnetite [2]. Among them, vanadium-titanium magnetite is the major raw material

W. Zhou · B. Xie (✉) · Z.-Q. Ke · J. Diao · W.-F. Tan · Y.-H. Liu · H.-Y. Li
College of Materials Science and Engineering, Chongqing University, Chongqing 400044, China
e-mail: bingxie@cqu.edu.cn

Z. Tao
Department of Science and Technology, Chongqing University of Education, Chongqing 400065, China

J.-Y. Hwang et al. (eds.), *8th International Symposium on High-Temperature Metallurgical Processing*, The Minerals, Metals & Materials Series,
DOI 10.1007/978-3-319-51340-9_14

to extract vanadium, and the most common process is schematically shown by the solid lines in Fig. 1 [3, 4]. To achieve a high extraction ratio of vanadium, the oxidization roasting of vanadium slag is decisive. The oxidation rate of vanadium is highly dependent upon spinel crystal size, the chemical components and phase compositions, of vanadium slag. Hence, it is of great significance to investigate the crystallization kinetics of spinels and the phase composition in vanadium slag.

More specifically, Chen and Yang [5] have found that a large spinel crystal size could promote the vanadium oxidation rate. According to the classic crystallization kinetic model reported by Turnbull [6, 7], Zhang et al. [8] developed a mathematical model which was used to describe the nucleation and growth kinetics of spinel crystals in vanadium slag, and an optimum temperature range of 1507–1557 K that benefits the precipitation of spinels was identified. The experimental study by Diao et al. [9] also found a similar behavior: the temperature range of 1473–1523 K was revealed as an optimum and the mean diameter of spinels could reach 23.5 μm. Therefore, it is reasonable for using the classic crystallization kinetic model to investigate the crystallization process of spinel crystals.

Presently, vanadium slags are traditionally sodium roasted followed by water leaching to extract vanadium, which requires a low CaO content in vanadium slags to avoid the formation of water-insoluble calcium vanadate [10]. However, the low CaO content in vanadium slags leads to a poor dephosphorization ability, and thus a heavy dephosphorization burden in the subsequent steelmaking process. At the same time there are some environmental issues during the sodium roasting-water

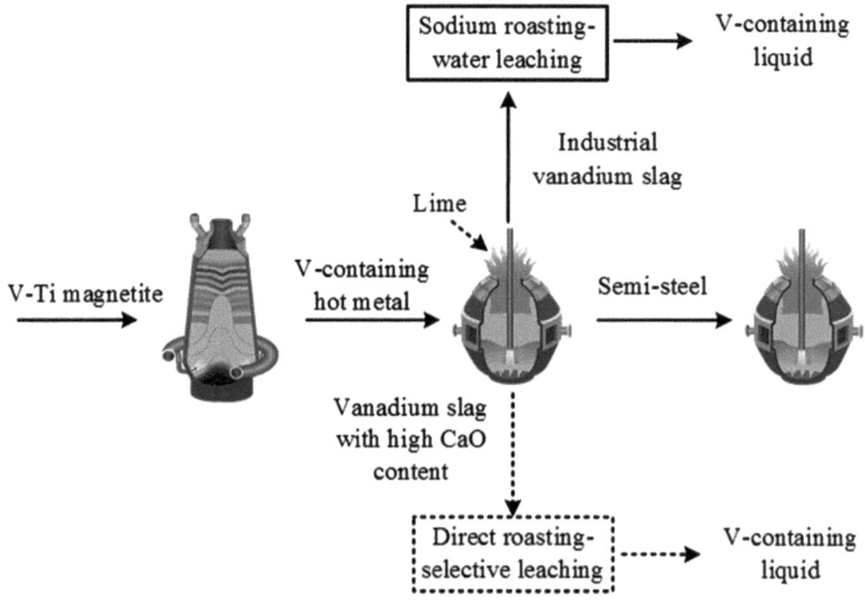

Fig. 1 Schematic of the vanadium extraction processes. *Solid lines* current industrial process, and *dotted lines* new process

leaching process [10], and consequently a lot of attention was paid to calcium roasting process to extract vanadium which is more environment-friendly [11–13]. In contrast to previous techniques, the calcium roasting technique allows high CaO contents in vanadium slags, which triggers the new idea of removing the phosphorus contained in hot metal by adding CaO into the converter during the production of vanadium slag. The novel process is schematically shown in Fig. 1 by dotted lines.

Due to the increase in CaO contents during slagging, the melting properties of vanadium slag will be changed. The hemisphere point temperature was found to be decreased significantly with increasing the CaO content, and the increase of the CaO content can reduce the precipitation temperature of spinels through thermodynamic calculations [4]. In addition, Zhou et al. [14] employed the CSD theory to analyze the non-isothermal crystallization kinetics of spinel crystals in vanadium slags containing high CaO content. However, the influence of P_2O_5 on crystallization kinetics of spinels and the phase composition of vanadium slag containing high CaO content was seldom reported and still not well defined, especially for phosphorus-containing phases in vanadium slag.

Based on above considerations, the aim of this work is to investigate the influence of P_2O_5 content on crystallization kinetics of spinels and the phase composition of vanadium slag containing high CaO content. This work was financially supported by National Natural Science Foundation of China (grant no. 51604047, no. 51474041 and no. 21105134), National Basic Research Program of China (973 Program, grant no. 2013CB632604), National Key Technology R&D Program (no. 2015BAB19B02) and Scientific and Technological Research Program of Chongqing Municipal Education Commission (grant no. KJ1601403).

Theoretical Calculation

The nucleation rate and growth rate can be expressed as Ref. [8].

$$I = \frac{N_V kT}{3\pi a^3 \eta} \exp\left[\frac{-16\pi\alpha^3\beta}{3T_r(\Delta T_r)^2}\right] \quad (1)$$

$$U = \frac{fkT}{3\pi a^2 \eta}\left[1 - \exp\left(\frac{-\beta \cdot \Delta T_r}{T_r}\right)\right] \quad (2)$$

where I is the nucleation rate per unit volume, U is growth rate, N_v is the number of molecules (or atoms) per unit volume, k is Boltzmann's constant, T is the absolute temperature, a is the lattice parameter of the crystal, η is viscosity, f is the fraction of acceptor sites in the crystal surface, α is the reduced crystal/liquid interfacial tension, β is the reduced molar heat of fusion, $T_r = T/T_m$ is the reduced temperature, T_m is the melting point of the crystal and $\Delta T_r = 1 - T_r$.

Table 1 Values of parameter used in calculation

Spinel type	*a* (Å)	T_m (°C)	α	β	*f*
FeV_2O_4	8.543	1750	1/3	1	$0.2\Delta T_r$
Fe_2TiO_4	8.378	1395			

The main spinel phases of vanadium slag in Pan-steel are MgV_2O_4, $Fe_xV_{3-x}O_4$, $Mn_xV_{3-x}O_4$ ($1 \leq x \leq 3$) and Fe_2TiO_4 spinels. Diao et al. [9] studies showed that V and Ti are concentrated in the spinel phases. Nevertheless, since the concentrations of Mg and Mn are fairly low to be compared with Fe, and the similarity of Fe spinels and (Mg, Mn) spinels is relatively high, only the Fe spinels are taken into consideration in the present study, namely FeV_2O_4 and Fe_2TiO_4. The parameters of structure and melting point of these spinel crystals are given in Table 1 [14]. As previously described, the viscosity of vanadium slag was calculated by the National Physical Laboratory (NPL) model [15].

Experimental

Preparation of Slag Samples

Initial materials were chemicals with purities of $FeC_2O_4{\cdot}2H_2O$ ($\geq$98.0%), SiO_2 ($\geq$99.0%), V_2O_3 ($\geq$99.0%), TiO_2 ($\geq$99.0%), $MnCO_3$ ($\geq$99.0%), MgO ($\geq$98.0%), CaO ($\geq$98.0%), $3CaO{\cdot}P_2O_5$ ($\geq$98.0%) supplied by Chengdu Kelong Chemical Co. Ltd. Reagent of CaO was calcined at 1373 K for 12 h before weighing and mixing. Due to the corundum crucible could be corroded during melting process, thus, Al_2O_3 (3 wt%) is not added into the mixture of initial materials. The mixture of initial materials was placed in 10 kW box-type electric resistance furnace with proportional-integral-differential (PID) controller (±2 K) for heating.

Experimental Method

In order to investigate the phase composition of vanadium slag with high CaO content, basicity and P_2O_5 content are considered as two variables. The chemical compositions of the slags are given in Table 2 (Ca1–Ca3). 200 g mixture was charged into a corundum crucible (inner diameter, 46 mm; height, 120 mm) which was then placed in another graphite crucible (inner diameter, 53 mm; height, 140 mm). They were placed in the furnace at 1073 K and heated to 1373 K at the rate of 5 K/min to allow the reagents of $FeC_2O_4{\cdot}2H_2O$ and $MnCO_3$ to decompose to FeO (totally assumed as FeO) and MnO. Then the mixture was heated to 1773 K (1500 °C) and held at 1773 K for 30 min to ensure homogeneity. The mixture was then cooled down to 1473 K, holding at 1473 K for 1 h, and then further cooled to

Table 2 Chemical composition of slag samples, wt%

	FeO	SiO_2	CaO	V_2O_3	TiO_2	MgO	MnO	P_2O_5	R (CaO/SiO_2)
Ca1	30.00	13.48	17.52	13.00	10.00	3.00	8.00	2.00	1.30
Ca2	29.00	13.00	26.00	11.00	8.00	3.50	5.50	2.00	2.00
Ca3	30.00	12.00	18.00	13.00	10.00	3.00	8.00	3.00	1.50
P1	30.00	12.80	19.20	13.00	10.00	3.00	8.00	1.00	1.50
P2	30.00	12.40	18.60	13.00	10.00	3.00	8.00	2.00	1.50
P3	30.00	12.00	18.00	13.00	10.00	3.00	8.00	3.00	1.50

about 1173 K at the rate of 5 K/min to ensure complete precipitation of minerals in the slag sample. Finally, the sample was removed and air-cooled.

The chemical compositions of P1, P2 and P3 are used for calculating the influence of P_2O_5 content on crystallization kinetics of spinels in vanadium slag with high CaO content.

Results and Discussion

Influence of P_2O_5 Content on Nucleation and Growth Rates of Spinels

The nucleation and growth rates of FeV_2O_4 and Fe_2TiO_4 spinels are shown in Figs. 2 and 3, respectively. The calculated results indicate that both the nucleation and the growth rates have a peak value. This is reasonable because both of them are greatly influenced by two contradictory factors, i.e., the degree of undercooling and the diffusive mass transfer in vanadium slag. It can be seen that with the increase in the P_2O_5 content from 1 to 3 wt%, the nucleation and growth rates of spinels decrease by less than half an order of magnitude. In other words, a lower P_2O_5 content benefits the nucleation and growth of spinels. However, the P_2O_5 content has little effect on the optimum temperature ranges for nucleation and growth of FeV_2O_4 and Fe_2TiO_4 spinels. The optimum temperature ranges for nucleation and growth of FeV_2O_4 spinels are 1023–1223 and 1425–1625 K, respectively. For Fe_2TiO_4 spinels, they are 856–1056 and 1206–1406 K, respectively (Fig. 3).

However, it is still difficult to give an accurate optimum temperature range for the crystallization of spinels. Therefore, it is necessary to combine the nucleation and growth rates. Zhang et al. [8] defined the crystallisation rate r as

$$r = \frac{\pi}{3} I U^3 \tag{3}$$

And the total crystallization rate r_{total} is given by

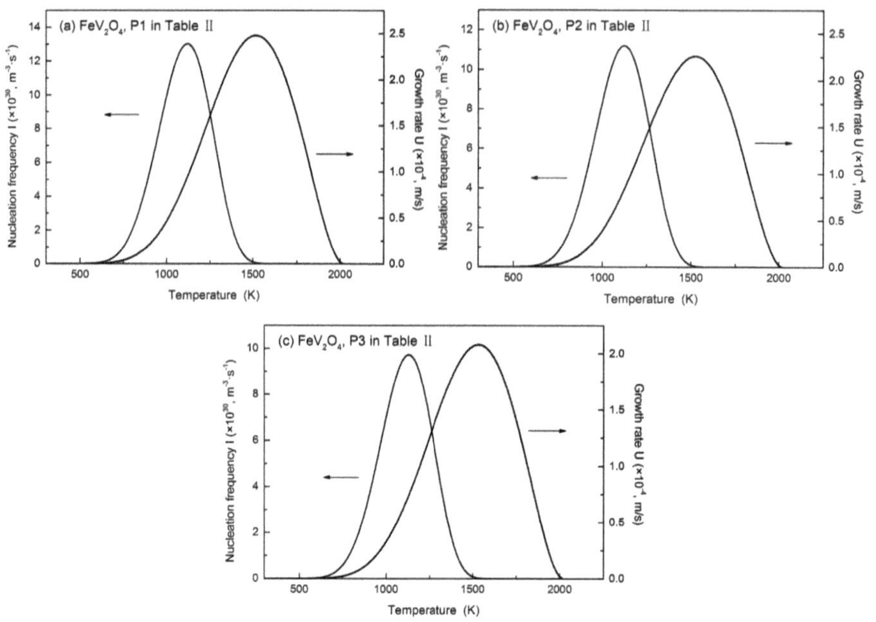

Fig. 2 Nucleation and growth rates of FeV_2O_4 **a** P1, **b** P2, **c** P3 in Table II

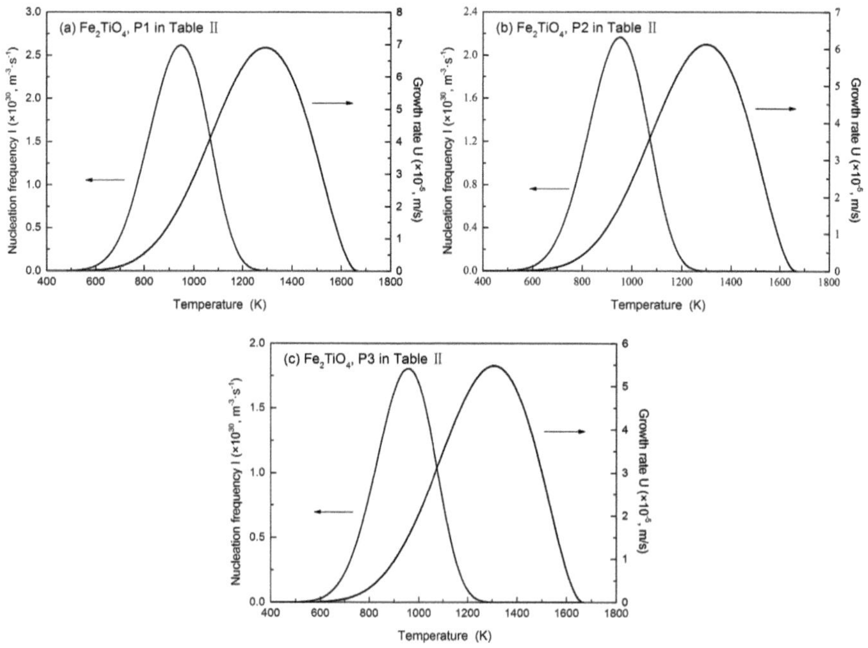

Fig. 3 Nucleation and growth rates of Fe_2TiO_4 **a** P1, **b** P2, **c** P3 in Table II

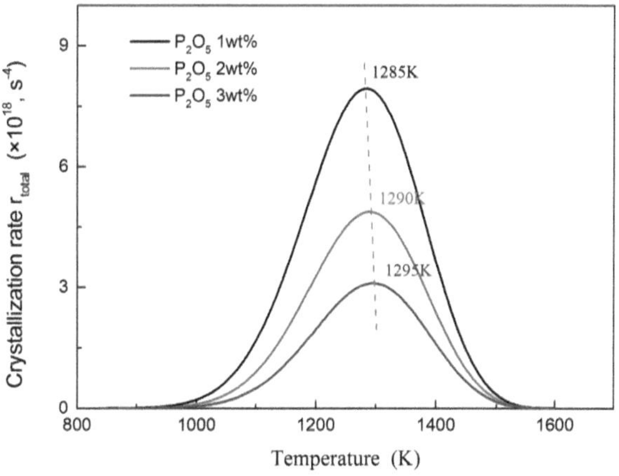

Fig. 4 Effect of P_2O_5 content on the crystallization rate of the total spinels

$$r_{\text{total}} = \sum_i \omega_i r_i \tag{4}$$

where ω_i is the weight percentage of a given crystal.

The effect of P_2O_5 content on the crystallization rate of the total spinels was investigated in the mass fraction range of 1–3 wt%. As shown in Fig. 4, with the increase of P_2O_5 content, the crystallization rate r_{total} decrease obviously. Moreover, the peak temperature for crystallization of spinels decrease 5 K with P_2O_5 content increasing 1 wt%. This suggests that the P_2O_5 content has disadvantageous influence on the crystallization of spinels in vanadium slag.

Mineralogical Phases in Vanadium Slag with High CaO Content

The mineralogical phases of the vanadium slags were investigated by X-ray diffraction (XRD, RIGAKUD/MAX 2500PC). The X-ray diffraction patterns shown in Fig. 5 indicate that the main spinel phases consist of $Fe_xV_{3-x}O_4$ (x = 1 or 2), Mg (V, Ti)$_2O_4$, MnV_2O_4 and Fe_2TiO_4 in vanadium slag with the basicity range of 1.3–2.0. The spinel phases in vanadium slag with high CaO content are consistent with the case in the vanadium slag with low CaO content [3, 11]. However, with 17.52 wt% content of CaO, perovskite ($CaTiO_3$) and akermanite ($Ca_2MgSi_2O_7$) occurs in the silicate phases, which agrees with previously reported results [4]. While the basicity of vanadium slag is 1.3, Fe_2SiO_4 transform to $CaFeSiO_4$. With further increase of basicity, $2CaO{\cdot}SiO_2$ could form in the silicate phases. Moreover, the phosphorus-containing phase will transform from $Ca_3(PO_4)_2$ to $Ca_3(PO_4)_2$-$n$$Ca_2SiO_4$ (C_3P-nC_2S), and the value of n increase from 1 to 6 with the increase in basicity from 1.5 to 2.0. In short, the CaO content has limited influence on the

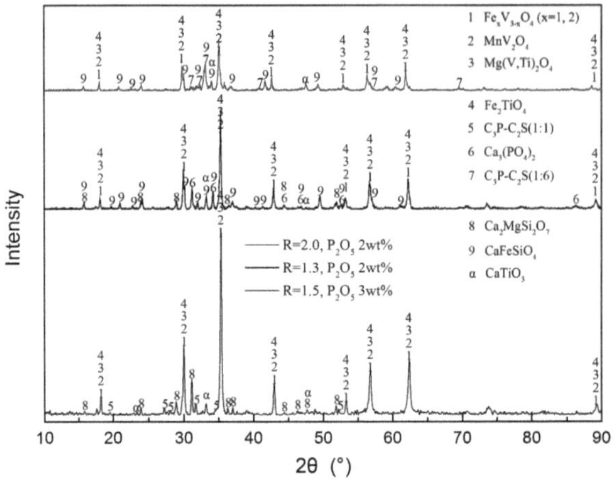

Fig. 5 X-ray diffraction patterns of the vanadium slag samples

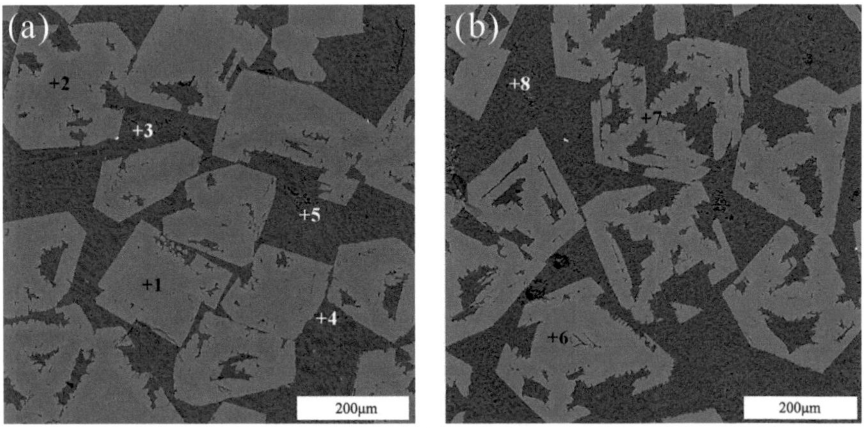

Fig. 6 Representative SEM images of the slags (R = 1.3) **a** Near the centre **b** Near the surface

composition of spinel phases, however, it has important influence on the composition of silicate phases (Fig. 5).

To investigate the evolution of mineralogical morphology of vanadium-containing phases in vanadium slags with high CaO content, SEM/EDS characterization was conducted. SEM images and EDS results are shown in Fig. 6, Table 3 and Fig. 7, respectively. The results indicate that V is concentrated in the white spinel phases, while Ca, Si and P are concentrated in the dark grey silicate phases. Fe, Mn, Mg, Al and O coexist in both phases. As shown in Fig. 6, near the surface of the sample, the morphology of spinel crystals is dendrite-like forms,

Table 3 Spot analysis of the slag as shown in the representative SEM images, wt%

Point	Phase	O	Mg	Al	Si	P	Ca	Ti	V	Mn	Fe
1	Spinel	29.26	3.22	7.83				7.60	16.44	6.03	29.61
2	Spinel	27.51	3.40	8.34				7.09	19.11	5.79	28.75
3	Silicates	44.67	2.10		14.95	1.89	19.40			4.50	12.48
4	Silicates	45.42	0.74	9.03	15.60		22.39			1.11	5.72
5	Silicates	44.27		7.80	14.47	1.46	24.79	0.95		1.33	4.93
6	Spinel	26.02	3.09	7.50				8.69	15.56	6.39	32.74
7	Spinel	27.27	2.75	7.60				7.85	16.09	6.49	31.95
8	Silicates	46.68		6.34	15.97		15.41	2.87		3.08	9.65

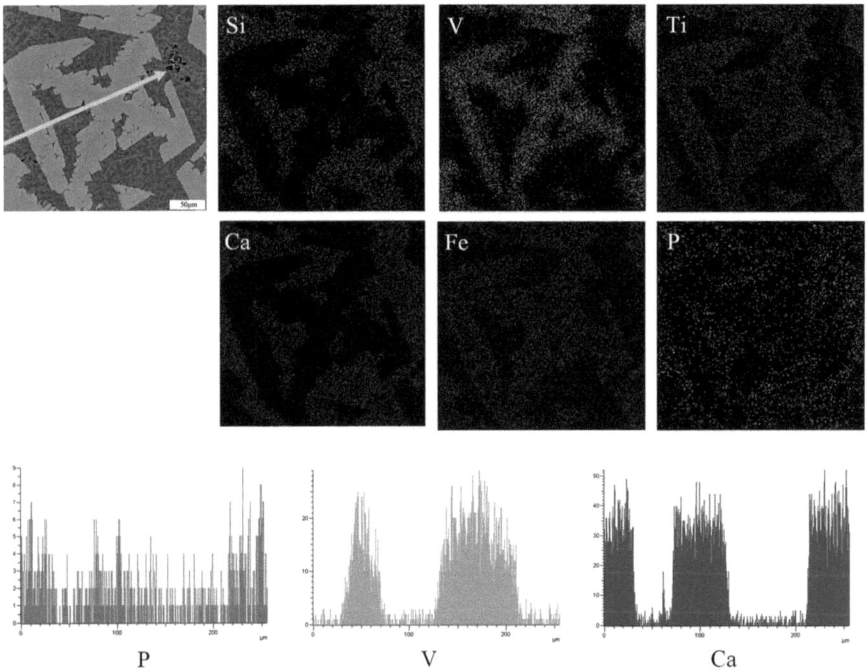

Fig. 7 Map and line scanning results of spinel crystal in vanadium slag sample (R = 1.3)

however, the morphology of spinel crystals is particle-like forms near the centre of the sample. It could be deduced that the morphology of spinel crystals in vanadium slag with high CaO content changes with the decrease of temperature, from dendrite-like to particle-like forms.

Conclusions

(1) With the increase of P_2O_5 content, the crystallization rate r_{total} of spinels in vanadium slag decrease obviously. Moreover, the peak temperature for crystallization of spinels decrease 5 K with P_2O_5 content increasing 1 wt%.
(2) Vanadium is concentrated in the white spinel phases, while Ca, Si and P are concentrated in the dark grey silicate phases. Fe, Mn, Mg, Al and O coexist in both phases.
(3) The main spinel phases consist of $Fe_xV_{3-x}O_4$ (x = 1 or 2), $Mg(V, Ti)_2O_4$, MnV_2O_4 and Fe_2TiO_4 in vanadium slag with the basicity range of 1.3–2.0, i.e. the CaO content has limited influence on the composition of spinel phases.
(4) $CaTiO_3$, $CaFeSiO_4$ and $Ca_2MgSi_2O_7$ phases will occur in the silicate phases while the basicity of vanadium slag is higher than 1.3.
(5) With the increase of basicity from 1.3 to 1.5, the phosphorus-containing phase will transform from $Ca_3(PO_4)_2$ to $Ca_3(PO_4)_2$-$n$$Ca_2SiO_4$ (C_3P-$n$$C_2S$), and the value of n increase from 1 to 6 with the increase in basicity from 1.5 to 2.0.

References

1. R.R. Moskalyk, A.M. Alfantazi, Processing of vanadium: a review. Miner. Eng. **16**, 793–805 (2003)
2. S.Z. Yang, *Vanadium metallurgy* (Metallurgical Industry Press, Beijing, 2010)
3. J. Diao, Y. Qiao, X. Zhang, C.Q. Ji, B. Xie, Growth mechanisms of spinel crystals in vanadium slag under different heat treatment conditions. CrystEngComm **17**, 7300–7305 (2015)
4. H.X. Fang, H.Y. Li, T. Zhang, B.S. Liu, X. Bing, Influence of CaO on existence form of vanadium-containing phase in vanadium slag. ISIJ Int. **55**, 200–206 (2015)
5. D.H. Chen, S.D. Yang, Systematic evaluation of the quality of vanadium slag. Hebei Yejin **15**, 19–23 (1993)
6. D. Turnbull, Formation of crystal nuclei in liquid metals. J. Appl. Phys. **21**, 1022–1028 (1950)
7. D. Turnbull, Under what conditions can a glass be formed. Contemp. Phys. **10**, 473–488 (1969)
8. X. Zhang, B. Xie, J. Diao, X.J. Li, Nucleation and growth kinetics of spinel crystals in vanadium slag. Ironmaking Steelmaking **39**, 147–154 (2012)
9. J. Diao, B. Xie, C. Ji, X. Guo, Y. Wang, X.J. Li, Growth of spinel crystals in vanadium slag and their characterization. Cryst. Res. Technol. **44**, 707–712 (2009)
10. H.A. Barolin, trans. by C.L. Wang, *Oxidation of Vanadium slag* (Metallurgical Industry Press, Beijing, 1982)
11. S.X. Li, Mechanism research on oxidation roasting and leaching process of high calcium low grade vanadium slag. Ph.D. thesis, Chongqing University, 2011
12. H.Y. Li, K. Wang, W.H. Hua, Z. Yang, W. Zhou, B. Xie, Selective leaching of vanadium in calcification-roasted vanadium slag by ammonium carbonate. Hydrometallurgy **160**, 18–25 (2016)

13. X.J. Li, The effect of roasting parameters in calcium roasting on the leaching rate of vanadium. J. Process. Eng. **12**, 56–57 (2012)
14. W. Zhou et al., Non-isothermal crystallization kinetics of spinels in vanadium slag with high CaO content. JOM **68**, 2520–2524 (2016)
15. K.C. Mills, S. Sridhar, Viscosities of ironmaking and steelmaking slags. Ironmaking Steelmaking **26**, 262–268 (1999)

Thermal Analysis Kinetics of the Solid-State Reduction of Nickel Laterite Ores by Carbon

Xueming Lv, Xuewei Lv, Lunwei Wang and Jie Qiu

Abstract Non-isothermal multiple scanning rate analysis method was used to analyse the solid-state reduction of nickel laterite by carbon. According to the TG curves, the activation energies were calculated and the most probable mechanism functions were determined by the Malek method. The results show that the process can be divided into three stages according to the reaction degree (α): 0–40%, 40–70%, 70–100%, respectively. In the stage of $0 \leq \alpha \leq 0.40$, the most probable mechanism function fits the chemical reaction order function. In the stage of $0.40 \leq \alpha \leq 0.70$ and $0.70 \leq \alpha \leq 1.0$, the process may not be exactly described by one mechanism function.

Keywords Nickel laterite · Solid-state reduction · Thermal analysis kinetics · Activation energy

Introduction

Nickel sulfide ores and nickel laterite ores are the two main resources for extracting nickel metal. In recent years, there has been an increased focus on the utilization of low-grade nickel laterite ore, along with a growing demand for stainless steel and a declining supply of sulfide ores. Therefore, in the long term, it will be necessary to optimise nickel extraction from nickel laterite ore [1–5].

Several investigations of nickel laterite ore reduction have been performed. For example, Jankovic et al. studied reduction kinetics of nickel oxide with hydrogen under non-isothermal conditions, the activation energies were 96.4 kJ/mol, then the most probable mechanism function was determined to be $f(\alpha) = \alpha^{0.63}(1 - \alpha)^{1.39}$ [6, 7]. Jungah et al. studied reduction kinetics of nanocrystalline Fe_2O_3–NiO by solid

X. Lv · X. Lv (✉) · L. Wang · J. Qiu
College of Materials Science and Engineering, Chongqing University,
Chongqing 400044, China
e-mail: lvxuewei@163.com

J.-Y. Hwang et al. (eds.), *8th International Symposium on High-Temperature Metallurgical Processing*, The Minerals, Metals & Materials Series,
DOI 10.1007/978-3-319-51340-9_15

carbon and the results showed that reduction rate improved constantly with increasing temperature, the reduction rate was controlled by Boudouard reaction [8].

Since the previous study mainly confined to reduction of pure nickel oxide, there is a few study on reduction of nickel laterite ores and rarely used carbon as reducing agent. In this paper, solid-state reduction of nickel laterite was determined, the activation energies were calculated and the most probable mechanism function was determined by the Malek method.

Thermal Analysis Kinetics Theory

The activation energies were calculated usually by integral and differential methods. FWO method is the most commonly used integral method, besides, Starink method and KAS method are the most commonly used differential methods. The equation of FWO method is derived as

$$\lg \beta = \lg\left(\frac{AE}{RG(\alpha)}\right) - 2.315 - 0.4567\frac{E}{RT} \tag{1}$$

The equation of Starink method is derived as

$$\ln\left(\frac{\beta}{T^{1.8}}\right) = \text{constant} - 1.0037\frac{E}{RT} \tag{2}$$

The equation of KAS method is derived as

$$\ln\left(\frac{\beta}{T^{2}}\right) = \ln\frac{AR}{E} - \frac{E}{RT} \tag{3}$$

The most probable mechanism function was determined by the Malek method, and the Malek method is used to determine $f(\alpha)$ by $y(\alpha)$ which is defined.

$$y(\alpha) = \left(\frac{T}{T_{0.5}}\right)^2 \frac{\left(\frac{d\alpha}{dt}\right)}{\left(\frac{d\alpha}{dt}\right)_{0.5}} = \frac{f(\alpha)\,.\,G(\alpha)}{f(0.5)\,.\,G(0.5)} \tag{4}$$

In this paper, the most probable mechanism function was derived by the standard curve of $y(\alpha) - \alpha$. When the experimental curve and the standard curve coincide, the corresponding equation $f(\alpha)$ of the standard curve is the most probable mechanism function.

Experimental

The purity of carbon powder used in this experiment was 99.99% and the particle size was in 200–300 mesh.

Nickel laterite ore contains a large amount of water, with a moisture content on a dry basis of up to 50%, in the form of adsorbed and crystallised water [9]. The XRD experiments were carried out using D/max-3C (Co Kα) under the scanning rate of 4° min^{-1}. The XRD patterns shown in Fig. 1 reveals that the nickel laterite ores are typical saprolitic ore, containing dominant amounts of Mg silicate and Mg–Fe silicate minerals, and in nickel laterite ores ores, Ni occurs mainly in Mg silicates [10]. The XRD patterns shown in Fig. 2 reveals that the adsorbed and crystallised water mainly in $CaAl_2Si_2O_8(H_2O)_4$ and hydroxyl groups mainly in $(Mg, Al)_3(Si, Fe)_2O_5(OH)_4$ and $Mg_3Si_2O_5(OH)_4$. Drying was performed at 1100 °C in a corundum crucible within a muffle furnace for 2 h. Dried nickel laterite ores were ground to a 200 mesh-size using a laboratory-scale ball mill to mix with pure carbon powder to be homogeneous by a ratio of 1–1.2 used for TG and DTG analysis.

The chemical composition of nickel laterite ore was analysed by AAS method which is shown in Table 1.

The reduction behavior of the prepared samples was investigated using two different techniques, tg technique to measure the total weight loss of carbon and oxygen as a function of time and effluent gas analysis method for the outlet gases. The TG analysis was carried out using SetsysEvoTG-DTA1750 with three kinds of

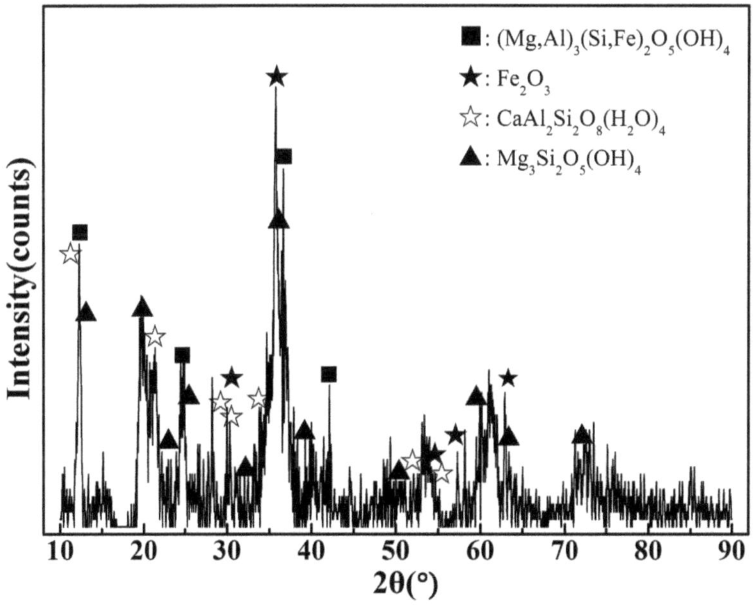

Fig. 1 X-ray diffraction pattern of nickel laterite

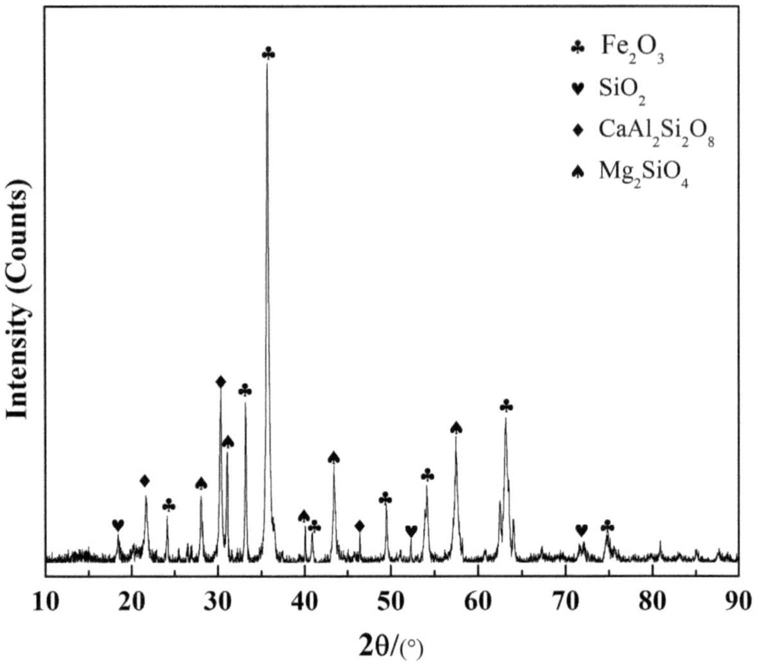

Fig. 2 X-ray diffraction pattern of nickel laterite at 1100 °C

Table 1 Chemical composition of nickel laterite/(wt%)

Ni	TFe	FeO	Cr_2O_3	SiO_2	Al_2O_3	CaO	MgO	P	S
1.814	17.87	0.44	0.51	34.97	4.75	1.54	13.5	0.005	0.064

heating rate of 10 K/min, 15 K/min, 20 K/min under Ar atmosphere at a constant rate of 20 mL min^{-1}. The concentration of CO and CO_2 in the outlet gas were carried out using advanced quadrupole mass spectrometer (TILON LC-D200, AMETEK LLC., USA). Pure argon gas was passed over the compact at a constant rate of 20 mL min^{-1}. The degree of reduction and total weight loss of carbon and oxygen were calculated by the oxygen and carbon balance based on the CO and CO_2 concentrations obtained from gas analysis.

Results and Discussion

TG Analysis

TG was used to determine the thermal properties of nickel laterite ore, as shown in Fig. 3. It can be observed that weight loss is greatly affected by reduction temperature. Increasing reduction temperature from 700 to 1400 °C causes the increase of the mass loss. In the initial stages of reduction, weight loss rate is slow, then the rate speeds up gradually and reach the maximum at about 1232 °C which can seen clearly from DTG curves in Fig. 4, until 1400 °C, the process of reduction comes to an end, the weight of nickel laterite no longer changes. Besides, the heating speed has little influence on the weight loss ratio of nickel laterite ores.

Activation Energy Results and Discussion

The reduction fraction (α) of reduction of nickel laterite is defined as

$$\alpha = \frac{m_0 - m_t}{m_0 - m_\infty} \times 100\% \quad (5)$$

where, m_0 is the initial mass of sample, m_t represents the mass of the sample at arbitrary time t, m_∞ is the mass of the sample at the end of the process. The reaction fraction and DTG curves as a function of reduction temperature are shown in Fig. 4, in which the solid lines express the DTG curves while the dotted lines express the reaction fraction curves, in which, the reduction degree has no significant change as the reduction temperature increased, besides, the reduction rate is the highest at early stages (α = 0.3).

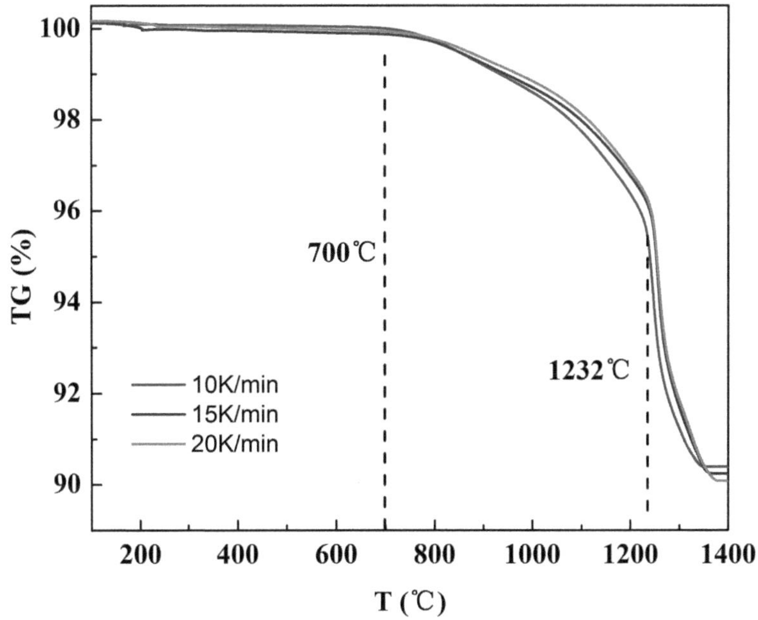

Fig. 3 TG curves of nickel laterite ore-carbon mixture at three heating rates

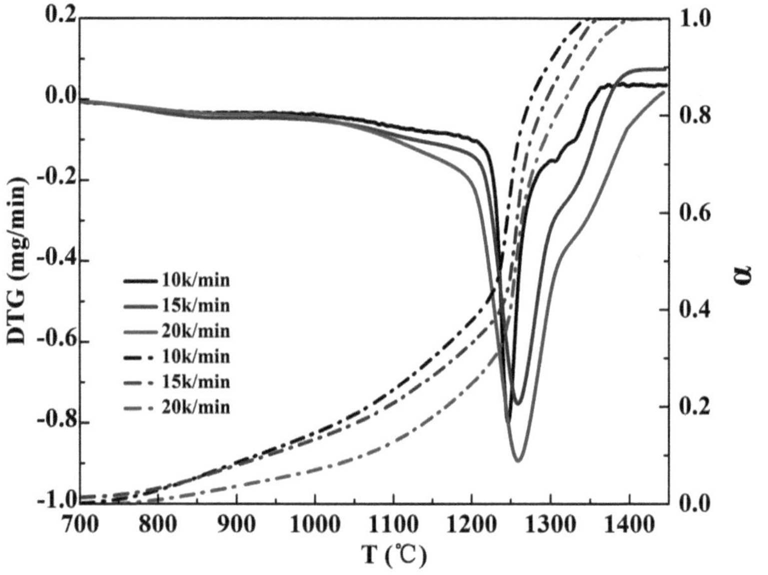

Fig. 4 Conversion fraction and DTG-temperature curves of nickel laterite ore-carbon mixture

The *E* values were determined by FWO method, Starink method and KAS method are shown in Fig. 5 and Table 2.

The reduction process is divided into three parts according to the *E* values, besides, the temperature corresponding to the three parts is shown in Fig. 5. In the range of $0.10 \leq \alpha \leq 0.40$ and $0.70 \leq \alpha \leq 0.90$, the activation energies are the constant value. Until for $0.40 \leq \alpha \leq 0.70$, the *E* values have a great change, the values increased with increasing α values.

The obtained activation energy is in good agreement with *E* values which were determined by Malek et al. [11] and Liao et al. [12], respectively. According to the research of Vyazovkin [13], if the activation energy is independent on α values, the reaction is single-step reaction, however, the reaction is multi-step reaction if the activation energy varies with the change of α values. From Fig. 5, in the wide range of $0.10 \leq \alpha \leq 0.40$ and $0.70 \leq \alpha \leq 0.90$, the proof of existing of single-step reaction is evident, however, in the range of $0.40 \leq \alpha \leq 0.70$ designated on the existence of parallel independent reaction.

In this study, the possible reactions are as follows,

$$2Fe_2O_3 + 3C = 4Fe + 3CO_2 \tag{6}$$

$$2NiO + C = 2Ni + CO_2 \tag{7}$$

$$CO_2 + C = 2CO \tag{8}$$

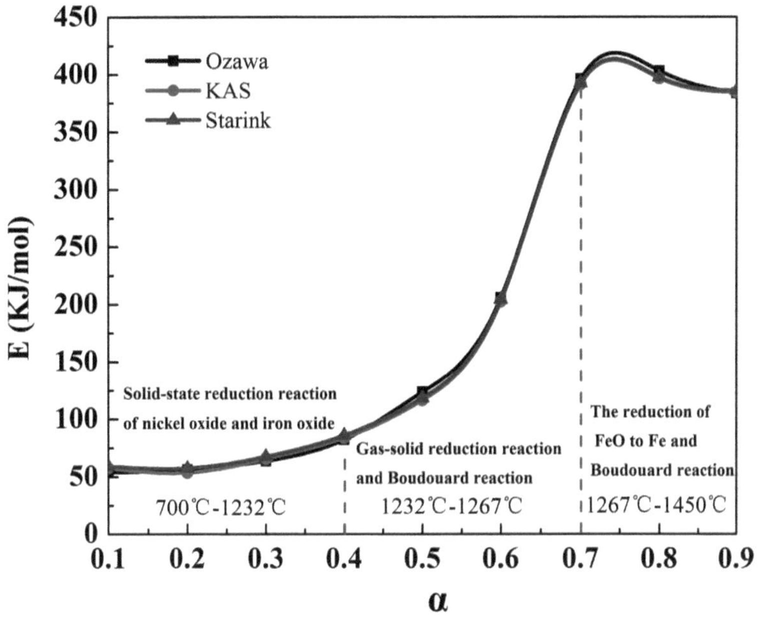

Fig. 5 Relationship between conversion fraction and activation energies

Table 2 Calculated activation energies through different methods

α	Eα/(KJ/mol)		
	FWO	Starink	KAS
0.10	53.93	57.47	59.28
0.20	56.31	53.49	57.50
0.30	63.94	65.56	67.58
0.40	82.26	83.95	86.00
0.50	123.79	116.55	118.53
0.60	205.92	202.37	204.09
0.70	396.20	391.38	392.46
0.80	403.01	396.61	397.71
0.90	383.95	384.59	386.03

$$Fe_2O_3 + 3CO = 2Fe + 3CO_2 \tag{9}$$

It is reported that NiO was firstly reduced, and then the metallic Ni could use metallic Fe as the carriers to form Fe–Ni alloy to separate from the gangue minerals [14]. At the beginning of the reduction process, the nickel laterite powders and pure carbon powders are close to each other, the preferential reduction of nickel oxide to metallic nickel can occur as Eq. (7), and the reduction of iron oxide to metallic iron takes place as Eq. (6). Then, the reaction between iron oxides and carbon monoxide generated by Boudouard reaction takes place as Eq. (9).

It is well known that the reduction reactions of iron oxides proceed in a stepwise manner (Fe_2O_3-Fe_3O_4-FeO-Fe). To detect the reduction situation of iron oxides in nickel laterite, XRD were used to study the reduction product at 1000, 1250, 1300 and 1400 °C which is shown in Fig. 6.

In the range of $0.10 \leq \alpha \leq 0.40$, the reduction reaction has a low activation energy and a fast speed, and Fig. 6 shows that the reduction product of iron oxides is Fe_3O_4, this suggests that Fe_2O_3 is mainly reduced into Fe_3O_4 by carbon, the possible reaction is listed as follows,

$$3Fe_2O_3 + 1/2C = 2Fe_3O_4 + 1/2CO_2 \tag{10}$$

In the unreacted-core shrinking model shown in Fig. 7a, the nickel laterite powders and pure carbon powders are close to each other in the initial stage, then the oxygen of Fe_2O_3 are removed to generate Fe_3O_4 particle and CO_2.

The connection of the rate-controlling step of reduction rate to the activation energy of iron oxide was investigated by Nasr [15] (Table 3). In this stage, the *E* values are almost within 55–67 kJ/mol, hence, the rate-controlling step in this stage is performed as inner chemical reaction.

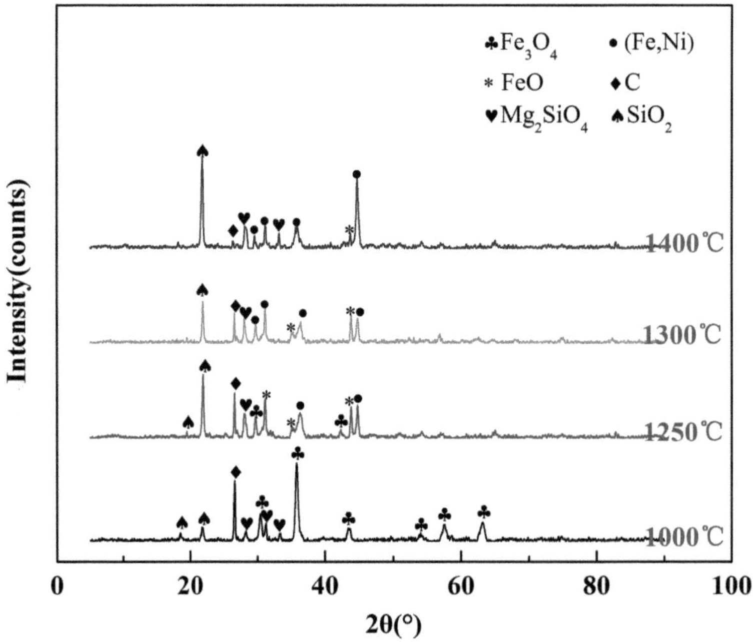

Fig. 6 X-ray diffraction pattern of nickel laterite ore-carbon mixture at 1000, 1250, 1300 and 1400 °C

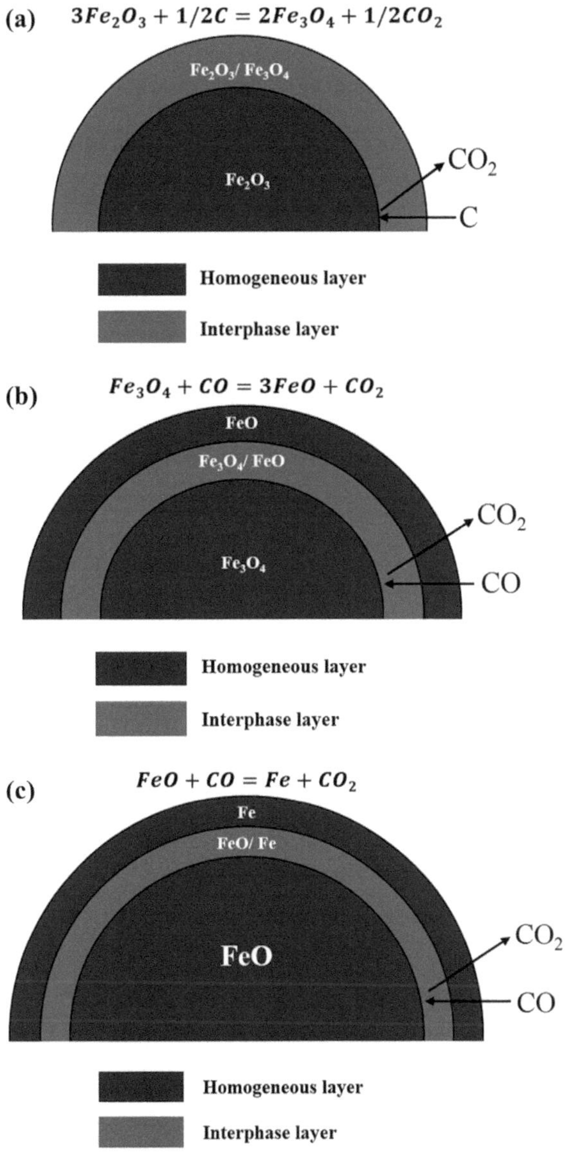

Fig. 7 Unreacted-core shrinking model for Fe_2O_3 reduction

Table 3 Relationship of rate-controlling step and activation energy of iron oxide

E/(kJ/mol)	Rate-controlling step
8–16	Inner gas diffusion
29–42	Inner gas diffusion and interface chemical reaction mixed
55–67	Interface chemical reaction
>80	Solid diffusion

In the range of $0.40 \leq \alpha \leq 0.70$, Fig. 6 shows that the reduction product of iron oxides are mostly FeO and a small part of Fe–Ni, this suggests that Fe_3O_4 is reduced into FeO and Fe–Ni, the possible reactions are listed as follows,

$$Fe_3O_4 + CO = 3FeO + CO_2 \tag{11}$$

$$FeO + CO = Fe + CO_2 \tag{12}$$

As Fig. 7b shows that CO diffuses across the FeO layer to reach the Fe_3O_4–FeO interphase layer, and CO reacts with Fe_3O_4 to generate CO_2. In this stage, the E values have a great change, the E values increased with increasing α values. In the initial stage ($\alpha = 0.4$), the E values are almost 80 kJ/mol, hence, the rate-controlling step is performed as solid diffusion according to the study of Nasr [15].

In the range of $0.70 \leq \alpha \leq 0.90$, Fig. 6 shows that the reduction products of iron oxides are mostly Fe–Ni and a small part of FeO at 1300 °C and iron oxides are all reduced into Fe–Ni at 1400 °C, the possible reaction is given as follows,

$$FeO + CO = Fe + CO_2 \tag{12}$$

As Fig. 7c shows that CO diffuses across the Fe layer to reach the FeO–Fe interphase layer, and CO reacts with FeO to generate CO_2.

In the range of 1100–1350 °C, the variation of CO and CO_2 in the outlet gas is shown in Fig. 8. It can be observed that the rate of formation of CO_2 is greater than that of CO particularly in the range of $0.50 \leq \alpha \leq 0.90$ and then slows down. It is reasonable to accept that Boudouard reaction is the major controlling step in this stage.

Model Function Results

Based on the Malek analysis, the standard curves were first obtained for the forty-one model functions. Figures 9, 10 and 11 show that the experimental data points are both almost have the same trend.

In the range of $0.10 \leq \alpha \leq 0.40$, the mechanism function is chemical reaction order function (No. 37) shown as follows:

$$f(\alpha) = (1 - \alpha)^{-1} - 1 \tag{13}$$

$$G(\alpha) = (1 - \alpha)^2 \tag{14}$$

In the range of $0.40 \leq \alpha \leq 0.70$ and $0.70 \leq \alpha \leq 0.90$, the shapes of the experimental curves vary somewhat with the standard curves, this indicates that the mechanism of the process under investigation may not be exactly described by one mechanism function.

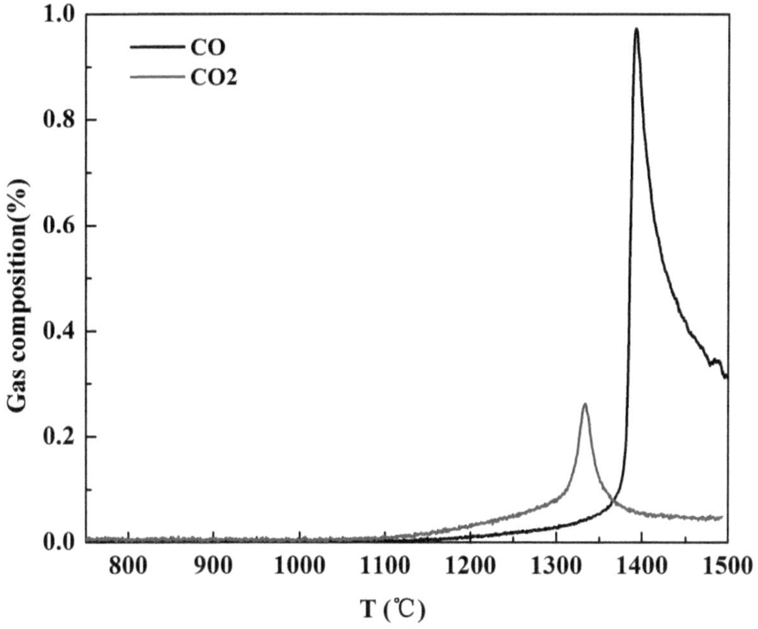

Fig. 8 Variation of gas composition (CO and CO_2) as a function of temperature

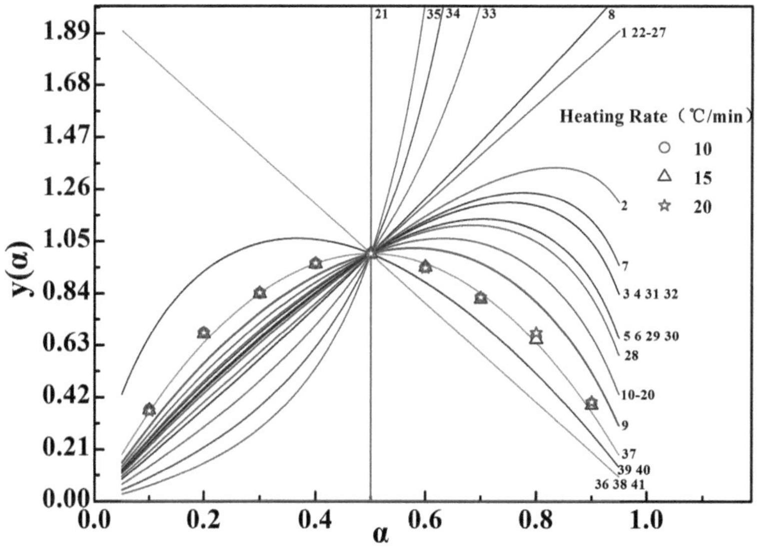

Fig. 9 The standard y(α)-α curves and the actual y(α)-α curves of solid-state reduction in the range of $0.10 \leq \alpha \leq 0.40$

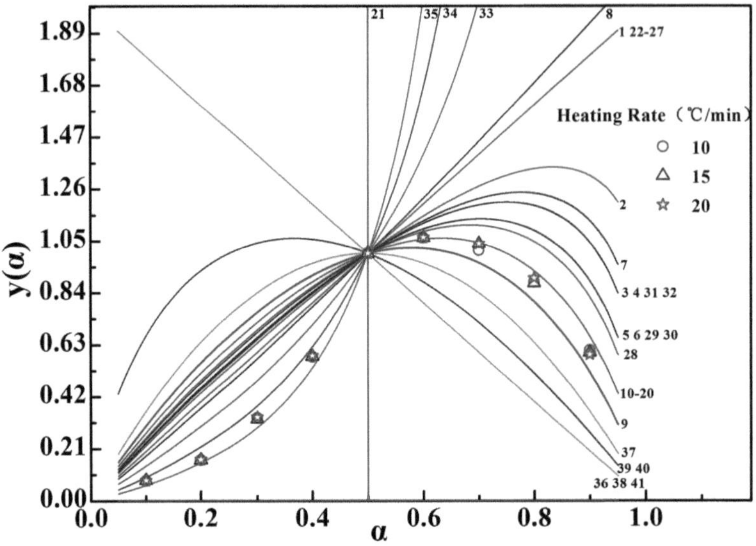

Fig. 10 The standard y(α)-α curves and the actual y(α)-α curves of solid-state reduction in the range of $0.40 \leq \alpha \leq 0.70$

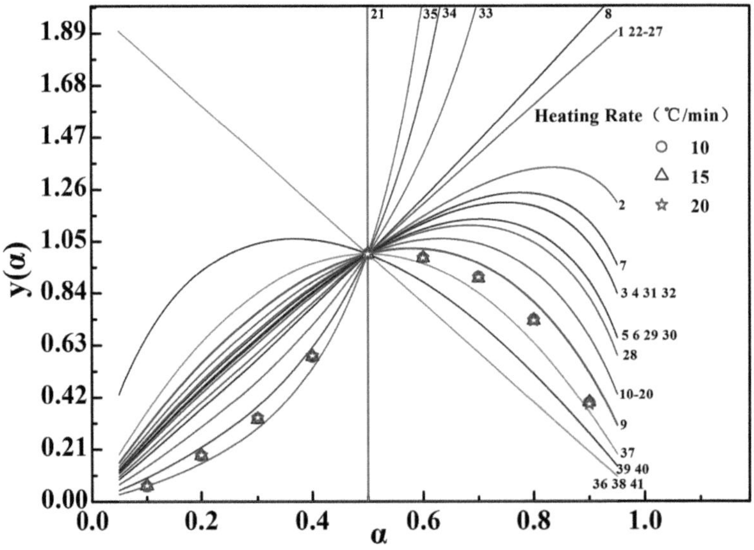

Fig. 11 The standard y(α)-α curves and the actual y(α)-α curves of solid-state reduction in the range of $0.70 \leq \alpha \leq 0.90$

Conclusions

(1) The weight loss ratio of nickel laterite ores is mainly related to the heating temperature; the heating speed has little influence.
(2) In the range of $0.10 \leq \alpha \leq 0.40$, the average E values are 59.45 kJ/mol. In the range of $0.40 \leq \alpha \leq 0.70$, the average E values are 136.77 kJ/mol and in the range of $0.70 \leq \alpha \leq 0.90$, the average E values are 383.95 kJ/mol.
(3) In the range of $0.10 \leq \alpha \leq 0.40$, solid-state reduction reaction of nickel oxide and iron oxide of Fe_2O_3 to Fe_3O_4 mainly happened. In the range of $0.40 \leq \alpha \leq 0.70$, gas-solid reduction reaction of Fe_3O_4 to FeO and Boudouard reaction mainly happened. In the range of $0.70 \leq \alpha \leq 0.90$, the reduction reaction of FeO to Fe and Boudouard reaction mainly happened.
(4) In the range of $0.10 \leq \alpha \leq 0.40$, the experimental data points are both targeting the chemical reaction order function. In the range of $0.40 \leq \alpha$ 0.70 and $0.70 \leq \alpha \leq 0.90$, the results showed that the process may not be exactly described by one mechanism function.

Acknowledgements The authors are especially grateful to The National Natural Science Foundation of China (No. 51234010).

References

1. X.M. Li et al., Technology of treating nickel-bearing laterite. Ferro-Alloys **4**, 24–28 (2007)
2. X.Y. Guo et al., The statue quo and prospect of the technologies of nickel laterites. Metal. Mater. Metall. Eng. **2**, 3–9 (2009)
3. J.B. Chen et al., Status quo of nickel mineral resource of our country and countermeasures. Express Inf. Mining Ind. **8**, 1–3 (2006)
4. Y.P. Zhang et al., Characteristics of laterite resource and analysis on its pyrometallurgy process. Ferro-Alloys **6**, 18–21 (2007)
5. S.W. Zhang, Status quo of nickel resources, production and consumption. World Nonferrous Metals **8**, 9–14 (2006)
6. B. Jankovic, B. Adnadevic, S. Mentus, The kinetic analysis of non-isothermal nickel oxide reduction in hydrogen atmosphere using the invariant kinetics parameters method. Thermochim. Acta **456**, 48–55 (2007)
7. B. Jankovic, B. Adnadevic, S. Mentus, The kinetic study of temperature-programmed reduction of nickel oxide in hydrogen atmosphere. Chem. Eng. Sci. **63**, 567–575 (2008)
8. J. Kim, G. Dodibiba, H. Tanno, Calcination of low-grade laterite for concentration of Ni by magnetic separation. Miner. Eng. **23**, 282–288 (2010)
9. D.A. Silva et al., Thermodynamic aspects of the roasting processes in the pre-treatment of nickelferrous garnierites. Mineral Process. Extr. Rev. Int. J. **9**, 97–104 (1992)
10. P. Wu et al., Prediction of the thermodynamic properties and phase diagrams of silicate systems-evaluation of the FeO-MgO-SiO_2 system. Trans. Iron Steel Inst. Jpn. **33**, 26–35 (1993)
11. J.M. Criado et al., Application of the master plots in kinetic analysis of non-isothermal data. Thermochim. Acta **147**, 377–385 (1989)

12. J.L. Liao et al., Influence of TiO_2 and basicity on viscosity of Ti bearing slag. Ironmaking Steelmaking **39**, 133–139 (2013)
13. S. Vyazovkin et al., An approach to the solution of the inverse kinetic problem in the case of complex process. Part III. Parallel independent reactions. Thermochim. Acta **197**, 41–51 (1992)
14. Y. Zhang et al., Reduction of Fe and Ni in Fe-Ni-O systems. J. Min. Metall. **49**, 13–20 (2013)
15. M.I. Nasr et al., Effect of nickel oxide doping on the kinetics and mechanism of iron oxide reduction. Trans. Iron Steel Inst. Jpn. **35**, 1043 (1995)

Evaluation of High Temperature Refractory Corrosion by Liquid Al_2O_3–Fe_2O_3–MgO–SiO_2

Christoph Sagadin, Stefan Luidold, Christine Wenzl and Christoph Wagner

Abstract Corrosion mechanisms have been investigated between MgO refractory substrates and synthetic FeNi slags. The materials taken into consideration comprised a simple synthetically mixed slag with specific oxides of slags from a ferroalloy producer. The MgO refractory substrates with the slag specimens on it were heated in a hot stage microscope to two different characteristic temperatures, 1350 and 1650 °C. The experiments proceeded under a controlled gas atmosphere that simulates the relevant process conditions. The corrosion mechanisms of each system were determined by SEM analyses. The obtained results showed that slag corrosion is dominating with a pronounced partial dissolution of refractory. It was also observed that iron oxide present in the slag diffused into the coarse refractory grains forming the relative low melting magnesia wuestite. Finally, the comparison of these findings with those predicted by thermodynamic calculation (FactSage) indicated the corrosion mechanisms and draw implications for improving the refractory performance and lifetime.

Keywords Ferronickel slag · Refractory corrosion · MgO–SiO_2–FeO system

C. Sagadin (✉) · S. Luidold
CD Laboratory for Extractive Metallurgy of Technological Metals,
Montanuniversitaet Leoben/Nonferrous Metallurgy,
Franz-Josef-Strasse 18, 8700 Leoben, Austria
e-mail: christoph.sagadin@unileoben.ac.at

S. Luidold
e-mail: stefan.luidold@unileoben.ac.at

C. Wenzl · C. Wagner
RHI AG, Wienerbergstrasse 9, 1100 Vienna, Austria
e-mail: christine.wenzl@rhi-ag.com

C. Wagner
e-mail: christoph.wagner@rhi-ag.com

J.-Y. Hwang et al. (eds.), *8th International Symposium on High-Temperature Metallurgical Processing*, The Minerals, Metals & Materials Series,
DOI 10.1007/978-3-319-51340-9_16

Introduction

The pyrometallurgical manufacturing of ferronickel in high temperature electric furnaces stresses the refractory linings by molten metal and/or slag. Chemical attack and penetration at high temperatures are the main factors for refractory corrosion. Reduced production efficiency and furnace shut-downs are the results of inferior refractory. Therefore, the aim is to select adequate furnace linings for the individual processes to provide homogeneous abrasion and predictable refractory life for scheduled maintenance and high furnace availability. Taking these facts into account and in order to improve ferroalloy processes along the production route continuous optimization of refractory materials, which is a decisive factor for furnace performance, is inevitable. For the identification of corrosion mechanisms different lab scale experiments are used [1, 2].

The presented work investigates synthetically ferronickel slags (typical slag composition based on the system $FeO–SiO_2–MgO–Al_2O_3$ [3–5]) in contact with MgO refractory substrate on a practical (corrosion test) and a theoretical basis (thermodynamic calculations with FactSage [6]).

Methods and Materials

Hot Stage Microscopy (HSM) and SEM Analysis

The corrosion tests were carried out in a HSM (EMI-201; Hesse Instruments Germany), by following experiment procedure: A synthetic slag sample cylinder with 3 mm diameter and 3 mm in height was prepared, placed on a plate of MgO refractory substrate (14 × 14 × 2 mm) and inserted in the horizontally tube furnace. The heating rate of the furnace was fixed at 10 K/min to a temperature of 1350 or 1650 °C, which remained constant for 60 min. The higher value was defined by the maximum reachable temperature of the HSM and the lower by the solidus temperature of the slag due to the assumption that the presence of any liquid phase will significantly enhance the chemical attack. To reflect process near conditions a mixture (Linde HiQ) of 60 % CO and 40 % CO_2 (corresponds to $p(O_2) = 1.94 \times 10^{-7}$ bar at 1650 °C) served as atmosphere.

After this treatment the samples were cut in cross sections, embedded in resin and grinded to produce a metallographic specimen for the following analyses by scanning electron microscope.

Materials

A. Synthetically produced slag

The pure oxide mixtures for a synthetic slag similar to real ones (i.e. composed of 48% SiO_2, 28.8% Fe_2O_3, 19.2% MgO and 4% Al_2O_3) were homogenized in a

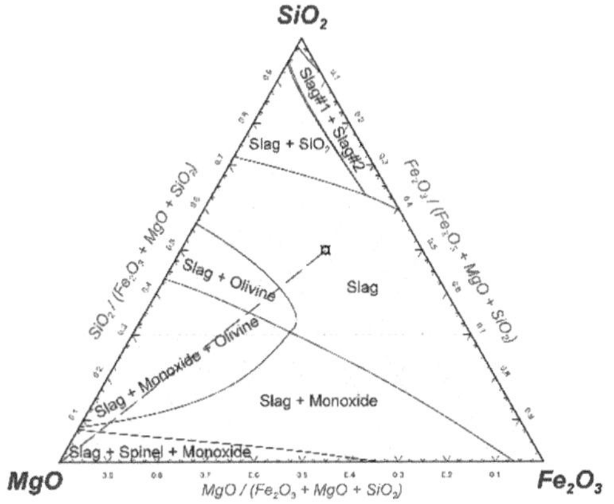

Fig. 1 Quaternary system at 1650 °C with a fixed Al_2O_3 content of 4% in CO/CO_2 atmosphere (60/40); the marker represents the slag composition and the dashed line varying mixtures of slag and refractory

Table 1 Chemical composition of slag and refractory

Chemical analysis	[wt.-%]							
	MgO	CaO	SiO_2	Fe_2O_3	Al_2O_3	BD	AP	CTE
Synthetical slag	17.6	–	49.8	29	4	–	–	–
Refractory substrate	>99.3	<0.35	<0.35	<0.12	<0.25	2.30	35	13

Chemical analysis -total Fe calculated as Fe_2O_3; non accredited laboratory procedure
BD bulk density [g/cm^3]; *AP* apparent porosity [wt.-%]; *CTE* coefficient of thermal expansion [10^{-6}/K]

swing mill and melted in a graphite crucible. Afterwards the slag was solidified and grinded in the swing mill to achieve a homogenized powder. The composition of the synthetic slag is illustrated in the quaternary $MgO–Fe_2O_3–SiO_2–Al_2O_3$ system at 1650 °C in equilibrium with the CO/CO_2 gas mixture by Fig. 1 which was compiled with FactSage. The chemical analysis of the slag is shown in Table 1.

B. Refractory substrate

A high quality magnesium oxide plate (14 × 14 × 2 mm^3) served as refractory. Its chemical analysis is shown in Table 1.

Practical Corrosion Tests and SEM/EDS Analysis

The mineralogical investigation of the prepared samples was carried out by SEM/EDS. This allows the determination of chemical reactions and penetration at the slag/refractory interface as well as identification of new phases, resulting from slag and refractory interactions. The EDS mapping provided a good qualitative detection of the main elements at the interface.

SiO_2–MgO–FeO_x–Al_2O_3 System (Synthetic Slag on Refractory Substrate) at 1650 °C

Figure 2 illustrates the interface between slag and refractory, which shows the infiltration of the slag into the refractory. The influenced zone in this figure is approximately 250–300 µm deep, however depending on the open porosity of substrate the infiltration depth can vary.

Figure 3 shows the spots analysed over the infiltration depth from the slag to pure refractory. The corresponding data of their analyses are summarized in Table 2.

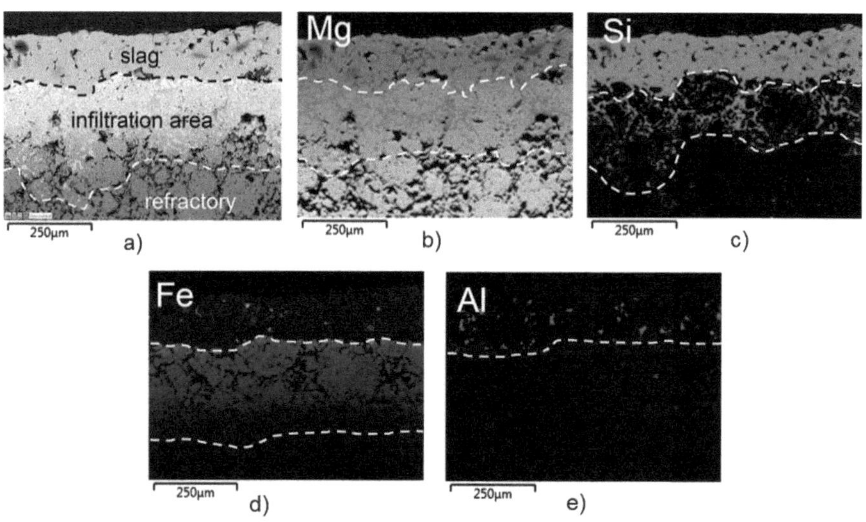

Fig. 2 SEM/EDS mapping of the interface between refractory material and molten slag: **a** overview with Mg, Si, Fe and Al; **b** Mg; **c** Si; **d** Fe; **e** Al

Fig. 3 Analyzed spots of the slag/refractory interface (1650 °C) based on EDS analysis: Mg *orange*, Si *green*, Fe *blue*, Al *red*

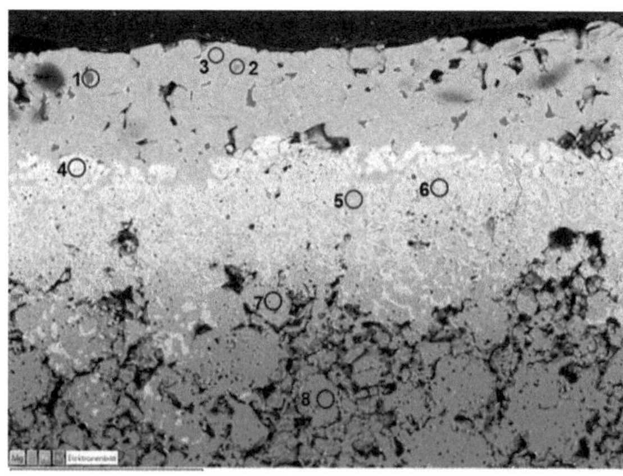

Table 2 Element concentrations at analyzed spots (1–8)

Element [mol-%]	EDX spot							
	1	2	3	4	5	6	7	8
Mg	29.2	23.8	62.7	83.6	62.6	85.2	90.8	98.9
Si	0.6	0.4	33.3	–	33.7	–	–	–
Fe	7.3	74.2	4.0	15.8	3.7	13.9	7.7	1.1
Al	62.9	1.6	–	0.6	0	0.9	1.5	–
(Mg, Fe)/Si ratio	56.7	243	2	–	2	–	–	–
Mg/Fe	4	0.3	15.7	5.3	17.1	6.1	11.7	88.3

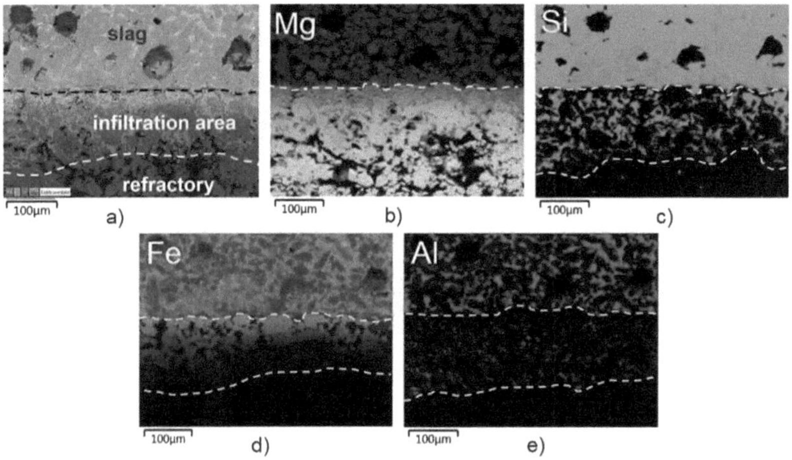

Fig. 4 Element distribution at the interface between slag/refractory: **a** over-view with Mg, Si, Fe and Al; **b** Mg; **c** Si; **d** Fe; **e** Al

SiO_2–MgO–FeO_x–Al_2O_3 System (Synthetic Slag on Refractory Substrate) at 1350 °C

The interface between slag and refractory at an experimental temperature of 1350 °C is pictured in Fig. 4. The thickness of the infiltration depth amounts to approximately 150 µm, which is about 100 µm less than at 1650 °C. The temperature was too low for a complete melting of the slag.

An analysis of a series of spots (Fig. 5) distributed over the infiltration area were additionally performed to identify the elements taking part in the corrosion between synthetic slag and refractory at 1350 °C. The corresponding data of the spot investigation are summarized in Table 3.

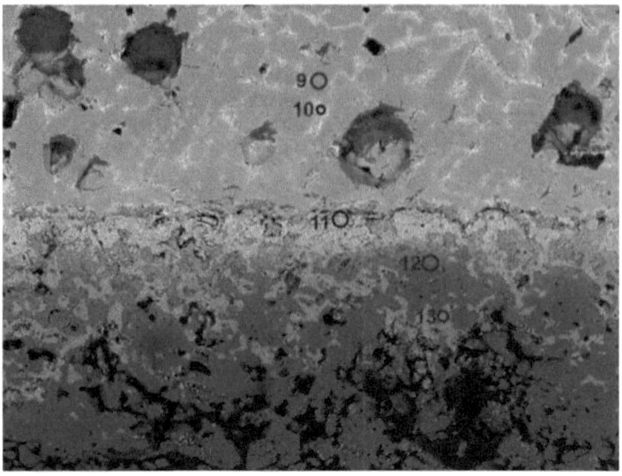

Fig. 5 Analyzed spots of the slag/refractory interface (1350 °C) based on EDS analysis: Mg orange, Si green, Fe blue, Al red

Table 3 Element concentrations at analyzed spots (9–13)

Element [mol-%]	EDX spot				
	9	10	11	12	13
Mg	34.2	13.3	83.1	87.4	61.3
Si	53.2	44.8	0.3	8.2	32.7
Fe	12.3	28.3	15.6	2.6	3.2
Al	0.3	13.5	1.0	1.6	1.2
(Mg, Fe)/Si-ratio	0.87	0.93	329.0	11.0	1.9
Mg/Fe	2.78	0.47	5.3	33.6	19.2

Thermodynamic Calculations

In addition to the practical experiments some thermodynamic calculations were carried out by FactSage software. Figure 6 shows the diagram with varying portions of synthetic ferronickel slag (SiO_2–MgO–Fe_2O_3–Al_2O_3 system) and the pure refractory (MgO). The calculation implied the same atmosphere which was used in practical experiments.

Results and Discussion

The melting trials and thermodynamic calculations of the slag/refractory interactions show in this case good coincidences. The refractory matrix fines (<100 µm) in the infiltrated area consist of Mg, Si and O. Large MgO particles of the substrate in the infiltrated layer are not dissolved. Slag first infiltrates along any available pores in the refractory and then reacts with the MgO. One newly formed phase is olivine which emerges in the pores of the MgO substrate.

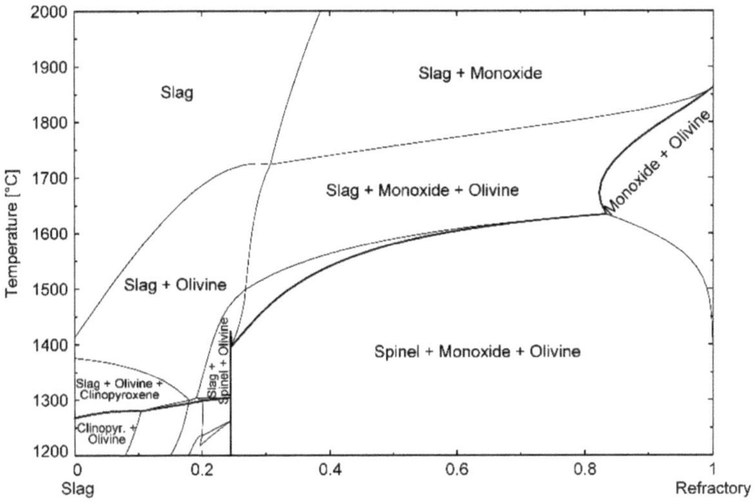

Fig. 6 Stable phases predicted by thermodynamic simulation as a function of refractory/slag ratio and temperature in CO/CO_2 atmosphere (60/40)

Spots 4, 6 (1650 °C) and 11 (1350 °C) represent different MgO grains which have an increased content of Fe. Silicate or alumina rich phases were not present, therefore these spot analyses represent Mg-wuestite ((Mg, Fe)O). Because of a high iron oxide content in the slag the MgO was highly enriched with Fe oxide reaching a Mg/Fe ratio of 6.1 in the MgO grain in spot 6. This Mg-wuestite phase corresponds to the phase monoxide (Fig. 4.1) of the thermodynamic FactSage database FToxid. Deeper in the refractory the Fe content in the MgO grains decreases as indicated by spots 7 and 8.

The mappings of synthetic slag visualizes different element concentrations: areas composed of Mg (yellow), Si (green), Fe (blue) and Al (pink).

At 1650 °C the effect of Mg-wuestite formation can be seen in Fig. 2.2d because of the enrichment of blue Fe in MgO grains. Figures 2.2c and 2.4c identify high Si concentrations in the slag and of the refractory pores. The spot analyses of 3 and 5 exhibit (Mg, Fe)/Si ratios of approximately 2 that coincides with the stoichiometry of olivine. Also spot 13 shows olivine phases in deeper refractory zones at 1350 °C. The theoretical calculation also conforms the formation of olivine. In Spot 1 increased contents of Al could be localized. Its Al/Mg ratio of about 2 indicate magnesiumaluminat ($MgAl_2O_4$), which also emerges in the FactSage calculation at MgO enriched slags. Also Fe-rich zones (Spot 2) appeared in the slag which can be interpreted as iron spinels due to the presence of Al.

The mapping at 1350 °C shows the occurrence of similar phases as at 1650 °C. The Fe content in the slag is high and the Mg-wuestite starts to form MgO which can be seen in spot 11. With decreasing Fe content and rising Mg content in spot 9 olivine phases were built. Spot 10 indicates an Al enriched zone which converts to magnesiumaluminate at 1650 °C.

Conclusion

For the investigation of corrosion mechanisms of refractories by liquid slags, it is vital to analyse the microstructure of the individual phases. The present study combines two investigation methods, namely hot stage microscopy and thermodynamic calculations. The former represents the practical analysis and was applied to investigate the corrosion mechanisms between magnesia refractory and synthetic ferroalloy slag. The thermodynamical calculations served as comparison for the practical results and verification of the formed phases and showed very good agreement.

The investigation of the refractory substrate was performed at two temperatures which show different infiltration and corrosion levels. The thermodynamic calculations are a suitable tool for predicting the sequence of phase transformations in equilibrium and therefore supports the selection and optimization of refractories.

Acknowledgements The financial support by the Austrian Federal Ministry of Science, Research and Economy and the National Foundation for Research, Technology and Development is gratefully acknowledged.

References

1. D. Gregurek et al., Refractory wear mechanisms in the nonferrous metal industry: testing and modeling results. JOM **65**(11), 1622–1630 (2013)
2. W.E. Lee, S. Zhang, Melt corrosion of oxide and oxide–carbon refractories. Int. Mater. Rev. **44** (3), 77–104 (2013)
3. S. Chen, E. Jak, P.C. Hayes, Phase equilibria in the olivine primary phase field of the system MgO–"FeO"–SiO_2 in equilibrium with metallic iron (ISIJ International, Japan, 2005), pp. 1095–1100
4. S. Chen, E. Jak, P. Hayes, Phase equilibria in the cristobalite, tridymite and pyroxene primary phase fileds in the MgO–FeO–SiO_2 system in equilibrium with metallic iron (ISIJ International, Japan, 2005), pp. 791–797
5. F. Crundwell et al., *Extractive metallurgy of nickel, cobalt and platinum group metals* (The Boulevard, Elsevier Professional, 2011)
6. C.W. Bale et al., FactSage thermochemical software and databases—recent developments. Calphad **33**(2), 295–311 (2009)

Thermodynamic Calculation on the Reactivity Between Slag and Ti-Stabilized Stainless Steel

Zhuo Chen, Kun-peng Xu, Sheng-ping He and Qian Wang

Abstract During the continuous casting (CC) process of Ti-Stabilized Stainless Steel, the nozzle clogging or even the crusts composed of TiN, Al_2O_3, Ti_xO_y and the complex inclusions $MgO{\cdot}Al_2O_3$, Al_2O_3 is often occurred. The main objective of the present study is to research the reaction between refining slag or mold flux and Ti-Stabilized Stainless Steel by thermodynamic calculation. A suitable refining slag composition was sought to absorb Al_2O_3, Ti_xO_y inclusions, and the reactions between different mold flux and steel was calculated and compared. It was found that a suitable refining slag was difficult to adjust aluminum and titanium to meet requirement simultaneously. The target refined slag is composited of 3–8.7% SiO_2, 56–59% CaO, 33–38% Al_2O_3 and 4% MgO with low melting point and high absorption capacity of inclusions for in the refining of the 321 stainless steel. SiO_2, Fe_2O_3, Na_2O and B_2O_3 of mold flux would react with titanium in the steel. The CaO–SiO_2 based mold flux had a strong reactivity than CaO–SiO_2–Al_2O_3 and CaO–Al_2O_3 based mold flux.

Keywords Ti-stabilized steel · Slag-steel reaction · Thermodynamics

Introduction

Ti-stabilization stainless steel such as 321 stainless steel has an excellent corrosion resistance in corrosive medium environment. Therefore, this steel is widely applied in petrochemical, transportation, chemical and light industries. However, clogging of the submerged entry nozzle (SEN) is a long term problem because of the high titanium of Ti-stabilization stainless steel during the continuous casting (CC).

Z. Chen · K. Xu · S. He · Q. Wang (✉)
College of Materials Science and Engineering,
Chongqing University, Chongqing 400044, China
e-mail: q_wang@cqu.edu.cn

J.-Y. Hwang et al. (eds.), *8th International Symposium on High-Temperature Metallurgical Processing*, The Minerals, Metals & Materials Series,
DOI 10.1007/978-3-319-51340-9_17

Previous studies [1, 2] researched that the clogging materials were composed of TiN, Al_2O_3, Ti_xO_y and the complex inclusions nucleated with TiN. Hence, the clogging deteriorates the properties of mold flux and will worsen the slab quality [3]. To improve the steel cleanliness and ensure the smooth running of the continuous casting, on the one hand, alumina and titanium oxides, solid deoxidation inclusions, have to be transformed into a liquid complex inclusion so that they could more easily aggregate and dissolve into the slag phase [4, 5]. On the other hand, during the continuous casting of Ti-stabilization stainless steel, several important components of mold flux are easily reduced by titanium in the steel. The reduction changes the initial components and the lubrication properties of mold flux. As a consequence, the performance of mold flux may be deteriorated, which will even cause sticker breakouts and surface defects on the slabs.

As is well known, refining slag with low oxidizability and high basicity also possesses function for secondary metallurgy of Al killed steel [6]. Therefore it is very hard to prevent the generation of high melting point inclusions. Moreover, the inclusions give rise to the formation of clogging in the SEN for Ti-stabilization stainless steel. Several researchers proposed that MnO or Fe_2O_3 [7, 8] could be added into mold flux to the oxidizability which is stronger than that of SiO_2, could be added to the mold flux to prevent SiO_2 from reduction because of its stronger oxidizability. Also, there is a new proposal to solve the problem fundamentally by reducing the content of reactive compositions such as SiO_2 and MnO [9].

The main purpose of the current study is to improve the cast ability of Ti-stabilization stainless steel. Since reactions between components of refining slag or mold flux and titanium in molten steel have not been studied systematically so far. Thermodynamic calculation of the equilibrium between molten steel and refining slag was investigated, and the reactivity between the titanium in molten steel and mold flux with different basicity were calculated and compared with assistance of the thermodynamic software FactSage 6.3. The composition of refining slag for Ti-stabilization stainless steel was preliminary designed to dissolve inclusions deoxidation inclusions into the slag phase.

The Reactivity Between Refining Slag and Steel

During the tapping process, molten steel would be well mixed by the tremendous kinetic energy generated by steel flow. Refining slag fully contacts with liquid steel and deoxidation type inclusions and it lasts for about 60–90 min from tapping to LF finish. Therefore, the progress can be settled as an equilibrium state for the stage of secondary metallurgy. Given that the silicon content of the steel was 0.73 wt%, The MgO and TiO_2 content of the refining slag was assumed to be 4 and 1.5 wt% respectively. The chemical compositions of Ti-Stabilized Stainless Steel and final

Table 1 The chemical compositions of Ti-stabilized stainless steel, wt%

C	Si	Mn	P	S	Cr	Ni	Ti	Al	N	Fe
0.03	0.73	0.99	0.028	0.003	17.31	9.35	0.3	0.019	0.127	71.11

Table 2 The chemical compositions of final refining slag, wt%

Al_2O_3	CaO	MgO	SiO_2	TiO_2
9.5	65.0	4.0	20.0	1.5

refining slag are given in Tables 1 and 2 respectively. In these reactions, T is the expression of temperature (K). The equilibrium reaction equations related silicon are as follows [10, 11]:

$$[\mathrm{Si}] + 2[\mathrm{O}] = (\mathrm{SiO_2})_{\mathrm{slag}} \tag{1}$$

$$\Delta G_1^o = -581,900 + 221.8T\,\mathrm{J/mol}$$

$$3[\mathrm{Si}] + 2(\mathrm{Al_2O_3})_{\mathrm{slag}} = 4[\mathrm{Al}] + 3(\mathrm{SiO_2})_{\mathrm{slag}} \tag{2}$$

$$\Delta G_2^o = 658,200 - 107.1T\,\mathrm{J/mol}$$

$$[\mathrm{Si}] + 2(\mathrm{CaO})_{\mathrm{slag}} = 2[\mathrm{Ca}] + (\mathrm{SiO_2})_{\mathrm{slag}} \tag{3}$$

$$\Delta G_3^o = 708,500 - 75.6T\,\mathrm{J/mol}$$

$$[\mathrm{Si}] + 2(\mathrm{MgO})_{\mathrm{slag}} = 2[\mathrm{Mg}] + (\mathrm{SiO_2})_{\mathrm{slag}} \tag{4}$$

$$\Delta G_4^o = 875,300 - 255T\,\mathrm{J/mol}$$

$$[\mathrm{Ti}] + (\mathrm{SiO_2})_{\mathrm{slag}} = [\mathrm{Si}] + (\mathrm{TiO_2})_{\mathrm{slag}} \tag{5}$$

$$\Delta G_4^o = 875{,}300 - 255T\,\mathrm{J/mol}$$

In the case of the SiO_2–CaO–Al_2O_3–4%MgO–1.5% TiO_2 refining slag and 0.73% silicon liquid steel, isolines of dissolved mass fractions of aluminum, the activities of dissolved oxygen a[O], calcium a[Ca] and magnesium a[Mg] were calculated using FactSage 6.3 in the 'Equilib' module. Meanwhile the temperature and pressure were set as 1873 K and 1 atm respectively.

The Reactivity Between Different Mold Flux and Steel

Three based mold fluxes (CaO–SiO_2–Al_2O_3, CaO–SiO_2 and CaO–SiO_2–Al_2O_3) were considered to calculate the reactivity between different mold flux and steel, and were numbered as No. 1, No. 2 and No. 3. The reactions Gibbs free energy

Table 3 The basicity and properties of basic mold fluxes

Basic mold fluxes	R = CaO/SiO_2	$\eta_{1300°C}$, Pa s	T_{m}, °C
1	1.97	0.126	1148
2	1.18	0.132	1110
3	6.1	0.126	1160

between components of mold flux with titanium in molten steel were calculated under pouring temperature to preliminary compare and estimate the reactivity. Chemical components of 321 stainless steel and the basicity and properties of basic mold flux are given in Tables 1 and 3 respectively. Assume that the initial mould fluxes content of TiO_2 in 4% of the total mold flux.

Since in continuous casting the consumption of mold fluxes is about 0.5 kg per ton of molten steel, which means that the update speed of molten steel is far more than the mold flux, the contents of mold flux components were regarded as the measuring unit in the thermodynamics calculation. The activity of components in the liquid steel was taken as 1 mass% substance while the activity of components in mold flux was taken as pure substance. The reaction equations are as follows:

$$[Ti] + (SiO_2) = [Si] + (TiO_2) \quad \Delta G^{\theta} = -101,050 + 7.3T\,J/mol$$

$$3/2[Ti] + (Al_2O_3) = 2[Al] + 3/2(TiO_2) \quad \Delta G^{\theta} = 189,408.3 - 50.6T\,J/mol$$

$$1/2[Ti] + (MgO) = [Mg] + 1/2(TiO_2) \quad \Delta G^{\theta} = 392,811.5 - 127T\,J/mol$$

$$3/2[Ti] + (Fe_2O_3) = 2[Fe] + 3/2(TiO_2) \quad \Delta G^{\theta} = -531,226.5 - 67.49T\,J/mol$$

$$1/2[Ti] + (Na_2O) = 2[Na] + 1/2(TiO_2) \quad \Delta G^{\theta} = 73,400 - 100.94T\,J/mol$$

$$3/2[Ti] + (B_2O_3) = 2[B] + 3/2(TiO_2) \quad \Delta G^{\theta} = -48,126.3 - 14.3T\,J/mol$$

$$1/2[Ti] + (CaO) = [Ca] + 1/2(TiO_2) \quad \Delta G^{\theta} = 296,565 - 36.4T\,J/mol$$

Table 4 Interaction coefficients of components in molten steel, e_i^k (1873 K)

B\K	C	Si	Mn	Al	P	S	Cr	Ni
Al	0.091	0.0056	–	0.045	–	0.03	–	–
Si	0.18	0.11	0.002	0.058	0.11	0.056	−0.003	0.005
Mg	−0.15 [13]	−0.09 [14]	–	−0.12	–	−1.38	–	–
Ca	0.34	−0.097	–	−0.07	–	–	–	–
B	0.22	0.078	–	–	–	0.048	–	–
Mn	−0.0761				−0.0038	−0.0522		

Table 5 The thermodynamic calculated data of the mould fluxes-steel system, J mol^{-1}

Number	ΔG_{SiO_2}	$\Delta G_{Al_2O_3}$	$\Delta G_{Fe_2O_3}$	ΔG_{MgO}	ΔG_{Na_2O}	$\Delta G_{B_2O_3}$	ΔG_{CaO}
No. 1	1.20E+04	5.00E+04	/	/	/	−6.77E+05	7.41E+04
No. 2	−1.66E+04	8.17E+04	−8.36E+05	4.95E+04	−3.21E+05	/	5.53E+04
No. 3	2.08E+04	2.42E+04	/	5.80E+03	/	−8.34E+05	7.54E+04

The activity of melt a_i is expressed as Eq. (6), while the activity of slag a_j is calculated by the thermodynamics software Factsage 6.3.

$$a_i = w_i \cdot f_i \tag{6}$$

where w_i is the concentration of melt component which is listed in Table 1; f_i is the interaction coefficient which can be obtained by

$$\lg f_i = \sum e_i^k [\%\mathrm{K}] \tag{7}$$

where K is the interaction coefficient and is listed in Table 4.

The interaction coefficient f_i in 1823 K would be calculated by combining Eqs. (7), (8) and K.

$$e_{i(T)}^k = \frac{1873}{T} e_{i(1873)}^k \tag{8}$$

The thermodynamic calculated data of the mould fluxes-steel system were shown in Table 5.

Result and Discussion

The restrain area of aluminum oxidation reaction is shown in Fig. 1, the grey area and the calculated results for isolines are shown in Fig. 2. The aluminum and titanium content in the 321 stainless steel ranged from 0.015 to 0.025% and 0.25 to

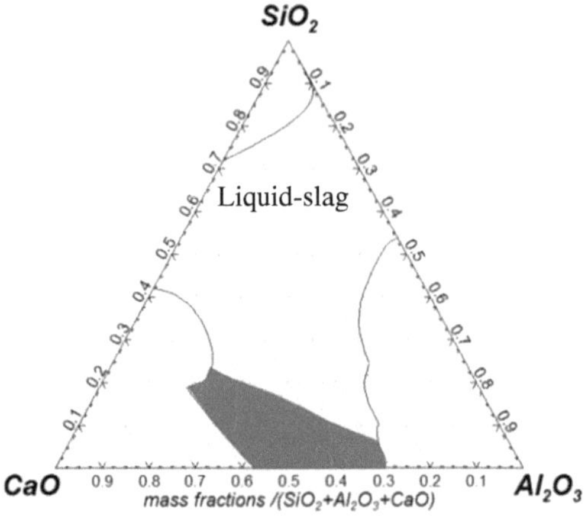

Fig. 1 The restrain area of aluminum oxidation reaction

0.35% for the requirement of the steel grade. Also, the ratio of CaO/Al_2O_3 was in the range of 1.5–1.8 which would be effective for absorbing non-metallic inclusions, as mentioned by Yoon et al. [12]. In the case of the SiO_2–CaO–Al_2O_3–4% MgO–1.5% TiO_2 refining slag and 0.73% silicon liquid steel, the target refining slag should be controlled to satisfy the following composition if aluminum meets the requirement: 17.1–22% SiO_2, 47.4–52% CaO, 27.8–33.1% Al_2O_3, 4% MgO and 1.5% TiO_2, as shown in gray zone of Fig. 2. The liquid complex inclusion was existed when the refining slag was used. The a[O], a[Ca] and a[Mg] is equal to $1.5–2.5 \times 10^{-4}$, $4–15 \times 10^{-8}$, $a_{[Mg]} = 2.2–3.5 \times 10^{-5}$ respectively. Also, the refining slag and molten steel containing 0.015–0.025% aluminum, 0.73% silicon and 0.004–0.01% titanium will reach a thermodynamic equilibrium. Similarly the target refining slag should be satisfied the following composition if titanium meets the requirement: 3–8.7% SiO_2, 56–59% CaO, 33–38% Al_2O_3, 4% MgO and 1.5% TiO_2, as shown in green zone of Fig. 2. The a[O], a[Ca] and a[Mg] is equal to $1.5–2.5 \times 10^{-4}$, $4–15 \times 10^{-8}$, $a_{[Mg]} = 2.2–3.5 \times 10^{-5}$ respectively. Also, the refining slag and molten steel containing 0.73% silicon and 0.1–0.5% titanium will reach a thermodynamic equilibrium. Considering titanium is absolutely necessary element for 321 Stainless Steel, so the target refined slag composition is 3–8.7% SiO_2, 56–59% CaO, 33–38% Al_2O_3, 4% MgO and 1.5% TiO_2.

Table 4 reveals that SiO_2 in mold flux No. 2 could react with titanium in the steel but not in No. 1 and No. 3, while the reactivity of SiO_2 are smaller. Since Fe_2O_3, B_2O_3, and Na_2O have a stronger reactivity than SiO_2, they would react with titanium even in the case of content is very few. To the contrary, neither CaF_2 nor MgO would not react with titanium in molten steel. Therefore, the reactivity of three basic mold fluxes are stronger than B_2O_3, SiO_2 in mold flux and the strong-to-weak sequence of the reactivity with titanium is No. 2, No. 3, No. 1.

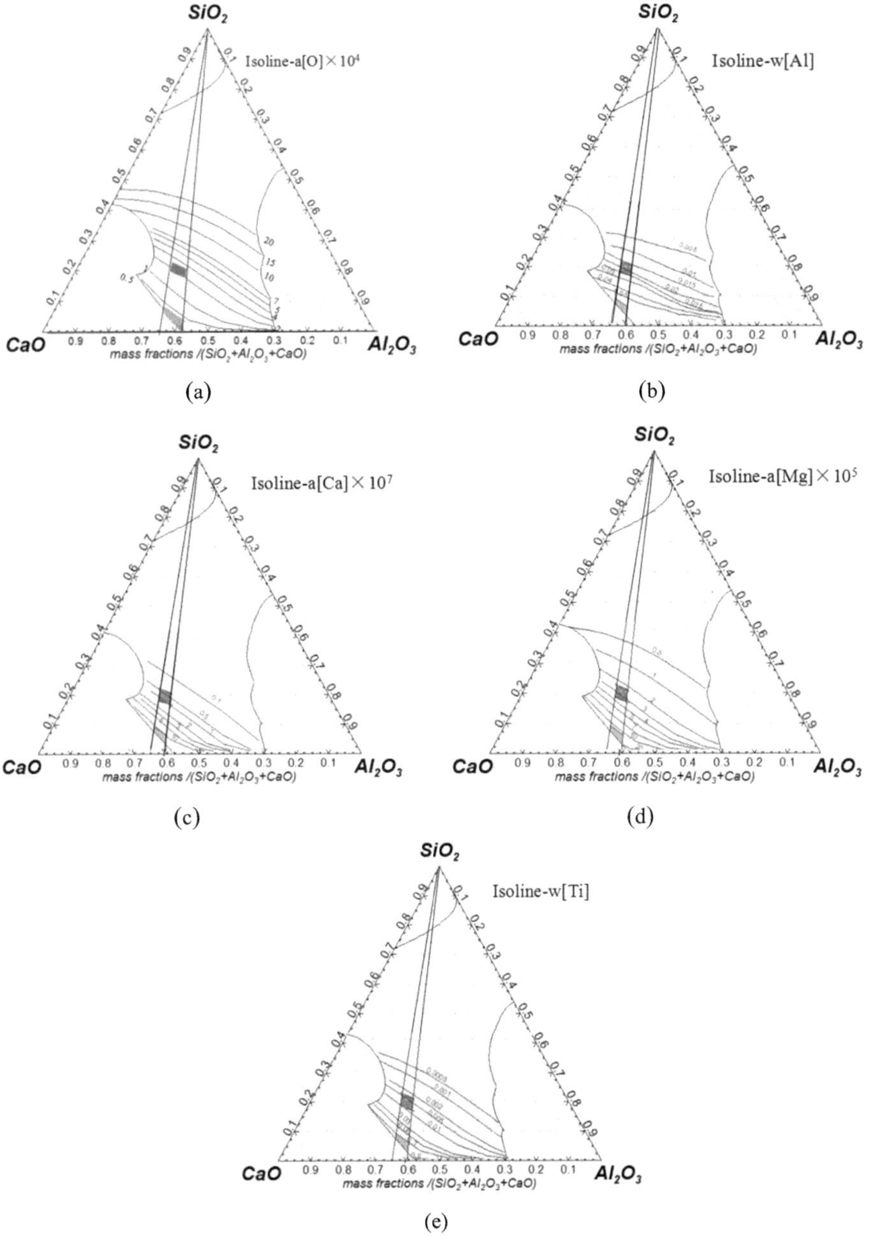

Fig. 2 The result of thermodynamic equilibrium between SiO_2–CaO–Al_2O_3–4% MgO–1.5% TiO_2 slag and molten steel with isolines of **a** a[O], **b** [wt, %Al], **c** a[Ca], **d** a[Mg], **e** [wt, %Ti] at 1873 K

Conclusion

The following conclusions can be drawn from the present work:

(1) The target refined slag is composited of 3–8.7% SiO_2, 56–59% CaO, 33–38% Al_2O_3 and 4% MgO and has low melting point and high absorption capacity inclusions for the refining of the 321 stainless steel.
(2) Fe_2O_3, Na_2O and B_2O_3 of mold fluxes would react with titanium in the steel. However, SiO_2 would react with CaO–SiO_2–Al_2O_3 and CaO–Al_2O_3 based mold fluxes but not CaO–SiO_2 based mold flux.
(3) The reaction of CaO, CaF_2, MgO, Li_2O, BaO with titanium in the steel not happened.
(4) The CaO–SiO_2 based mold flux has a strong reactivity than CaO–SiO_2–Al_2O_3 and CaO–Al_2O_3 based mold fluxes.

Acknowledgements The authors greatly appreciate the supports from the Key Program of National Natural Science Foundation of China (No. U1660204)

References

1. R. Qiang, L. Ju-Cang, X. Bin, The inclusion formation mechanism of the 409L stainless steel. Mater. Res. Innovations **19**(S9), 378 (2015)
2. R.C. Nunnlngton, N. Sutcliffe, The steelmaking and casting of Ti stabilized stainless steel, in *59th Electric Furnace Conference and 19th Process Technology Conference*
3. S.K. Choudhary, A.J. Khan, Nozzle clogging during continuous slab casting at Tata Steel. Steel Times Int. **24**(5), 21 (2000)
4. S. Abdelaziz, G. Megahed, I. El-Mahallawi, H. Ahmed, Ironmaking Steelmaking **36**(6), 432–441 (2009)
5. L.E.K. Holappa, A.S. Helle, J. Mater. Process. Technol. **53**, 177–186 (1995)
6. J. Yang, X. Wang, M. Jiang, W. Wang, J. Iron. Steel Res. Int. **8**(7), 8–14 (2011)
7. Q. Wang, J.H. Chi, Study on properties of mould fluxes for continuous casting of E2 steel. Sichuan Metall. **3**, 40–46 (1991)
8. S.P. He, Q. Wang, J.H. Zeng, Performance control of mould fluxes for high-Al steel. J. Iron Steel Res. **21**(12), 59–62 (2009)
9. Baoshan Iron and Steel, China, CN200710042540.6, 2009.10.14
10. H. Suito, R. Inoue, ISIJ Int. **36**(5), 528–536 (1996)
11. X. Zhang, Q. Han, D. Chen, Metall. Trans. B **22B**(6), 918–921 (1991)
12. B.-H. Yoon, K.-H. Heo, J.-S. Kim, H.-S. Sohn, Ironmaking Steelmaking **29**(3), 215–218 (2002)
13. G.K. Sigworth, J.F. Elliott, The thermodynamics of liquid dilute iron alloys. Metal Sci. **8**, 298 (1974)
14. H. Ohta, H. Suito, Activities in CaO-MgO-Al_2O_3 slags and deoxidation equilibria of Al, Mg and Ca. ISIJ Int. **36**(8), 983 (1996)

Part IV
Alloys and Materials Preparation

Development of a Novel, Low-Cost Titanium Extraction Process for Bulk or Powder Applications

James Cox, Chanaka De Alwis, Benjamin Kohler, Mike Lewis, Matthew Call, Julia Kluck, Amelinda Olson and Marc Snyderman

Abstract UTRS Inc. has developed a novel two-step method to extract commercially viable titanium alloys (patent pending no. 20130164167 [1]). This approach has proven its potential to produce a variety of titanium alloys (Ti-6Al-4V, Ti_xAl_y, etc.,) and has demonstrated a number of advantages over the current commercial titanium production process. Additionally, the UTRS Process operates without producing CO_2 and eliminates the use of $TiCl_4$. Step 1: Aluminothermic reduction makes available many types of ores that could otherwise not be used. Due to the aluminothermic reduction technique and the composition of the ore, metal extracted from this step will inherently contain non-titanium elements. Step 2: Electro-refining removes undesirable elements from the thermally extracted titanium. This refined metal has a dendritic morphology with a wool-like appearance that is referred to as "titanium wool". This new morphology of titanium allows it to have a diversity of end product types from titanium powder to bulk titanium.

Keywords Titanium · Alloy · Titanium-bearing · Ores · Ti-6Al-4V · Aluminothermic · Titanium wool · Electro-refining

Introduction

Titanium provides a combination of unique properties that cannot be found in any other single structural metal. Its strength-to-weight ratio (120 kNm/kg) is much higher than the most commonly used structural metal, steel (38 kNm/kg) or stainless steel 304 (77 kNm/kg) [2]. Its resistance to corrosion is superior to any other structural metal (steel, magnesium, aluminum). Although these properties make titanium the obvious choice as a structural metal, its use is limited due to the high cost associated with the production (~ $9.00/kg, Ti sponge) [3, 4]. Despite this

J. Cox (✉) · C. De Alwis · B. Kohler · M. Lewis · M. Call · J. Kluck · A. Olson · M. Snyderman
Universal Technical Resource Services (UTRS) Inc, Butte, USA
e-mail: jcox@utrs.com

J.-Y. Hwang et al. (eds.), *8th International Symposium on High-Temperature Metallurgical Processing*, The Minerals, Metals & Materials Series,
DOI 10.1007/978-3-319-51340-9_18

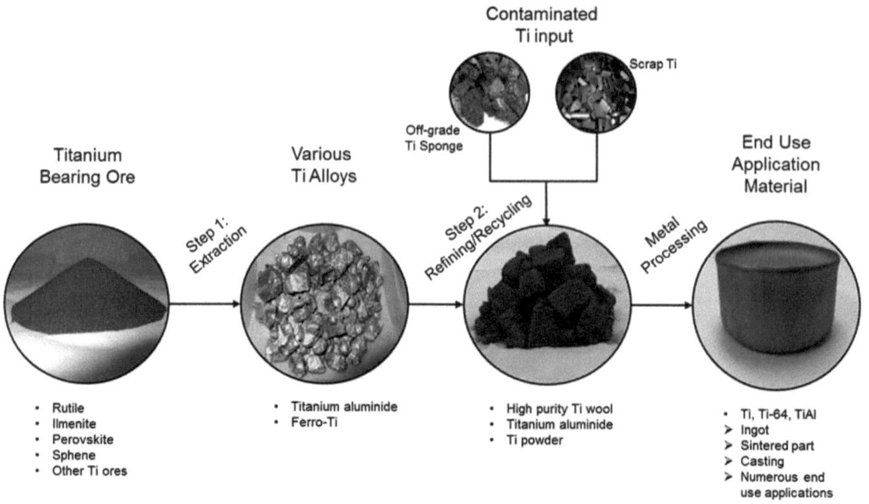

Fig. 1 Various stages of the UTRS titanium extraction process

high cost, titanium and its alloys are the only choice in many industries (e.g. medical implants, aerospace, sports equipment, etc.). UTRS has developed a novel production method to reduce the titanium alloy cost by 40–50%. The underlying technology is accomplished in two major processing steps: (1) metal extraction and (2) electro-refining. The electro-refined product has a dendritic morphology that we refer to as titanium wool. This high purity titanium wool can be pressed without the aid of a binder into ingots for bulk titanium, near-net shaped parts or can be ground into titanium powder for additive manufacturing applications.

The UTRS Titanium Extraction Process is illustrated in Fig. 1.

Methodology

Extraction of Titanium Metal from Ores

In step 1 of the UTRS Titanium Process, titanium metal is extracted by reducing titanium bearing ores using aluminum as a reducing agent. The resulting composition of the metal can be tuned by using a variety of titanium bearing ores [(e.g., Rutile (natural TiO_2), Ilmenite ($FeTiO_3$), Perovskite ($CaTiO_3$), and Sphene ($CaTiSiO_5$)].

Historically, the affinity of oxygen to titanium and Ti–Al alloys has prevented the usefulness of this technique for producing commercial titanium [5, 6]. Additionally, other metal impurities such as iron (Fe), silicon (Si), etc., can find their way to the metal-phase during the aluminothermic process.

Fig. 2 Cross section image of extraction product showing good metal/slag separation

Although, the reaction between the titanium bearing ore (e.g. TiO_2) and Al is exothermic, it does not generate sufficient heat to drive the reaction to completion. UTRS has been successful in developing a process that enables the use of aluminum as a viable reducing agent for titanium extraction. In part, by using a combination of external heating and fluxing agents, the UTRS method enables the reaction to proceed to completion in a controlled manner and promotes good metal/slag separation. Figure 2 is a digital image of the UTRS extracted product.

The UTRS Process brings to light the following benefits of using Al as a reducing agent for titanium metal extraction:

- Aluminum is inexpensive relative to other commonly used reducing agents
- Aluminum has a wide operable temperature range that is conducive for titanium metal extraction (melting point of 660 °C to a boiling point of 2470 °C)
- Aluminothermic reduction makes available many types of titanium bearing ores that could otherwise not be used by current commercial titanium process
- Eliminates the need for $TiCl_4$ in the process and the process does not produce CO_2

The UTRS approach to using aluminothermic extraction of titanium removes much of the oxygen from TiO_2 ores and prevents or significantly reduces impurities such as calcium and rare-earth elements from entering the metal. However, depending on the composition of the ore, the metal extracted during this step will inherently contain non-titanium elements. These undesirable elements are removed during the electro-refining step described herein.

Refining of Extracted Metal

In the previous section, it was demonstrated that titanium can be thermo-chemically extracted from titanium bearing ores. However, the extracted titanium contains elevated levels of other elements that are too high for many practical applications. Therefore, in step 2 of the UTRS Process, the extracted metal is refined

Table 1 Calculated EMF values for selected reactions at 750 °C (HSC Chemistry 7, Outotec, Finland)

Reaction	EMF versus Cl^-/Cl (V)
Ti^{2+}/Ti	−1.82
Al^{3+}/Al	−1.74
Ti^{3+}/Ti	−1.77
Ti^{4+}/Ti	−1.47
Si^{4+}/Si	−1.37
Fe^{2+}/Fe	−1.13
Ni^{2+}/Ni	0.79

electrochemically to remove undesired elements. The electro-refining of metals depends on their ability to selectively dissolve (oxidize) at an anode (positive electrode) and to selectively deposit (reduce) at a cathode (negative electrode). The selectivity of these two processes is governed by the sequence of electrochemical potentials (electromotive force—EMF). The potential, or EMF, represents the energy that is required to dissolve or deposit the metal. Metals with more positive potentials are difficult to oxidize and metals with more negative potentials are difficult to reduce. The calculated EMF values for selected metals are shown in Table 1.

As shown in Table 1, Al may not be completely removed from titanium alloys due to similarities in dissolution potentials of Al and Ti. However, we believe that a partial separation can be achieved due to different activity coefficients of Al and Ti in Ti–Al alloys [7]. Contrarily, metals such as iron (Fe), silicon (Si), etc., can effectively be eliminated from the extracted titanium metal by the electro-refining process since there is a wider potential gap between these elements and titanium. Therefore, this paper will focus on achieving a lower level of aluminum concentrations ($\sim$5.0 wt%) in the refined titanium wool. Additionally, the oxygen concentration can be significantly reduced by electro-refining. However, as mentioned in the experimental section, the harvesting and handling of the titanium metal was performed in normal atmosphere where contamination with oxygen could not be avoided.

Experimental

Analysis

All analytical work was out-sourced and performed at external laboratories. Data were collected by IGF to determine concentrations of: O, C, N and H; and by ICP-OES for the remaining elements as presented in all subsequent tables.

Metal Extraction

The following is a typical example of an extraction experiment using the UTRS Titanium Process. Data from analysis of the titanium ore and metal product from a representative extraction experiment are presented in Table 2.

A titanium metal product was extracted from a titanium bearing ore as follows. The ore concentrate was ground in a ball mill to pass through a 200 mesh standard sieve (<74 μm). A chemical blend was prepared by mixing stoichiometric ratios of the ground ore concentrate (399.4 g) and aluminum powder (179.9 g), and calcium fluoride (260.3 g) in a jar mill for 30 min to achieve a uniform blend. The chemical blend was then charged into a cylindrical graphite crucible and placed in a containment housing within an induction furnace. The containment housing was closed and purged with argon gas to displace oxygen from the system. A continuous flow of inert gas was maintained throughout the remainder of the extraction operation. A temperature profile consisting of a ramp rate of 10 °C/min towards ~1725 °C and a soak time of ~20 min was used to heat the blend and control the reaction. The induction furnace was then cooled to room temperature and the reaction products were removed and prepared for analysis. See patent [1] for additional information.

Electro-refining

A quartz crucible (8000 ml) was used as a reaction vessel. A circular molybdenum plate was placed parallel to the bottom of the reaction vessel and used as the cathode. The extracted product was placed inside a 500 ml nickel anode basket which had been perforated on the side wall and suspended above the cathode. The reaction vessel was filled with 7500 g of an alkali halide salt blend of sodium chloride (NaCl) and potassium chloride (KCl) at a 44:56 NaCl:KCl respective weight percent ratio. The quartz crucible was placed inside a secondary container

Table 2 Analytical data of Ti ore and extracted metal from UTRS titanium process values are listed as wt% except where noted

Elements	Ti ore	Extracted metal example
Ti	59.7	73.7
Al	29 ppm	11.4
O	39.9	11.0
C	27 ppm	1.10
H	500 ppm	12 ppm
N	50 ppm	13 ppm
Si	<100 ppm	70 ppm
Ca	70 ppm	1.50
Fe	120 ppm	740 ppm
Total	99.7	98.8

made from Inconel and then lowered into a closed-end Inconel tube. The Inconel tube was sealed off with an air tight lid which had air tight ports for electrode connections. The entire assembly was lowered into a "Mellen" vertical crucible furnace and heated to 850 °C. Argon was circulated through the Inconel tube during the experiments. Electrochemical experiments were performed using a potentiostat/galvanostat "PAR 4000" by Princeton Applied Research. Potentials were measured in reference to a titanium rod in constant voltage experiments.

After each experiment, the quartz crucible was removed from the Inconel tube and broken to harvest the refined product. The refined wool was recovered by washing with distilled water to remove salt. Complete removal of the salt was established by monitoring the conductivity of discarded water. However, due to equipment limitations, harvesting and handling operations were performed in normal atmosphere and therefore, it was not possible to avoid oxygen contamination.

Results and Discussion

The refining was performed at various potentials to investigate the partial separation of Al and Ti. Initially, compositions of Ti-10Al and Ti-13Al were tested as synthetic anode feed materials for the refining. These alloys were prepared by melting Al and Ti according to the desired ratio in a Vacuum Arc Remelt (VAR) furnace. The final Ti and Al composition, determined by ICP-OES, is given in Table 3. Data indicate that partial separation between Al and Ti can be achieved. Aluminum content in the refined products was 2.0–3.0 wt% when electrolysis was performed at 0.4 V. As expected, an elevated level of Al (7.45 wt%) was observed after refining at 0.8 V. The data confirm that final Al content in titanium wool can be controlled by adjusting the refining voltage.

Figure 3 presents the voltage dependence with the time during the refining process for synthetic Ti-10Al. The refining was performed at a constant current of 1.0A. The data suggest that dissolution takes place via three distinct stages. The first stage is up to ~5000 s, the second stage is between 5000 and 30,000 s, and the third stage is after 30,000 s. During the initial stage, the dissolution is more difficult than in the second and third stage. Though hypothetical, it is possible that initial oxidation must break through a passivated layer (e.g., oxide layer) in the first stage.

Table 3 Composition (wt% by ICP-OES) of extracted and refined product at constant voltage

Voltage (V)	Elements	Before refining	After refining
0.4	Al	10.0	3.07
	Ti	90.0	95.5
0.4	Al	13.0	1.98
	Ti	81.0	96.1
0.8	Al	13.0	7.45
	Ti	81.0	88.8

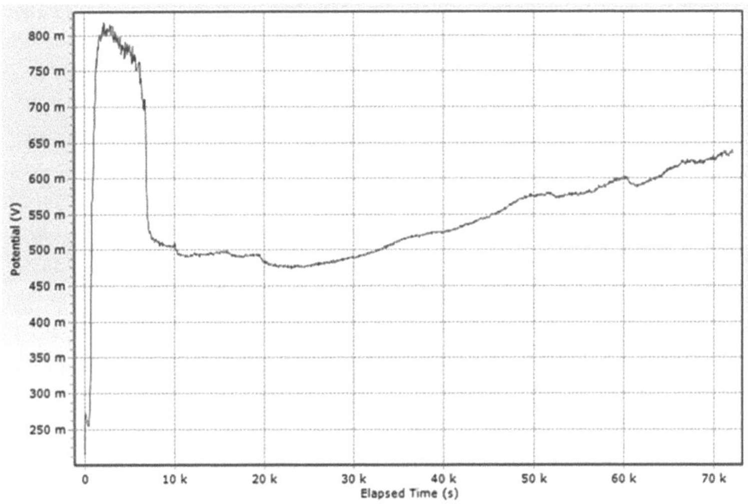

Fig. 3 Time-voltage profile of VAR Ti-10Al alloy during refining at a constant current of 1.0 A

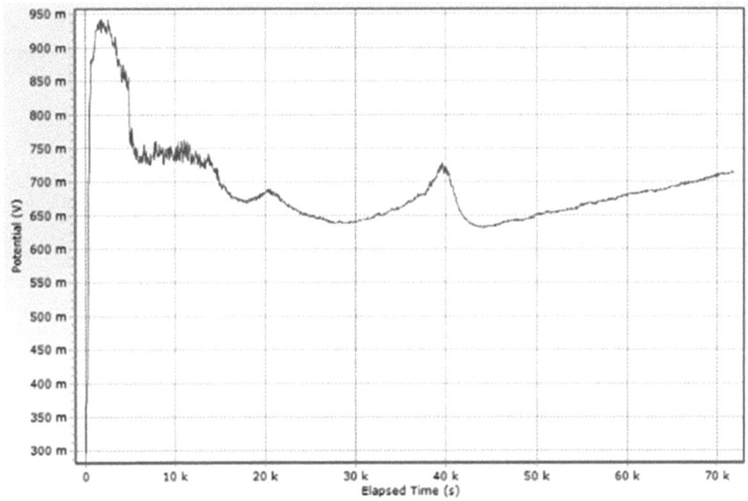

Fig. 4 Time-voltage profile of extracted metal 1 during refining at a constant current of 1.0 A

In the second stage, oxidation of loosely bound titanium occurs and thereafter, in the third stage, the oxidation of hard bound titanium takes place (as aluminum concentration increases).

Figure 4 presents the time-voltage profile during the refining of extracted metal produced during step 1 of the UTRS Process. This curve demonstrates a similar trend to that of synthetic Ti-10Al alloy (see Fig. 3), where the dissolution takes place via three distinct stages. Initially, the dissolution is more difficult than in second and third stages. NOTE: The time scale for each of these stages in Figs. 3 and 4 is not comparable due to differences in size and geometry of the anode feed. However, it is noteworthy that the third stage begins later than in the case of synthetic Ti-10Al alloy. Therefore, if the steady increase in voltage reflects the concurrent dissolution of Al, then less Al is expected in the final product of the extracted metal than in the case of synthetic Ti-10Al alloy.

The final composition of the refined product at constant current is given in Table 4. As can be seen from the data shown in Table 4, a better separation of Ti from Al was achieved for extracted metals than for the synthetic Ti-10Al alloy. A 5–10 wt% reduction in Al content was demonstrated for extracted metals. Only 3.0 wt% reduction in Al content was achieved for synthetic Ti-10Al alloy under similar refining conditions. These results demonstrate that a 5.0% Al level in refined titanium alloy can be easily achieved.

Encouraged by the data in Table 4, we performed another refining test on materials that had been through the refining process one previous time (i.e., double refining). Refining was carried out at a constant current of 3.0A. The final analysis is presented in Table 5. The data shows that Al can be removed from extracted metal and pure titanium metal (98.95%) can be obtained by double refining of extracted titanium-aluminum metal.

Table 4 Composition (wt% by ICP-OES) of refined product at a constant current

Anode Feed	Current(A)	Elements	Before refining	After refining
Synthetic alloy	1.0	Al	10.0	6.7
		Ti	90.0	91.1
Extracted metal 1	1.0	Al	13.0	2.5
		Ti	74.0	94.7
Extracted metal 2	1.0	Al	8.0	3.0
		Ti	76.0	91.4
Extracted metal 3	6.0	Al	6.41	3.0
		Ti	80.6	95.4

Table 5 Composition (wt% by ICP-OES) of double refined product at a constant current

Elements	Initial composition of extracted metal	Composition after double refining
Ti	76.7	98.95
Al	9.65	0.20
O	12.25	0.74
Si	0.18	0.09
Ca	0.24	ND
Fe	0.08	ND
Total	99.09	99.98

ND not detected

Table 6 ICP-OES and IGF analysis of the final alloy product

Elements	Composition (wt%)		
	Extracted metal	Refined metal	Melted metal
Ti	78.6	94.3	88.5
Al	9.9	4.6	6.3
V	0.0	0.0	3.8
O	9.8	0.5	0.6
C	0.8	100 ppm	NA
H	12 ppm	320 ppm	NA
N	650 ppm	48 ppm	NA
Other	0.8	0.6	0.7

NA not analyzed

It was also demonstrated that refined metal can be converted to Ti-6Al-4V alloy. The conversion was achieved by adding Al-V master alloy to the refined metal. A comprehensive economic analysis showed that production cost will be almost 50% less than today's commercial production method of Ti-6Al-4V alloy (Economic study was performed by an independent engineering firm). Details of the economic analysis is beyond the scope of this article but will be presented elsewhere. ICP-OES and IGF analysis of the final alloy is presented in Table 6.

Summary

The UTRS Titanium Extraction Process has been proven to have potential as a low cost method to produce titanium and titanium alloys. During this work it was demonstrated that titanium alloys can be extracted and refined into a commercially viable product. Furthermore, it was demonstrated that pure titanium can be obtained by double refining extracted titanium alloy. It was estimated that production cost for UTRS Titanium Extraction Process will be almost 40–50% less than today's commercial production method of Ti-6Al-4V alloy.

References

1. J.R. Cox, C.L. De Alwis, B.A. Kohler, M.G. Lewis, System and method for extraction and refining of titanium. US Patent Application no. 20130164167, 21 Dec 2012
2. J.F. Klausner, Innovating the future of titanium production at U.S. Department of Energy Advanced Research Projects Agency-Energy (ARPA-E). Paper presented at Titanium 2015, Orlando, Florida, 5 Oct 2015
3. M. Qian, F.H. Froes (eds.), *Titanium Powder Metallurgy Science, Technology and application* (Elsevier Inc., Amsterdam, 2015), 37

4. E.R. Poulsen, J.A. Hall, Extractive metallurgy of titanium: a review of the state of the art and evolving production techniques. J. Metals 60–65 (1983)
5. G.W. Fletcher, Aluminothermic process. Patent no. US4169722, 2 Oct 1979
6. J.-C. Stoephasius, B. Friedrich, J. Hammerschmidt, A new processing route for titanium alloys by aluminothermic reduction of titanium and refining by ESR. Paper presented at the 10th World Conference on Titanium, Hamburg, Germany, 13–18 July 2003
7. G.H. Meier, F.S. Pettit, S. Hu, Oxidation behavior of titanium aluminides. J. De Physique **IV**(3), 395–402 (1993)

Evolution of Non-metallic Inclusions in Solid Fe–Al–Ti–N Alloy During Heating

Hiroyuki Matsuura, Wonjin Choi and Gen Kamimura

Abstract Control of non-metallic inclusions in terms of morphology, number density or composition is essential to produce steel products having the desired mechanical properties. Though many studies have been conducted to clarify various properties and evolution mechanisms of inclusions at steelmaking temperature, it has been reported that the states of inclusions in final products are significantly different from those in crude steels for some steel grades. The present study has focused on the evolution of oxide and nitride inclusions in solid Fe–Al–Ti–N alloys during heating to increase the understanding of inclusion evolution in a simplified metal system. Compositional and morphological change of initial inclusions, as well as precipitation and growth of new inclusions have been observed even when the alloy was in a solid state. This behavior has been observed due to the change of thermodynamic stability and solubility of existing phases in inclusions.

Keywords Non-metallic inclusion · Heat treatment · Ti-alloyed steel

Introduction

Titanium-alloyed steels have been widely used for various products such as automobile sheets, or heavy plates for ship construction. The added titanium forms various types of non-metallic inclusions in steels such as oxide, nitride, carbide or sulfide. Titanium has a strong affinity with oxygen and thus molten steel is firstly deoxidized by adding Al before Ti alloying for economical reason. Therefore, Al deoxidation and Ti addition is a common process for Ti-alloyed steel.

H. Matsuura (✉) · G. Kamimura
Graduate School of Frontier Sciences, The University of Tokyo, 5-1-5 Kashiwanoha, Kashiwa, Chiba 277-8561, Japan
e-mail: matsuura@k.u-tokyo.ac.jp

W. Choi
School of Advanced Material Engineering, Yonsei University, 262 Seongsanno, Seodaemun-Gu, Seoul 120-749, South Korea

J.-Y. Hwang et al. (eds.), *8th International Symposium on High-Temperature Metallurgical Processing*, The Minerals, Metals & Materials Series,
DOI 10.1007/978-3-319-51340-9_19

Some of the Ti-based inclusions formed improve steel microstructure during solidification or heat treatment, by prevention of austenite grain growth and assisting a finer ferrite microstructure formation [1]. Therefore, many studies have been conducted so far to clarify various properties and evolution mechanisms of inclusions at steelmaking temperature. On the contrary, it has been well known that properties of inclusions in final products are significantly different from those in molten steel or solidified steel slab for some steel grades, which clearly demonstrates that inclusions may change during the reheating process.

Therefore, the evolution and growth of inclusions in solid steel are very important phenomena to understand the role of inclusions on the creation of steel microstructure. Therefore, many researchers have studied the formation of inclusions during solidification [2], estimated the effect of inclusions on the solidification microstructure [3] and clarified the effect of inclusions on the microstructure in solid steel during heating [4].

In the present study, the evolution of oxide and nitride inclusions in solid Fe–Al–Ti–N alloys during heating has been focused on to achieve a deeper understanding of inclusion evolution in a simplified metal system [5, 6].

Experimental

About 100 g of electrolytic iron were melted in an Al_2O_3 Tammann tube (O.D. 30 mm, I.D. 24 mm, height: 150 mm) by an induction furnace at 1873 K in Ar or Ar–N_2 atmosphere. After melting, aluminum and titanium were added in sequence with an interval of 2 min. After holding 3 min, the crucible with the melt was quenched by water. The compositions of the obtained specimens were analyzed by ICP-OES for soluble Al and Ti, and by combustion analyzer (LECO TC600 and CSLS 600) for total oxygen, nitrogen and sulfur. Sulfur was mainly introduced from the electrolytic iron, which contained 0.002 mass% of sulfur. The compositions of the prepared six samples are shown in Table 1.

A cylindrical piece of the alloy (O.D. 24 mm, height: from 13 to 15 mm) was machined from the produced alloy and the piece was encapsulated into a low carbon steel ampoule (O.D. 32 mm, I.D. 24 mm, height: 20 mm) by Ar

Table 1 Composition of samples in mass%

Sample	Soluble Al	Soluble Ti	Total O	Total N	Total S
F–A	0.0000	0.277	0.0006	0.0004	0.0015
F–B	0.0064	0.036	0.0060	0.0010	0.0020
S–A	0.0274	0.089	0.0005	0.0005	0.0017
S–B	0.0392	0.093	0.0007	0.0058	0.0012
S–C	0.0003	0.109	0.0012	0.0013	0.0020
S–D	0.0009	0.123	0.0010	0.0055	0.0017
K–A	0.0077	0.0251	0.0023	0.0097	Not analyzed

arc-welding at high purity Ar atmosphere. Or a cubic piece 5 mm on a side was machined to seal it in a quartz tube ampoule (I.D. 8 mm, length 30 mm) at evacuated condition. The ampoule was heated at predetermined conditions in an electric furnace. As-cast and heated specimens were mirror-polished by SiC papers and diamond suspensions to analyze by FE-SEM (JEOL JSM-7001FA) and EDS (JEOL JED-2300). For microstructure observation, the specimens were etched by methanol-5 vol.% nitric acid etching solution and observed by an optical digital microscope (KEYENCE VHX-1000).

Results and Discussion

The typical photomicrographs in as-cast and heated samples are shown in Fig. 1 [6]. The grain refinement was observed by heating at 1473 K for 3 h in all samples. Table 2 shows the number density of inclusions in as-cast and heated samples [6]. The number density increased by heating for all samples and this increment would affect the evolution of the microstructure.

To quantitatively analyze the microstructure and evaluate the effect of inclusions on the evolution of microstructure, ASTM grain size number was calculated by using the ASTM comparative chart method [7]. An increase of ASTM grain size number means a decrease of average grain size. In as-cast samples, ASTM grain size number was in the range from 1.9 to 2.5. Figure 2a shows the relationship between ASTM grain size number and the number of TiN-based inclusions in the inner layer of all of the samples and Fig. 2b shows the change in grain size number by heating with regard to the variation of the number of TiN-based inclusions in a unit area [6]. In the case of as-cast samples, ASTM grain size number increased with increasing number of TiN-based inclusion. On the other hand, the change in the grain size number increased linearly with increased change in the number of TiN-based inclusions in the heated samples. From these results, it is concluded that the formation of TiN-based inclusions becomes a decisive factor for grain refinement during solidification as well as heating. Comparing to the above results, the effect of TiS-based inclusions on the evolution of microstructure was not observed.

Sample K–A was used to further investigate the evolution of TiN-based inclusions during heating and several heating patterns were employed to observe inclusion formation during the initial stage of the heating process. Figure 3 shows the size distribution of TiN-based inclusions in the as-cast sample as well as heated samples by two patterns, one heated at 1573 K for 10 min before water quenching and the other heated at 1573 K for 10 min followed by graduate cooling to 1273 K (−10 K/min) and water quenching. Comparing to the as-cast sample, the average size of inclusions in the sample annealed at 1573 K for 10 min increased, which would be due to the dissolution of fine TiN particles and the growth of coarse TiN particles by Ostwald ripening. The slow cooling of annealed sample to 1273 K resulted in the decrease of the average size of inclusions because of the precipitation and growth of fine TiN particles of less than 1 μm during cooling.

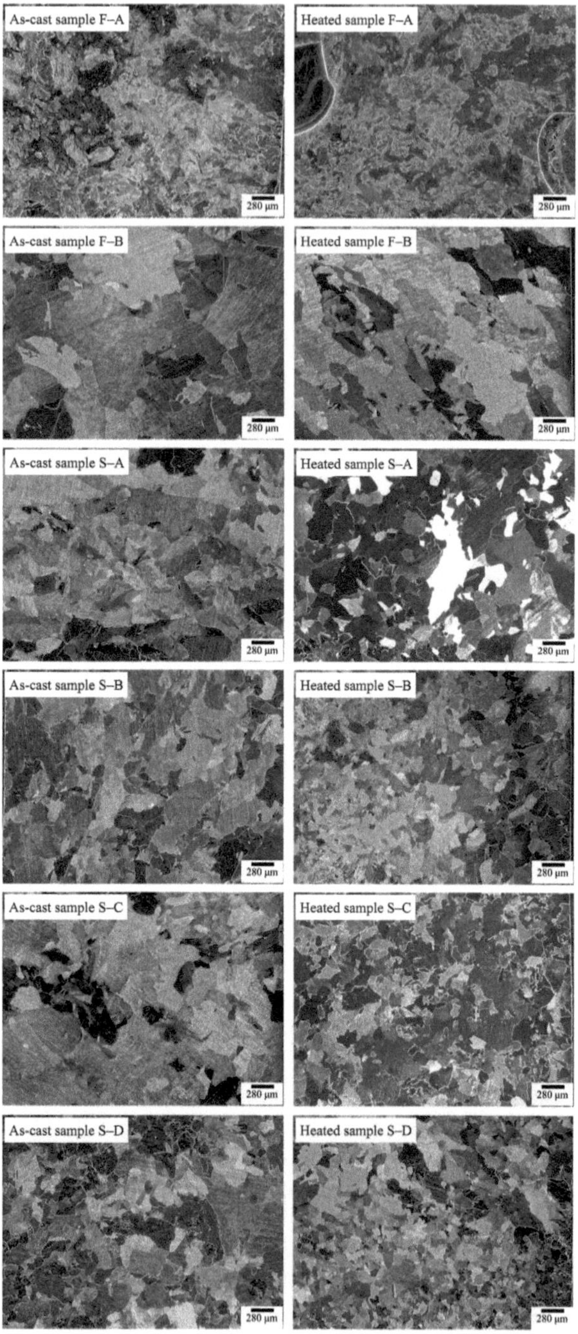

Fig. 1 Microstructure of the inner layer of as-cast and heated F–A, F–B, S–A, S–B, S–C and S–D samples [6]

Table 2 Number density of inclusions (per 1 mm^2) [6]

Sample	As-cast	Heated
F–A	500	570
F–B	190	490
S–A	470	3940
S–B	1530	2140
S–C	410	3550
S–D	680	2150

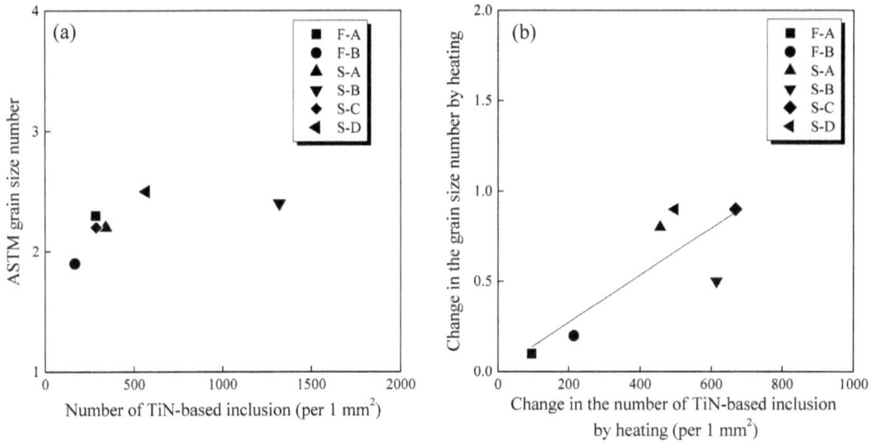

Fig. 2 Effect of TiN-based inclusions on the grain size in **a** as-cast and **b** heated samples [6]

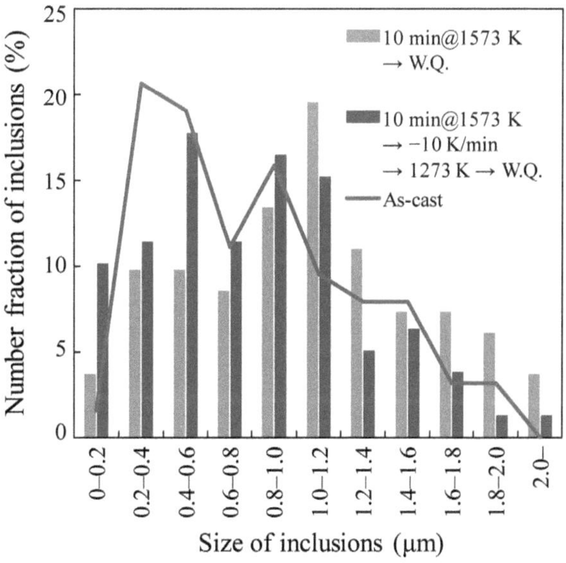

Fig. 3 Size distribution of TiN-based inclusions in as-cast and heated sample K–A

From the above observations, it was found that the dissolution, evolution and growth (rearrangement) of TiN inclusion are initiated from the beginning of heating and such behavior of inclusions affects the change in the microstructure significantly.

Conclusions

The change of inclusion composition, number density and size distribution during heating of solid Fe–Al–Ti–N alloys was observed and the effect of such change on the change of the microstructure of alloys was studied. TiN inclusions changed by heating at a range of temperatures over 1273 K due to the change of its thermodynamic stability and solubility product. Precipitation of fine TiN particle during slow cooling as well as the coarsening of TiN during heating at constant temperature, due to the effect of Ostwald ripening were observed. Microstructure of alloys became finer by heating because of the increase of TiN-based inclusion density. Though TiS-based inclusions also increased by heating, TiS-based inclusions did not exert the ability to assist the formation of a finer microstructure.

References

1. H. Fujimura, S. Tsuge, Y. Komizo, T. Nishizawa, Effect of oxide composition on solidification structure of Ti added ferritic stainless steel. Tetsu-to-Hagane **87**(11), 707–712 (2001)
2. S.-C. Park, I.-H. Jung, K.-S. Oh, H.-G. Lee, Effect of Al on the evolution of non-metallic inclusions in the Mn-Si-Ti-Mg deoxidized steel during solidification: experiments and thermodynamic calculations. ISIJ Int. **44**(6), 1016–1023 (2004)
3. J.H. Park, Effect of inclusions on the solidification structures of ferritic stainless steel: computational and experimental study of inclusion evolution. CALPHAD **35**(4), 455–462 (2011)
4. D. Yu, D.P. Dunne, T. Chandra, F.J. Barbaro, Austenite grain coarsening and formation of intragranular ferrite in HSLA steels deoxidized with Ti and Al. Mater. Trans. **37**(10), 1554–1560 (1996)
5. W. Choi, H. Matsuura, F. Tsukihashi, Effect of nitrogen on the formation and evolution of non-metallic inclusions in Fe–Al–Ti–N alloy. ISIJ Int. **53**(11), 2007–2012 (2013)
6. W. Choi, H. Matsuura, F. Tsukihashi, Effect of non-metallic inclusions in Fe-Al-Ti-O-N-S alloy on grain size. Metall. Mater. Trans. B **47B**(3), 1851–1857 (2016)
7. ASTM Standard E112-10, Standard test methods for determining average grain size. ASTM International, West Conshohocken, PA (2010). doi:10.1520/E0112-10, www.astm.org

Preparation of Low-Carbon Ti_2O_3 by Carbonthermal Reduction of the Mixture of Titanium Dioxide and Activated Carbon Under Vacuum Condition

Kejia Liu, Yaowu Wang, Yuezhong Di, Jianping Peng, Xinzhong Deng, Naixiang Feng and Yi Zhang

Abstract Low-carbon Ti_2O_3 was prepared by carbonthermal reduction of the mixture of titanium dioxide and activated carbon at vacuum condition. The kinetics and phase evolution of Ti_2O_3 formation process were investigated through X-ray diffraction and gravimetric analysis. The only intermediate phase detected is Ti_3O_5. The reduction time was reduced sharply to 4 h when temperature was raised to 1300 °C. The best molar ratio of activated carbon and TiO_2 is 0.6–1. At this condition, the content of Ti_2O_3 in slag is up to 95.63% and the content of the carbon element is below 0.004%, which meets the requirements for the further preparation of metal Ti by FFC Cambridge process. At the same time, the microstructure of Ti_2O_3 generated presented porosity, which would promote the electro-deoxidation process extremely.

Keywords Carbonthermal reduction · Vacuum · Ti_2O_3

Introduction

Metal titanium has been prepared successfully through direct electro-deoxidation method, also called FFC Cambridge process, which was a novel molten salt electrochemical technique developed by Fray et al. [1, 2]. In the process, TiO_2 pellets

K. Liu · Y. Wang (✉) · Y. Di · J. Peng · X. Deng · N. Feng
Department of Metallurgy, Northeasten University, Shenyang, People's Republic of China
e-mail: wangyw@smm.neu.edu.cn

K. Liu
e-mail: kejia1919@foxmail.com

Y. Zhang
Institute of Process Engineering, Chinese Academy of Sciences, Beijing, People's Republic of China

J.-Y. Hwang et al. (eds.), *8th International Symposium on High-Temperature Metallurgical Processing*, The Minerals, Metals & Materials Series,
DOI 10.1007/978-3-319-51340-9_20

Table 1 List of reactions taking place during electrochemical reduction of TiO_2

Stage	Cathode reactions		
1		$5TiO_2 + Ca^{2+} + 2e^- = Ti_4O_7 + CaTiO_3$	(1)
		$4Ti_4O_7 + Ca^{2+} + 2e^- = 5Ti_3O_5 + CaTiO_3$	(2)
		$3Ti_3O_5 + Ca^{2+} + 2e^- = 4Ti_2O_3 + CaTiO_3$	(3)
		$2Ti_2O_3 + Ca^{2+} + 2e^- = 3TiO + CaTiO_3$	(4)
	or	$Ti_2O_3 + 2e^- = 3TiO + O^{2-}$	(5)
		$CaTiO_3 + 2e^- = TiO + Ca^{2+} + 2O^{2-}$	(6)
2		$CaTiO_3 + TiO = CaTi_2O_4$	(7)
	or	$CaTi_2O_4 + 2e^- = 2TiO + Ca^{2+} + 2O^{2-}$	(8)
		$2CaTiO_3 + Ti_2O_3 + 2e^- = 2CaTi_2O_4 + O^{2-}$	(9)
3		$CaTi_2O_4 + Ca^{2+} + 2e^- = 2TiO + 2CaO$	(10)
		$TiO + 2(1 - \delta)e^- = Ti[O]_\delta + (1 - \delta)O^{2-}$	(11)

and graphite rods serve as cathode and anode respectively, molten calcium chloride is employed as electrolyte, and a potential below the decomposition potential of $CaCl_2$ molten salt but above that of TiO_2 is applied. Based on oxygen ionization mechanism, ionised oxygen in TiO_2 pellets at cathode passes into electrolyte under the driving of the potential, and metal Ti is left. At the same time, O^{2-} entering into electrolyte discharges at graphite anode and evolves CO or CO_2 gas. According to Schwandt' study [3, 4], the typical reduction pathway of TiO_2 pellets is summarized in Table 1. Obviously, the formation of calcium titanate during electrolysis made the cathode reactions become rather complicated and brought the current efficiency down. For purpose of avoiding the formation of Ca compounds ($CaTiO_3$, $CaTi_2O_4$), Ti_2O_3 is a better choice as precursor for electro-deoxidation [5].

Generally Ti_2O_3 is prepared by means of thermal reduction of titanium dioxide. The common reductants are hydrogen gas [6] and titanium powder [7]. Hydrogen gas is flammable gas, explosion risk could easily happen in the experimental operation. And Ti powder is not suitable for present study obviously. Recently Zr was used as reductant to prepare Ti_2O_3 through a phase separation method [8]. But the reduction efficient was very low.

In present study, the activated carbon was used as reductant to prepare Ti_2O_3. And it was ever used to prepare TiC powder through carbothermal reduction [9–13]. Ti_2O_3 is the lowest oxide phase before forming TiC_xO_y and TiC when carbon content in raw materials is enough [9, 10]. Through controlling the proportion of activated carbon, low-carbon Ti_2O_3 can be prepared. The kinetics and phase evolution of Ti_2O_3 formation process were investigated using X-ray diffraction (X'Pert Pro MPDDY2094, PANalytical B.V with CuKα radiation) and gravimetric analysis and the microstructure by SEM (S-4800, Hitachi). Impurity C in final production was analysed by instrument of high-frequency infrared of Carbon and Sulphur (LECO, CS230).

Experimental

Chemically-pure TiO_2 (98%, 10022728, Sinopharm) and analytically-pure activated carbon (10006619, Sinopharm) were used as raw material and reductant respectively. TiO_2 was mixed with activated carbon according to different molar ratios. The mixtures were rotated at least for 10 min by a universal high-speed smashing machine (28,000 r/min, HC-150, HuangCheng). And then they were pressed into pellets(10 ± 0.5 g) of 30 mm diameter under the axial pressure of 40 MPa. The pellets were put into vacuum tube furnace (GSL-1700, HEFEI KEJING MATERIALS TECILNOLOGY CO.LTD.). When system pressure reached 5 Pa, the furnace began to heat up at 10 °C/min and held on at target temperature for different times. After reduction finished, the vacuum pump was switched off and highly-pure Ar gas was pumped into the furnace at 5 L/s. Until the temperature was below 50 °C, samples were drawn out.

The phase evolution was investigated using X-ray diffraction (X'Pert Pro MPDDY2094, PANalytical B.V with CuKα radiation). The microstructure of samples were analysed by SEM (S-4800, Hitachi). The content of Ti_2O_3 in final production was tested through gravimetric analysis. And impurity C in final production was tested by instrument of high-frequency infrared of carbon and Sulphur (LECO, CS230). The contents of Ti_2O_3 in slags were tested through gravimetric analysis. The theoretical maximum weight gain is 11.11% when Ti_2O_3 is oxidated into TiO_2. The content of Ti_2O_3 (including the Ti_2O_3 in Ti_3O_5) in slag was obtained by the ratio of the observed relative weight gain ($\Delta m/m_0$) to its theoretical maximum weight gain (11.11%) according to Eq. (1):

$$\alpha = \frac{\Delta m}{m_0 \times 0.1111} \tag{1}$$

Results and Discussions

Thermodynamic Analysis

According to Refs. [8–12], carbothermal reduction of titanium dioxide proceeds through successive steps. The main reactions during carbothermal reduction are shown in Table 2 [9–13]. According to the data in Ref. [14], the Gibbs free energy changes were calculated. Because carbon monoxide is the unique gas phase, the partial pressure of carbon monoxide is approximately equal to system pressure. And in present experiment, the system pressure is about 1–50 Pa. The ΔG-T curves of reaction (a)–(d) at 5 Pa are shown in Fig. 1. From the ΔG-T curves, the lowest reaction temperatures of Reaction a–d at 5 Pa in theory are 703, 790, 1017, and 798 °C successively. However the temperature during experiment needs to be far higher than that in theory, or carbothermal reduction reactions will proceed very

Table 2 Main reactions in reduction process

No.	Reactions	$\Delta G (J/mol)$ $(T = 773 - 1773\,K)$
a	$TiO_2(s) + 1/3C(s) = 1/3Ti_3O_5(s) + 1/3CO(g)$	$\Delta G = 91167 - 65.993\,T + 8.314 \times T \times \ln(P_{CO}/P^{\theta})^{1/3}$
b	$1/3Ti_3O_5(s) + 1/6C(s) = 1/2Ti_2O_3(s) + 1/6CO(g)$	$\Delta G = 41583 - 25.411\,T + 8.314 \times T \times \ln(P_{CO}/P^{\theta})^{1/6}$
c	$1/2Ti_2O_3(s) + 1/2C(s) = TiO(s) + 1/2CO(g)$	$\Delta G = 179250 - 97.835\,T + 8.314 \times T \times \ln(P_{CO}/P^{\theta})^{1/2}$
d	$1/2Ti_2O_3(s) + 5/2C(s) = TiC(s) + 3/2CO(g)$	$\Delta G = 394650 - 245.15\,T + 8.314 \times T \times \ln(P_{CO}/P^{\theta})^{3/2}$

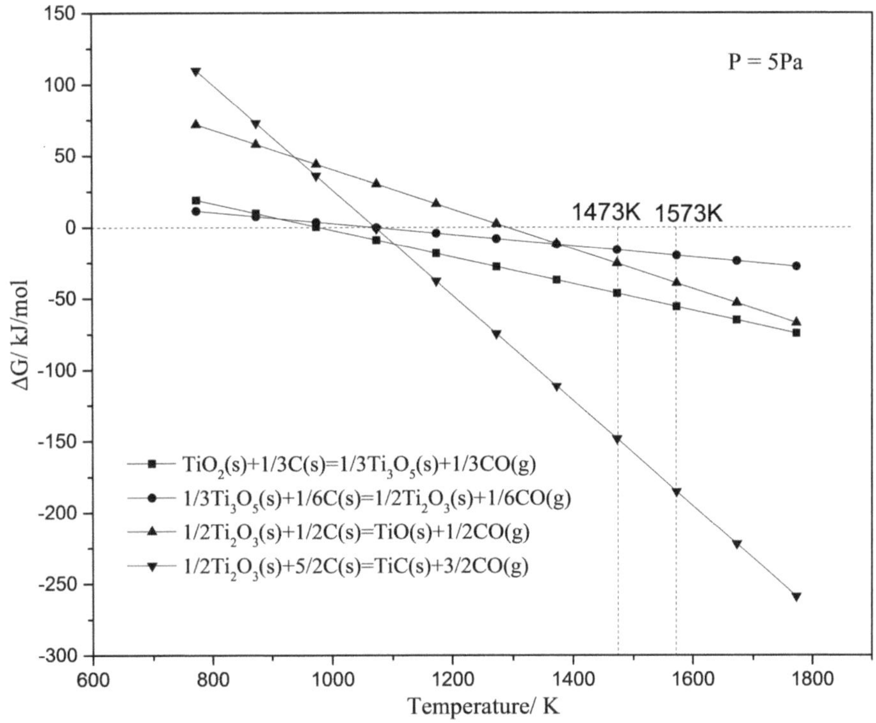

Fig. 1 ΔG-T curves of reaction a–d at 5 Pa

slowly. In present study, the experimental temperatures were 1200 and 1300 °C respectively. As shown in Fig. 1, at these temperatures, the formation of TiC is far more favorable than TiO. So if activated carbon is excess in mixture or the mixture is not homogeneous enough, some carbon-containing compounds may form, which will lead to the increase of C element in production dramatically. So an accurate proportion of activated carbon in mixture and good material-mixing process are crucial importance to prepare low-carbon Ti_2O_3. And according to Reaction (a) and (b), the molar ratio of C to TiO_2 in theory is 0.5: 1 for preparing Ti_2O_3. But in practice, the amount of activated carbon would be excessive slightly because of the weight loss of activated carbon on drying.

Reduction Process

Figure 2 shows the relationship curves between mass loss (wt%) of samples and time at 1200 and 1300 °C. From that, 24 h or more time was needed to reach reduction endpoint at 1200 °C, while just 3 h was needed at 1300 °C. Based on Reaction (12)–(14), the theoretical mass loss for Ti_3O_5 is 11.11%, for Ti_2O_3 is

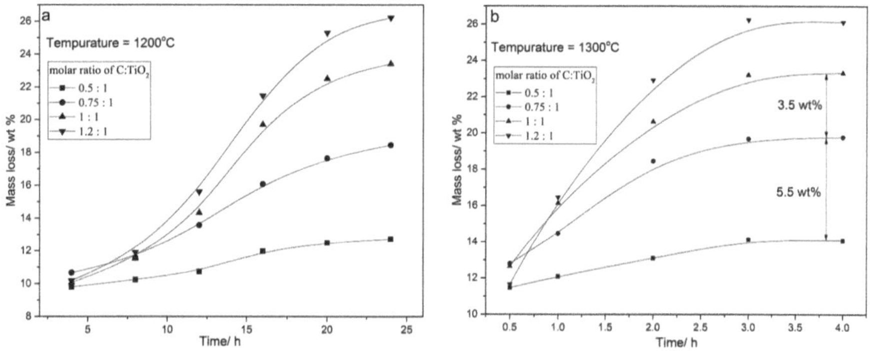

Fig. 2 Relationship curves between mass loss (wt%) of samples and time at 1200 °C (**a**) and 1300 °C (**b**)

16.28% and for TiC is 48.28%. From Fig. 2b, when the molar ratio of activated carbon and titanium dioxide is 0.5:1, the mass loss should have been 16.28% according to Reaction (13), but the mass loss in practice was just 14.25% (including some volatile in activated carbon), which indicated that the effective C in activated carbon is below 87.5%. When the molar ratio of activated carbon and TiO_2 increased to 0.75:1, the mass loss increased by 5.5 wt%, and when the molar ratio increased to 1:1, the mass loss just increased by 3.5 wt%, which indicated that some C was left in samples. Figure 3 shows the X-ray diffraction patterns of partially reduced samples at 1300 °C for different time when activated carbon: titanium dioxide = 0.75:1. The main phase was Ti_3O_5 after 0.5 h. As time went, the diffraction peaks of Ti_3O_5 weakened and that of Ti_2O_3 heightened. After 2 h, Ti_2O_3 was the only phase observed. However, little Ti–O–C phase was detected after 4 h, which was always detected in preparing TiC [9–17]. It indicated that activated carbon was excessive for preparing Ti_2O_3 when activated carbon: titanium dioxide is 0.75:1.

$$3TiO_2(s) + C(s) = Ti_3O_5(s) + CO(g) \tag{12}$$

$$2TiO_2(s) + C(s) = Ti_2O_3(s) + CO(g) \tag{13}$$

$$TiO_2(s) + 3C(s) = TiC(s) + 2CO(g) \tag{14}$$

Carbon Element Content

As discussed above, the ideal molar ratio of activated carbon and titanium dioxide for preparing Ti_2O_3 is between 0.5:1 and 0.75:1. The mixtures corresponding to activated carbon: titanium dioxide = 0.5:1, 0.55:1, 0.6:1, 0.65:1, 0.7:1 respectively

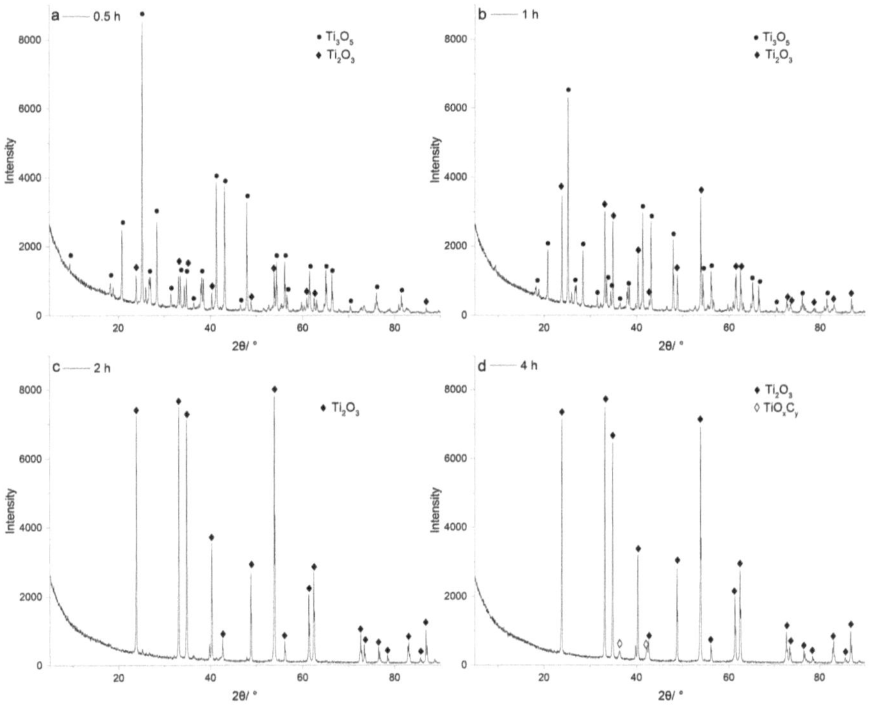

Fig. 3 X-ray diffraction patterns of partially reduced samples for different times at 1300 °C when activated carbon: titanium dioxide = 0.75:1

were heated to 1300 °C for 4 h in vacuum. And Fig. 4 shows their X-ray diffraction patterns. Obviously, as proportion of activated carbon increasing in raw material, Ti_3O_5 was replaced by Ti_2O_3 in slags and continuous solid solution TiO_xC_y was never detected. From Fig. 4d, e, only Ti_2O_3 was detected. However, according to the contents of carbon element in slags as shown in Table 3, the content of carbon element was up to 0.226 wt% in slag when activated carbon: titanium dioxide = 0.7:1, which means little TiO_xO_y generated. Obviously it cannot meet the requirement for further preparation of metal Ti. The content of carbon element in reduced slag was still up to 70 ppm when activated carbon: titanium dioxide = 0.65:1. Though it could be reduced by using more homogeneous mixture through extending mixing time or changing mixing blending manner such as ball-milling [9, 10, 15], it was still not suitable to prepare Ti_2O_3 for this study. When activated carbon: titanium dioxide = 0.6:1, the content of C in slag is below 0.004%. Though the slag contained little Ti_3O_5, it may be the best molar rate for preparing Ti_2O_3. Table 4 shows the content of carbon element in slags which were reduced from raw materials with different mixing times. From that, the mixing time, or the uniformity degree of mixtures is of crucial importance for the content of carbon element in reduced slags.

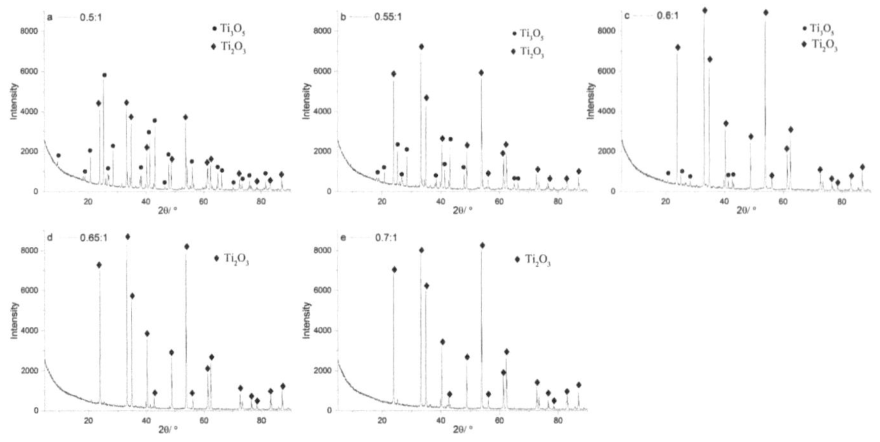

Fig. 4 X-ray diffraction patterns of fully reduced samples from raw material of activated carbon: titanium dioxide = 0.5:1 (**a**), 0.55:1 (**b**), 0.6:1 (**c**), 0.65:1 (**d**), 0.7:1 (**e**)

Table 3 Contents of carbon element and Ti_2O_3 in slags reduced from raw materials with different molar ratio

Molar ratio ($C:TiO_2$)	0.5:1	0.55:1	0.6:1	0.65:1	0.7:1	0.75:1
Content of C (%)	<0.004	<0.004	<0.004	0.007	0.226	0.515
Content of Ti_2O_3 (%)	78.71	88.71	95.63	102.12		

Table 4 Content of carbon element in slags reduced from raw materials with different mixing times

Molar ratio ($C:TiO_2$)		0.61:1	0.62:1	0.63:1	0.64:1
		Content of carbon element (%)			
Mixing time	3 min	0.081	0.088	0.161	0.263
	10 min	<0.004	<0.004	<0.004	<0.004

Content of Ti_2O_3 in Slags

The contents of Ti_2O_3 in slags were shown in Table 3. When activated carbon: titanium dioxide = 0.65:1, the contents of Ti_2O_3 in slags was greater than 100%. It could be attributed to two factors. First is the generation of TiO_xC_y in slags. From Table 3, the content of carbon element in reduced slag was 0.007% when titanium dioxide = 0.65:1. Obviously the influence of C element is very slight. Second is that titanium sesquioxide is a non-stoichiometry compound and the composition ranges from $TiO_{1.49}$ to $TiO_{1.51}$. Though the content of Ti_2O_3 in slag when titanium dioxide = 0.6:1 is 95.63%, Ti(IV) has been rather little, which can avoid calcium titanate largely. At the same time, the content of carbon element is little enough for

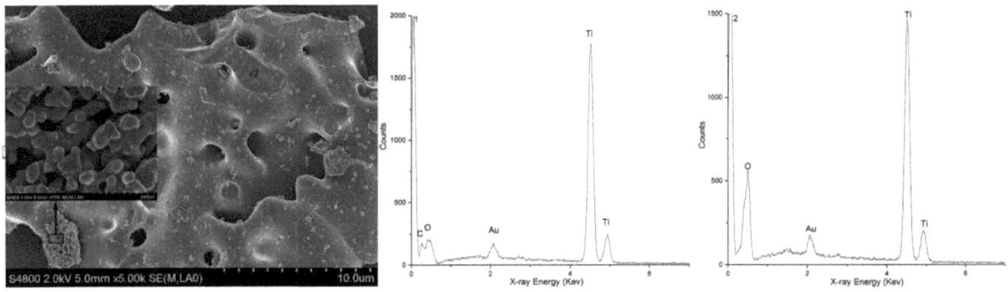

Fig. 5 SEM and EDS of the slag reduced from the mixture of titanium dioxide = 0.7:1

preparing metal titanium. Of course, the purity of Ti_2O_3 can be higher through well mixing and more accurate proportion of activated carbon in mixture.

Figure 5 shows the microstructure of the slag reduced from the mixture of activated carbon: titanium dioxide = 0.7:1. Porous structure was observed, which was attributed to the emitting of CO gas. Those cavities would be extremely favorable for preparing metal titanium through FFC process. At the same time, some little particles (about 100 nm) were observed. Energy Dispersive Spectrometer in Fig. 5 shows that main elements contained in them were Ti, O and C.

Conclusions

Ti_2O_3 was prepared successfully through carbonthermal reduction of the mixture of titanium dioxide and activated carbon under vacuum condition. Reduction temperature has a great influence on reduction rate and 24 h or more time is needed to reach reduction endpoint at 1200 °C, while just 3 h is needed at 1300 °C. Proportion of activated carbon in mixture need to be rather accurate and the mixing process is also very important or continuous solid solution TiO_xC_y cloud be generated, which will increase the content of carbon element in slag. In present study, the best molar ratio of activated carbon and TiO_2 is 0.6 to 1. At this proportion, the content of carbon element in slag is below 0.004% and the Ti_2O_3 content is 95.63%. The Ti_2O_3 generated is porous, which will be extremely favorable for preparing metal titanium through FFC process.

Acknowledgements The authors sincerely acknowledge the financial support of the State Key Development Program for Basic Research of China (973 Program, Grant No. 2013CB632606), the National Natural Science Foundation of China for the financial support of the project (51674076) and the National Natural Science Foundation of China (51304044).

References

1. D.J. Fray, T.W. Farthing, G.Z. Chen, Removal of oxygen from metal oxides and solid solutions by electrolysis in a fused salt. International patent NO. WO 9964638 (1999)
2. G.Z. Chen, D.J. Fray, T.W. Farthing, Direct electrochemical reduction of titanium dioxide to titanium in molten calcium chloride. Nature **407**, 361–364 (2000)
3. C. Schwandt, D.J. Fray, Determination of the kinetic pathway in the electrochemical reduction of titanium dioxide in molten calcium chloride. Electrochim. Acta **51**, 66–76 (2005)
4. C. Schwandt, D.T.L. Alexander, D.J. Fray, The electro-deoxidation of porous titanium dioxide precursors in molten calcium chloride under cathodic potential control. Electrochim. Acta **54**, 3819–3829 (2009)
5. R. Bhagat, D. Dye, S.L. Raghunathan, R.J. Talling, D. Inman, B.K. Jackson, K.K. Rao, R. J. Dashwood, In situ synchrotron diffraction of the electrochemical reduction pathway of TiO_2. Acta Mater. **58**, 5057–5062 (2010)
6. Yuan Wei, Liao Haida, An investigation on the optimal technological conditions for preparing titanium(III)oxide from titanium(IV)oxide. J. Beijing Univ. Chem. Technol. **27**(3), 79–82 (1999)
7. Y. Lu, Y. Matsuda, K. Sagara, L. Hao, T. Otomitsu, H. Yoshida, Fabrication and thermoelectric properties of Magneli phases by adding Ti into TiO_2. Adv. Mater. Res. **1291**, 415–417 (2012)
8. A. Kitada, G. Hasegawa, Y. Kobayashi, K. Kanamori, K. Nakanishi, H. Kageyama, Selective preparation of macroporous monoliths of conductive titanium oxides Ti_nO_{2n-1} (n = 2, 3, 4, 6). J. Am. Chem. Soc. **134**, 10894–10898 (2012)
9. S. Wei, B.Q. Xu, H.Y. Sun, J.X. Song, H.L. Wan, Y.N. Dai, Preparation of TiC powders by carbothermal reduction method in vacuum. Trans. Nonferrous Met. Soc. China **21**, 185–190 (2011)
10. W. Sen, H. Sun, B. Yang, B. Xu, W. Ma, D. Liu, Y. Dai, Preparation of titanium carbide powders by carbothermal reduction of titania/charcoal at vacuum condition. Int. J. Refract. Metals Hard Mater. **28**, 628–632 (2010)
11. B. Xu, B. Yang, J.P. He, W. Seng, Y. Dai, D. Liu, Study on carbothermic reduction of titanium dioxide in vacuum, Paper presented at proceedings of the 9th vacuum metallurgy and surface engineering conference, pp. 499–505 (2009)
12. R. Koc, Kinetics and phase evolution during carbothermal synthesis of titanium carbide from ultrafine titania/carbon mixture. J. Mater. Sci. **33**, 1049–1055 (1998)
13. P. Lefort, A. Maitre, P. Tristant, Influence of the grain size on the reactivity of TiO_2/C mixtures. J. Alloy. Compd. **302**, 287–298 (2000)
14. A. Roine, HSC Chemistry, version 6.0, Outokumpu Research Oy, Pori, Finland
15. Y.-C. Woo, H.-J. Kang, D.J. Kim, Formation of TiC particle during carbothermal reduction of TiO_2. J. Eur. Ceram. Soc. **27**, 719–722 (2007)
16. T.L. Lekanova, YuI Ryabkov, O.A. Sevbo, Particle-size effect on the rate of carbon reduction of TiO_2. Inorg. Mater. **39**(7), 715–719 (2003)
17. A. Maitre, D. Tetard, P. Lefort, Role of some technological parameters during carburizing titanium dioxide. J. Eur. Ceram. Soc. **20**, 15–22 (2000)

Pyrometallurgical Studies for Manganese Extraction Using Turkish Ore Reserves

Ender Keskinkilic

Abstract Pyrometallurgical studies conducted for Mn extraction using Turkish Mn-ore reserves were reviewed. Turkish Mn-ores are low-grade ones having ~30% Mn. The most important Mn ore reserves in Turkey are in the Denizli-Tavas region, where more than 2 million tons of proven reserves are reported. Mining rights to the Denizli-Tavas manganese belongs to Eregli Iron and Steel Works Co. (ERDEMIR). The ore is usually charged to the blast furnaces of ERDEMIR during hot metal production. So far, nearly 20 Master's thesis studies have been conducted for the analysis of manganese extraction in Turkey. In the scope of the present work, only pyrometallurgical and related activities (i.e. ore beneficiation, calcination, etc.) relevent to the treatment of a specific ore were examined.

Keywords Manganese · Ferromanganese · Silicomanganese

Introduction

Manganese is an important element since it increases the strength, toughness and hardenability of ferrous products, especially steel. It is also used as deoxidizer and desulphurizer in steelmaking. Most of the manganese used in the iron and steel industry are in the form of ferromanganese and silicomanganese. Turkey has some low-grade manganese ores having ~30% Mn. However, there is no ferromanganese/silicomanganese production plant in Turkey. Therefore, the demand of the Turkish iron and steel industry for manganese ferroalloys has been met by imports. The most important Mn-ore reserve is present in the Western Anatolia region, namely Denizli-Tavas. More than 2 million tons of proven reserves are reported in that area. In the last two decades, pyrometallurgical and

E. Keskinkilic (✉)
Department of Metallurgical and Materials Engineering, Atilim University, Incek, Ankara 06836, Turkey
e-mail: ender.keskinkilic@atilim.edu.tr

J.-Y. Hwang et al. (eds.), *8th International Symposium on High-Temperature Metallurgical Processing*, The Minerals, Metals & Materials Series,
DOI 10.1007/978-3-319-51340-9_21

hydrometallurgical studies were performed to treat this ore and similar reserves in Turkey. In the scope of this work, only pyrometallurgical and pyrometallurgical related (i.e. ore beneficiation, calcination, etc.) activities were reviewed. In this paper, important results obtained from these studies were summarized and outlined to create a question mark in the reader's mind whether industrial production of manganese ferroalloys from those ores is possible together with the method that could be used. The results presented should hopefully give alternative ideas related to the use of these valuable deposits in iron and steelmaking. Mining rights for the Denizli-Tavas manganese ore belongs to Eregli Iron and Steel Works Co. (ERDEMIR). The ore is usually charged to the blast furnaces of ERDEMIR during hot metal production. Therefore, various alternative uses should be considered and applied such as a mini blast furnace (and mini basic oxygen furnace) process for high/medium carbon ferromanganese production in the steelmaking company and the use of the ore in steelmaking converters as a manganese source. The first alternative obviously decreases the amount of ferromanganese imports, due to production from domestic sources. Selection of a blast furnace method is related to both high electricity prices in Turkey and relatively low ore grades. The second alternative is rather challenging and could be a topic of another paper.

Literature Review

So far, nearly 20 Master's thesis studies have been conducted for the analysis of manganese extraction in Turkey. In the scope of present work, only pyrometallurgical and pyrometallurgical related (i.e. ore beneficiation, calcination, etc.) activities were examined. Among these, the ones dealing with treatment of a specific ore were reviewed. In other words, literature about hydrometallurgical studies and studies including treatment of manganese containing chemicals were outside of the authors' interest. Thirteen studies have been performed for pyrometallurgical manganese extraction using Turkish manganese ore reserves. Eleven of them are related to the treatment of the biggest manganese ore reserves in Turkey, which are present in the Western Anatolia region, namely Denizli-Tavas. This is a carbonate type ore body having 30–31% Mn. Chemical composition of the Denizli-Tavas manganese ore is tabulated in Table 1. In this region, more than 2 million tons of proven reserves are reported. Only two studies are related to treatment of other reserves present in Central and Eastern Anatolia, namely Karaman and Erzincan, respectively. Those are very small ore reserves that are not suitable for industrial scale treatment.

Table 1 Chemical composition of Denizli-Tavas manganese ore (wt%)

Mn	Fe	SiO_2	CaO	MgO	Al_2O_3
30.34	5.56	18.72	7.32	2.41	3.52

Pyrometallurgical studies conducted for the treatment of manganese ores of Turkey may be grouped into three categories: (1) Ferromanganese/Silicomanganese production, (2) Carbothermic reduction, (3) Pretreatment operations for increasing Mn content of the ore.

Six studies have been performed in the first category. The first pyrometallurgical study was conducted by Celik and Bilgen [1]. According to their study, Denizli-Tavas manganese ore was smelted using electrical and induction furnaces. The ore was mixed with charcoal and limestone, and smelted at 1440 °C for two hours. It was reported that an iron-manganese alloy with 67% Mn was obtained with 60% manganese recovery.

Emeksiz [2] investigated the production of high carbon ferromanganese from Denizli-Tavas manganese ore. The ore was first calcined and then smelted in graphite crucibles in an induction furnace at 1600 °C. Four small graphite crucibles inside a large graphite crucible were placed in the furnace. The smaller graphite crucibles had outer diameters of 60 mm, inner diameters of 50 mm and depths of 73 mm. The charges for smelting were contained in the smaller graphite crucibles. Smelting time, charge basicity (i.e. CaO/SiO_2), coke to ore weight ratio and CaF_2 addition were the main experimental variables and the effect of each variable on manganese recovery were studied. Two hours smelting time together with coke to ore ratio less than 0.2 were reported to yield optimum results. The main problem encountered was the poor metal-slag separation. The reason of this problem was considered to be inadequate metal depth in the crucibles. This result was more pronounced when excess coke was added to the charge such that coke to ore weight ratio was greater than 0.3.

Imer [3] studied the production of high carbon ferromanganese from the same ore with the same experimental variables. First smelting experiments were conducted in an Al_2O_3 crucible contained in a graphite crucible in an induction furnace. The Al_2O_3 crucible was found to have interacted with the liquid slag that formed during the experiment and the liquid slag was found to have leaked through the Al_2O_3 crucible and to have collected at the bottom of the outer graphite crucible. A similar experiment conducted in an electronically controlled muffle furnace led to the same problems mentioned above. Similar difficulties arose in an experiment conducted with an MgO crucible. As a result of these findings, smelting experiments were decided to be conducted in graphite crucibles. With cylindrical graphite crucibles the leakage problem was overcome, but the metal-slag separation was not good as in the case of study performed by Emeksiz [2]. A conical shaped graphite crucible shown in Fig. 1 was considered to be helpful in obtaining good metal-slag separation. It was found that metal-slag separation was satisfactory in the experiment conducted with such a crucible design. Imer [3] performed a series of experiments with these crucibles using an electronically controlled muffle furnace. A schematic drawing of the alloy-slag system obtained at the end of the experiments is presented in Fig. 1.

According to Imer's findings [3], the manganese concentration and the weight of the metal increased with time in the experiments conducted for 1, 2 and 4 h. The manganese content of the metal increased with increasing basicity in the

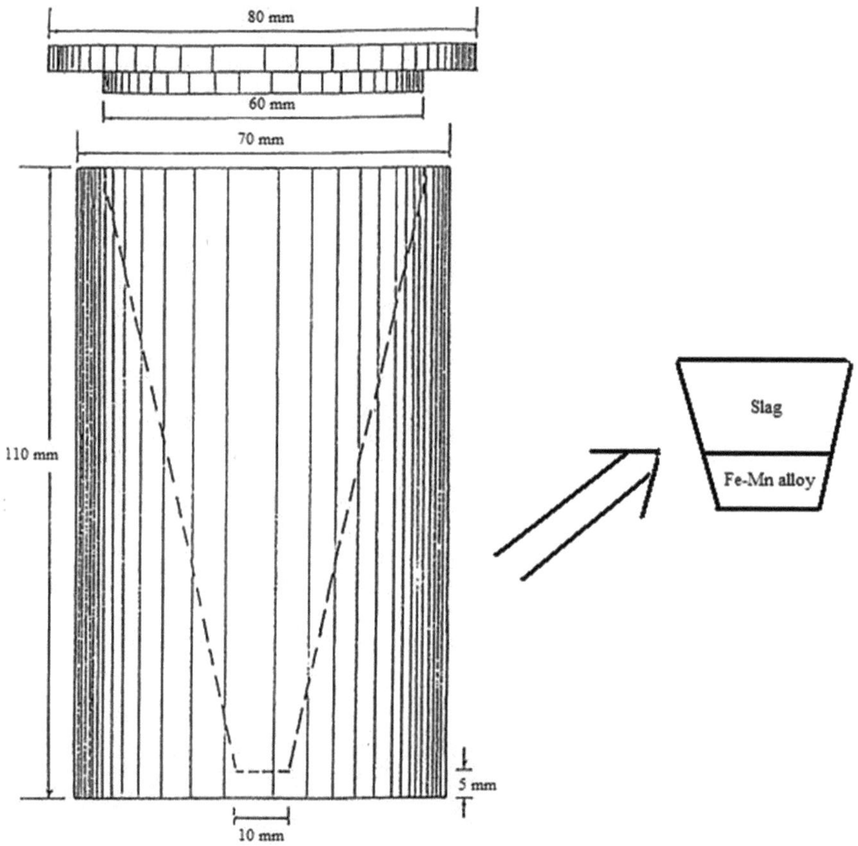

Fig. 1 Design of the graphite crucible used in smelting experiments and schematical drawing of the Fe–Mn alloy-slag system obtained at the bottom of the crucible

experiments conducted with CaO/SiO_2 ratios of 1, 1.5 and 2.0. CaF_2 addition was found to help obtain good metal-slag separation using additions of 5 and 10% of the weight of the (ore + coke + lime + scrap iron) mixture. The manganese content of the ferromanganese produced was very close to standard ferromanganese [4]. These aforementioned two studies showed that this ore could be used for the production of ferromanganese. However, some pretreatment processes were advised to upgrade the manganese content of the ore. The study conducted by Akarca [5] had this target and it could be counted in the third category previously indicated. He studied the possibility of concentration of Denizli-Tavas region manganese ore by magnetic concentration and gravity separation methods. It was observed that a concentrate containing 39.46% Mn can be produced by magnetic separation with 89.07% manganese recovery and a concentrate of 45.34% Mn with 62.37% manganese recovery can be produced by gravity separation. Since Denizli-Tavas manganese ore is a carbonate type, calcination is, of course, a simple way to increase manganese content. In that respect, Keceli [6] studied the mineralogy of the ore. He reported that the manganese carbonate (rhodochrosite) and manganese hydroxide (manganite) were found to be most abundant, 21 and 16% of the ore by weight, respectively (Table 2). According to experiments conducted, it was proposed to carry out calcination at 900 °C, and the calcination time (with a good stirring) should be a minimum of 1 h, and the particle size of ore should be about 1 mm. With those conditions satisfied, the manganese content of the ore could be increased to above 35%.

Ari [7] conducted a set of experiments using a laboratory scale arc furnace to determine the feasibility of ferromanganese and silicomanganese production using Denizli-Tavas manganese ore. He found that the ore is suitable for the production of high carbon ferromanganese and silicomanganese. When a charge containing 13.69% fixed carbon and 7.69% CaO was smelted, an alloy with 85.08% Mn and 1.28% Si was obtained with 82.50% manganese recovery. The smelting period was indicated as 30 min. In relation to silicomanganese production, a charge containing 15.60% fixed carbon and SiO_2 sand/ore wt ratio of 0.28 yielded a silicomanganese alloy having 69.7% Mn and 18.2% Si. The smelting period was noted as half an hour and manganese recovery was reported as 70.7%. He also included an economic analysis showing that industrial scale ferromanganese production from Denizli-Tavas manganese ore is feasible.

Keskinkilic [8] investigated silicomanganese production from Denizli-Tavas manganese ore using electronically controlled muffle furnace. Experiments were conducted in the same design of graphite crucibles as used in the previous study [3]. Results of the experiments performed with SiO_2 sand addition indicated that the charge having basicity of 0.58 and smelted at 1700 °C for a 1.5 h smelting period was the optimum conditions for silicomanganese smelting. Variations of manganese and silicon recoveries with charge basicity are shown in Fig. 2. Both manganese and silicon recoveries were found to increase as charge basicity increased from 0.39 to 0.58. Further increase in charge basicity gave rise to a decrease in both manganese and silicon recoveries. Maximum manganese and

Table 2 The mineralogical composition of Denizli-Tavas manganese ore

Mineral	Weight, %[a]
Rhodochrosite	21.0
Manganite	16.0
Psilomelane	12.0
Pyrolusite	5.0
Bixbyite + Hausmannite + Braunite	3.0
Calcite	13.1
Quartz	18.7
Pyrite	2.8
Al_2O_3	3.5
MgO	2.4

[a]Total is 97.5%. Rest, which is 2.5%, is the sum of other elements or oxides. These are not indicated here

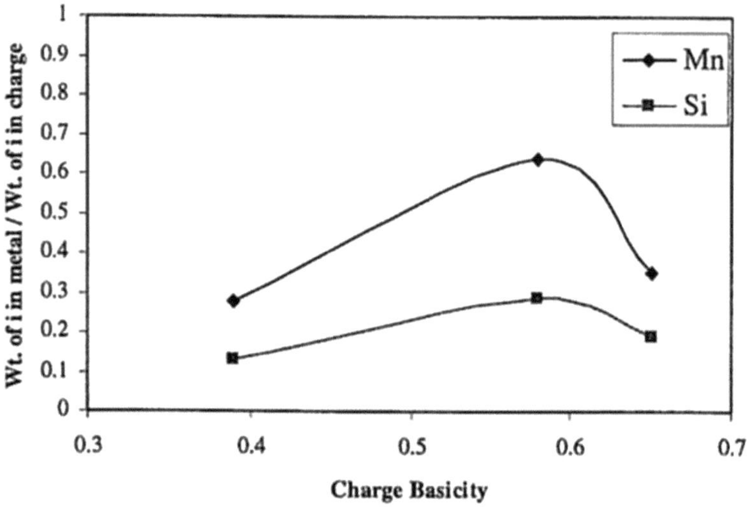

Fig. 2 Variations of manganese and silicon recoveries with charge basicity (Temperature: 1700 °C, Time: 1.5 h, Active carbon/(ore + SiO_2 sand): 0.26)

silicon recoveries were obtained at the charge basicity value of 0.58. At this point, manganese and silicon recoveries were found as 64 and 29%, respectively. Both manganese and silicon recoveries were expected to be higher. These relatively small recoveries may be attributed to the large mass losses in the experiments, which were also reported in the work conducted by Ari [7]. Such losses were most probably related to increased vapor pressure of manganese at increased temperature. Regardless of the losses encountered in the laboratory scale set up, it was finally concluded that production of silicomanganese [9] from Denizli-Tavas manganese ore is possible.

In the literature, the most recent study in relation to manganese ferroalloy production from a domestic ore was conducted by Cardakli [10]. He used a manganese ore sample (ore grade was about 42%) taken from a small reserve present in Erzincan. Experiments conducted using the same smelting system revealed that a high carbon ferromanganese alloy having manganese content more than 75% could be produced using Erzincan manganese ore. Manganese recovery during the smelting experiments was reported as about 60–80%.

The remaining five studies present in the literature were related to carbothermic reduction of manganese oxides using Turkish manganese ores. Aytis [11] investigated the reduction mechanism of manganese oxides. With a series of experiments using Denizli-Tavas manganese ore, he first outlined the well-known reduction sequence of manganese oxides starting from MnO_2 until MnO. Then, he conducted a number of experiments between 1100 and 1350 °C under argon atmosphere using graphite as the reducing agent. It was indicated that the reduction was found to occur in two stages: (1) The first stage includes the rapid reduction of higher oxides of manganese and iron to MnO and FeO. (2) The second stage is slower during which MnO and FeO are reduced to mixed carbide of iron and manganese. Kalfaoglu [12] studied the solid state reduction behavior of Denizli-Tavas manganese ore to produce ferromanganese, by using pure carbon in the form of graphite. Weight loss data was collected with respect to time as the reactions took place in the furnace. Experiments were done at 1100, 1150, 1200, 1250, and 1300 °C, and a constant amount of carbon and calcined manganese ore were used for each experiment. Reactions were conducted under argon atmosphere. The weight loss was converted to percent reduction, R%, using the formula given below.

$$R\% = \frac{Mass\ of\ oxygen\ removed}{Mass\ of\ removable\ oxygen} \times 100 \tag{1}$$

Mass of oxygen removed was calculated from the mass of CO removed and which was assumed to be equal to the weight loss. Mass of removable oxygen was taken as the oxygen present in the reducible oxides in the system. Iron oxides (assumed as Fe_2O_3 in stoichiometric calculations) and manganese oxides (assumed as Mn_2O_3 in stoichiometric calculations) were considered to be the reducible oxides. Therefore, all other oxides were assumed to be non-reducible in the calculation of R%. It was found that approximately 60% reduction of the manganese ore took place at and below 1200 °C. At and above 1250 °C, it was reported that this reduction efficiency increased to approximately 90%. It was concluded from his study that it is possible to produce high carbon ferromanganese alloy from the Denizli-Tavas manganese ore by solid state reduction with carbon. According to the weight loss data and Energy Dispersive Analysis of X-Rays (EDAX), the overall mechanism was observed to take place in three stages: (1) Reduction of higher oxides to lower oxides. (2) Nucleation of a metal phase. (3) Growth of the metal phase together with the reduction of MnO to Mn [12].

Akil [13] aimed to find the optimum conditions to produce a charge material for ferromanganese production or steelmaking with high content of iron and manganese carbides from Denizli-Tavas manganese ore by carbothermic reduction. The experiments conducted using a horizontal tube furnace under argon atmosphere revealed that solid state reduction performed at 1250 °C for 4 h with a charge containing 100% stoichiometric active carbon and 5% CaO yielded about 84% reduction of the calcined ore. This percentage reduction value was again calculated from Eq. (1). Stoichiometric active carbon was defined as the necessary carbon for the chemical reactions shown as (2) and (3). It was indicated that the ore product obtained this way had higher manganese content (~53%), so it could be used for ferromanganese production or as an additive in the production of high carbon steels.

$$7Mn_2O_3 + 27C = 2Mn_7C_3 + 21CO \tag{2}$$

$$3Fe_2O_3 + 11C = 2Fe_3C + 9CO \tag{3}$$

Cil [14] studied the effect of mechanical activation on the extent of solid state reduction of manganese oxides. Some of the Denizli-Tavas ore samples were subjected to planetary milling for different milling times and the reduction degree of those samples were compared with the ones reduced without mechanical activation. It was found that the extent of reduction for non-activated manganese ore at 1100 and 1200 °C for 1 h were reported as 34.29 and 55.11%, respectively. On the other hand, the reduction degrees of the activated (60 min) manganese ore at the same reduction conditions were found as 41.70 and 70.43%, respectively. Therefore, it was concluded that mechanical activation was beneficial to increase the extent of reduction of manganese oxides.

Aranci [15] investigated the effect of carbonized tea plant wastes on the production of ferromanganese using the carbothermic method from ferrous manganese oxide ore subjected to mechanical activation. The manganese ore obtained from Karaman was mixed with Elazig-Keban region magnetite ore and the carbonization product obtained from tea wastes from the Karadeniz CAYKUR plants. The whole mixture was subjected to mechanical activation for 10–120 h. The product was pelletized and then subjected to reduction at 1000, 1100, 1200, 1300 and 1400 °C for various periods ranging from 30 to 120 min. It was concluded that the mechanical activation process influenced the degree of reduction positively. It was also indicated that carbonized tea plant's wastes could be possibly used in the carbothermic reduction.

Conclusions

According to laboratory-scale pyrometallurgical studies using ore from Turkish Mn reserves, the following main conclusions were drawn:

- Denizli-Tavas manganese ore (or other similar ore bodies in Turkey) is suitable for ferromanganese production. However, some pretreatment operations are advised to increase the ore grade. Mechanical activation improves recovery.
- Smelting of silicomanganese from Denizli-Tavas manganese ore (or other similar ore bodies in Turkey) is possible.
- Solid-state carbothermic reduction studies revealed that a product of higher manganese content was obtained. Therefore, it could be used for ferromanganese production or as an additive in the production of high carbon steels.

References

1. O. Celik, N. Bilgen, Denizli-Tavas Mangan Cevherlerinin Pirometalurjik Yöntemle Değerlendirilmesi, Report, General Directorate of Mineral Research and Exploration, MTA, Report No. 502 (1990)
2. F. Emeksiz, in *Smelting of High Carbon Ferromanganese from Denizli-Tavas Manganese Ore*. MS Thesis, Middle East Technical University (1991)

3. S.T. Imer, in *Smelting of High Carbon Ferromanganese from Denizli-Tavas Manganese Ore*. MS Thesis, Middle East Technical University (1997)
4. American Society for Testing and Materials (ASTM), “Annual Book of ASTM Standards,” Volume 01.02, “Ferrous Castings; Ferroalloys,” 59–61 (1993)
5. S.S. Akarca, *Beneficiation of Denizli-Tavas Region Manganese Ores*. MS Thesis, Middle East Technical University (2000)
6. H. Keceli, in *Calcination of Denizli Tavas Manganese Ore*. MS Thesis, Middle East Technical University (1997)
7. M.E. Ari, in *Tavas Manganez Cevherlerinden Laboratuvar Tipi Ark Fırınında Ferromanganez/Ferrosilikomanganez Üretim Koşullarının Belirlenmesi*. MS Thesis, Istanbul Technical University (1996)
8. E. Keskinkilic, in *Smelting of Silicomanganese from Denizli-Tavas Manganese Ore*. MS Thesis, Middle East Technical University (2001)
9. American Society for Testing Materials (ASTM), Annual book of ASTM standards, Vol. 01.02, “Ferrous Castings; Ferroalloys,” 256–257 (1993)
10. I.S. Cardakli, in *Production of High Carbon Ferromanganese from a Manganese Ore Located in Erzincan*. MS Thesis, Middle East Technical University (2010)
11. C. Aytis, in *Production of Ferromanganese by Carbothermic Reduction*. MS Thesis, Yildiz Technical University (1997)
12. F. Kalfaoglu, in *Solid State Carbothermic Reduction of Denizli-Tavas Manganese Ore*. MS Thesis, Middle East Technical University (1997)
13. C. Akil, in *Optimization of Conditions to Produce Manganese and Iron Carbides from Denizli-Tavas Manganese Ore by Solid State Reduction*. MS Thesis, Middle East Technical University (2007)
14. G. Cil, in *Effect of Activation on Ferromanganese Production from Denizli-Tavas Manganese Ore by Carbothermal Reduction*. MS Thesis, Sakarya University (2013)
15. E. Aranci, in *Effect of Carbonized Tea Plant Wastes to Production of Ferromanganese by Carbothermal Method from Ferrous Manganese Oxide Ores Subjected to Mechanical Activation*. MS Thesis, Firat University (2014)

Trace Elements Behavior During the Oxidation of Liquid SiMn Alloy

Yan Ma, Ida Kero, Sarel Johannes Francois Gates and Gabriella Tranell

Abstract During the tapping and casting of manganese ferroalloys, fumes are generated and released to the working environment. The fumes are mostly composed of metallic oxides generated due to reaction between high temperature molten metal and ambient oxygen in the air. From an environment perspective, it is important to limit these emissions. However, to do so, it is important to understand fuming mechanisms and kinetics as well as fume compositions, i.e. element behavior during the oxidation process of liquid manganese ferroalloys. Silicomanganese alloys are composed of a minimum of 65 wt% Mn and 17 wt% Si, 2 wt% C as well as minor and trace elements. These elements include Fe, Mg, Al and other elements originating from ores and reductants. Trace elements behavior in fume from liquid silicomanganese generated under an impinging air jet in the temperature range of 1450–1700 °C has been investigated in this work. The thermodynamic and kinetic conditions governing the generation mechanisms are discussed.

Keywords Silicomanganese · Oxidation · Fume · Trace elements

Introduction

Fugitive emissions, originating from various sources in the metallurgical industry is receiving increased attention worldwide. The aerosols produced in most metallurgical processes may be harmful and the inhalation and exposure to high levels of

Y. Ma (✉) · S.J.F. Gates · G. Tranell
Norwegian University of Science and Technology (NTNU),
Alfred Getz vei 2, NO-7491, Trondheim, Norway
e-mail: yan.ma@ntnu.no

I. Kero
SINTEF Materials and Chemistry, Alfred Getz vei 2, NO-7465, Trondheim, Norway

S.J.F. Gates
University of Pretoria, Private Bag X20 Hatfield, Pretoria 0028, South Africa

J.-Y. Hwang et al. (eds.), *8th International Symposium on High-Temperature Metallurgical Processing*, The Minerals, Metals & Materials Series,
DOI 10.1007/978-3-319-51340-9_22

particles has been linked to respiratory and cardiovascular syndromes [1–3]. Characteristic properties of the fume particles, especially particle size and chemical compositions, influence their impact on human health.

Fumes are generated and released to the working environment during the production of manganese ferroalloys. The significant fume generation during tapping and casting of both FeMn and SiMn results from the relatively high vapour pressure of manganese. The fumes are mostly composed of metallic oxides and generated in a reaction between high temperature molten metal and ambient oxygen in the air [4]. There is also a certain amount of trace elements existing in the fume, originating from the tapped alloy, for example heavy metals. Heavy metals are carried into the process as trace elements in the raw materials [5]. The heavy metals get redistributed to metal, slag or fume during the smelting process. Part of the heavy metals will escape as metal vapor. This process depends heavily on the type of ferroalloy produced and the temperature of the smelting process. As soon as the metal vapors exit the reactor/furnace, the temperature drops and the metal vapors may be condensed and finally collected as fume.

In previous publications, the elemental distribution of both major and trace elements in the fume from the metallurgical grade silicon (MG–Si) refining process has been reported [6, 7]. The predominant trace elements in the silica fume are Pb, Mg and Zn [6]. Kero et al. [8] performed elemental analysis of fumes collected in a SiMn alloy production plant, using EDX coupled to an electron microscope. The major elements found were Si, Mn, and O, and minor elements detected were Mg, Ca, Al, and K, while trace elements were Na, Fe, Zn, Cu, and Cl. However, the behavior of trace elements from SiMn fume were not further studied in detail.

The aim of this study was to characterize the trace elements of fume from a liquid SiMn alloy, kept at varying temperature under an impinging air jet. The objective is to increase the understanding of trace elements behavior during active oxidation and fume formation from liquid SiMn. The knowledge will support higher accuracy in emission reporting of trace elements from an environmental protection prospective. It will also provide more information on trace element contents, which is important in the re-processing/reuse of the dust.

Theory

There have been very few papers regarding the fume generation mechanisms in manganese ferroalloy production process published [4, 9, 10]. The general mechanisms for fume formation from liquid metal have been described by Gray [11] and Gonser [12], as shown in Fig. 1. Metal alloy droplet expulsion, as shown in mechanisms (a) and (b), occur under the conditions of turbulent flows. Unfractionated evaporation of the liquid alloy occurs at high temperature, as illustrated in mechanisms (c) and (d), the metal vapor comes into contact with oxygen directly or after condensation to form an oxide. The presence of trace elements within the alloy may result in fractionated vaporization and subsequent

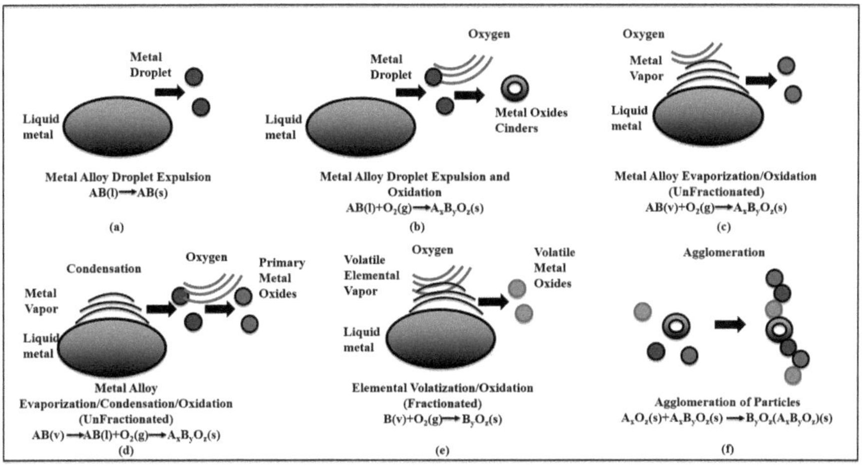

Fig. 1 Mechanisms of fume formation from a liquid metal [12]

oxidation as illustrated in mechanism (e). Agglomeration of particles occurs when the alloy oxide particles coalesce, forming larger particles.

In the studied system, mechanisms (a) and (b) should not happen as laminar flow conditions apply at the melt surface in the experiments without bubbling of gases through the melt. Unfractionated evaporation and subsequent oxidation in mechanisms (c) and (d) are also not likely to happen, as major elements display very different evaporation/oxidation behavior. Mechanisms (e) and (f) are likely to occur in the studied system. Formation mechanism (e) which is fractionated evaporation and subsequent oxidation will be further discussed here.

Næss et al. [6] used the "boiling-point-model" developed by Myrhaug and Tveit [13] and "oxidation potential of elements" to explain the element distribution through the Si refining process. According to the boiling point model, elements with boiling temperatures higher than the process temperature will preferentially stay in a condensed phases, while elements with boiling temperatures below the process temperature will go off as fume or gas. If the elements are more stable as oxides or other compounds (such as in slag), the boiling temperature of the compound will be applicable. In the use of oxidation potential of elements, elements with higher Gibbs free energy of oxidation than silicon have two possible paths: silicon or fume. The elements with lower Gibbs free energy of oxide formation will oxidize and report to the slag or gas (if the oxide is volatile at the process temperature). However, these two explanations are both based on an ideal description of the system; the elemental behaviors will depend on solute-solvent interactions.

Elements with low boiling temperatures and respectively high vapor pressure will preferentially report to the fume, e.g. Hg, S, K, Na, Zn, Mg, Li, Pb [14]. Elements that may form high volatile oxides may also report to the fume, such as Si, Al [15, 16]. Elements with higher Gibbs free energy of oxidation than Mn are expected to stay in elemental state (as vapor or in liquid), e.g. Hg, Pb, Cu, As, Ni, Co, S, K, Fe, Zn, Na and Cr [17]. If vaporized, however, elemental aerosols may form and subsequently oxidize away from the liquid surface.

Experimental

The experimental set-up is shown in Fig. 2, the same as used in the study by Kero et al. [18]. A standard grade SiMn alloy and dry air was used for the investigation. The experiments were carried out at 6 different temperatures between 1450 and 1700 °C. After the experiments, the fume was collected at three sites; the so-called transition tube, the cooler and the filter. The majority of fume was collected from the filter. The deposition temperatures at the three sites were different; there was a significant temperature drop inside the transition tube, from about 600 °C at the transition tube entry point to 200 °C at the point of connection to cooler. In the cooler, the temperature dropped further to 25 °C which was also the constant temperature of the filter. The use of the graphite crucible has led to formation of CO(g), as reported in our previous study [9]. The presence of CO will affect the chemistry and potentially the kinetics of the fuming process.

The fumes collected from the three different sites at six different temperatures were characterized by inductively coupled plasma mass spectrometry (ICP-MS). A total of 18 fume samples were collected from the experiments. The fume samples were dissolved in 0.5 mL concentrated HF, 0.5 mL concentrated HNO_3, and

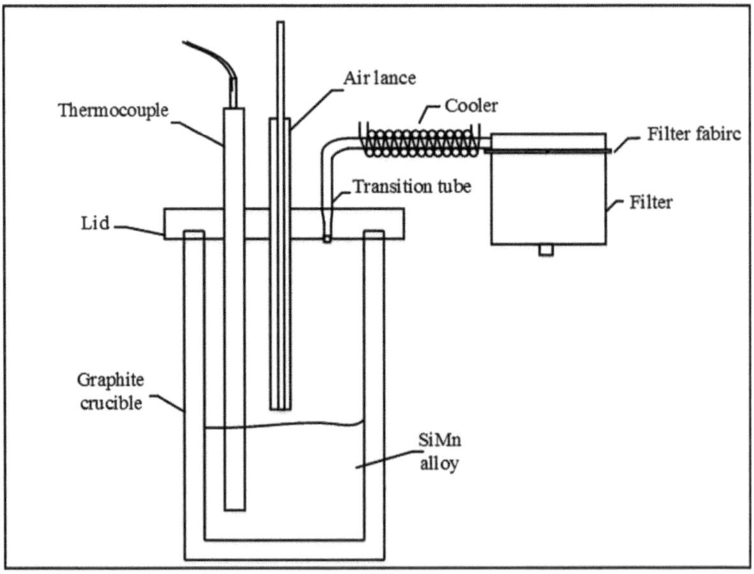

Fig. 2 Experimental setup

0.5 mL concentrated H_2SO_4, and then digested in an Ultra Clave prior to analysis. Sample of the SiMn alloy used for the experiments was also measured by the same method. An instrumental detection limit (IDL) for each element was determined by the blanks and by the concentration of the element. If the detected results were below its respective IDL value, the results are considered unusable. Each sample was analyzed three times, and a relative standard derivation (RSD) value from ICP-MS analysis was also obtained. An analysis with an RSD value above 10% was considered unreliable.

Results and Discussion

While 32 elements were included in the ICP-MS analysis, only 14 elements (in addition to Si and Mn) results were considered reliable. The other elements were either present in concentrations below the IDL value or the RSD value >10%. The concentration of Si and Mn varied from 10 to 15 wt% and from 45 to 55 wt% respectively. The set of trace elements that can be used for discussion are Al, As, Cr, Co, Cu, Fe, Pb, Li, Mg, Ni, K, Na, S and Zn. The 14 elements are classified into four groups, according to their chemical properties, as seen in Table 1.

Figure 3a illustrates the total amount of trace elements in filter fumes generated at different temperatures. It is shown that the weight percentage of trace elements increases with temperature. Some elements are present at percent level concentrations, such as Mg, S, K, Na, Fe, Al and Zn. As, Cu, Ni, Cr, Co, Li and Pb have ppm-level concentrations, as shown in Fig. 3b. The contribution of Mg increases particularly fast with increasing temperature (4 wt% at 1700 °C), as may be expected from the strong coupling between temperature and Mg vapor pressure. Sulphur has consistently high concentration (between 1 and 2 wt%) with a notable exception at 1600 °C. The alkali metals Na and K are also present at significant levels in the fume, while Fe only makes a contribution below 1650 °C. Aluminum only makes a significant contribution above 1650 °C.

Three groups of trace elements, alkali metals, toxic metals and other metals, will be discussed in detail here, as they represent the majority of elements that are present in the fume.

Figure 4 illustrates the concentration of elements of three element groups at three different sites with different temperatures, in the transition tube, the cooler and the filter. It is clear that the highest concentration of all trace elements can be found in the transition tube, i.e. the deposition site with the highest temperature, where the temperature also drops rapidly. The transition tube was located directly above the

Table 1 Trace elements detected in fume

Groups	Elements
Alkali metals	Li, Na, K, Mg
Toxic metals	Pb, As
Other metals	Zn, Cu, Ni, Cr, Co, Al, Fe
Non-metals	S

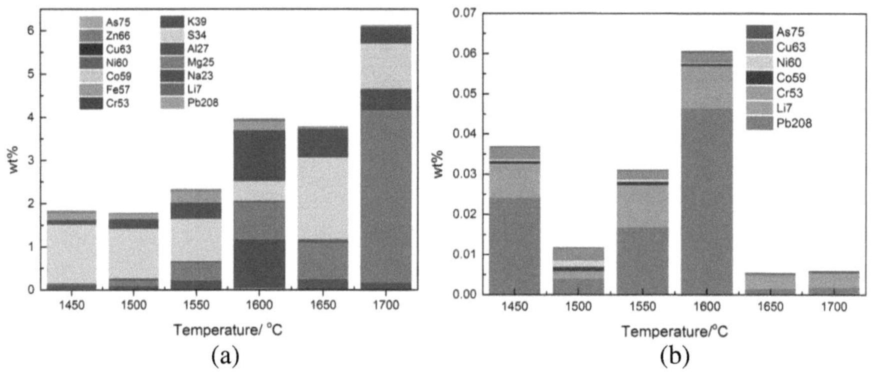

Fig. 3 Weight percentage of trace elements in filter fume **a** total amount of trace elements, **b** trace elements with concentration below 0.05 wt%

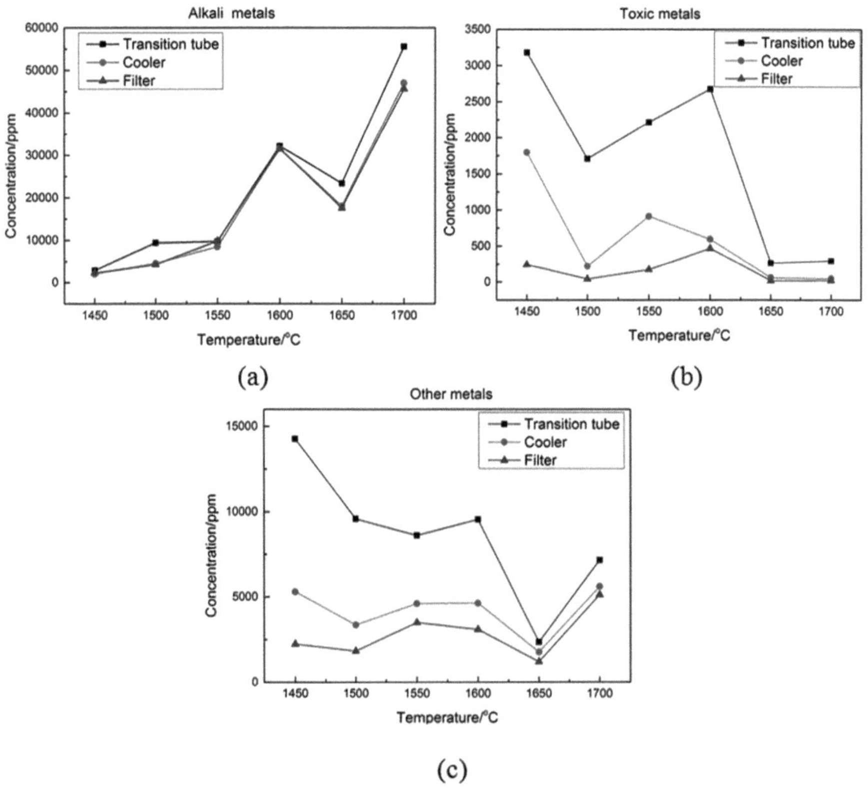

Fig. 4 Concentration of elements from three groups at three different positions

reaction crucible and metal vapor/oxides were indeed expected to condense there. As shown in Fig. 4a, the total concentration of alkali metals gradually increases with fume generation temperatures. In accordance with the overall trend, the fume deposited in the transition tube has the highest concentration of alkali metals, while there is no significant concentration difference between cooler fume and filter fume which can be attributed to the similar temperatures of the cooler and filter sites. The concentration of alkali metals in fume generated at 1650 °C appears to be slightly lower than at 1600 °C. This general trend differs from that of both the toxic elements group and that of the "other metals" group, as illustrated in Fig. 4b, c, both decreasing in concentration in the transition tube fume with increasing temperature, while element concentrations in the cooler and filter fumes are relatively similar at different temperatures (with some exceptions, see Fig. 4b at 1450 and 1550 °C). The behavior of trace elements in all fumes from the experiments at 1650 °C follows a different trend than at other temperatures, and it is difficult to assess without several parallel samples from identical experiments.

The concentrations of individual elements from filter fume are plotted in groups in Fig. 5. As illustrated in Fig. 5a, concentrations of alkali metals Mg, K and Na are at the same orders of magnitude except for at 1700 °C where the concentration of Mg is very high. Concentration of Li is generally very low, compared to the other alkali metals. The boiling points of the alkalis are low compared to the experimental temperatures and thus promote their vaporisation from the melt. The concentration of Pb is one or two orders of magnitude larger than As, as shown in Fig. 5b. The highest concentration of Pb appears at 1600 °C. Zn concentration behaves the same way as Pb, but is one order of magnitude larger. As shown in Fig. 5c, the concentrations of Fe, Al and Zn are significant, while the concentrations of more noble elements Cr, Co, Ni and Cu are much lower. Fe has the highest concentration at 1550 °C and the lowest concentration at 1700 °C. Al vapor pressure starts to become significant at 1700 °C.

Concentrations of Mg, Li, Na, S, K and Pb in the raw material-SiMn alloy were all below the IDL value. Concentration of Zn was only 1 ppm in the alloy, which was three or four orders of magnitude smaller than that in the fume. It means these elements get up-concentrated in the fume. The results confirmed that these elements expected to be in the fume have relatively high measured concentrations, due to their low boiling points and high vapor pressures. Considering their oxidation potentials, Mg and Li will be further oxidized after evaporation, while Na, S, K, Pb and Zn are expected to get oxidized to a lesser extent.

Concentration of As in the fume was one order of magnitude smaller than that in the alloy, which may be explained by two competing properties: high vapor pressure and negative derivation from ideality in both Si and Mn [19]. As explained in the theory section, As was, a priori, expected to remain in its elemental state.

Concentration of Al in the alloy was also below the detection limit, but Al was detected in significant concentrations in the fume at 1700 °C, which may be explained by relatively high vapour pressure at 1700 °C and the formation of the gaseous oxide Al_2O.

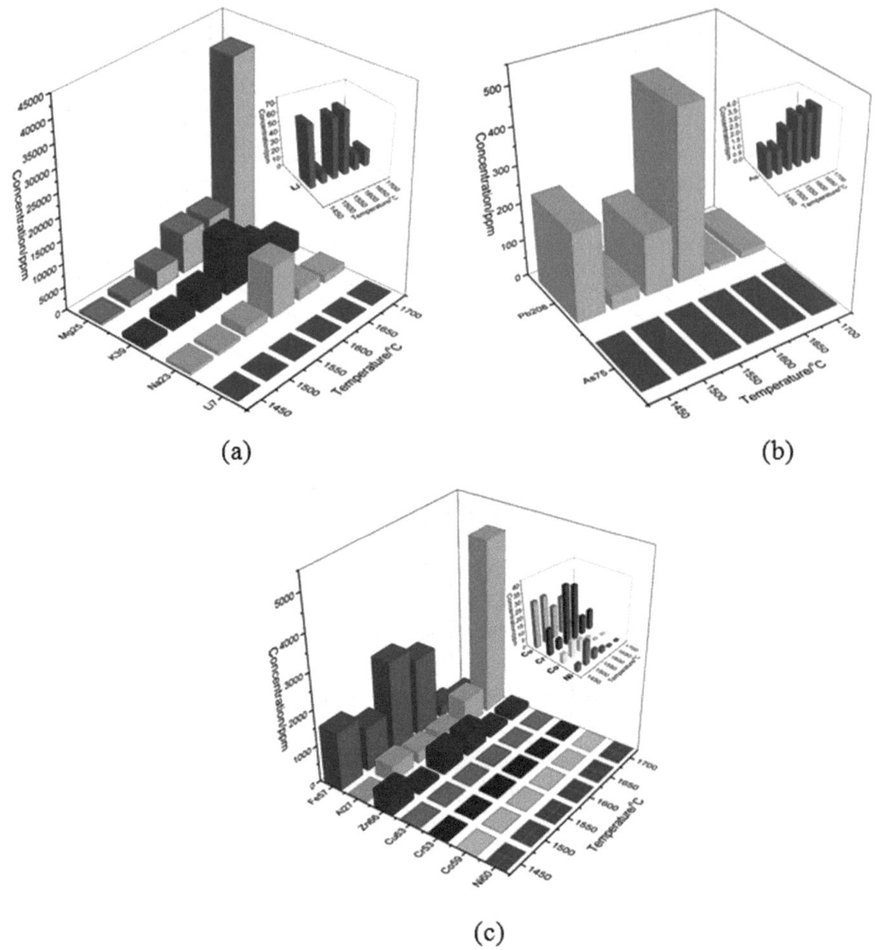

Fig. 5 Concentration of elements from filter fume

The concentrations of Fe, Cu, Co, Cr and Ni in the alloy were several orders of magnitude larger than that in the fume, which may be explained by their low vapor pressures and negative deviations from ideality in both Si and Mn [19]. Cr shows positive derivation in Mn and negative derivation in Si [19], its relatively low concentration in the fume may be explained by the low vapor pressure and lower oxidation potential than Mn.

Conclusions

The concentration and distribution of different trace elements in fume, resulting from the oxidation of liquid silicomanganese under an impinging air jet at temperatures between 1450 and 1700 °C have been studied.

Some elements are present at relatively high concentrations in fume, such as Mg, S, K, Na, Fe, Al and Zn, while As, Cu, Ni, Co, Cr, Li and Pb have ppm level concentrations. The highest concentration of all trace elements can be found at the first temperature-drop zone for the off-gas. The total concentration of alkali metals gradually increases with fume generation temperature, while the toxic metals and 'other metals' decrease in concentration with increasing temperature.

The distribution of elements in the filter fume under elevated temperatures is generally in accordance with the boiling-point-model. Elements that get up-concentrated in the fume include Mg, Li, Na, S, K, Pb, Zn and Al. Elements which have relatively lower concentrations in the fume compared to the alloy include Fe, Cu, Cr, Co and Ni. The concentration of As was one order of magnitude smaller in the fume than in the alloy.

Acknowledgements These studies were funded by Norwegian Research Council (NFR) through the project "DeMaskUs", grant number 245216.

References

1. Å. Hobbesland, H. Kjuus, D.S. Thelle, Mortality from nonmalignant respiratory diseases among male workers in Norwegian ferroalloy plants. Scand. J. Work Environ. Health **23**, 342–350 (1997)
2. A.D. Kappos et al., Health effects of particles in ambient air. Int. J. Hyg. Environ. Health **207**, 399–407 (2004)
3. H. Kjuus, A. Andersen, S. Langård, K.E. Knudsen, Cancer incidence among workers in the Norwegian ferroalloy industry. Br. J. Ind. Med. **43**, 227–236 (1986)
4. S. Rencken, in Analysis of secondary fume generation and capture during the tapping of high carbon ferromanganese. Report, University of Petoria, (2013)
5. European Environment Agency, EMEP and EEA emission inventory guidebook. Report (2013)
6. M.K. Næss et al., Element distribution in silicon refining: thermodynamic model and industrial measurements. JOM **66**(11), 2343–2354 (2013)
7. M.K. Næss, I. Kero, G. Tranell, A new method for estimation of emissions and sources of measurements error in the silicon refining process. JOM **65**(8), 997–1006 (2013)
8. I. Kero, M.K. Naess, G. Tranell, Particle size distributions of particulate emissions from the ferroalloy industry evaluated by electrical low pressure impactor (ELPI). J. Occup. Environ. Hygiene **12**(1), 37–44 (2015)
9. I. Kero et al. Fume formation from liquid ferromanganese. Paper presented at the 2015 sustainable industrial processing summit, Antalya, Turkey, 4–9 October 2015, pp. 135–146
10. P. Cowx et al. The use of fine water sprays to suppress fume emissions when casting refined ferromanganese. Paper presentated at the Infacon XIV. Kyiv, Ukraine, 1–4 June 2015, pp. 81–90

11. C. Gray, in *Fume Formation in Electric Arc Welding*. Ph.D. thesis, University of Bradford (1980)
12. M. Gonser, H. Theodore, *Arc Welding Health Effects, Fume Formation Mechanisms, and Characterization Methods* (INTECH Open Access Publisher, USA, 2011), p. 305
13. E.H. Myrhaug, H. Tveit, Material balances of trace elements in the ferrosilicon and silicon processes. Paper presented at the 58th electric furnace conference and 17th process technology conference proceedings, iron and steel society, Orlando, US, Nov. 12–15, pp. 591–604 (2000)
14. E.T. Turkdogan, P. Grieveson, L.S. Darken, Enhancement of diffusion-limited rates of vaporization of metals. J. Phys. Chem. **67**(8), 1647–1654 (1963)
15. M.K. Næss et al., Parameters affecting the rate and product of liquid silicon oxidation. Oxid. Met. **82**(5), 395–413 (2014)
16. M. Ratto et al., Oxidation of metals with highly reactive vapors: extension of Wagner theory. Metall. Mater. Trans. B **32**(5), 903–911 (2001)
17. Outotec, HSC. 7.1, (2016)
18. I. Kero, G. Tranell, Active oxidation and fume formation from liquid SiMn. Paper presented at TMS Annual Meeting. Nashville, USA, pp. 77–84 (2016)
19. T.B. Massalski, *Binary Alloy Phase Diagrams* (Materials Park, ASM International, Ohio, 1990)

Effect of Microalloy Elements V and Mg on Organization at High Heat Input Welding Shipbuilding Structure Steel

Wang Yan, Han Yihua, Zhu Liguang, Zhang Qingjun, Wang Shuoming and Zhang Caijun

Abstract An industrial test on 210 t converter-LF-CC production line on high heat input welding shipbuilding structure steel DH36 has been taken in a factory in north China, the effect of adding microalloy elements V and Mg on casting billet microstructure has been researched. It was found that when the composition of V was between 0.03 and 0.05% and the composition of Mg was between 0.002 and 0.005%, a large number of acicular ferrite precipitate in austenitic crystal in as-cast structure and cross into a net structure. The inclusions size which induced intra-granular ferrite are mainly less than 4 μm, especially those between 2 and 3 μm account for 50.80%, in these inclusions the composite inclusions mixed with Mg accounted for more than 60%.

Keywords High heat input welding shipbuilding structure steel · Microalloy elements · Acicular ferrite · Inclusions · Structure

The national natural science fund project (51574106). Hebei province natural science foundation of key funds (E2016209396). Hebei province natural science fund project (E2017209223). Hebei province High-level personnel of institutions of higher learning scientific research projects (GCC2014030).

W. Yan · H. Yihua · Z. Liguang (✉) · W. Shuoming · Z. Caijun
Hebei High Quality Steel Continuous Casting Engineering Technology Research Center, College of Metallurgy and Energy, North China University of Science and Technology, Tangshan 063009, China
e-mail: zhulg@ncst.edu.cn

Z. Qingjun
Analysis and Testing Center, North China University of Science and Technology, Tangshan 063009, China

J.-Y. Hwang et al. (eds.), *8th International Symposium on High-Temperature Metallurgical Processing*, The Minerals, Metals & Materials Series,
DOI 10.1007/978-3-319-51340-9_23

Introduction

Using traditional methods to enhance the strength of steel plate will result in a seriously decline in the impact of the steel plate toughness and welding performance. Increasing the toughness of high-heat input welding HAZ through refining the grain size by oxide metallurgy [1, 2] and microalloy technology is an important research direction in steelmaking. The third generation of oxide metallurgy technology improves the HAZ [3, 4] mainly by using deoxidizer Ca and Mg to obtain deoxidizing inclusions. Both its high temperature stability and dispersion distribution are better, the inclusions size is more reasonable, and the performance of the HAZ can be greatly improved. Aimed at adding microalloy elements Mg and V, the influence of Mg and V on inclusions in high heat input welding shipbuilding steel was researched.

Experimental Materials and Procedure

The sample was acquired from a 210 t converter-LF-CC and 40 mm hot rolled plate production line in a factory in northern China. The type, size and distribution of the inclusions were observed by metallographic microscope and S-4800 field emission scanning electron microscope.

Metallographic Observation

The influence of microalloy elements V and Mg on the structure of casting billet has been researched. The as-cast structure is austenitic and it is found that a large number of acicular ferrite precipitate in austenitic crystal in as-cast structure and crosse into a net structure with the composition of V and B between 0.03–0.05% and 0.002–0.005%, respectively, as shown in Fig. 1. A large number of acicular ferrite is induced by inclusions in austenitic crystalline, the grain is refined effectively, and the as-cast structure is optimized, providing a congenital basis for refining grains in rolling and welding.

Cross-section microstructure of 40 mm thick steel is shown in Fig. 2. By V–Mg microalloying, the rolling state structure mainly include ferrite, pearlite, a little tertiary cementite and granular carbide. The size of grain is small, and the average size of grain is about 11 level.

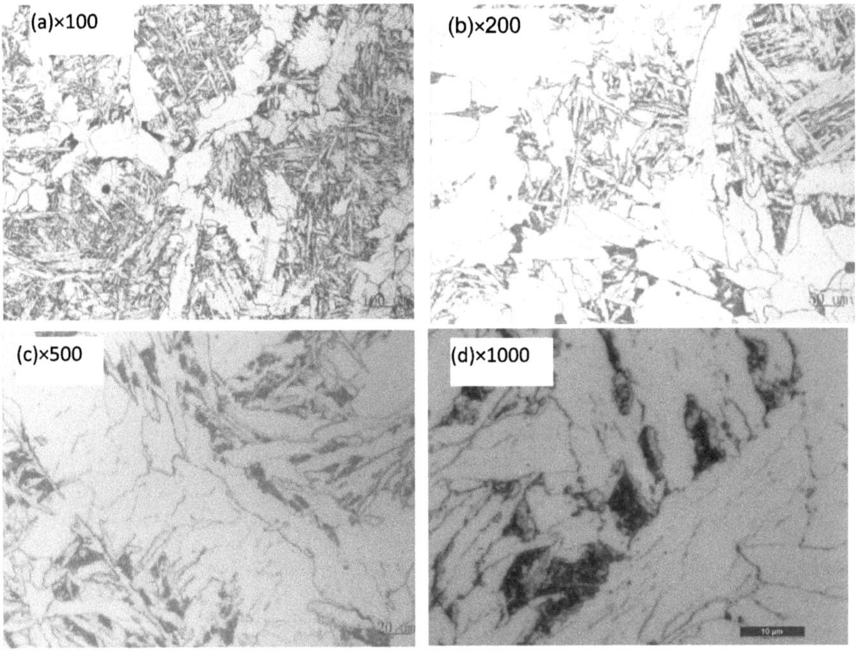

Fig. 1 As-cast structure chart under different magnification

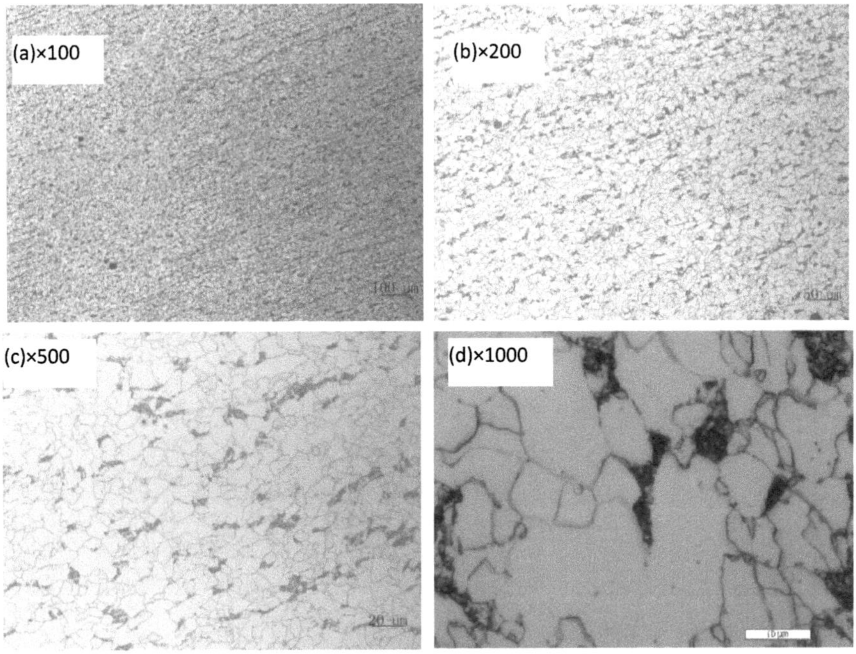

Fig. 2 Cross section structure diagram under different magnification

Inclusions Particle Size Distribution

420 inclusions of 15 metallographic samples separated from three locations in as-cast, 200 inclusions from cross section and 200 inclusions from lengthwise section of 6 rolled metallographic samples were selected to analyze particle size of all the inclusions. The statistical results of inclusions size distribution was shown in Figs. 3 and 4.

Figure 3 shows that the particle size distribution of inclusions in as-cast has two peaks. The content of size of 3–4, 10–15 μm is about 18.45 and 12.8%, respectively. The total content of size of 9–10 and 15–20 μm is only 2.4%.

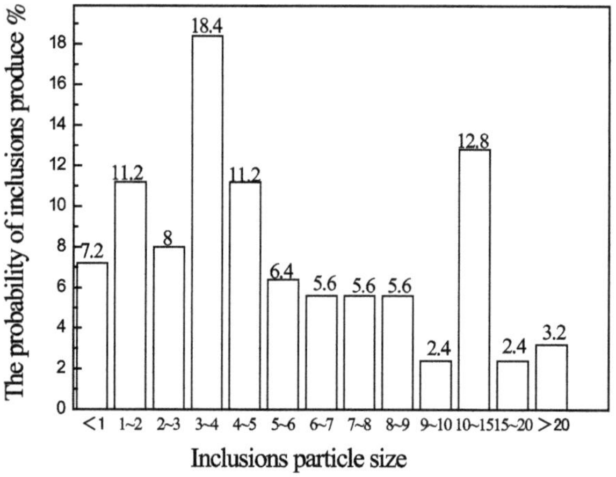

Fig. 3 Percentage for different sizes of as-cast inclusions

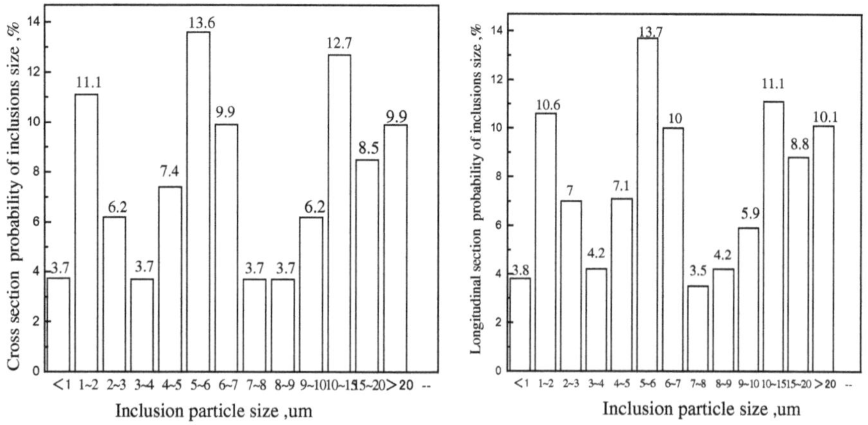

Fig. 4 Percentage for different sizes of inclusions

Figure 4 shows that the size distribution of inclusions in longitudinal section is the same as that in cross section. The content of inclusions with size of 5–6 μm in cross sections is about 13.6, and 13.7% in longitudinal section. The content of inclusions with size of 10–15 μm in cross sections and longitudinal section is about 11.1 and 10.6%, respectively. Comparing Fig. 3 with Fig. 4, it is obvious that large-sized inclusions in rolling state is notably higher than that in as-cast. Especially those with size more than 20 microns, is only about 3.2% in as-cast, 10.1% in cross section, and 9.9% in longitudinal section in rolling state. It illustrates that the inclusions in V–Mg microalloying ship plate steel had obvious deformation in the rolling process.

Related Elements Statistics in Inclusions

Composite inclusions induce IGF and the inclusions energy spectrum are shown in Fig. 5 and Table 1.

The proportion of elements included in the inclusions in different structure are shown in Fig. 6.

Figure 6a shows that the proportion of oxygen mixed in inclusions in as-cast is the largest, up to 86.9%, that of Al and Mg are 84.4 and 82.9% respectively. Although the chemical reaction of Mg alloying is violent, the homogeneity decrease. In a large range of Mg content, the inclusions with Mg distributed uniformly in the steel, giving a full play to the microalloying of Mg, and promoting acicular ferrite formation in austenitic crystalline. As shown in Fig. 6b, the content of Mg in longitudinal section and cross section of rolling states are 90 and 88% respectively. The content of Al in inclusions decrease from 84.4% in as-cast to 76

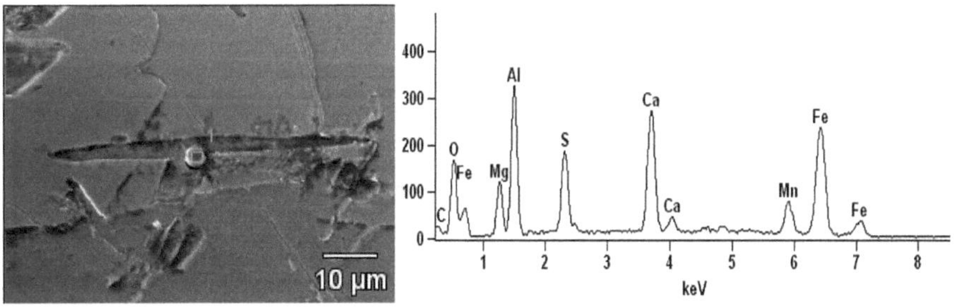

Fig. 5 Inclusions morphology and energy spectrum

Table 1 Inclusions composition table

Element	O	Mg	Al	S	Ca	Mn	Fe
Content	25.38	6.02	14.12	6.11	11.69	6.97	29.72

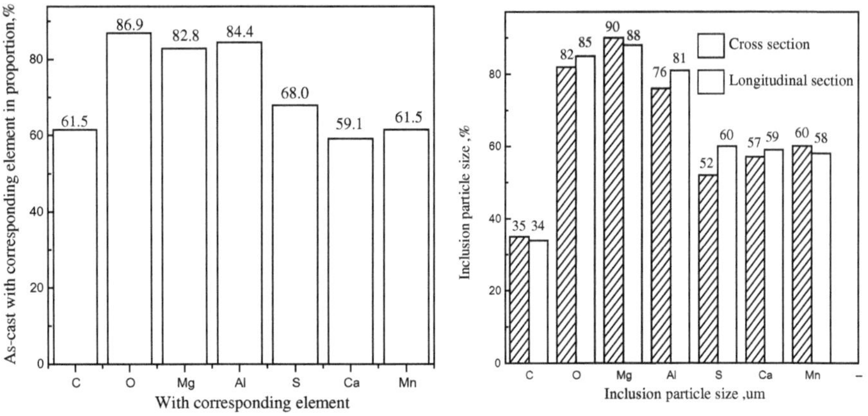

(a)The ratio of element of as-cast inclusions (b) The ratio of element of rolling inclusions

Fig. 6 The proportion of the corresponding elements in the inclusions

Table 2 Average content of each element in the inclusion (wt%)

		C	O	Mg	Al	S	Ca	Mn
As-cast		24.55	36.41	19.71	16.86	9.68	8.56	17.40
Rolling state	Cross	13.24	34.39	24.68	13.04	7.79	10.29	17.54
	Longitudinal	12.55	35.21	23.01	16.10	9.02	10.30	17.01

and 75% in rolling state. The content of V in inclusions to effectively induce acicular ferrite in austenite intracrystalline is very little. The effect of V mainly reflects in pinning austenite grain boundary. The precipitation of VN in austenite can inhibit the growth of austenitic recrystallization grain, and also increase the core of ferrite nucleation, which both promote ferrite grain refinement [5, 6].

The average content of each element in inclusions (wt%) is shown in Table 2. Table 2 illustrates that average content of oxygen in inclusions is the highest. The content of Mg is 24% in rolling state, higher than that of 19.71% in as-cast. The average content of aluminum is 16% similarly. The average content of C decreases from 24.55% in as-cast to 13% in rolling state, which infer that the carbide inclusion is partly dissolved in the rolling state.

Analysis of Inclusions Type

The statistical result of inclusions type in as-cast and rolling state is shown in Table 3. It can be seen in Table 3 that the inclusions in as-cast and rolling are mostly Mg–Al–O–CaS–MnS and Mg–Al composite inclusions, and some CaS–MnS, MgO and Al_2O_3 composite inclusions.

Table 3 Percent of different inclusions/%

Type	As-cast state	Rolling state	Type	As-cast state	Rolling state	Type	As-cast state	Rolling state
$MgO–Al_2O_3–CaS–MnS$	26.6	20.6	$MgO–Al_2O_3–MnO$	3.9	5.8	$MgO–Al_2O_3$	14.8	15.0
$Al_2O_3–CaS$	0.8	1.2	CaS–MnS	0.8		MgO–MnO	2.3	2.3
MgO–CaS–MnS	0.8		MgO–CaO–MnO	0.8		$Al_2O_3–MnO$	0.8	
$MgO–Al_2O_3–MnS$	3.1	1.2	$Al_2O_3–MnS$	0.8		$MnO–SiO_2$		2.3
$MgO–Al_2O_3–CaS$	0.8		MgS–CaS–MnS	3.9	4.6	MgO–CaO		3.5
MgO–CaS–MnS	0.8		CaS–MnS		1.2	MgO–MgC		2.3
CaO–MnO–CaS–MnS		1.2	$Al_2O_3–CaO–MnO$		1.2	Al_2O_3	4.7	3.5
MgC–MgS–CaS–MnS		1.2	MgS–MnS		3.5	MnS	6.2	
$MgO–Al_2O_3–CaO–MnO$		1.2	$Al_2O_3–MnO$	4.7	2.3	MgO	3.1	8.1
$MgO–Al_2O_3–CaO$	2.3		MgO–CaS	2.3	1.2	CaO		1.2
$MgO–Al_2O_3–CaS$	11.7	10.3	CaS–MnS	0.8	1.2			

Relationship of Inclusions Size and Induce IGF

Relationship of inclusions size and inducing rate of IGF are shown in Fig. 7. It can be seen that the inductions induced by inclusions with size of 2–3 μm are the most, up to 50.80%. The induction rate declined with the size increasing. Inclusions with Mg in induced IGF is more than 40%.

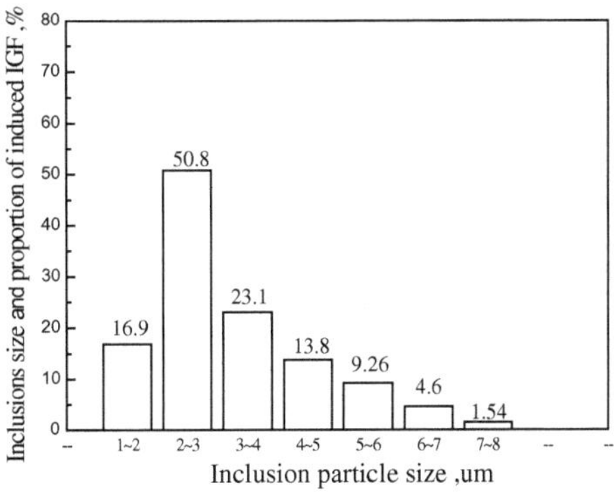

Fig. 7 Relationship between inclusion size and induced rate

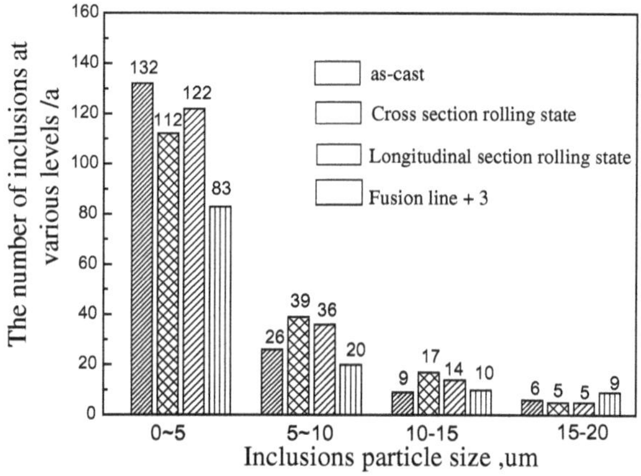

Fig. 8 Inclusions size distribution from as-cast state to welding

Inclusions size distribution from as-cast state to welding state is shown in Fig. 8. The content of inclusions with particle size of less than 5 μm decreases, while those of size larger than 5 μm in rolling state increases. The total number decreases when welding.

Conclusions

(1) Inclusions which induced IGF in V and Mg micro alloying shipbuilding steel are mainly Mg–Al–O–CaS–MnS and Mg–Al composite inclusions.
(2) The number of oxygen inclusions is the most in as-cas, up to 86.9%. The number of Aluminium mixed in as-cast is about 84.4%, and Mg inclusions is about 82.8%. Although the chemical reaction of Mg alloying is violent, the homogeneity decrease. In a large range of Mg content, the inclusions with Mg distributed uniformly in the steel, giving a full play to the microalloying of Mg, and promoting acicular ferrite formation in austenitic crystalline.
(3) The size of inclusions inducing IGF are mainly less than 4 μm, especially 2–3 μm, up to 50.80%. The content of induced inclusions with Mg is more than 60%. Inclusions size statistics found that inclusion particle size is given a priority to 3–4 μm in as-cast, up to 18.45% of total inclusions. Grain size in rolling state is 5–6 μm mostly, up to 13.7%.

References

1. J.I. Takamura, S. Mizoguhci, Metallurgy of oxides in steels, in *Proceedings of The Six International Iron and Steel Congress*, 1990, Nagoya, ISIJ, pp. 591–597
2. S. Mizoguchi, J.I.T. Akamura, Control of oxides as inoculants -metallurgy of oxides in steels 2, in *Proceedings of The Six International Iron and Steel Congress*, 1990, Nagoya, ISIJ, pp. 598–604

3. Y. Jian, S. Jian Guo, Z. Kai. Progress in the technological development of oxide metallurgy for manufacturing steel plates with excellet HAZ toughness. Baosteel Tech. Res. **2**(4), 1–16 (2008)
4. Y. Jian, Z. Kai, W. Rui Zhi et al., Technology for improving the toughness of heat affected zones with strong deoxidizers. Baosteel Tech. Res. **4**(4), 1–6 (2010)
5. Liu Qingyou, Mao Xinping, Lin Zhenyuan et al., Study on metallurgical characteristics of V-N microalloyed strip produced by CSP. Iron Steel **40**(12), 4–8 (2005)
6. F. Ishikawa., T. Takahashi, T. Ochi, Intragranular ferrite nucleation in medium-carbon vanadium steels. Metall. Mater. Trans. A. **25**(5), 929–936 (1994)

Part V
Extraction and Recovery of Metals

Effect of Carbon Reductant on the Formation of Copper Doped Titanium Oxycarbonitride by Carbothermal Reduction and Nitridation

Yong Jing Hui, Noor Izah Shoparwe, Sheikh Abdul Rezan, Norlia Baharun, Srimala Sreekantan and M.N. Ahmad Fauzi

Abstract In this paper, the effect of carbon reductant on carbothermal reduction and nitridation of copper-doped titanium oxide (Cu–TiO_2) was investigated. Reduction was performed under H_2–N_2 (50 vol.% N_2) gas mixtures with a temperature range of 1100–1300 °C for 1–3 h respectively. Three different carbon sources were investigated in the reaction which were graphite, activated carbon (AC) and carbon nanotube (CNT). The highest extent of reduction of 99.9% was obtained at 1300 °C for 180 min using 1:4 carbon to oxide molar ratio with CNT. Extent of nitridation was 71.7% while extent of carburization was 28.2% for this sample. Elemental analysis indicated an oxycarbonitride phase with a composition of Cu–$TiO_{0.002}C_{0.28}N_{0.717}$. With CNT and AC as a reductant, the formation of Cu–$TiO_xC_yN_z$ occurred at 1100 °C after 180 min. Increasing the reduction temperature, time and molar ratio, the TiN and TiC contents in the synthesized Cu–$TiO_xC_yN_z$ are increased.

Keywords Cu–Titanium oxycarbonitride · Cu–Titania · Carbothermal reduction · Nitridation · Hydrogen · Nitrogen · Activated carbon · CNT

Introduction

Titanium oxycarbonitride ($TiO_xC_yN_z$) is a solid solution of titanium oxide, carbide, and nitride [1]. The phases of TiO_x, TiC_y and TiN_z are a NaCl type crystallographic structure. It is through this structure, substitution of O, N, C occurs easily by solid-solid diffusion at 1100–1500 °C [2]. By doping with various metals into TiO_2 matrix, the rate of conversion to $TiO_xC_yN_z$ will be affected. Iron chloride is the

Y.J. Hui · N.I. Shoparwe · S.A. Rezan (✉) · N. Baharun · S. Sreekantan · M.N. Ahmad Fauzi
School of Materials and Mineral Resources Engineering, Engineering Campus,
Universiti Sains Malaysia, 14300 Nibong Tebal, Pulau Pinang, Malaysia
e-mail: srsheikh@usm.my

J.-Y. Hwang et al. (eds.), *8th International Symposium on High-Temperature Metallurgical Processing*, The Minerals, Metals & Materials Series,
DOI 10.1007/978-3-319-51340-9_24

most common catalyst used by many researchers to enhance the formation of $TiO_xC_yN_z$ [3–5]. Doping with transition metals such as Copper (Cu) to enhance the formation of $TiO_xC_yN_z$ was rarely reported in the literature. Furthermore, Cu–$TiO_xC_yN_z$ is a promising material for use in photo catalytic systems due to the oxycarbonitride structure which has a lower optical bandgap than rutile [6].

Besides photocatalytic applications, studies by Rezan et al. and Adipuri et al. [7, 8] have shown that $TiO_xC_yN_z$ can be chlorinated at low temperature. The low temperature chlorination of $TiO_xC_yN_z$ is attractive as it will lower energy cost and reduce the chlorination of unwanted impurities [8, 9]. This paper investigates the carbothermal reduction and nitridation of Cu–TiO_2 to Cu–$TiO_xC_yN_z$ in H_2–N_2 atmosphere at different temperature, time, carbon source and carbon molar ratio.

Experimental

Cu–TiO_2 (99.9 + %, <200 nm) and CNT (<110 nm), AC (<25 μm), graphite (<20 μm) were wet mixed with addition of 0.3 wt% carboxymethylcellulose as a binder. Cu–TiO_2 was prepared by hydrothermal process [10]. The mixture was dried and pressed into pellets (400 MPa). The Cu–TiO_2-carbon mixture had a titania-to-carbon molar ratio that varied from 1:2 to 1:4. The pellets were 8 mm in diameter and with 13 mm thickness, weighing about 1.0 g. Reduction of Cu–TiO_2 by graphite (G), CNT & AC in H_2–N_2 gas atmosphere was studied in a laboratory fixed bed reactor in a horizontal tube electric furnace. Experimental set-up and gas system were described elsewhere [3]. The gases were of super high purity. The outlet gas was analysed by gas chromatography to obtain volume concentration of CO, CO_2 and CH_4. The total gas flow rate was maintained at 2.0 L/min. The reduced pellets were analysed by XRD (Bruker D8-advance). Surface area of the carbon source was measured by BET method (Autosorb-IQ). Oxygen content in the reduced samples was determined using LECO analyses (TC-436DR). Final carbon and nitrogen analysis was done by Perkin Elmer Series II, model 2400. The extent of reduction (X) was defined as a fraction of oxygen in titanium dioxide removed in the course of reduction, and was calculated from gas composition and oxygen and carbon contents of reduced samples. The extent of nitridation (X_N) was defined as a fraction of titanium which was present as nitride (TiN) in the reduced sample. Extent of carburization (X_C) was defined as a fraction of titanium which was present as carbide (TiC) in the reduced sample. In Cu–$TiO_xC_yN_z$ phase, the stoichiometric values for x, y, z were obtained from calculations of X, X_N and X_C. Thermodynamic simulations was done by Outokompu HSC Chemistry v5.11 software [11].

Results

Effect of Temperature, Time and Carbon Source

Figures 1 and 2 presents the reduction curves obtained in the carbothermal reduction of Cu–TiO_2 in a 50 vol.% H_2–50 vol.% N_2 gas mixture at 1100 and 1300 °C with respect to carbon source and time. The molar carbon ratio was 1:4. From the figure, it was clearly evident that CNT had a higher degree of reduction compared to AC and graphite. At 1100 °C after 180 min, the X for CNT was 85.7% compared to AC which was 41%. The XRD analysis at 85.7% reduction with CNT showed the main phases were $TiO_xC_yN_z$ while Ti_3O_5 was a minor phase. On the other hand, at 41% reduction with AC, the major phases were Ti_3O_5 with minor phases being TiO_2 and $TiO_xC_yN_z$. The detection of TiO_2 phase indicated that the reduction was slower with AC at 1100 °C.

At 1300 °C, the X curves changed significantly. Both CNT and graphite had a high X close to 100%. For CNT, the X was 99% after 180 min whereas for graphite it was 93.5%. AC had the lowest X of 85.7% among the 3 carbon sources at 1300 °C. Extending the reduction time at this temperature to more than 300 min brought about total conversion of Cu–TiO_2 into Cu–$TiO_xC_yN_z$ for all 3 carbon sources. As reduction temperature increased from 1100 to 1300 °C, the reaction was accelerated indicating a chemical controlled reaction from the kinetic curve. For CNT, the X increased from 85.7 to 99% which was an overall increase of about 13%. Full conversion from Cu–TiO_2 into Cu–$TiO_xC_yN_z$ using CNT occurred within 30 min based on the kinetic data in Fig. 2. After full conversion occurred, X became slower

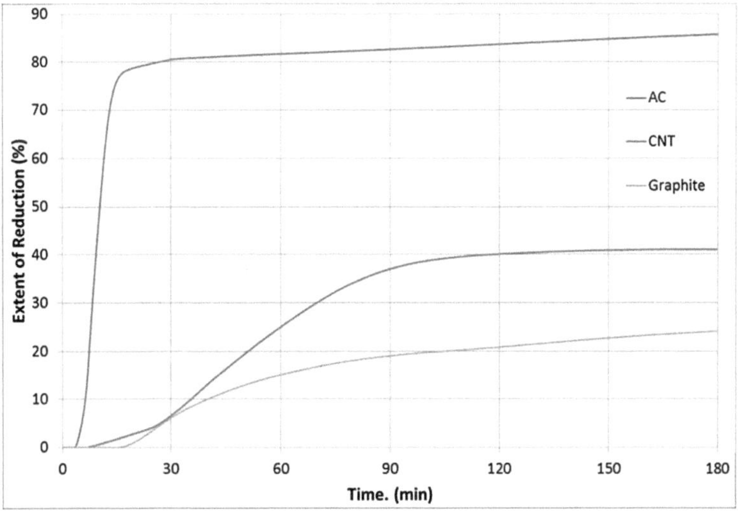

Fig. 1 Effect of time and carbon source on extent of reduction in 50 vol.% H_2–50 vol.% N_2 gas mixture at 1100 °C

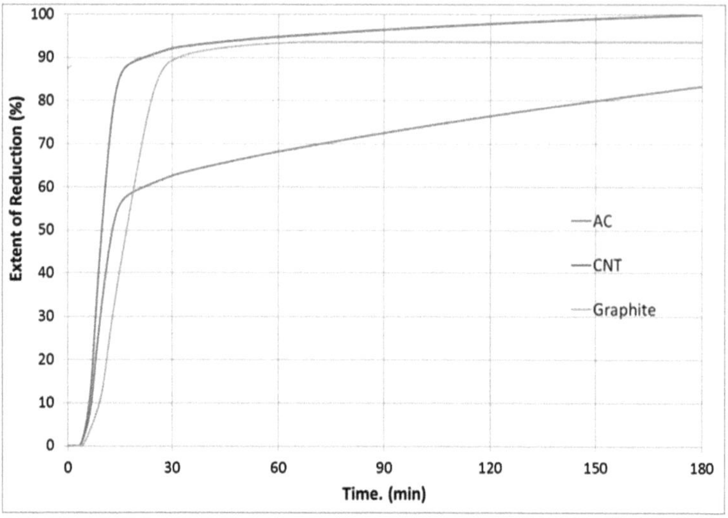

Fig. 2 Effect of time and carbon source on extent of reduction in 50 vol.% H_2–50 vol.% N_2 gas mixture at 1300 °C

as shown in Fig. 2. Graphite (G) reductant showed a similar behaviour with CNT. On the other hand, the X curve for AC at 1300 °C indicated that the reaction was still increasing after 180 min. The phase development during carbothermal reduction was examined in the next section.

Progress of Conversion

Table 1 shows the Design of Experiment (DOE) parameters for the reduction of Cu–TiO_2 at different temperature, time, molar ratio and carbon source. From the table, XRD analysis was performed at different time for Cu–TiO_2 reduced by CNT and AC at 1100 and 1300 °C. The phases identified with their respective extent of reduction (X), nitridation (X_N) and carburization (X_C) are listed in Table 2. With the use of CNT at 1100 °C, the phase identified by LECO was Cu–$TiO_{0.40}C_{0.14}N_{0.46}$ at 60 min. As reduction time progresses, the phase of Cu–$TiO_{0.28}C_{0.04}N_{0.68}$ was identified and listed as sample #15 in Table 2. The X and X_N was increasing with a decrease in X_C. This trend change in X, X_N and X_C was similar to results obtained by Sheikh A. Rezan et al. [2] using graphite reduction of titania. No formation of suboxides was detected at 60 and 180 min using CNT at 1100 °C.

On the other hand, using AC at 1100 °C, the X was significantly affected with increasing time. After 60 min, the phase identified by XRD was Ti_3O_5, TiO_2 and carbon. Rietveld Refinement analysis was performed on the XRD data and the quantitative amount was listed in Table 2 under sample #5. As the reduction time

Table 1 Design of experiment (DOE) parameters for reduction of Cu–TiO_2 in 50 vol.% H_2–50 vol.% N_2 gas mixture at different temperature and time

Sample #	Temp. (°C)	Time (min)	Molar ratio	Carbon type #
1	1100	60	1:2	AC
2	1300	60	1:2	AC
3	1100	180	1:2	AC
4	1300	180	1:2	AC
5	1100	60	1:4	AC
6	1300	60	1:4	AC
7	1100	180	1:4	AC
8	1300	180	1:4	AC
9	1100	60	1:2	CNT
10	1300	60	1:2	CNT
11	1100	180	1:2	CNT
12	1300	180	1:2	CNT
13	1100	60	1:4	CNT
14	1300	60	1:4	CNT
15	1100	180	1:4	CNT
16	1300	180	1:4	CNT
17	1200	120	1:3	G
18	1200	120	1:3	G
19	1200	120	1:3	G

CNT carbon nanotubes, *AC* activated carbon and *G* graphite

Table 2 Summary of X, X_N, X_C and phases identified by XRD

Sample	X_o	X_N	X_C	Phase (wt%)
1	19.29	1.46	–	31.4% Ti_3O_5, 68.6% graphite
2	46.43	19.49	–	47.4% $TiO_xC_yN_{0.20}$, 52.7% Ti_3O_5
3	32.31	2.92	–	99.5% Ti_3O_5, 0.5% graphite
4	41.12	11.80	–	21.5% $TiO_xC_yN_{0\cdot118}$, 49.9% Ti_3O_5, 28.6% TiO_2
5	28.19	16.48	–	87.5% Ti_3O_5, 11.7% anatase, 0.9% graphite
6	69.01	36.46	1.56	$TiO_{0.62}C_{0.02}N_{0.36}$
7	41.03	5.57	–	17.6% $TiO_xC_yN_{0.06}$, 81.9% Ti_3O_5, 0.5% C
8	83.38	15.25	51.51	$TiO_{0.33}C_{0.52}N_{0.15}$
9	53.42	24.88	–	35.8% $TiO_xC_yN_{0.25}$, 64.2% Ti_3O_5
10	65.97	74.19	–	$TiO_xC_yN_{0.74}$
11	61.47	55.99	–	64.5% $TiO_xC_yN_{0.56}$, 35.4% Ti_3O_5
12	63.63	61.47	–	76% $TiO_xC_yN_{0.62}$, 24% Ti_3O_5
13	80.02	46.22	13.81	$TiO_{0.40}C_{0.14}N_{0.46}$
14	92.81	70.22	15.40	$TiO_{0.15}C_{0.15}N_{0.70}$
15	85.73	67.70	3.77	$TiO_{0.28}C_{0.04}N_{0.68}$
16	99.90	71.68	28.13	$TiO_{0.002}C_{0.28}N_{0.717}$
17	90.93	70.88	10.99	$TiO_{0.18}C_{0.11}N_{0.71}$
18	91.22	69.84	12.60	$TiO_{0.18}C_{0.12}N_{0.70}$
19	91.57	70.48	12.65	$TiO_{0.17}C_{0.13}N_{0.70}$

increased to 180 min, the phases identified were Cu–$TiO_xC_yN_{0.06}$, Ti_3O_5 and carbon. The removal of TiO_2 phase indicated nitridation and carburization had started to occur in this sample. The quantitative analysis was listed in Table 2 as sample #7. Increasing the reduction temperature to 1300 °C affected phase changes significantly. For CNT, at a reduction time of 60 min, all suboxides have been converted to Cu–$TiO_xC_yN_z$. At 60 min, the phase of Cu–$TiO_{0.15}C_{0.15}N_{0.70}$ was identified by LECO with an X of 92.8%. As the reduction time increased to 180 min, the phase identified were Cu–$TiO_{0.002}C_{0.28}N_{0.717}$ with an X of 99%. From the XRD and LECO data, X_C was increasing from 1100 to 1300 °C while X_N increase was small.

At 1300 °C after 180 min with CNT, the overall increase in X_C was about 400% while the X_N only improved by 6%. The significant increase in X_C illustrates the influence of reduction temperature and time to control the degree of carburization. Similar reduction studies with graphite by Rezan et al. [2] have shown that identical X_N and X_C values were obtained above 1500 °C. By changing the carbon source to AC at 1300 °C, X was significantly affected compared to 1100 °C. The formation of Ti_3O_5 was non-existent based on XRD analysis shown in sample 6 and 8 in Table 2. The phase identified after 60 min reduction time with AC was Cu–$TiO_{0.62}C_{0.02}N_{0.36}$ (sample #6) which had an X of 69%. As the reduction time increased to 180 min, the phase identified was Cu–$TiO_{0.33}C_{0.52}N_{0.15}$ (sample #8) which had an X of 83.4%. Again, similar trend with CNT was observed in AC although the overall increase in X, X_N and X_C was lower. The kinetic curve for AC shown in Fig. 2 indicates the reduction was not completed at 180 min like CNT and graphite. In general, the TiC content in the titanium oxycarbonitride was always lower than TiN which was reflected in the X_N and X_C values.

Design of Experiment (DOE) Analysis

DOE analysis was performed on reduction data obtained in Table 2 using a factorial model. In order to differentiate the different carbon sources used in this study, BET surface area was chosen. Table 3 lists the BET surface area for the different carbon sources. For CNT, the surface area was smaller than AC due to pores in the nanotubes being occupied by metal catalyst particles [12]. Analyses of Variance (ANOVA) done by DOE are given in Table 4. From the ANOVA analysis of the data, standard deviation and correlation regression factor (R-squared) was acceptable to 95% confidence interval level. This indicated the data was good and fitted well by the factorial model.

Figure 3 shows the DOE perturbation plot of the factors used with respect to X. From the Figure, it can be concluded that increases in reduction temperature, reduction time and TiO_2: C molar ratio will increase X. On the other hand, a high carbon surface area will lower the X, and carbon surface area had the most significant influence among all the experimental factors based on its gradient change.

Table 3 Summary of BET surface properties for graphite, AC and CNT

Properties	Graphite	Activated carbon	CNT
BET surface area (m^2/g)	4.463	817.578	155.643
Pore volume (cm^3/g)	0.002	0.405	0.076
Average pore diameter (nm)	3.537	2.027	3.537

Table 4 Tabulated statistical data for ANOVA analysis of experimental data

Response	Standard deviation	R-squared
Extent of reduction	6.25	0.98
Extent of nitridation	3.22	0.99
Carbon consumed	6.69	0.96

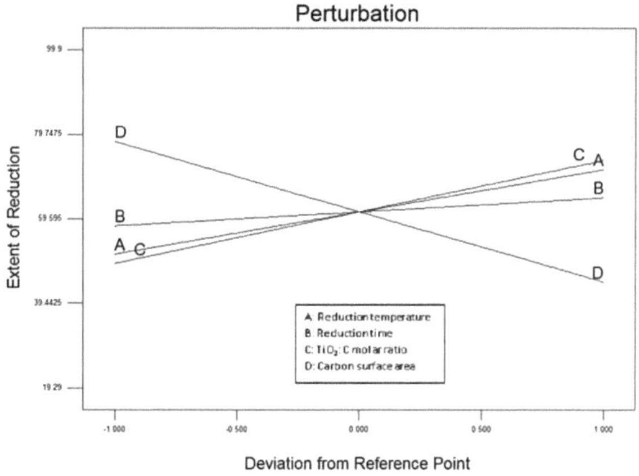

Fig. 3 DOE perturbation plot for X with respect to reduction temperature, reduction time, molar ratio and carbon surface area

This observed trend was supported by work done by White et al. [1] which showed that the reactivity of carbon was inversely proportional to its surface area.

Figure 4 shows the DOE perturbation plot for X_N. Reduction temperature, reduction time and molar ratio affected X_N positively. When these factors were increased, X_N will increase proportionately. From the Figure, an increase in carbon surface area will lower the X_N. However, reduction temperature had a lower significant impact compared to carbon surface area. The gradient for the perturbation plot for both factors were compared and carbon surface area had a higher negative gradient. From the literature, work done by Jha and Yoon [5] showed that X_N was affected substantially by particle size and surface area.

Comparing CNT and AC, it was found CNT had the highest impact on X, X_N and X_C. This was shown by the data in Table 2. Figure 5 shows a 3D surface plot of X with respect to reduction temperature and time. In the Figure, the interaction of reduction temperature and time indicates that the highest X can be achieved at 1300 °C for 180 min when the molar ratio was 1:4 with CNT carbon source. The lowest X was achieved when the temperature was at 1100 °C at 60 min. In general, CNT carbon source gave a high degree of reduction with increasing degree of

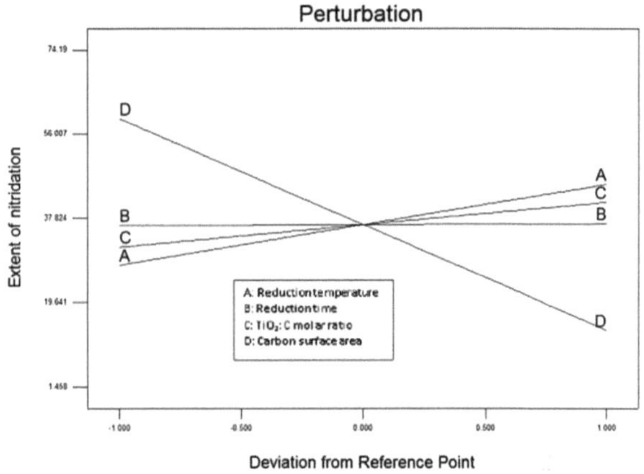

Fig. 4 DOE perturbation plot for X_N with respect to reduction temperature, reduction time, molar ratio and carbon surface area

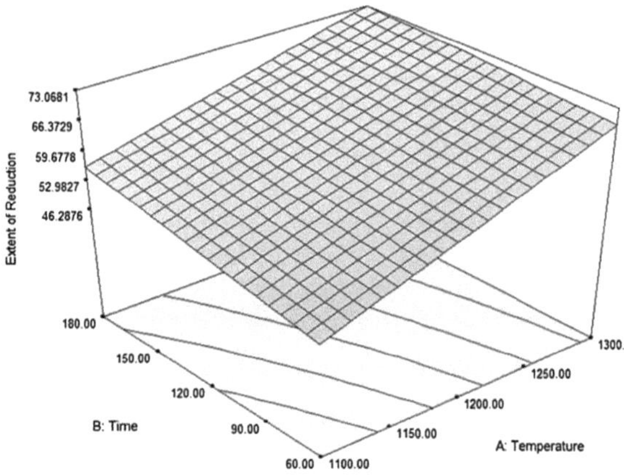

Fig. 5 3D surface plot of X with respect to time and temperature. Cu–TiO_2 to carbon molar ratio of 1:4 with CNT

nitridation and carburization. If graphite carbon source was used, X_N and X_C were lower. The high degree of reduction, nitridation and carburization was attributed to metal catalyst used for CNT synthesis. XRD analysis of the CNT powder used in this study indicated the presence of Mg catalyst [13].

Discussion

In synthesis of copper doped titanium oxycarbide and oxycarbonitride by carbothermal reduction, titania was reduced through the Magneli phases to Ti_3O_5 and Ti_2O_3, and then to TiO_x–TiC_y or TiO–TiN_z solid solution [14]. However, Ti_2O_3 was not detected by XRD in the process of formation of titanium oxycarbonitride [2]. The formation of Ti_2O_3 was detected when N_2 gas atmosphere was absent [15]. When reduction was performed in 100 vol. % N_2 atmosphere, only Ti_3O_5 phase was the intermediate suboxide detected between 1100 and 1300 °C [16]. The slow transformation of $Ti_3O_5 \rightarrow Ti_2O_3$ in H_2 rich gas atmosphere shows the importance of nitridation in converting Ti_3O_5 to $TiO_xC_yN_z$ instead of TiO_xC_{1-x}. Studies by Rezan et al. [2] have shown the slow conversion of $Ti_3O_5 \rightarrow Ti_2O_3$ can be achieved by addition of high concentration of Carbon Monoxide (CO) gas in H_2–N_2 gas atmosphere.

Reduction of TiO_2 to Ti_3O_5 can be presented by Reaction (1):

$$3TiO_2 + C = Ti_3O_5 + CO, \quad \Delta G_1^\circ = 279.154 - 0.202142T(\text{kJ/mol}) \tag{1}$$

Titanium oxycarbonitride (solid TiO–TiC–TiN solution) formation proceeds through Reactions (2)–(4):

$$Ti_3O_5 + 2C = 3[TiO] + 2CO, \quad \Delta G_2^\circ = 599.026 - 0.325492T(\text{kJ/mol}) \tag{2}$$

$$\begin{aligned} Ti_3O_5 + 8C = 3[TiC] + 5CO, \\ \Delta G_3^\circ = 1279.102 - 0.794764T(\text{kJ/mol})(\text{kJ/mol}) \end{aligned} \tag{3}$$

$$Ti_3O_5 + 5C + 1.5N_2 = 3[TiN] + 5CO, \Delta G_4^\circ = 836.368 - 0.560122T(\text{kJ/mol}) \tag{4}$$

The standard Gibbs free energy for Reactions (1)–(4) was calculated using NIST-JANAF Thermochemical Tables in the temperature range of 1027–1527 °C [17]. The equilibrium partial pressure of CO (P_{CO}) for these reactions was presented in Fig. 6. From the equilibrium CO partial pressure, the most favourable Reaction was (1) followed by Reaction (4) due to high partial pressure needed to sustain the reactions. During carbothermal reduction and nitridation, the TiN fraction was high compared to TiC fraction. This was evident with reduction of Cu–TiO_2 with CNT.

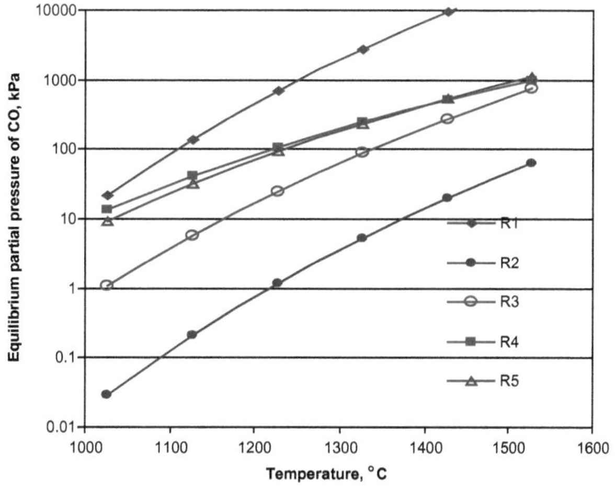

Fig. 6 Equilibrium partial pressure of CO for reactions (1)–(5)

X_C at 1300 °C after 180 min were 28% compared to X_N which was 72%. This explains why TiN is the major component in titanium oxycarbonitride. Formation of TiC by Reaction (3) is thermodynamically more favourable than formation of TiO by Reaction (2) based on equilibrium partial pressure shown in Fig. 6. The low TiC_y fraction in Cu–$TiO_xC_yN_z$ can be explained by diffusion constraint of carbon transfer during the carburization process. Since the conversion of TiO → TiC_y is a solid-solid process, high temperature and time was needed. Studies by Sharifah A. et al. have shown $FeCl_3$ catalyst can enhanced this process of carburization [3]. The presence of Copper has also been linked to a similar role as $FeCl_3$ catalyst in promoting the Boudouard Reaction (9).

Reactions (2)–(4) can be summarised as Reaction (5):

$$Ti_3O_5 + (5-3x+3y)C + (3/2)zN_2 = 3(x[TiO] + y[TiC] + z[TiN])(x) + (5-3x)CO,$$
$$\Delta G_5 = x\Delta G_2^\circ + y\Delta G_3^\circ + z\Delta G_4^\circ + 3RT\ln\left[(a_{TiO})^x(a_{TiC})^y(a_{TiN})^z(P_{CO})^{(5/3-x)}/(P_{N2})^{z/2})\right] \tag{5}$$

where a_i represents the activity of TiO, TiC and TiN in titanium oxycarbonitride.

The activities of the species are proportional to their composition measured experimentally by LECO through Raoult law [18]. The equilibrium partial pressure of CO for Reaction (5) can be calculated for an ideal TiO–TiC–TiN solution containing 15 mol% TiO, 10 mol% TiC and 75 mol% TiN ($TiO_{0.15}C_{0.10}N_{0.75}$), and the partial pressure of nitrogen 50 kPa, as also presented in Fig. 7. The partial pressure of CO from Reaction (5) is close to that from Reaction (4). Formation of titanium oxycarbonitride is spontaneous when CO partial pressure is less than 10 kPa for Reaction (4).

In the final stage of reduction when titanium oxides are consumed, the reduction progresses slowly with conversion of TiO and TiC to TiN in the solid solution:

$$[TiO] + C + 1/2N_2 = [TiN] + CO, \quad \Delta G_6^\circ = 79.114 - 0.078210T(kJ) \tag{6}$$

$$[TiO] + [TiC] + N_2 = 2[TiN] + CO, \quad \Delta G_7^\circ = -68.464 + 0.000004T(kJ) \tag{7}$$

The formation of TiN_z and TiC_y was a slow conversion process at this stage. Since carbon is in excess inside the system, both carburization and nitridation occur. The latter occurring faster below 1400 °C while carburization is accelerated above 1400 °C.

Carbothermal reduction of solid oxides proceeds through the gas phase mainly through the evolution of CO, CO_2 and CH_4. In inert atmosphere, reduction of titanium oxide is reduced by CO with formation of CO_2. In inert atmosphere, CH_4 will not be generated.

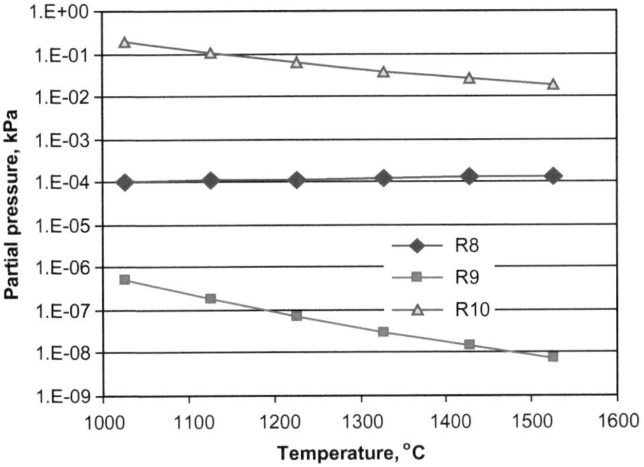

Fig. 7 Equilibrium partial pressure of CO_2 and CH_4 for reactions (8)–(10) calculated for an ideal $TiO_{0.15}C_{0.10}N_{0.75}$ solution with partial pressures of H_2, CO and N_2 at 50, 1 and 50 kPa respectively

$$\begin{aligned} &Ti_3O_5 + (5 - 3x + 6y)CO + (1.5z)N_2 = 3(x[TiO] + y[TiC] + z[TiN]) \\ &+ (5 - 3x + 3y)CO_2, \Delta G_8 = x(269.200 + 0.015668T) \\ &+ y(-40.202 + 0.569876T) + z(11.803 + 0.292778T) \\ &+ 3RT \ln \frac{(a_{TiO})^x (a_{TiC})^y (a_{TiN})^z P_{CO_2}^{5/3-x+y}}{P_{CO}^{5/3-x+2y} P_{N_2}^{z/2}} \end{aligned} \tag{8}$$

CO_2 is then converted to CO by Boudouard reaction (9):

$$CO_2 + C = 2CO, \quad \Delta G_9^\circ = -21.118 + 0.189648T(\text{kJ}) \tag{9}$$

In reduction in hydrogen containing gas atmosphere, titanium oxides can be reduced by methane which is formed by Reaction (11) by hydrogen with carbon:

$$C + 2H_2 = CH_4, \quad \Delta G_{10}^\circ = -92.503 + 0.111638T(\text{kJ/mol}) \tag{10}$$

$$\begin{aligned} &Ti_3O_5 + (5 - 3x + 3y)CH_4 + (1.5z)N_2 \\ &= 3(x[TiO] + y[TiC] + z[TiN]) + (5 - 3x)CO + 2(5 - 3x + 3y)H_2 \end{aligned} \tag{11}$$

Figure 8 shows the effect of varying the N_2 partial pressure (P_{N_2}) on the equilibrium P_{CO}. From the Figure, as P_{N_2} increases, the P_{CO} also increases which makes Reaction (8) more favourable with increasing temperature.

Figure 9 is the Temperature-Pressure-Pressure (TPP) diagram simulated in Outokumpu HSC Chemistry software. In this simulation, the partial pressure of oxygen (P_{O_2}) was set to be 1×10^{-20}atm The low P_{O_2} was needed for the conversion of $TiO_2 \rightarrow TiN$. During carbothermal reduction and nitridation of

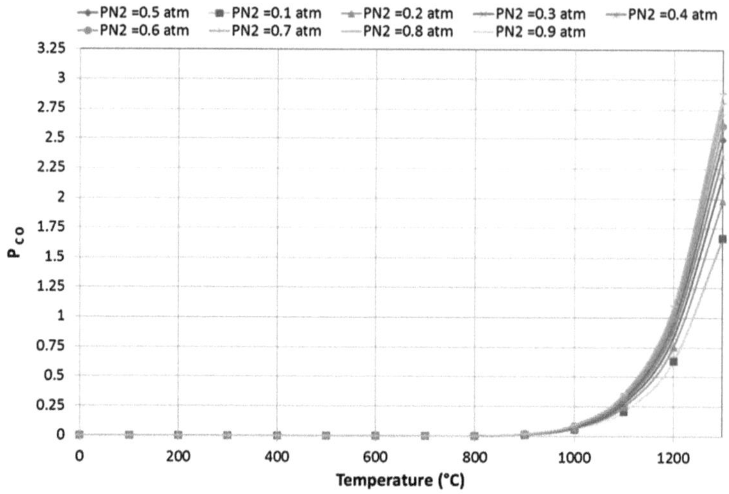

Fig. 8 Partial pressure of CO at different temperature for different P_{N_2}

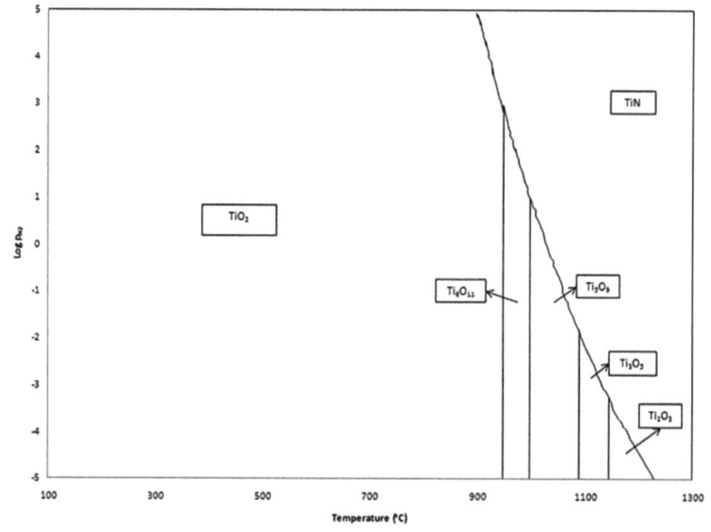

Fig. 9 Predominance diagram for Ti–O–N system at constant oxygen partial pressure

Cu–TiO_2, the equilibrium reaction P_{CO} needed to be high like in Reaction (3–5) shown in Fig. 6. To achieve this equilibrium pressure for the conversion into oxycarbonitride, P_{O_2} was in the region of 1×10^{-20}atm. From Fig. 9, when the P_{N_2} was high, Cu–TiO_2 can be transform directly to Cu–TiN at lower temperatures, without going through the formation of suboxides. Since the experimental thermodynamic data for $TiO_xC_yN_z$ was not available in Outokumpu HSC Chemistry software database, ideal conditions are assumed to be present for the formation of Cu–TiN. In reality, conditions are non-ideal and formation of Cu–$TiO_xC_yN_z$ was proven by XRD and LECO analysis. At low P_{N_2}, the conversion of Cu–TiO_2 → Cu–$TiO_xC_yN_z$ was not favourable. From this simulation, it was critical to maintain a high P_{N_2} of more than 50 kPa (50 vol. %) to enable the conversion to oxycarbonitride.

Conclusions

Carbothermal reduction of Cu–TiO_2 to Cu–$TiO_xC_yN_z$ was studied in H_2–N_2 gas atmospheres. Increasing temperature and time accelerated the reduction. By using a more reactive carbon source such as CNT as compared to AC, the X, X_N and X_C were more significant. With CNT and hydrogen-nitrogen gas atmosphere, the conversion of Cu–TiO_2 into Cu–$TiO_xC_yN_z$ at 1300 °C was completed in 60 min; it decreased to 20 min at 1300 °C. By changing the carbon source to AC, the reduction was not completed at 1100 °C after 180 min. At 1300 °C, the complete conversion to Cu–$TiO_xC_yN_z$ occurred after 120 min although X was only 83% which was lower than 99% with CNT present. Increasing reduction temperature or time increased TiN_z fraction in the synthesized oxycarbonitride. Increasing temperature resulted in formation of Cu–$TiO_xC_yN_z$ with higher TiC_y content. The utilization of CNT was a promising carbon source to enhance nitridation and carburization at lower temperatures.

Acknowledgements The authors wish to thank Universiti Sains Malaysia (USM) and Ministry of Higher Education (MOHE) of Malaysia for supporting this work. This research was supported by three grants, MOHE Fundamental Research Grant Scheme (FRGS) grants no: 203/PBAHAN/6071230, FRGS grant no: 203/PBAHAN/6071261 and USM Research University Individual (RUI) Grant no: 1001/PBAHAN/814273. Special thanks to USM technician Mr. Shahrul and Mr. Syafiq for supporting the author's experimental work.

References

1. G.V. White, K.J.D. Mackenzie, J.H. Johnston, Carbothermal synthesis of titanium nitride: part 1. J. Mater. Sci. **27**(16), 4287–4293 (1992)
2. S.A. Rezan, G. Zhang, O. Ostrovski, Carbothermal reduction and nitridation of titanium dioxide in a H_2–N_2 gas mixture. J. Am. Ceram. Soc. **94**(11), 3804–3811 (2011)
3. A.S.S. Sharifah, H.S. Saidin, N. Baharun, S.A. Rezan, H. Hashim, Microstructural study of reduced malaysian ilmenite by carbothermal reduction and nitridation in nitrogen atmosphere. Adv. Mater. Res. **858**, 265–271 (2014)
4. S.K. Gupta, V. Rajarumar, P. Grieveson, Reduction behavior of ilmenite with carbon at 1240 °C. Metall. Mater. Trans. B **26B**(2), 401–404 (1995)

5. A. Jha, S.J. Yoon, Formation of titanium carbonitride phases via the reduction of TiO_2 with carbon in the presence of nitrogen. J. Mater. Sci. **34**, 307–322 (1999)
6. K. Maeda, K. Domen, Oxynitride materials for solar water splitting. MRS Bull. **36**(01), 25–31 (2011)
7. S.A. Rezan, A. Adipuri, G. Zhang, O. Ostrovski. Carbothermal reduction and nitridation of ilmenite concentrates and chlorination of the reduced samples. in *XXV International Mineral Processing Congress (IMPC 2010)* (Brisbane, Australia, 2010)
8. A. Adipuri, G. Zhang, O. Ostrovski, Chlorination of titanium oxycarbonitride produced by carbothermal nitridation of rutile. Ind. Eng. Chem. Res. **48**(2), 779–787 (2008)
9. A. Adipuri, Chlorination of titanium oxycarbide and oxycarbonitride, in *School of Material Science and Engineering*, Phd Thesis (University New South Wales: Sydney, Australia, 2009), p. 177
10. C.W. Lai, S. Sreekantan, Visible light photoelectrochemical response of copper deposited titanium dioxide nanotubes. Eur. Phys. J. Appl. Phys. **59**(2), 20402 (2012)
11. A. Roine, Outokumpu HSC chemistry v5.11, in *Chemical Reaction and Equilibrium Software with Extensive Thermochemical Database* (2002), Outokompu
12. Y.J. Hui, Synthesis of copper doped titanium oxycarbonitride from copper doped titanium oxide by carbothermal reduction in H_2–N_2 Atmosphere, in *School Of Materials And Mineral Resources Engineering* (Universiti Sains Malaysia, Nibong Tebal, Penang, 2014), p. 106
13. W.-W. Liu, A. Aziz, S.-P. Chai, A.R. Mohamed, U. Hashim, Synthesis of single-walled carbon nanotubes: effects of active metals, catalyst supports, and metal loading percentage. J. Nanomater. **2013**, 8 (2013)
14. S. A. Rezan, G. Zhang, O. Ostrovski, Synthesis of titanium oxycarbonitride by carbothermal reduction of titania innitrogen containing gas mixtures. in *Proceeding of 11th World Conference Titanium*, 2007. Kyoto, Japan
15. M.A.R. Dewan, G. Zhang, O. Ostrovski, Carbothermal reduction of titania in different gas atmospheres. Metall. Mater. Trans. B **40**(1), 62–69 (2009)
16. S.A. Rezan, G. Zhang, O. Ostrovski, Effect of gas atmosphere on carbothermal reduction and nitridation of titanium dioxide. Metall. Mater. Trans. B: Process Metall. Mater. Process. Sci. **43**(1), 73–81 (2012)
17. M.W.Chase, in *NIST-JANAF Thermochemical Tables*. (Journal of Physical Chemistry Reference Data, Washington DC, 1988)
18. C.B. Alcock, in *Principles of Pyrometallurgy* (Academic Press, Canada, 1988), p. 348

Cohering Behavior of Iron Ore Pellet Powder in Kiln by a Novel Natural Stacking Method

Xiaoliang Liu, Yongbin Yang, Yan Zhang, Qian Li, Bin Xu and Tao Jiang

Abstract The cohering behavior of iron ore pellet powder in kiln cannot be reflected entirely through the conventional briquetting method, so a novel natural stacking method was proposed in this paper. The results showed that in the two methods the cohering strength of powders from different iron ore pellets (numbered A, B and C) rose with the increase of roasting time and temperature and presented the same result, i.e., C > B > A. Comparatively, the natural stacking powders obtained much lower cohering strengths under the same roasting condition. The structure analyses of ring from the kiln tail and cohering briquette by the natural stacking method indicated that the occurrence of crystallized particles of them were extremely similar. Therefore, the initial information of pellet powder's cohering behavior could be provided by the natural stacking method, improving the methods to study the actual ringing process and mechanism of pellet powders in kiln.

Keywords Kiln ringing · Cohering behavior · Iron ore pellet powder · Briquetting method · Natural stacking method

Introduction

Currently, grate-kiln process, one of the most important processes, has been extensively used to produce oxidized pellets in China due to a series of advantages including simplified operation, easy control and large production capacity, etc. The main obstacle, i.e., ringing in rotary kiln, however, is apt to occur in the production process, especially when there are no appropriate preventive measures or handling without delay. Once ringing occurs, it will result in serious yield reduction and even

X. Liu · Y. Yang (✉) · Y. Zhang (✉) · Q. Li · B. Xu · T. Jiang
School of Minerals Processing and Bioengineering, Central South University, Changsha 410083, Hunan, People's Republic of China
e-mail: ybyangcsu@126.com

Y. Zhang
e-mail: zhangyancsu@126.com

J.-Y. Hwang et al. (eds.), *8th International Symposium on High-Temperature Metallurgical Processing*, The Minerals, Metals & Materials Series,
DOI 10.1007/978-3-319-51340-9_25

stop production of oxidized pellets as well as plenty of human maintenance. Severe cases can even damage the kiln liner and the refractory of rotating annular cooler [1–4].

During the actual production process in kiln, the iron ore pellet powders firstly adsorb onto the kiln wall and are stationary relative to the kiln wall, then under relatively high temperatures, a certain consolidation of the pellet powders is caused, and namely, the initial ringing occurs. Afterwards, the new pellet powders will cohere onto the formed ring. The initial ring thickens and spreads generally and finally develops into the common ring problem. The ringing degree in different locations of the kiln varies because of the differences of roasting conditions. So studying the cohering behavior of pellet powder under different roasting conditions will provide valuable theoretical foundation to better understand and effectively control and prevent ringing in kiln [5–10].

The conventional briquetting method has been used to research most of the cohering behaviors of pellet powders in kiln. Namely, under certain pressures in a mould, the powders are firstly made into briquettes which have fixed definite shape and mechanical strength. Then during roasting the briquettes consolidate, and finally the consolidation strength of cooled briquettes are tested and used to measure the cohering characteristics of powders [11, 12]. In the initial stage of kiln ringing, however, nearly no pressure is exerted on the powder. In addition, the cohering characteristic of powder themselves and that between powder and kiln wall are most relevant to ringing. As a consequence, the original state of pellet powder before ringing cannot be presented by the briquetting method, and hence, the actual ringing of pellet powder will not be completely shown by it.

On the basis of the above, comparing with the briquetting method, a new natural stacking method [13] is adopted to investigate the cohering behavior of pellet powder to reveal the ringing mechanism in its entirety.

Experimental

Raw Materials

The iron ore pellet powders (including three different types which were numbered A, B and C, respectively) were all provided from an Iron & Steel Group of China. The chemical compositions of them are shown in Table 1.

Table 1 Chemical compositions of the iron ore pellet powders/wt%

Iron ore pellet powder	Fe	FeO	Fe_2O_3	SiO_2	Al_2O_3	CaO	MgO	MnO	K_2O	Na_2O
A	65.88	4.67	88.82	2.06	0.42	0.82	0.32	0.11	0.030	0.10
B	65.29	4.86	87.87	2.50	0.76	0.75	0.32	0.11	0.035	0.10
C	65.33	5.72	86.97	2.46	0.73	0.86	0.32	0.11	0.035	0.10

It is indicated from Table 1 that all the pellet powders were oxidized sufficiently because the content of Fe_2O_3 are all relatively high (>85%). Additionally, according to the content of FeO, the degree of oxidation for powder A was slightly higher than that for powder B, but both of them were much higher than powder C's. As for the gangue like SiO_2, Al_2O_3, CaO and MgO, etc., their contents in powder B and C were higher than in powder A.

Methods

As seen from Fig. 1, the horizontal tube furnace was used to conduct the roasting tests, and the roasting temperature was calibrated before each test.

For the natural stacking method, the iron ore pellet powder was naturally stacked in an inside diameter 10 mm × depth 10 mm alumina crucible to be just full and parallel to the feed opening. Then 8 crucibles were closely put together in the center of a length 60 mm × width 30 mm × depth 20 mm porcelain boat. After that, the porcelain boat was placed into the center of the refractory furnace tube where was directly beneath the thermocouple to make sure the temperature profile of these crucibles were the same. After oxidative roasting under an air atmosphere and at a certain temperature, the strength of cooled cohering briquettes was determined using the compression strength testing machine (ZQYC-10000NIII). The point to emphasize here is that the cohering briquettes are easily removed from the molds after a minute or less of vibration owning to the shrinkage property of powder during the roasting.

Comparatively, the briquetting method was that firstly the pellet powder was mixed well with water whose addition was around 7–8% (mass fraction). Then the wet materials were compressed into Φ10 × 10 mm briquettes through an inside diameter 10 mm × depth 50 mm cylindrical mould. In the same way, the cohering strength of cooled briquettes was obtained as the natural stacking method.

The ringing degree can be measured by the cohering briquette strength, and the higher of the strength, ringing in kiln occurs more easily and seriously.

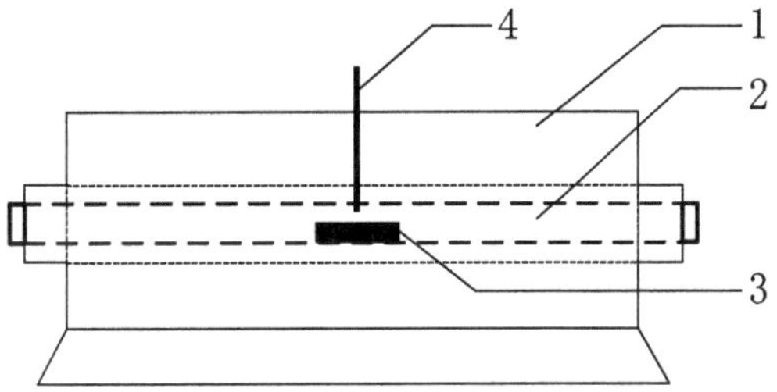

Fig. 1 Schematic diagram of the experimental facility for oxidative roasting. *1* Horizontal tube furnace; *2* refractory furnace tube; *3* porcelain boat; *4* thermocouple

Results and Discussion

Effect of Roasting Time on Cohering Behavior of Powders

Under the air atmosphere and at the roasting temperature of 1250 °C, the effects of roasting time on cohering behavior of powders (A, B and C) were investigated by the briquetting method and the natural stacking method. The results were presented in Fig. 2.

Using the briquetting method, it is indicated from Fig. 2a that the variation of cohering strength of different powders with roasting time was similar to each other. At the early stage of roasting within around 10 min, the cohering strength all increased dramatically from 250–500 N/Briquette to 900–1100 N/Briquette. Then with the increase of time to 30 min, the cohering strength experienced very slow growth to 1050–1275 N/Briquette. When the natural stacking method was used, Fig. 2b shows that similar cohering behavior of the powders was also presented. All the cohering strength rose with the increase of time, but during the roasting of within 30 min it was always relatively fast, increasing noticeably from 0–80 N/Briquette to 450–630 N/Briquette.

Comparatively, by the two methods, the cohering strength of powders were both in the sequence of C > B > A. This mainly lies in the differences of oxidation degree among the powders. The more insufficiently oxidized the powders, the larger the content of FeO. But FeO can react with SiO_2 to generate a series of eutectic mixtures in the form of $4FeO{\cdot}SiO_2$, $2FeO{\cdot}SiO_2$ and $FeO{\cdot}SiO_2$, etc., whose fusion temperature are in the range of 1138–1180 °C. Moreover, other dystectic component can be dissolved by the eutectic mixtures, thus decreasing the fusion temperature of powders. In other words, the higher the oxidation degree of powders, the less likely the cohering, i.e., ringing of powders. This is in agreement with the results of Table 1 that the oxidation degree of powders followed in the order of

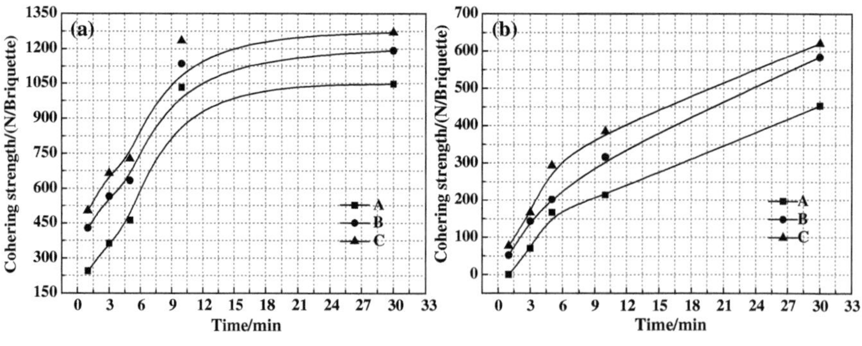

Fig. 2 Effect of roasting time on cohering strength of powders by **a** the briquetting method and **b** the natural stacking method

C > B > A. So during the practical firing in kiln, the primary reason for ringing in kiln is the undesirable effect of preheating.

Furthermore, the cohering of compacted powders could be finished basically within around 20 min whilst it seems that the natural stacking powders continued to cohere in a relatively long time (>30 min). And the complete cohering strengths of compacted powders were obviously higher than those of natural stacking powders after roasting 30 min. As for the powders in a natural stacking state, it is normal for them to first shrink naturally in a certain time. Then the cohering of powders occurs on the particles which approach each other to continuously increase their own cohering strength. In comparison with the compacted powders, a firmly cohering could not occur on the natural stacking powders in a relatively short time, but the cohering degree was still not negligible. This is actually the initial ringing of powders, and the initial ringing degrees varied with different powders.

Effect of Roasting Temperature on Cohering Behavior of Powders

Likewise, under the air atmosphere and after roasting 10 min, the effects of roasting temperature on cohering behavior of powders were researched by the two methods, and the results were indicated in Fig. 3.

As shown from Fig. 3a, the cohering behaviors of compacted powders resembled each other, and with the increase of temperature the cohering strengths of different powders all rose constantly. Specifically, as the temperature went up from 1200 to 1250 °C, the strengths increased from 100–400 N/Briquette to 950–1200 N/Briquette. There was a slow increase of cohering strength from 1250 to 1280 °C, but it became fast again like the initial stage when the temperature exceeded 1280 °C. Similarly, it can be seen from Fig. 3b that rising from 1200 to

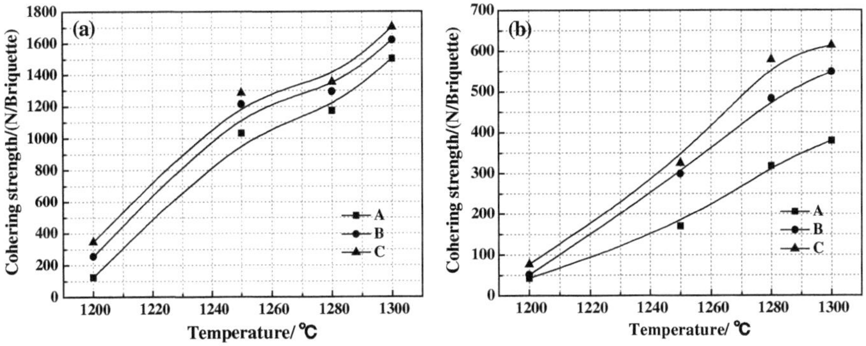

Fig. 3 Effect of roasting temperature on cohering strength of powders by **a** the briquetting method and **b** the natural stacking method

1300 °C the natural stacking powders cohered to raise their own strengths nearly linearly. At 1250 °C, the cohering strength could amount to around 200–350 N/Briquette which was much lower than that obtained by the briquetting method but could not be neglected in light of initial ringing.

Also, the cohering behaviors of powders by the two methods were both similar with the results obtained in section "Effect of Roasting Time on Cohering Behavior of Powders", namely, the cohering strengths of powders followed in the succession of C > B > A. Due to under pressure, the growth rates and values of cohering strengths in the briquetting method are both much larger comparing with the natural stacking method. Nevertheless, ringing in kiln possibly starts with the unremarkable cohering of powders which are in a natural stacking state and can stick to the kiln wall.

Structure Analysis of the Ring Formed in Kiln and the Cohering Briquette Formed in Laboratory

Based on the above analysis, we know that the cohering behavior of powder in kiln varies with the contact state of particles with each other. When the pellet powders just come into the kiln from the tail, less pressure from the pellets is exerted on the powders. So in the kiln tail powders are more likely to exist in a natural state, and thus the initial ringing of powder can be reflected better. In this section, the microstructures of ring formed in the kiln tail and cohering briquette formed by the natural stacking method in laboratory were primarily studied to reveal the actual ringing process and its mechanism more fully. The relevant results can be seen from Figs. 4 and 5.

The structures of ring formed in the kiln tail are shown in Fig. 4. It can be indicated from Fig. 4a that there was no metallic luster and no entrained pellets, and the ring was presented in a layered structure. In addition, as seen from the SEM

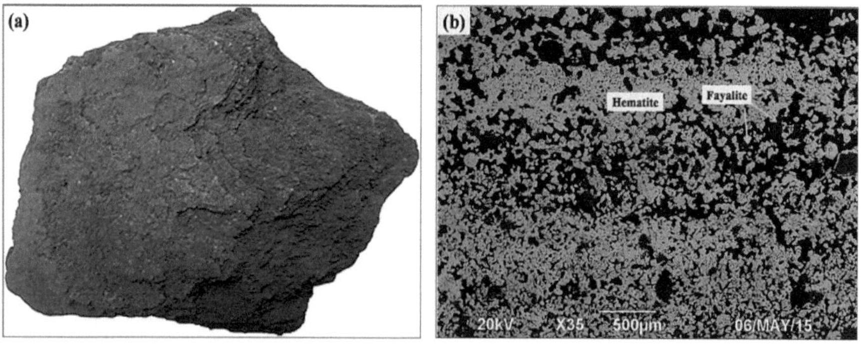

Fig. 4 Structures of the ring formed in the kiln tail: **a** macrostructure and **b** SEM image

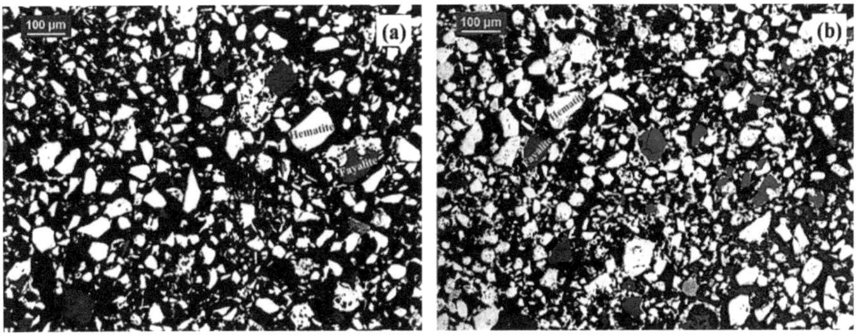

Fig. 5 Microstructures of **a** the loose layer of ring and **b** the cohering briquette obtained by the natural stacking method

image of the ring (see Fig. 4b) that the layered structure included a compact layer and a loose layer which distributed alternatively. In the loose layer, the crystalline particles of hematite (Fe_2O_3) and fayalite ($2FeO{\cdot}SiO_2$) were mainly dissipated separately in single particles. In contrast, particles of hematite and fayalite in the compact layer interconnected with each other and the boundaries between them became rather obscure. It had also been shown by other researchers that the inlayer structure of ring was rather compact while the out layer was relatively loose [14, 15]. This is largely because that the powder first cohered in a relatively loose state, and under the pressure from rolling pellets the structure of the initially formed ring would become compact. It is the approximately natural stacking state of powder that the loose layer structure forms in the outside part of ring. That is to say, at the early stage the powder rings probably under no external pressure.

By means of the optical microscopy, microstructures of the ring's loose layer and the cohering briquette obtained after roasting 10 min at 1250 °C by the natural stacking method were further analyzed, which are presented in the following Fig. 5.

It is clearly demonstrated from Fig. 5 that their microstructures were extremely similar and the crystalline particles (mainly hematite and fayalite) were both distributed almost individually. This further manifests that the natural stacking method can provide information about the actual ringing of powders in kiln in the initial stage.

To sum up, the ringing mechanism could be improved using the new natural stacking method. Namely, the iron ore pellet powders in a loose and natural accumulation state firstly adsorb onto the kiln wall and then cohere in a certain time to generate an initial ring which presents a loose structure and certain cohering strength. After that, under the continuous pressing of the pellets, the formed loose ring is pressed to be compact gradually. As a result, new pellet powders will constantly cohere onto the formed compact ring which spreads and thickens. Ultimately, a ring whose inner structure is compact and outer structure is loose is developed, and the cohering strength of the whole ring is rather high, namely, the common ring problem occurs.

Therefore, the conventional briquetting method reflects the cohering behavior of powder particles under certain pressures whereas the natural stacking method provides information on the cohering behavior of powders in a loose or natural stacking state, i.e., the initial ringing. More importantly, the natural stacking method not only supplies a novel research technique for studying the cohering characteristics of powders, but also more comprehensively, associating with the briquetting method, elucidates and improves the actual ringing behavior and mechanism of powders in kiln.

Conclusions

(1) Using the briquetting method and the natural stacking method, the cohering strength of different iron ore pellet powders all increased with the increase of roasting time and temperature. And under the same roasting conditions, the cohering strength showed the same result, i.e., C > B > A. In comparison with the briquetting method, the natural stacking powder presented much lower cohering strength, but it just means the occurrence of initial ringing.
(2) According to the structure analyses of the ring in the kiln tail and the cohering briquette obtained by the natural stacking method, it is found that the occurrence of crystallized particles were extremely similar, confirming the validity and applicability of the natural stacking method.
(3) The natural stacking method is novel and feasible, and can provide the initial cohering information of pellet powders in kiln. In association with the briquetting method, the actual ringing process and mechanism of powders in kiln can be reflected more fully and effectively by the natural stacking method.

Acknowledgements This work was financially supported by National Natural Science Foundation of China (Grant No. 51574284), National Science Foundation for Distinguished Young Scholars of China (Grant No. 51504293) and Science and Technology Program of Yunnan (No. 2013IB020).

References

1. D. Gong, H. Jiang, Prospects of the development of grate-kiln for oxidized pellets production in China. Sinter. Pelletizing **2**, 17–23 (1988)
2. S. Chen, About the grate-kiln pelletizing process and production. Sinter. Pelletizing **31**(4), 11–13 (2006)
3. L. Zhang, Technological progress of grate-kiln oxide pellet production system of Tieying Company. Sinter. Pelletizing **35**(4), 40–43 (2010)
4. Y. Xu, Development and application on grate-kiln pellet process. China Metall. **15**(4), 17–20 (2005)
5. J. Fu, G. Bai, Mechanism of ringing in grate-kiln. in *Proceedings of the 2004 China Annual Meeting on the Technology Exchange of Sintering and Pelletizing*, pp. 35–38

6. X. Yang et al., Mechanism study and countermeasures on ring formation of rotary kiln used to produce oxidized pellets in KISC. Sinter. Pelletizing **34**(3), 21–24 (2009)
7. J. Li et al., Analysis and prevention of kiln-ringing of Hangang oxidized pelletizing production line. Sinter. Pelletizing **32**(3), 55–58 (2007)
8. H. Zhang, Causes for ringing and lumping of grate-kiln for oxidized pellets and precautions. Metal Mine **7**, 59–61 (2005)
9. J. Wang, X. He, G. Sun, Analysis of rotary kiln ringing and improvement measures. Henan Metall. **2**, 37–39 (2012)
10. J. Liu et al., Study on ring-forming cause of Baosteel oxidized pellet rotary kiln. Sinte. Pelletizing **5**, 31–35 (2013)
11. X. Fan et al., Study of mechanism on ring formation in grate-kiln of acid pellet. Iron Steel **43** (3), 15–20 (2008)
12. T. Jiang et al., Forming mechanism of rings in rotary-kiln for oxidized pellet. J. Iron. Steel Res. Int. **16**, 292–297 (2009)
13. Y. Yang et al., A measurement method for the ringing characteristics of scrap powders in oxidized pellet kiln. China Patent CN201410630761.5 (2015)
14. J. Si et al., Characteristic of microstructure for ring-forming material of oxidized pellet rotary kiln. Res. Iron Steel **1**, 5–8 (2014)
15. J. Nie et al., Characteristics of the pellet rotary kiln ring. J. Wuhan Univ. Sci. Technol. **33**(5), 527–530 (2010)

Direct-to-Blister Copper Smelting with the ISASMELT™ Process

Paul Voigt, Alistair Burrows, Michael Somerville and Chunlin Chen

Abstract To date, many types of primary concentrate have been smelted using the ISASMELT™ technology, with annual smelting of copper concentrate alone exceeding 7 million tonnes. Most of this material could best be described as chalcopyrite concentrate, owing to the predominance of copper attributable to this single mineral. Chalcopyrite concentrates lend themselves readily to smelting with a fayalite slag, and producing a matte containing 55–65 wt% Cu. However, the ISASMELT™ technology also stands ready to meet future copper demand through the treatment of new types of concentrates containing relatively low iron and high silica levels. Various slag systems have been considered for direct-to-blister smelting. They have been evaluated computationally, using the MPE thermodynamic modelling package, and practically, at pilot scale. The lime-silica-iron oxide slag was found to have the lowest slag make, lower dissolved copper and hence lowest copper losses. Based on the outcomes of the modelling and pilot scale testwork a future smelter flowsheet has been defined, comprising a single ISASMELT™ direct-to-blister furnace, an electric slag cleaning furnace, and appropriately designed anode furnaces.

Keywords Single-stage copper making · Concentrates · Blister copper · TSL smelting · MPE · Direct copper · ISASMELT

P. Voigt (✉)
Glencore Technology, Level 10, 160 Ann Street, Brisbane, QLD 4000, Australia
e-mail: paul.voigt@glencore.com.au

A. Burrows
Kazzinc Ltd., 1 Promyshlennaya Street, Ust-Kamenogorsk 070003, Kazakhstan
e-mail: aburrows@kazzinc.com

M. Somerville · C. Chen
CSIRO Mineral Resources, Private Bag 10, Clayton South, VIC 3169, Australia
e-mail: Michael.Somerville@csiro.au

C. Chen
e-mail: Chunlin.Chen@csiro.au

J.-Y. Hwang et al. (eds.), *8th International Symposium on High-Temperature Metallurgical Processing*, The Minerals, Metals & Materials Series,
DOI 10.1007/978-3-319-51340-9_26

Introduction

The copper ISASMELT™ process has evolved over more than a quarter of a century and is considered mainstream, with single furnaces producing more than 340,000 tpa of copper in matte. Over the last 30 years the ISASMELT™ technology has been developed into a process suited to various roles in the non-ferrous smelting industry including primary and secondary copper smelting, and primary and secondary lead smelting. Furnace capacities have increased more than 30-fold from the first demonstration plant, to the current 230 tph at Kansanshi's copper ISASMELT™ furnace [1]. More than 340,000 tonnes per annum of cathode copper equivalent have been produced in the Vedanta ISASMELT™ furnace in India. There is a realistic expectation for still higher throughputs in the future.

The rate of adoption of the ISASMELT™ furnace has been highest in the primary copper smelting industry with eleven furnaces in operation today. Success in this field has been due to the high rates of oxygen transfer that are possible for a submerged lance operating with an iron silicate slag. This permits furnaces of small footprints to be built with a high smelting capacity. Specially optimised furnaces are then used downstream for the purpose of settling a matte-slag mixture, and providing a reservoir of matte as a buffer between the ISASMELT™ furnace and the downstream converting furnace(s). The performance and generic operating statistics of such plants have been documented previously [2, 3].

One of the ISASMELT™ technology applications that did not find a licensee has been the direct-to-blister copper smelting process. First developed in 1996 [4] the direct-to-blister ISASMELT™ furnace was envisaged as a future development for the copper industry that would provide benefits of reduced cost, reduced complexity and faster construction of greenfield smelters.

Direct-to-blister smelting is a process that becomes most economical when the total slag fall is low as a proportion of the copper produced. A low slag fall increases the "first pass" yield of blister copper and reduces the amount of slag that requires reduction. Typically it is the amount of iron in a copper concentrate that most influences the slag fall, so concentrates containing large amounts of chalcocite (Cu_2S) and bornite (Cu_5FeS_4) are more amenable to direct-to-blister smelting than concentrates containing chalcopyrite ($CuFeS_2$) as the sole copper-bearing mineral. For this reason there are relatively few potential new smelters ideally suited to direct-to-blister smelting.

Initial Challenges of Direct-to-Blister Smelting in TSL Furnace

Direct-to-blister smelting is well established in the copper industry via the Flash process that uses a suspension smelting mechanism. Unlike suspension smelting, where oxygen transfer to the concentrate occurs in the gas phase, the TSL technology relies on oxygen transfer through the molten slag phase. In the case of copper smelting this is achieved principally by the exchange reaction between the Fe^{3+}/Fe^{2+} couple. A major challenge for making a robust direct-to-blister process using a TSL furnace is therefore to identify a slag system that:

1. Has good oxygen transfer in the molten slag
2. Stays molten at the prevailing conditions of pO_2 and pS_2
3. Has a liquidus region tolerant of various oxides that may exist in the copper concentrate.

In its original conception [4], the direct-to-blister ISASMELT™ process was to use the calcium ferrite slag system. From a 20-year history (at that time) in the Mitsubishi C-Furnace, it was already well-known that the calcium ferrite slag was an excellent transporter of oxygen and stable in the conditions of blister copper production. The high solubility of magnetite (Fe_3O_4) in this slag makes it especially attractive. However, despite successful trials at the pilot scale showing promise that this slag system would be suitable, a practical problem existed that was yet to become apparent.

In 2002 Mopani Copper Mines approached the then Xstrata Technology requesting assistance for the rebuild of the Mufulira Copper Smelter using ISASMELT™ technology. Mufulira concentrate was to be a major input to the planned smelter. At that time Mufulira concentrate had the following characteristics as shown in Table 1.

Initially, the high proportion of bornite in the Mufulira concentrate seemed to invite an introduction of the ISASMELT™ direct-to-blister technology to the world. On closer inspection of the data, that did not seem like a wise choice, and was not pursued. The main impediment was that the concentrate contains a high proportion of silica among the oxide minerals in comparison with the iron contained in the sulphides. This combination of high silica and low iron prevented the use of the familiar calcium-ferrite slag system, and would have required the use of a ferrous-calcium-silicate slag instead. While there is plenty of work (for example [5, 6]) defining the liquidus boundary of this slag system, practical experience on ISASMELT™ pilot furnaces gave a strong indication that such a slag system would

Table 1 Mufulira smelter concentrate analysis

Mineral	Units	Value
Chalcopyrite	%w/w	23.5
Bornite	%w/w	56.0
Chalcocite	%w/w	4.0
Pyrite	%w/w	1.5
Microcline	%w/w	0.05
Talc	%w/w	1.5
Silica	%w/w	9.0
Kaolinite	%w/w	0.45
Dolomite	%w/w	2.8
Alumina	%w/w	1.2

be prone to sudden uncontrollable foaming when operated at a pO_2 of 10^{-6} atm, which for a commercial furnace was considered an intolerable risk.

An ISASMELT™ direct-to-blister furnace was not chosen for the Mufulira Copper Smelter, which history has shown to be a good decision. During the design and commissioning processes, the tolerance and flexibility of the technology offers significant advantages to deal with any deviations in the feed composition that may occur between the time of initial design and the later smelter operation. The change in concentrate proportions treated by the Mufulira Copper Smelter subsequent to the technology selection, documented elsewhere [3], could only have been tolerated by a conventional primary copper ISASMELT™ furnace.

Slag for the Next Evolution of Direct-to-Blister Smelting

Thwarted efforts always provide salutary lessons, and during the development of a direct-to-blister ISASMELT™ furnace there have been numerous lessons learned. The principal lesson was that it is realistic to expect a lot of non-sulphide gangue minerals in the concentrate supplied to a smelter, so it is helpful to develop a slag system tolerant of them. Work has been done to define a slag system that will be tolerant of a large amount of silica, without creating the conditions that are associated with uncontrollable foaming slag in an ISASMELT™ furnace [7].

The research has led to a small region of slag, defined in Fig. 1 in the SiO_2–CaO–FeO_x ternary diagram, as being suitable for the proposed process. This slag has been shown to have a good compromise between low viscosity, low slag fall, and low copper solubility with both thermodynamic modelling using the CSIRO MPE package and experimentally [7, 8]. It is believed that a slag system of this

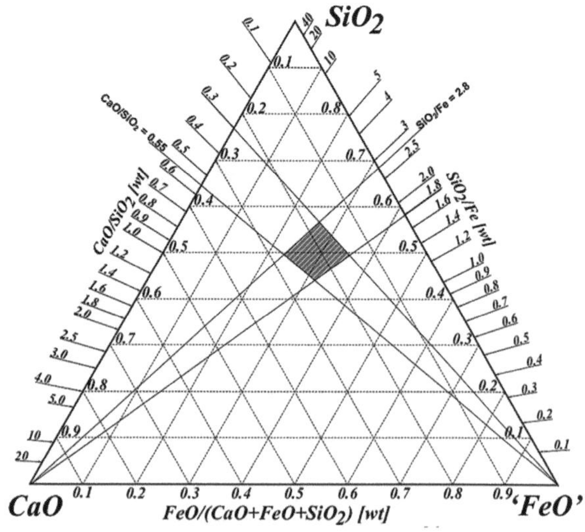

Fig. 1 Intended slag chemistry for ISASMELT™ direct-to-blister copper smelting

composition is likely to be applicable to concentrates where direct-to-blister smelting is a realistic and attractive option.

Recent testwork [7] provided great confidence that the slag will transfer oxygen at sufficiently rapid rates to equilibrate with the blister copper layer, in the same way that slag-matte equilibrium has been easily achieved in primary copper ISASMELT™ furnaces in the past. Blister containing around 1.2%S may be expected when operating in this regime.

Where to Next?

Glencore plc has been a major miner and producer of copper in Africa for many years. Aside from Mopani Copper Mines, mentioned earlier, there are other operations across the copper-belts of both Zambia and the Democratic Republic of the Congo (DRC). In addition to the current operations development is ongoing for the future mine projects and processing plants that are possible in this area.

During the past few years development of existing deposits with underlying sulphides in the DRC has been under consideration. Testwork has shown that the likely concentrate may have the mineralogy indicated in Table 2.

Like Mufulira concentrate a decade ago, this type of concentrate has a large proportion of bornite and chalcocite minerals compared with the amount of chalcopyrite. It invites the application of direct-to-blister smelting. This time the ISASMELT™ technology will be ready for it.

An ISASMELT™ furnace will fit into the direct-to-blister smelter as shown schematically in Fig. 2. In the particular case of the DRC deposit, where the carrolite ($CuCo_2S_4$) content is significant, provision is made for a possible second stage of electrothermic reduction, depending on the future requirement for cobalt recovery. Other deposits around the copperbelt are also likely to be suitable for this technology.

Table 2 Sulphide concentrate analysis

Mineral	Units	Value
Chalcopyrite	%w/w	24.5
Bornite	%w/w	40.0
Carrolite	%w/w	3.7
Chalcocite	%w/w	18.0
Cubanite	%w/w	5.2
Pyrite	%w/w	0.4
Quartz	%w/w	1.0
Talc	%w/w	2.1
Amphibole	%w/w	0.2
Mica	%w/w	0.2
Chlorite	%w/w	0.5
Magnesite	%w/w	1.2
Dolomite	%w/w	2.1

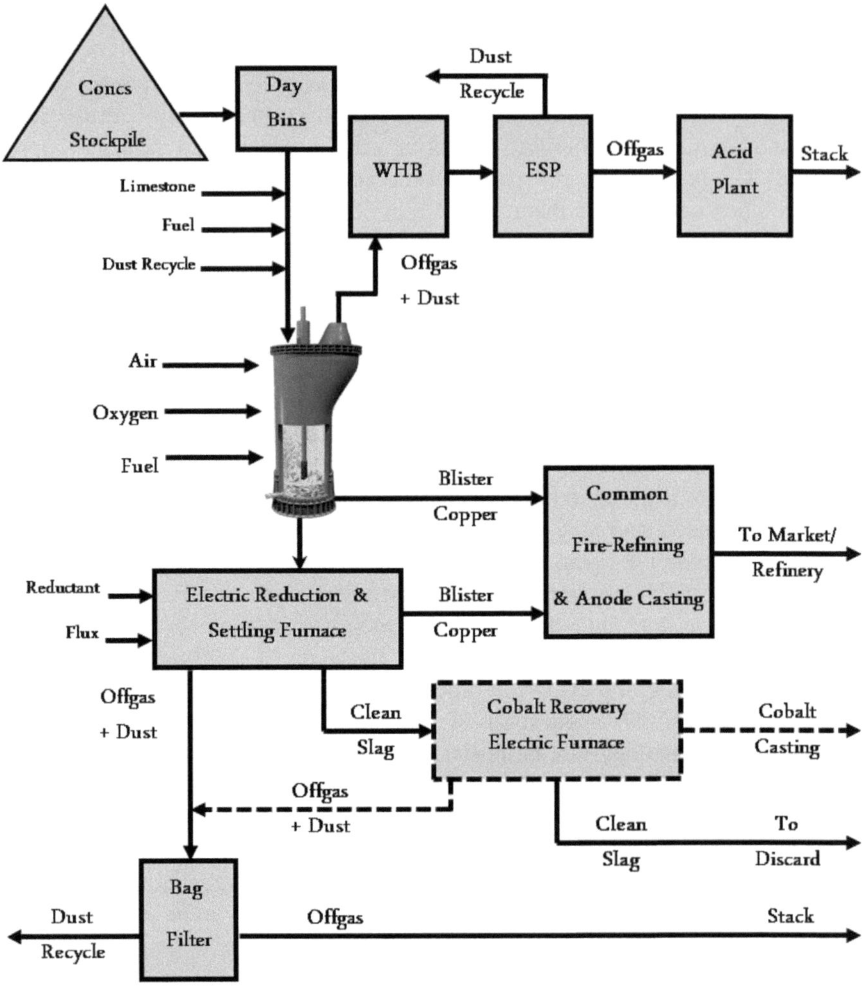

Fig. 2 Direct-to-blister smelter flowsheet, based on ISASMELT^{TM} technology

Conclusions

After three decades of continuous development, the ISASMELT^{TM} technology is ready for the next step in the copper industry and will in future take on direct-to-blister smelting. Fundamental work has been done, pilot testing is complete and flowsheet development is underway.

Acknowledgements The authors would like to thank their respective companies (Glencore Technology, Kazzinc Ltd., and CSIRO Mineral Resources) for permission to publish this paper.

References

1. D. de Vries, S. Hunt, N. Dyussekenov, M. Milovanov, H.M. Chikashi, Design, commissioning and operation of an ISASMELT™ furnace at Kansanshi copper smelter. in *Proceedings of "Copper 2016" Conference*, Kobe, Nov 2016
2. A. Burrows, G.R.F. Alvear-Flores, P.J. Mackey, B. Das, E. Herrera, ISASMELT™—smelting more with less. in *Proceedings of "COM 2015" Conference*, Torstein Utigard Memorial Symposium, Paper No. 9163, CIM 2015
3. A.S. Burrows, P.J. Partington, P.H. Mascrenhas, ISASMELT™ at Mufulira—increased flexibility on the zambian copperbelt. in *Proceedings of the Fray International Symposium*, ed. by F. Kongoli, vol. 1, pp. 217–226
4. J.S. Edwards, S. Jahanshahi, Copper converting. United States Patent, No. 5,888,270, 30 Mar 1999
5. S. Nikolic, P.C. Hayes, E. Jak, Phase equilibria in ferrous calcium silicate slags part III: copper saturated slag at 1250 °C and 1300 °C at an oxygen partial pressure of 10^{-6} atm. Metall. Trans. B **39B**(2), 200–209 (2008)
6. S. Nikolic, P.C. Hayes, E. Jak, Phase equilibria in ferrous calcium silicate slags part IV: liquidus temperatures and solubility of copper in "Cu_2O"-FeO-Fe_2O_3-CaO-SiO_2 slags at 1250 °C and 1300 °C at an oxygen partial pressure of 10^{-6} atm. Metall. Trans. B **39B**(2), 210–217 (2008)
7. M.A. Somerville, C. Chen, G. Alvear, S. Nikolic, Fluxing strategies for the direct-to-blister smelting of high-silica and low-iron copper concentrates. Paper presented at Molten 2016, Seattle, Washington, USA, 22 May 2016
8. L. Zhang, S. Jahanshahi, S. Sun, C. Chen, B. Bourke, S. Wright, M. Somerville, CSIRO's multiphase reaction models and their industrial applications. JOM **54**(11), 51–56 (2002)

Improving Separation of Cu–Fe from Copper Slag by Modification of the Molten Slag

Zheng-qi Guo, De-qing Zhu, Jian Pan and Feng Zhang

Abstract Iron mainly exists as the magnetite and fayalite, while copper is primarily dispersed in the form of fine matte particles in the copper slag, which is difficult to recover by conventional separation techniques. In this paper, the feasibility of an integrated technological route to improve beneficiation of copper and iron components from copper slag through the modification of molten slag was investigated, and the results shows that: the copper grade of rougher copper concentrate was increased from 6.43 to 11.04%, iron recovery from magnetic separation was increased significantly from 32.40 to 63.36%, and other evaluation indexes were changed slightly, in comparison with unmodified copper slag. Meanwhile, the grains of the target minerals in the modified slag grew considerably to a mean size >50 μm after slow cooling. Ultimately, the beneficiation efficiency of copper and iron was improved because of the ease with which the target minerals were liberated.

Keywords Copper slag · Mineral phase modification · Recovery · Particles

Introduction

Copper slag is a waste during pyrometallurgical production of copper from copper sulfide concentrates. Generally, 2.0–3.0 tons of copper slag are generated per ton of copper production, depending on the properties of raw materials and operating conditions [1–4]. It is estimated that approximately 12 million tons of copper slag is produced every year, and the amount of accumulated waster copper slag reached 140 million tons by 2015 in China.

The copper slag contains about 35–45 wt% iron and 0.5–1.5 wt% copper, which can be recovered if treated properly. Unfortunately, more than 80% of the waste

---

Z. Guo (✉) · D. Zhu · J. Pan · F. Zhang
School of Minerals Processing and Bioengineering, Central South University, Changsha 410083, China
e-mail: gzqcsu@126.com

J.-Y. Hwang et al. (eds.), *8th International Symposium on High-Temperature Metallurgical Processing*, The Minerals, Metals & Materials Series,
DOI 10.1007/978-3-319-51340-9_27

slag generated in China is discarded at dump sites, which not only occupies plenty of land, but also wastes many valuable metals resource and potentially causes severe damages to the ecological environment [2–4]. Thus, it is extremely urgent to recovery the iron and copper from the copper slag in the viewpoint of comprehensive utilization of mineral resources and elimination of environmental problems. In recent years, many researchers have attempted to extract the copper and iron from the slag and some new methods were proposed. The methods are roughly classified into: (i) physical separation process, (ii) pyrometallurgical and (iii) hydrometallurgical processes. These processed, included acid leaching to recovery the Cu [5], roasting followed by electrowinning or leaching [6], froth flotation separation to recycle ferrous metals [7, 8], reduction or direct reduction-wet magnetic separation process to recovery iron [9], and flocculation-magnetic separation to recover iron as magnetite [10]. However, it is worth noting that the mentioned processes for upgrading the iron from waste slag are not effective, due to the fact that iron constituents mainly exist in the form of fayalite (Fe_2SiO_4).

Therefore, in order to overcome the disadvantages of traditional processes, such as low metal recovery, high temperature and energy consumption, an innovative technology has been developed to transform fayalite to magnetite and to induce the grain growth of copper matte and magnetite by modifying the molten copper slag. In the new process, the properties of slag can be tailored to the requirements of the targeted application through modification of the copper slag and a heat treatment system. As a result, it is effective to improve recovery and separation of copper and iron from the copper slag; meanwhile, the latent heat of the molten slag can be effectively utilized and the energy consumption reduced.

Materials and Methods

Materials

Copper Slag

The copper slag used in the experiments was a byproduct generated during the cleaning of electric furnace slag and refining of copper in Tongling Non-ferrous Metals Group Holding Co., Ltd. (Tongling city, China). As shown in Table 1, the main chemical compositions contained in the waste slag were 42.85 wt% FeO and 31.11 wt% SiO_2, and the content of Cu reached 0.80 wt%.

Chemical phase analysis of the sample was used to measure the quantitative distribution characteristic of iron and copper in the copper slag, and the results are

Table 1 Chemical compositions of copper slag (wt%)

TFe	FeO	SiO_2	CaO	MgO	Al_2O_3	Cu	Pb	S	P	LOI
39.10	42.85	31.11	2.44	1.79	2.29	0.80	0.13	0.45	0.10	0.88

shown in Tables 2 and 3, respectively. Iron mainly exists in the form of fayalite and magnetite, which will result in relatively low recovery of iron by the traditional magnetic separation process, and copper mainly occurs in the form of copper sulfide and metallic copper, while the combined copper oxide fraction is also as high as 17.50%.

Flux

CaO of analytical grade was used as flux in this work to adjust the binary basicity (ratio of CaO/SiO_2) of the mixture, and its particle size was less than 74 μm.

Compound Additive

A compound additive containing hematite, sulfide and a viscosity reducer was produced by Central South University, in a powder form with over 80% passing 43 μm.

Experimental Methods

Modification of Copper Slag

In the study, the copper slag was re-melted in order to simulate the conditions of hot slag discharging from an electric furnace during normal production. First, the slag was crushed to a particle size below 0.15 mm in diameter by a hammer and pulverizer. Then, the crushed copper slag was uniformly mixed with a certain quantity of additives (burnt lime and additive compound) according to the required amount in an alumina crucible. After that, the mixed raw material was charged into an electric resistance furnace, and the charged mixture was subjected to a modification reaction at a temperature of 1350 °C for 120 min under the N_2 atmosphere. Then the sample was cooled slowly to an ending point temperature in the furnace at a fixed cooling rate. Once the slow cooling stage was concluded, the samples were taken out of the furnace and cooled to ambient temperature under the protection of nitrogen for the subsequent froth flotation and magnetic separation.

Table 2 The distribution of iron in associated minerals (wt%)

Mineral	Ferric sulfate	Magnetite	Iron sulfide	Hematite	Fayalite	Fe_{total}
Content	0.41	12.85	0.01	3.75	22.08	39.10
Fraction	1.05	32.86	0.03	9.59	56.47	100.00

Table 3 The distribution of copper in associated minerals (wt%)

Mineral	Copper oxide	Metallic copper	Copper sulfide	Combined copper oxide	Cu_{total}
Content	0.02	0.18	0.46	0.14	0.80
Fraction	2.50	22.50	57.50	17.50	100.00

In this research, direct one stage flotation (rougher) and magnetic separation (rougher) has been used. The experimental results were evaluated by modification efficiency and phase transformation degree.

Mechanism of Modification

X-ray diffraction (XRD) patterns were recorded by a diffractometer (RIGAKU D/Max 2500, Rigaku Co., Tokyo, Japan). Microstructures of slag were performed by Leica DMLP optical microscopy, FEI Quata-200 scanning electron microscope (FEI Company, GG Eindhoven, The Netherlands) and EDAX32 genesis spectrometer (Ametek Inc., Paoli, Co., USA). Scanning electron microscope (SEM) images were recorded in backscatter electron modes operating in low vacuum mode at 0.5 Torr and 20 keV.

Results and Discussion

Phase Alteration Through Modification of the Molten Slag

In the paper, the purpose of the burnt lime and additive compound is to convert the iron-bearing and copper-bearing minerals into the desired phases (magnetite and copper sulfide phase) within the copper slag, meanwhile, the proper heating treatment system is key to growth of the valuable mineral particles.

The results of the phase alteration by the addition of the additive compound and the heat treatment system in the modification process are shown in Fig. 1. As can be seen from that, with an increase of binary basicity, the iron recovery significantly increased, yet it is unprofitable to copper recovery. In contrast, it seems that the additive compound plays a distinct role in the modification process. This is due to the fact that the presence of calcium oxide in the sample benefited the decomposition of $2FeO{\cdot}SiO_2$ through replacing the "FeO", which led to an increase of the activity of free "FeO" in the slag and promoted more magnetite generation. However, with more magnetite generated and precipitated in the modification and slow cooling process, the viscosity of the copper slag was increased sharply, which had extremely negative influence on aggregation and growth of the particles of the copper bearing minerals, leading to more fine copper minerals formation and increased losses.

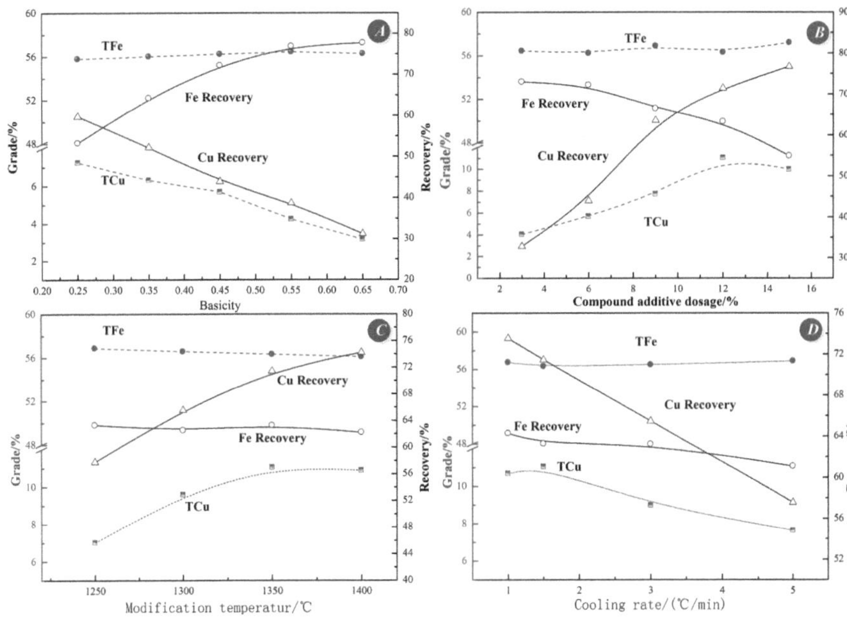

Fig. 1 The results of the phase alteration by the modification of the molten slag. **a** Effects of binary basicity on recovery of copper and iron (modifying at 1350 °C for 120 min with 6% of compound additive and cooling at 1.5 °C/min from 1350 to 900 °C). **b** Effects of additive compound dosage on recovery of copper and iron (0.45 of binary basicity, modifying at 1350 °C for 120 min and cooling at 1.5 °C/min from 1350 to 900 °C). **c** Effects of slow cooling rate on recovery of copper and iron (0.45 of binary basicity, blending with 12% of compound additive, modifying at 1350 °C for 120 min and cooling down to 900 °C). **d** Effects of end point temperature of slow cooling on the recovery of copper and iron (0.45 of binary basicity, blending with 12% of compound additive, modifying at 1350 °C for 120 min and cooling at 1.5 °C/min)

In addition, with increasing temperature, bonded silicate anions were broken, which caused a decrease in the viscosity and an increase in fluidity of the molten slag, which contributed to collision and aggregation of matte particles. Therefore, higher temperature is beneficial to modification of the molten slag. As a result, the copper recovery increased significantly.

It also can be seen from Fig. 1d, the cooling rate has significant influence on the iron and copper recovery. The lower the cooling rate, the higher the metals recovery rate. A possible explanation for this is that slow cooling of slag from its molten state to solidification ensures full growth for the particles of valuable mineral and increases the size of the particles, thereby creating ideal mineralogical conditions for subsequent flotation and magnetic separation.

Comparison of Separation Indexes Between Modified Slag and Unmodified Slag

Based on the above results, a modified slag was obtained under the conditions of 0.45 of binary basicity, blending with 12% of additive compound, modifying at 1350 °C for 120 min and cooling at 1.5 °C/min from 1350 to 900 °C. The conditions of flotation and magnetic separation process for the unmodified copper slag is similar to modified copper slag, and the results are compared with each other as shown in Fig. 2.

As seen in Fig. 2, the iron recovery from flotation tailings of two types of slag, presents a great difference in the magnetic separation process. The iron recovery was increased significantly from 32.40 to 63.26%, while the iron grade of the magnetite concentrates was similar. The improved magnetic separation efficiency of the copper slag is a result of the simultaneous increase of the content and grain size of magnetite in the slag during the modification, as the fayalite had been broken and converted into magnetite and hedenbergite.

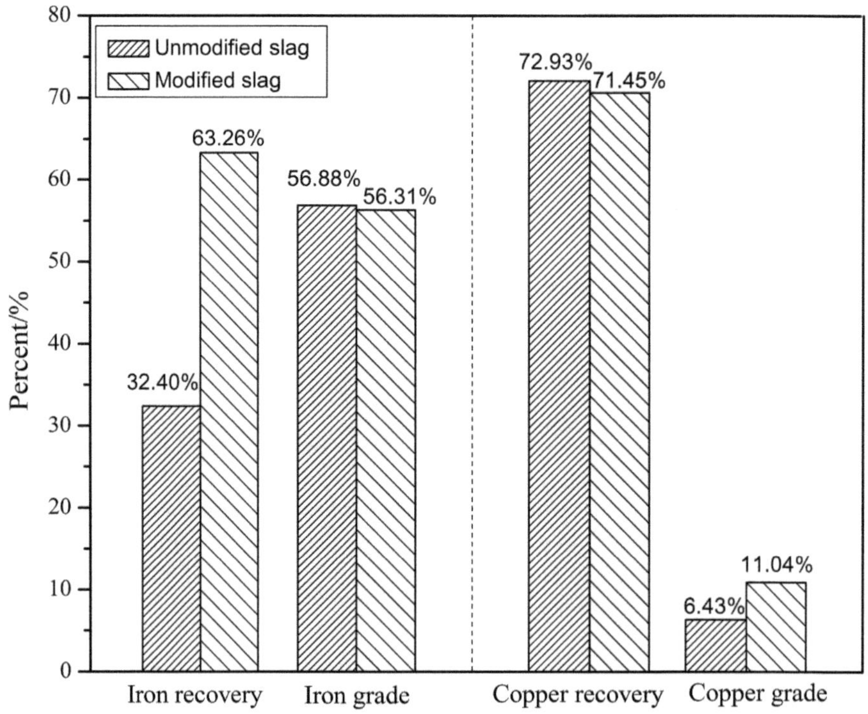

Fig. 2 Comparison of beneficiation indexes of copper and iron between two kinds of slags

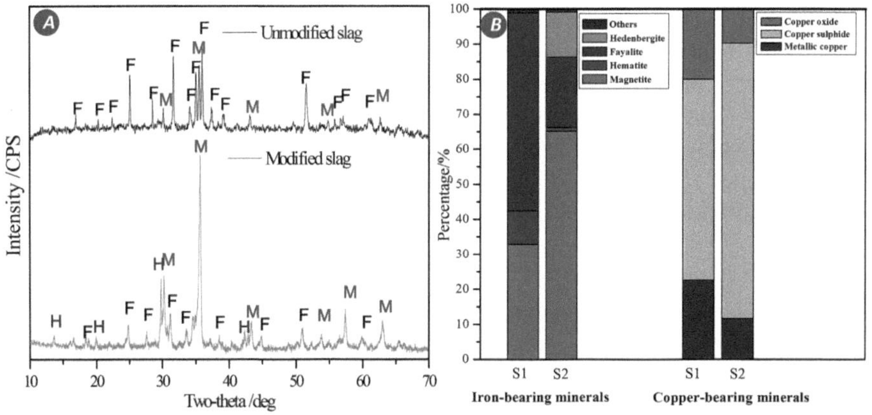

Fig. 3 Phase reconstruction of the copper slag. **a** XRD patterns of the copper slag, **b** comparison between the simulated mineralogy of two types of slags. 0.45 of binary basicity, blending with 12% of compound additive, modifying at 1350 °C for 120 min and cooling at 1.5 °C/min down to 900 °C

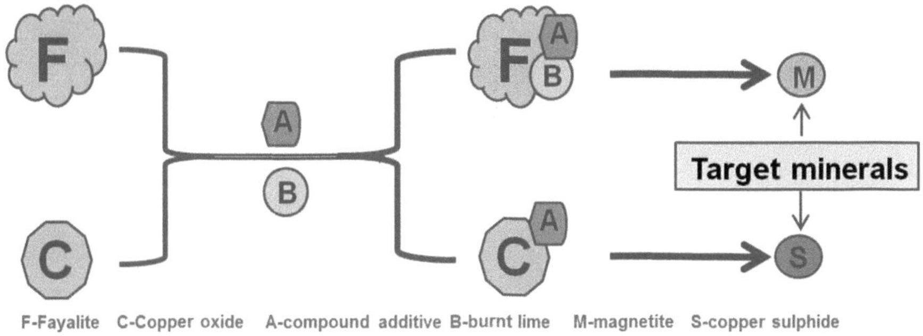

Fig. 4 Illustration for phase modification mechanism of copper slag in modification process

Intensification Mechanisms of Modification of Copper Slag on Cu and Fe Separation

Phase Transformation

The comparison of experiment results of the phase transformations obtained with modified and unmodified slag are displayed in Fig. 3.

As seen from Fig. 3a, in the unmodified slag, the major minerals constituents are fayalite (Fe_2SiO_4) and magnetite (Fe_3O_4). With the modification process, the peak intensity of magnetite increased, and that of fayalite decreased significantly. Simultaneously, hedenbergite, as a new phase, was detected in the modified slag. When the binary basicity is low (less than 1), the CaO is reactive with fayalite

(Fe_2SiO_4) to form hedenbergite ($CaFeSi_2O_6$) instead of wollastonite ($CaSiO_3$), which is not helpful with recovery of the iron.

Moreover, it also can be seen from Fig. 3b that the percentage of magnetite in iron-bearing minerals was increased from 32.86 to 65.12% by the modification process, which is easily enriched by magnetic separation, and the percentage of metallic copper and copper sulfide in copper-bearing minerals was simultaneously improved noticeably from 80.00 to 90.33%. Hence, the results above further

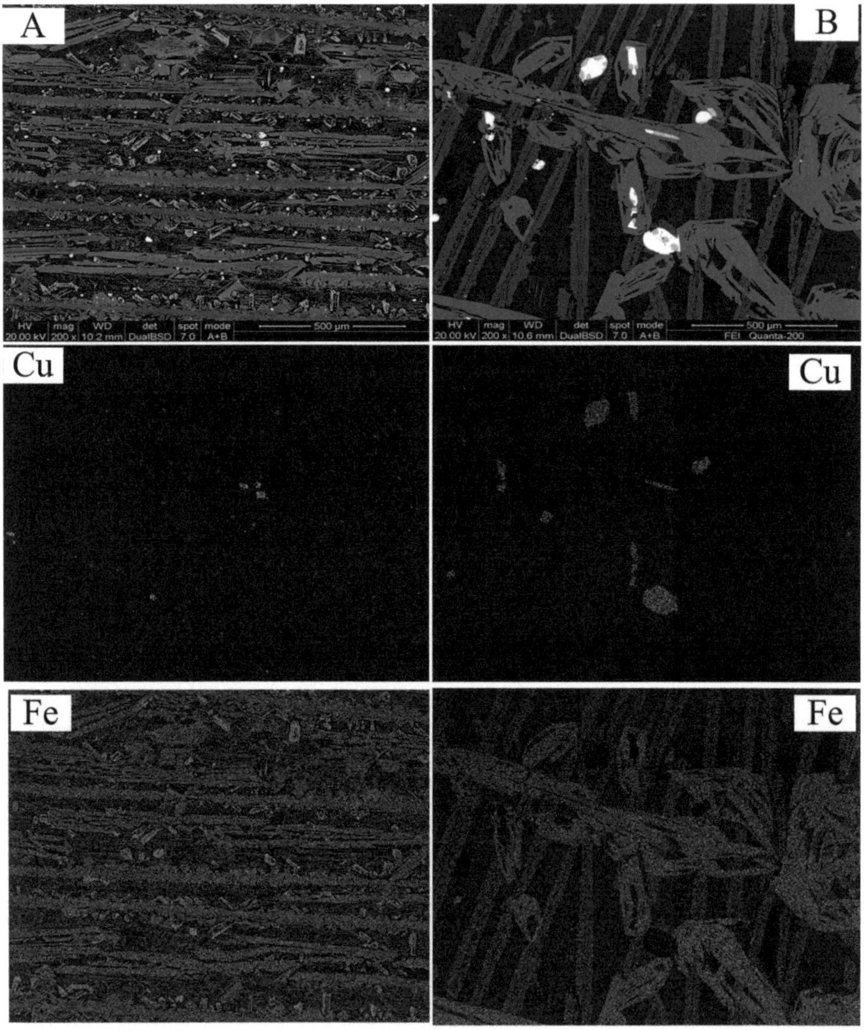

Fig. 5 Microstructure and corresponding elemental surface scanning images of copper slag. **a** Unmodified slag, **b** modified slag. 0.45 of binary basicity, blending with 12% of additive compound, modifying at 1350 °C for 120 min and cooling at 1.5 °C/min down to 900 °C

indicate the recycling potential of iron and copper may be enhanced by post-process additions to the high temperature slag to change its composition.

According to the results of thermodynamics analysis, XRD and phase analysis, the detailed phase reconstruction mechanism of copper slag with adding CaO and compound additives is schematically illustrated in Fig. 4.

Growth of Valuable Mineral Particles

To further reveal the mechanism of the modification on the improvement of the mineralogy and the grain size of the valuable mineral particles, the microstructures of the slag were studied using an optical microscope, scanning electron microscope and energy dispersive spectrometer (SEM-EDS). The results are shown in Fig. 5.

As seen from Fig. 5, the matte particles are finely disseminated throughout the slag and surrounded by magnetite and fayalite. What is worse, the particle size of the copper matte particle is relatively fine, i.e. only around 10 μm. This fine dissemination of the minerals responds very poorly to flotation and magnetic separation, due to insufficient mineral liberation from the gangue material through grinding.

On the contrary, in the modified slag, with the addition of the additive compound, the particles of copper matte and magnetite grew significantly to a size >50 μm. Moreover, Fig. 5 shows the microstructure and corresponding elemental surface scanning images of copper slag, which further confirmed that most of matte and magnetite particles in the modified slag have aggregated together [11].

And the illustration for growth of target minerals particles is shown in Fig. 6, which displays the behaviors of target mineral particles growth in the modification process.

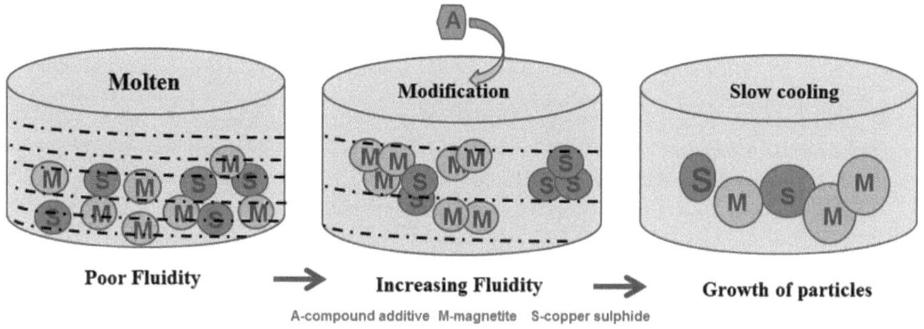

Fig. 6 Illustration for growth of target minerals particles through improved fluidity of molten slag

Conclusions

(1) Elemental iron in the copper slag mainly existed in the form of fayalite and magnetite, and the copper was present in sulfide form with fine size, as well as being closely embedded in the fayalite matrix, revealing that traditional physical beneficiation process may result in poor valuable metal recovery.
(2) The physical beneficiation studies showed that compared with unmodified slag, copper grade of the rougher copper concentrate was increased from 6.43 to 11.04%; meanwhile, the copper recovery was kept about 72%, while the corresponding iron recovery was increased significantly from 32.40 to 63.36%, and the total iron grade of magnetite concentrate was similar under the optimum conditions.
(3) The synergy of burnt lime and compound additives drive the mineral phase modification of the copper slag through converting more of the iron into magnetite and copper into both matte and metallic copper.

References

1. B. Gorai, R. Jana, M. Premchand, Characteristics and utilization of copper slag—a review. Resour. Conserv. Recycl. **39**, 299–313 (2003)
2. H. Shen, E. Forssberg, An overview of recovery of metals from slags. Waste Manag. **23**, 933–949 (2003)
3. C. Shi, C. Meyer, A. Behnood, Utilization of copper slag in cement and concrete. Resour. Conserv. Recycl. **52**, 1115–1120 (2008)
4. I. Alpa, H. Deveci, H. Sungun, Utilization of flotation wastes of copper slag as raw material in cement production. J. Hazard. Mater. **159**, 390–395 (2008)
5. M.M. Antonijevica, M.D. Dimitrijevic, Z.O. Stevanovic, S.M. Serbula, G.D. Bogdanovic, Investigation of the possibility of copper recovery from the flotation tailings by acid leaching. J. Hazard. Mater. **158**, 23–34 (2008)
6. H.S. Altundogan, F. Tumen, Metal recovery from copper converter slag by roasting with ferric sulphate. Hydrometallurgy **44**, 261–267 (1997)
7. D. Urosevic, M. Dimitrijevic, Z. Jankovic, D. Antic, Recovery of copper from copper slag and copper slag floatation tailings by oxidative leaching. Physicochemical Probl. Mineral Process. **51**, 71–82 (2015)
8. I. Muravyov, V. Fomchenko, V.A. Usoltsev, F.K. Vasilyev, Leaching of copper and zinc from copper converter slag flotation tailings using H_2SO_4 and biologically generated $Fe_2(SO_4)_3$. Hydrometallurgy **2012**(119), 40–46 (2009)
9. X.L. Zhou, D.Q. Zhu, J. Pan, T.J. Wu, Utilization of waste copper slag to produce directly reduced Iron for weathering resistant steel. Int. ISIJ. **55**, 1347–1352 (2015)
10. J. Zheng, K. Xiao, The research on using dispersant agent to iron recovery in copper slag. Nonferrous Metals Sci. Eng. **6**, 71–73 (2011) (in Chinese)
11. D. Durincka, F. Engströmb, S. Arnouta, J. Heulensa, P.T. Jonesa, B. Björkmanb, B. Blanpaina, P. Wollants, Hot stage processing of metallurgical slags. Resour. Conserv. Recycl. **52**, 1121–1131 (2008)

Phase Transformation of High Calcium Type Tin-, Iron-Bearing Tailings During Magnetizing Roasting Process

Zijian Su, Yuanbo Zhang, Yingming Chen, Bingbing Liu, Guanghui Li and Tao Jiang

Abstract High calcium type tin, iron-bearing tailing is recognized as one of typically complex iron ore resources in China. A process of magnetizing roasting process followed by magnetic separation was used to utilize the tailings in our previous paper. Under the optimal conditions, qualified magnetic concentrates were obtained and could be used for direct reduced iron production. In this study, the phase transformation of these tailings during the magnetizing roasting process was investigated. And the reaction behavior between calcium oxide and iron oxide was revealed by using XRD, TG-DSC, VSM, etc. The results indicated that calcium ferrite was easily formed under $CO–CO_2$ atmosphere, which has an unfavorable effect on the separation of iron and tin oxides from the tailings.

Keywords High calcium type tin-, iron-bearing tailings · Magnetizing roasting · $CO–CO_2$ atmosphere · Calcium ferrite · Phase transformation

Introduction

High calcium type tin, iron-bearing tailing is recognized as one of typically complex iron ore resources in China. In Gejiu of Yunnan province, the reserve of this tailings accounted for more than 250 million tons with about 0.4 million tons of tin metal and 40 million tons of iron metal [1, 2]. There are more than 50 kinds of Sn-bearing minerals in nature, and cassiterite (SnO_2) is the only one with economic value among them [3]. In the tin-bearing resource of Gejiu, more than 95 wt% of tin are existed in the form of cassiterite. Cassiterite is always fine grained dissemination in gangue minerals and iron oxide, ore-grinding process is necessary to realize the

Z. Su · Y. Zhang (✉) · Y. Chen · B. Liu · G. Li · T. Jiang
School of Minerals Processing and Bioengineering, Central South University, Changsha 410083, Hunan, China
e-mail: zybcsu@126.com

J.-Y. Hwang et al. (eds.), *8th International Symposium on High-Temperature Metallurgical Processing*, The Minerals, Metals & Materials Series,
DOI 10.1007/978-3-319-51340-9_28

monomer separation of cassiterite. But cassiterite is a brittle mineral, and is easily over-grounding during the mineral processing. In addition, the iron oxides have detrimental effects on the flotation process, which aggravates the separation and recovery of cassiterite. Thus, the recovery of tin during the mineral processing was only about 30 wt%. So far, little attention has been paid to the recovery of iron-bearing minerals, and they are discharged as solid wastes [4, 5]. It was known that the key problem to utilize the tin, iron-bearing tailings is to realize the effective separation of tin oxide from iron oxide. After multistage beneficiation, the tin content remained in the iron concentrates is required to be no more than 0.08 wt%, if it is used as iron-making materials. Additionally, the iron content in the cassiterite concentrates for smelting should not be higher than 15 wt%, because the reduction of iron oxides would consume extra reducing agents, and the formation of Fe–Sn alloy (called "hardhead") during the smelting reduced the recovery of metallic tin [6, 7].

The main mineral compositions in the high calcium type tin, iron-bearing tailings, as reported, were goethite, dolomite and calcite. Goethite is a kind of weak magnetic mineral that can't be recovered effectively by traditional separation methods, and magnetizing roasting is the only effective process to obtain magnetite concentrate from this kind of resource. Our previous studies indicated that high qualified magnetic concentrates was obtained under the optimal conditions [8, 9]. During the magnetizing roasting process, roasting temperature has the most significant effect on the separation and recovery of iron and tin. The study indicated that $CaSnO_3$ was easily formed under $CO/(CO + CO_2)$ atmosphere and it was beneficial for the separation of iron and tin.

In this study, the high calcium type tin, iron-bearing tailings were used as raw materials, and the effect of roasting temperature on the phase transformation process during the magnetizing roasting was investigated by using TG-DSC, XRD, VSM, SEM-EDS, etc.

Experimental

Materials

The high calcium type tin, iron-bearing tailings used in this study were the same as our previous researches, which were obtained from Yunnan province of China. The total iron grade (TFe) of the tailings 40.15 wt%, and the content of tin was 1.13 wt%. The main impurity compositions included 14.75 wt% CaO, 5.72 wt% MgO and 1.74 wt% SiO_2. The mineralogical characteristics of the tailings indicated that the iron existed as goethite and the main gangue minerals are calcite and dolomite [8, 9]. The loss of ignition of the tailings at 900 °C was as high as 15.67 wt%, which was attributed to the dehydration of goethite and the decomposition of carbonate.

Method

All the magnetizing roasting tests were conducted in a horizontal resistance furnace, the schematic diagram of experimental equipment has been reported previously [8, 9]. The main experimental procedure included balling, drying, preheating, magnetizing roasting and cooling. The tailings were first mixed with 0.5 wt% bentonite and balled into 8 mm diameter green pellets in a Φ1000 mm disk pelletizer, then dried in an oven at 105 °C for 4 h. After that, dried pellets were weighed and put into a corundum crucible, and then roasted in a horizontal furnace. The total flow rate of the inlet gas was fixed at 4.0 L/min. After roasting at given temperatures for a period, the pellets were cooled to room temperature under N_2 atmosphere. Based on our previous study, the $CO/(CO + CO_2)$ was fixed on 5.0 vol.% and the roasting time was 60 min [8, 9]. The roasted products were subjected to wet-grinding in a ball mill (XMQΦ240*90) for 10 min, the grinding concentration was fixed at 50 wt%. After that, the magnetic separation was conducted in a magnetic separator (XCGS-73, Davis magnetic tube) with a magnetic field intensity of 0.1 T. Then the magnetic and non-magnetic materials were dried and weighted precisely.

Characterization

XRF (XRF, Axios mAX, PANalytical Co., Ltd., in Holland) was used for determining the iron and tin grade of the sample. The mineral constituents of the roasted samples were identified by X-ray diffraction (XRD, D/max 2550PC, Japan Rigaku Co., Ltd.) with the step of 0.005° at 10° min^{-1} in ranging from 10° to 80°. The changes of the sample magnetism were investigated by VSM (Vibration sample magnetometer, BHV-50HTI, Riken Keiki, in Japan). The morphological change of the sample was detected using an environmental scanning electron microscope (ESEM; FEI QUANTA 200; FEI, Eindhoven, The Netherlands) equipped with an EDAX energy dispersive x-ray spectroscopy (EDS) detector (EDAX Inc., Mahwah, NJ).

Results and Discussion

Effect of Roasting Temperature on the Separation of Iron and Tin

Table 1 presents the effect of roasting temperature on the iron grade and recovery of iron in the magnetic concentrate, and the tin content in the concentrate was investigated as well. The results indicated that the roasting temperature has obvious effect on the recovery of iron and the separation of tin. The iron recovery of the

Table 1 Effect of roasting temperature on the separation of iron and tin

Roasting temperature/°C	Evaluation indices of magnetic concentrate		
	Iron grade/wt%	Recovery of iron/wt%	Tin content/wt%
600	65.2	76.1	0.47
700	65.8	82.5	0.46
800	66.9	87.8	0.31
850	67.0	93.3	0.13
900	66.8	83.5	0.18
1000	65.2	70.2	0.24

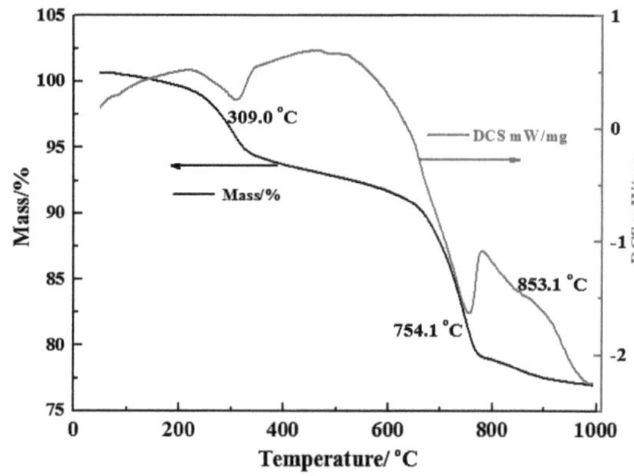

Fig. 1 TG-DSC analysis of the tailings under N_2 atmosphere

magnetic concentrate increased as the roasting temperature increases from 600 to 850 °C, meanwhile the tin content decreased sharply from 0.47 to 0.13 wt%. The optimal roasting temperature was 850 °C, and the iron recovery decreased significantly when the temperature was higher than 850 °C.

Phase Transformation of the Tailings During the Magnetizing Roasting Process

TG-DSC analysis The TG-DSC curve of the high calcium type tin, iron-bearing tailings under 100 vol.% N_2 atmosphere is presented in Fig. 1. The first endothermic peak at 309.0 °C in Fig. 1 was ascribed to the dehydration reaction associated with goethite ($2FeOOH = Fe_2O_3 + H_2O$) with the weight loss of 6.1 wt% [2]. As the increasing of the temperature, a distinct and sharp exothermic peak was found from about 500 to 754.1 °C, which was attributed to the thermal decomposition of dolomite ($CaMg(CO_3)_2 = CaCO_3 + MgCO_3$; $MgCO_3 = MgO + CO_2$) [2]. With further increase of the temperature, a weak exothermic peak was found at 853.1 °C, and it was deemed as the decomposition of calcite ($CaCO_3 = CaO + CO_2$).

XRD analysis Fig. 2 demonstrates the XRD patterns of the roasted products under the temperatures varying from 300 to 1000 °C, and the results of the raw material is shown as well. Figure 2b indicated that the diffraction peaks of goethite almost vanished at 300 °C, and goethite was converted to hematite and magnetite. As the roasting temperature increased from 300 to 700 °C, the peaks of dolomite decreased and calcite increased synchronously, which was attributed to the thermal decomposition of dolomite as presented in Fig. 1. The diffraction peaks of magnetite were enhanced at 800 °C, because higher roasting temperature promoted the reduction process of Fe_2O_3 to Fe_3O_4. With the temperature increasing from 800 to 1000 °C, the diffraction peak intensity of calcite decreased, while that of the lime

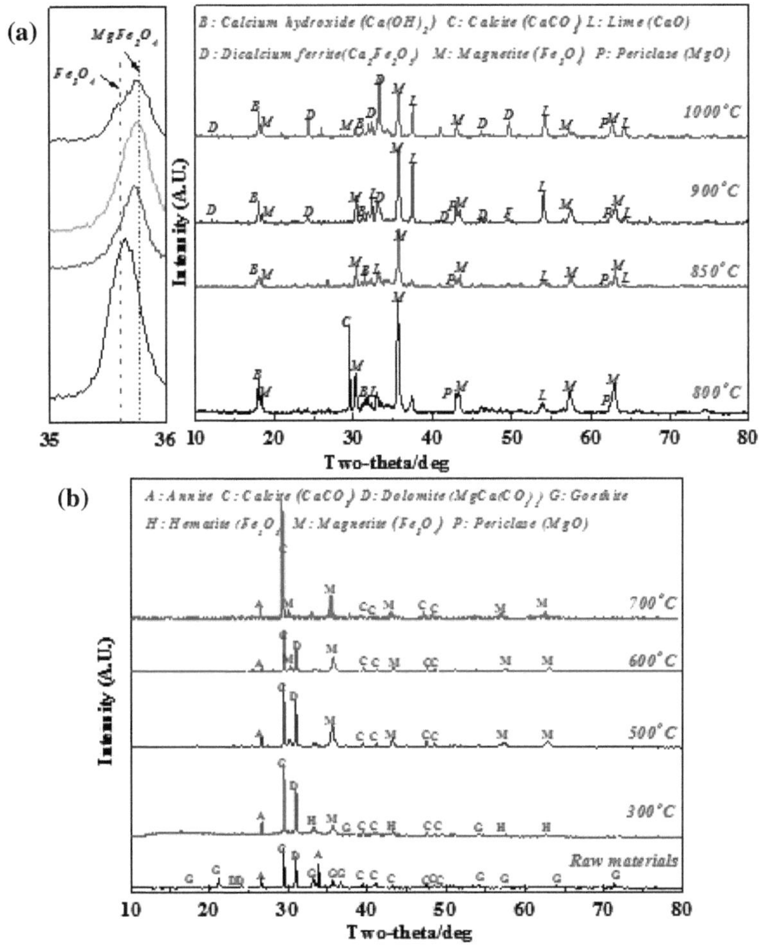

Fig. 2 XRD patterns of magnetizing roasting products at different roasting temperature

increased significantly, indicating that higher roasting temperature promoted the decomposition of $CaCO_3$. Besides, Fig. 2a illustrated that the diffraction peaks of dicalcium ferrite ($Ca_2Fe_2O_5$) increased sharply as the temperature increased higher than 850–1000 °C. The results above indicated that the $Ca_2Fe_2O_5$ was formed under CO–CO_2 atmosphere, and the increase of temperature promoted the reaction between the iron oxide and CaO. In addition, it was found by careful observation in Fig. 2a that the main diffraction peaks of Fe_3O_4 gradually shifted to $MgFe_2O_4$ as the increasing of roasting temperature, meanwhile periclase decreased and vanished at 1000 °C. It is known that magnesioferrite ($MgFe_2O_4$) has the same spinel structure as magnetite. Thus, parts of Mg^{2+} would enter into the lattice of Fe_3O_4 and replace Fe^{2+} to form $MgFe_2O_4$, and this process was enhanced by increasing the temperature.

VSM analysis In order to further study the effect of roasting temperature on the phase transformation of tailings, the magnetic hysteresis loops of the roasted products were analyzed by VSM and the results are shown in Fig. 3. The saturation magnetization (Ms) of raw tailings was only 0.51 emu/g in Fig. 3, so it was almost impossible to recover the iron oxides by magnetic separation process. The Ms value of the roasted products markedly increased as the roasting temperature, and value reached 53.07 emu/g when the temperature was 850 °C. But exorbitant temperature could result in the formation of $Ca_2Fe_2O_5$ and $MgFe_2O_4$, which decreased of Ms value and was unfavorable for the iron recovery. The changing rule of Ms is in accordance with the results given in Fig. 3 and Table 1.

SEM-EDS analysis Fig. 4 shows the morphologies of the raw material and the roasted products at 850 °C. It can be observed in Fig. 4 that obvious cracks were formed on the particle surface of the roasted products, especially on the interface between different minerals. It was known that the volume of the main minerals changed significantly during the magnetic roasting process as follows: 6FeOOH

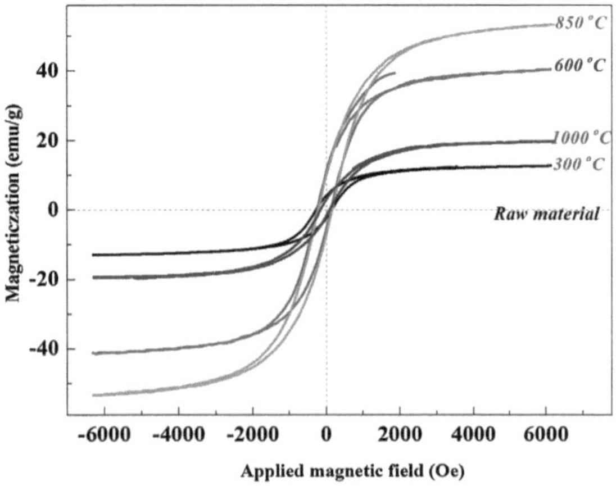

Fig. 3 Magnetic hysteresis loops of magnetizing roasted products at different temperature

Fig. 4 Morphologies of the raw material and the roasted products at 850 °C

(100%) → $3Fe_2O_3$ (108.7%) → $2Fe_3O_4$ (142.2%), $CaCO_3$ (100%) → CaO (30.3%) and $CaMg(CO_3)_2$ (100%) → CaO + MgO (58.1%) [Data obtained from ICSD Patterns]. The volume changes and cracks formed during the roasting process were beneficial for the liberation of iron and tin. Subsequent wet-grinding enhanced the liberation of the embedded minerals, which promoted the separation and recovery of iron oxides during the magnetic separating process. In order to further investigate the phase transformation of cassiterite during the roasting process, cross-section image of the roasted sample at 850 °C was analyzed by SEM-EDS, and the results are shown in Fig. 5. It can be observed from Fig. 5 that the Ca–Sn compounds were formed on the outside surface of the CaO or MgO particles. Our previous studies have indicated that the reactions between CaO and SnO_2 were promoted by the CO–CO_2 atmosphere above 800 °C, hence, $CaSnO_3$ and Ca_2SnO_4 were easily formed under magnetizing roasting conditions. Then the newly formed Ca–Sn compounds were entered into the nonmagnetic materials, and the non-magnetic materials were separated during the magnetic separation process. In this way, the tin content of magnetic concentrate was decreased as the increase of the roasting temperature.

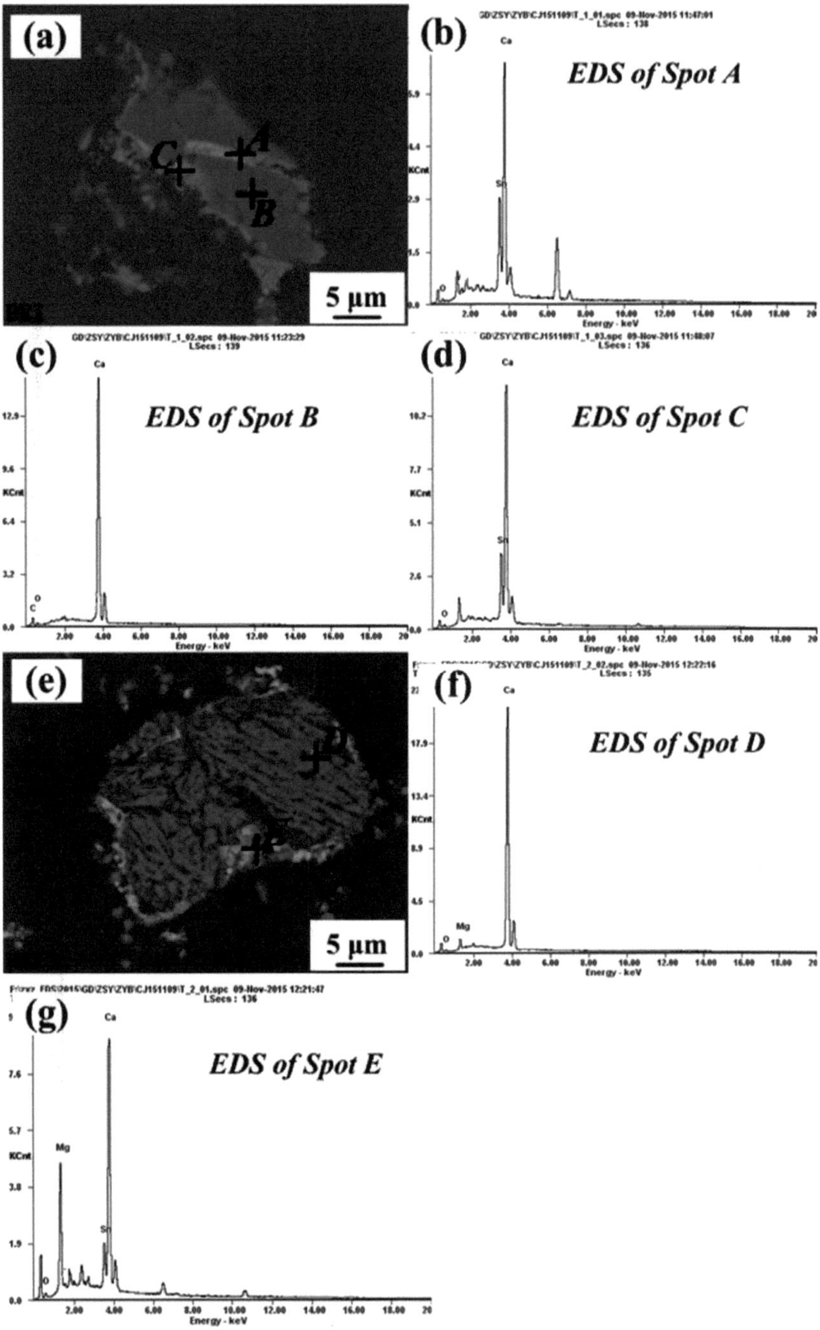

Fig. 5 SEM-EDS of cross-section images of the roasted sample at 850 °C

Conclusions

(1) The phase transformation process of high calcium type tin, iron-bearing tailing was mainly affected by the roasting temperature. The optimal roasting temperature for tin separation and iron recovery was 850 °C.
(2) From 300 to 850 °C, the decomposition reactions of goethite, dolomite and calcite were beneficial for the liberation of iron and tin. However, exorbitant temperature above 850 °C could result in the formation of $Ca_2Fe_2O_5$ and $MgFe_2O_4$, which decreased the recovery of iron. During the magnetizing roasting process, the reactions between CaO and SnO_2 were promoted by the CO–CO_2 atmosphere, which was beneficial to reduce the tin content in the magnetic materials.

Acknowledgements The authors would express their heartful thanks to National Natural Science Foundation of China (No. 51574283 and No. 51234008), the Natural Science Foundation of Hunan Province, China (No. 2016JJ2143), Co-Innovation Center for Clean and Efficient Utilization of Strategic Metal Mineral Resources, and Hunan Provincial Innovation Foundation for Postgraduate (CX2015B054).

References

1. Y.H. Jin, Research on reutilization and on process mineralogy of old tailings of YunXi. Min. Metall. **7**, 36–39 (1998) (in Chinese)
2. X. Tong et al., Recovering tin and iron from veintin tailings in Yunnan tin group by roasting-cohesion-magnetic separation technology. Chin. J. Nonferrous Met. **21**, 1676–1704 (2011) (in Chinese)
3. P.A. Wright, *Extractive Metallurgy of Tin*, 2nd Rev. edn. (Elsevier Scientific Publishing Company Press, New York, 1982), pp. 26–32
4. W.Q. Qin et al., Electro-flotation and collision-attachment mechanism of fine cassiterite. Trans. Nonferrous Met. Soc. China **22**, 917–924 (2012)
5. P.P. Wang et al., Solution chemistry and utilization of alkyl hydroxamic acid in flotation of fine cassiterite. Trans. Nonferrous Met. Soc. China **23**, 1789–1796 (2013)
6. Y.B. Zhang et al., Tin and zinc separation from tin, zinc bearing complex iron ores by selective reduction process. Ironmak Steelmak **38**, 613–619 (2011)
7. Y.B. Zhang et al., Reduction behavior of tin-bearing iron concentrate pellets using diverse coals as reducers. Int. J. Miner. Process. **110–111**, 109–116 (2012)
8. Z.J. Su et al., Selective separation and recovery of iron and tin from high calcium type tin- and iron-bearing tailings using magnetizing roasting followed by magnetic separation. Sep. Sci. Technol. **51**, 1900–1912 (2016)
9. J. Chen, et al., Research on recovering iron oxide from the iron, tin-bearing tailings. in *7th International Symposium on High-Temperature Metallurgical Processing*, (2016), pp. 395–402

Sensitivity of Contactless Ultrasound Processing to Variations of the Free Surface of the Melt with Induction Heating

G. Djambazov, V. Bojarevics, D. Shevchenko, D. Burnard, W. Griffiths and K.A. Pericleous

Abstract Contactless ultrasound treatment can be useful for high-temperature or reactive alloys in the liquid state, where contact with an immersed vibrating probe is undesirable. The alternating component of the Lorentz force can generate sound pressure levels leading to cavitation of gas bubbles in the melt, through acoustic resonance. Resonance is a function of the speed of sound in the liquid, the shape of the volume containing it and surrounding boundary conditions. Induction forces applied to a crucible lead to bulk stirring and in general deform the free surface, whose precise shape may influence the resonance conditions. This effect is investigated here by multi-physics computer modelling. Calculated results for aluminium melts are compared with experimental data and conclusions are drawn as to the sensitivity to process parameters and the reliability of this type of ultrasound metal processing.

Keywords Ultrasound treatment · Liquid metal · Electromagnetic induction heating · Acoustic resonance · Cavitation

Introduction

Treating metal melts with ultrasound prior to casting can improve the quality of the final products by enhancing fine grain structure [1], removing unwanted gases, or ensuring even dispersal of nanoparticle additives [2]. Using an axially vibrating immersed 'sonotrode' probe is the standard practice for low-temperature alloys, e.g. Al or Mg [3].

Materials melting at higher temperatures, e.g. steels, nickel-based super-alloys or reactive γ-TiAl, would benefit of the same technique [4], but then probe erosion

G. Djambazov · V. Bojarevics · K.A. Pericleous (✉)
University of Greenwich, London, UK
e-mail: K.Pericleous@gre.ac.uk

D. Shevchenko · D. Burnard · W. Griffiths
University of Birmingham, Birmingham, UK

J.-Y. Hwang et al. (eds.), *8th International Symposium on High-Temperature Metallurgical Processing*, The Minerals, Metals & Materials Series,
DOI 10.1007/978-3-319-51340-9_29

becomes a serious problem. This leads to contamination of the treated volume and to high costs due to frequent replacement of exotic material probes.

Cavitation resulting from the presence of gas bubbles in the melt is thought to be the main mechanism through which ultrasound treatment helps improve the final quality [3]. The cavitating mode is established when high-amplitude pressure oscillations (resulting from the intense vibration of the probe) interact with microscopic gas bubbles in the liquid metal. The bubbles themselves appear as dissolved gases, which come out of solution during degassing, or may have been brought into the system attached to clusters of particles, or entrained oxides.

Non-contact vibration of liquid metal samples has been attempted using a combination of static and AC magnetic fields [5–8], with some evidence of intermittent cavitation. However, the effect of a strong static field on cavitation dynamics is not well understood and it may prove impractical for industry. Electromagnetic induction with AC current at half of the target *resonant* acoustic frequency has been shown as a viable alternative [9]. It can be applied with a dedicated induction coil lowered onto the melt surface [10] or even by tuning the main induction melting coil [11]. The acoustic excitation in both cases is a result of the oscillating Lorentz force active near the metal surface or the vessel walls. (This force is caused by the interaction of the driving magnetic field with the induced current in the metal melt.) The induction coils do not touch the melt; hence, the sound generation is *contactless*. Unlike the immersed probe, where the amplitude of the vibration can be adjusted as needed, the strength of the naturally occurring Lorentz force is not sufficient to cause cavitating acoustic waves. As stated earlier, laboratory tests [5] have shown that it can be achieved by adding strong DC magnetic fields from superconducting magnets. However, in industrial settings, tuning the coil or coils to exploit resonance seems to be a much simpler alternative.

In this brief communication, the acoustic effect of the free surface shape of the aluminium melt in a crucible is considered. The shape changes in two cases of contactless sound generation (top coil or side coil) affecting resonance are studied and simulation results are compared to experimental measurements.

Computational Model

When heating a charge of molten metal by induction in a crucible, the radial Lorentz force pushes inwards the melt creating a dome-like free surface. If then a second inductor [9] is lowered onto the free surface, its Lorentz force pushes back a depression at the top of the dome (Fig. 1). The liquid volume is not static, but continuous recirculation develops inside, due to transverse gradients (curl) of the Lorentz forces. The observed time-averaged shape where the metal is pushed away from the vessel walls, is a result of the balance of gravity, Lorentz, and inertial forces [12].

In order to take into account the dynamic inertial forces, the liquid flow in the melt is simulated by solving the Navier-Stokes equations with the k-ω turbulence

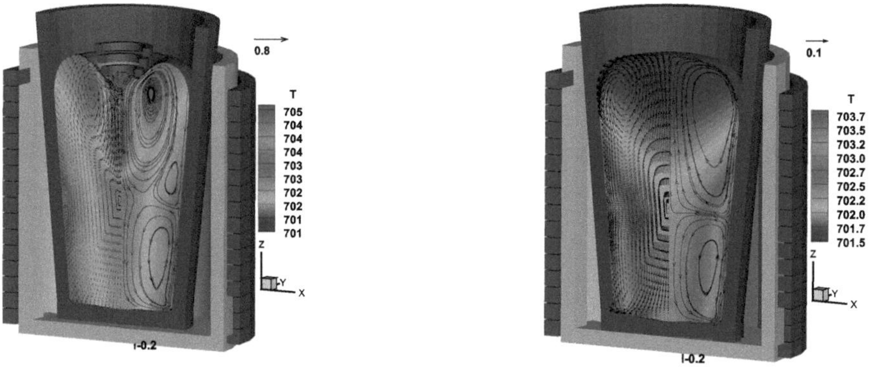

Fig. 1 Typical shapes of induction melted dome: with top and side coil (*left*) and with side coil only (*right*)

model. As shown in Fig. 1, mixing of the melt and its temperature distribution can be computed at the same time [9, 12].

Given the quasi-steady shape of the melt and associated Lorentz force field, a second stage simulation is needed to determine the internal acoustic waves. These are driven by the time-dependent component of the Lorentz force and arise from the 'skin layer' near the surface of the melt. At typical ultrasound processing AC frequencies (about 10 kHz) the electromagnetic skin layer in the melt is in the order of millimetres which is many times smaller than the radius of the crucible. Then, if x (m) is the inward distance from the surface, this time-dependent Lorentz force (N/m^3) can be expressed [9] as:

$$\tilde{F} = \frac{1}{2\mu\delta} B_0^2 e^{-2\frac{x}{\delta}} \sqrt{2} \cos\left(2\omega t - 2\frac{x}{\delta} + \frac{\pi}{4}\right) \tag{1}$$

where B_0 (T) is the amplitude of the magnetic field just outside the surface, μ (H/m) is the magnetic permeability, ω (s^{-1}) is the rotational frequency of the AC current and the skin thickness (m) is defined as $\delta = \sqrt{\frac{2}{\mu\sigma\omega}}$ with σ (S/m) being the electrical conductivity of the liquid. This time-dependent force, on its own, creates acoustic waves typically of around 2000 Pa magnitude for B_0 being in the order of 0.1 T in aluminium; not sufficiently strong for the onset of cavitation and therefore acoustic resonance conditions must be sought after.

For simple shapes (e.g. cylindrical vessels) the natural frequencies that would lead to resonance can be determined analytically [9]. For the general case of curved shapes (as in Fig. 1) direct numerical simulation of the sound field can be performed as described below.

The 3-dimensional volume of the melt is discretised in a stepwise manner into hexahedral cells whose size is chosen such that (a) at least 20 cells are contained within the wavelength of the chosen driving frequency and (b) the geometry of the vessel and the shape of the melt dome are adequately resolved. A high-accuracy numerical scheme, optimised for acoustic wave propagation is then applied for the duration of the chosen simulation interval (which must cover several thousand cycles in order to obtain reliable Fast Fourier Transform (FFT) frequency analyses).

The acoustic simulation algorithm is based on finite-difference approximations of the mass continuity equation and of the simplified momentum equations (neglecting convection or viscous terms) describing the acoustic fluid motion:

$$\frac{\partial p}{\partial t} + \rho c^2 \frac{\partial \tilde{v}_\imath}{\partial x_i} = 0 \tag{2}$$

$$\rho \frac{\partial \tilde{v}_\imath}{\partial t} + \frac{\partial p}{\partial x_i} = \tilde{F} \tag{3}$$

Here p (Pa) and $\tilde{v}_\imath$ (m/s) denote the acoustic perturbations of the pressure and of the velocity vector components, ρ (kg/m^3) is the fluid density, c (m/s) is the speed of sound in the liquid metal (e.g. 4700 m/s for aluminium [3]), t is time and, according to the index summation convention, the terms containing repeated indexes are summed for the three spatial directions x_i, i = 1, 2, 3.

Equations (2) and (3) are discretised on a fully staggered computational mesh (pressure—in the cell centre; velocities—on the cell faces, in the middle of the time steps, Fig. 2, right) where the approximating 4-th order polynomials are specially adjusted (optimised) for accurate long-time simulation of sinusoidal waves [13]. Zero-velocity boundary conditions are imposed on the solid walls of the vessel, while at the free surface, zero-pressure boundary conditions are assumed.

Simulated signals are extracted from several chosen points within the melt volume. A continuous increase in time of the sound amplitude then indicates resonance conditions. FFT analysis of the recorded signals shows the driving frequency (in this case 17 kHz) together with the neighbouring natural frequencies excited by it (Fig. 3).

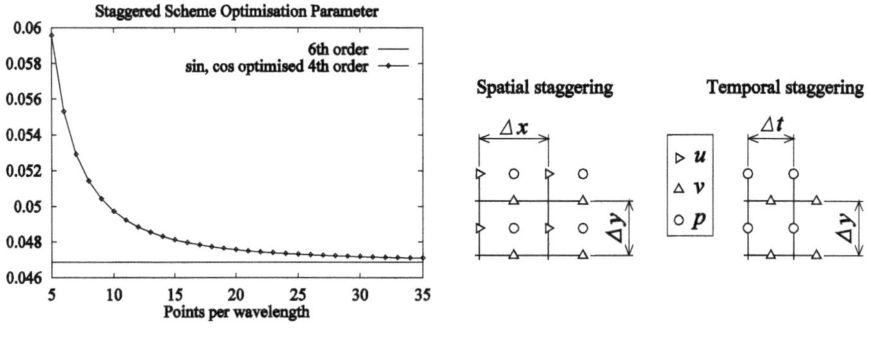

Fig. 2 *Left* dependence of the optimal value of the acoustic numerical scheme parameter on the grid spacing [13]; *right* diagram of the spatial-temporal grid-point staggering of pressure (p) and velocities (u, v)

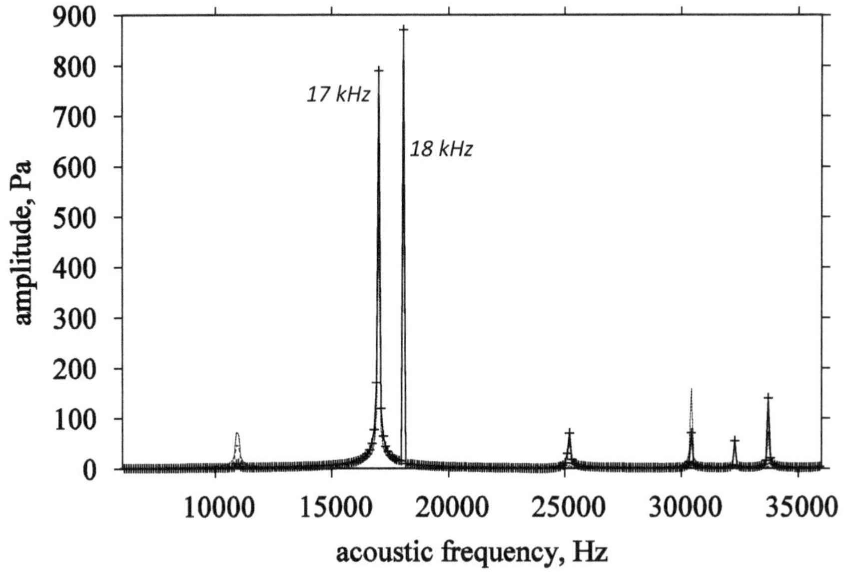

Fig. 3 Spectrum of simulation with a driving frequency of 17 kHz AC, indicating no resonance, but revealing a strong neighbouring peak at 18.06 kHz

Retuning the power supply frequency to one of the indicated natural modes is a safe way of achieving resonance. Recalling that for this system the electrical frequency is always half of the Lorentz force frequency, a second acoustic simulation with the new frequency can be used to confirm the development of resonance. To ensure that the accuracy of the algorithm is maintained for the new frequency, either the acoustic simulation optimisation parameter should be adjusted (Fig. 2, left), or the mesh spacing can be altered to follow the 20-cells rule. This is because the acoustic simulator is most accurate for the chosen optimisation frequency, i.e. cells per wavelength, and its performance deteriorates continuously for higher frequencies.

Experimental Validation

The top-coil inductor [10] was installed onto an induction furnace with an electrically non-conducting clay-graphite crucible (Fig. 4) and numerous experiments were carried out to investigate its performance. A high-frequency microphone was positioned above the crucible and the ultrasound emitted to the air from the melt surface was digitally recorded. FFT analysis of these records was then carried out and presented as the spectrogram in Fig. 5 with frequency normalised with respect to the driving frequency as the ordinate and time as the abscissa. In this form, the figure reveals how the spectrum changes continuously in time. The driving

Fig. 4 Overhead view of top coil being lowered onto the melt

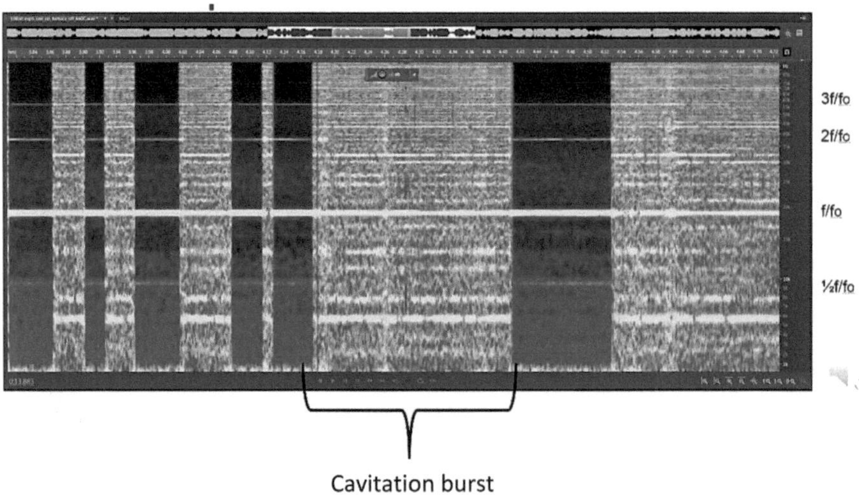

Fig. 5 Measured sound emission from aluminium containing crucible with the top coil inductor operating at near-resonant conditions. Cavitation bursts appear as broad-spectrum bands at several time intervals

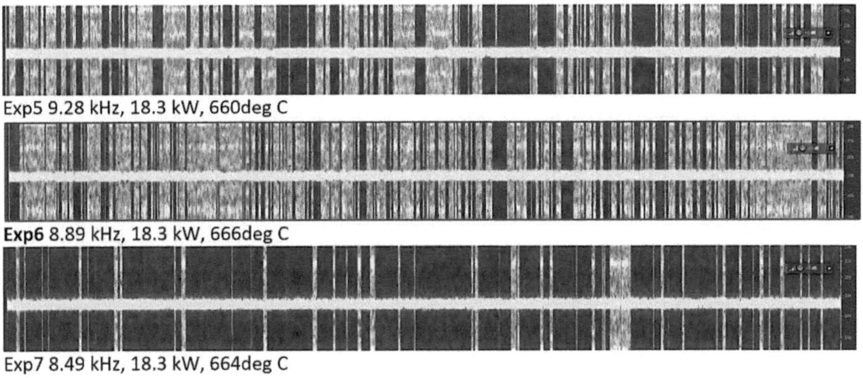

Fig. 6 Spectrograms from top-coil experiments at three different AC frequencies showing the evolution of the spectrum of the recorded signals: horizontal axis—time; vertical axis—frequency; light colour—sound intensity

frequency then appears as a thick horizontal line while the vertical lines mark the occurrence of broadband noise, which is an indicator of ongoing cavitation [14] and which can be continuous, or in bursts.

Three successive experiments are shown in Fig. 6, with coil frequency ranging from 9.28 to 8.49 kHz. Experiment '#6' at 8.89 kHz shows the most intense cavitation, therefore one would expect that frequency to be the nearest of the three to a resonant natural mode.

Indeed, when a simulation of experiment 7 at 8.49 kHz electrical frequency was performed, its FFT analysis revealed a natural frequency at 18.06 kHz (or 9.03 electrical) (Fig. 3). A second simulation using this as the driving frequency confirms the resonance, which lies very close to experiment #6 producing an almost continuous cavitation.

With satisfactory agreement between experimentally observed cavitation and simulations the model can be used to study the effect of different geometries and driving AC currents.

Effect of Dome Shape

Lowering an auxiliary induction coil from the top onto the melt may not always be practical—it may be more convenient to heat the metal with a side coil and choose the frequency such that resonance and cavitation are established at the same time. The problem parameters of the process being size and shape of vessel, AC frequency and magnitude.

With the resonant frequency determined for the top coil (18.06 kHz) supplied to the cylindrical side inductor, (the volume of metal being the same but the top coil removed) the dome shape shown in Fig. 1 (right) is established. The FFT analysis

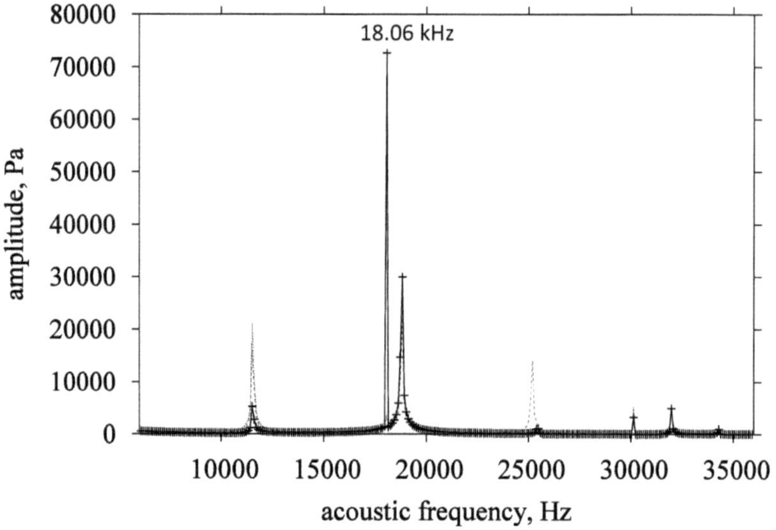

Fig. 7 Simulated spectrum with side coil excitation at 18.06 kHz

of the acoustic simulation result shows the targeted natural frequency has shifted by 740 Hz to 18.8 kHz (Fig. 7), which is confirmed by a subsequent run of the model. The proximity of the driving and the natural frequency appears as a strong signal beat (Fig. 8), which alone would be sufficient to initiate cavitation in the melt in the presence of gas bubbles.

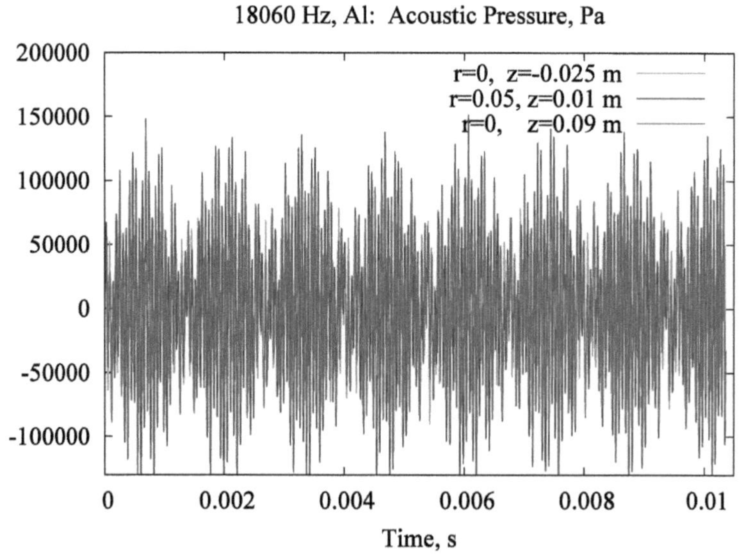

Fig. 8 Simulated acoustic signals with side coil excitation at 18.06 kHz

Conclusions

We have demonstrated that ultrasonic waves strong enough to produce cavitation in a crucible containing liquid metal can be generated using a contactless electromagnetic technique. This technique, originally based on a theoretical concept was proven experimentally in the treatment of liquid aluminium. The experiments were used to provide validation for the accompanying computational model, which couples fluid flow, heat transfer, electromagnetic interaction and soundfield simulations. It was found that resonant conditions are necessary to produce soundwaves of sufficient strength for cavitation, which means process geometrical details become important for a successful implementation. Although the experiments so far are conducted in aluminium, since the technique is contactless future experiments will investigate its use in steels, nickel alloys and titanium.

Acknowledgements The authors acknowledge financial support from the ExoMet Project (co-funded by the European Commission, contract FP7-NMP3-LA-2012-280421, by the European Space Agency and by the individual partner organizations).

References

1. T.V. Atamanenko et al., Criteria of grain refinement induced by ultrasonic melt treatment of aluminum alloys containing Zr and Ti. Metall. Mater. Trans. A **41**(8), 2056–2066 (2010)
2. R. Casati, M. Vedani, Metal matrix composites reinforced by nano-particles—a review. Metals **4**, 65–83 (2014)
3. G.I. Eskin, D.G. Eskin, *Ultrasonic treatment of light alloy melts* (CRC Press, Boca Raton, FL, 2015), pp. 1–346
4. C. Ruirun et al., A novel method for grain refinement and microstructure modification in TiAl alloy by ultrasonic vibration. Mat. Sci. Engg. A **653**, 23–26 (2016)
5. I. Kaldre et al., Nanoparticle dispersion in liquid metals by electromagnetically induced acoustic cavitation. Acta Mater. **118**, 253–259 (2016)
6. C. Vives, Crystallization of aluminium alloys in the presence of cavitation phenomena induced by a vibrating electromagnetic pressure. J. Crystal Growth **158**, 118–127 (1996)
7. A. Radjai, K. Miwa, Structural refinement of gray iron by electromagnetic vibrations. Metall. Mater. Trans. A **33**(9), 3025–3030 (2002)
8. I. Grants, G. Gerbeth, A. Bojarevics, Contactless magnetic excitation of acoustic cavitation in liquid metals. J. Appl. Phys. **117**, 204901 (2015)
9. V. Bojarevics, G.S. Djambazov, K.A. Pericleous, Contactless ultrasound generation in a crucible. Metall. Mater. Trans. A **46**, 2884–2892 (2015)
10. D.J. Jarvis, K. Pericleous, V. Bojarevics, C. Lehnert, Manufacturing of a metal component or a metal matrix composite component involving contactless induction of high-frequency vibrations. Patent EP3038771A1. www.google.com/patents/WO2015028065A1 (2013)
11. G. Djambazov, V. Bojarevics, B. Lebon, K. Pericleous, Contactless acoustic wave generation in a melt by electromagnetic induction. in *Light Metals*, ed. by J. Grandfield (Wiley, Hoboken, NJ, USA, 2014), pp. 1379–1382

12. V. Bojarevics et al., The development and experimental validation of a numerical model of an induction skull melting furnace. Metall. Mat. Trans. B **35**(4), 785–803 (2004)
13. G.S. Djambazov, C.H. Lai, K.A. Pericleous, Staggered-mesh computation for aerodynamic sound. AIAA J. **38**(1), 16–21 (2000)
14. I. Tzanakis et al., Effect of input power and temperature on the cavitation intensity during the ultrasonic treatment of molten aluminium. Trans. Indian Inst. Met. **68**(6), 1023–1026 (2015)

Extraction of Zinc from Willemite by Sodium Salt Roasting and Ammonia-Leaching Process

Xudong Liu, Ganghua Fu, Yufeng Guo, Tao Jiang, Wei Chen and Yujia Tan

Abstract Willemite is usually found in high-silicon oxide zinc ore and zinc metallurgical secondary resources but without effective extraction method yet. In this study, zinc was extracted from willemite by sodium salt roasting and ammonia-leaching process based on the idea of zinc phase reconstruction. The effects of Na_2CO_3 dosage, roasting temperature and roasting time on phase transformation and zinc leaching rate were studied, respectively. The results show that Zn_2SiO_4 was transformed into ZnO or Na_2ZnSiO_4 during roasting process. After roasting at 800 °C for 2.5 h at ratio of $n(Na_2CO_3)/n(Zn_2SiO_4) = 2.5:1$, willemite was thoroughly transformed into ZnO. The zinc leaching rate of the roasted sample was 97.62% in $NH_4Cl–NH_3–H_2O$ leaching system, which increased by 91.57% compared with zinc leaching rate of 6.05% when willemite was directly leached in $NH_4Cl–NH_3–H_2O$ system.

Keywords Zinc · Willemite · Na_2CO_3 · Ammonia-leaching

Introduction

According to the Industrial Statistics of China, China is the largest producer and distributor of zinc all over the world, and approximately 6.28 million tons of refined zinc was consumed in 2015. Several types of zinc oxide ores have been found in nature, such as smithsonite, willemite, hemimorphite and zincite [1]. Apart from the natural zinc resources such as zinc sulfide ore and zinc oxide ore, there are also a lot of oxidized zinc resources storaged in the secondary slag of lead (Pb) metallurgy

X. Liu · G. Fu (✉) · Y. Guo (✉) · T. Jiang · W. Chen · Y. Tan
School of Mineral Processing and Bioengineering, Central South University, Changsha 410083, Hunan, China
e-mail: fugh@csu.edu.cn

Y. Guo
e-mail: guo.yf@126.com

J.-Y. Hwang et al. (eds.), *8th International Symposium on High-Temperature Metallurgical Processing*, The Minerals, Metals & Materials Series,
DOI 10.1007/978-3-319-51340-9_30

and zinc metallurgy. Obviously, utilization of the secondary resources of zinc can make up for the lack of zinc.

The utilization status of willemite-bearing zinc resources at home and abroad were summarized from its physical and chemical characters. Normally, zinc oxide ore belongs to a low-grade and refractory zinc resource for its high content of silicon and alkaline gangue. Among the oxidized zinc resources, hemimorphite and willemite are the most refractory for their stable structures and higher silicon contents [2]. It has a long process and serious equipment corrosion using hydro-metallurgical methods. The two main problems in the traditionally pyrometallurgy process are low reduction efficiency and high energy consumption. New processing technologies, such as microwave, vacuum metallurgical technology, aluminium bath melting method have the advantages of high zinc recovering rate and more environmental protection, However, they are hard to be commercialized in industry due to the large equipment investment and the immature technology [3–6]. What's more, during the processes of zinc extraction from the willemite-bearing zinc resources by strong acid or strong alkaline, a large number of impurities are also dissolved in the solution, which is great harmful for purifying of solution and zinc extraction [7]. In addition, the leaching process requires strict control of the experimental conditions with great technical difficulty. Thus, many scientists have studied the alkali leaching for high-silicon contained oxidized zinc resources [8]. Z.W. Zhao leached hemimorphite in NaOH solution by mechanical activation method and found that silicon was dissolved into the solution in form of $H_2SiO_4{}^{2-}$ without the formation of silica gel while hemimorphite was dissolved in the form of $Zn(OH)_4^{2+}$ [9]. This method was featured by high zinc leaching rate but its drawbacks are also obvious: poor selectivity, poor quality of products and difficulty to the subsequent leachate purification, leachate regeneration and zinc recycling [10]. Leaching zinc in ammonia solution from oxidized zinc resources has the comprehensive advantages of high selectivity, fewer impurities and simplified purification. But the zinc leaching rate of willimete by ammonia leaching process is much lower than that of other oxidized zinc ore [10]. Z.Y. Liu studied the leaching mechanism of hemimorphite and willemite pure minerals in ammonia system and found that the zinc leaching efficiency of hemimorphite could reach as high as 95% under the condition of high liquid-solid ratio, and β-Zn_2SiO_4 was much easier to be leached than α-Zn_2SiO_4 regardless of crystallinity. In addition, the precipitation rate of dissolved silica in the leaching solution was significantly slow, which lead to difficulties of willemite leaching in the NH_3–$(NH_4)_2SO_4$–H_2O system, especially at a low liquid/solid ratio [11].

Previous researches indicated that sodium salt has positive effect on mineral separation. However, few experimental investigation of sodium salt roasting and ammonia-leaching process was reported. In order to improve the utilization of willemite-bearing zinc recourses, this paper developed a novel process to treat willemite by roasting with Na_2CO_3 followed by ammonia leaching, and good results were obtained. In the process, willemite was transformed into zinc oxide, and then zinc oxide was leached by ammonia.

Experimental

Raw Materials

Willemite samples used in this study was the product after roasting process from natural hemimorphite ($Zn_4Si_2O_7(OH)_2$), which was obtained from Lanping Oxidized Zinc Ore, Yunnan Province of China. Hemimorphite sample was crushed and screened to collect the −74 μm size, and then sample was roasted at temperature of 1300 °C for 2.5 h. The roasting reaction equation is as follows.

$$Zn_4Si_2O_7(OH)_2(s) = 2Zn_2SiO_4(s) + H_2O(g) \uparrow \tag{1}$$

The XRD pattern of willemite sample is shown in Fig. 1. The main chemical composition of willemite sample is listed in Table 1.

Results in Fig. 1 and Table 1 show that hemimorphite sample was completely dehydrated and transformed into willemite (JCPDF NO 83-2270) after roasting.

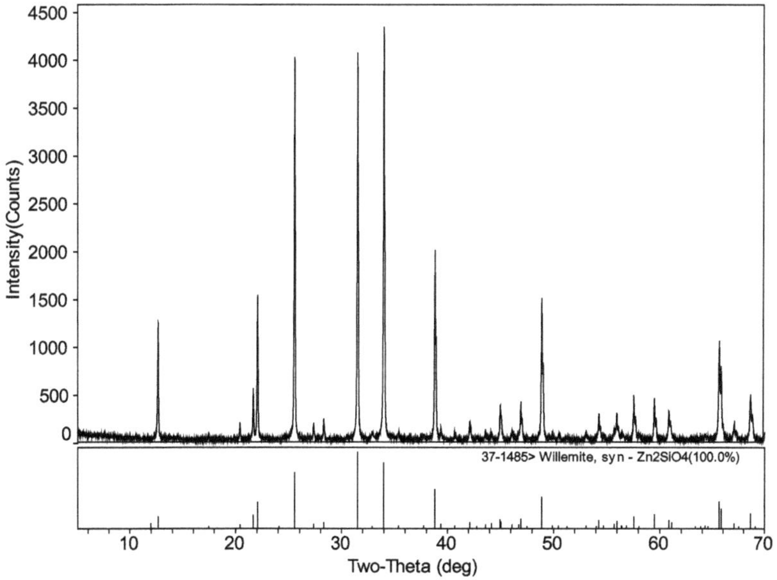

Fig. 1 XRD pattern of willemite sample

Table 1 The main chemical composition of willemite sample

$\sum Zn$	$\sum Fe$	SiO_2	CaO	MgO	Al_2O_3
57.89	0.34	26.02	0.10	0.033	0.095

The zinc element content of willemite sample is 57.89 wt%, which is very close to theoretical content (58.34 wt%) of willemite.

Experimental Method

(1) Roasting Experiment: First, the total weight of 10 g of willemite sample powder and analytical Na_2CO_3 powder were mixed by a certain percentage, then the mixture was loaded in a corundum boat and roasted in a horizontal tube furnace at a target reaction temperature.
(2) Leaching Experiment: The roasted samples are ground into particles with size <74 μm. Leaching experiment was conducted in a 250 mL beaker heated in a water bath at a target temperature. The leaching procedure initiated by pouring 100 mL of required concentration ammonia leaching agent into a 250 mL beaker with stirring. Then the beaker was heated to the experimental temperature. When the required reaction time was up, the slurry was filtered by a vacuum suction filter, and then solution and leaching residue were taken for analysis. Each group of analysis is repeated for three times and the average value is taken as the final data.

Analysis Methods

Zinc content: EDTA Complexometric Titration (GB/T 8151.1-2012).
Chemical phase analysis: X-ray Diffraction (Rigaku-TTRJII, Cu target, Kα, $\lambda = 0.15406$ nm)

Results and Discussion

Effects of Na_2CO_3 additive dosage, roasting temperature and roasting time on leaching rate of zinc were studied, respectively. Willemite samples and sodium-transformed roasting samples were leached in $NH_4Cl–NH_3{\cdot}H_2O$ leaching system under the conditions of total ammonia concentration of 6 mol/L, mole ratio of NH_4Cl to $NH_3{\cdot}H_2O$ of 1:1, leaching temperature of 40 °C, the solid-liquid ratio of 20 g/L, the stirring speed of 350 r/min and the leaching time of 3 h. A blank comparative experiment referring to Dr. Z.Y. Liu's thesis [14] under the above leaching conditions was conducted and the zinc leaching rate of willemite leached in $NH_4Cl–NH_3{\cdot}H_2O$ leaching system was 6.02%.

Effect of Na_2CO_3 Dosage on Zinc Leaching Rate

The influence of the dosage of Na_2CO_3 additive on zinc leaching rate of the roasted sample roasting at 800 °C for 2 h is very significant as shown in Fig. 2. As shown in Fig. 2, with the dosage of Na_2CO_3 increasing from 0.5:1 to 2.75:1, zinc leaching rate increases from 49.23 to 90.06%. The curve shows a linear relationship when the dosage of Na_2CO_3 additive increases from 0.5:1 to 2.5:1 (molar ratio). While continuing to increase the dosage of Na_2CO_3, zinc leaching rate increases slowly, which indicates that the dosage of Na_2CO_3 has reached a sufficient level and it is not a restriction factor of willemite roasting transformation anymore. XRD patterns of sodium-transformed roasting samples (molar ratio of 0.5:1 and 2.5:1 roasted at 800 °C for 2 h) are shown in Fig. 3. As seen in Fig. 3, there are diffraction peaks of ZnO and Zn_2SiO_4 in roasted ore sample (molar ratio of 0.5:1 and 2.5:1, roasted at 800 °C for 2 h) while diffraction peaks of Zn_2SiO_4 are not found in XRD pattern (molar ratio of 2.5:1, roasted at 800 °C for 2 h). It can be explained that the most willemite had been transformed into ZnO during sodium-transformed roasting process under the condition of molar ratio of 0.5:1 and roasting temperature of 800 °C. Obviously, 0.5:1 ratio of Na_2CO_3 is insufficient for the phase reconstruction of willemite during roasting process.

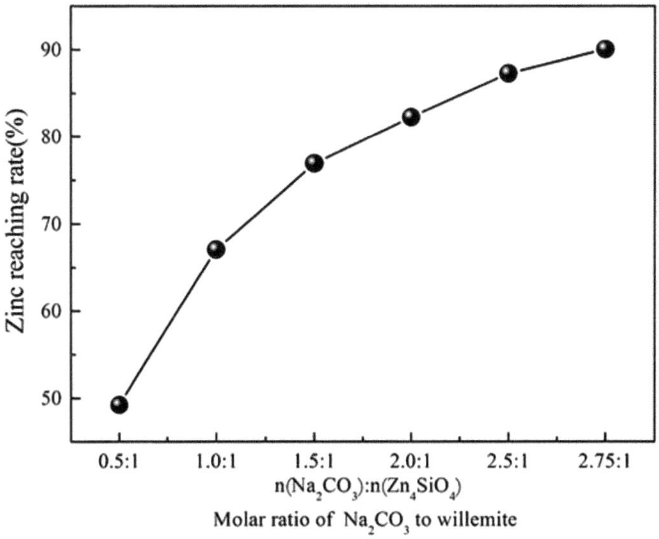

Fig. 2 Influence of Na_2CO_3/willemite (molar ratio) on zinc leaching rate

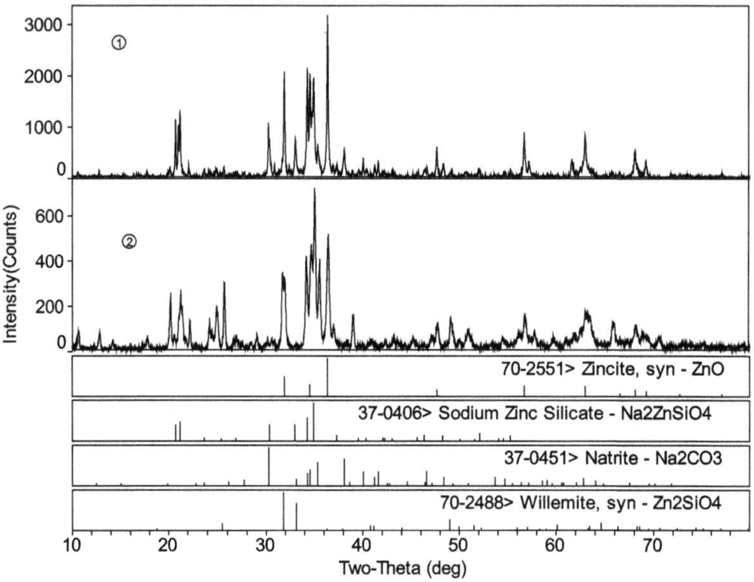

Fig. 3 XRD pattern of roasted sample. (① molar ratio of 2.5:1;② molar ratio of 0.5:1)

Effect of Roasting Temperature on Zinc Leaching Rate

The mixed willemite and Na_2CO_3 samples (Na_2CO_3: willemite = 2.5:1 molar ration) were roasted at the temperature of 650, 700, 750, 800, 850 and 900 °C for 2 h, respectively. The effect of roasting temperature on zinc leaching rate is shown in Fig. 4. As seen from Fig. 4, the promoting effect of temperature on zinc leaching rate is obvious. With the roasting temperature increasing from 650 to 850 °C, the zinc leaching rate increases sharply from 52.62 to 93.70%, while continuing to increase the temperature, the zinc leaching rate decrease. The XRD patterns of sodium-translated roasting samples are shown in Fig. 5 As seen from Fig. 5, diffraction peaks of ZnO and Zn_2SiO_4 exist in roasted samples below 800 °C while diffraction peaks of Zn_2SiO_4 are not found in roasted sample which was roasted at 850 °C). It can be explained that all willemite has been transformed into ZnO during sodium—transformed roasting process at 850 °C.

Effect of Roasting Time on Zinc Leaching Rate

The effect of roasting time on zinc leaching rate of willemite and Na_2CO_3 samples under the conditions of n(Na_2CO_3): n(willemite) = 2.5:1 roasted at 800 °C is shown in Fig. 6. As shown in Fig. 6, the influence of roasting time on zinc leaching rate is significant. The XRD patterns of sodium-transformed roasting samples under different roasting time are shown in Fig. 7. As seen in Fig. 7, ZnO, Na_2ZnSiO_4 and

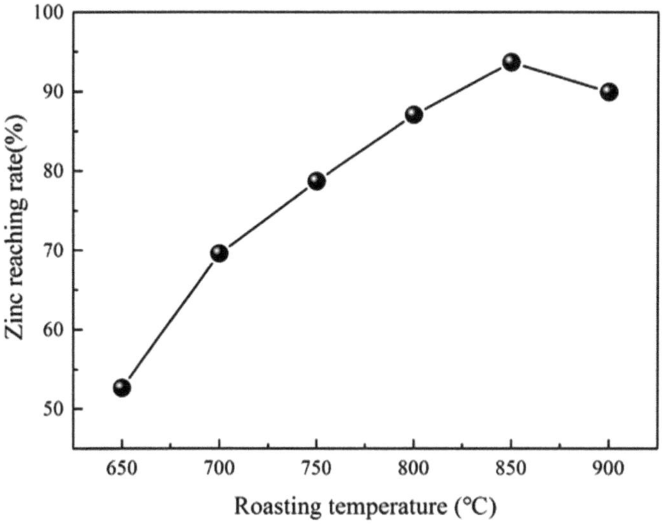

Fig. 4 Influence of roasting temperature on zinc leaching rate

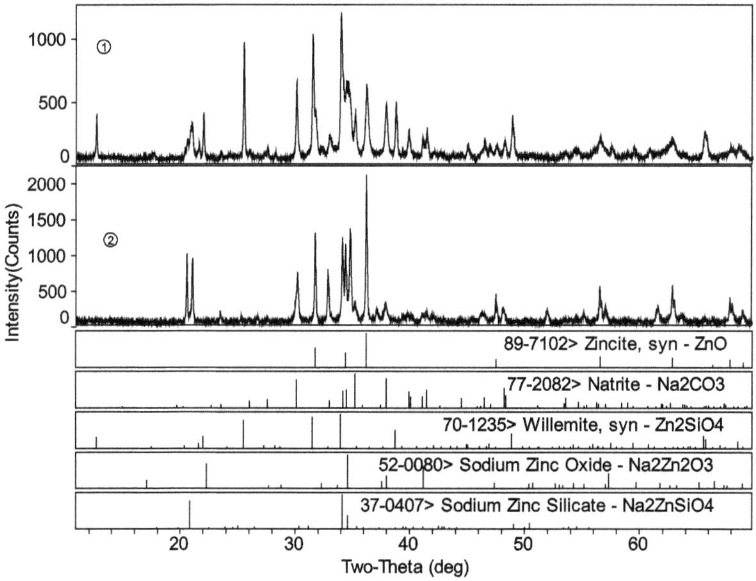

Fig. 5 XRD pattern of roasted sample. (① 650 °C for 2 h; ② 850 °C for 2 h)

Na_2CO_3 exist when the sample was roasted for 0.5 h, while diffraction peaks of Na_2CO_3 disappear when roasted sample was roasted for 2.5 h. It can be explained that most willemite has been transformed into ZnO after sodium-transformed roasting for 2.5 h and that may account for the increase of zinc leaching rate.

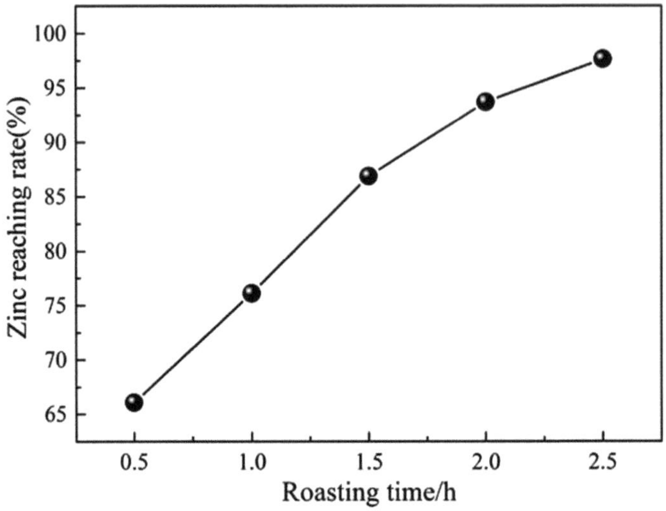

Fig. 6 Influence of roasting time on zinc leaching rate

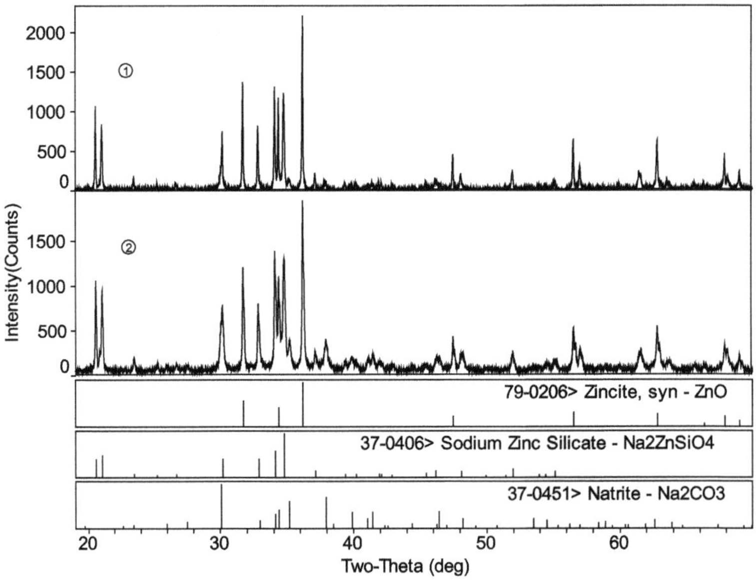

Fig. 7 XRD pattern of roasted sample. (① 800 °C for 2.5 h; ② 800 °C for 0.5 h)

Conclusions

(1) Under different conditions of sodium salt roasting, willemite (Zn_2SiO_4) can react with Na_2CO_3 and be transformed into Na_2ZnSiO_4 or ZnO due to different Na_2CO_3 dosage, roasting temperature and roasting time. An appropriate increase of Na_2CO_3 dosage, roasting temperature or roasting time can promote the transformation at the reaction and lead to an obvious increase of zinc leaching rate.

(2) Compared with the raw willemite, the sodium-transformed roasting samples are much easier to be leached in $NH_4Cl–NH_3·H_2O$ leaching system. The zinc leaching rate reached as high as 97.62% when willemite was roasted under the condition of Na_2CO_3/willemite ratio of 2.5:1, roasting temperature of 800 °C and roasting time of 2.5 h while the zinc leaching rate of raw willemite was only 6.02% under the same leaching condition.

Acknowledgements The authors are grateful to Major State Basic Research Development Program of China (973 Program Grant No. 2014CB643403).

References

1. Y.M. Zhang, J. Li, Influence of ultrasonic irradiation on ammonia leaching of zinc from low-grade oxide zinc ore. Chin. J. Nonferrous Met. **19**(5), 961–966 (2009)
2. A. Chen, Z. Zhao, D. Xu et al., Effect of sodium silicate on the solubility of zinc oxide of the $Na_2O–ZnO–H_2O$ equilibrium system. Hydrometallurgy **136**(4), 46–50 (2013)
3. W.Q. Qin, Mineralogical characterization of blast furnace slag quenched by water. Chin. J. Nonferrous Met. **24**(1), 262–268 (2014)
4. X.Z. Guo, B.H. Zhang, H.B. Yang et al., Melting-reduction of low-grade zinc oxide ores. Min. Metall. Eng. **23**(2), 57–60 (2003)
5. G. Strohmeier, J.E. Bonestell, Steel works residues and the Waelz kiln treatment of electric arc furnace dust. Iron Steel Engineer **73**(4) (1996)
6. L.Z. Xiong, Q.Y. Chen, Z.L. Yin et al., Preparation of metal zinc from hemimorphite by vacuum carbothermic reduction with CaF_2, as catalyst. Trans. Nonferrous Met. Soc. China **22** (3), 694–699 (2012)
7. Z.Y. Liu, Z.H. Liu, Leachingmechanism of willemitein $(NH_4)_2SO_4$-$NH_3·H_2O$ system. Trans. Nonferrous Met. Soc. China **21**(11), 2929–2935 (2011)
8. H.S Xu, C Wei, Sulfuric acid leaching of zinc silicate ore under pressure. Hydrometallurgy **105**, 186–190 (2010)
9. Z.W. Zhao, X.J. Jia, A.L. Chen et al., Leaching zinc oxide ore in alkaline solution. J. Central South Univ. **41**(1), 39–43 (2010)
10. Z.W. Zhao, L. Shuang, A.L. Chen et al., Alkali leaching of refractory high silica zinc oxide ore by mechanical activation. Zhongnan Daxue Xuebao **41**(4), 1246–1250 (2010)
11. Z. Liu, Z. Liu, Q. Li et al., Dissolution behavior of willemite in the $(NH_4)_2SO_4–NH_3–H_2O$ system. Hydrometallurgy **125–126**, 50–54 (2012)

Part VI
Ironmaking and Steelmaking

Influence of Ruhrstahl Heraeus Refining Process on Aluminum Consumption in Interstitial-Free Steel Smelting Process

Siyuan Zhang, Yanping Bao, Chaojie Zhang, Lu Lin and Wei Xiao

Abstract In order to decrease high aluminum consumption in Interstitial-Free (IF) steel production in a steel plant, Ruhrstahl Heraeus (RH) process data from 2428 heats were analyzed by statistical methods. The results showed that the blowing oxygen quantity and oxygen activity, before the RH and before deoxidization, have positive correlations with aluminum consumption, especially the blowing oxygen quantity. Waiting time before refining also has a positive correlation with aluminum consumption when it exceeds 7 min. On the contrary, steel temperature before RH has a significant negative correlation with aluminum consumption, when it is between 1600 and 1630 °C. According to the statistical results, the RH refining process was optimized when the steel temperature before RH was controlled between 1630 and 1645 °C, waiting time was controlled at less than 7 min, blowing oxygen quantity was controlled to less than 100 m^3, and the oxygen activity before RH was controlled between 0.050–0.075%. After optimization, the reduction of aluminum consumption was remarkable. Average aluminum consumption decreased from 1.45 to 1.17 kg/t steel.

Keywords Interstitial-free steel · Aluminum consumption · Mathematical statistics · Al killed steel

Introduction

IF steel is a kind of low carbon Al killed steel, which is the abbreviation for Interstitial-Free steel. It has excellent deep drawability and is widely used in the automobile industry, as well as in household appliances [1]. With the improvement of the quality requirement for IF steel, cleanness of the steel becomes more and more important. Thus one of the main challenges is to control non-metallic

S. Zhang · Y. Bao (✉) · C. Zhang · L. Lin · W. Xiao
State Key Laboratory of Advanced Metallurgy, University of Science and Technology Beijing, Xueyuan Road 30, 100083 Beijing, China
e-mail: baoyp@ustb.edu.cn

J.-Y. Hwang et al. (eds.), *8th International Symposium on High-Temperature Metallurgical Processing*, The Minerals, Metals & Materials Series,
DOI 10.1007/978-3-319-51340-9_31

inclusions. The morphology, number, size, composition and distribution of inclusions not only influence the quality of the steel, but also lead to the occurrence of nozzle clogging in the continuous casting process [2–5]. A large amount of aluminum alloy is added to steel in the IF steel smelting process, producing a lot of Al_2O_3 non-metallic inclusions [6]. High aluminum consumption causes the amount of non-metallic inclusions in steel liquid to be increased, which not only affects the quality of the steel, but also increases the production costs. Therefore, it is important to improve the cleanness of the steel and reduce the production costs by decreasing aluminum consumption.

In recent years, metallurgists have performed a great deal of research on Al-killed steel. Yang Shufeng et al. studied the modification of $MgO{\cdot}Al_2O_3$ spinel inclusions in Al-killed steel by Ca-treatment by industrial trials and thermodynamic calculations, finding $MgO{\cdot}Al_2O_3$ inclusions would transform into liquid complex inclusions when the content of Ca in the liquid steel reached a certain value [7]. Guo Yuming et al. analyzed and contrasted the number, shape and composition of non-metallic inclusions in the samples taken at different periods in the IF steel smelting process. The change of inclusions and Ca-treatment on the inclusions were studied. The results showed that Ca-treatment is effective in improving the cleanness of steel and modifying the inclusions [8]. Li et al. [9] analyzed the effect of Al_2O_3 on nozzle clogging by analyzing the composition and morphology of each layer deposited on the nozzle during IF steel casting and quantified the effects. Deng et al. [10] studied the evolution mechanism of inclusions in Al-killed alloyed steel during secondary the refining process by industrial experiments and thermodynamic calculations.

The research above has mostly concentrated in reducing the effect of the generated inclusions on the cleanness of steel, or only describe the state and harmfulness of inclusions. Few metallurgists have made a detailed analysis of the refining process, which affects aluminum consumption. In this paper, RH process data from 2428 heats in 2015 from IF steel production in a steel plant were analyzed by statistical methods. The influence of different RH refining process parameters on aluminum consumption was defined, and the actual refining process was optimized. Finally, the achieved reduction in aluminum consumption was significant.

Research Methods

Pure aluminum particles are used in the deoxidization process and mostly added into steel liquid 3 min after the end of the decarbonization step. Firstly, RH process data from 2428 heats were collected, and aluminum consumption was calculated. Secondly, the correlations of the main RH process parameters and aluminum consumption were tested by using SPSS statistical software and the statistical significance of different process parameters on aluminum consumption were identified. Finally, influences of the different process parameters on aluminum

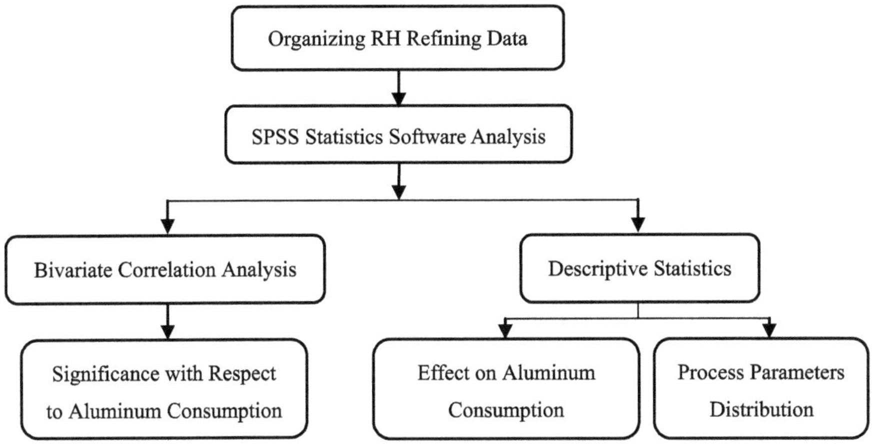

Fig. 1 Flow chart of statistical analysis

consumption were defined by descriptive statistics, which were provided to the actual production facility as optimization measures.

The correlations between different process parameters and aluminum consumption were tested by bivariate correlation analysis. Pearson correlation coefficients were used to measure the linear relationship. The closer to 1 or −1 the correlation coefficients are, the stronger the relevance is. They represent positive and negative correlations respectively. Significance tests were conducted with a two-sided test. Decisions were made according to concomitant probabilities, which were calculated by a two-sided test. When the concomitant probability was more than the significance level (0.01), the linear relationship between two variables was not significant. The correlation was not obvious. The process of the statistical analysis is shown in Fig. 1.

Research Results and Discussions

Data from 1919 heats were selected from RH process data originating from 2428 heats by filtering out data when the oxygen activity before the RH was less than 0.03% and other abnormal conditions. The correlations between the main process parameters and aluminum consumption were tested, as shown in Table 1.

Table 1 shows significant results between the aluminum consumption and the oxygen blowing quantity, waiting time before refining, steel temperature before the RH, and oxygen activity before RH and deoxidization, which are each less than the significance level (0.01), and which shows that aluminum consumption has a correlation with these process parameters. Correlation coefficients between aluminum consumption and blowing oxygen quantity, oxygen activity before RH and

Table 1 Correlation test results between main RH process parameters and aluminum consumption

Process parameter	Waiting time before refining	Time to low vacuum time	Minimum vacuum degree	Steel temperature before RH	Blowing oxygen quantity	Oxygen activity before RH	Oxygen activity before deoxidization
Correlation coefficient	0.175	0.041	−0.027	−0.464	0.716	0.064	0.148
Significant result	0.000	0.071	0.235	0.000	0.000	0.005	0.000

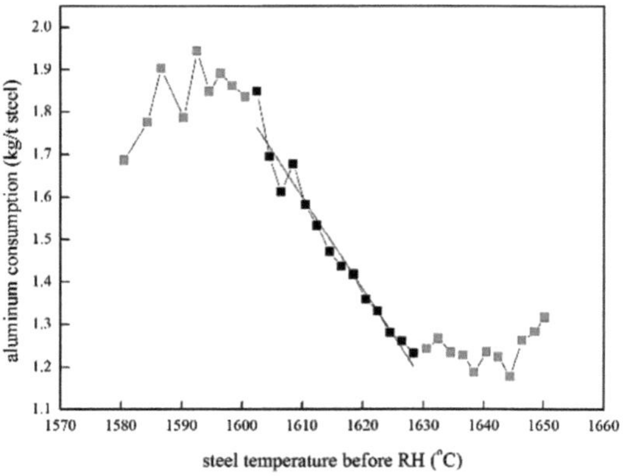

Fig. 2 Relationship between aluminum consumption and the steel temperature before the RH

deoxidization, waiting time before refining, are more than 0 respectively, which indicates that aluminum consumption has a positive correlation with these process parameters. On the contrary, the correlation coefficient between aluminum consumption and steel temperature before RH is less than 0, which indicates that aluminum consumption has a negative correlation with this process parameter. Specific effects of different process parameters on aluminum consumption will be analyzed by descriptive statistics in the following sections.

Effect of Steel Temperature Before the RH Process on Aluminum Consumption

Table 1 shows that the relationship between the steel temperature before the RH process and the aluminum consumption has passed a significance test. The correlation coefficient is −0.464 indicating a strong correlation. Aluminum consumption for different steel temperatures before the RH process is shown in Fig. 2.

Figure 2 shows clearly that the aluminum consumption and steel temperature before the RH have a linear relationship when the steel temperature before the RH is between 1600 and 1630 °C. Aluminum consumption increases with an increase

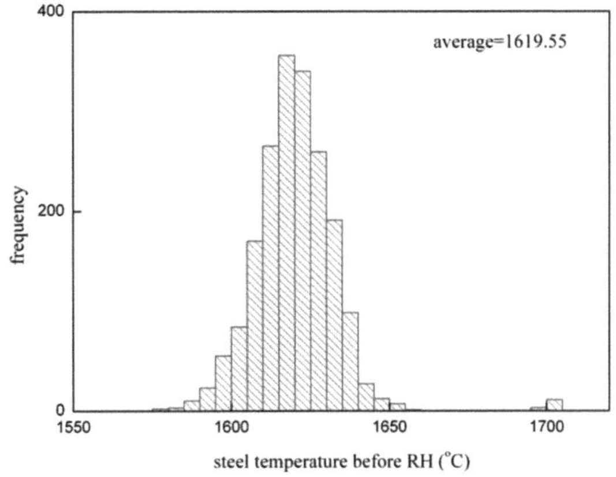

Fig. 3 Distribution histogram of steel temperature before the RH

in steel temperature before the RH. Aluminum consumption will decrease by about 0.217 kg/t steel, when the steel temperature before the RH increases about 10 °C. Aluminum consumption is maintained between 1.80 and 1.95 kg/t steel or 1.1–1.3 kg/t steel when the steel temperature before RH is lower than 1600 °C or higher than 1630 °C.

The main reason for molten steel temperature increasing is by the thermite reaction [11]. With the increase of steel temperature before the RH, aluminum consumption for heating will be decreased. When steel temperature before the RH reaches a certain value, there is no need to add extra aluminum to heat the steel liquid to meet the requirements of the next working process [12]. In order to reduce aluminum consumption, the steel temperature before the RH should be kept above 1630 °C. But too high a temperature of liquid steel would increase the burden on the converter and reduce the service life of the lining, thus it is recommended to control the steel temperature before the RH between 1630 and 1645 °C. In order to know the distribution of steel temperature before the RH in the steel plant, the distribution histogram of steel temperature before the RH is shown in Fig. 3.

Figure 3 shows that the average steel temperature before the RH is 1619 °C. Percentage of values in the range of 1630–1645 °C is only about 20%, i.e. the proportion is on the low side. Therefore, increased temperature has great potential to reduce aluminum consumption by controlling the steel temperature before the RH to between 1630 and 1645 °C.

Effect of Waiting Time Before Refining on Aluminum Consumption

Table 1 shows that the relationship between the waiting time before refining and aluminum consumption has passed the significance test, with a correlation coefficient of 0.175. The average aluminum consumption for different waiting times before refining is shown in Fig. 4.

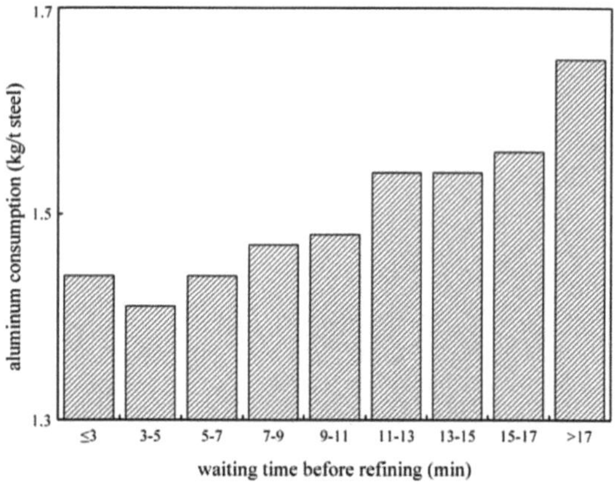

Fig. 4 Average aluminum consumption for different waiting times before refining

Figure 4 shows aluminum consumption changes little, remaining around 1.4 kg/t steel, when waiting time before refining is less than 7 min. When the waiting time before refining is more than 7 min, aluminum consumption increases with the increase of waiting time before refining. The rise is more obvious when the time is more than 10 min. This is mainly because the longer the waiting time before refining, the greater the temperature drop in the liquid steel will be. Reduced steel temperature before the RH, contributes to high aluminum consumption as shown previously. Therefore, it is very important to control the waiting time before refining to less than 7 min in order to have optimum aluminum consumption.

Effect of Blowing Oxygen Quantity on Aluminum Consumption

Table 1 shows that the relationship between oxygen blowing quantity and aluminum consumption has passed the significance test, with a correlation coefficient of 0.716 indicating a strong correlation. The average aluminum consumption for different oxygen blowing quantities is shown in Fig. 5.

Figure 5 shows that the aluminum consumption increases with an increase of the oxygen blowing quantity. Aluminum consumption increases rapidly when oxygen blowing is more than 100 m^3 and exceeds the average aluminum consumption of 1.45 kg/t steel. This effect is mainly because the free oxygen activity increases with the increase of the amount of oxygen blowing, which leads to an increased consumption of aluminum for deoxidization. The correlation between the oxygen blowing quantity and aluminum consumption is strong, thus the specific factors affecting the amount of oxygen blowing should be made clear.

Oxygen blowing in the RH is mainly used for heating and decarburization [13], thus the process parameters effecting the oxygen blowing quantity may include the steel temperature before the RH, waiting time before refining and oxygen activity

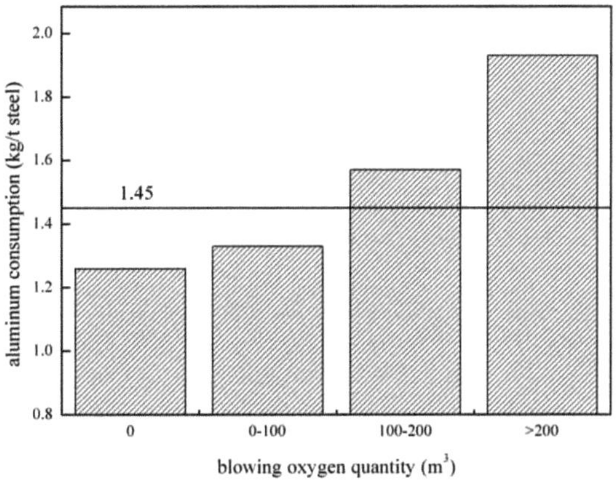

Fig. 5 Average aluminum consumption under different blowing oxygen quantity

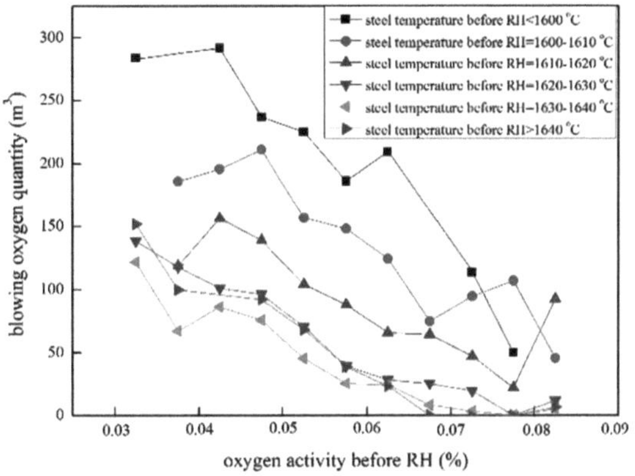

Fig. 6 Relationship between oxygen blowing quantity and oxygen activity before the RH

before the RH. Data from 1389 heats were selected by filtering out the data that had a waiting time before refining of more than 7 min. Under conditions of different steel temperatures before the RH, the relationship between oxygen blowing quantity and oxygen activity before the RH was determined and is shown in Fig. 6.

Figure 6 shows that the oxygen blowing quantity decreases with the increase of steel temperature and oxygen activity before RH. When steel temperature before the RH is less than 1620 °C, almost all heats will be blown with oxygen. When steel temperature before the RH is more than 1630 °C, the oxygen blowing quantity of almost all heats was less than 100 m^3, except the heats whose oxygen activity before the RH was less than 0.030%. The oxygen blowing quantity under different conditions was shown in Table 2.

Table 2 shows the conditions that lead to the oxygen blowing quantity more than 100 m^3. Therefore, smelting process parameters should be changed according to the influence of steel temperature and oxygen activity before the RH on oxygen blowing quantity to control the required amount to less than 100 m^3.

Table 2 Oxygen blowing quantity under different conditions of steel temperature and Oxygen activity before RH

Oxygen activity before the RH	Steel temperature before the RH			
	Less than 1600 °C	1600–1610 °C	1610–1620 °C	1620–1630 °C
Less than 0.075%	More than 100 m^3	–	–	–
Less than 0.065%	–	More than 100 m^3	–	–
Less than 0.050%	–	–	More than 100 m^3	–
Less than 0.045%	–	–	–	More than 100 m^3

Table 3 Correlation test results between oxygen activity and aluminum consumption

Process parameter	Oxygen activity before RH	Oxygen activity before deoxidization
Correlation coefficient	0.541	0.644
Significant result	0.000	0.000

Effect of Oxygen Activity Before RH and Deoxidization on Aluminum Consumption

Table 1 shows the relationship between oxygen activity before the RH and the aluminum consumption, and the relationship between the oxygen activity before deoxidization and aluminum consumption have both passed the statistical significance test. However, the correlations are not strong because of oxygen blowing. Data from 580 heats were selected by filtering out data where oxygen blowing was used. Correlations between oxygen activity and aluminum consumption were tested by bivariate correlation analysis, as shown in Table 3.

Table 3 shows that both the relationship between oxygen activity before the RH and the aluminum consumption, and the relationship between oxygen activity before deoxidization and aluminum consumption, have passed the statistical significance test. Correlation coefficients of 0.541 and 0.644 respectively, indicated strong correlations. The aluminum consumption for different oxygen activities before the RH and deoxidization is shown in Fig. 7.

Figure 7 shows that the aluminum consumption increases with the increase of oxygen activity before the RH and deoxidization. This is mainly because aluminum consumption for deoxidizing increases with the increase of oxygen activity before the RH and deoxidization, which both contribute to high aluminum consumption. When aluminum consumption is 1.45 kg/t steel, oxygen activity before the RH and deoxidization is 0.075 and 0.045% respectively. Correlation between the oxygen

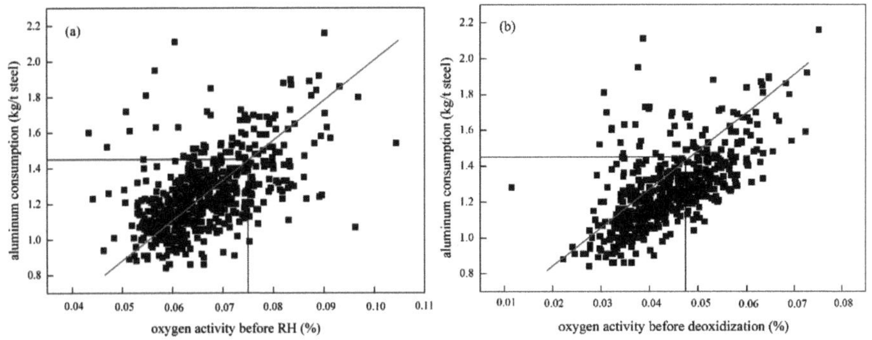

Fig. 7 Aluminum consumption for different oxygen activities before the RH and deoxidization, (**a**) different oxygen activity before the RH, (**b**) different oxygen activity before deoxidization

Table 4 Correlation test result between oxygen activity before the RH and oxygen activity before deoxidization

Process parameter	Oxygen activity before RH
Correlation coefficient	0.704
Significant result	0.000

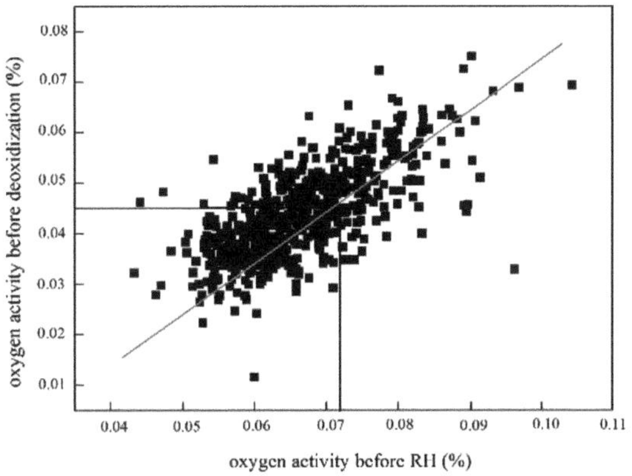

Fig. 8 Oxygen activity before deoxidization under different oxygen activity before RH

activity before the RH and oxygen activity before deoxidization was tested, as shown in Table 4.

Table 4 shows the relationship between oxygen activity before the RH and oxygen activity before deoxidization has passed the statistical significance test, where a correlation coefficient of 0.704 indicates a strong correlation. Oxygen activity before deoxidization under different oxygen activity before the RH is shown in Fig. 8.

Figure 8 shows that the oxygen activity before deoxidization increases with the increase of oxygen activity before the RH. When oxygen activity before deoxidization is 0.045%, oxygen activity before the RH is 0.075%, and aluminum consumption is 1.45 kg/t steel. Decreasing oxygen activity before the RH can contribute to the decrease of oxygen activity before deoxidization. Therefore, the oxygen activity before the RH should be controlled to between 0.050 and 0.075%.

Comparison of Results Before and After Process Optimization

According to the process influences described above, a 300 ton RH refining process for IF steel production at a steel plant were optimized: steel temperature before the RH was controlled between 1630 and 1645 °C, waiting time before refining was controlled to less than 7 min, oxygen blowing quantity was controlled to less than 100 m^3, and oxygen activity before the RH was controlled between 0.050 and 0.075%. Results showing the aluminum consumption before and after the process optimization are compared, as shown in Fig. 9.

Figure 9 shows that the aluminum consumption before process optimization was concentrated in the range of 1–1.8 kg/t steel. After optimization, the aluminum consumption of every heat was less than 1.4 kg/t steel. The percentage of the heats that used less than 1.4 kg/t steel was greatly improved. The average aluminum consumption decreased from 1.45 to 1.17 kg/t steel, i.e. a remarkable decrease of 19.3%.

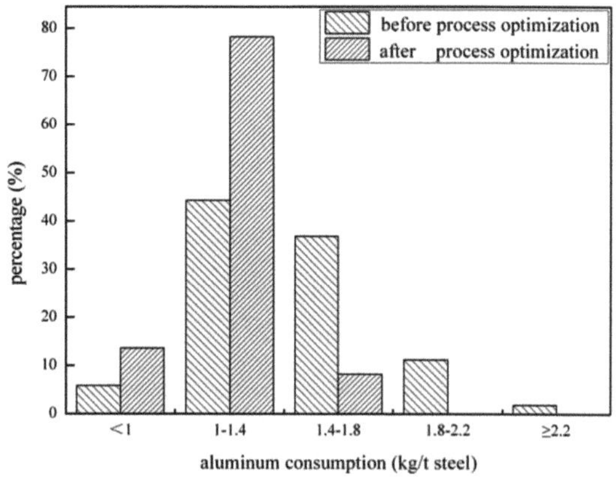

Fig. 9 Comparison of aluminum consumption before and after process optimization

Conclusions

1. Statistical methods have identified which of the RH refining process parameters affect aluminum consumption in IF steel production. The statistically significant parameters include: the oxygen blowing quantity, oxygen activity before the RH and deoxidization, waiting time before refining, and the steel temperature before the RH.
2. Oxygen blowing quantity, oxygen activity before the RH and deoxidization, each have positive correlations with aluminum consumption. Waiting time before refining also has a positive correlation with aluminum consumption when it exceeds 7 min.
3. Steel temperature before the RH has a significant negative correlation with aluminum consumption, when it is between 1600 and 1630 °C.
4. The effect of blowing oxygen quantity on aluminum consumption is the most important parameter.
5. The effects of steel temperature before the RH, oxygen activity before the RH and deoxidization on aluminum consumption are moderate and the effect of waiting time before refining on aluminum consumption is minimal.
6. Process optimization measures were put forward. Steel temperature before the RH was controlled between 1630 and 1645 °C, waiting time before refining was controlled to less than 7 min, oxygen blowing quantity was controlled to less than 100 m^3, and oxygen activity before the RH was controlled between 0.050 and 0.075%. After optimization, the average aluminum consumption decreased from 1.45 to 1.17 kg/t steel, i.e. a decrease of 19.3%.

Acknowledgements This work was supported by the Natural Science Foundation of China (No. 51404022).

References

1. M. Longauerová, M. Federová, S. Longauer, Distribution of microalloying elements and impurities in surface zone of CC IF steel. Ironmaking Steelmaking **36**, 176–185 (2009)
2. B. Coletti, S. Vantilt, B. Blanpain, Observation of calcium aluminate inclusions at interfaces between Ca-treated, Al-killed steels and slags. Metall. Mater. Trans. B **34**, 533–538 (2003)
3. E. Zinngrebe, V.H. Corrie, H. Visser, Inclusion population evolution in Ti-alloyed Al-killed steel during secondary steelmaking process. ISIJ Int. **52**, 52–61 (2012)
4. S. Abdelaziz, G. Megahed, I. El-Mahallawi, Control of Ca addition for improved cleanness of low C, Al killed steel. Ironmaking Steelmaking **36**, 432–441 (2009)
5. Y. Vermeulen, B. Coletti, B. Blanpain, Material evaluation to prevent nozzle clogging during continuous casting of Al killed steels. ISIJ Int. **42**, 1234–1240 (2002)
6. D. Liu, H. Lei, T.L Wang, Thermodynamics of deoxidation equilibrium with Al in liquid iron at 1873 K. J. Mater. Metall. **14**, 96–99 (2015)
7. S. Yang, J. Li, Z. Wang, Modification of MgO·Al_2O_3 spinel inclusions in Al-killed steel by Ca-treatment. Int. J. Min. Metall. Mater. **18**, 18–23 (2011)

8. Y.M Guo, Z.B Xu, H.T Wang, Research on inclusions in Al-killed steel during LF refining. Iron Steel Vanadium Titanium **28**, 14–18 (2007)
9. X. Li, Y.P Bao, M. Wang, Effect of aluminum consumption and oxygen potential at final deoxidation on nozzle clogging of IF steel, Chin. J. Eng. **37**, 700–705 (2015)
10. Z. Deng, M. Zhu, Evolution mechanism of non-metallic inclusions in Al-killed alloyed steel during secondary refining process. ISIJ Int. **53**, 450–458 (2013)
11. X.C Li, P. He, F. Wang, Study on temperature drop in RH refining process. Res. Iron Steel **38**, 13–15 (2010)
12. G.M. Zhai, X.L. Liu, Production practice of IF steel in RH furnace. Steelmaking **25**, 5–7 (2009)
13. Y. Wu, H.W Ni, H. Zhang, Development of liquid temperature prediction model in RH-MFB refining process. Special Steel **35**, 8–10 (2014)

Formation Mechanisms of Inclusions in Spring Steels

Sha Lv, Zongze Huang, Zan Yao, Xiaodong Ma, Geoff Wang, Zhouhua Jiang, Jin Zou and Baojun Zhao

Abstract 55SiCr steel is one of the important spring steels widely used in automobile manufacturing. The presence of micro inclusions in the steel can significantly influence a number of spring properties such as fatigue strength. It is important to understand the mechanism of the inclusion formation so that size and proportion of the inclusions can be controlled. Inclusions in Si-killed 55SiCr spring steel have been analysed by scanning electronic microscopy and electron probe X-ray microanalysis. The chemistry of the inclusions observed in 55SiCr steels can be represented by the SiO_2–Al_2O_3–CaO–MgO system. It was found that the compositions of the inclusions cover a wide composition range, which indicates that these inclusions were formed from different reactions. The possible mechanisms of the inclusion formation are discussed. It would be possible to characterize and manage the inclusions in the spring steels deoxidized by Si.

Keywords Formation mechanism · Spring steel · Inclusions · Si deoxidization

Introduction

55SiCr spring steel is widely applied as a suspension spring wire in automobiles, due to the excellent mechanical properties, such as high strength, fatigue and impact resistance. It has been proven that non-metallic inclusions can cause serious crack

S. Lv · X. Ma · G. Wang · J. Zou · B. Zhao (✉)
The University of Queensland, Brisbane, Australia
e-mail: baojun@uq.edu.au

Z. Huang · Z. Yao
Baoshan Iron and Steel Co. Ltd, Shanghai, China

Z. Jiang
Northeastern University, Shenyang, China

J.-Y. Hwang et al. (eds.), *8th International Symposium on High-Temperature Metallurgical Processing*, The Minerals, Metals & Materials Series,
DOI 10.1007/978-3-319-51340-9_32

initiation during the working and therefore can obviously affect the in-service life of the spring [1]. As a result, it is widely acknowledged that inclusions should be controlled to achieve a small size and excellent plasticity. It was suggested that the primary factor affecting the plasticity of inclusions is the melting point, which is closely related to the chemical composition [2, 3]. Alumina is one of the most common inclusions with poor plasticity. Therefore, inclusions like alumina with high melting point and high hardness should be avoided.

It was suggested that there are two types of SiO_2–containing inclusions in final wire products: one is the deoxidation product and the other is the material precipitated during casting [4]. The inclusions may be deformable if they contain about 20 wt pct Al_2O_3. To achieve this optimal Al_2O_3 content in inclusions, slag with low basicity near 1.0 and Al_2O_3 of about 8 wt pct is preferred [5]. MnO–SiO_2–Al_2O_3 inclusions are believed to be deoxidation products [6]. However, there are controversies on the formation mechanism of CaO–SiO_2–Al_2O_3–(MgO) inclusions. Some researchers proposed that CaO–SiO_2–Al_2O_3–(MgO) inclusions come from the entrapped slag particles [7]. Whereas other studies suggested that these MgO-containing inclusions originate from interaction of the molten steel reacted with top slag in a ladle [8, 9]. Alumina content in CaO–SiO_2–Al_2O_3–MgO inclusions is usually higher than that in the top slag. Moreover, CaO–SiO_2–Al_2O_3–(MgO) inclusions always contain a small amount of MgO. The CaO–SiO_2–Al_2O_3–(MgO) system inclusions are considered as contributing factors to wire breakage, because of their poor deformability during cold drawing. Therefore, classification of formation mechanisms for the CaO–SiO_2–Al_2O_3–(MgO) system inclusions in spring steel should be further specified.

This paper systematically investigated the inclusions present in the spring steel rod and wires collected from production practices. Different from previous studies, attention was paid by the present authors to the origin and evolution of CaO–SiO_2–Al_2O_3–MgO inclusions. The formation mechanism of CaO–SiO_2–Al_2O_3–MgO inclusions is proposed on the basis of the experimental findings.

Experimental

The spring steel used in the present study mainly consists of 0.55% carbon, 1.5% Si, 0.7% Mn and 0.7% Cr. The steelmaking process was characterized by: (i) Electric Arc Furnace (EAF) steelmaking → (ii) Ladle Furnace (LF) refining → (iii) Vacuum degassing → (iv) Bloom casting → (v) Heat-treatment → (vi) Cold drawling. The molten steel was transferred from the EAF to the LF station. Low basicity slag (CaO/SiO_2 1.0) was used in the LF refining together with argon bottom blowing at a proper stirring energy to prevent exposure of liquid steel to the air. At the end of the LF refining, the ladle was transported to vacuum degassing and then poured into a tundish.

To investigate the origins of $CaO–SiO_2–Al_2O_3–MgO$ inclusions in wire rods, solid steel samples had been taken after the hot rolling (BP1), heat treatment (BS2) and cold drawing processes (BS3). Then the samples were observed under the optical microscope, electron probe X-ray microanalysis (EPMA) and SEM, to implement a thorough study of the composition of the inclusions. The as-received steel samples were ground and mirror polished for inclusion inspection under the EPMA. Therefore, important information about inclusions including the number, size, chemistry, and morphology during rolling was determined. Inclusions of less than 1 μm were not taken into account during the study.

Results and Discussions

Morphology and Composition of Inclusions

From the EPMA results, it can be seen that inclusions mainly contain CaO, Al_2O_3, MgO and SiO_2. The inclusions can be classified into four types according to the EPMA results, as shown in Fig. 1 and Table 1. The first kind of inclusions are $Al_2O_3–SiO_2–CaO$ system and contained a very high level of SiO_2 with an almost

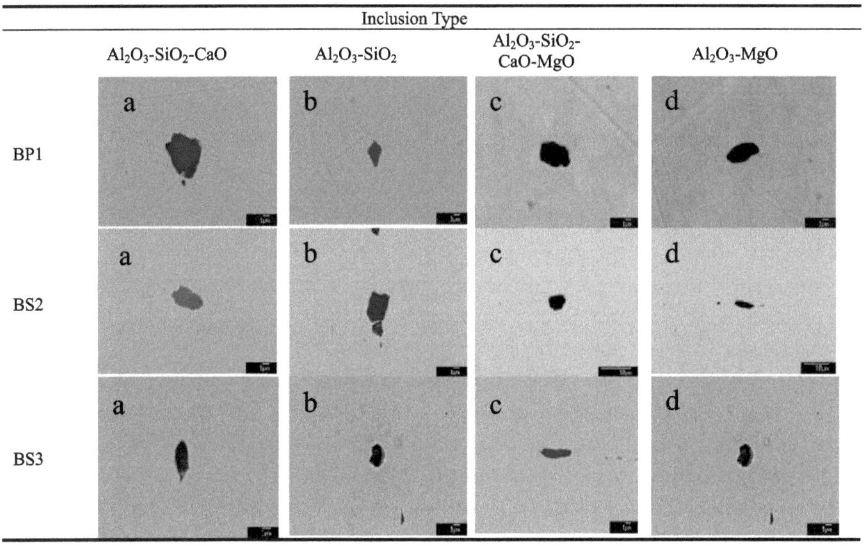

Fig. 1 Typical inclusions in steel samples deoxidized by Si

Table 1 Chemical compositions of typical inclusions in steel samples deoxidized by Si

			Al_2O_3	SiO_2	CaO	FeO	Cr_2O_3	MgO	S	Mn	TiO_2
Al_2O_3–SiO_2–CaO	BP1-a	wt%	20.1	32.5	39.5	2.0	0.0	5.9	0.0	0.1	0.0
		mol%	12.2	33.4	43.6	1.7	0.0	9.0	0.0	0.1	0.0
	BS2-a	wt%	33.5	42.1	16.6	6.7	0.2	0.3	0.0	0.6	0.0
		mol%	25.2	47.4	19.8	6.2	0.1	0.6	0.0	0.7	0.0
	BS3-a	wt%	31.4	30.7	20.1	7.9	0.2	7.0	0.7	1.5	0.5
		mol%	22.7	35.2	22.8	7.4	0.1	8.4	1.2	1.8	0.4
Al_2O_3–SiO_2	BP1-b	wt%	41.1	34.7	3.4	7.0	0.5	0.9	3.5	8.6	0.3
		mol%	29.7	39.6	3.9	4.7	0.2	1.6	7.6	12.4	0.3
	BS2-b	wt%	48.6	36.0	11.8	0.2	0.9	0.0	0.4	1.0	1.0
		mol%	36.6	46.3	12.7	0.1	1.8	0.0	0.6	0.9	1.0
	BS3-b	wt%	42.3	33.8	8.4	4.3	0.2	5.5	2.7	2.8	0.1
		mol%	31.3	37.7	10.4	2.7	0.1	11.0	3.2	3.5	0.1
Al_2O_3–SiO_2–CaO–MgO	BP1-c	wt%	37.7	32.1	13.8	4.1	0.0	11.1	0.0	0.6	0.6
		mol%	24.6	35.5	16.4	3.9	0.0	18.4	0.0	0.8	0.5
	BS2-c	wt%	37.8	27.8	17.3	3.9	0.2	10.6	0.0	2.2	0.2
		mol%	28.3	30.1	20.1	3.5	0.1	14.7	0.1	2.7	0.4
	BS3-c	wt%	10.2	37.4	36.6	3.4	0.1	9.2	1.5	1.6	0.0
		mol%	5.8	36.1	37.8	2.7	0.0	13.3	2.6	1.7	0.0
Al_2O_3–MgO	BP1-d	wt%	68.5	5.9	1.6	5.0	0.1	11.7	1.0	5.7	0.4
		mol%	51.7	7.6	2.2	5.4	0.1	22.2	2.4	8.0	0.4
	BS2-d	wt%	31.1	51.7	0.2	3.0	0.0	10.3	0.0	3.4	0.3
		mol%	19.8	56.0	0.2	2.9	0.0	16.6	0.0	4.0	0.5
	BS3-d	wt%	44.0	6.5	2.8	7.4	0.5	35.8	0.1	2.7	0.2
		mol%	32.7	7.7	3.7	8.1	0.3	43.1	0.3	3.7	0.4

semi-spherical shape. The second type of inclusions are Al_2O_3–SiO_2 and contain a much higher Al_2O_3 content. The third type are CaO–SiO_2–Al_2O_3–MgO system. The last type of inclusions are spinels. It is widely accepted that SiO_2 is the product of deoxidation, Al_2O_3 may come from ferroalloys, which usually contain Al as an impurity. And it may also come from the refractories.

Figure 2 shows the morphology of inclusions in rolled products. CaO–SiO_2–Al_2O_3–MgO system inclusions changed from a spherical shape to elongated shape after rolling as shown in Fig. 2a, b. Distinctively, CaO–SiO_2–Al_2O_3 system inclusions can be easily elongated during rolling and can present as a long strip shape in rolled products as shown in Fig. 2c, d. The chemical compositions of CaO–SiO_2–Al_2O_3–MgO and MgO–SiO_2–Al_2O_3 system inclusions are homogeneous. Discontinuous dot-linked shaped inclusions and eye-like shaped inclusions are detected in rolled products and demonstrated in Figs. 3 and 4 (Tables 2 and 3), respectively. Point analysis indicates compositions of these inclusions are not homogeneous. Meanwhile, some very large CaO–SiO_2 inclusions over 10 μm were also observed, as shown in Fig. 5.

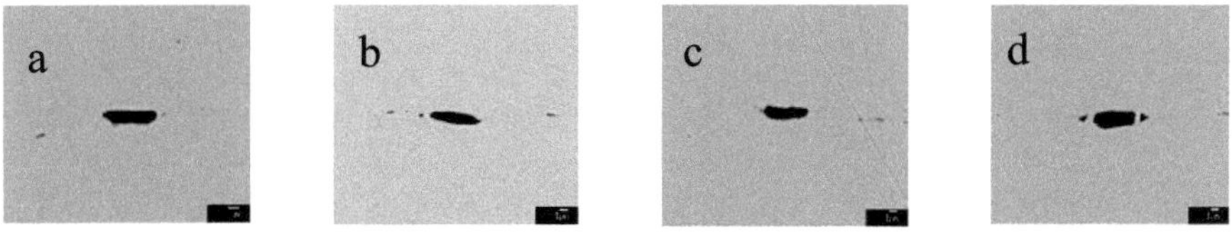

Fig. 2 Morphology of $CaO–SiO_2–Al_2O_3–MgO$ and $MgO–SiO_2–Al_2O_3$ inclusions along the rolling direction

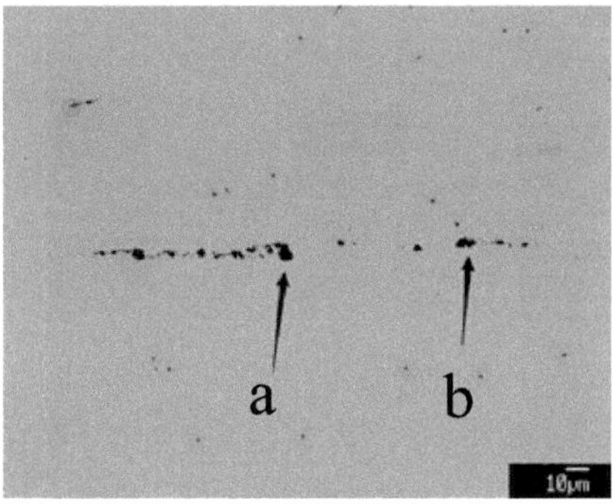

Fig. 3 Discontinuous dot-linked shape along the rolling direction

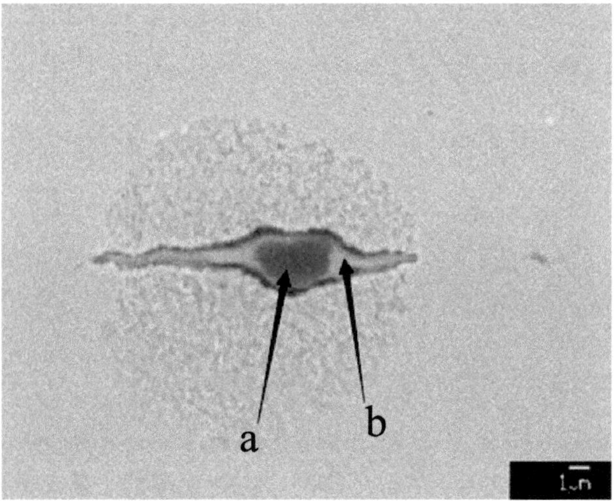

Fig. 4 Eye-like dual-phase inclusions along the rolling direction

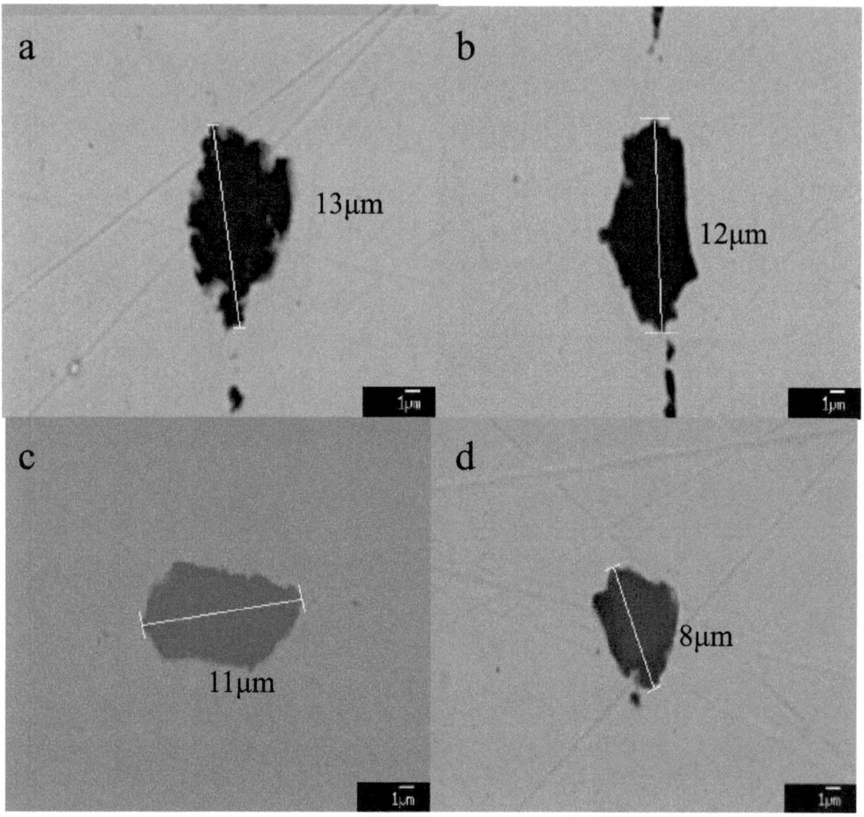

Fig. 5 Large size CaO–SiO_2 inclusions detected in the steel samples

Table 2 Chemical compositions of dual-phase inclusions in discontinuous dot-linked shape along the rolling direction

		Al_2O_3	SiO_2	CaO	FeO	Cr_2O_3	MgO	S	Mn	TiO_2
a	wt%	40.1	23.8	10.6	2.8	0.1	19.0	0.1	1.1	0.3
	mol%	25.9	26.1	12.5	2.6	0.1	31.1	0.1	1.3	0.3
b	wt%	15.7	39.9	20.0	8.4	0.2	12.6	0.0	1.0	0.6
	mol%	9.4	40.7	21.9	7.1	0.1	19.2	0.1	1.1	0.5

Table 3 Chemical compositions of dual-phase inclusions along the rolling direction

		Al_2O_3	SiO_2	CaO	FeO	Cr_2O_3	MgO	S	Mn	TiO_2
a	wt%	55.8	12.7	20.0	6.3	0.1	2.6	1.7	0.4	0.3
	mol%	41.1	15.8	26.8	6.6	0.0	4.8	4.0	0.5	0.3
b	wt%	22.3	1.0	6.9	18.6	0.8	7.7	18.7	23.9	0.0
	mol%	13.3	1.0	7.5	15.7	0.3	11.7	24.3	26.3	0.0

Number of Typical Inclusions in Steel

The statistic results of inclusion distribution and size are plotted in Fig. 6. Average size and density of inclusions per unit area are lower in BS2 than in BP1. This could be explained by the crystallization of inclusions. After cold drawing, the number density of the inclusions increases and average size decreases greatly, because brittle inclusions of Al_2O_3–MgO–SiO_2–CaO are broken into small pieces, due to the weak plastic deformation capability in the drawing process. Almost 80% of all inclusion are less than 5 μm in BS3. And in steel BS3, almost no inclusions larger than 10 μm can be found.

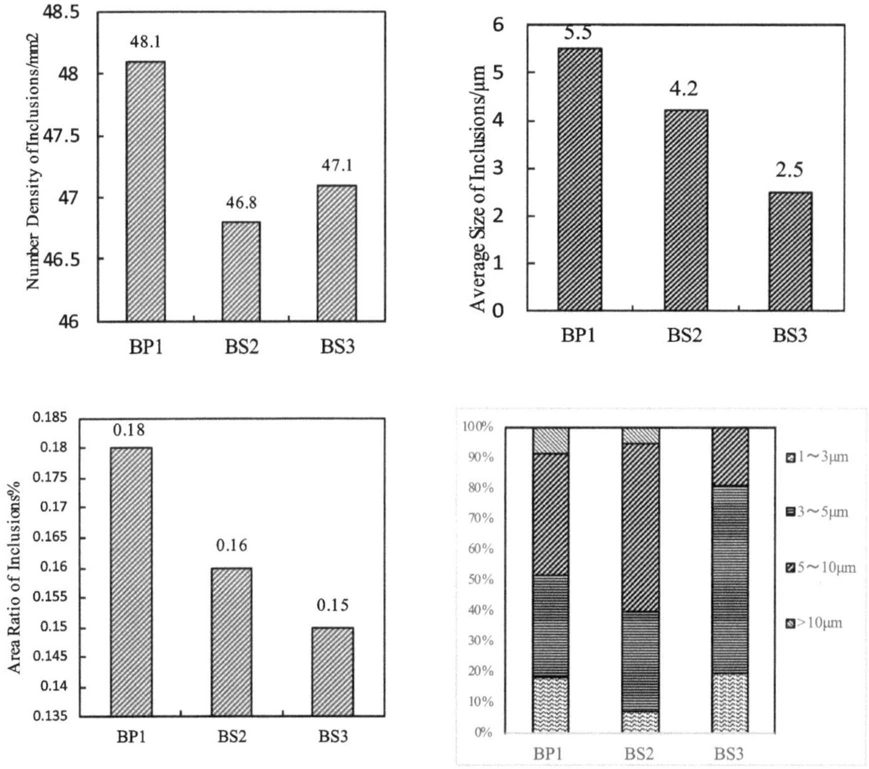

Fig. 6 Inclusions size and number in three samples

Change of Inclusions Composition in Three Samples

Inclusions Composition Changes

In order to confirm the formation of inclusions, the ratio of each kind of inclusions was studied by examining hundreds of oxide inclusions after the hot rolling (BP1), heat treatment (BS2) and cold drawing processes (BS3). The number of inclusions analysed in three samples are 105, 115 and 106 respectively. From the statistical results shown in Fig. 7, we can see that the ratio of Al_2O_3–SiO_2–CaO inclusions decreases from 73 to 59% after heat-treatment, while it increases to 79% after cold drawing. Furthermore, some Al_2O_3–MgO inclusions are found in wire rods, but few of them exist after heat-treatment and cold drawing.

Transformation of the average composition of the compound oxides during different stages is shown in Fig. 8. The compositions of the inclusions in steel are

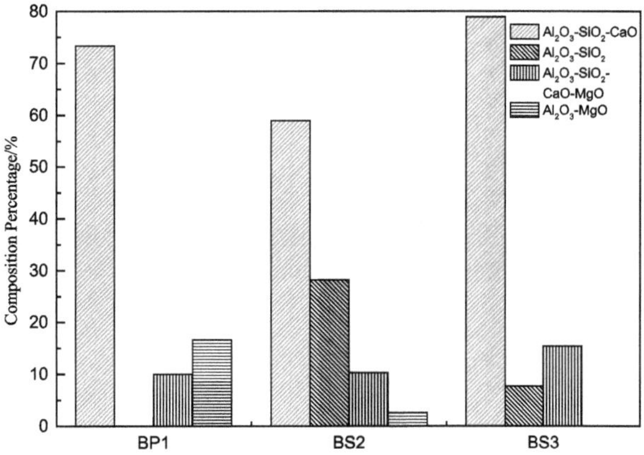

Fig. 7 Ratio of each kind of inclusions at different stages

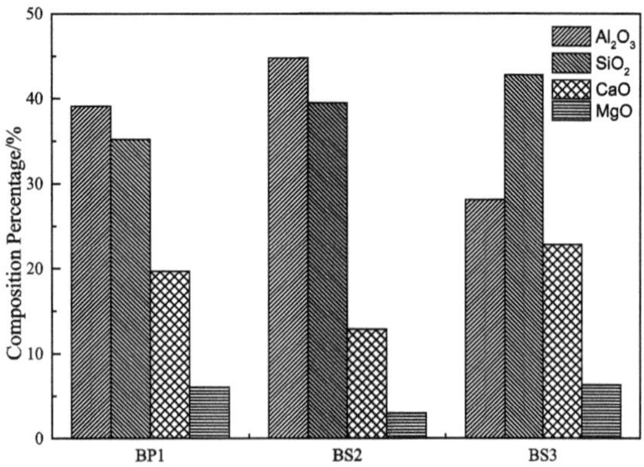

Fig. 8 Average compositions of CaO–SiO_2–Al_2O_3–MgO inclusions

primarily Al_2O_3: 39.1%, SiO_2: 35.2%, CaO: 19.7%, and MgO: 6.1% in BP1. Al_2O_3 and SiO_2 contents in inclusions increases gradually, while the CaO and MgO content decreases gradually in BS2. For BS3, the content of SiO_2 in inclusions keeps increasing gradually, while the content of Al_2O_3 reaches the lowest level and CaO and MgO slightly increases.

The typical line analysis of Al_2O_3–SiO_2–CaO inclusion is shown in Fig. 9. It can be seen that this inclusion is single phase and a much higher level of SiO_2 exists in the core of this inclusion, which is almost equal to the sum of the content of Al_2O_3 and CaO. Compositions of CaO–SiO_2–Al_2O_3 and Al_2O_3–SiO_2 inclusions analysed in BP1, BS2 and BS3 are projected onto the Al_2O_3–SiO_2–CaO phase diagram as presented in Fig. 10. As can be seen in Fig. 10, the melting point of most CaO–SiO_2–Al_2O_3 system inclusions are above 1500 °C. From Fig. 10 we could find that there are no significant changes in the composition of CaO–SiO_2–Al_2O_3 inclusions after cold drawing and heat-treatment. Unlike Al_2O_3–SiO_2 inclusions produced in deoxidation, CaO–SiO_2–Al_2O_3 inclusions always contained a small amount of CaO. Composition of CaO–SiO_2 system inclusions was similar to the top slag. Moreover, from Fig. 5, it is shown that most of the CaO–SiO_2 inclusions were over 10 μm. Therefore, CaO–SiO_2 system inclusions were concluded to be entrapped slag particles. By comparison, CaO–SiO_2–Al_2O_3 inclusions always have a higher CaO content than the CaO–SiO_2 inclusions, and more Al_2O_3 than the top slag. In addition, the size of the CaO–SiO_2–Al_2O_3 inclusions is larger than the Al_2O_3–SiO_2 inclusions, but smaller than the CaO–SiO_2 inclusions. If a line is drawn to link the average composition of Al_2O_3–SiO_2 inclusions and the top slag, it can be seen that most of the CaO–SiO_2–Al_2O_3 inclusions are concentrated in the middle of this line. Hence, a reasonable explanation can be proposed that the CaO–SiO_2–Al_2O_3 inclusions come from the coalescence between Al_2O_3–SiO_2 inclusions and entrapped slag particles.

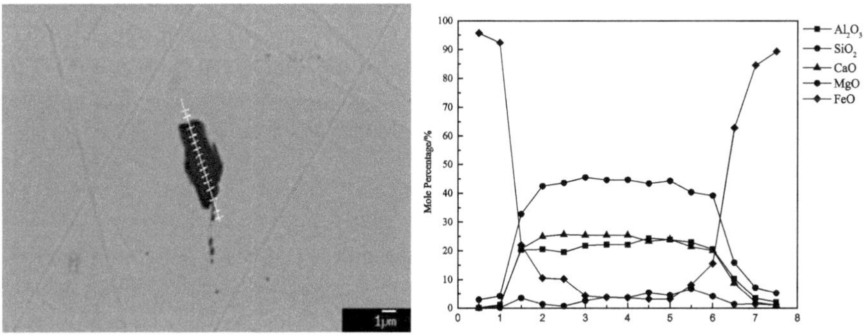

Fig. 9 Analysis location and values for an Al_2O_3–SiO_2–CaO inclusion (0.5 μm interval)

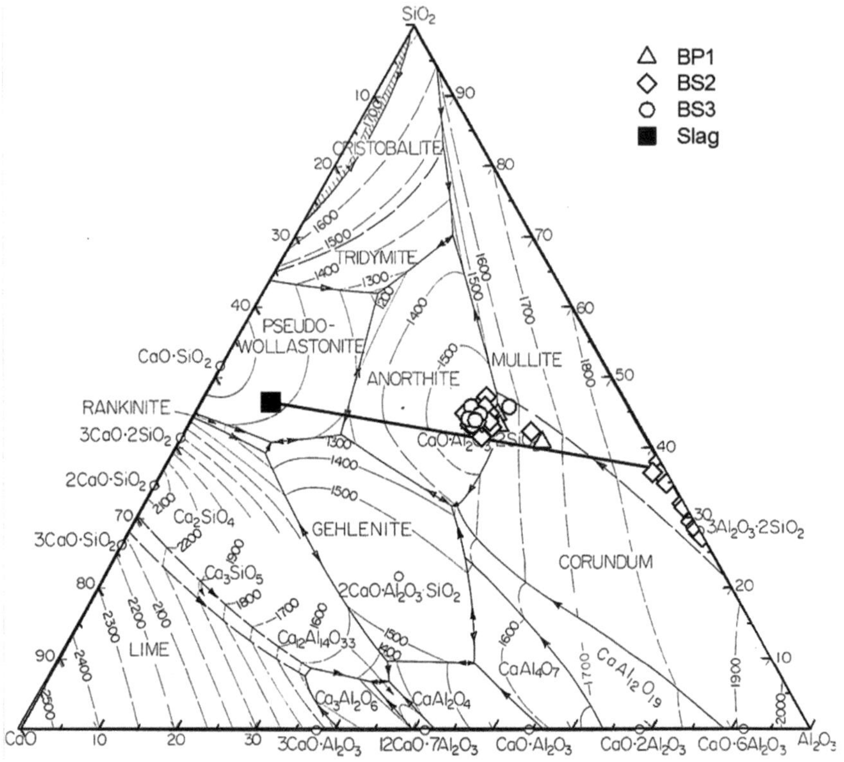

Fig. 10 Composition distribution of CaO–SiO_2–Al_2O_3 inclusions detected in BP1, BS2 and BS3

Origin and Evolution of SiO_2–Al_2O_3 System Inclusions

The average composition of the inclusions in the steel samples taken from BP1, BS2 and BS3 are shown in Fig. 8. It is of interest that the contents of SiO_2 and Al_2O_3 in the inclusions are always very high. As for the high SiO_2 content, it is mainly owing to the deoxidization when FeSi alloys were added, as given by Eq. (1).

$$[\mathrm{Si}] + 2[\mathrm{O}] = (\mathrm{SiO_2})^{\mathrm{inclusion}} \quad (1)$$

Actually, an increase in the content of alumina is partly due to aluminum contamination (Eq. 2) when the ferrosilicon alloys were added as a deoxidizer or as an alloying element, because the commercial grade ferrosilicon alloys have aluminum as an impurity.

$$2[\mathrm{Al}]^{FeSi} + 3[O] = (Al_2O_3)^{inclusions} \quad (2)$$

Origin and Evolution of $CaO–SiO_2–Al_2O_3$ System Inclusions

With the addition of Fe–Si alloy-containing low levels of aluminium, these reactive metals would quickly dissolve and react with dissolved oxygen. As a result, Al_2O_3–SiO_2 inclusions would be largely produced. At the same time, owing to the strong agitation of the liquid steel during tapping, emulsification of the added slag in liquid steel could take place and therefore result in the formation of exogenous inclusions.

Due to the violent agitation of the flowing steel, the emulsified slag particles and Al_2O_3–SiO_2 inclusions can frequently collide and coalescence with each other, contributing to the change of slag particle composition. If the sizes of slag particles were large, changes of compositions would be negligible. These small-sized emulsified slag particles would be changed into CaO–SiO_2–Al_2O_3-inclusions, while large-sized ones would remain essentially as CaO–SiO_2-inclusions, which accounts for why the former was always detected in smaller sizes, while the latter were found in larger sizes during the experiments.

Conclusions

In current study, formation of CaO–SiO_2–Al_2O_3–MgO complex inclusions were examined and reevaluated in a more detailed way. By analyzing the types, sizes, compositions, and morphology of inclusions, continuous changes of inclusions at different stages from casting to cold drawing can be depicted. A mechanism was proposed to explain the formation of CaO–SiO_2–Al_2O_3 system inclusions. Important findings can be summarized as following:

(1) Inclusions discovered in the wire rods can be divided into four kinds, that are, Al_2O_3–SiO_2–CaO, Al_2O_3–SiO_2, Al_2O_3–SiO_2–CaO–MgO and Al_2O_3–MgO inclusions. CaO–SiO_2–Al_2O_3 system inclusions can be easily elongated during rolling and can present as a long strip shape in rolled products. Some very large CaO–SiO_2 inclusions over 10 μm were observed.

(2) CaO–SiO_2–Al_2O_3-inclusions were the reaction product of Al_2O_3–SiO_2 inclusions and CaO–SiO_2 inclusions. Al_2O_3–SiO_2 inclusions were likely produced during Si deoxidation and CaO–SiO_2 inclusions mainly came from the slag.

(3) Formation of CaO–SiO_2–Al_2O_3–(MgO) inclusions involved two important steps. First, emulsion of slag occurred, resulting in the entrapment of slag particles. Collisions and coalescence between these exogenous particles and Al_2O_3–SiO_2 inclusions happened because of the violent flow of molten steel. Second, CaO–SiO_2 inclusions reacted with molten steel, causing the increase of Al_2O_3 in the inclusions and resulting in the evolution of CaO–SiO_2 into CaO–SiO_2–Al_2O_3 system inclusions.

Acknowledgements The authors would like to thank: (1) financial support from Baosteel through Baosteel-Australia Joint Research and Development Centre; (2) China Scholarship Council (CSC) and The University of Queensland for providing a scholarship for Sha Lv; (3) Mr. Ron Rasch and Ms. Ying Yu of the Centre for Microscopy and Microanalysis (CMM) at the University of Queensland for technical support for EPMA and SEM analyses.

References

1. Q.Y. Wang, C. Bathias, N. Kawagoishi, Q. Chen, Effect of inclusion on subsurface crack initiation and gigacycle fatigue strength. Int. J. Fatigue **24**(12), 1269–1274 (2002)
2. H.L. Yang, J.S. Ye, X.L. Wu, Y.S. Peng, Y. Fang, X.B. Zhao, Effect of top slag with low basicity on transformation control of inclusions in spring steel deoxidized by Si and Mn. ISIJ Int. **56**(1), 108–115 (2016)
3. M. Jiang, X. Wang, B. Chen, W. Wang, Laboratory study on evolution mechanisms of non-metallic inclusions in high strength alloyed steel refined by high basicity slag. ISIJ Int. **50** (1), 95–104 (2010)
4. S. Feichtinger, S.K. Michelic, Y.B. Kang, C. Bernhard, In Situ observation of the dissolution of SiO_2 particles in $CaO–Al_2O_3–SiO_2$ slags and mathematical analysis of its dissolution pattern. J. Am. Ceram. Soc. **97**, 316–325 (2014)
5. J.W. Kim, S.K. Kim, D.S. Kim, Y.D. Lee, P.K. Yang, Formation mechanism of Ca-Si-Al-Mg-Ti-O inclusions in type 304 Stainless Steel. ISIJ Int. **36**(Suppl), S140–S143 (1996)
6. D.-H. Woo, Y.-B. Kang, H.-G. Lee, Thermodynamic study of MnO-SiO_2-Al_2O_3 slag system: liquidus lines and activities of MnO at 1823 K. Metall. Mater. Trans. B **33**(6), 915–920 (2002)
7. K. Wang, M. Jiang, X. Wang, Y. Wang, H. Zhao, Z. Cao, Formation mechanism of CaO-SiO_2-Al_2O_3-(MgO) inclusions in Si-Mn-killed steel with limited aluminum content during the low basicity slag refining. Metall. Mater. Trans. B **47**(1), 282–290 (2016)
8. C.-W. Seo, S.-H. Kim, S.-K. Jo, M.-O. Suk, S.-M. Byun, Modification and minimization of spinel ($Al_2O_3 \cdot xMgO$) inclusions formed in Ti-added steel melts. Metall. Mater. Trans. B **41**(4), 790–797 (2010)
9. H. Todoroki, K. Mizuno, Effect of silica in slag on inclusion compositions in 304 stainless steel deoxidized with aluminum. ISIJ Int. **44**(8), 1350–1357 (2004)

Investigation on Coal Combustion Behaviors in the Oxygen Blast Furnace

Zhenfeng Zhou, Yuanyuan Zhang, Guang Wang, Jingsong Wang and Qingguo Xue

Abstract The oxygen blast furnace is a new technology for ironmaking, where the oxygen and CO_2-stripped blast furnace top gas are blown into the blast furnace replacing the conventional hot air. The oxygen blast furnace could reduce the consumption of coke and emission of carbon dioxide. In this chapter, a three-dimensional numerical model was developed to simulate the lance-blowpipe-tuyere-raceway region of a blast furnace, and the flow and combustion behaviors of coal particles were investigated based on an oxygen blast furnace. Compared with the traditional blast furnace, the combustion characteristics in the raceway of the oxygen blast furnace were different. In addition to the combustion of coal and coke, the combustion of large amounts of reducing gas was included. The coal combustion of different oxygen content (70, 80, 90, 100%) were investigated. The simulation results indicated that the coal burnout was only 21.64% at an oxygen content of 70%, a decrease of 51.52%, compared with a traditional blast furnace. The coal burnout was increased with the increase of oxygen content. When the oxygen content is 100%, the coal burnout was 90.38%, an increase of 17.22%.

Keywords Raceway · Oxygen blast furnace · Burnout

Z. Zhou (✉) · Y. Zhang · G. Wang · J. Wang · Q. Xue
State Key Laboratory of Advanced Metallurgy, University of Science and Technology Beijing, Beijing 100083, People's Republic of China
e-mail: zhouzhenfeng1989@sina.com

Y. Zhang
e-mail: 15841693946@163.com

G. Wang
e-mail: wangguang@ustb.edu.cn

J. Wang
e-mail: wangjingsong@ustb.edu.cn

Q. Xue
e-mail: xueqingguo@ustb.edu.cn

J.-Y. Hwang et al. (eds.), *8th International Symposium on High-Temperature Metallurgical Processing*, The Minerals, Metals & Materials Series,
DOI 10.1007/978-3-319-51340-9_33

Introduction

Blast furnaces are still the main facilities for hot metal production due to their superiority in productivity and heat utilization [1–3]. Under the pressure of resource scarcity and environmental degradation, researchers have to develop a series of new blast furnace ironmaking technologies. To overcome this problem, during the operation of a modern blast furnace, as much coal as possible is injected in order to partially replace coke. However, to improve the coal ratio, the main problem is to obtain sufficient combustion of the coal in the raceway region [4]. The most effective way to solve this problem is oxygen enrichment. The oxygen blast furnace (OBF) is a new ironmaking process to solve this problem. The Full Oxygen Blast Furnace (FOBF) proposed by Qin [5] is characterized by operation using pure oxygen instead of the hot blast and the recycling the top gas after removal of CO_2.

Compared to the Traditional Blast Furnace (TBF), the blast parameters of the OBF, and the coal combustion behavior are very different; therefore, is very different. Therefore, the coal combustion behavior under OBF conditions has been investigated.

Due to the high temperature and harsh environment of the blast furnace, it is difficult and expensive to carry out a trial of PCI directly in an industrial blast furnace. Furthermore, due to the limitations of many conditions, the results may be not reliable [6]. The development of computational fluid dynamics (CFD) and its application in coal combustion provide a new research method for blast furnace PCI investigation. Many studies have been carried out on PCI in blast furnace using CFD. Wijayanta et al. [7, 8] investigated the injection of pulverized biochar into the blast furnace using CFD. Shen et al. [9–12] carried out many studies on the flow and combustion behavior of pulverized coal in blast furnace using CFD. Shen validated his models using a pilot-scale test rig. Many CFD investigations on coal combustion have demonstrated the reliability of this approach.

In this chapter, a numerical model was developed to simulate the lance-blowpipe-tuyere-raceway of an OBF using CFD. The effect of arrangement of oxygen and recycling gas on coal combustion is investigated. Furthermore, the effect of oxygen content is examined.

Model Description

The gas-particle flow and coal combustion in the tuyere and raceway region were calculated based on the framework of the software package ANSYS–FLUENT. The mathematical formulation is described below.

Basic Equations

In this model, the gas phase is treated with an Eulerian frame and described by the steady-state Reynolds-averaged Navier–Stokes equations closed by a k–ε turbulence model [13]. The governing equations for the gas phase are summarized in Table 1.

Particles of pulverized coal are treated as discrete phase, modeled using the Lagrangian method, where the trajectories of the discrete particles are determined by integrating Newton's second law of motion. The drag force (f_D) and turbulence dispersion are included. Full coupling of mass, momentum, and energy of particles with the gas phase is implemented. The change of particle temperature is governed by three physical processes: convective heat transfer, latent heat transfer associated with mass transfer, and radiative heat transfer. The governing equations for the particle phase are summarized in Table 2.

Combustion Process

Coal combustion is regarded as a multistage overlapping process: (1) preheating, (2) the devolatilization of raw coal, (3) the volatiles combustion, and (4) the oxidation/gasification of the residual char.

(1) Release of volatiles

The devolatilization process releases volatiles ($C_\alpha H_\beta O_\gamma N_\delta$) and char (C(s)). The release of volatiles is simulated using the so-called two-competing-reactions model [14]. A pair of reactions with different rates (k_1, k_2) and volatile yields (α_1, α_2) compete to pyrolyse the raw coal.

Table 1 Governing equations for the gas phase

Mass	$\nabla \cdot (\rho U) = \sum_{np} \dot{m}$
Momentum	$\nabla \cdot (\rho UU) - \nabla \cdot \left[(\mu + \mu_t)\left(\nabla U + (\nabla U)^T\right)\right] = -\nabla\left(P + \frac{2}{3}\rho k\right) + \sum_{np} f_D$
Energy	$\nabla \cdot \left[\rho UH - \left(\frac{\lambda}{C_P} + \frac{\mu_t}{\sigma_H}\right)\right] = \sum_{np} q$
Gas species	$\nabla \cdot \left[\rho UY_i - \left(\Gamma_i + \frac{\mu_t}{\sigma_{Yi}}\right)\right] = W_i$
Turbulent kinetic energy	$\nabla \cdot \left[\rho Uk - \left(\mu + \frac{\mu_t}{\sigma_k}\right)\nabla k\right] = P_k - \rho\varepsilon$
Turbulent dissipation rate	$\nabla \cdot \left[\rho U\varepsilon - \left(\mu + \frac{\mu_t}{\sigma_\varepsilon}\right)\nabla\varepsilon\right] = \frac{\varepsilon}{k}(C_1 P_k - C_2\rho\varepsilon)$

Table 2 Governing equations for the particle phase

Mass	$\frac{dm_p}{dt} = -\dot{m}$
Momentum	$m_p \frac{du_p}{dt} = -f_D$ $-f_D = \frac{1}{8}\pi d_p^2 \rho C_D \lvert U - U_p \rvert (U - U_p)$
Energy	$m_p C_p \frac{dT_P}{dt} = -q$ $\pi d_p \lambda N_\mu (T_g - T_p) + \sum \frac{dm_p}{dt} H_{reac} + A_p \varepsilon_p \left(\pi I - \sigma_B T_p^4\right)$

$$\text{Raw coal} \xrightarrow{k_1} \alpha_1 VM_1 + (1 - \alpha_1) C_1 \qquad \text{(Low Temperature)}$$

$$\text{Raw coal} \xrightarrow{k_2} \alpha_2 VM_2 + (1 - \alpha_2) C_2 \qquad \text{(High Temperature)}$$

The rate constants k_1 and k_2 are given in the Arrhenius format [15]:

$$k = A \exp(-E/T) \tag{1}$$

where A represents the Arrhenius rate constant, E represents the activation energy. A and E are $3.7 \times 10^5\ \mathrm{s}^{-1}$ and 18,000 K at low temperature, and $1.46 \times 10^{13}\ \mathrm{s}^{-1}$ and 30,189 K at high temperature.

The rate of volatiles production is given by:

$$\frac{dw(VM)}{dt} = (\alpha_1 k_1 + \alpha_2 k_2) C_0 \tag{2}$$

(2) Gas combustion

The volatiles are treated as one fuel gas ($C_\alpha H_\beta O_\gamma N_\delta$), and its combustion is simplistically represented by two overall reactions as follows:

$$\left(C_\alpha H_\beta O_\gamma N_\delta\right) + aO_2 \rightarrow bCO + cH_2O + dN_2 \cdots R_1$$
$$CO + 0.5O_2 \rightarrow CO_2 \cdots R_2$$

For the gas combustion, in addition to the combustion of volatiles, the combustion of hydrogen is included:

$$H_2 + 0.5O_2 \rightarrow H_2O \cdots R_3$$

In turbulent flow, the gas reaction mechanisms are represented by the finite rate/eddy dissipation model.

(3) Oxidation/gasification of the residual char

For the char reactions, the heterogeneous surface reactions model is used. The solid reactions include the following processes.

$$C(s) + 0.5O_2 \rightarrow CO \cdots R_4$$
$$C(s) + CO_2 \rightarrow 2CO \cdots R_5$$
$$C(s) + H_2O \rightarrow CO + H_2 \cdots R_6$$

The reaction rate is expressed as follows:

$$R_{j,r} = A_p \eta_r Y_j P \frac{k_r D_{0,r}}{D_{0,r} + k_r} \tag{3}$$

where A_p and η_r indicate the particle surface area [m^2] and effectiveness factor, respectively. Y_j is the mass fraction of species j. P represents the partial pressure [Pa]. The diffusion rate coefficient, $D_{0,\ r}$, for reaction r is computed as:

$$D_{0,r} = C_{j,r} \frac{\left[\left(T_p + T_g\right)/2\right]^{0.75}}{d_p} \tag{4}$$

where $C_{j,r}$ is the molar concentration of species j in reaction r. d_p represents particle diameter [m].

Geometry and Operating Condition

An oxygen blast furnace of 120 m^3 will be built at a steel company in China based on a TBF of 120 m^3. The main operational parameters of the TBF and OBF are summarized in Table 3.

The TBF of 120 m^3 was dissected in 2007. The results of the blast furnace dissection show that the depth of the raceway was about 700 mm. The detailed parameters of this model are shown in Fig. 1. The main focus was the flow and combustion behavior of pulverized coal in the coal plume formed, which plays a significant role in determining the permeability of the bird's nest. The raceway was designed as a tube of 700 mm long with a divergence angle of 3° based on the work of others [9]. This geometric setting could avoid the formation of flow recirculation, so that this simulation mainly focuses on coal flow and combustion behaviors of pulverized coal in the coal plume formed. The tuyere and raceway of the TBF and OBF are designed to be the same.

Table 3 Blast parameters: (a) TBF; (b) OBF

(a)

Blast volume (Nm^3/h)	Blast temperature (K)	Volume (m^3)	Coal ratio (kg/t)
19766.78	1473	120	150

(b)

Oxygen		Recycling top gas						
Volume (Nm^3/t)	Content (%)	Volume (Nm^3/t)	Temperature (°C)	CO (%	H_2 (%)	CO_2 (%)	N_2 (%)	H_2O (%)
258	90	350	900	64.32	14.62	4.5	15.94	0.62

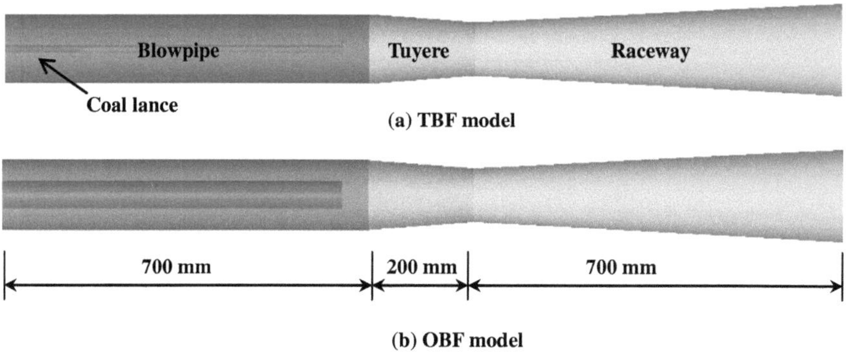

Fig. 1 Geometry of the model

Results and Discussion

Effect of Arrangement of Oxygen and Recycling of Top Gas

Figure 2 shows the arrangement of oxygen and recycling gas. The oxygen, recycling gas, and coal are arranged coaxially. The coal is arranged in the center. For case 1, the recycling gas is arranged on the outside. For case 2, the oxygen is arranged on the outside. For case 1 and 2, the velocity of oxygen and recycling gas is the same.

Figure 3 shows the effect of the arrangement of oxygen and recycling gas on coal burnout. The coal burnout is 73.16% under TBF. For case 1, the coal burnout is 53.42%, with a decrease of 19.74%. For case 2, the coal burnout is 88.03%, with an increase of 14.87%. The reasons causing the aforementioned phenomena are as follows: For case 2, the coal particles contact with the recycling gas directly. However, due to the limitation of diffusion, the coal particles cannot full contact with the oxygen. Therefore, the coal burnout is lower due to the lack of oxygen. For case 1, the coal particles contact with the oxygen directly. After devolatilization, the

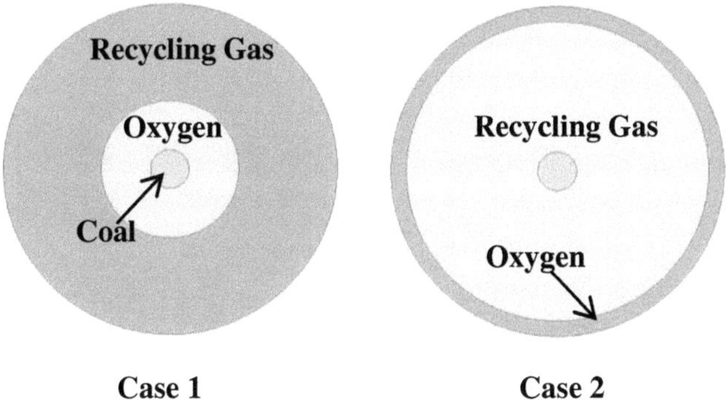

Fig. 2 Arrangement of oxygen and recycling gas

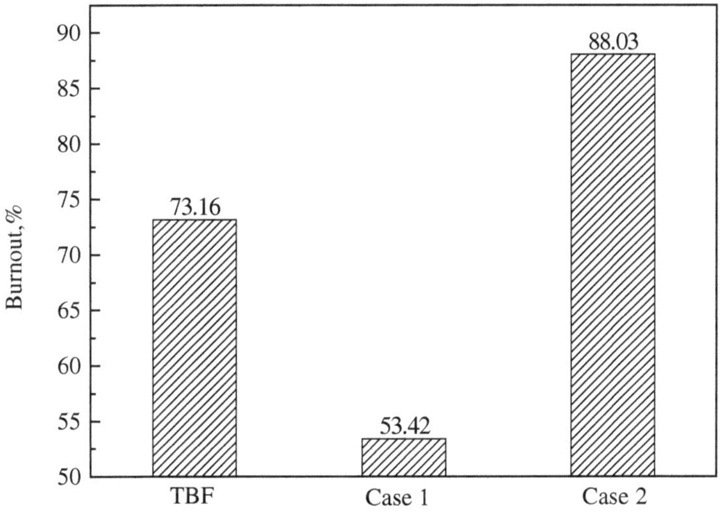

Fig. 3 Effect of arrangement of oxygen and recycling top gas on coal burnout

coal particles rapidly react with large amounts of oxygen; therefore, the burnout increases significantly.

Figure 4 shows the coal burnout at different positions and under different arrangements. At less than 0.7 m, the burnout in the TBF is higher than that of case 1 and 2. After at the distance of 0.7 m, the burnout of the TBF is lower than that of case 1 and higher than that of case 2. This is because the hot blast temperature is 1473 K and higher than that of oxygen and recycling gas in the OBF. In the first half part of the raceway, the coal particles mainly release volatiles. This is mainly affected by the temperature. After devolatilization, it is mainly the combustion of the volatiles and char. This is mainly affected by the oxygen. Therefore, for case 1,

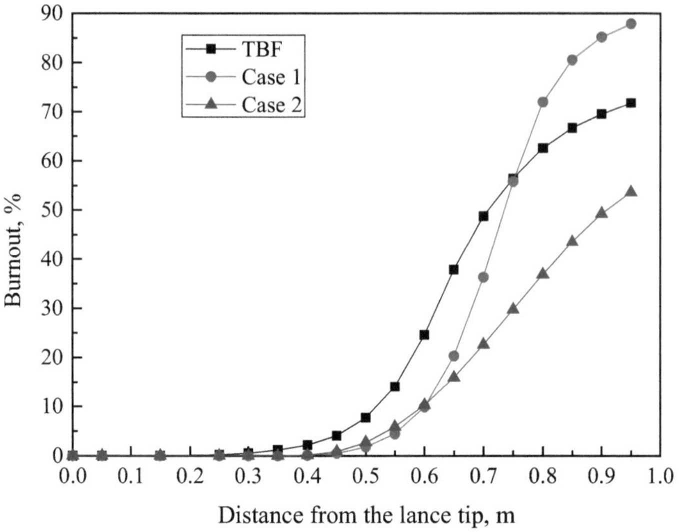

Fig. 4 Coal burnout at different positions under different arrangements

before the distance of 0.5 m, the burnout is lower, and rapidly increases after the distance of 0.5 m. At the end of the raceway, the burnout increases slowly due to the lack of oxygen. For case 2, the burnout is lower due to the lack of oxygen around the coal particles.

Effect of Velocity Ratio of Oxygen and Recycling Gas

Figure 5 shows the effect of velocity ratio of oxygen and recycling gas on coal burnout. The coal burnout slightly increases when the velocity ratio increases or decreases. When the velocity ratio is 0.5, the coal burnout is 89.65%, with an increase of 1.77%. This is because the effect of gas flow on coal particles decreases, and the residence time of coal particles extends. When the velocity ratio is 1.5, the coal burnout is 89.04%, with an increase of 1.16%. This may be because the velocity of recycling gas decreases, and the recycling gas combusts more completely, releasing more heat. This is advantageous to coal combustion.

Effect of Oxygen Content

Table 4 shows the blast parameters of different OBFs of 70, 80 and 100% O_2. Figure 6 shows the effect of oxygen content on coal burnout. The coal burnout

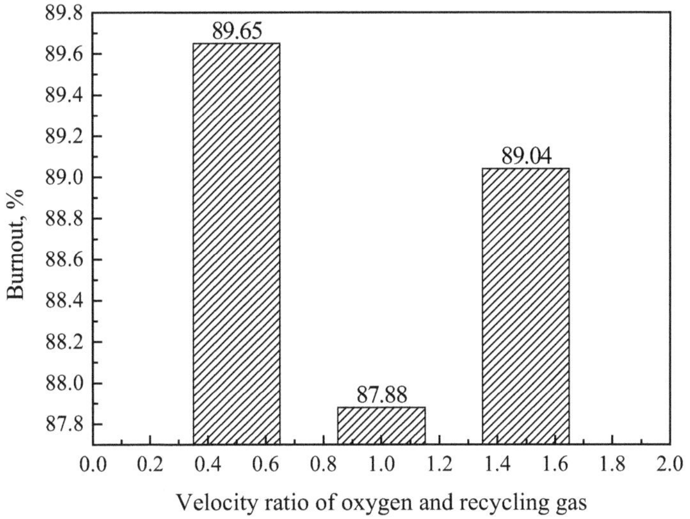

Fig. 5 Effect of velocity ratio of oxygen and recycling gas on coal burnout

Table 4 Blast parameters of different OBFs: (a) 70% O_2; (b) 80% O_2; (c) 100% O_2

Oxygen		Recycled top gas						
Volume (Nm^3/t)	Content (%)	Volume (Nm^3/t)	Temperature (°C)	CO (%)	H_2 (%)	CO_2(%)	N_2 (%)	H_2O (%)
(a)								
366	70	280	900	52.58	10.81	4.5	31.49	0.62
(b)								
306	80	300	900	57.87	12.59	4.5	24.42	0.62
(c)								
228	100	365	900	74.08	17.41	4.5	3.39	0.62

increases with the increase of oxygen content significantly. When oxygen content is 70%, the coal burnout is only 21.64%, with a decrease of 51.52%. When oxygen content is 80%, the coal burnout is 71.63%, with a decrease of 1.53%. When oxygen content is 90%, the coal burnout is 88.03%, with an increase of 14.87%. When oxygen content is 100%, the coal burnout is 90.38%, with an increase of 17.22%. The reasons causing aforementioned phenomena are as follows: (i) the room-temperature oxygen and nitrogen limits the release of volatiles, and (ii) the whole combustion is delayed. This is very disadvantageous to coal combustion and the effect is more obvious at lower oxygen content. As shown from Table 4, with the increase of oxygen content, the volume and CO content in the recycled gas both increases; these are both advantageous for optimal coal combustion.

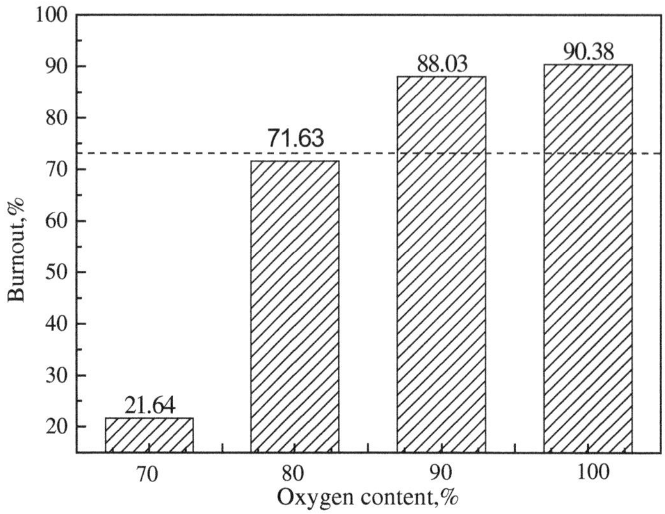

Fig. 6 Effect of oxygen content on coal burnout

Conclusions

The coal combustion behavior in the oxygen blast furnace was investigated. The main conclusions are as follows:

1. The effect of the geometry of the injection arrangement for oxygen and recycle gas on coal burnout is significant. For case 2 that the oxygen is arranged on the outside, the coal burnout is 53.42%, with a decrease of 19.74%. For case 1 that the recycling gas is arranged on the outside, the coal burnout is 88.03%, with an increase of 14.87%.
2. The velocity ratio of oxygen and recycled gas has some effect on coal burnout. However, the effect is small. When the velocity ratio is 0.5, the burnout is 89.65%; when the velocity ratio is 1, the burnout is 88.03%; when the velocity ratio is 1.5, the burnout is 89.04%;
3. The coal burnout increases significantly with the increase of oxygen content. When the oxygen content is 70 and 80%, the burnout is lower than that of the TBF. Particularly, when the oxygen content is 70%, the burnout is only 21.64%. When oxygen content is 90 and 100%, the burnout is higher than that of the TBF.

Nomenclature

C_1, C_2	Turbulent model constants
C_D	Drag coefficient
C_p	Particle heat capacity, $Jkg^{-1}\ k^{-1}$

f_D	Drag force from a particle, N
H	Enthalpy, Jkg^{-1}
h_g	Heat transfer coefficient
H_{reac}	Reaction heat, Jkg^{-1}
I	Radiation intensity, $Wm^{-2}\ s^{-2}$
m	Mass transfer rate from a particle, kgs^{-1}
n_p	Particle number per unit volume, m^{-3}
N_u	Nusselt number
U	Mean (true) velocity of gas, ms^{-1}
W_i	Reaction rate of species i (per unit volume), $kgm^{-3}\ s^{-1}$
Y_i	Mass fraction of species i
ε	Turbulent dissipation rate, $m^2\ s^{-3}$
ε_p	Particle emissivity
λ	Thermal conductivity, $Wm^{-1}\ k^{-1}$
σ_k, σ	Turbulence model constant
σ_B	Stefan–Boltzmann constant
ρ	Density, kgm^{-3}
μ	Dynamic viscosity, Pa s
μ_t	Turbulent viscosity, Pa s
Γ_i	Molecular diffusivity of species i, $kgm^{-1}\ s^{-1}$

References

1. M. Geerdes, R. Van Laar, R. Vaynshteyn, Low-cost hot metal: the future of blast furnace ironmaking. Iron Steel Technol. **8**, 51 (2011)
2. W.-H. Chen, S.-W. Du, C.-H. Tsai, Z.-Y. Wang, Torrefied biomasses in a drop tube furnace to evaluate their utility in blast furnaces. Bioresour. Technol. **111**, 433–438 (2012)
3. H. Suopajärvi, E. Pongrácz, T. Fabritius, Bioreducer use in Finnish blast furnace ironmaking–analysis of CO_2 emission reduction potential and mitigation cost. Appl. Energy **124**, 82–93 (2014)
4. J.-I. Yagi, Mathematical modeling of the flow of four fluids in a packed bed. ISIJ Int. **33**, 619–639 (1993)
5. Q. Minshen, Z. Yuting, Blast furnace operation with full oxygen blast. Ironmaking Steelmaking **15**, 287–292 (1988)
6. A. Majeski, A. Runstedtler, J. D'alessio, N. Macfadyen, Injection of pulverized coal and natural gas into blast furnaces for iron-making: lance positioning and design. ISIJ Int. **55**(7), 1377–1383 (2015)
7. A.T. Wijayanta, M.S. Alam, K. Nakaso, J. Fukai, K. Kunitomo, M. Shimizu, Combustibility of biochar injected into the raceway of a blast furnace. Fuel Process. Technol. **117**, 53–59 (2014)
8. A.T. Wijayanta, M.S. Alam, K. Nakaso, J. Fukai, K. Kunitomo, M. Shimizu, Numerical study on pulverized biochar injection in blast furnace. ISIJ Int. **54**, 1521–1529 (2014)
9. Y. Shen, D. Maldonado, B. Guo, A. Yu, P. Austin, P. Zulli, Computational fluid dynamics study of pulverized coal combustion in blast furnace raceway. Ind. Eng. Chem. Res. **48**, 10314–10323 (2009)

10. Y. Shen, B. Guo, A. Yu, P. Zulli, A three-dimensional numerical study of the combustion of coal blends in blast furnace. Fuel **88**, 255–263 (2009)
11. Y. Shen, A. Yu, P. Austin, P. Zulli, CFD study of in-furnace phenomena of pulverised coal injection in blast furnace: Effects of operating conditions. Powder Technol. **223**, 27–38 (2012)
12. Y. Shen, A. Yu, P. Austin, P. Zulli, Modelling in-furnace phenomena of pulverized coal injection in ironmaking blast furnace: Effect of coke bed porosities. Miner. Eng. **33**, 54–65 (2012)
13. B. E. Launder, D. B. Spalding, Lectures in mathematical models of turbulence. (1972)
14. S.K. Ubhayakar, D.B. Stickler, C.W. Von Rosenberg, R.E. Gannon, Rapid devolatilization of pulverized coal in hot combustion gases. In Paper Presented at the 16th Symposium (International) on Combustion, (Pittsburgh, 1976), p. 427
15. Y. Shen, B. Guo, A. Yu, D. Maldonado, P. Austin, P. Zulli, Three-dimensional modelling of coal combustion in blast furnace. ISIJ Int. **48**, 777–786 (2008)

Inclusion Control with Ca Treatment to Improve Castability of Low Carbon Aluminum-Killed Steel

S. Sun, S. Waterfall, N. Strobl, D. Liao and D. Holdridge

Abstract High melting point inclusions, such as spinel, alumina and Al_2O_3 rich calcium-aluminates, have a strong tendency to agglomerate on the tundish stopper rod and Submerged Entry Nozzle (SEN) during the steel casting process, and eventually to cause SEN clogging. Calcium treatment by wire injection during steel refining during Ladle Metallurgy (LM) is commonly used to modify these troublesome inclusions to lower melting point inclusions. However, too much Ca injection (Ca over-treatment) can cause operational issue (e.g. slide gate plate erosion), as well as increase Ca cost. Therefore, it is desirable to add an appropriate amount of Ca to avoid either under- or over-treatment. Based on inclusion analysis with an ASPEX® computer controlled scanning electron microscope (ASPEX) instrument, Al_2O_3 and Al_2O_3 rich calcium-aluminates were identified as a major cause of SEN clogging at ArcelorMittal Dofasco Electric Arc Furnace (EAF) steelmaking stream. A Ca treatment index was developed to improve the inclusion modification. The improved inclusion control has reduced SEN clogging and enhanced steel quality.

Keywords $CaO–Al_2O_3$ · Inclusions · Ca treatment · SEN clogging · Castability

Introduction

Nozzle clogging has been one of the main problems in continuous casting. Non-metallic inclusions formed due to de-oxidation and/or re-oxidation play a major role causing the clogging. In production of low carbon aluminum-killed

S. Sun (✉)
ArcelorMittal Global R&D—Hamilton, ArcelorMittal Dofasco Inc., Burlington St. E., Box 2460 1330, Hamilton, ON L8N 3J5, Canada
e-mail: Stanley.sun@arcelormittal.com

S. Waterfall · N. Strobl · D. Liao · D. Holdridge
Steelmaking Technology, ArcelorMittal Dofasco Inc., Burlington St. E., Box 2460 1330, Hamilton, ON L8N 3J5, Canada

J.-Y. Hwang et al. (eds.), *8th International Symposium on High-Temperature Metallurgical Processing*, The Minerals, Metals & Materials Series,
DOI 10.1007/978-3-319-51340-9_34

(LCAK) steel, alumina inclusions have a tendency to agglomerate and stick to the surface of SEN walls to form a deposit layer during casting, and eventually clog the SEN. A common practice to minimize the clogging is by Ca treatment. Ca is injected into fully killed steel to modify solid alumina inclusions to liquid or partial liquid calcium-aluminate inclusions (such as C12A5 in Fig. 1 [1]) to prevent/minimize the clogging. However, improper control of Ca treatment could result in either *under* (insufficient addition of Ca) or *over* (excessive addition of Ca) modification of alumina. Both situations may not prevent the clogging, and could also introduce other problems. Under-modification results in formation of CA6 and CA2 inclusions (Fig. 1), which are more detrimental than ordinary alumina to SEN clogging due to increased inclusion volume and lower density [2–4]. On the other hand, an over-modification could generate solid CaS and cause clogging [5], while high CaO inclusions can result in severe erosion of alumina graphite stopper [6] and slide gate plate.

Theoretically, the Ca injection should be controlled just enough to modify alumina to near C12A7 for liquid inclusions. In practice, however, it is a challenge to achieve this mainly due to the unknown alumina content in steel. Industry practices were usually developed based on plant experience. Setting a target Ca content is one common practice, such as controlling to 10–12 ppm Ca in steel [7], or adding 0.18 kg/t of calcium into steel [8] for the treatment. Other common practices include adding Ca to achieve a set Total-Ca/Total-Al or Total-Ca/Total-O

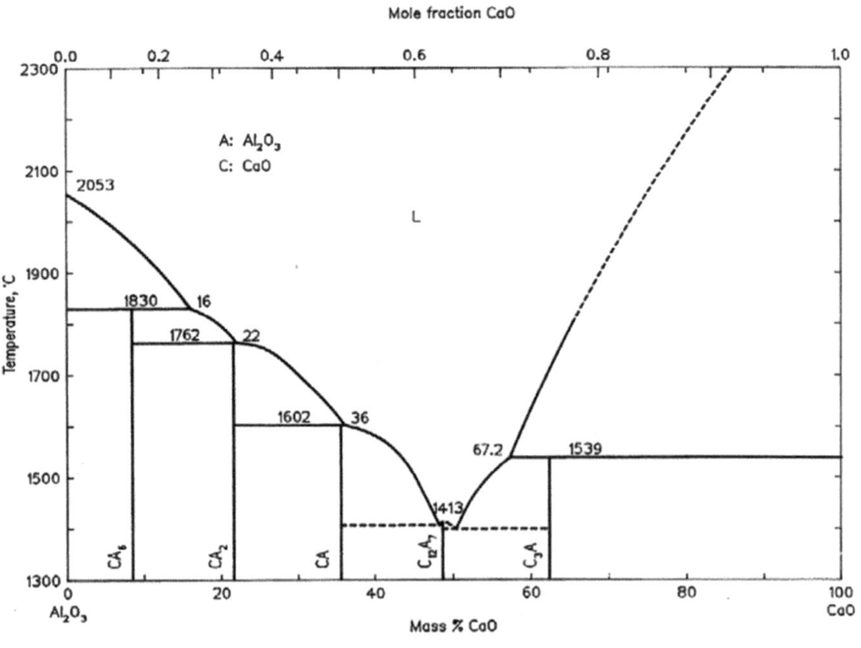

Fig. 1 CaO–Al_2O_3 phase diagram [1]

ratio [6], or Ca/Al ratio [9]. Those practices have limited success in improving the castability, due to inability to detect alumina inclusion content in steel for an effective modification of the inclusions. A new Ca treatment practice was developed at ArcelorMittal Dofasco with a Ca treatment index to improve the inclusion modification. The improved inclusion control has reduced SEN clogging and improved steel quality.

LCAK Steel Production and SEN Clogging Definition

The ArcelorMittal Dofasco steelmaking shop has both an oxygen steelmaking and an EAF steelmaking stream. The EAF stream consists of a 165 tonne twin-shell EAF, twin ladle metallurgy (LM) facility and a single strand conventional (216 mm) slab caster. LCAK is a major steel grade groups produced in the EAF stream. Major element ranges of chemistry are shown in Table 1. Calcium is injected at the LM facility towards the end of refining for good castability.

During casting of LCAK steels, SEN clogging was observed, especially when casting the first heat in a tundish. Although it is impossible to have a direct check on the SEN for clogging during continuous casting, clogging causes operational issues, such as mold level fluctuation, stopper rod bumping, or even a cast sequence break. These issues were observed and recorded by caster operators. Any heat with these operation issues was considered as a "Clogging Heat" in this work, while those without the casting issues were considered as "Good Heats" for comparison.

Effect of Inclusions on the Clogging

In order to investigate the effect of inclusions on the clogging, an inclusion baseline study was conducted during 2013–14. Inclusion samples (lollypop samples) from the LM (during steel refining) and the tundish (during casting) were taken and analyzed for inclusion characterization and evolution. An ASPEX instrument for automated inclusion analysis was used to measure the inclusion content. Inclusions data from tundish samples (after 70 tonne steel cast) were used in this work for investigation of effect of inclusions on the clogging.

In the baseline study, a total of 85 heats were sampled for the inclusion analysis. There was evidence of clogging in 22 heats (Clogging Heats), while the other 63 were "Good Heats" without clogging.

Table 1 Major element range of most LCAK grades

Element	C	Mn	Si	S	P	Al
wt%	0.03–0.08	0.15–0.30	<0.04	<0.01	<0.01	0.025–0.05

(a) **Inclusion comparison between "Clogging" and "Good" Heats**

In general, six types of inclusions were observed in this type of steel due to strong de-oxidation, de-S and Ti addition at the LM. In particular, *alumina, spinel, Total CA* (calcium aluminates comprising CA6, CA2, CA, C12A7 and C3A), *MnS, Total CaS* (CaS and CaS containing small amount of MnS, etc.) and *Ti-based* (TiN, TiS, TiO_2, TiO_2–Al_2O_3, TiO_2–Al_2O_3–CaO). Figure 2 is a comparison of inclusion area fraction (a ratio of total inclusion cross section area to total sample scanning area) between clogging and good heats. The detailed types of calcium aluminates are shown in Fig. 3. It can be seen that the clogging heats have a higher content of Al_2O_3, CA6 and CA2, especially Al_2O_3 and CA6, but fewer CA and CA3 inclusions. The contribution of the high melting point inclusions (Al_2O_3 and CA6) to the clogging are consistent with observations from other investigators [2–4]. The results also show that complete modification of alumina may not be necessary. Specifically, good castability can be achieved with only partial modification of alumina. To improve castability, Al_2O_3 and CA6 inclusions must be avoided/minimized.

(b) **Evaluation of existing Ca treatment practice for inclusion modification**

The LCAK grades are partially Ca-treated grade. The existing Ca injection practice aims to partially convert Al_2O_3 to lower melting point inclusions C12A7 and CA (Fig. 1). The higher content of Al_2O_3 and CA6 in the clogging heats indicated an ineffective modification of alumina, although complete modification is not necessary.

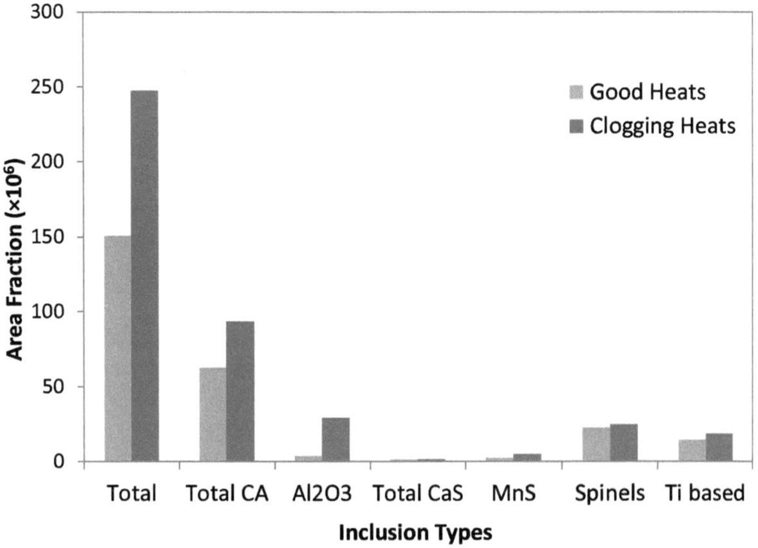

Fig. 2 Comparison of general inclusions between good and clogging heats

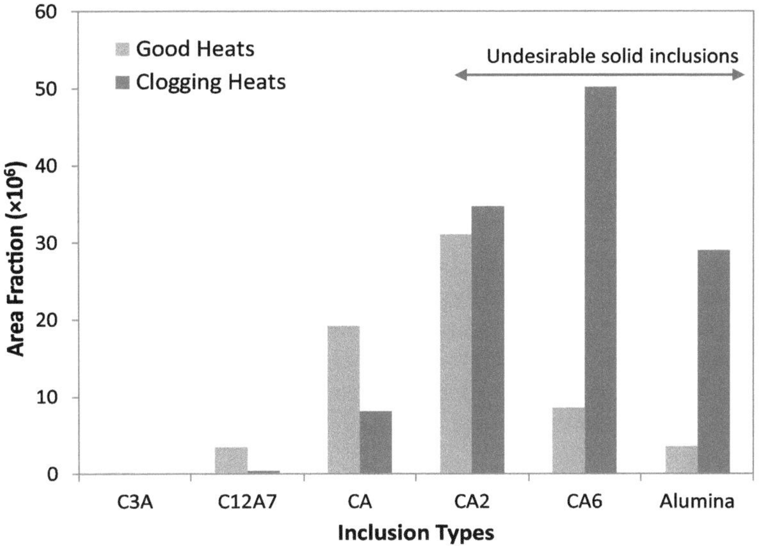

Fig. 3 Comparison of $CaO–Al_2O_3$ inclusions between good and clogging heats

Although different mechanisms of alumina modification with Ca are reported in literature, stoichiometrically the reaction may be expressed as;

$$x[\mathrm{Ca}] + \left(y + \frac{x}{3}\right)\mathrm{Al_2O_3} = (\mathrm{CaO})_x(\mathrm{Al_2O_3})_y + \frac{2x}{3}[\mathrm{Al}] \quad (1)$$

The final product, $(CaO)_x(Al_2O_3)_y$, of the reaction depends on the relative amount of [Ca] and (Al_2O_3) in steel before the reaction; i.e. the effectiveness of alumina modification depends on a proper [Ca] to (Al_2O_3) ratio. Where [Ca] is calcium content and (Al_2O_3) should be Al_2O_3 content in steel before reaction. In Ca treatment practice, the Ca content is the target of Ca injection which can be checked after injection by chemical analysis. However, the Al_2O_3 content is usually unknown. The existing Ca treatment practice is essentially an injection of a fixed amount of Ca for every heat without any knowledge of Al_2O_3 content. This practice results in a variable range of Ca to Al_2O_3 ratio due to variable and unknown quantity of Al_2O_3 in the steel. When steel has a high Al_2O_3 content, the fixed amount of Ca is not enough to modify Al_2O_3, which results in an ineffective modification of Al_2O_3, leaving a large content of Al_2O_3 and CA6 in the steel (Fig. 3).

Development of Ca Treatment Index—A Parameter for Evaluation of the Ca to Al_2O_3 Ratio

For effective modification of Al_2O_3 it is necessary to have an appropriate Ca to Al_2O_3 ratio, and the Al_2O_3 content in steel must be known. Although there is no technology currently available to determine the Al_2O_3 content, Al_2O_3 can be indirectly estimated.

A close look at the chemical analysis of production samples, two values of Al content are reported by OES analysis, i.e. total Al content ($[Al]_{total}$) and soluble Al content ($[Al]_{sol}$). Because Al is mainly used for de-oxidation purposes in LCAK steel, other than alloying, there is no other significant reaction in the steel to consume Al. The product of Al reaction in the steel should be mainly Al_2O_3. Therefore, the difference between the $[Al]_{total}$ and $[Al]_{sol}$ should be mostly the Al_2O_3 in the steel.

$$(Al_2O_3) = f(\Delta[Al]) \tag{2}$$

where

$$\Delta[Al] = [Al]_{total} - [Al]_{sol} \tag{3}$$

Therefore, the [Ca] to (Al_2O_3) ratio may be estimated from the delta Al as a [Ca] to Δ[Al] ratio, or what we are calling the Ca *treatment index*.

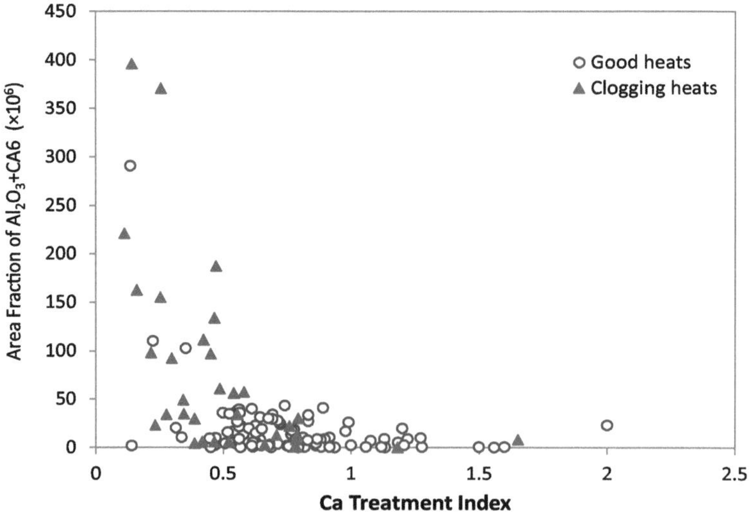

Fig. 4 Effect of the Ca treatment index on the area fraction of Al_2O_3 and CA6 inclusions for both good and clogging heats

$$\text{Ca}\,\textit{treatment index} = [\text{Ca}]/\Delta[\text{Al}] \quad (4)$$

A higher Ca treatment index value means more Ca is available to modify Al_2O_3 inclusions, and thus produce fewer Al_2O_3 and CA6 inclusions in the steel. Figure 4 shows the effect of the index ([Ca]/Δ[Al]) on the area fraction of Al_2O_3 and CA6 inclusions. Most of the clogging heats have a low value of the Ca treatment index (<0.6), an indication of insufficient Ca for alumina modification, and thus shows a high content of Al_2O_3 and CA6 inclusions to cause clogging issues. While in the good heats, a higher Ca treatment index results in sufficient modification of alumina to minimize clogging issues with fewer Al_2O_3 and CA6 inclusions. It may therefore be desirable to control the index to >0.6 to minimize clogging by reducing the Al_2O_3 and CA6 content in steel.

Improvement in Castability and Slab Quality

With the Ca treatment index, it is possible to improve steel castability with a better control of Ca injection. One example of using the Ca treatment index is to improve castability of the first heat of a new tundish.

(a) **Inclusions evaluation for first heat of a new tundish**

It is always a challenge to cast the first heat of a new tundish, due to a higher frequency of clogging issues than other heats. A comparison of inclusions showed that first heats have a higher content of Al_2O_3 and CA6 (Fig. 5), which is a result of lower values of the Ca treatment index (Fig. 6), likely leading to ineffective modification of Al_2O_3.

The lower value of the Ca treatment index and higher area fraction of bad inclusions (Al_2O_3 and CA6) may be due to re-oxidation of steel. When casting the first heat, liquid tundish slag is not yet formed to cover steel in the tundish, and re-oxidation of steel when in contact with air causes Ca fade (oxidation) and generation of new Al_2O_3.

(b) **Increase Ca for the first heats to minimize the clogging**

To compensate for the new formation of Al_2O_3 due to re-oxidation, it is desirable to control the Ca treatment index to greater than 0.6 (Fig. 4) with an increase in Ca injection at LM. Based on new aim Ca treatment index, the Ca addition was increased by an average 0.8 kg/heat for the first heats. To evaluate its effectiveness after the practice change, tundish samples from first heats of 14 caster sequences were analyzed for inclusion evaluation. In addition, slabs from more than 250 first heats made after the practice change were assessed to determine the affect on quality.

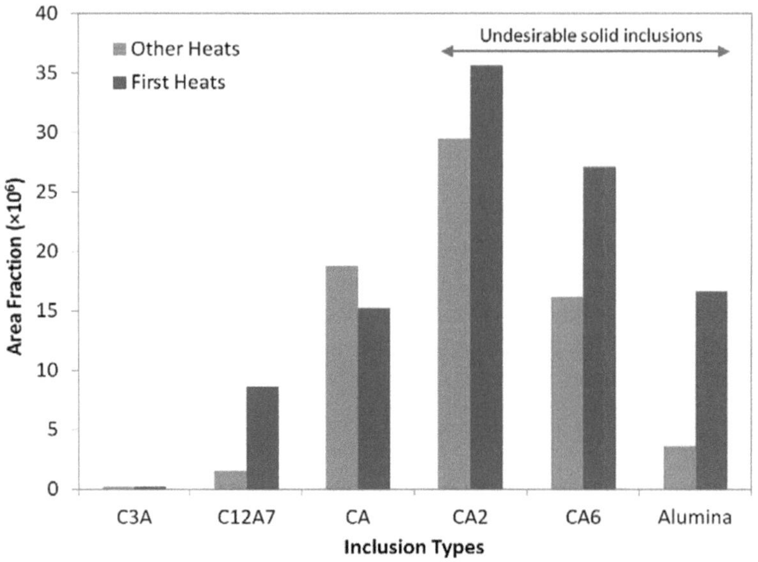

Fig. 5 Comparison of CaO–Al_2O_3 inclusions between the first heats and other heats

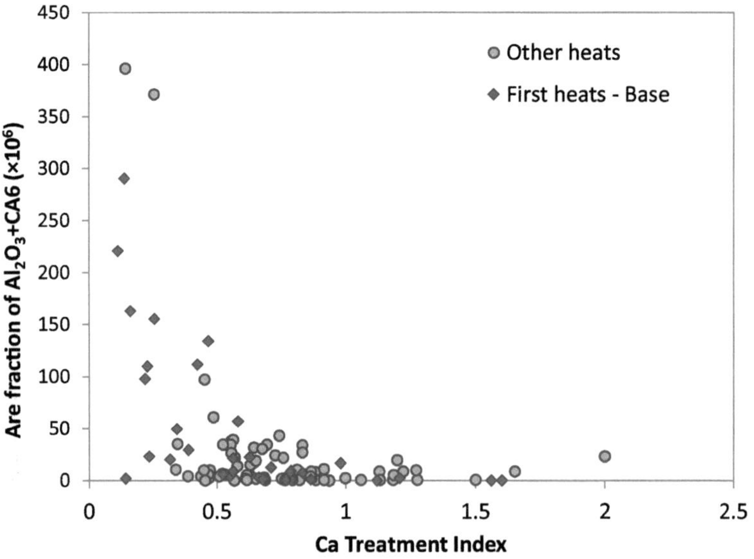

Fig. 6 Effect of the Ca treatment index on the area fraction (AF) of Al_2O_3 and CA6 inclusions for the first and other heats

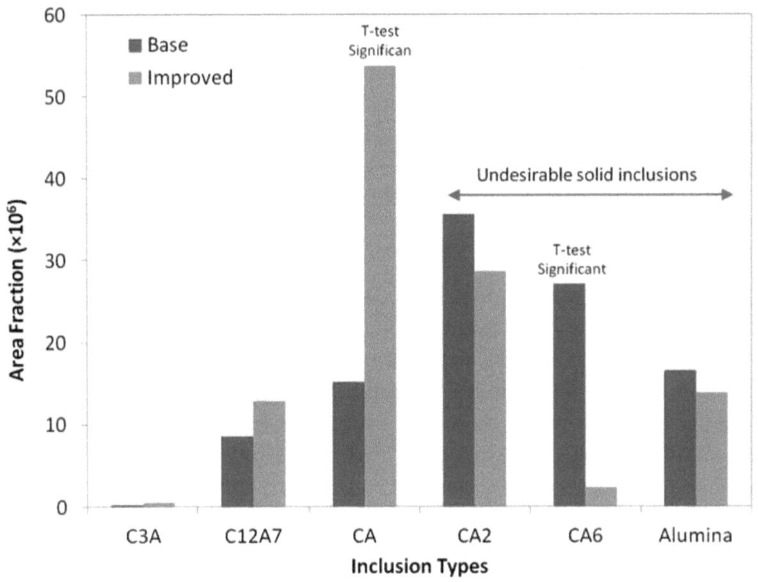

Fig. 7 Comparison of $CaO–Al_2O_3$ between the base and improved period

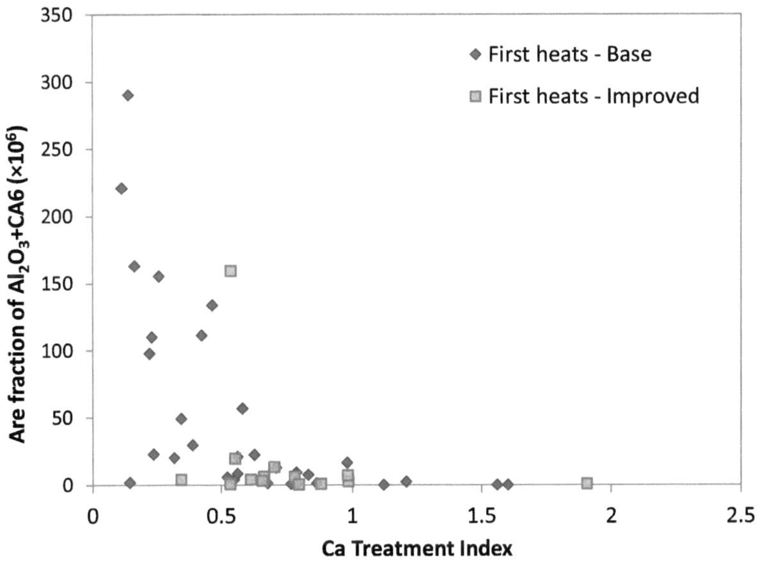

Fig. 8 Effect of the Ca treatment index on the area fraction of Al_2O_3 and CA6 inclusions

(c) **Improved Al_2O_3 inclusion modification**

Figure 7 is a comparison of calcium-aluminate inclusions for the first heats made between the baseline (Base—before the change) and the improved (after the change) periods. With the small increase of Ca content, most of the CA6 inclusions were converted to CA inclusions (Fig. 7), due to a more favorable Ca treatment index value (Fig. 8).

(d) **Improved castability and slab quality**

Alumina and CA6 are the major cause of SEN clogging issues, such as mold level fluctuation, excessive tundish stopper rod movement or stopper rod bumps. Those issues cause delays or slowdowns in casting, and slab quality down grades. With the reduction in those undesirable inclusions by the improved control of Ca injection, the clogging issues on the first heat of a new tundish were significantly reduced. Results included a reduction in average casting delay by 1.6 min/heat and a 45% reduction in slab quality downgrades.

Conclusions

- High content of Al_2O_3 and CA6 inclusions in the steel is the major cause for SEN clogging in cast LCAK steel at the ArcelorMittal Dofasco EAF stream. The Al_2O_3 and CA6 inclusions are a result of insufficient modification of the Al_2O_3 inclusions by Ca.
- A Ca treatment index was developed based on an estimation of Al_2O_3 content in steel and subsequent practice changes to tune the amount of Ca injection.
- Improvement in inclusion modification resulted in improved castability; with benefits of reducing casting delay time by 1.6 min/heat and a 45% reduction in slab quality downgrades on the first heat of a new tundish.

Acknowledgements The authors would like to thank L. Sun, L. Stewart and other members of Product Characterization Group, Global R&D Hamilton, for their contributions in inclusion analysis. The authors would also like to thank ArcelorMittal Dofasco EAF steelmaking management and operators for their support and contributions to the success of this project. Our special thanks to B. Farrand for his discussions and advice during the project and preparation of this manuscript.

References

1. VDEh (ed.), *Slag Atlas* (Verlag Stahleisen GmbH, Dusseldorf, 1995), p. 39
2. G.M. Faulring, J.W. Farrell, D.C. Hilty, Steel flow through nozzle: influence of calcium. in *AIME Electric Furnace Conference, Detroit, USA*, (1979), pp. 219–229

3. S.N. Sing, Mechanism of alumina buildup in tundish nozzles during continuous casting of aluminum-killed steels. Metall. Trans. **5**(10), 2165–2178 (1974)
4. S. Dawson, Tundish nozzle blockage during the continuous casting of aluminum-killed steel. in *1990 Steelmaking Conference Proceedings*, pp. 15–31
5. D. Janke, Z. Ma, P. Valentin, A. Heinen, Improvement of castability and quality of continuously casting steel. ISIJ Int. **40**(1), 31–39 (2000)
6. Y. Kusano et al., Calcium treatment technologies for special steel bars and wire rods. ISIJ Int. **36**(Supplement), S77–S80 (1996)
7. F. Fuhr et al., Relationship between nozzle deposits and inclusion composition in the continuous casting of steels. in *ISSTech 2003 Conference Proceedings*, pp. 165–175
8. K. Ahlborg et al., Inclusion in aluminum-killed steel with varying calcium additions. in *ISSTech 2003 Conference Proceedings*, pp. 177–194
9. S.R. Story et al., Inclusion analysis to predict casting behavior. in *ISSTech 2003 Conference Proceedings*, pp. 129–139

High Temperature Mineralization Mechanism of Granules During Iron Ore Sintering Process

Wei Lv, Xiaohui Fan, Min Gan, Xuling Chen, Zhiyun Ji, Yang Zhou, Guojing Wang and Qiang Li

Abstract Mineralization during sintering is a process to make part of the raw material melt after serial complex chemical reactions at high temperature. However, high temperature mineralization behavior of the adhesion layer and nucleus particles in the granules during the sintering process has been studied. Research findings reveal that solid-phase reactions between iron ores and fluxes in adhesion layer occurred first, of which the product would induct the generation of the initial liquid phase. By the assimilation of the initial liquid phase, fluxes which served as nucleus particles in granules, such as limestone and dolomite, could dissolve in the liquid phase and extend its amount as increasing temperature, whereas iron ore nuclear particles remained as unfused ores for their insufficient mineralization. Therefore, the mineralization proceeding for sintering was achieved: the iron ores in fine fraction (−0.5 mm) reacted with all fractions of fluxes to form the melt zone, while the coarse iron ores (+0.5 mm) acted as unfused ores, which formed the final sinters together with the melt zone.

Keywords Iron ore sintering · Mineralization proceeding · Granule · Adhesion fines · Nucleus particles

Introduction

Mineralization during sintering is a process to make part of the raw material melt after serial complex chemical reactions at high temperature. After condensation and consolidation of melt, raw mixtures are consolidated into agglomerates with good strength to satisfy the requirement of ironmaking in blast furnace [1]. The mineralization in sintering process impacts on the structure and mineral composition of sinters. Therefore the important indexes, such as production, quality and energy

W. Lv · X. Fan (✉) · M. Gan · X. Chen · Z. Ji · Y. Zhou · G. Wang · Q. Li
School of Minerals Processing and Bioengineering, Central South University, Peace Building, No. 252, Changsha 410083, Hunan, People's Republic of China
e-mail: csufanxiaohui@126.com

J.-Y. Hwang et al. (eds.), *8th International Symposium on High-Temperature Metallurgical Processing*, The Minerals, Metals & Materials Series,
DOI 10.1007/978-3-319-51340-9_35

consumption of sinter, are depended on the high temperature mineralization behavior of raw materials [2]. Most researches of mineralization mainly focused on the ability of a single kind of iron ores to take part in mineralization, as well as the influence of chemical components [3–5].

Temptation to evaluate the mineralization ability of iron ores through researching their reaction behaviors has been done. Among the mineralization properties, assimilation capacity which represented the reaction ability of iron ores won more favors. Loo, C.E. from BHP Corp. Researched the assimilation through measuring the formation of calcium ferrite in the roasting process [6]. The method is that the raw materials are ground into −0.125 mm, then blended, briquetted and roasted in a heating furnace on the condition of simulating sintering temperature and atmosphere, at last, calcium ferrite in roasted agglomerate was tested to characterize assimilation properties. Wu, S.L. et al. from China researched assimilation capacity like this, the iron ore powder cake placed on the upside of CaO cake were sent into the furnace, and the temperature when the cakes appeared a cycle of product were measured to evaluate the assimilation capacity [7]. Besides, some other researchers tried to reveal the relationship between the chemical compositions and sinter strength by studying the influence of SiO_2, Al_2O_3, MgO and basicity on mineralization [8–13]. All above research work have promoted the understanding of mineralization in the sintering process, but it's still not easy to distinguish the mineralization ability among various grains of raw materials. This paper will study the mineralization behavior of iron ores and fluxes with different grain sizes in granules, and aims at revealing the mineralization mechanism during sintering.

Materials and Methods

Properties of Raw Materials

Iron ores and fluxes involving limestone, dolomite and quicklime were utilized to produce sinter. The chemical compositions of materials are given in Table 1. Two representative types of iron ores, hematite and limonite, were selected in this investigation. Hematite came from Brazil which had a high content of Fe, with few impurities except for 3.78% SiO_2 and 0.77% Al_2O_3. And the limonite was from Austria, which had higher SiO_2 and Al_2O_3 contents than that in hematite. Lime was the product of limestone after calcinations, which had a high CaO content of 89.66%. Limestone and dolomite belonged to carbonate, and limestone contained 52.31% CaO, and dolomite contained 33.08% CaO and 17.35% MgO. The size characteristics of raw materials are shown in Table 2. The contents of fine particles with −0.5 mm of hematite and limonite were 40.30 and 50.07% respectively. The sizes of limestone and dolomite were mainly −3 mm, which accounted for nearly 80%.

Table 1 Chemical compositions of materials/wt%

Material types	TFe	FeO	CaO	MgO	SiO_2	Al_2O_3	LOI
Hematite	64.94	0.86	0.03	0.01	3.78	0.77	1.67
Limonite	57.31	1.05	0.17	0.13	6.82	2.02	8.46
Quicklime	0.15	0.08	89.66	0.87	2.30	0.38	5.63
Limestone	0.20	0.10	52.31	1.45	1.66	0.52	42.95
Dolomite	0.27	0.22	33.08	17.35	1.88	0.66	45.27

Table 2 Size distributions of materials

Material types	Size distributions (mm)/%							Average diameter/mm
	+8	5–8	3–5	1–3	0.5–1	0.25–0.5	−0.25	
Hematite	13.20	24.40	8.10	8.60	5.40	4.40	35.90	3.24
Limonite	10.22	12.58	10.67	7.78	8.68	5.38	44.69	2.41
Quicklime	0.00	0.04	1.68	52.36	11.62	14.38	19.91	1.31
Limestone	0.00	2.13	19.69	27.56	16.05	10.84	23.73	1.70
Dolomite	0.00	0.68	21.02	21.54	18.07	10.61	28.07	1.56

Methods

The micro-sintering was used to research the physicochemical reactions at high temperature. The micro-sintering adopted a horizontal furnace with controllable temperature-rising rate and atmosphere. In order to simulate sintering process exactly, this experiment divided the high temperature process into preheating belt, reaction belt, fusion belt, solidification belt and sinter belt. On the basis of physicochemical characteristics of each belt and temperature curve of the actual sintering layer, the micro-sintering used the heating program and atmosphere in Table 3 to simulate the sintering process.

Mineralization testing process included ore matching, agglomeration, roasting and mineralogical analysis. Iron ores and fluxes (dolomite, limestone, and

Table 3 Heating program and atmosphere of micro-sintering

Sintering belts	Temperature/°C	Time to destination temperature/min	Atmosphere
Drying-preheating belts	60 → 700	1	N_2
Reaction belts	700 → 1200	1	$CO:O_2:CO_2 = 1:1:5$
Melt belts	1200 → 1300	Rate of 10 °C/min	$CO:O_2:CO_2 = 1:1:5$
Solidification belts	Tmax → 1000	2	Air
Sinter belts	1000 → 700	1	Air

quicklime) were mixed to a feed, which then was used to produce sinter with SiO_2 4.7%, basicity (CaO/SiO_2) 2.0 and MgO 2.0%, the proportions of each material were shown in Table 4. The principle of ore matching for adhesion layer was screening out the fine powder (−0.5 mm) in iron ores and fluxes, and their proportions see Table 4. Added various kinds of nucleus particles in different size (0.5–1, 1–2, 2–3, 3–5 mm) into adhesive powder for simulating the structure of granules. After the ore matching and blending, the mixtures were compacted into cylinders with 30 × 25 mm size of the punching machine, using 300 kg/cm^2 pressure for 1 min. The cylinders were heated to 1300 °C according to the parameters in Table 3. After the agglomerates were cooled down, they were made into polished sections for optical microscope (Leica DMRXP) and its image analysis software (Qwin), together with the SEM (JMS-56600LV), to observe the mineralogical characteristics of sintering mixtures. The experimental flow is shown in Fig. 1.

Table 4 Materials matching for sintering and adhesion layer/wt%

Schemes	Hematite	Limonite	Quicklime	Limestone	Dolomite
Proportions for sintering	58	26	4	5	7
Proportions for adhesion layer	54.44	30.88	3.25	4.10	6.42

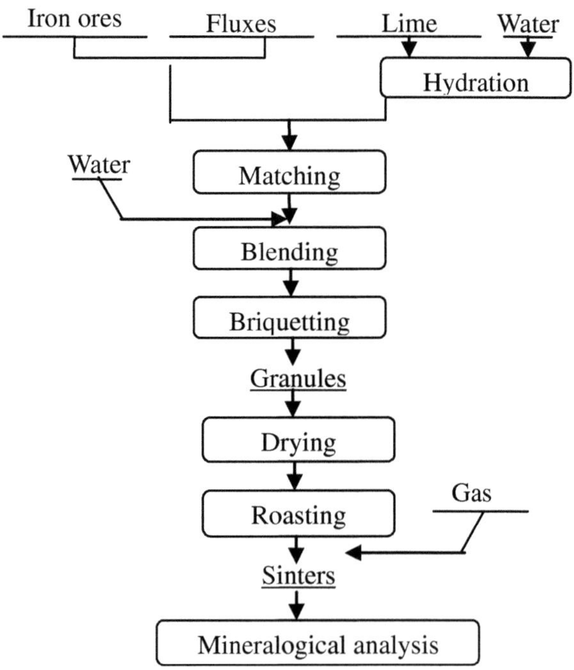

Fig. 1 Experimental flow chart of micro-sintering

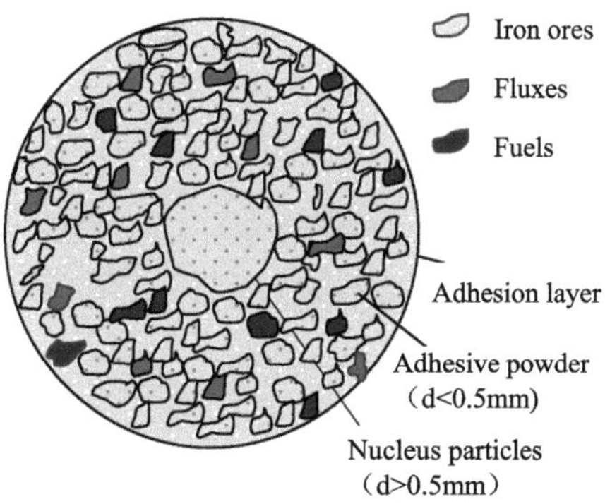

Fig. 2 Structural model of granule

Results and Discussion

Mineralization Behavior of Adhesion Layer

The mixtures were granulated to granules before sintering. The previous study [14] of authors has confirmed that the granule consisted of adhesion layer and nucleus particles. The particles of −0.5 mm served as the adhesive fines, while the particles of +0.5 mm served as the nuclei, of which the typical structure is shown in Fig. 2.

Fine ores and fluxes in adhesion layer possessed large specific surface areas, thus the contact areas between the particles were correspondingly sufficient. Therefore the chemical reactions take place in the adhesion layer first. Mineralization behavior of adhesion layer is shown in Fig. 3.

When the temperature ranged within 1100–1150 °C, solid-phase reaction had been occurred. However, obvious phenomenon was hardly observed due to the slow reaction speed. After increasing the temperature to 1200 °C, a large number of calcium ferrite (abbreviate as CF) generated by solid-phase reactions. As the temperature increased further to 1225–1250 °C, the liquid phase generated obviously and the holes began to shrink. The liquid content was developed when the temperature increased to 1300 °C, and the main mineral constituents were CF, magnetite and hematite, without unfused fluxes, which proved that all the iron ores and fluxes in adhesion layer could participate in mineralization completely.

Mineralization Behavior of Flux Nucleus Particles

When nucleus particles of lime and dolomite (2–3, 3–5 mm) were added to the adhesion layer, the mineralization behaviors are shown in Fig. 4. From Fig. 4a, c it was found that the lime and dolomite of the fraction of 2–3 mm could mineralize completely, and there were no residual nucleus particles. From Fig. 4b, d, it could be found that most of fluxes in the fraction of 3–5 mm participate in the

(a)1100°C; (b)1150 °C; (c)1200 °C; (d)1225 °C; (e)1250 °C; (f)1300 °C
H—Fe_2O_3, M—Fe_3O_4, CF—Calcium Ferrite, P—Pole

Fig. 3 Effect of temperature on mineralization of adhesion layer

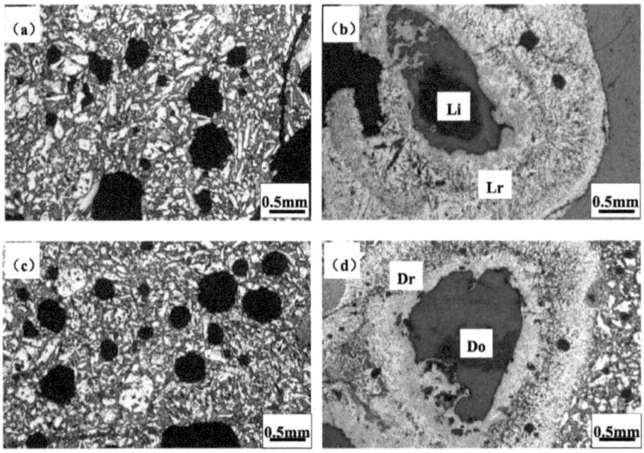

(a) Limestone of 2~3mm,(b)Limestone of 3~5mm,(c)Dolomite of 2~3mm,(d)Dolomite of 3~5mm;
Li—Limestone, Do—Dolomite, Lr—Products from reaction of Li, Dr—Products from reaction of Do

Fig. 4 Mineralization behavior of different size of flux particles

mineralization, and only a few tiny nucleus remained. On account of the fact that the fraction of fluxes used in sintering were all controlled under 5 mm (exactly the fraction of −3 mm accounting for more than 80%), fluxes, such as lime and dolomite, could take part in mineralization mostly.

Mineralization Behavior of Iron Ore Nucleus Particles

Iron ore nucleus particles of hematite or limonite with the size of 0.5–1 and 1–2 mm accounting for 25% (wt%) in mixture were added into adhesion layer (75%) for the micro-sintering, and compared with the scheme without nucleus particles. The microstructure of sinter roasted at 1300 °C is shown in Fig. 5.

The structure of adhesion layer without nucleus particles is shown in Fig. 5a, and it was hard to find out the unreacted ores after roasting. Figure 5b is the image of sinter with 0.5–1 mm nucleus particles of hematite, in which some remained unfused ores still existed. Figure 5c shows that the nucleus particles with a fraction of 1–2 mm could hardly be mineralized in the sintering process. In the same way, for limonite, liquid could penetrate into the interior of the unreacted core due to the developed porosity resulting from the removing of hydrate water, but the nucleus particles of limonite were still difficult to participate in mineralization (Fig. 5d). Therefore, whether hematite or limonite, when coarse iron ores (+0.5 mm) were added to mixtures, they would remain as unfused ores.

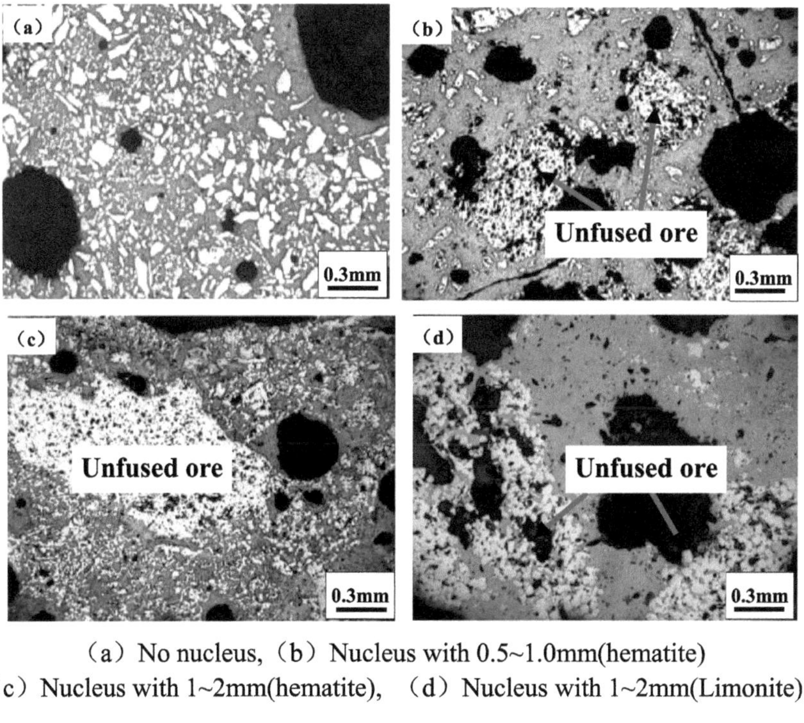

Fig. 5 Mineralization behavior of different size of iron ore particles

Therefore, the structure of sinter consisted of melt zone and unfused ores. It can be concluded that the unfused ores were surrounded by molten liquid. In the sintering process, all the fluxes participated in the mineralization, while the grain boundary for iron ores was 0.5 mm. Specifically, fine iron ores (−0.5 mm) took part in mineralization and formed the melting zone, while coarse iron ores (+0.5 mm) remained as unfused ores.

Condensation of the Liquid Phase and the Structures of Sinter

The initial liquid phase generated in adhesion layer first, and nucleus particles of fluxes melt into the liquid phase, which formed the melt zone of sinter after cooling and condensation. The microstructures of melt zone are shown in Fig. 6. There were mainly 3 kinds of structures, corrosion structure of magnetite and CF, eutectic structure of CF and silicate, and pilotaxitic texture of hematite and CF. The major part of the melting zone was the corrosion structure of magnetite and CF, accounting for 80–90%. Only a small area was characterized by the pilotaxitic texture of hematite and CF, the eutectic structure of CF and silicate.

Surface scanning analysis by SEM was used to research the element distribution in melt zone, as shown in Fig. 7. It is shown that Fe mainly existed in iron oxide, secondly in CF, then in silicate; Ca mainly existed in CF and silicate; Si mainly existed in silicate, as well as a few in CF; the distribution of Al was relatively uniform, while its content in CF was slightly higher than that in iron oxide; Mg distributed relatively evenly.

EDS was used to analyze the distribution of chemical compositions in melt zone, and the results are shown in Table 5. It can be seen that there were some difference between the compositions of each field, which proved that the melt zone was non-homogeneous. Under the assumption that all the fine iron ores (−0.5 mm) and fluxes were taken part in mineralization, the theoretical compositions of melt zone could be figured out. Compared the theoretical compositions with the average value

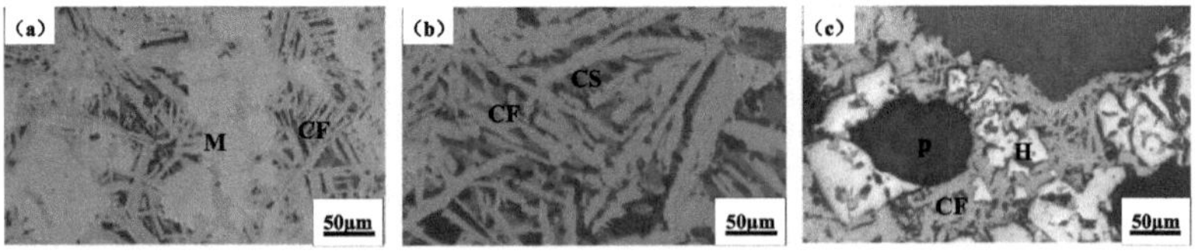

(a) Corrosion structure of magnetite and CF, (b) Eutectic structure of CF and silicate, (c) Pilotaxitic texture of hematite and CF;H—hematite, M—magnetite, CF—calcium ferrite, CS—silicate, P—pole

Fig. 6 Microstructures of melt zone

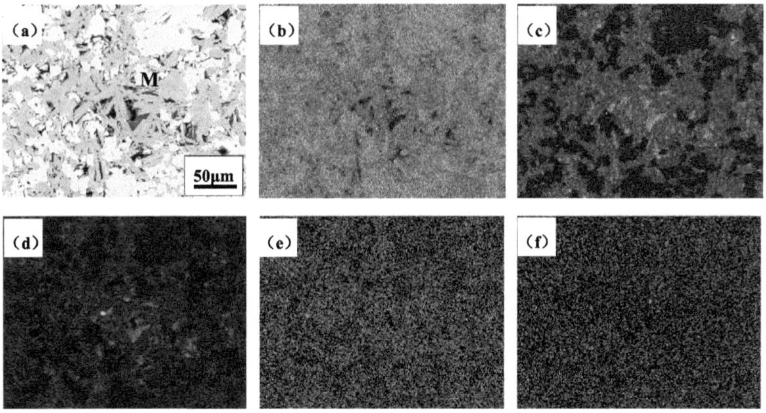

(a) Figure of SEM; (b) Distribution of Fe; (c) Distribution of Ca; (d) Distribution of Si; (e) Distribution of Al; (f) Distribution of Mg

Fig. 7 Element distribution of melt zone in sinter

Table 5 Chemical compositions of melt zone in sinter/%

Composition	TFe	SiO_2	CaO	MgO	Al_2O_3
Region 1	50.87	5.75	17.75	3.88	1.40
Region 2	48.66	6.13	19.45	3.65	1.38
Region 3	51.15	4.66	15.54	3.05	1.55
Region 4	54.48	4.51	16.37	2.95	1.40
Region 5	50.65	4.78	15.12	3.12	1.36
Region 6	53.24	4.48	14.55	3.01	1.44
Average of testing	51.51	5.05	16.46	3.28	1.42
Theoretical value	51.13	4.96	16.09	3.23	1.43

of the testing ones, it can be found out that they were basically the same. The accuracy of sintering mineralization model could be further verified as well.

Mineralization Proceed of Granules During Sintering

From the results finding introduced, the mineralization path of granules in the sintering process can be schematically shown in Fig. 8. (1) Solid-phase (Fig. 8a, b) reaction took place in the adhesion layer of granule. Fine iron ores (−0.5 mm) with large specific surface area reacted with fine fluxes (−0.5 mm), which formed the low melting point material. Because CaO in flux had more chances for the touch

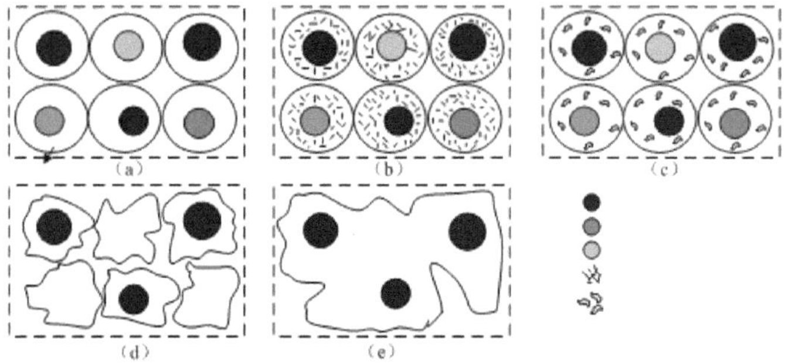

(a) granule; (b) Solid reaction; (c) Generation of primary liquid phase; (d) Extension of liquid phase; (e) Convergence of liquid

Fig. 8 Schematic of mineralization processes of granules

with Fe_2O_3, the main product of solid-phase reaction was CF. (2) Primary liquid phase (Fig. 8b, c). When the temperature increased to the melting temperature of low melting point material, the CF that existed in adhesion layer began to melt to primary liquid phase. The lowest temperature for the generation of liquid phase in the CF system was the eutectic point of $CaO{\cdot}2Fe_2O_3$ and $2CaO{\cdot}Fe_2O_3$, that is 1206 °C. Therefore, 1200 °C is proximately the temperature to form the primary liquid phase. (3) Extension of liquid phase (Fig. 8c, d). The fluxes which served as nucleus particles could dissolve in the primary liquid phase, which made the liquid phase get an extension. While the iron ores which served as nucleus particles could not take part in the mineralization because of the limitation of dynamical condition. As a result, the coarse iron ores (+0.5 mm) remained as unfused ores. (4) Liquid phase aggregation and crystallization (Fig. 8d, e). As temperature continued to rise, the liquidity of liquid phase got improved. And the liquid phases between the granules began to coverage, and the holes started to shrink. As temperature dropped from the peak, liquid phase began to condense gradually. As a result, crystals separated out successively and bonded the unfused ores together to form the final sinter.

Conclusions

1. Solid-phase reaction takes place first in the adhesion layer of granules. Fine iron ores (−0.5 mm) with large specific surface area react with fine fluxes (−0.5 mm), which creates the low melting point material-CF. All the iron ores and fluxes in adhesion layer enable to participate in mineralization completely.

2. The flux which serves as nucleus particles can dissolve in the primary liquid phase, which makes the liquid phase extended. However, the iron ores which serve as nucleus particles cannot take part in the mineralization. As a result, the coarse iron ores (+0.5 mm) remain as unfused ores.
3. The sinter comprises melt zone and unfused ores. Fine iron ores (−0.5 mm) take part in mineralization and form the melt zone, and coarse iron ores (+0.5 mm) remain as unfused ores.

Acknowledgements The research is financially supported by National Natural Science Foundation of China (No.51474237, U1660206), the Postdoctoral Science Foundation (No. 2013M540639), the Natural Science Foundation of Hunan Province in China (No. 2015JJ3164), Hunan Provincial Co-Innovation Center for Clean and Efficient Utilization of Strategic Metal Mineral Resources and Innovation Driven Plan of Central South University (No. 2015CX005), Hunan Provincial Innovation Foundation for Postgraduate (CX2016B054).

References

1. D.F. Ball, *Agglomeration of Iron Ores* (Heinemann Educational, 1973), p. 125
2. P.R. Dawson, Recent developments in iron ore sintering. Part 2. Research studies on sintering and sinter quality. Ironmaking Steelmaking **20**, 137–143 (1993)
3. L.H. Hsieh, J.A. Whiteman, Effect of raw material composition on the mineral phases in lime-fluxed iron ore sinter. ISIJ Int. **33**, 462–473 (1993)
4. H.G. Li et al., Melting characteristics of iron ore fine during sintering process. J. Iron Steel Res. **18**, 11–15 (2011)
5. S. Machida et al., Experimental evaluation of chemical composition and viscosity of melts during iron ore sintering. ISIJ Int. **45**, 513–521 (2005)
6. D. Debrincat, C.E. Loo, Effect of iron ore particle assimilation on sinter structure. ISIJ Int. **44**, 1308–1317 (2004)
7. D. Oliveira, Sintering properties and optimal blending schemes of iron ores. J. Iron Steel Res. **19**, 1–5 (2012)
8. J.H. Park et al., Effect of Al_2O_3, SiO_2 and MgO on the formation of calcium ferrites in sinter using X-ray diffraction method. J. Korean Inst. Met. Mater. **40**, 811–816 (2002)
9. U.S. Yadav et al., Influence of magnesia on sintering characteristics of iron ore. Ironmaking Steelmaking **29**, 91–95 (2002)
10. X.H. Fan et al., Influence of flux additives on iron ore oxidized pellets. J. Central South Univ. Technol. **17**, 732–737 (2010)
11. L.H. Hsieh, Effect of raw material composition on the sintering properties. ISIJ Int. **45**, 551–559 (2005)
12. G.L. Zhang, S.L. Wu, S.G. Chen, Influence of gangue existing states in iron ores on the formation and flow of liquid phase during sintering. Int. J. Miner. Metall. Mater. **21**, 962–968 (2014)
13. N.V.Y. Scarlett et al., Reaction sequences in the formation of silico-ferrites of calcium and aluminum in iron ore sinter. Metall. Mater. Trans. B **35**, 929–936 (2004)
14. M. Gan et al., Optimising method for improving granulation effectiveness of iron ore sintering mixture. Ironmaking Steelmaking **42**, 351–357 (2015)

Reduction Behaviors of Sinter Made from Magnetite Concentrates in Reducing Process Simulated COREX Shaft Furnace

Benjing Shi, Deqing Zhu, Jian Pan and Yuxiao Xue

Abstract COREX smelting reduction process which is freed of metallurgical coke provides a sustainable developing way for ironmaking industry. However, the shortage and high cost of pellets limit its development in China. Meanwhile, iron ore sinter has a great output in China. This work studied the reduction behaviors of sinter made from magnetite concentrates during the reducing process of COREX shaft furnace and compared the RI and RDI of sinter with BF process. The results showed that the reduction degree of sinter increased obviously during the temperature zone from 600 to 850 °C, while the reduction degradation presented rapidly aggravate in the temperature zone from 425 to 550 °C in COREX shaft furnace, and the increase of metallic iron content leads to the degradation of sinter got decreased slightly during the stage of temperature above 650 °C. Sinter got higher RI than BF process but also got higher RDI in low reduction temperature of 550 °C, which means sinter would have heavier disintegration in COREX shaft furnace.

Keywords COREX · Sinter · Reduction behaviors · Degradation

Introduction

COREX smelting reduction technology is a new process that uses coal to reduce iron ore at high temperatures and obtain liquid iron. Mainly employing iron lumps, pellets and non-coking coal, COREX is the earliest established commercially and the only one smelting reduction process that realized industrial production [1, 2]. Due to that 80–90% of fuel requirement is non-coking coal, the COREX process has greatly reduced dependence on high-quality coking coal resources compare with traditional iron-making process in blast furnace [3, 4]. The availability of wide

B. Shi (✉) · D. Zhu · J. Pan · Y. Xue
School of Minerals Processing and Bioengineering, Central South University, Changsha 410083, China
e-mail: shi_benjing@126.com

J.-Y. Hwang et al. (eds.), *8th International Symposium on High-Temperature Metallurgical Processing*, The Minerals, Metals & Materials Series,
DOI 10.1007/978-3-319-51340-9_36

variety of non-coking coals and elimination of harmful environmental emission have led COREX to the most promising alternative route to blast furnaces [5, 6]. However, the production shortage and higher prices of pellets, which have became one important factor that increases the cost of hot metal, restricts the development of COREX process in China. On the other hand, with excess production capacity and relative lower price, sinter has great application potential for COREX process in China. But due to the proportion of reducing gas which is hydrogen-enriched in shaft furnace of COREX process is quite higher than that in BF process, the reduction behaviors of sinter in COREX process would be different with in BF.

The reduction behaviors of sinter during reduction process are mainly reduction and reduction degradation, especially the latter would strongly affect the permeability of burden and the production of the shaft furnace. Intensive researches on reduction degradation of sinter in traditional blast furnace have been made by many researchers, and people have got a comprehensive understanding of the objective [7–10]. It is well known that the main factor of the disintegration is volumetric expansion of the iron oxide phase during reducing from hematite to magnetite, and various phases composing sinter have different effects on disintegration [11, 12]. However, studies on reduction behavior of sinter in COREX process were rarely reported. The relevant researches have shown that, furnace burden has higher reduction degree and heavier disintegration degree at reduction condition containing H_2 [12–15]. In this work, the reduction behaviors of sinter made from magnetite concentrates were studied. The RDI and RI of sinter samples were tested under various reduction time, temperature and gas composition simulated COREX shaft furnace. Besides, comparisons of reduction degree and disintegration degree between COREX process and BF process for sinter were made and results were analyzed.

Raw Materials and Experimental

Raw Materials

The sinter samples used in this work were made from several kinds of Chinese domestic magnetite concentrates, and sintering process were conducted in a sintering pot of φ180 mm × 700 mm. The product sinter was obtained at condition that the coke ratio and moisture were appropriate and return fines got balance. Table 1 shows the chemical composition of sinter samples. The main iron-oxide phases of the sinter samples are hematite and magnetite according to the XRD

Table 1 Chemical composition of sinter samples (mass percent, %)

Fe_{total}	FeO	SiO_2	CaO	MgO	Al_2O_3	TiO_2	S
56.92	10.63	5.98	10.55	1.32	1.40	1.05	0.040

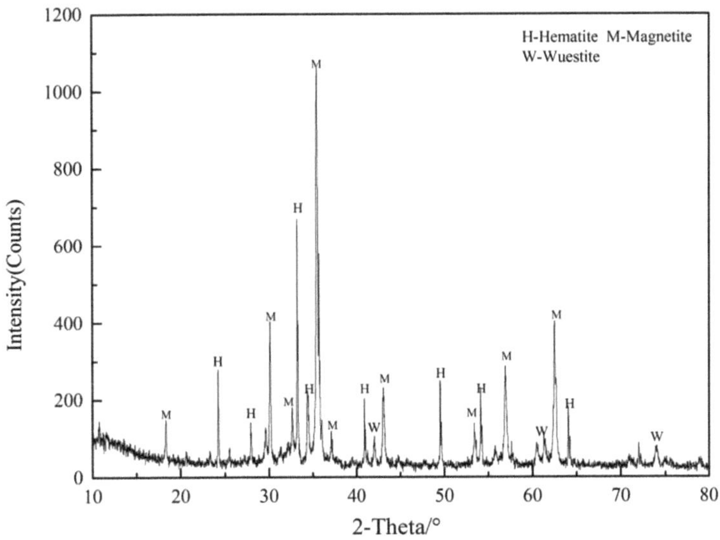

Fig. 1 X-ray diffraction inspection of the sinter samples

pattern as showed in Fig. 1. Before reduction experiments, the sinter samples were crushed and sieved to 10–12.5 mm.

Experimental Methods

All the reduction experiments were conducted in a static gas-solid reaction apparatus as Fig. 2 show. The size of reduction tube is φ75 mm × 800 mm, and it could resist the temperature above 900 °C. The total weight of samples for each test was 500 g.

As for the tests of the reduction behaviors, the sinter samples were put into the reactor tube and heated from air temperature with a heating rate of 2.5 °C under shielding with pure N_2 of 5 L/min. When the temperature got increased to 250 °C, the reducing gas was introduced into the reactor tube and the N_2 was ceased. In order to simulate the composition of reducing gas in COREX shaft furnace, the reducing gas composition in the tests changed gradually from 35% CO, 50% CO_2 and 15% H_2 to 72% CO, 8% CO_2 and 20% H_2 during the next 240 min, which is different from the standard reducing atmosphere (30% CO + 70% N_2) in the traditional BF process. The thing to note here is that, the heating was interrupted when the temperature of sinter samples reached 375, 425, 450, 475, 500, 525, 575, 625, 700, 750, 800 and 850 °C respectively. At the meantime, the reducing gas was replaced by N_2 again to cool down the samples to below 100 °C. The changes of temperature and reducing gas composition are shown in Fig. 3.

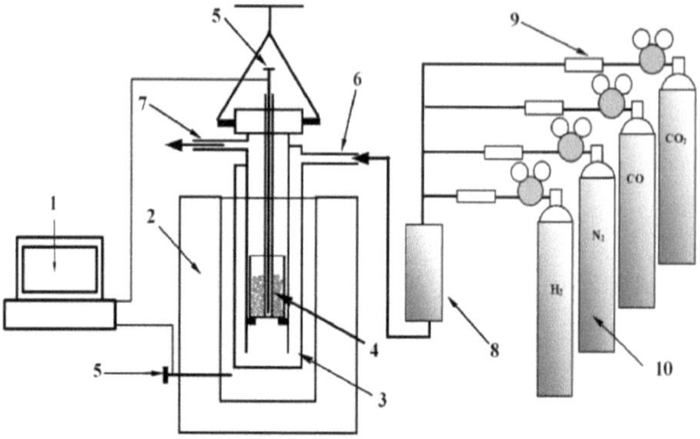

1-Temprature controller, 2-Heater, 3-Reaction tube, 4-Sinter samples, 5-Thermocouple, 6-Gas inlet, 7-Gas outlet, 8-Mix tank, 9-Flowmeter, 10-Gas cylinder

Fig. 2 Schematics of experimental apparatus for reduction experiments

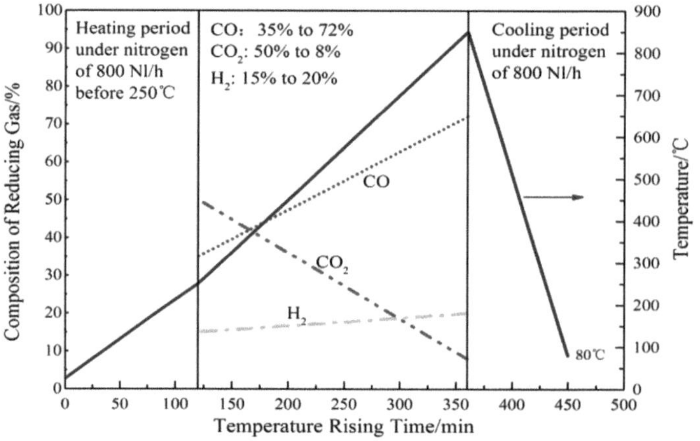

Fig. 3 Changes of reaction temperature and reducing gas composition

After the sinter samples were cooled down to room temperature, the reduction degree was calculated through the change in weight before and after the reduction process refers to GB/T13241-91. Then the sinter samples after reduction were charged into a tumbling drum (φ130 mm × 200 mm) and rotated for 10 min at a speed of 30 rpm. Then the samples were sieved with 6.3, 3.15 and 0.5 mm mesh respectively to determine the particle size distribution of the sinter samples after tumbling. The RDI was calculated by the equation followed: $RDI_{+6.3} = \frac{M_1}{M} \times 100\,\%$, $RDI_{+3.15} = \frac{M_2}{M} \times 100\,\%$, $RDI_{-0.5} = \frac{M_3}{M} \times 100\,\%$, where M was the weight of sinter samples after reduction, M1 and M2 were the weight of particles above 6.3 and 3.15 mm respectively after tumbling and M3 was the weight of particles smaller than 0.5 mm. Additionally, the sinter samples of different reduction stages were chemically analyzed by chemical titration, the mineral composition and microstructure were investigated by means of X-ray diffraction (XRD) and microscope analysis, using a diffractometer (RIGAKU D/Max 2500, Rigaku Co., Tokyo, Japan) and Leica DMLP optical microscopy respectively. Besides, to compare the reduction behavior of sinter in different reduction processes, the RI and RDI of sinter in traditional blast furnace was determined according GB\T13241-91 and GB\T13242-91 respectively.

Results and Discussion

Reduction Behaviors of Sinter in COREX Shaft Furnace

Figure 4 shows the fluctuation of RDI and reduction degree of sinter samples during the reducing process simulated COREX shaft furnace. Generally speaking, along with moving down of burden in furnace the temperature and reducing gas concentration increase gradually. It can be clearly seen from this figure that the reduction degradation degree of sinter increases first then decreases and the reduction degree shows an upward tendency in the whole reducing process.

For reduction behaviors of sinter at condition of simulating COREX furnace, the reduction process could be divided into three stages. The first stage is the duration of temperature below 425 °C, in this stage the sinter was almost no reduced and the reduction degradation degree was also relatively slight. The second stage is the duration of temperature from 425 to 550 °C, in this stage the reduction degree of sinter shown obvious rising even though the absolute value was still not high (below 10%), but in the meantime the reduction degradation degree of sinter samples occurred a sharp increase with high degradation rate. The $RDI_{+6.3}$ and $RDI_{+3.15}$ decreased 58.37 and 31.56% respectively during the second stage, occupied 83.19 and 84.49% of the whole degradation ratio respectively in the reduction process. The third stage is the temperature zone of 550–850 °C, and in this step the reduction degradation degree decreased slightly while the reduction degree increased 83.94% and occupied ratio of 90.45% for the whole reduction degree during the reduction process.

In summary, it could be concluded that the sinter samples mainly disintegrated in the second stage during which the temperature increased from 425 to 550 °C despite the reduction degree of the sinter remained relatively level. Meanwhile, the reduction of sinter samples mainly occurred in the third stage of temperature increased from 550 to 850 °C.

As Fig. 5 shows, the content of FeO reach the peak at around 650 °C and then decline along with reduction. The content of metallic iron (MFe) increases obviously from around 625 °C and the same with the metallization ratio (MR).

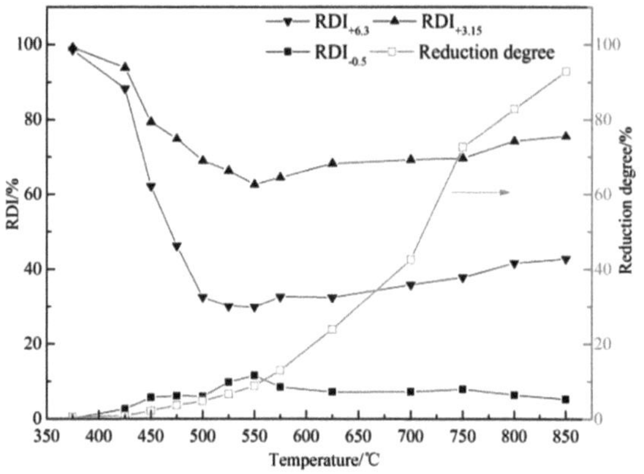

Fig. 4 RDI and reduction degree during reduction process

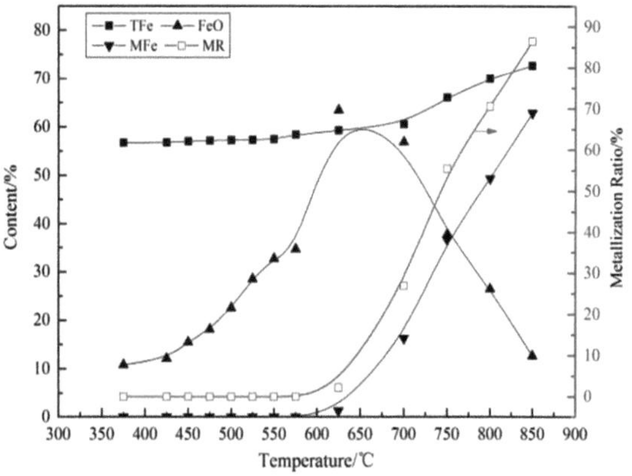

Fig. 5 The TFe, FeO, MFe and metallization ratio of sinter during reduction

Former researches have indicated that when the reduction degree reached a certain value the increase rate of the reduction degradation degree would decrease, and this value was considered to be around 11% at which the hematite was totally reduced to magnetite [16, 17]. Slightly different with those results, the degradation degree of sinter samples was reduced as the increase of metallization ratio. This may be due to the increase of metallic iron content leads to the drum strength of sinter got improved after reduction.

Mineralogical Phases and Structure Changes During Reduction

Figure 6 shows the X-ray diffraction inspection of sinter samples that reduced in different reduction stages.

In the interval of temperature 375–550 °C, the reducing reaction was mainly the transformation of hematite to magnetite, as shown in Fig. 5 there was little hematite existing in sinter samples when the temperature reach 550 °C. The magnetite was reduced to be Fe_xO gradually in duration of temperature increase from 550 to 650 °C and then the wuestite (Fe_xO) was reduced to metallic iron when the temperature is above 700 °C. Combine with the reduction behaviors of sinter in COREX shaft furnace as above results shows, it could be concluded that the degradation of sinter mainly take place during the stage in which hematite was reduced to magnetite.

Figure 7 shows the microstructure of sinter samples during reduction process. The symbols “H”, “M”, “P”, “R”, “S”, “W” and “CF” indicate the phases of hematite, magnetite, pore, resin, slag, wuestite and multi-component calcium ferrite respectively. As the Fig. 7a shows, the microstructure has no obvious change of the sinter in reducing stage of 375 °C under the present circumstance. Figure 7b shows the cross section near the surface of sinter samples in reducing stage of 450 °C. It can be clearly seen the hematite in the outer layer of sinter has been reduced to magnetite and many cracks are observed not only in the new formed magnetite area but also CF phase near the particle surface. The reducing reaction has not occurred in internal part of sinter.

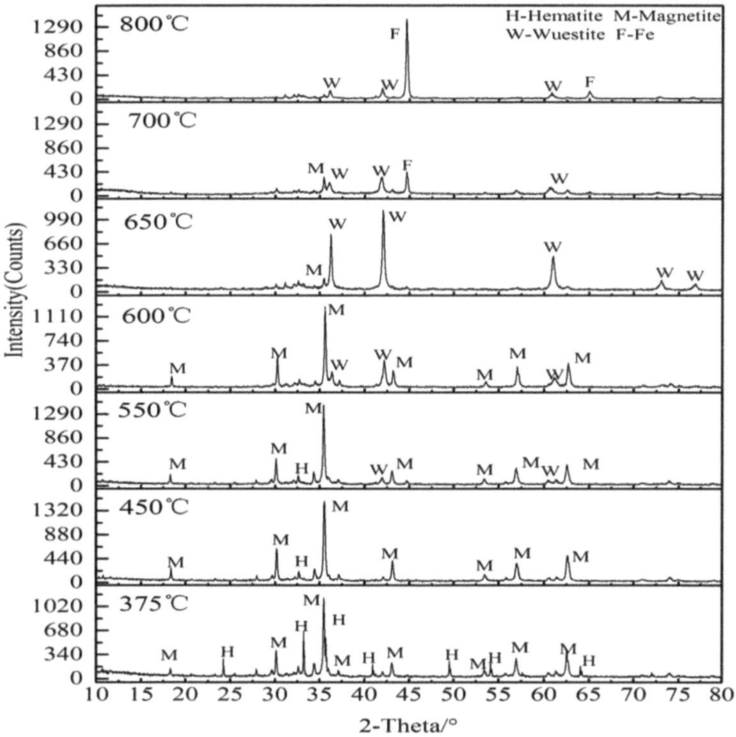

Fig. 6 X-ray diffraction inspection of sinter during reduction process

Figure 7c indicates the microstructure of sinter samples reduced in the stage of 550 °C. Reduction reaction proceeded to a greater degree inside the sinter particles with increasing reducing time. Compared with former stage, more hematite has transformed into magnetite in the meantime longer crack formed and cross linking with each other. The degradation of sinter reach peak at this reduction stage. This result corresponds well to some former relevant researched about the degradation mechanism of sinter in blast furnace, and the main reason was attributed to the volume expansion produced in the reduction process from hematite to magnetite, especially by the skeletal hematite [8–11]. When the reduction proceed to the stage of 700 °C as Fig. 7d shows, metallic iron could be observed in many areas in the sinter. It seem like reduction occur gradually along the thick cracks formed in degradation because the metallic iron was generated earliest near cracks during the process reducing form outer to inside.

Low Temperature Reduction Comparison of Sinter in COREX and BF Process

Compared with traditional blast furnace process, the reduction conditions in COREX shaft furnace are very different. Therefore, the reduction degradation degree and reduction degree of sinter samples at reduction temperature of 550 °C in

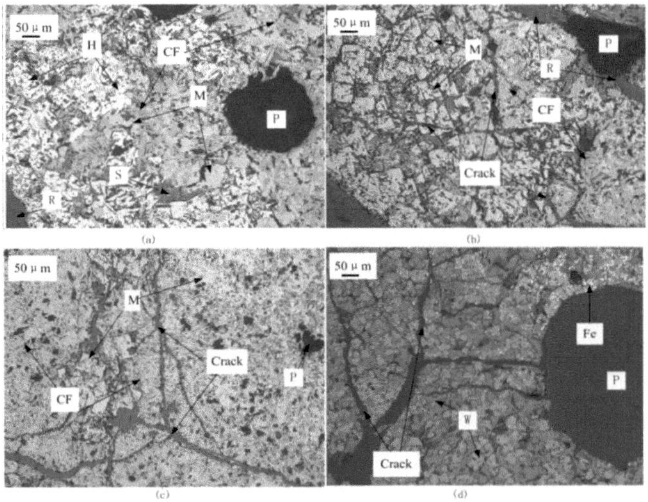

a-375℃, b-450℃, c-550℃, d-700℃, H-hematite, M-magnetite, P-pore, R-resin, S-slag, W- wuestite, CF- multi-component calcium ferrite

Fig. 7 Microstructure of sinter during reduction process traditional blast furnace and COREX process are compared in Fig. 8. Obviously the reduction degradation degree and reduction degree in COREX process were both extended. The $RDI_{+6.3}$ and $RDI_{+3.15}$ decreased by 19.55 and 12.52% respectively, and $RDI_{-0.5}$ increased by 2.66% in the COREX process, which means more fines particles generated during reduction. Meanwhile, the reduction degree raised 4.4%, almost doubled to that in BF process.

According the knowledge of thermodynamics, the reducing ability of CO in stronger than that of H_2 below 810 °C. However, the molecule radius of H_2 is smaller which means its diffusion coefficient is higher than that of CO. Therefore the reduction rate of iron oxide by H_2 is higher than that by CO even below 810 °C [14]. So the H_2 contained in the reducing gas of COREX process promoted reduction of sinter even at 550 °C, as the result of reduction degree shown above. On the other hand, the reduction rate of sinter in COREX process increased nearly doubled to BF process, which caused the inner expansion stress in sinter samples increased. Consequently, the sinter tended to crack further and then the reduction degradation degree of sinter got higher in COREX process than that in BF process.

In summary, sinter has higher reduction degree in the COREX shaft furnace. However, its degradation degree is more serious than in BF at the same time, the small particles and fine powder would both possess a great influence on the furnace permeability. Consequently, the measures decreasing the degradation degree of sinter in COREX process should be developed and adopted in order to realize largely applying in this promising ironmaking process for sinter.

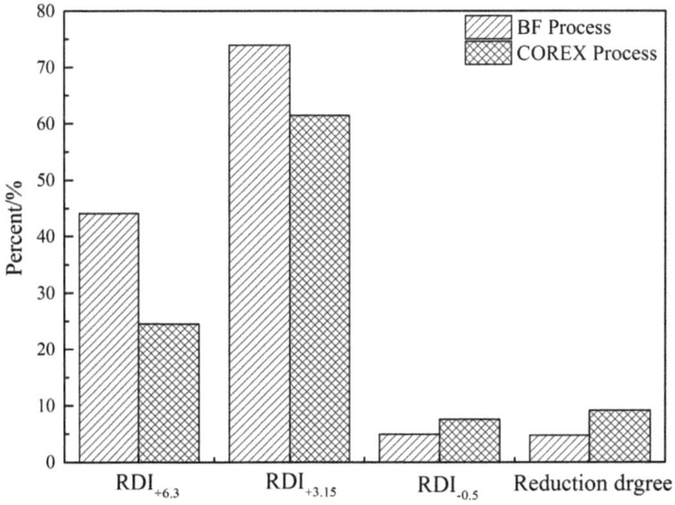

Fig. 8 Reduction degradation degree and reduction degree of sinter samples in different processes

Conclusions

1. For reduction behaviors of sinter at condition of simulating COREX furnace, the reduction process could be divided into three stages. The sinter samples mainly disintegrated in the second stage of 425–550 °C, and the reduction of sinter samples mainly occurred in the third stage of temperature increased from 550 to 850 °C.
2. The hematite transformed to magnetite in the interval of temperature 375–550 °C, many long and cross linking cracks occurred in sinter during this process and lead to serious reduction degradation. The increase of metallic iron content leads to the degradation of sinter got decreased slightly during the stage of temperature above 650 °C.
3. Due to the high proportion of H_2 contained by reducing gas in COREX shaft furnace, sinter has higher reduction degree and more serious degradation degree at low reduction temperature, which caused by faster rate of reduction from hematite to magnetite.

References

1. Y.S. Zhou, Review of current development of coal-based smelting reduction ironmaking process. Iron Steel **40**(11), 1–8 (2005)
2. D.U. Rui-Ling, W.U. Keng, W.L. Zhan et al., Establishment and application of rist operating line for COREX shaft furnace. Chin. J. Process Eng. **13**(1), 94–98 (2013)
3. P.P. Kumar, D. Gupta, T.K. Naha et al., Factors affecting fuel rate in Corex process. Ironmaking Steelmaking **33**(4), 293–298 (2013)
4. Q.H. Liu, W.U. Keng, W.L. Zhan et al., High temperature performance of coke formed from COREX coal. J. Northeast. Univ. **35**(8), 1155–1159 (2014)
5. Y. Qu, Z. Zou, Y. Xiao, A comprehensive static model for COREX process. ISIJ Int. **52**(12), 2186–2193 (2012)
6. L.I. Hai-Feng, H. Zhou, T. Zhang et al., Influence of screw design on burden descending velocity and particle segregation in COREX shaft furnace. J. Iron. Steel Res. Int. **23**(6), 516–524 (2016)
7. M. Asada, M. Shima, Y. Omori, Measurement of macro strain in the course of reduction of the skeletal hematite in sinter (sinter). Phys. Rev. C **30**(3), 1776–1778 (1984)
8. C.E. Loo, N.J. Bristow, Study on low temperature reduction degradation mechanism of sinter. Sinter. Pelletizing **20**(3), 25 (1995)
9. C.E. Loo, N.J. Bristow, Study on low temperature reduction degradation mechanism of sinter (continuation). Sinter. Pelletizing **4**, 23–28 (1995)
10. H.P. Pimenta, V. Seshadri, Influence of Al_2O_3 and TiO_2 degradation behaviour of sinter and hematite at low temperatures on reduction. Ironmaking Steelmaking **29**(3), 175–179 (2002)
11. R. Nakajima, T. Sumigama, K. Wakimoto et al., Reduction degradation behavior of sinter in the blast furnace shaft (blast furnace phenomena). Tetsu to Hagane **73**, 1964–1971 (1987)
12. T. Murakami, Y. Kamiya, T. Kodaira et al., Reduction disintegration behavior of iron ore sinter under high H_2 and H_2O conditions. ISIJ Int. **52**(8), 1447–1453 (2012)
13. S. Wu, X. Liu, J. Wu, Reduction disintegration behavior of lump ore in COREX shaft furnace. ISIJ Int. **55**(8), 1608–1616 (2015)

14. W.U. Sheng-Li, X.Q. Liu, Q. Zhou et al., Low temperature reduction degradation characteristics of sinter, pellet and lump ore. J. Iron Steel Res. **18**(8), 20–24 (2011)
15. M. Lin, X. Jiang, Q.J. Gao et al., Effect of hydrogen addition on low temperature metallurgical property of sinter. J. Iron Steel Res. **19**(4), 6–10 (2012)
16. N. Taguchi, T. Otomo, Y. Omori, Immersion treatment of sinter into calcium halide solution. ISIJ Int. **30**(4), 281–289 (1990)
17. H.P. Pimenta, V. Seshadri, Characterisation of structure of iron ore sinter and its behaviour during reduction at low temperatures. Ironmaking Steelmaking **29**(3), 169–174 (2002)

Part VII
Treatment and Recycling of Slag/Wastes

Introduction of Matte Droplets in Copper Smelting Slag

Xiangfeng Cheng, Zhixiang Cui, Leonel Contreras, Mao Chen, Anh Nguyen and Baojun Zhao

Abstract Smelting is an important process in copper production to produce matte and slag. The slag, containing about 1–12% Cu, can be further processed to recover the copper contaminated, but recycle of the slag concentrate reduces overall productivity and efficiency of the operation. The best solution is to minimize the copper loss to the slag. The entrained matte droplets in copper smelting slag may originate from a few sources. The present study aims to investigate possible copper matte loss mechanisms by visualization of room temperature experiments and high-temperature examination. It is believed that the gas bubbles from the matte phase disperse the matte droplets into the slag causing entrained matte in slag. Analyses of quenched slags from industrial operation and laboratory experiments confirmed that matte droplets in the slag are extensively associated with the gas bubbles. Water model is used to simulate the procedure and the bubble behaviors at the liquid-liquid interface are observed experimentally with a high-speed camera.

Keywords Copper loss · Gas bubble · Matte droplet · Settlement

Introduction

Copper loss is one of the significant factors which limits the overall metal recovery in primary and secondary copper production. In the smelting process, the copper loss in slag can be high up to 12 wt% Cu. Further operations in the slag cleaning process are expensive and also reduce the overall productivity and efficiency, therefore minimizing the copper loss to slag is essential in copper smelting

X. Cheng · M. Chen · A. Nguyen · B. Zhao (✉)
The University of Queensland, Brisbane, Australia
e-mail: baojun@uq.edu.au

Z. Cui
Dongying Fangyuan Nonferrous Metals, Dongying, China

L. Contreras
National Copper Corporation of Chile, Santiago, Chile

J.-Y. Hwang et al. (eds.), *8th International Symposium on High-Temperature Metallurgical Processing*, The Minerals, Metals & Materials Series,
DOI 10.1007/978-3-319-51340-9_37

processes. It is necessary to acquire a comprehensive understanding of mechanisms of copper losses to the slags. The loss of copper in smelting slags can be divided into chemically dissolved copper and mechanically entrained matte droplets. The chemically dissolved copper is closely related to slag compositions, operating conditions (temperature and Po_2), etc. The mechanically entrained matte droplets are the matte droplets that are not settled to the matte layer and remained in the slag layer [1].

The sources of the mechanically entrained copper may include

- Unreacted copper concentrate droplets
- Very fine dispersions of copper or copper sulphide precipitated from slag when their solubility in the slag is decreased
- Refining and converting slags returned to the smelting furnace contain significant amounts of finely dispersed matte droplets
- Attachment of matte or metal droplets to solid magnetite particles present in the slag phase
- Gas bubbles from the matte phase dispersing the matte into the slag as they cross the matte-slag interface

Among all the sources, researchers observed that the gas bubbles may play a significant role in the matte entrainments [1]. The study of mechanical entrapment was focused on the bubble behaviors in the melt bath. Several theories of copper losses by the entrained matte have been proposed [2]. Kemiha et al. [3] studied the gas bubble passage and dynamic behavior of the interface experimentally with a high-speed camera and numerically by volume-of-fluid approach. In the present study, the copper loss entrapments were examined by the as-received industrial slags and the copper entrainments by bubbles were investigated by high-temperature and cold model experiments.

Analysis of BBF and TC Slags

In order to identify the sources of matte entrapments, the as-received slag samples from the bottom blown smelting furnace (BBF) and Teniente Converter (TC) were analysed by Scanning Electron Microscope (SEM). From the SEM images of the BBF slags and TC slags analysed, the statistical information of the free matte droplets and the matte droplets attached to gas bubbles and spinel particles were reported.

Through the analysis of quenched BBF and TC slags, the volume proportions of free matte droplets, matte droplets attached to bubbles and matte attached to spinel particles in the slag phase are shown in Fig. 1a, b.

From Fig. 1a, b, it can be seen that a large number of matte droplets are attached to gas bubbles in the slag phase, accounting for 14.8 and 11.2% of total volumes in BBF and TC slag samples, respectively. However, only 0.2 and 0.4% of matte

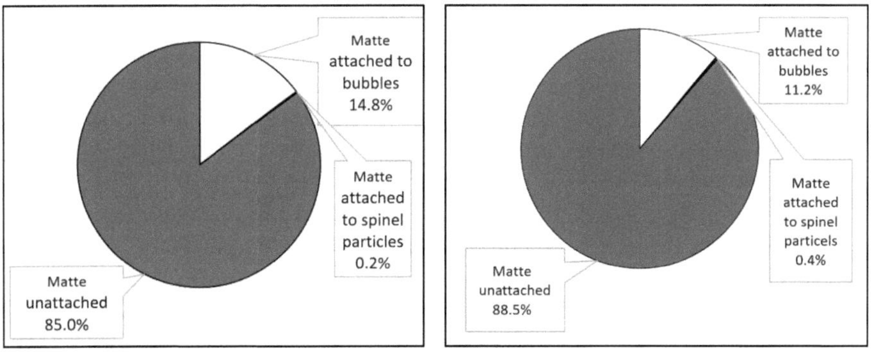

(a) (b)

Fig. 1 The proportions of matte droplets with/without attachments in BBF slags (**a**) and TC slags (**b**)

Table 1 Mean diameter of matte droplets and gas bubbles in BBF and TC slags (unit: μm)

Sample source	Matte droplets	Matte droplets attached to bubbles	Matte droplets attached to spinel	Bubble size	Bubbles attached to matte droplets
BBF slags	55.6	176	22.2	130	147
TC slags	88.5	309	32.0	168	256

droplets volume are attached to spinel particles. The statistical results from the as-received slag samples from two different smelting technologies indicate that gas bubbles play a vital role in the copper matte entrapment in the slags. It should be kept in mind that the slag samples were collected from the slag tapping hole where some of the matte droplets have already separated from the gas bubbles.

The sizes of the gas bubbles and matte droplets in the slag phase are also compared. The analysis results for BBF slags and TC slags are shown in Table 1.

It can be seen from Table 1 that, the mean diameter of the matte droplets is 55.6 and 88.5 μm respectively in BBF and TC slags. Accordingly, the mean diameter of the matte droplets attached to gas bubbles and spinel particles in TC slags are much larger than those in BBF slags. In both slags the average size of the matte droplets attached to bubbles is over three times larger than the total average and those attached to spinel particles. The small matte droplets attached to the spinel may represent partially reacted copper concentrates.

The mean diameter of gas bubbles is 168 μm in the TC slags, which is larger than 130 μm in the BBF slags. Particularly, the mean diameter of the bubbles attached to matte droplets (256 μm) in TC slags is almost double of those in BBF slags (147 μm). The bubbles with large sizes can easily carry large matte droplets in the slag phase, which may explain the differences in the matte droplet sizes between the BBF and TC slags.

The sizes of total bubbles and matte droplets in slag phase vary in broad ranges. Supposing all the matte droplets are free living in slags irrespective of other effects including the attachment to bubbles and spinel particles, the statistics for the size distribution of matte droplets and the study on sedimentation velocity are of great importance as the settlement time of the matte droplets in slag layer is strongly related to the copper losses. The size distribution of the matte droplets was obtained from SEM images, therefore the sedimentation velocity of each matte droplet can be obtained. The terminal velocity of a fluid particle in creeping flow can be achieved by the Hadamard-Rybczynski equation [4]:

$$U_T = \frac{2}{3}\frac{gr^2\Delta\rho}{\mu}\left(\frac{1+k}{2+3k}\right)$$

where u_T is the sedimentation velocity; r is the radius of droplet; $\Delta\rho$ is the density difference; μ is the viscosity of continuous phase; $\kappa = \mu_p/\mu$ is viscosity ratio of bubble, droplet or particle to continuous phase. The viscosities of the BBF slags and TC slags can be calculated using FactSage software from their compositions and operating temperatures [5]. The average viscosities of BBF slags and TC slags are 0.180 and 0.124 Pa s respectively. The densities of slags and mattes in both BBF and TC are assumed to be the same, which are $\rho_{\text{slag}} = 3.5 \times 10^3\ \text{kg/m}^3$ and $\rho_{\text{matte}} = 4.7 \times 10^3\ \text{kg/m}^3$. The calculations of the sedimentation velocity of matte droplets with different sizes in both BBF slags (Fig. 2a) and TC slags (Fig. 2b) give a general overview of the effects of matte size on matte entrapment process.

From Fig. 2a, b, it can be seen that nearly 50% of the matte droplets in BBF slags are smaller than 20 μm. Assuming the matte droplet needs to cross 0.4 m thick slag phase to be back to the matte layer, the settlement time based on the calculated sedimentation velocity can be obtained from the figure. The results show that 82% matte droplets are smaller than 100 μm in BBF slags, which will cost more than 4 h to be settled down from the top to the bottom of the stagnant slag

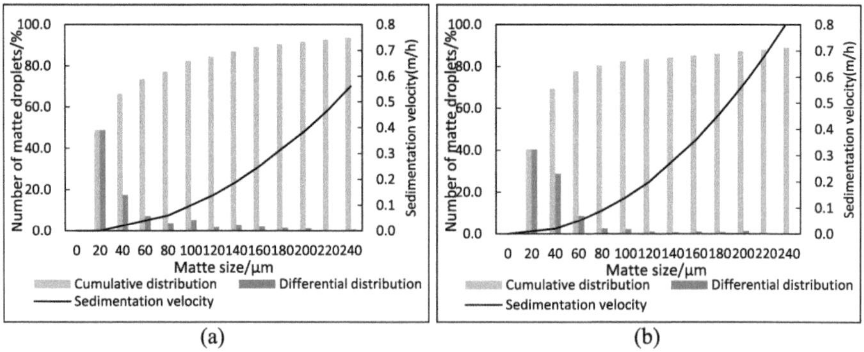

Fig. 2 Sedimentation velocity of matte droplets with different size in BBF slags (**a**) and TC slags (**b**)

phase. Similar results are obtained for the TC slags although the required settling time is relatively shorter. The settling time of the matte droplets in both BBF and TC is usually shorter than 0.5 h. This means that most of the matte droplets would not have enough time to be settled to the matte layer before being tapped out. It is therefore more important to understand and reduce the raise of the matte droplets attached to gas bubbles [6].

High Temperature Experiments of Synthetic Slag

To verify the attachment phenomenon between gas bubbles and matte droplets, high-temperature experiments were conducted in vertical tube furnace (Fig. 3a) at 1200 °C using alumina crucibles (inner dimensions: 2 cm diameter, 5 cm height) [7]. The samples were prepared in three layers (Fig. 3b): slag layer at the top, matte layer in the middle and mixture of Cu_2O and FeS at the bottom. The bottom layer with Cu_2O and FeS mixture was to generate SO_2 bubbles in the melts to simulate the reactions in the bath smelting furnaces. The temperature was programmed in three different periods with different heating rates: 600–900 °C, 5 °C/min; 900–1200 °C, 10 °C/min; 1200 °C for 5 min. After holding at 1200 °C for 5 mins, the sample descended slowly to the cool end of the furnace and fast-cooled by the argon gas. The samples were then sectioned and mounted for SEM analysis. A typical microstructure is shown in Fig. 4. It can be clearly seen from the figure

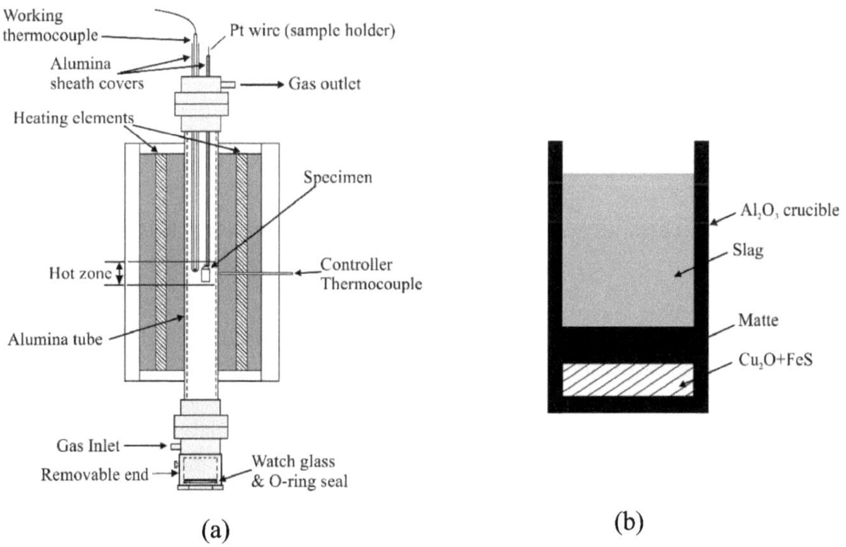

Fig. 3 Schematic of the tube furnace used in the high temperature experiment (**a**) and the sample with three layers (**b**)

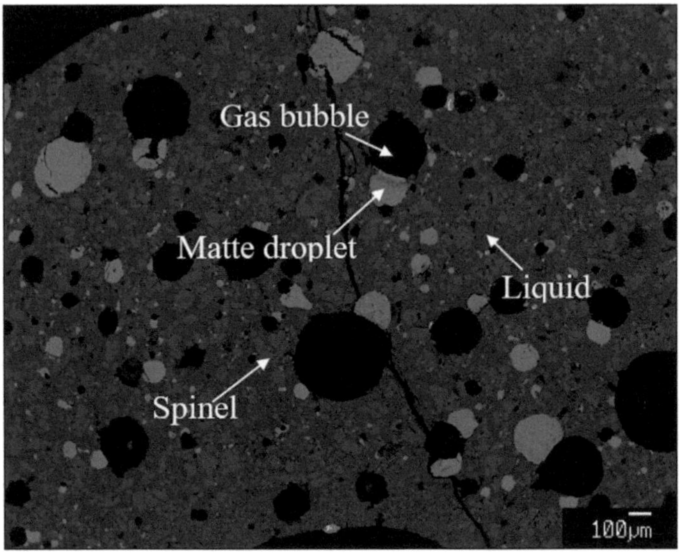

Fig. 4 Typical microstructure of the high temperature experiment sample

that many matte droplets were attached to the bubbles and suspended in the slag layer.

The statistics information of the matte droplets attached to the and the free-matte droplets in the high temperature samples are shown in Fig. 5. It can be seen that more than 90% matte droplets were attached to gas bubbles, which is far greater than those in the BBF and TC slags. The higher proportion of matte attachment indicates that in the smelting furnaces, a large number of matte droplets will be detached from the gas bubbles during rising. The results also demonstrate that gas bubbles indeed play a significant role in copper matte entrapment mechanism.

The Cold Model Experiments

In order to visualize the bubble behaviours and their abilities of carrying matte droplets, simulated room temperature experiments were carried out. In the cold model experiments, based on the kinematic similarity, silicone oils with different viscosities were used to simulate the slag phase and water was used to simulate the matte phase. The gas injection was controlled manually by adjusting air flowrate to produce expected bubble generation frequency. Basically, the main units of the experimental setup shown in Fig. 6 consisted of a high speed camera, a light source, and one computer for displaying. A high-speed camera (FASTCAM SA1.1) was placed in front of a glass cube (5 cm × 5 cm × 45 cm) to capture the gas bubbles and water droplets behavior in silicone oil phase. The frame rate was set as 125 fps

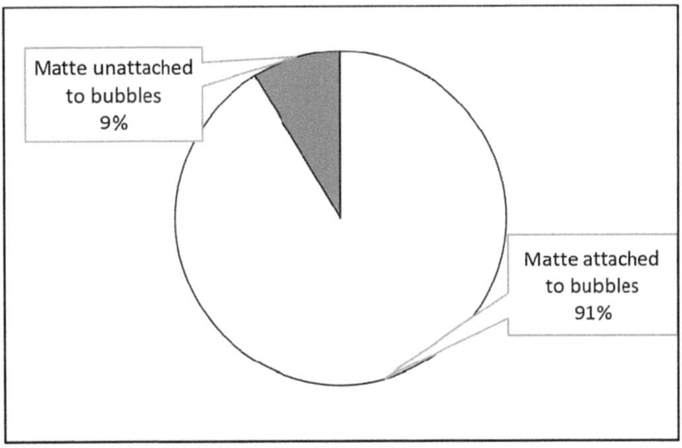

Fig. 5 Matte attachment volume proportion in high-temperature lab experiments

Fig. 6 Setup of the model experiments

and the resolution of the camera was 1024 × 1024 pixels. A light source (XGY-*II*) was placed at the opposite direction of the high-speed camera separated by a plastic board to disperse the strong light and illuminate liquids in the glass cube.

Figure 7 shows the images of the whole process of one bubble passing through the water–silicone oil interface captured by the high-speed camera. The viscosity of the silicone oil is 0.010 Pa s. It can be seen that when t = 0.133 s, the bubble started

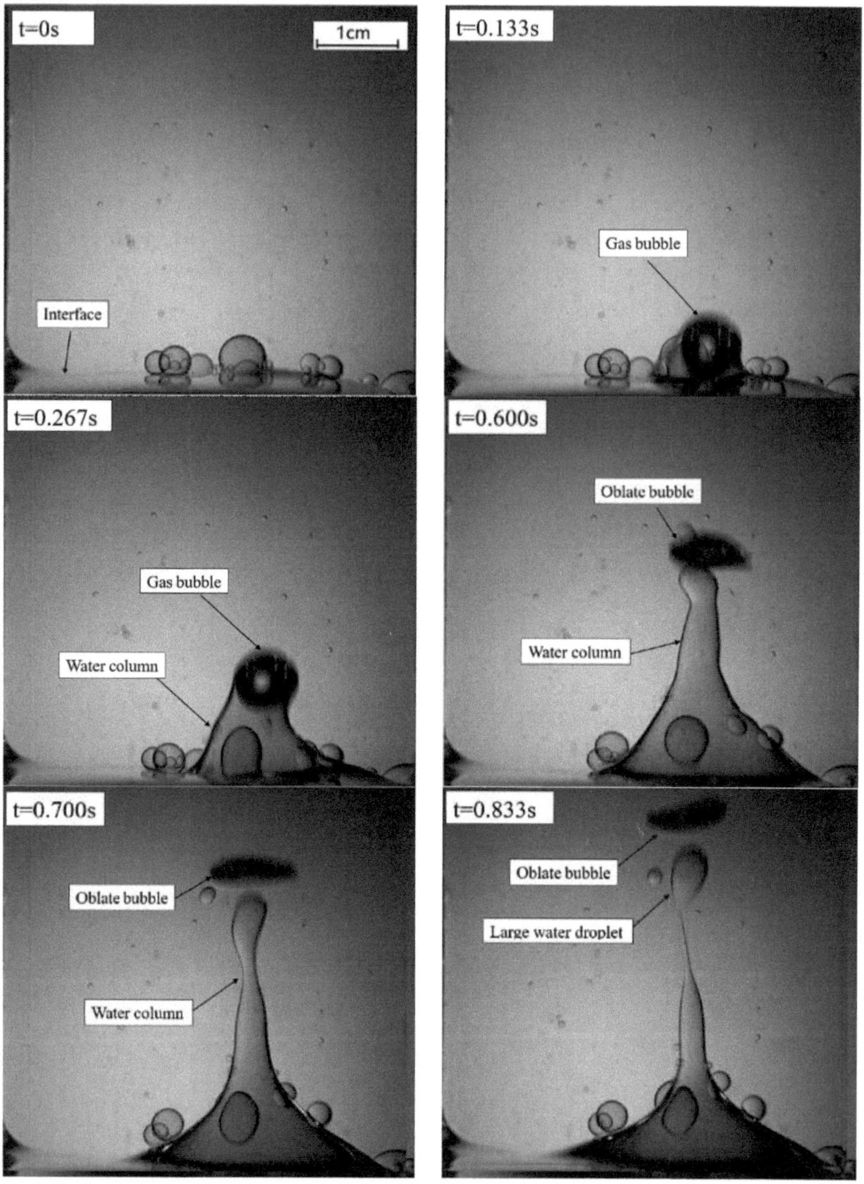

Fig. 7 The bubble behaviour at the water–silicone oil interface

to cross the water–silicone oil interface as a spherical shape. At t = 0.267 s, the bubble kept rising and dragged by the interfacial tension. The bubble was in the silicone oil layer, but coated by a thin water film, and a water column behind bubble started to form at t = 0.267 s. The water column was extended to a certain length due to its momentum and surface tension. The bubble was seriously stretched to be an oblate shape after being detached from water column at t = 0.600 s. Later, the water column began to be scattered into a large water droplet and some small droplets which continued to rise due to bubble wake at t = 0.700 s and 0.833 s.

In all cold model experiments, the water columns dragging the gas bubbles were scattered into small droplets after being detached the gas bubbles. From the present observation, it shows even a single bubble crossing the water–silicone oil interface will bring up several water droplets into the silicone oil layers. The droplets will go on rising for a certain distance before they start to descend. This phenomenon can also be confirmed by the observations in BBF and TC slags and high temperature lab slags. The reactions in the crucible in high temperature experiments can be considered to take place near the interface as the limited height of 5 cm. These scattered droplets are one of the important sources to cause matte droplets entrapment in slag phase. It is also assumed that the water film on the gases bubbles will fascinate the water droplets, which may be brought up by other bubbles, to form a large droplet attached to the bubble. The large droplet volume and the height it can rise are associated with the bubble wake, which is a function of the bubble rising velocity, the bubble radius, the continuous liquid phase viscosity and density [4, 8, 9]. The bubble wake effects on the entrapment of water droplets will be simulated and quantified afterward.

Conclusions

In the present study, the effects of bubble on the copper losses to the smelting slags were investigated through statistical analysis of the as-received bottom blown smelting furnace slags and Teniente Converter slags, high temperature lab experiments and cold model experiments. The statistical investigations of the industrial slags show that high proportion of matte droplets volume was attached to bubbles so that the gas bubbles play a significant role in copper matte entrapment mechanisms. The high temperature lab experiments simulating the bubble attachment with the matte droplets suspended in the slag layer demonstrate that the matte droplets are indeed brought up by the bubbles. The cold model experiments confirm that when passing through the water–silicone oil layers, the rising bubbles will drag the water columns, and this inevitably generates additional water droplets scattered in the silicone oil layer. The water film covering the bubble will be easily shattered into a large water droplet in the bubble wake area behind the rising bubble.

Acknowledgements The authors would like to thank National Copper Corporation of Chile (Codelco), Dongying Fangyuan Nonferrous Metals (Fangyuan) and Australia Research Council for financial support through the ARC Linkage program. The authors also would like to thank China Scholarship Council (CSC) and The University of Queensland to provide Xiangfeng Cheng scholarship.

References

1. D. Poggi, R. Minto, W.G. Davenport, Mechanisms of metal entrapment in slags. J. Met. **21** (11), 40–45 (1969)
2. R. Minto, W.G. Davenport, Entrapment and flotation of matte in molten slags. Inst. Min. Metall. Trans. Sect. C **81**, C36–C42 (1972)
3. M. Kemiha, E. Olmos, W. Fei, S. Poncin, H.Z. Li, Passage of a gas bubble through a liquid-liquid interface. Ind. Eng. Chem. Res. **46**(19), 6099–6104 (2007)
4. Roland Clift, J.R. Grace, M.E. Weber, *Bubbles, drops, and particles* (Dover Publications, INC., New York, NY, 2005), pp. 30–34
5. C. Bale, E. Bélisle, P. Chartrand, S. Degterov, G. Eriksson, K. Hack, I.H. Jung, Y.B. Kang, J. Melançon, A. Pelton, FactSage thermochemical software and databases—recent developments. Calphad **33**, 295–311 (2009)
6. E. De Wilde, I. Bellemans, L. Zheng, M. Campforts, M. Guo, B. Blanpain, N. Moelans, K. Verbeken, Origin and sedimentation of Cu-droplets sticking to spinel solids in pyrometallurgical slags. Mater. Sci. Technol. (2016). doi:10.1080/02670836.2016.1151998
7. M. Chen, B. Zhao, Phase equilibrium Studies of "Cu_2O"–SiO_2–Al_2O_3 system in equilibrium with metallic copper. J. Am. Ceram. Soc. **96**(11), 3631–3636 (2013)
8. P. George, A.V. Nguyen, G.J. Jameson, Assessment of true flotation and entrainment in the flotation of submicron particles by fine bubbles. Miner. Eng. **17**(7–8), 847–853 (2004)
9. G.K. Bachelor, F.R.S., *An Introduction to Fluid Dynamics* (Cambridge, UK: Cambridge University Press, 1967), pp. 378–380

Dissolution Behavior of Fe from Glassy Oxides Phase in Steelmaking Slag

Shohei Koizumi, Xu Gao, Shigeru Ueda and Shin-ya Kitamura

Abstract Steelmaking slag was confirmed to be a useful fertilizer for paddy growth as it efficiently supplies Ca and Si to soil water. In addition, supplement of Fe from slag fertilizer has been expected to suppress the H_2S generation in paddy soil, especially for the degraded paddy soil. However, on the basis of our previous research, the Fe dissolution ability changed with slag brands, and even some brands could dissolve Fe under paddy soil pH, the supplying ability were not high. To improve Fe dissolution, we found that among various kinds of Fe-rich mineralogical phases in slag, the $CaO–SiO_2–FeO$ glassy phase showed highest water solubility of Fe. Because the $CaO–SiO_2–FeO$ glassy phase was a quenched molten phase, its dissolution behavior may depend on chemical composition, and this is important for controlling slag composition to produce fertilizer which can easily release Fe. Therefore, in this research, the $CaO–SiO_2–FeO$ glassy phase of various CaO/SiO_2 ratio, Fe^{2+}/T–Fe ratios, and FeO contents were synthesized, and the dissolution behaviors of Fe as well as other cations from these synthetic glassy phases into an aqueous solution at pH of 5 were investigated. The result showed that the dissolution ratio of Fe increased significantly by increasing Fe^{2+}/T–Fe ratio. An increase in FeO content of glassy phase also increased the dissolution of Fe. In addition, the Fe dissolution ratio showed a local maximum with increasing CaO/SiO_2 ratio.

Keywords Steelmaking slag · Dissolution · Glassy phase · Fertilizer

S. Koizumi (✉)
Graduate School of Engineering, Tohoku University,
2-1-1, Katahira, Aoba-ku, Sendai 980-8577, Japan
e-mail: shohei.koizumi.q5@dc.tohoku.ac.jp

X. Gao · S. Ueda · S. Kitamura
Institute of Multidisciplinary Research for Advanced Materials,
Tohoku University, 2-1-1, Katahira, Aoba-ku, Sendai 980-8577, Japan

J.-Y. Hwang et al. (eds.), *8th International Symposium on High-Temperature Metallurgical Processing*, The Minerals, Metals & Materials Series,
DOI 10.1007/978-3-319-51340-9_38

Introduction

In Japan, degradation of paddy fields proceeded because of the increase in fertilizer price and the decrease of rice price. To restore the paddy soil, the supply of Ca, Si, and Fe into the soil water is important. Shortage of Ca leads to the decrease in soil pH and thus the paddy soil becomes acidic. Shortage of SiO_2 declines photosynthesis of paddy and that of Fe results in the accumulation of H_2S which is extremely harmful for paddy root. To suppress the soil degradation, application of fertilizer made of steelmaking slag has been focused, because the steelmaking slag is mainly composed of CaO, SiO_2, and FeO which are the desired elements to modify the degraded paddy field. Before applying the fertilizer made of steelmaking slag to the actual degraded paddy, the dissolution behavior of each element and its effect on the yield should be confirmed. For this purpose, leaching tests, soil incubation tests, and field tests have been conducted using the commercial fertilizers made of steelmaking slag [1–5]. The positive effects for paddy growth and yield by the supplement of Ca and Si on soil were clarified [6, 7].

On the other hand, the dissolution ratio of Fe from the fertilizer made of steelmaking slag was found to be low and vary with its brands. To clarify this phenomenon, the mineralogical phases that found in the commercial fertilizers made of steelmaking slag which contained iron oxides were synthesized and their dissolution behaviors were studied [8]. The results showed that, the glassy phase which mainly consisted of CaO–SiO_2–FeO showed higher Fe solubility than other Fe rich phases such as calcium ferrite, magnesio-wustite, etc. Therefore, to improve the soluble Fe from slag fertilizer to paddy soil environment, the CaO–SiO_2–FeO glassy phase holds the key. However, because the CaO–SiO_2–FeO glassy phase is a liquid phase at steelmaking temperature, its composition can vary by the refining and solidification conditions and these variations may affect the dissolution behavior when it is applied as fertilizer. Therefore, in this study, the dissolution behavior of the CaO–SiO_2–FeO glassy phase as a function of chemical composition was studied.

Experimental Method

The glassy phase with different CaO/SiO_2, Fe^{2+}/T–Fe ratios, and FeO content were synthesized. The composition of each synthetic glassy phase is shown in Table 1. For the raw materials, $CaCO_3$ was sintered at 1373 K to make CaO, and Fe_2O_3 was reduced at 1723 K by metallic iron to make FeO. For SiO_2 and Fe_2O_3, commercial regents were used. To synthesize the glassy phases with various CaO/SiO_2 ratios and FeO contents, prepared CaO, SiO_2, and FeO were mixed and melted in a Fe crucible at 1723 K under Ar atmosphere. After melting, the melt was quenched by water. To change the F^{e2+}/T–Fe ratios, CO and CO_2 gas mixture was used. In addition, to synthesize the glassy phase contained Fe_2O_3, the mixture of reagents

Table 1 Synthesized glassy phase composition

	Composition (mass%)				Basicity, mass%CaO/mass%SiO_2
	CaO	SiO_2	FeO	Fe_2O_3	
A	19.0	51.0	30.0	0.0	0.37
B	28.0	42.0	30.0	0.0	0.67
C	35.0	35.0	30.0	0.0	1.00
D	26.7	39.9	0.0	33.3	0.67
E	32.0	48.0	20.0	0.0	0.67
F	24.1	35.9	40.0	0.0	0.67

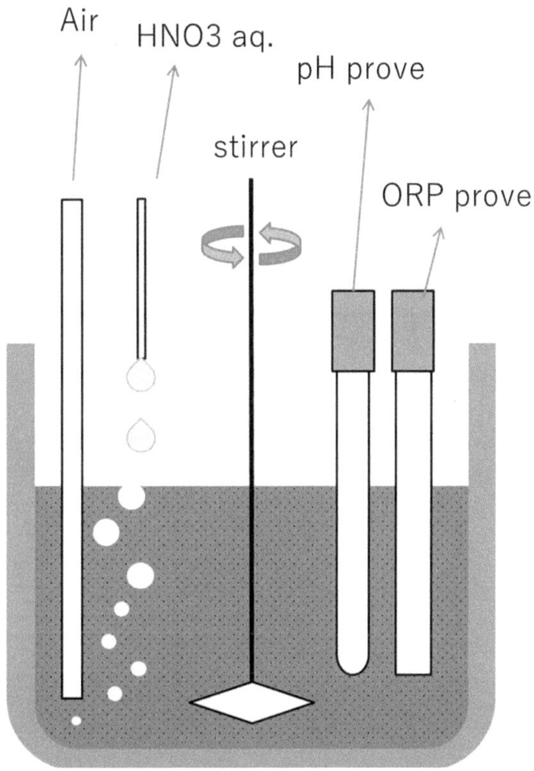

Fig. 1 Experimental setup for leaching test

was melted in a Pt crucible under air atmosphere. The crystalline in each synthesized phase was confirmed by XRD.

After synthesis, the dissolution behavior of each phase was investigated using an experimental setup as shown Fig. 1. In the leaching experiment, air-saturated water was used and the target pH was selected as 5 to simulate the initial stage after irrigation in actual paddy fields [1]. For the experimental procedure, firstly, 400 mL of deionized water was poured into a Teflon container and kept at 298 K using a thermostatic bath. After air bubbling for about 30 min, 1 g of synthesized glassy slag powder (particle size <53 μm) was dropped into the water and the stirring by mechanical stirrer was started. At the same time, dilute nitric acid was continuously added to control the pH to 5. After the inlet of synthetic glassy phase and the start of pH control, the time was recorded as 0 min and then the sampling of solution was repeated at certain time intervals. The sampled solution was analyzed by ICP-AES. After a given time, the leaching test was stopped and the residue were collected and analyzed by XRD.

Result and Discussion

As an example, the leaching behavior of Fe from 1 g of glassy phase B in water is shown in Fig. 2. Concentration of Fe in solution increased with time and reached steady state after 120 min. In addition, in other experiments, Fe showed similar dissolution behavior. Therefore, the dissolution ratios of Fe for all experiments were calculated using the following calculation. Here, D_{Fe} is the dissolution ratio of Fe from 1 g of glassy phase, mg/g; C_{Fe} is the content of Fe in solution that analyzed by ICP-AES after leaching for 120 min, mg/L; V_{water} is the water volume, L; and V_{HNO_3} is the volume of the added dilute nitric acid to keep pH at 5, L.

$$D_{Fe} = C_{Fe} \times (V_{water} + V_{HNO_3})$$

The dissolution ratios of Fe from the glassy phase of various CaO/SiO_2 ratios (slag A, slag B, slag C) after leaching for 120 min are shown in Fig. 3. The Fe dissolved little from low-basicity glassy slag, and an increase in basicity improved the dissolution ratio. The dissolution ratios of Fe from the glassy phase with varied the Fe valence (slag B, slag D) are compared in Fig. 4. When iron oxide was changed from FeO to Fe_2O_3, the dissolution ratio of Fe decreased about 98%.

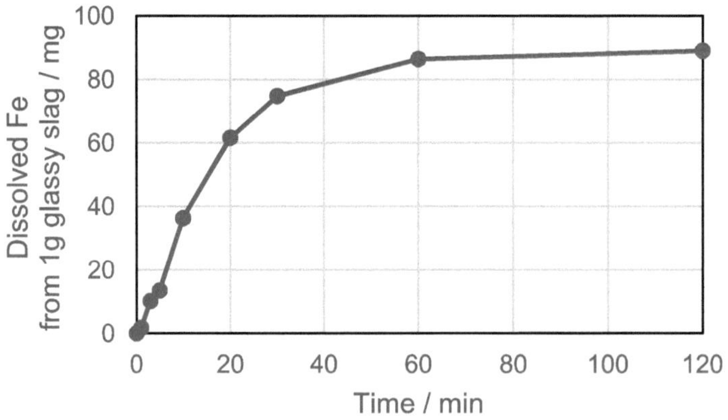

Fig. 2 Leaching behavior of Fe in water from glassy phase B

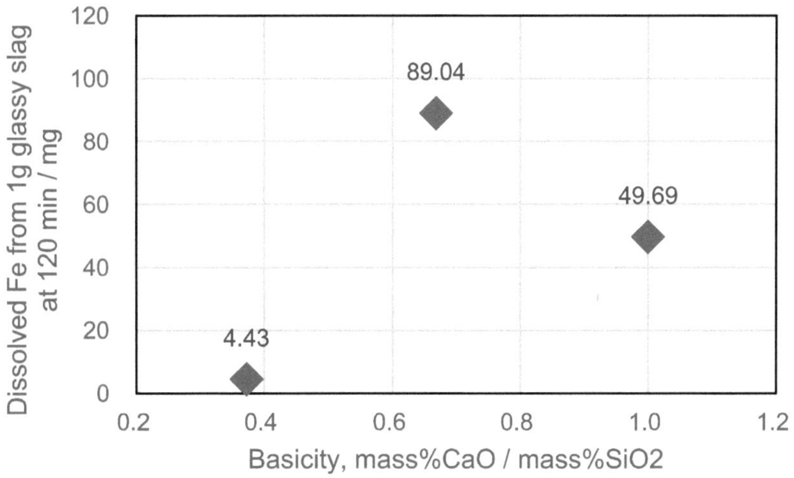

Fig. 3 The dissolved Fe from glassy phase with varied basicity at 120 min

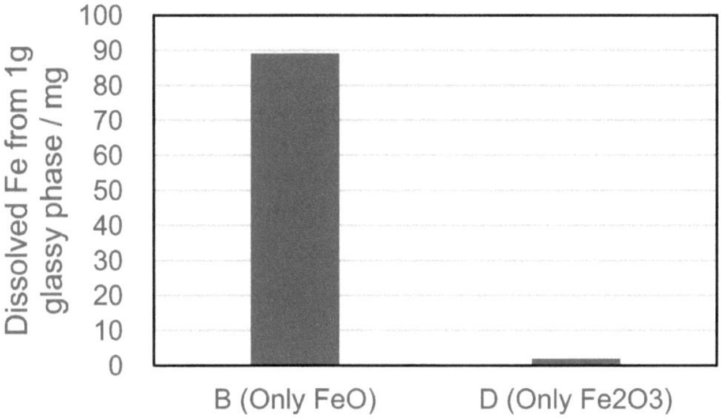

Fig. 4 The dissolved Fe from glassy phase with varied the Fe

In addition, the change in the dissolution ratios of Fe from the glassy phase of different FeO contents (slag B, slag E, slag F) are shown in Fig. 5. With an increase in FeO content, the dissolution ratio of Fe was increased. In summary, the dissolution ratio of Fe from the CaO–SiO_2–FeO glassy phase increased with increasing the Fe^{2+}/T–Fe ratio and FeO contents, and showed a local maximum by the increase in CaO/SiO_2 ratio.

To understand the above experimental phenomenon, the change in the glass structure by chemical composition was considered. For the CaO–SiO_2–FeO glassy phase, the network structure should be made by SiO_2, and basic oxides, such as CaO and FeO, modify the SiO_2 network structure and produces non-bridging oxygen. In addition, the Fe_2O_3 works as a network former. Therefore, with an increase in Fe^{2+}/T–Fe ratio, FeO content, and CaO/SiO_2 ratio, network modifier

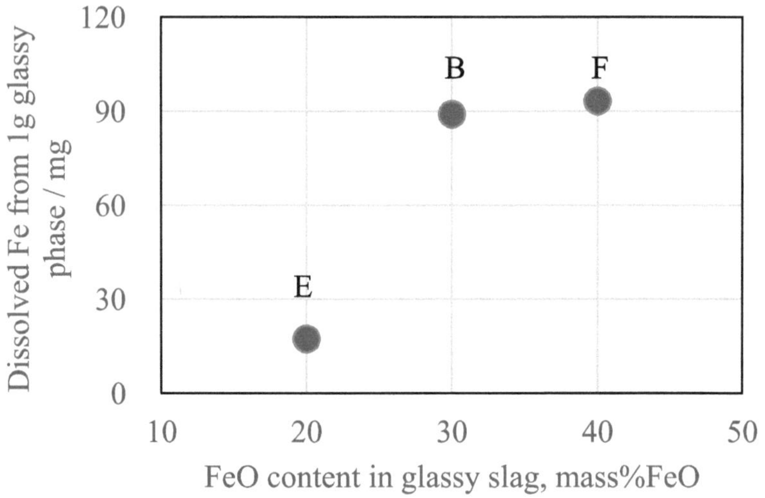

Fig. 5 The dissolved Fe from glassy phase with varied the Fe valence at 120 min

increased and stability of the glassy phase was weakened. However, when the CaO/SiO_2 ratio increased to 0.67 or more, the dissolved Fe started to decrease. It was known that the coordination number of Fe partially changes from 6 to 4 by the increase in the CaO/SiO_2 ratio over certain value [9]. Therefore, this trend would be caused by the existence of a part of Fe^{2+} as tetrahedron which is a relatively stable structural unit. However, further study is necessary to confirm the coordination number of Fe in the glassy phase.

Conclusions

To improve the Fe dissolution from the fertilizer made of steelmaking slag, the dependency of the dissolution behavior of $CaO–SiO_2–FeO$ glassy phase on its chemical composition was studied. In conclusions, an increase in the $Fe^{2+}/T–Fe$ ratio of $CaO–SiO_2–FeO$ glassy phase dramatically improved the dissolution of Fe into the aqueous solution of pH 5 controlled by nitric acid. The FeO content also increased the dissolution ratio of Fe. In addition, with an increase CaO/SiO_2 ratio, the dissolution ratio of Fe showed a local maximum. Therefore, to improve the soluble Fe from slag fertilizer to paddy soil environment, it is effective to bring the composition of $CaO–SiO_2–FeO$ glassy phase to $CaO/SiO_2 = 0.67$, $Fe^{2+}/T–Fe = 1$, and higher FeO content as possible.

References

1. N. Maruoka, M. Okubo, H. Shibata, X. Gao, T. Ito, S. Kitamura, Improvement of desalted paddy soil by the application of fertilizer made of steelmaking slag (recovery of a paddy field damaged by the tsunami using fertilizer made of steelmaking slag—1). Tetsu-to-Hagane **101**(8), 445–456 (2015)
2. M. Okubo, N. Maruoka, H. Shibata, X. Gao, T. Ito, S. Kitamura, Long-term dissolution characteristics of various fertilizers made of steelmaking slag in a desalted paddy soil environment (recovery of a paddy field damaged by the tsunami using fertilizer made of steelmaking slag—2). Tetsu-to-Hagane **101**(8), 457–464 (2015)
3. S. Kitamura, T. Ito, Final report of the ISIJ innovative program for advanced technology "Recovery of paddy fields damaged by the tsunami by the used steelmaking slag". Feramu **20**(11), 545–551 (2015)
4. X. Gao, N. Maruoka, S.J. Kim, S. Ueda, S. Kitamura, Dissolution behavior of nutrient elements from fertilizer made of steelmaking slag under the environment of rice field after irrigation. J Sustain Metal. **1**, 304–313 (2015)
5. X. Gao, T. Ito, H. Nasukawa, S. Kitamura, Application of fertilizer made of steelmaking slag in the recovery of paddy fields damaged by the Tsunami of 2011. ISIJ Int. **56**(6), 1103–1110 (2016)
6. T. Ito, K. Nasu, M. Saito, S. Kitamura, Productivity improvement of saline paddy soils caused by seawater inflow with steelmaking slag fertilizer. *CAMP-ISIJ* **27**, 322 (2014)
7. X. Gao, M. Okubo, N. Maruoka, H. Shibata, T. Ito, S. Kitamura, Production and utilisation of iron and steelmaking slag in Japan and the application of steelmaking slag for the recovery of paddy fields damaged by Tsunami. Trans. Inst. Min. Metall. C **124**(2), 116–124 (2015)
8. X. Gao, S. Kim, H. Minoshima, T. Ito, S. Kitamura, Effect of the mineralogical phases in steelmaking slag fertilizer on the dissolution behavior of nutritious elements and rice growth. *J. Sustain. Metall.* 1–12 (2016)
9. S. Sumita, T. Miori, K. Morinaga, T. Yanagase, Viscosity of slag melts containing Fe_2O_3. J. Jpn. Inst. Met. **44**(1), 94–95 (1980)

Penetration Depth of Microwave in Tire Rubber

Yuzhe Zhang, Jiann-Yang Hwang, Zhiwei Peng, Matthew Andriese, Bowen Li, Xiaodi Huang, Xinli Wang and Xin Yan

Abstract Using a TM_{0n0} cavity system as the main reactor, an integrated process was developed to find the best condition for waste tire rubber to absorb microwaves. The penetration depth was the main character for determining the best condition. The penetration depth is defined as the distance from the surface into the materials at which the traveling wave power drops to from its value at the surface. The waste tire powder was irradiated by two different microwave frequencies, including 915 and 2466 MHz. The study was conducted with same powder size and atmospheric condition. The effects of temperature and microwave frequency were analyzed based on the experimental data. The main trends are higher the temperature, the absorption will be better and the higher the frequency, the better the absorption will be.

Keywords Waste tire · Rubber · Microwave · Penetration depth

Y. Zhang (✉) · J.-Y. Hwang · Z. Peng · M. Andriese · B. Li · X. Huang · X. Wang · X. Yan
Michigan Technological University, Houghton, MI 49931, USA
e-mail: yuzhez@mtu.edu

Y. Zhang · J.-Y. Hwang
Futianbao Environmental Technology Protection Corporation, Xi'an, Shaanxi, China

Z. Peng
School of Minerals Processing and Bioengineering, Central South University, Changsha 410083, Hunan, China

B. Li · X. Huang
Advanced Materials R&D Center of WISCO, Beijing 102211, China

J.-Y. Hwang
Tongchang Environmental Technology Company, Yangzhou, Jiangsu, China

J.-Y. Hwang
Evergreen Technology, Inc., Hancock, MI 49930, USA

J.-Y. Hwang et al. (eds.), *8th International Symposium on High-Temperature Metallurgical Processing*, The Minerals, Metals & Materials Series,
DOI 10.1007/978-3-319-51340-9_39

Introduction

The disposal of waste tires all over the world has become a serious economic and environment problem. The number of waste tire keeps increasing by about 2% each year. Since the waste tire is excessively resistant to any forms of degradation, such as biological, chemical degradation, the landfilled waste tires will hold the valuable land for very long time. In addition, waste tries can cause health problems and fire hazards. Methods for recycling of waste tires by chemical or mechanical means are also limited. For example, incineration can utilize the energy of waste tires, but it will produce large amounts of SO_2, NO_x and other toxic emissions.

Pyrolysis is an alternative way with high potential to recycle the valuable contents of waste tires. Waste tires are inhomogeneous, they have various compositions based on the tire design, applications, weather, grade, manufacturer, and other issues. Pyrolysis is a way to process the tire rubber to a few essential products. In this paper, the microwave will be used for the pyrolysis of waste tire.

Microwave Working Principles

Microwave heating is considered a more advanced heating method than the conventional heating [1]. The reason is that materials absorb microwave energy can be absorbed at molecular level by materials. There are four loss mechanisms: conduction, hysteresis, resonance which is also called magnetic loss, and dipolar polarization [2].

Some experimental work found the non-thermal effect of microwave field. The activation energy which is generated through orientation effects of molecules (entropy effect) may be lowered by the microwave field. This is because microwave has high acceleration of reaction rate. As a result, the microwave can also lower the storage energy which is the same as oscillatory energy of molecules (enthalpy effect). But this effect can't be confirmed so far. Unlike traditional synthesis, the microwave assisted synthesis can hardly control some experimental conditions like pressure and temperature. This situation becomes even worse without accurate and reliable temperature measurement techniques [3]. To ensure the existence of the non-thermal effect, the quantitative evaluations are necessary.

With the ability of the electric field to polarize the electric charge of the stuff where polarization cannot follow the changing electric fields, microwave can heat dielectric materials [4]. There are four important kinds of dielectric polarization: The first one is orientation polarization. This kind of polarization is resulted in by the reorientation of the constant dipoles because of the influencing electric field [5]. The second one is electron polarization [6]. This kind of polarization is because of the position changes of electrons around the nucleus. The third one is spatial charge polarization. This kind of polarization is found when materials consisting of free electrons whose allocation are limited by the surface of grain [7]. The fourth one is

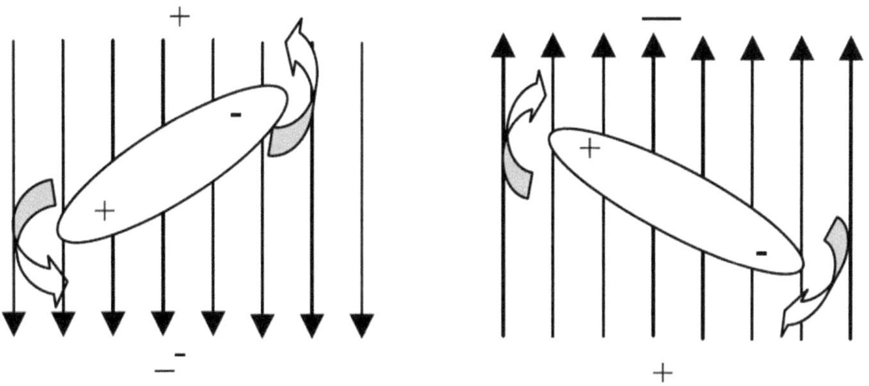

Fig. 1 Realignment of a dipole in an electromagnetic area [9]

atomic polarization. This kind of polarization is produced by the nucleus positional shifting because of the asymmetry allocation of the charge within the molecule [8]. Figure 1 shows how polar molecules reorient in a microwave field.

Orientation (dipole) polarization is the most vital mechanism in microwave frequency range because electron and atomic polarization require much more energy than the energy which can be produced at microwave frequency [9].

Another important microwave heating mechanism is ionic conduction. Due to their native charge, the ions of a solution move under microwave field and collide. The collisions will lead to transformation of kinetic energy to thermal energy. There are an increasing number of collisions with the increasing concentration of ions in solution so that the solution will be heated faster at higher concentration [10].

There are three kinds of materials which are classified based on the response to microwave heating: The first kind is conductors which prevent microwave penetrating and reflect microwaves [9]. The second kind is high loss or absorbing materials. This kind of material absorbs the microwave and then convert electromagnetic energy to heat.

Penetration Depth Principles

The penetration depth D_p is a parameter to evaluate the microwave absorption ability of materials. It is defined as the distance from the surface into the materials at which the traveling wave power drops to 1/e from its value at the surface. The penetration depth can be calculated from the dielectric loss factor, ε_r'', and the relative dielectric constant, ε_r:

$$D_{\mathrm{p}} = \frac{\lambda_0}{2\pi(2\varepsilon_r')^{\frac{1}{2}}}\left\{\left[1+\left(\frac{\varepsilon_r''}{\varepsilon_r'}\right)^2\right]^{\frac{1}{2}}-1\right\}^{-\frac{1}{2}} \quad (1)$$

where λ_0 is the wavelength of the microwave in free space.

The reflection loss calculation is also needed for this research. Since microwave heating is generally conducted in a metallic cavity, the reflection loss RL can be employed to further analyze the microwave absorption ability of sample. The equation is:

$$RL = 20\log\frac{\left|\sqrt{\frac{\mu_{\mathrm{r}}}{\varepsilon_{\mathrm{r}}}}\mathrm{tsnh}\left(j\frac{2\pi f}{c}\sqrt{\mu_{\mathrm{r}}\varepsilon_{\mathrm{r}}}d\right)-1\right|}{\left|\sqrt{\frac{\mu_{\mathrm{r}}}{\varepsilon_{\mathrm{r}}}}\mathrm{tsnh}\left(j\frac{2\pi f}{c}\sqrt{\mu_{\mathrm{r}}\varepsilon_{\mathrm{r}}}d\right)+1\right|} \quad (2)$$

where μ_{r} is complex relative permeability of sample and its value is assumed to be 1 in the calculation of reflection loss of coal and d is thickness of sample. The c is the velocity of microwave in free space.

Experiment

In this research, I use the same technical equipment as same as former researches which is cavity perturbation technique for measuring the penetration depth (D_{p}) of the tire rubber during pyrolysis at frequency of 915 and 2466 MHz. The experiment was performed under the stagnant air condition (Fig. 2).

The samples were prepared by cutting off rubber rods from the center of tire tread. These rods have a length of 12 mm and diameter of 7 mm.

Results and Discussion

Figure 3 shows that the penetration depth of tire rubber at 915 MHz keeps decreasing in the range from room temperature to 366 °C. It keeps increasing in the range of 366–415 °C before it decreases in the range of 415–444 °C. However, it increases when the temperature is increased to 444 °C until there is a substantial decrease in the range 500–642 °C. It increases firstly and then decreases after increasing in the range of 642–666 °C. The penetration depth of tire rubber at 2466 MHz behaves similarly to the tire rubber at 915 MHz from room temperature to 666°C. Only a difference happens where the penetration depth of tire rubber at 2466 MHz keeps decreasing in the range of 415–642 °C. The penetration depth of tire rubber at 915 MHz is higher than that at 2466 MHz.

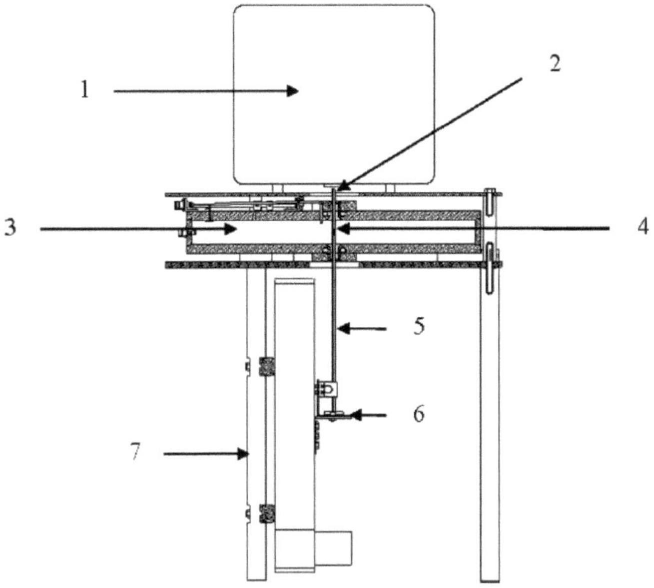

Fig. 2 Diagram of the TM_{0n0} cavity system. *1* High temperature resistance furnace, *2* Entry hole, *3* TM_{0n0} right cylindrical cavity, *4* Sample in quartz holder, *5* Quartz sample holder, *6* Sample stage on linear actuator, and *7* Support stand [10]

Figure 4 shows the relations between the temperature and sample density, volume and mass. Sample mass begins to decrease at temperature above 300 °C, experiences huge weight loss from 300 to 400 °C, and then slows down from 400 to 750 °C.

Tire is composed of primarily rubber and carbon black. Carbon black is a strong microwave absorbing material while rubber absorbs little microwave. The penetration depth is inversely proportional to the microwave absorption capability. Therefore, the results obtained could be related to the behavior of rubber and carbon black, respectively, in the tire.

The sharp decrease of penetration depth in the temperature range from room temperature to 366 °C means that the microwave absorption ability increases significantly in the temperature range. The absorption of microwave of a material generally increases with the temperature because of the increase of dieletric constant with the temperature. Therefore, both the rubber component and the carbon black component will contribute to the sharp decrease of penetration depth.

In the temperature range of 366–415 °C, the penetration depth increases. It is observed that there is a significant loss of weight and expansion in volume in the vicinity of the temperature range. There is a time lapse between the penetration depth measurement and weight and volume measurements during the experiments, since sample has to be taken out of the device at each time of weight and volume measurement. The weight loss and volume expansion is associated with the

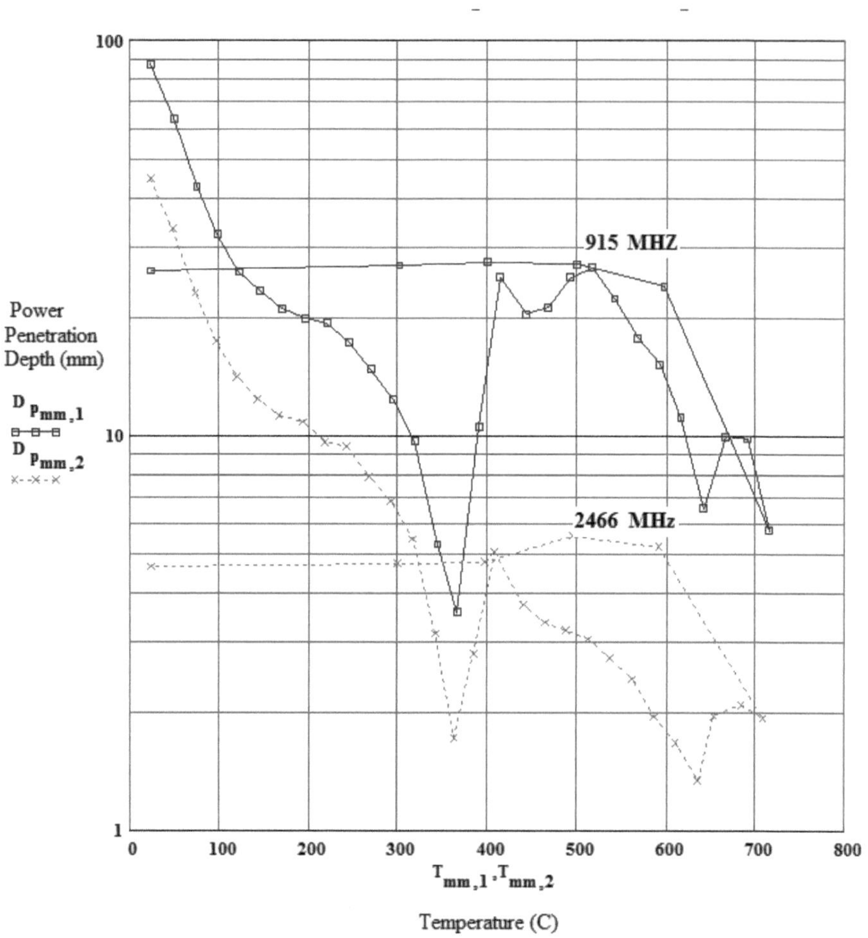

Fig. 3 Temperature dependences of log (penetration depth) under stagnant air. The Dpmm, 1 is log (penetration depth) at 915 MHz, the Dpmm, 2 is log (penetration depth) at 2466 MHz

pyrolysis of the rubber component because its volatiles in the rubber will escape while being heated. Carbon black component will not have such strong changes. Therefore, the phase transformations of the tire rubber during pyrolysis is attributed to the phenomenon.

There is not much changes of penetration depth from 415 °C to roughly 500 °C. Decrease of penetration depth is observed at temperature higher than 500 °C. This is probably due to the carbonization of the rubber component. Graphite structure begins to form. The pi bond of graphite provides the conductivity and hence yields a good increase of dielectric constant.

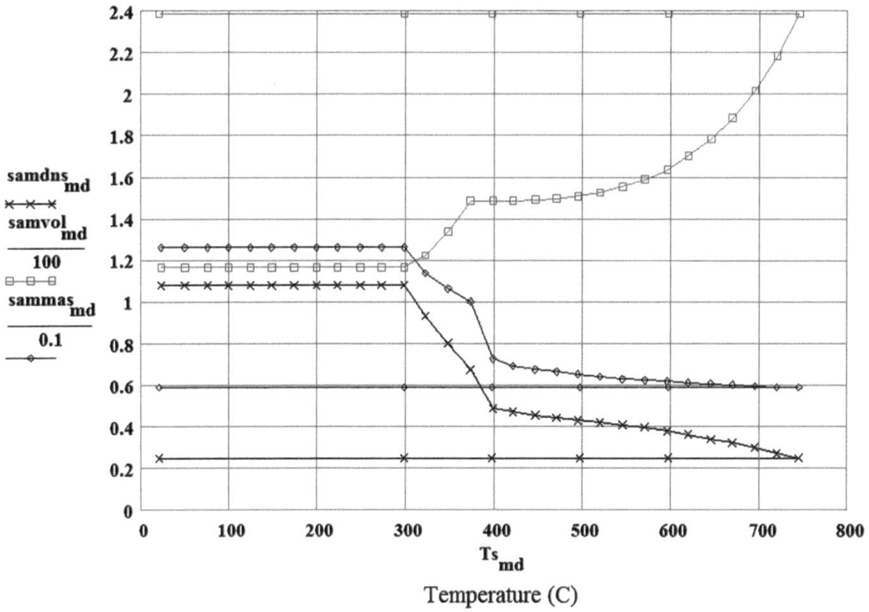

Fig. 4 Temperature dependences of sample density (gm/cc), sample volume (mm^3) and sample mass (gm) under UHP nitrogen. The $samdns_{md}$ is sample density, the $samvol_{md}$ is sample volume and the $sammas_{md}$ is sample mass

Conclusions

Waste tire rubber powders were irradiated with microwave in TM_{0n0} cavity system to find the appropriate conditions for waste tire to absorb microwaves. The system was run under various temperatures with stagnant air and two different frequencies of microwave, which are 915 and 2466 MHz. Experiments on this system have shown that from room temperature to 800 °C, penetration depth generally decreases with the rise of temperature. The penetration depth of tire rubber with 2466 MHz microwave are smaller than the penetration depth of tire rubber with 915 MHz microwave. The strong loss of volatiles at temperatures from 300 to 400 °C and carbonization above 500 °C during pyrolysis may have significant impacts on the results.

References

1. J.M. Lee, J.S. Lee, J.R. Kim, S.D. Kim, Pyrolysis of waste tires with partial oxidation in a fluidized-bed reactor. Energy **20**, 969–976 (1995)
2. H. Huang, L. Tang, Pyrolysis treatment of waste tire powder in a capacitively coupled RF plasma reactor. Energy Convers. Manage. **50**, 611–617 (2009)

3. X. Sun, Treatment of electric arc furnace dust by microwave heating, Master thesis, Michigan Technology University, 2006
4. Z. Peng, J.-Y. Hwang, J. Mouris, R. Hutcheon, X. Sun, Microwave absorption characteristics of conventionally heated nonstoichiometric ferrous oxide. Met. Mater. Soc. ASM Int. 2259–2263 (2011)
5. X. Dai, X. Yin, C. Wu, W. Zhang, Y. Chen, Pyrolysis of waste tires in a circulating fluidized-bed reactor. Energy **26**, 385–399 (2001)
6. M.C. Bignozzi*, F. Sandrolini, Tyre rubber waste recycling in self-compacting concrete. Cem. Concr. Res. **36**, 735–739 (2006)
7. V.K. Gupta, B. Gupta, A. Rastogi, S. Agarwal, A. Nayak, Pesticides removal from waste water by activated carbon prepared from waste rubber tire. Water Res. **45**, 4047–4055 (2011)
8. E.L.K. Mui, D.C.K. Ko, G. McKay, Production of active carbons from waste tyres—a review. Carbon **42**, 2789–2805 (2004)
9. M. Al-Harahsheh, S.W. Kingman*, Microwave-assisted leaching-a review. Hydrometallurgy **73**, 189–203 (2004)
10. Z. Peng, J.-Y. Hwang*, B.-G. Kim, J. Mouris, R. Hutcheon, Microwave absorption capability of high volatile bituminous coal during pyrolysis. Energy Fuels **26**, 5146–5151 (2012)
11. P. Ariyadejwanich, W. Tanthapanichakoon, K. Nakagawa, S.R. Mukai, H. Tamon, Preparation and characterization of mesoporous activated carbon from waste tires. Carbon **41**, 157–164 (2003)

Effect of TiO_2 on Thermophysical Properties and Structure of P-Bearing Steelmaking Slags

Zhanjun Wang, Zuotai Zhang, Mei Zhang and Min Guo

Abstract In this study, the viscous flow and crystallization behavior of $CaO–SiO_2–MgO–Al_2O_3–Fe_tO–P_2O_5–TiO_2$ steelmaking slags have been investigated over a wide range of temperatures. The results indicated that the viscosity of the slags slightly decreased with increasing TiO_2 content. The constructed non-isothermal continuous cooling transformation (CCT) diagrams revealed that the addition of TiO_2 lowered the crystallization temperatures. This can mainly be ascribed that the addition of TiO_2 promotes the formation of $[TiO_6]$-octahedra units, thereby $MgFe_2O_4–Mg_2TiO_4$ solid solution is formed. Moreover, the decreasing viscosity has a significant effect on the diffusion of ion units, such as Ca^{2+} and $[TiO_4]$-tetrahedra, from bulk melts to crystal-melt interface. The crystallization of $CaTiO_3$ and $CaSiTiO_5$ was therefore accelerated, which can improve the phosphorus content in P-enriched phase.

Keywords TiO_2 · Viscosity · Crystallization · Structure

Introduction

In 2015, the production of crude steel surpassed 800 million tons in China, possessing nearly 80–100 million tons of steelmaking slags, which can be used as construction materials, recovery of metals and so on [1, 2]. Among these methods, recycling the steelmaking slags into the iron- and steelmaking process was considered as the most effective way. However, it will inevitably result in the enrichment of phosphorus and further increase the burden in the following

Z. Wang (✉) · M. Zhang · M. Guo
School of Metallurgical and Ecological Engineering, University of Science and Technology Beijing, Beijing 100083, People's Republic of China
e-mail: wangzhanjun1990@163.com

Z. Zhang
School of Environmental Science and Engineering, South University of Science and Technology of China, Shenzhen 518055, People's Republic of China

J.-Y. Hwang et al. (eds.), *8th International Symposium on High-Temperature Metallurgical Processing*, The Minerals, Metals & Materials Series,
DOI 10.1007/978-3-319-51340-9_40

dephosphorization process. The selective crystallization and phase separation (SCPS) method has been proved to successfully enrich phosphorous into a certain phase, which can be separated and used as phosphorus fertilizer, and the residues may be recycled into iron- and steelmaking process [3].

Many studies have been conducted and considered $n(2CaO{\cdot}SiO_2–3CaO{\cdot}P_2O_5)$ solid solution as the efficient P-enriched phase [4]. A series of studies have been carried out to explore the P concentrating phase in steelmaking slags modified by Fe_2O_3, P_2O_5, Al_2O_3 and CaF_2 additives, where the essential structure of the slags has been systematically studied. Meanwhile, some studies showed that TiO_2 can also be used as an additive to improve P (or V) solid solubility in the P (or V)-enriched phase, during which process the viscosity and crystallization ability changed correspondingly. It is also well known that Ti^{4+} can act as a network former or a network modifier, therefore, the effect of TiO_2 on enrichment of phosphorus was strongly needed for systematical extrapolation to expand the previous experimental data.

The present study aims to clarify of the thermophysical properties of $CaO–SiO_2–Al_2O_3–MgO–Fe_tO–P_2O_5$ slags modified through TiO_2 from the viewpoint of kinetics and Raman structural analysis, including the viscous flow and crystallization behaviors.

Experimental

The raw materials were analytical grade CaO, SiO_2, MgO, Al_2O_3, Fe_2O_3, P_2O_5 and TiO_2. The samples were prepared at 1823 K under 0.3 L/min of Ar in a platinum crucible to obtain the homogeneous slags for two hours. The later quenched slags were examined by X-ray fluorescence (XRF), which was shown in Table 1.

The viscosity was measured using a calibrated Brookfield digital viscometer with a full-scale torque of 6.737×10^{-5} Nm with connecting suspension wires and spindle.

Differential scanning calorimeter (DSC) of the quenched samples was performed by a SETARAM instrument (*Setsys Evolution, S60/58341, SETARAM Instrumentation, France*).

After the DSC measurement, the pre-melted slags were heated to 1823 K and then cooled to the target temperature at a cooling rate of 10 K/min. Then the

Table 1 Analyses result of synthesized slag compositions (wt%)

Slag	$Fe^{3+}/\Sigma Fe$	TiO_2	CaO	SiO_2	MgO	Al_2O_3	Fe_tO	P_2O_5
1	0.79	0	37.81	21.91	6.54	4.85	19.47	9.42
2	0.78	3.11	35.51	20.24	6.67	4.85	20.11	9.51
3	0.77	4.86	34.89	19.46	6.81	4.72	19.87	9.39
4	0.79	6.69	33.28	18.98	6.91	4.91	19.69	9.54
5	0.77	9.67	32.26	17.84	6.52	4.69	19.36	9.66

precipitated crystalline phases were analyzed by X-ray diffraction (XRD) and scanning electronic microscopy (SEM).

A more quantitative analysis of structure can be obtained by using Raman spectra and MAS-NMR.

Result and Discussion

Viscosity Measurement

The effect of TiO_2 addition on the viscosity of the typical basic oxygen $CaO–SiO_2–Al_2O_3–MgO–Fe_tO–P_2O_5$ steelmaking slags at a fixed CaO/SiO_2 ratio of 1.8 is shown in Fig. 1. It can be noted that the slag viscosity decreases with increasing TiO_2 content.

DSC Measurement

The DSC investigation of the studied slags was inquired under non-isothermal conditions at five cooling rates and each separate peak suggested a discrete crystallization of different phases during cooling process. Figure 2 records the DSC curves of the slags with different TiO_2 contents during continuous cooling process at a cooling rate of 5, 10, 15, 20 and 25 K/min, respectively. During the continuous cooling process, slag 1 exhibits four DSC exothermic peaks, as shown in Fig. 2a. However, five peaks were detected for slags containing TiO_2, indicating the occurrence of a new crystallization event in the slag with TiO_2 additions. It can be

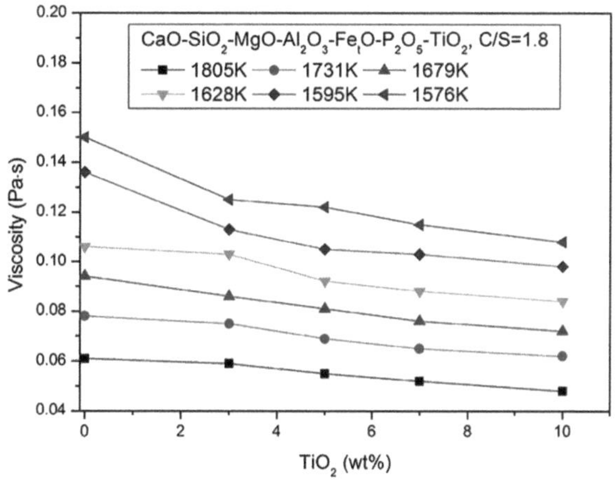

Fig. 1 Effect of TiO_2 on the viscosity of $CaO–SiO_2–MgO–Al_2O_3–Fe_tO–P_2O_5$ steelmaking slags at various temperatures

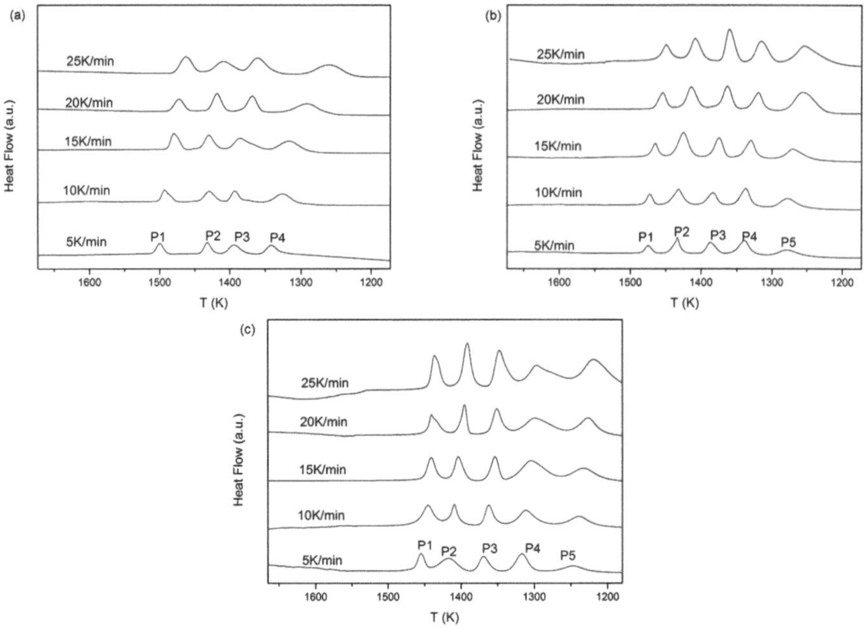

Fig. 2 DSC curves of non-isothermal crystallization of slags at various cooling rates: **a** Slag 1, **b** Slag 3, **c** Slag 5

seen that a significant shift occurred toward a lower temperature, i.e., the addition of TiO_2 lowered the crystallization temperatures.

Identification of Crystalline Phases

A series of heat treatment experiments were carried out to determine the sequence of crystalline phase precipitation, i.e., the slags were melted at 1823 K and then quenched at the target temperatures of the DSC curves. It can be seen from Fig. 3a that $MgFe_2O_4$ spinel became as the first generated crystal for slag 1 at 1473 K, whereas $MgFe_2O_4$ and nC_2S–C_3P co-precipitated at 1413 K, indicating that the exothermic peaks P1 and P2 represented the precipitation of spinel ($MgFe_2O_4$) and P-enrich phase (nC_2S–C_3P), respectively. With increasing TiO_2 content, it can be seen from Fig. 3b that the first crystalline phase is transformed to the mixture of $MgFe_2O_4$ and Mg_2TiO_4. The second exothermic peak P2 still signifies the formation of P-enriched phase. In addition, it can be seen from Fig. 3b $CaTiO_3$ and Ca_2TiSiO_4 can be detected for slag 3 quenched at 1173 K can be detected, which can also be interpreted by phase diagram of CaO–SiO_2–TiO_2 system.

Furthermore, SEM/EDS technique was employed to display the typical phase microscope observation of the crystals. It can be seen from Fig. 4a that, for slag 1,

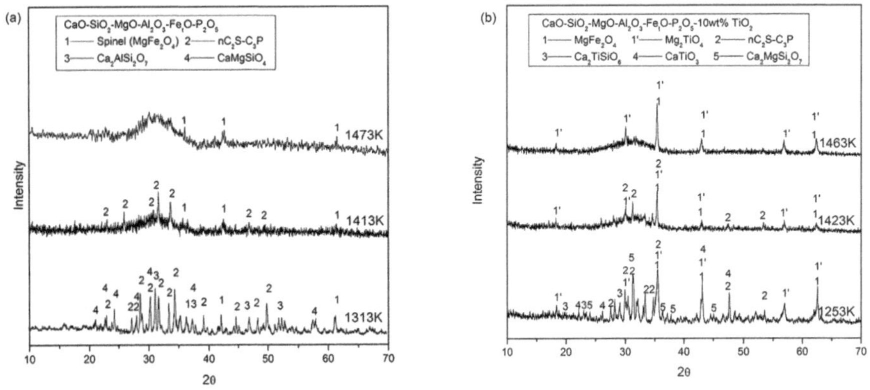

Fig. 3 XRD results of slag 1 and slag 3 quenched at corresponding temperatures

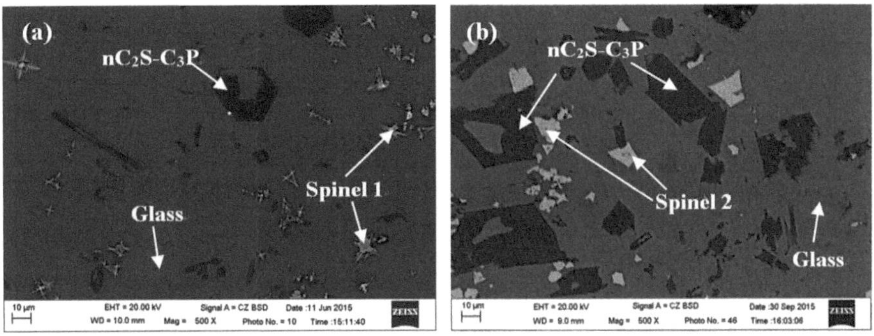

Fig. 4 BSE images of slags at the cooling rate of 10 K/min and quenched at corresponding temperatures: **a** Slag 1 (1413 K), **b** Slag 3 (1423 K)

the first precipitated granular pale crystal can be identified as spinel ($MgFe_2O_4$) and the following precipitated polygonal grey crystal can be determined as P-enriched phase (nC_2S–C_3P). As demonstrated in Fig. 4b, for slag 3, the primary crystal transformed to the white lumps. By combining the XRD results with the study of Harrison [5], the white lumps may be explained as the solid solution of $MgFe_2O_4$–Mg_2TiO_4.

Raman Analysis

Figure 5 exhibits the deconvoluted Raman spectra of slag 1, 3 and 5. The study assigned the bands at about 855, 920, 988 and 1050 cm^{-1} to the presence of Q^0, Q^1, Q^2 and Q^3 units (Q^n, n = NBO/Si, NBO is the non-bridging oxygen number), respectively. In addition, the peaks near 970 and 1100 cm^{-1} correspond to the

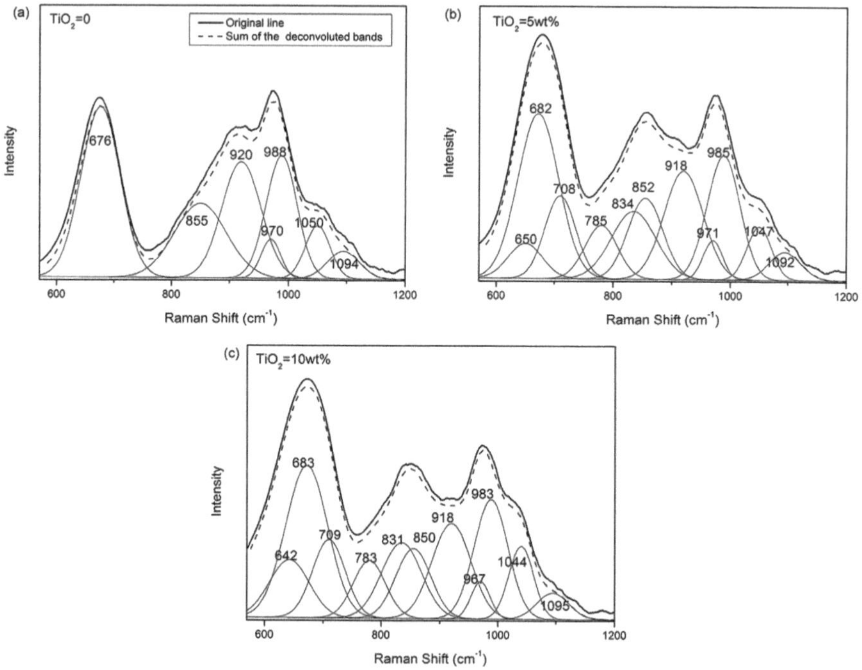

Fig. 5 Raman results of slag 1, slag 3 and slag 5 quenched at 1823 K

vibration of P–O–P and Si–O–P. The band near 680 cm^{-1} can be assigned to the tetrahedrally coordinated Fe^{3+}. By analogy with the results with others, the bands at about 790 and 845 cm^{-1} can be assigned to $Ti_2O_6^{4-}$ chain units and TiO_4^{4-} monomers (or Si–O–Ti) structural units. The band around 644 cm^{-1} could be ascribed to [TiO_6]-octahedra. Additionally, the band in the vicinity of 710 cm^{-1} probably is due to O–Ti–O or O–(Si,Ti)–O deformation in chain or sheet units or both [6].

With increasing TiO_2 content, the Q^2 and Q^3 bands gradually increase and the intensity of Q^0 and Q^1 bands decrease, which means the increasing degree of the polymerization (DOP). The competition between Ti^{4+} and Si^{4+} can be represented by the following reaction [6]:

$$\mathrm{Si-O-M-O-Si+Ti-O-Ti \rightarrow Si-O-Si+Ti-O-M-O-Ti} \quad (1)$$

$$\mathrm{Si-O-M+Ti-O-Ti \rightarrow Si-O-Ti+Ti-O-M} \quad (2)$$

However, viscosity is dependent on both the degree of polymerization and the strength of oxygen bridges. Therefore, the decreasing viscosity could not be explained by the increasing DOP. According to Zheng et al. [7], the strength of Ti–O bonds seems to be much weaker than Si–O bonds for the reason that Ti^{4+} cation is larger in size and smaller in electronegativity comparing with Si^{4+}, thus addition of TiO_2 to highly polymerized melts will decrease the viscosity.

From the viewpoint of structure, $[FeO_4]$-tetrahedra and $[TiO_6]$-octahedra can copolymerize to promote the formation of $MgFe_2O_4$–Mg_2TiO_4 solid solution, which has a lower crystallization temperature in thermodynamics. In addition, the $[TiO_6]$-octahedra structure unit is also the main component to build structure of $CaTiO_3$. Therefore, $CaTiO_3$ was gradually crystallized. Moreover, the appearance of Ti–O–Si band suggested that the $[TiO_4]$-tetrahedra units can copolymerize with $[SiO_4]$-tetrahedra units to enhance the crystallization of $CaSiTiO_5$.

As the chemical Eq. (3) shows as follows, TiO_2 would capture C_2S (Ca^{2+} and $[SiO_4]$ units) from the early precipitated solution (nC_2S–C_3P) to form $CaSiTiO_5$ and $CaTiO_3$, i.e., C_2S vanished gradually and improved the concentration of phosphorus in P-enriched phase.

$$\mathrm{nCa_2SiO_4 \cdot Ca_3(PO_4)_2 \; + \; TiO_2 \rightarrow CaSiTiO_5 + CaTiO_3 + n'Ca_2SiO_4 \cdot Ca_3(PO_4)_2} \tag{3}$$

Conclusions

1. The viscosity of the slags decreased with increasing TiO_2 content and the crystallization tendency also decreased with the addition of TiO_2.
2. The increasing TiO_2 content promotes the formation of $[TiO_6]$-octrahedra units, which are octahedrally coordinated in $MgFe_2O_4$-Mg_2TiO_4 solid solution and $CaTiO_3$. In addition, $[TiO_4]$-tetrahedra units can copolymerize with $[SiO_4]$-tetrahedra to form $CaSiTiO_4$.
3. The formation of $CaTiO_3$ and $CaSiTiO_4$ were beneficial to the enrichment of phosphorus in the P-enriched phase.

References

1. B. Das, S. Prakash, P. Reddy, V. Misra, An overview of utilization of slag and sludge from steel industries. Resour. Conserv. Recycl. **50**, 40–57 (2007)
2. L. Lin, Y.P. Bao, M. Wang, H.M. Zhou, Influence of Al_2O_3 modification on phosphorus enrichment in P-bearing steelmaking slag. Ironmaking Steelmaking **41**, 193–198 (2014)
3. Z.J. Wang, Q.F. Shu, S. Sridhar, M. Zhang, M. Guo, Z.T. Zhang, Effect of P_2O_5 and Fe_tO on the viscosity and slag structure in steelmaking slags. Metall. Mater. Trans. B **46**, 758–765 (2015)
4. L. Lin, Y.P. Bao, Q. Yang, M. Wang, W. Jiang, Effect of CaF_2 and SiO_2 modification on phosphorus utilization in P-bearing slag. Ironmaking Steelmaking **42**, 331–338 (2015)

5. R.J. Harrison, E.J. Palin, N. Perks, A computational model of cation ordering in the magnesioferrite-qandilite ($MgFe_2O_4$-Mg_2TiO_4) solid solution and its potential application to titanomagnetite (Fe_3O_4-Fe_2TiO_4). Am. Mineral. **98**, 698–708 (2013)
6. B.O. Mysen, The influence of TiO_2 on the structure and derivative properties of silicate melts. Am. Mineral. **65**, 1150–1165 (1980)
7. K. Zheng, J.L. Liao, X.D. Wang, Z.T. Zhang, Raman spectroscopic study of the structural properties of CaO-MgO-SiO_2-TiO_2 slags. J. Non-Cryst. Solids **10**, 209–215 (2013)

Analysis for Optimum Conditions for Recovery of Valuable Metals from E-waste Through Black Copper Smelting

M.A.H. Shuva, M.A. Rhamdhani, G.A. Brooks, S. Masood, M.A. Reuter and M. Firdaus

Abstract Declining grade of primary ores and resource efficiency have led us to process more alternative metal resources such as e-waste. One of the processing routes for extracting valuable metals from e-waste is through the black copper smelting. However, the underlying knowledge of the thermodynamics behaviour of the valuable metals contained in e-waste during smelting are limited which prevent us from developing an optimised process to recover all the metals. These different metals clearly will have different favourable conditions for their extraction. To illustrate this, the distribution behaviour of germanium (Ge) and palladium (Pd) between liquid copper and ferrous-calcium-silicate slag during black copper smelting was analysed. It was demonstrated that oxygen partial pressure and slag composition affect the partitioning of these metals to the copper phase and the favourable slag chemistry for recovering these metals is opposing. Considering the available thermodynamic data of these metals, an analysis for the optimum conditions is presented.

Keywords Resources efficiency · E-waste · Black copper · Metal recovery

Introduction

Electronic waste or e-waste can be defined as an end-of-life (EoL) electrical or electronic equipment, which have become obsolete. It is composed of variety of metallic and non-metallic substances and also considered as a hazardous waste as it contains harmful elements such as lead, mercury, and cadmium. The generation rate of e-waste is much higher compared to the general waste, and increases about 4–5% per annum [1]. The main reasons behind the growing rate of e-waste generation are

M.A.H. Shuva (✉) · M.A. Rhamdhani · G.A. Brooks · S. Masood · M. Firdaus
Swinburne University of Technology, Melbourne, Australia
e-mail: mshuva@swin.edu.au

M.A. Reuter
Helmholtz Institute Freiberg for Resource Technology, Freiberg, Germany

J.-Y. Hwang et al. (eds.), *8th International Symposium on High-Temperature Metallurgical Processing*, The Minerals, Metals & Materials Series,
DOI 10.1007/978-3-319-51340-9_41

rapidly changing of the technologies and designs and the shorter lifespan of available equipment.

E-waste contains more than 50 elements which include precious and hazardous metals. Landfilling of e-waste is one of the way of e-waste management in many developing countries which provides significant burden to the environment. At the same time, e-waste is an alternative resource for precious and strategic metals. The concentration of these valuable metals in e-waste is also much higher compared to their ores or mines. For example, the amount of gold extracted from one tonne of gold ore is around 5 g while 30 times of this quantity of gold can potentially be obtained from the same amount of weight of mobile phone waste. The metals composition and concentration vary from one type of e-waste to another and also change with time as the technology changes [1, 2]. Thus, inappropriate management of e-waste not only affects the environment but also results in the loss of valuable resources.

The metals recycling industries currently process e-waste using different advanced metallurgical route (mostly a combination of pyrometallurgy and hydrometallurgy processes). One of the routes include processing of e-waste through primary or secondary copper smelting followed by hydrometallurgy processes to recover precious metals. Noranda, at Quebec Canada, adds about 14% of e-waste to mined copper concentrates in the smelter [3]. The precious metal containing copper matte is upgraded in converters, and finally refined in an anode furnace and cast with a purity of 99.1%. The rest of the composition (0.9%) include the precious metals, including gold, silver, platinum and palladium, along with other recoverable metals such as selenium, tellurium, and nickel. These metals are then recovered from the by-product of subsequent electrorefining of anode copper. The smelter of Rönnskår in Sweden usually has two particular ways for treating the scrap. The waste containing high-grade copper is fed directly into a converting process while the low grade is fed into a Kaldo Furnace. It was reported that they process scrap, including e-waste, about 100,000 tonnes annually. A mixed copper alloy is produced in the Kaldo furnace which is then sent to copper converting for recovery of metals, such as Cu, Ag, Au, Pd, Ni, Se and Zn. The dust, containing Pb, Sb, In and Cd, are processed to recover the metals through other operations [4].

Umicore in Belgium uses integrated smelting and refining processes for the recovery of precious and other valuable metals from non–ferrous scrap, spent industrial catalysts, car exhaust catalysts and printed circuit boards [5, 6]. An IsaSmelt furnace is used to smelt the wastes in the first stage of the precious metals operations (PMO). The precious metals are separated into the copper bullion while the other metals concentrate into a slag which is the by-product of the PMO. Then leaching and electrowinning are performed to recover copper and lastly the precious metals are recovered in the precious metal refinery. The special metals residue like indium, selenium, and tellurium are generated in base metal operations which are further refined in the special metals refinery. Dowa Mining in Kosaka Japan uses an Outotec Top Submerged Lance (TSL) furnace to process approximately 150,000 tonnes of secondary copper feed annually. The operations combine, in different stages, both oxidation and reduction processes to produce a high-grade copper product and to recover valuable elements in refining process such as PGMs, Ni, Zn, and Pd. LS Nikko's Global Resources & Material (GRM) in Danyang, Korea also operate the process under reduced conditions to produce black copper and then refine it to segregate the precious metals during electrolytic refining for further processing [7].

In summary, there are different types of recycling technology and process routes practiced in industries for recycling of end-of-life consumer goods including e-waste. However, the existing processes applied in industry for recovering all valuable metals are far from optimised. To recover all valuable metals, understanding the underlying knowledge of thermodynamics relevant to each metal in e-waste is vital [8, 9]. It has been reported that there is a limited thermodynamics data and information available for most of the metals found in e-waste for conditions associated with non-primary routes of copper processing, although some information is available for primary copper processing operating conditions [10].

In this paper, an analysis for the optimum conditions of recovering valuable metals from e-waste through black copper smelting process is presented. Case study of thermodynamic behaviour of two elements (Ge and Pd) between copper and magnesia-saturated ferrous-calcium-silicate (FCS) slag are considered to illustrate the need of different slag systems. A possible process route for the metals is also discussed and presented.

Optimum Conditions for Germanium and Palladium Recovery as a Case Study

The authors have carried out investigations on the thermodynamic behaviour of Ge and Pd during reactions between FCS slag and copper. The composition of master alloys was Cu-3.9 mass pct Ge and Cu-5.9 mass pct Pd. Ferrous-calcium-silicate (FCS) slags were prepared with varying the ratio of Fe/SiO_2 from 0.8 to 1.30 and mass pct of CaO from 5 to 20%. The details of the experimental setup and preparation of master alloys and slags have been described elsewhere [11, 12]. The distribution ratio of each metal has been measured from equilibrium experiments in a vertical tube furnace in a controlled atmosphere using a CO–CO_2 gas mixture. The distribution of germanium, $L_{Ge}^{s/m}$, and palladium, $L_{Pd}^{s/m}$, in FeO_x–CaO–SiO_2–CaO–MgO and copper at equilibrium from the conducted experiments are shown in Fig. 1.

The effect of oxygen potential was investigated under equilibrium for both Ge and Pd at 1573 K. It can be observed from Fig. 1a that the partitioning of Ge and Pd to copper phase sharply increases with the decrease of oxygen partial pressure. More than 90 mass pct Ge and Pd distributed into the copper phase at $pO_2 = 10^{-10}$ atm. Therefore, it can be noted that more reducing condition is preferable to recover more Ge and Pd to the copper phase.

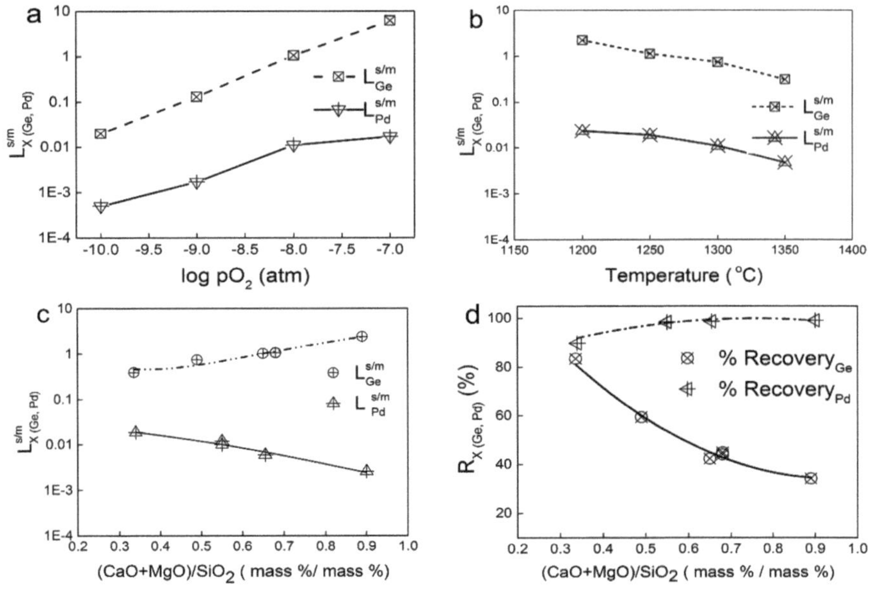

Fig. 1 The influence of, **a** oxygen potential at 1573 K, **b** temperature at $pO_2 = 10^{-8}$ atm, **c** (CaO + MgO)/SiO_2 (slag basicity) at 1573 K and $pO_2 = 10^{-8}$ atm, on the distribution of germanium and palladium in FeO_x–CaO–SiO_2–MgO and copper at equilibrium, **d** effect of (CaO + MgO)/SiO_2 (slag basicity) on Ge and Pd recovery through copper phase [11, 12]

The effect of temperature on the $L_{Ge}^{s/m}$ and $L_{Pd}^{s/m}$ at a fixed pO_2 is shown in Fig. 1b. It can be seen from Fig. 1b that both Ge and Pd mostly report to copper phase at higher temperatures. Operating at higher temperatures is good for slag-copper separation as it decreases the slag viscosity. However, at higher temperatures more copper is lost to the slag and gas phases, i.e. approximately 2% at 1623 K. Thus, it is recommended that the operating temperature should between 1573 and 1623 K (not beyond 1623 K) under reducing condition for good recovery of Ge and Pd.

The influence of slag basicity, represented by (CaO + MgO)/SiO_2 in mass pct, on the $L_{Ge}^{s/m}$ and $L_{Pd}^{s/m}$ at a fixed $pO_2 = 10^{-8}$ atm and temperature 1573 K, is illustrated in Fig. 1c. It can be clearly seen from this figure that the slag basicity has a contradictory effect on the distribution of Ge and Pd in FCS slag. As the basicity is increased, more Pd report to copper phase and in contrary more Ge report to slag phase. The recovery of Ge and Pd (at equilibrium) to liquid copper phase was also plotted against (CaO + MgO)/SiO_2 and presented in Fig. 1d. It was found that more than 80% of Ge and Pd recovered to the copper phase using FCS slag with low basicity. As the basicity is increased, the recovery of Ge decreases and recovery of Pd increases at a fixed $pO_2 = 10^{-8}$ atm and at 1573 K. The results suggest that the nature of Ge and Pd oxides in the FCS slag are acidic and basic, respectively. Therefore, by understanding the behaviour and nature of oxides of valuable

Table 1 Calculated electrostatic bond strength (z/a^2) of minor elements

Minor elements	Oxide	z/a^2	Character of oxide[a]	Preferred type of slag
Ag	Ag_2O	0.15	Basic	Basic
Pb	PbO	0.30		
Pd	PdO	0.39		
Co	CoO	0.43	Intermediate	Basic
Sn	SnO_2	0.46		
Bi	Bi_2O_3	0.51		
In	In_2O_3	0.62		
Sb	Sb_2O_3	0.64		Acid
Te	TeO_2	0.71		
Ga	Ga_2O_3	0.74		
Ge	GeO_2	1.07	Acid	Acid
Se	SeO_2	1.11		

[a]Characteristic of the nature of oxide as proposed by Gilchrist [13]

elements, one can potentially pre-determine the suitable type of slag that likely promotes their recovery into the copper phase.

The acid-basic behaviour of metal oxides can be predicted using a relationship suggested by Gilchrist [13] where he related the acid-basic characteristic with electrostatic bond strength (z/a^2) of metal oxide. In this case, z and a represent the charge of metal ion and sum of ionic radii of two ions in metal oxide, respectively. Table 1 shows the value of electrostatic bond strengths (z/a^2) for selected oxides of interest in e-waste. Low value of electrostatic bond strength suggests a basic nature of the oxide. With the increasing value of the electrostatic bond strength, the oxides behaviour gradually changes from basic to intermediate to acidic. The results of Ge and Pd oxides' behaviour in the FCS slag were found to be in good agreement with the calculated result from electrostatic bond strength.

The information of electrostatic bond strength can help understand the behaviour of oxide species in the slag system and provide insight on the selection of the best chemistry of slag for the black copper smelting process. This analysis suggest that basic slag is preferred for the recovery of Ag, Pb, Pd, Co, Sn, Bi, In; while acidic slag is suitable for Sb, Te, Ga, Ge, Se.

Analysis of the Conditions for Recovering Metals in Black Copper Smelting

Recovery of all the valuable metals from e-waste is quite a complex task as there are different and conflicting conditions that need to be sought for all these elements. One of the key points is the understanding of where will the elements report to during the process (i.e. dissolved to Cu, oxidised to slag or form gaseous phases to form metals/oxides dusts) and tune the operating conditions to achieve particular elements distribution which requires the understanding of detailed slag chemistry, thermodynamics and kinetics of the process.

Tuning the slag chemistry will allow maximum deportment of selected valuable elements to copper, for example by considering multi-stage operation using basic and acidic slag. Another consideration is the role of technology and reactor types. Certain technology and reactor may have limited operating conditions (temperature and oxygen partial pressure ranges), while others may have a wider and more versatile range (can be operated from reducing to oxidising conditions). Understanding the capabilities of different technologies is useful for developing feasible (from economical and technical) process route, e.g. it can inform the minimum number of reactors needed as well minimum operational requirements (such as appropriate refractory linings for operations with basic or acidic slag).

In general, multi-stage black copper smelting can be carried out through a number of process routes (reduction followed by oxidation; or vice versa) depending on the type and composition of feeds, as can be seen from the Fig. 2. The key points from the black copper smelting process are to:

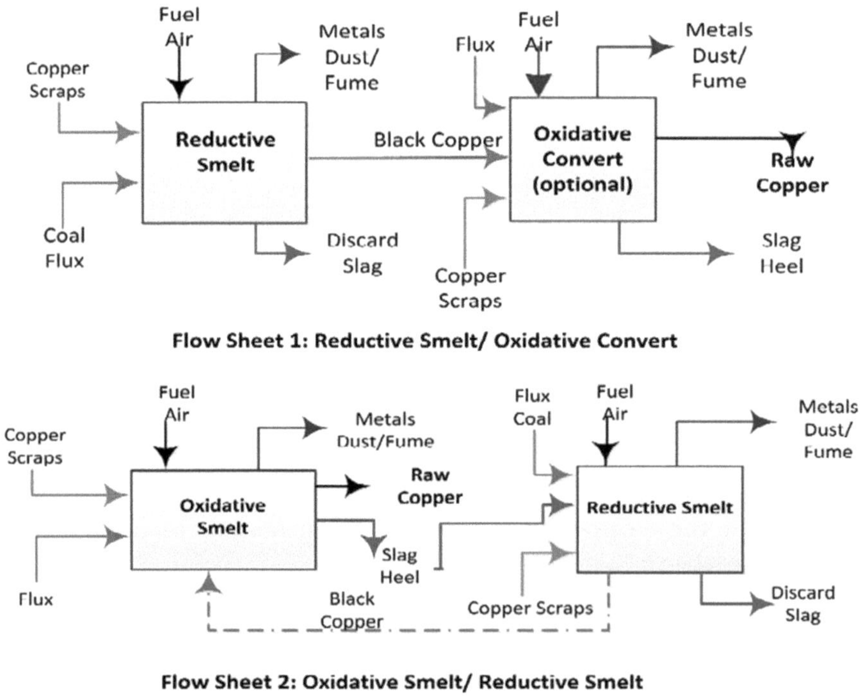

Fig. 2 Variation of black copper smelting process routes depending on the type/composition of feed [8]

- Use copper as the metal collector for valuable metals such as In, Sn, Ge, Pd, Au, Ag, and others.
- Separate the volatile valuable metals from copper/slag phases to the gaseous phase (Zn, Pb, Cu and some other valuable metals that have large partitioning to the gas phase) and condensed them to allow further processing for their recovery.
- Produce a clean slag (typically contain oxides of Fe, Mg, Si, Ca, Al) so that it can be discarded or re-use for other applications (such as concrete and road mix).

The recovery of valuable elements including Ge and Pd from e-waste is already practiced in industry. For example, Aurubis in Germany processes e-waste through black copper multi-stage smelting using Kayser Recycling System (KRS) where top submerged lance (TSL) technology is used to carry out reduction process and followed by the use of top blown rotary converter (TBRC) furnace for oxidation process [14]. Nevertheless, by considering the slag chemistry (tuning the acidity and basicity of the slag) the existing process routes can be modified for potential improved-recovery of both "acidic" and "basic" metals to copper phase.

One possible modification is to have a multi-stage black copper smelting that consists of two sub-process routes where in one of the route the secondary copper resources are processed using basic slag and while in the other using acid slag. This proposed route is explained in more details in Fig. 3.

A single versatile reactor can be used to carry out the reduction and followed by oxidation operations using basic slag (top left). In this "basic" operation, copper scraps, e-waste, and coke are treated with basic slag and black copper is produced. The suitable conditions are $pO_2 = 10^{-10}$ atm and temperature 1573 K for the reduction process. The precious metals and "basic" metals are partitioned into the copper phase (e.g. Au, Ag, Pd, In, Sn). The slag from this operation contains some

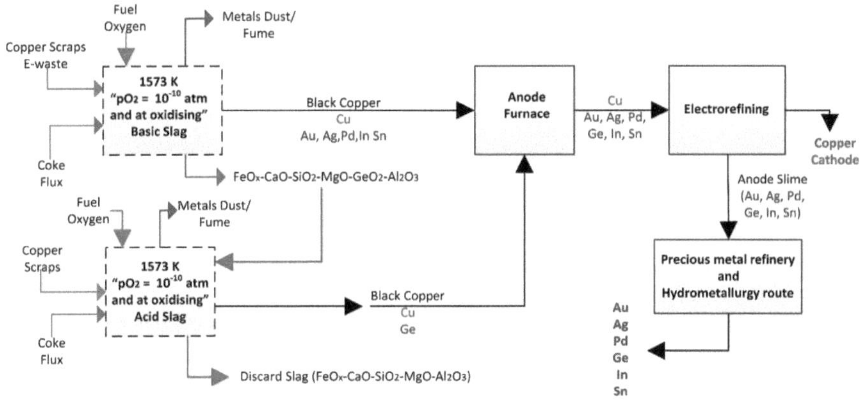

Fig. 3 A route of e-waste recycling through black copper smelting

valuable "acidic" elements, e.g. Ge, Se. Thus, the slags need to be re-used and processed further (reduction-oxidation reaction with copper scrap) in another reactor operating with an acid slag (bottom left). The suitable conditions for the reduction are $pO_2 = 10^{-10}$ atm and temperature 1573 K. By doing this, the valuable "acidic" metals can also be enriched into the copper phase. The copper from both "basic" and "acidic" operation is then treated in anode furnace to produce anode copper and followed by electrorefining for rectifying the purity of copper. The valuable metals in the black copper (for example Ag, Pd, Ge, Sn, In, and Au) mainly goes to anode slime, a by-product of the electrorefining, which is processed further through hydrometallurgical route for their recover. Other valuable metals partitioned the gaseous phases (Zn, Pb, Cu) can also be further processed for their recovery.

It should also be pointed out that different types of slag will attack refractory lining in different ways. Each reactor is designed with appropriate refractory linings capable to handle a particular slag system. In industry, operation using different slag systems (acidic and basic) in a single reactor is difficult to implement because it is difficult to find a suitable refractory. Slag chemistry affects technological issues such as wear of refractory as replacement cost of refractory lining is very high. One of the possible solution is using copper cooling panels could mitigate this issue, allowing a more robust and flexible operation which can be tuned to the feed mix.

Concluding Remarks

From the present work above, the optimised conditions for recovering the valuable metals Ge and Pd from e-waste using black copper route were found to be at $pO_2 = 10^{-10}$ atm and temperature around 1573 K (1300 °C). The suitable slag chemistry was found to be opposite for high recovery of Ge and Pd. It was also found that the recovery of Ge to the copper phase is less than 80% with an acid slag at $pO_2 = 10^{-8}$ atm and at 1573 K (1300 °C). On the other hand, more than 95% of Pd was recovered to the copper phase with a basic slag at the same oxygen potential pressure and temperature. A possible modification of black copper smelting operated using basic and acidic slag was also discussed. It is hoped that present study will give some insight about the importance of underlying thermodynamics knowledge of valuable metals and effect of slag chemistry for their maximum recovery through the black copper smelting route.

Acknowledgements The study is supported by the Swinburne University Postgraduate Research Award (SUPRA) and the *Wealth from Waste Research Cluster*, a collaborative program between the Australian CSIRO, Swinburne University of Technology, University of Technology Sydney, Monash University, University of Queensland, and Yale University.

References

1. R.M. Izatt, *Metal Sustainability: Global Challenges, Consequences, and Prospects* (John Wiley and Sons, UK, 2016)
2. G. Harvey, The potential of Urban mining, in *The Diplomate* (2016). http://thediplomat.com/2013/11the-potential-of-urban-miniing/. Accessed 30 July 2016
3. H. Veldbuizen, B. Sippel, Mining discarded electronics. Ind Environ **17**(3), 7–14 (1994)
4. L. Theo, Integrated recycling of non-ferrous metals at Boliden Ltd. Ronnskar smelter, in *IEEE International Symposium on Electronics & the Environment* (Oak Brok, IL, 1998), pp. 42–47
5. C. Hageluken, Recycling of electronic scrap at umicore's integrated metals smelter and refinery. World Metall.—Erzmetall **59**(3), 152–161 (2006)
6. C. Hagelüken, Recycling of electronic scrap at Umicore's integrated metals smelter and refinery, in *European Metallurgical Conference*, 2005
7. J. Wood, S. Creedy, R. Matusewicz, M. Reuter, Secondary copper processing using Outotec Ausmelt TSL technology, in *AusIMM MetPlant Conference* (Perth, Australia, 2011)
8. E. Worrell, M. Reuter, *Handbook of Recycling*, 1st edn. (Elsevier, UK, 2014), pp. 85–94
9. A. Khaliq, M.A. Rhamdhani, G.A. Brooks, S. Masood, Metal extraction processes for electronic waste and existing industrial routes: a review and Australian perspective. *Resources* **3**(1), 152–179 (2014)
10. M.A.H. Shuva, M.A. Rhamdhani, G.A. Brooks, S. Masood, M.A. Reuter, Thermodynamics data of valuable elements relevant to e-waste processing through primary and secondary copper production: a review. J. Clean. Prod. **131**, 795–809 (2016)
11. M.A.H. Shuva, M.A. Rhamdhani, G.A. Brooks, S. Masood, M.A. Reuter, Thermodynamics behaviour of germanium during equilibrium reactions between FeO_x–CaO–SiO_2–MgO slag and molten copper. Metall. Mater. Trans. B **47**(5), 2889–2903 (2016)
12. M.A.H. Shuva, M.A. Rhamdhani, G.A. Brooks, S. Masood, M.A. Reuter, Thermodynamics of palladium (Pd) and tantalum (Ta) relevant to secondary copper smelting. Metall. Mater. Trans. B (2016). doi:10.1007/s11663-016-0839-y
13. J.D. Gilchrist, *Extraction Metallurgy*, 3rd edn. (Pergamon Press, London, 1989), pp. 198–200
14. Aurubis (2016) www.Aurubis.com. Accessed 31 Aug 2016

Precipitation Behavior of $M_xTi_{3-x}O_5$ in the Titanium-Bearing Electric Furnace Slag

Fuqiang Zheng, Xiaoming Qu, Guanzhou Qiu, Yufeng Guo and Tao Jiang

Abstract $M_xTi_{3-x}O_5$ is the main titanium-bearing mineral in the titanium-bearing electric furnace slag (TEFS), which is produced by direct reduction-electric furnace smelting process with vanadium titanomagnetite concentrate in Panzhihua-Xichang region of China. In order to effectively utilize the titanium resource in TEFS, the thermomechanical analysis and conditions for the growing up of $M_xTi_{3-x}O_5$ grain were investigated, which emphasized on the effects of CaF_2 and cooling rate on the crystallization of $M_xTi_{3-x}O_5$ in the melting process. The mineral phase and morphology of titanium-bearing slag were determined by XRD and electron microscope technique, respectively. The results indicated that the average grain size of $M_xTi_{3-x}O_5$ grown up from 30 to 92 μm and the area fraction of $M_xTi_{3-x}O_5$ reached about 66% from 61% under the conditions of 5 wt% CaF_2 and cooling rate at 1 °C/min.

Keywords $M_xTi_{3-x}O_5$ · Titanium slag · Crystallization · Electric furnace

Introduction

Panzhihua-Xichang region of China is rich in vanadium titanomagnetite resources which account for 11.6% of vanadium and 35.17% of titanium all over the world [1]. The vanadium titanomagnetite concentrate, containing about 89% vanadium and 52% titanium of the vanadium titanomagnetite ore, is a main titanium-bearing product of vanadium titanomagnetite separation and is commonly used to produce iron and extract vanadium in the conventional blast furnace process and direct reduction-electric furnace smelting process [2, 3]. Because the direct reduction-electric furnace smelting process is more environmentally friendly than the process

F. Zheng · X. Qu · G. Qiu · Y. Guo (✉) · T. Jiang
School of Mineral Processing and Bioengineering, Central South University, Changsha 410083, Hunan, China
e-mail: guoyufengcsu@163.com

J.-Y. Hwang et al. (eds.), *8th International Symposium on High-Temperature Metallurgical Processing*, The Minerals, Metals & Materials Series,
DOI 10.1007/978-3-319-51340-9_42

of conventional blast furnace, it has been paid more attentions all over the word [4, 5]. However, there is no effective method to recover titanium from titanium-bearing electric furnace slag (TEFS) even though its TiO_2 grade (40–55 wt%) is higher than that of the titanium-bearing blast furnace slag. Thus, it is crucial to develop an effective method to extract titanium from TEFS for the utilization of the vanadium titanomagnetite concentrate and the development of direct reduction-electric furnace smelting process.

There are many methods [1, 6] develop or under development all over the world to utilize the titanium resource from TEFS, including preparation of nanostructured titanium dioxide, hydrochloric acid leaching of TEFS for titanium dioxide preparation. But too much impurities in TEFS lead to the high acid, energy consumption and the difficulty of acid leaching residue treatment. The selective enrichment method is a method to recover titanium resource from titanium-bearing slag in many studies as taken the considerations of large production scale, simple devices and environment protection [7, 8]. In the selective enrichment process, the titanium in molten slag is enriched into the target titanium-bearing mineral phase, which including perovskite, anosovite, rutile and TiC_xN_y. The perovskite is difficult to be utilized because the its titanium content is lower than the others. The rutile and TiC_xN_y crystals are difficult to grow up. The main titanium-bearing mineral in the TEFS is $M_xTi_{3-x}O_5$, therefore, $M_xTi_{3-x}O_5$ is the preferred target titanium-bearing mineral to realize the selective enrichment for TEFS. The enriched $M_xTi_{3-x}O_5$ should meet the requirement of beneficiation including crystal size and crystal morphology. The larger crystal size and more simple embedded features are benefit to the separation of minerals. Then, it is necessary to study the crystallization behavior of $M_xTi_{3-x}O_5$ in TEFS.

There are few studies on the crystallization behavior of $M_xTi_{3-x}O_5$. Wang X. etc. studied the crystallization behavior of anosovite by adding 3 wt% P_2O_5 into titanium-bearing blast furnace slag under reduction atmosphere, and the quantity of crystallization of anosovite reached 28% with particle size at 80–150 μm. Li J. etc. pointed out the perovskite could be transformed to anosovite by adding SiO_2 up to 42–50 wt% in the slags under the C/CO equilibrated atmospheres [9, 10]. The raw materials of these methods were titanium-bearing blast furnace slag containing 20–25 wt% TiO_2. The TiO_2 content of TEFS reached 40–60 wt%, which was much higher than that of titanium-bearing blast furnace slag. Therefore, it is important to study the precipitation behavior of $M_xTi_{3-x}O_5$ in the titanium-bearing electric furnace slag.

In the present study, the thermodynamic analysis of the selective enrichment of $M_xTi_{3-x}O_5$ with CaF_2 as additive was carried out. The effects of CaF_2 and cooling rate on the enrichment of $M_xTi_{3-x}O_5$ from TEFS was performed. The findings will provide a technical basis for the effective utilization of titanium resource from TEFS.

Experimental

Raw Materials

The titanium-bearing electric furnace slag used in this study was produced in the direct reduction-electric furnace smelting process with the vanadium titanomagnetite concentrate in Panzhihua-Xichang region of China. The chemical composition of TEFS is listed in Table 1. It can be seen that the TiO_2 content of TEFS is 47.35 wt%, and the major impurities are aluminum, silicon, magnesium, calcium-bearing diopside and spinel. The XRD pattern of TEFS are shown in Fig. 1. The major phases in the TEFS are $M_xTi_{3-x}O_5$, diopside and spinel. CaF_2 of analytical reagent (AR) grade was employed as additive.

Experimental Method

The melting experiments were carried out in a high temperature electric furnace (furnace cavity Φ200 × 250 mm) with the $MoSi_2$ heating element. 100 g TEFS mixed with a certain amount of CaF_2 and loaded in a graphite crucible (inner diameter: 50 mm, height: 100 mm) heated at 1773 K for 10 min. When the melting

Table 1 Chemical composition of the original TEFS (wt%)

Component	TiO_2	TFe	V_2O_5	Al_2O_3	SiO_2	MgO	CaO
wt%	47.35	3.16	0.49	12.10	16.08	9.88	9.97

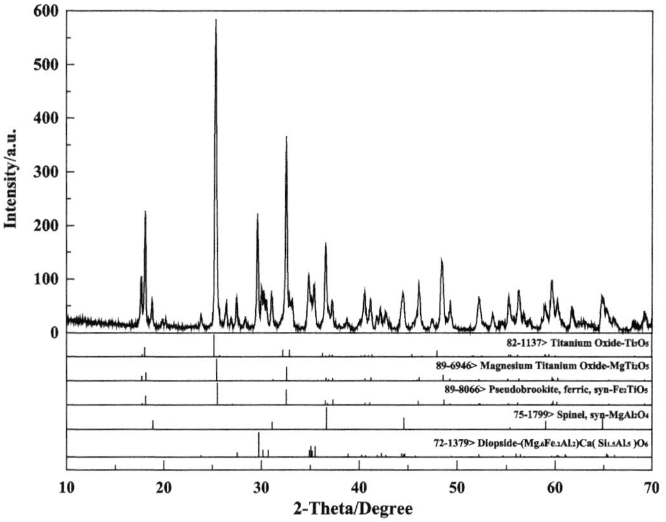

Fig. 1 XRD pattern of the TEFS

process was finished, the melting slag was cooled to room temperature at a certain cooling rate. The sample was under N_2 atmosphere with flow rate at 3 L/min over the melting and cooling process. Then the modified slag was prepared for analyzing the mineral phases and microstructure.

Analysis Methods

The chemical compositions of the TEFS was determined by X-ray fluoroscopy (XRF, Axios mAX, Holland PANalytical Co., Ltd).

The slag samples were mounted into epoxy resin, mechanically ground and polished. Then the polished surface was characterized by an electron microscope for microstructure.

The mineral phases of the TEFS and residue were characterized by X-ray diffraction (Cu K_a radiation, $\lambda = 0.154056$ nm, 40 kV, 250 mA, SCAN: 10.0/70.0/0.02/.15 s, D/max2550PC, Rigaku Co., Ltd., Japan).

Results and Discussion

Thermodynamic Analysis of Mineral Phases Composition

There are many factors that can affect the crystallization behavior of $M_xTi_{3\text{-}x}O_5$ in the slag, such as the composition of slag and the fluidity of the slag during the melting and cooling process. The TiO_2–CaO–SiO_2–MgO–Al_2O_3 phase diagram is shown in Fig. 2, which is drawn using the PhaseDigam module of Factsage 7.0 with slag datebase.

The Fig. 2 indicated that point A was the composition of TEFS, and the mineral phase of point A was consistent with the main mineral phase of TEFS. It is well known that a wide temperature range of liquid phase is benefit to the crystallization of minerals in the slag. It could be seen that the temperature ranges of liquid phase expanded when the CaO content increased only. According to Fig. 2, when CaO content of TEFS increased, the mineral phase would change along the line of A-CaO. When the CaO content of TEFS reached that of point B, the main titanium-bearing mineral phase would be perovskite. In order to make the temperature ranges of liquid phase as wider as possible without changing the main titanium-bearing mineral phase in the slag, the CaO content should not exceed the CaO content of point B (16.8 wt%).

The Analysis of Liquid Phase Mass Fraction and Viscosity in TEFS

The effects of CaF_2 dosage on the liquid phase mass fraction and viscosity are shown in Fig. 3. It indicated that adding CaF_2 could effectively increase the liquid

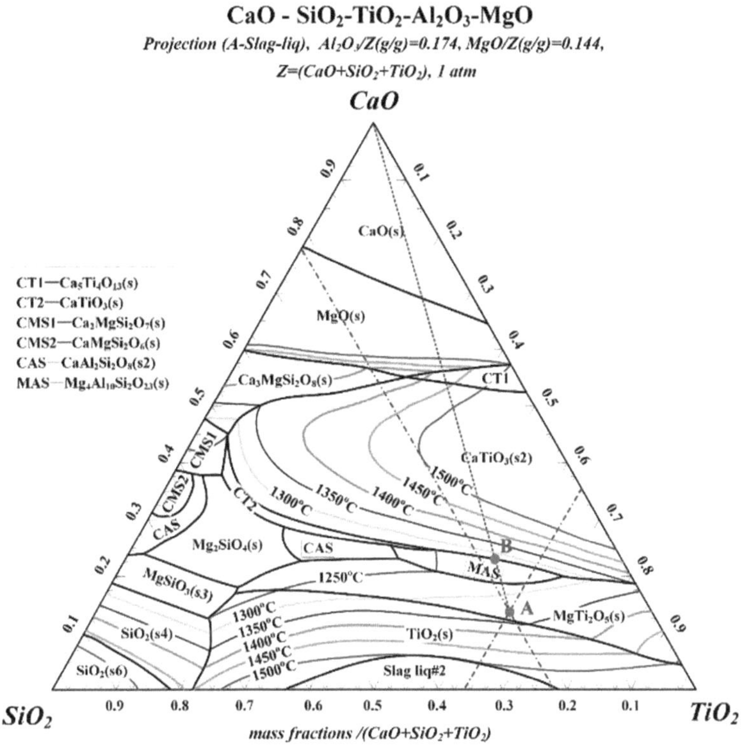

Fig. 2 The TiO_2–CaO–SiO_2–MgO–Al_2O_3 phase diagram

phase mass fraction of the slag in the temperature range of 1350–1140 °C during the cooling process. The viscosity decreased with the increasing of CaF_2 dosage. The low viscosity can reduce the resistance of the ion motion and promote the mineral crystallization in the molten slag. In the molten TEFS, the crystallization temperature of $M_xTi_{3-x}O_5$ is much high than that of silicate. The CaF_2 can effectively reduce the melting temperature and viscosity of the TEFS, which can promote the growth of the $M_xTi_{3-x}O_5$.

Effects of the CaF_2 on the Crystallization of $M_xTi_{3-x}O_5$

The effect of CaF_2 dosage on XRD patterns is shown in Fig. 4. It indicated that when the CaF_2 dosage increased from 0 to 5 wt%, the main mineral phases in the slag had no change, and the intensity of $M_xTi_{3-x}O_5$ peaks intensified slightly. The perovskite was detected in the titanium slag when the CaF_2 dosage reached 7 wt%, which was consistent with the thermodynamic analysis of mineral phases composition.

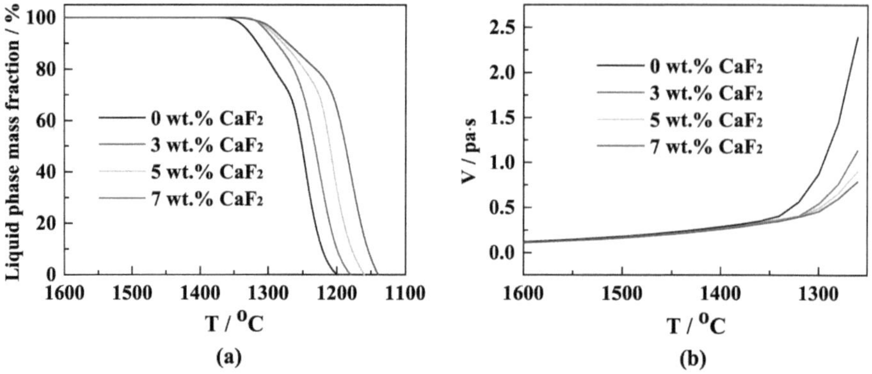

Fig. 3 The effects of CaF_2 dosage on the liquid phase mass fraction and viscosity

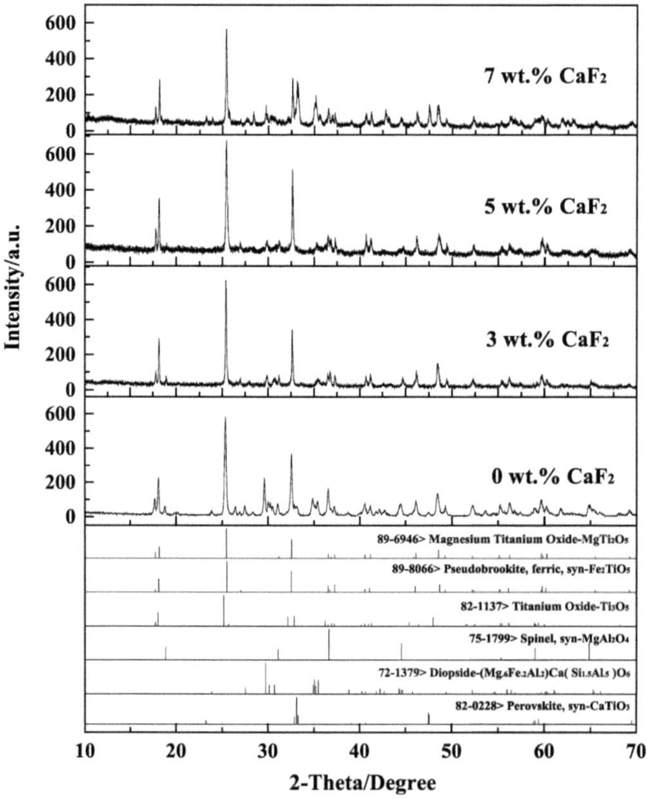

Fig. 4 The effect of CaF_2 dosage on XRD patterns

Fig. 5 The effects of CaF_2 dosage on the microstructures of titanium slags

The effects of CaF_2 dosage on the microstructures of titanium slags with a cooling rate at 10 °C/min is shown in Fig. 5. It indicated that the particle size of $M_xTi_{3-x}O_5$ increased with the increasing of CaF_2 dosage. When the CaF_2 dosage reached 7 wt%, the dendrite perovskite crystallized in the treated titanium slag. When the CaF_2 was added in TEFS, the shape of $M_xTi_{3-x}O_5$ changed from long strip to granular.

The effects of CaF_2 dosage on the particle size and area fraction of $M_xTi_{3-x}O_5$ is shown in Fig. 6, which is determined by analyzing of the grayscale image of microstructures using Image-Pro-Plus. It can be seen that when the CaF_2 dosage increased from 0 to 5 wt%, the average grain size and area fraction of $M_xTi_{3-x}O_5$ increased observably. However, when the CaF_2 dosage reached 7 wt%, the average grain size and area fraction of $M_xTi_{3-x}O_5$ decreased because of the crystallization of perovskite crystallite in the treated titanium slag. When the CaF_2 dosage was 5 wt%, the average grain size and area fraction of $M_xTi_{3-x}O_5$ reached 60 μm and 66%.

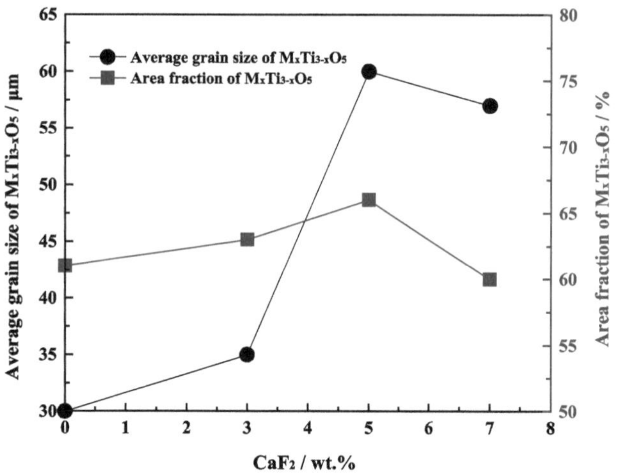

Fig. 6 The effects of CaF_2 dosage on the particle size and area fraction of $M_xTi_{3-x}O_5$

Effects of the Cooling Rate on the Crystallization of $M_xTi_{3-x}O_5$

The effects of cooling rate on the microstructures of titanium slags with a CaF_2 dosage at 5 wt% is shown in Fig. 7. It indicated that a slow cooling rate can effectively promote the crystallization of $M_xTi_{3-x}O_5$. The average grain size of $M_xTi_{3-x}O_5$ increased from 60 to 92 μm when the cooling rate reduced from 10 to 1 °C/min. The shape of $M_xTi_{3-x}O_5$ was granular, which provided a good base for the further grinding and separation. The cooling rate of the molten slag had no obviously effect on the area fraction of $M_xTi_{3-x}O_5$.

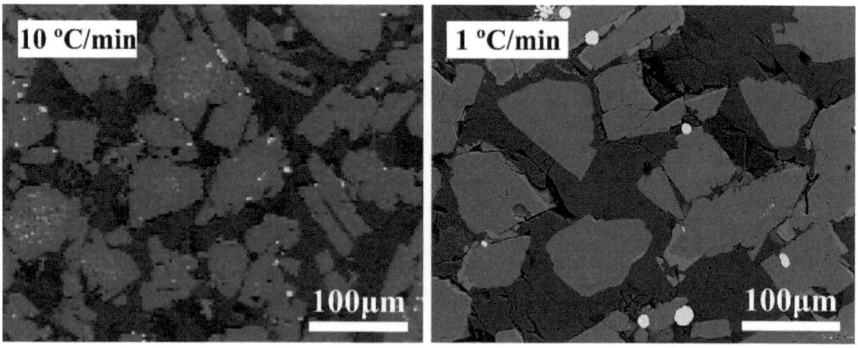

Fig. 7 The effects of cooling rate on the microstructures of titanium slags

Conclusions

(1) The major titanium-bearing mineral phase in the TEFS are $M_xTi_{3-x}O_5$. The CaF_2 can effectively reduce the melting temperature and viscosity of the TEFS, which can promote the growth of $M_xTi_{3-x}O_5$.

(2) The average grain size of $M_xTi_{3-x}O_5$ grew up from 30 μm to more than 92 μm and the area fraction of $M_xTi_{3-x}O_5$ reached about 61% from 66% under the conditions of adding 5 wt% CaF_2 and cooling rate at 1 °C/min.

References

1. F. Zheng, F. Chen, Y. Guo, T. Jiang, A.Y. Travyanov, G. Qiu, Kinetics of hydrochloric acid leaching of titanium from titanium-bearing electric furnace slag. JOM **68**, 1476–1484 (2016)
2. P.R. Taylor, S.A. Shuey, E.E. Vidal, J.C. Gomez, Extractive metallurgy of vanadium containing titaniferous magnetite ores: a review. Miner. Metall. Proc. **23**, 80–86 (2006)
3. S. Samanta, S. Mukherjee, R. Dey, Upgrading metals via direct reduction from poly-metallic titaniferous magnetite ore. JOM **67**, 467–476 (2015)
4. Y.F. Guo, M.J. Tang, T. Jiang, L.J. Qing, J.F. Zhou, Research on the slag phase type of vanadium-titanium magnetite in pre-reduction-electric furnace smelting, in *4th International Symposium on High-Temperature Metallurgical Processing, TMS Annual Meeting*, San Antonio, Texas, USA, 2013, pp. 87–94
5. G.G. Liu, Study on utilization technology of vanadium titanium magnetite based on the rotary hearth furnace direct reduction process. Appl. Mech. Mater. **217–219**, 441–444 (2012)
6. Y. Li, Y. Yue, Z.Q. Que, M. Zhang, M. Guo, Preparation and visible-light photocatalytic property of nanostructured Fe-doped TiO_2 from titanium containing electric furnace molten slag. Int. J. Min. Met. Mater. **20**, 1012–1020 (2013)
7. W. Mingyu, Z. Linnan, Z. Li, S. Zhitong, T. Ganfeng, Selective enrichment of TiO_2 and precipitation behavior of perovskite phase in titania bearing slag.pdf. Trans. Nonferrous Met. Soc. China **16**, 421–425 (2006)
8. J. Li, Z. Guo, J. Gao, Isothermal enriching perovskite phase from $CaO–TiO_2–SiO_2–Al_2O_3–MgO$ melt by super gravity. ISIJ Int. **54**, 743–749 (2014)
9. J. Li, Z. Zhang, M. Zhang, M. Guo, X. Wang, The influence of SiO_2 on the extraction of Ti element from Ti-bearing blast furnace slag. Steel Res. Int. **82**, 607–614 (2011)
10. J. Li, X. Tang, M. Guo, M. Zhang, Z. Zhang, X. Wang, Conditions for extracting anosovite from Panzhihua Ti-bearing blast furnace slag. J. Chin. Rere Earth Soc. **28**, 365–371 (2010)

Part VIII
Utilization of Complex Ores

Evaluation of Molybdenum Concentrates

Kagan Benzeşik, Seref Sonmez and Onuralp Yücel

Abstract The aim of this study is processing of domestic molybdenum concentrate to produce technical grade molybdenum trioxide may be the starting material for the production of various molybdenum products. In order to determine the optimum parameters for the total oxidizing roasting process of molybdenite concentrate to produce MoO_3 and purify the MoO_3 product by removing the copper content; first, MoS_2 was roasted by controlling reaction temperature and duration in a chamber type furnace. The highest Mo concentration rate was obtained as 56.6 wt% of Mo at 650 °C for the roasting duration of 45 min. Secondly, roasted product was leached with H_2SO_4 to remove copper. Leaching conditions were optimized by investigating the effects of different H_2SO_4 concentrations and S/L ratios. The minimum copper content was obtained as 0.13 wt% at leach residue. The raw materials and the products were characterized by using AAS (atomic absorption spectrometry) and XRD (X-Ray Diffraction) techniques.

Keywords Molybdenite · Roasting · Technical-grade molybdenum trioxide · Leaching

Introduction

Generally molybdenum metal is produced from its high grade sulphide concentrate through oxidizing roasting of molybdenite (MoS_2) in order to obtain molybdenum trioxide (MoO_3). Then purification of MoO_3 and hydrogen reduction of MoO_3 follows.

K. Benzeşik (✉) · S. Sonmez · O. Yücel
Metallurgical and Materials Engineering Department, Faculty of Chemistry and Metallurgy, Istanbul Technical University, 34469 İstanbul, Turkey
e-mail: kaganbenzesik@gmail.com

S. Sonmez
e-mail: ssonmez@itu.edu.tr

O. Yücel
e-mail: yucel@itu.edu.tr

J.-Y. Hwang et al. (eds.), *8th International Symposium on High-Temperature Metallurgical Processing*, The Minerals, Metals & Materials Series,
DOI 10.1007/978-3-319-51340-9_43

Molybdenum is widely used for the production of ferromolybdenum, which is required for the production of alloyed steel. Almost 80% of Mo, produced from molybdenum trioxide, is used for steel making in industries.

Usually, concentrates are roasted to obtain low copper and lead levels to produce a calcine that is essentially MoO_3 which contains low sulphur. Such calcines can be used directly in steel making, because liquid iron will reduce MoO_3 to metal in high yield.

Technical-grade MoO_3 is produced by roasting MoS_2 in air atmosphere in a multiple-hearth furnace. The roasted MoO_3 product usually has <0.1% sulphur content. Technical-grade MoO_3 typically contains 85–90 wt% MoO_3, the balance is silica with some Fe_2O_3 and Al_2O_3.

In some cases, additional hydrometallurgical processing is needed to produce technical-grade MoO_3. In steel making, using molybdenum trioxide with high copper content has some harmful effects on the mechanical properties of the alloyed steel produced. Therefore, the up-gradation of molybdenum trioxide is required in steel industry. The up-graded molybdenum trioxide must have less than 0.5 wt% Cu content for ferromolybdenum production [1–4].

Experimental Procedure

The raw material which was used in total oxidizing roasting experiments, is local molybdenite concentrate which has 0.40 wt% Cu and 0.05 wt% Mo content. The XRD pattern of MoS_2 concentrate is shown in Fig. 1.

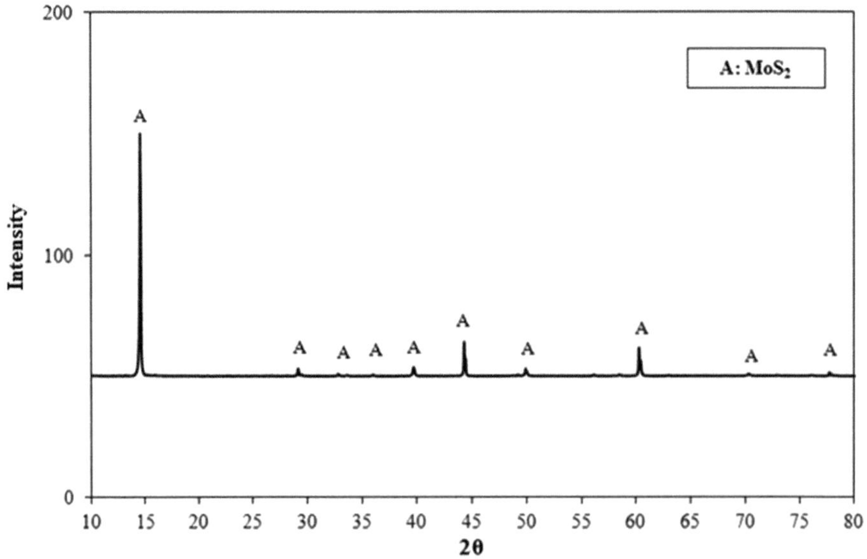

Fig. 1 XRD pattern of MoS_2 concentrate

Table 1 Chemical analysis of MoS_2 concentrate

Mo	52.03
S	39.19
C	0.05
Fe	2.94
Cu	2.56
Pb	0.08
Si	1.30
Al	0.45
Ca	0.60
Mg	0.17

MoS_2 concentrate was analyzed by using chemical analysis and AAS. Chemical analysis of MoS_2 concentrate is given in Table 1.

Present study was conducted in two main stages: First, MoS_2 was roasted by controlling parameters such as reaction temperature and duration in a chamber type furnace. To make a homogeneous roasting, alumina boats which have large surface area, were used. The samples were put in the alumina boats as thin layers. Samples were roasted at 600, 625, 650 °C with 15, 30, 45, 60, 90 and 120 min.

In the second experimental sets, roasted products were leached by H_2SO_4 to remove copper. Leaching conditions were optimized by investigating the effects of different H_2SO_4 concentrations and S/L ratio. Leaching experiments were done with Merck quality 95–98% H_2SO_4 with the duration of 15 min, at room temperature and 400 rpm mixing rate. All the samples were 10 g. In order to investigate the effects of different H_2SO_4 concentrations and S/L ratios, 10 g of samples were leached with 0.2, 0.4, 0.6 M acid concentrations with 1/5, 1/2 and 1/1 solid-liquid ratios.

Results and Discussion

In the first experimental set, effect of roasting temperature and duration on molybdenum trioxide recovery were investigated. In Fig. 2, the change of sulphur content at the concentrate depending on the roasting temperature and duration, is given.

As it is seen from Fig. 2, the sulphur content of the concentrates decreases rapidly till 40 min of roasting. It can be said that roasting at 650 °C with the duration of 45 min has given the closest results to the standard technical-grade MoO_3. As the roasting temperature and duration decrease, the sulphur amount increases.

The chemical analysis of the samples which were roasted at different temperatures and durations, is given in Table 2.

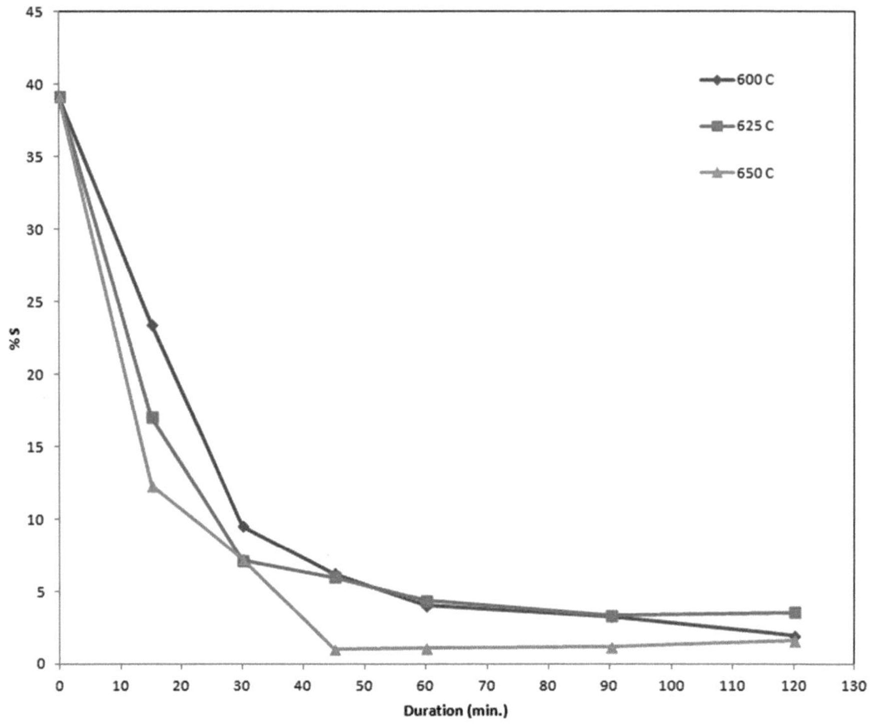

Fig. 2 The change of sulphur content at the concentrate

Table 2 The chemical analysis of the roasted samples

Duration (min)	Temperature (°C)	Concentration (%)		
		Cu	Fe	Mo
15	600	2.84	3.60	46.51
45		3.02	3.65	50.82
60		2.89	3.96	48.11
15	625	2.81	3.41	50.19
45		2.96	3.51	48.79
60		2.97	3.81	49.36
15	650	2.31	3.28	58.10
45		2.62	3.49	56.58
60		2.77	3.72	52.38

The optimum Mo concentration rate was obtained as 56.58 wt% at 650 °C for the roasting duration of 45 min.

In the second experimental sets, roasted product was leached by H_2SO_4 to remove copper. Leaching conditions were optimized by investigating the effects of different H_2SO_4 concentrations and S/L ratio.

The metal concentrations in H_2SO_4 solutions are given in Table 3.

Table 3 The metal concentrations in H_2SO_4 solutions

Molarity	S/L ratio	Metal concentrations in H_2SO_4 solution (ppm)		
		Cu	Fe	Mo
0.2	1/5	566.75	125.15	2200.00
	1/2	483.25	106.70	1650.00
	1/1	1123.00	231.85	3770.00
0.4	1/5	789.50	171.85	1274.00
	1/2	794.50	198.50	1450.00
	1/1	717.50	140.95	1116.00
0.6	1/5	762.25	190.50	1787.50
	1/2	903.00	224.25	2090.00
	1/1	866.70	212.70	1650.00

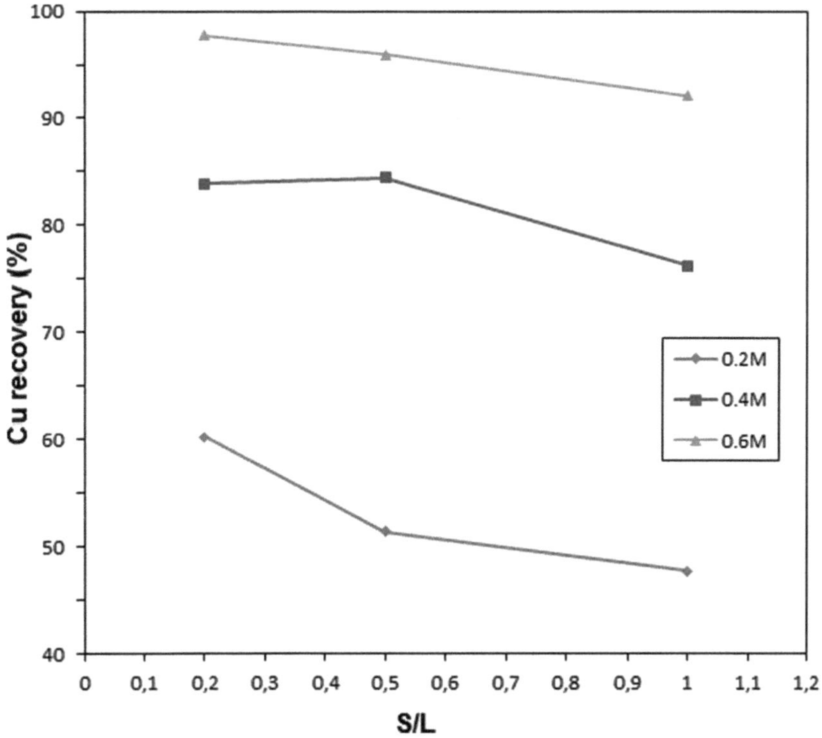

Fig. 3 The effects of different H_2SO_4 concentrations and S/L ratios on Cu recovery from the solution

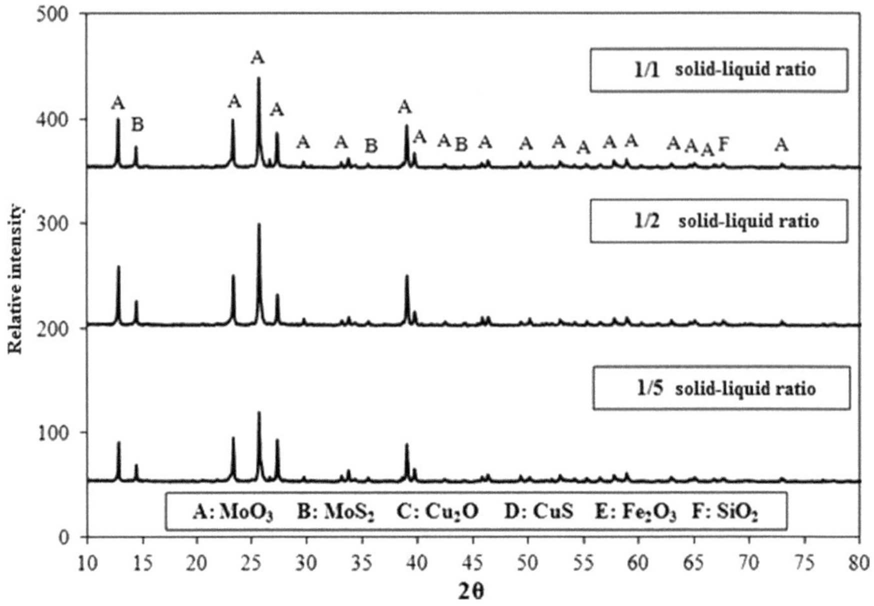

Fig. 4 XRD patterns of leached samples with 0.6 M H_2SO_4 solution

The concentration values were put in Formula (1) in order to calculate the metal recovery from H_2SO_4 solutions.

$$\text{Metal recovery}\ (\%) = \frac{\text{concentration (ppm)} \times \text{stock (ml)} \times 100}{10^6 \times \text{metal mass in the samples (g)} \times \text{metal ratio}} \quad (1)$$

In Fig. 3, the effects of different H_2SO_4 concentrations and S/L ratios on Cu recovery from the solution are given.

As it is seen from Fig. 4, almost 96% of Cu was removed from calcine by using 0.6 M H_2SO_4 solution at 1/5 solid-liquid ratio [5, 6].

Figure 4 presents the XRD patterns of leached samples with 0.6 M H_2SO_4 solution.

Figure 4 shows that, the main phases of leaching residue are MoO_3. There are not seen any copper compounds rather the roasted samples. Because the majority of the copper was solution treated.

Conclusions

Based on the results of present study of roasting molybdenite concentrate at different temperatures and durations, leaching of roasted samples with different H_2SO_4 concentrations at different solid-liquid ratios, following conclusions can be given.

(1) Optimum Mo recovery was determined as 56.58% Mo and 1.05% S with the roasting temperature of 650 °C and duration of 45 min.
(2) Leaching with 0.6 M H_2SO_4 solution at 1/5 solid-liquid ratio ended up with 95.92% Cu recovery from the H_2SO_4 solution. These parameters gave the best results as 0.13% Cu content in the leaching residue.

References

1. S.C. Grund, K. Hanusch, H.J. Breuning, H.U. Wolf, *Ullmann's Encyclopedia of Industrial Chemistry* (ABD, Wiley-VCH Verlag GmbH & Co. KGaA.O, Weinheim, NY, 2006)
2. C.K. Gupta, *Extractive Metallurgy of Molybdenum* (CRC Press, USA, 1992)
3. Url-1. http://www.imoa.info
4. F. Habashi, in *Handbook of Extractive Metallurgy, Volume III: Precious Metals, Refractory Metals, Scattered Metals, Radioactive Metals, Rare Earth Metals*, (Wiley-VCH, Weinheim, 1997), pp. 1361–1402
5. J. Lee, B. Kim, M.K. Jha, J. Jeong, Leaching of impurities for the up-gradation of molybdenum oxide and cementation of copper by scrap iron. Int. J. Miner. Process. **88**, 7–12 (2008)
6. J.R. Kumar, J.Y. Lee, H.S. Jeon, J.S. Kim, Up-gradation of MoO_3 and separation of copper, iron, zinc from roasted molybdenum ore by a leaching process. Braz. J. Chem. Eng. **30**, 391–397 (2013)

Intensification of Gold Leaching from a Multi-refractory Gold Concentrate by the Two-Stage Roasting—Alkaline Sulfide Washing—Cyanidation Process

Qian Li, Yan Zhang, Xishan Li, Bin Xu, Yongbin Yang, Tao Jiang and Hongwei Li

Abstract The cyanide leaching rate of gold from a multi-refractory gold concentrate (18.05 g/t Au) containing 1.62% As, 20.76% S, 6.95% C and 1.31% Sb was only 1.33%. The enhancement of gold leaching was studied mainly by the pretreatment of oxidative roasting and alkaline sulfide washing. Results showed that after the two-stage roasting (i.e., I stage of 500 °C, 1 h and II stage of 700 °C, 2 h) most of sulfur (93.84%) and carbon (98.27%) were removed, but the gold leaching rate of calcine just reached 71.33% because the residual arsenic (0.63%) and antimony (1.15%) were still high. After alkaline sulfide (i.e., NaOH and Na_2S) washing, the residual arsenic and antimony was further decreased to 0.21 and 0.65%, respectively. The gold leaching rate of the treated calcine by cyanidation could increase to 93.52% substantially. Also, the optimized particle size, NaCN concentration and leaching time were −0.045 mm over 85%, 0.1% and 36 h, respectively.

Keywords Refractory gold concentrate · Roasting · Alkaline sulfide · Cyanidation

Introduction

A gold deposit whose proven reserve beyond 40 tons has been discovered in Yunnan province of China, but its gold concentrate by floatation is multi-refractory due to bearing detrimental impurities of S, As, Sb and C, etc. [1–3]. Most of gold is associated with sulfide minerals like pyrite, arsenopyrite and stibnite, etc. and is generally encapsulated in the matrixes of associated sulfide minerals in the form of ultrafine inclusions. Also, extra consumption of the lixiviant (i.e., cyanide) and the

Q. Li · Y. Zhang (✉) · X. Li · B. Xu · Y. Yang · T. Jiang · H. Li
School of Minerals Processing and Bioengineering, Central South University, Changsha 410083, Hunan, People's Republic of China
e-mail: zhangyancsu@126.com

Q. Li
e-mail: zhangyancsu@126.com

J.-Y. Hwang et al. (eds.), *8th International Symposium on High-Temperature Metallurgical Processing*, The Minerals, Metals & Materials Series,
DOI 10.1007/978-3-319-51340-9_44

oxidant (i.e., oxygen) can be caused by the associated sulfide minerals. Particularly, it has been found that stibnite decomposes easily in alkaline cyanide solutions, and the decomposition products like thioantimonites can result in significant consumption of oxygen and subsequent side reaction with cyanide as well as possible passivation of gold surfaces [4–6]. The presence of carbonaceous matters is also unfavorable for gold leaching because of the preg-robbing effect [7]. Consequently, the gold leaching from the concentrate by direct cyanidation is greatly restricted.

Pretreatments have to be considered and various available methods have been applied to deal with these refractory gold ores in China. At present, oxidative roasting has been playing a dominant role in treating carbonaceous or sulfide gold ores, which presents a strong adaptability to most types of refractory gold ores like those. So, oxidative roasting points out the direction on treating the carbonaceous sulfide gold ores [1].

Furthermore, an alkaline sodium sulfide solution is a universal solvent for the majority of antimony compounds. Most of other metals present low solubility in this solution, but arsenic is not included in this rule [5, 6, 8]. It was indicated that the gold leaching rate from a gossan type gold ore could be improved from 80 to 87–95% by the pretreatment of sulfidization (i.e., Na_2S solution washing) prior to cyanidation [9].

In this paper, pretreatments of oxidative roasting and alkaline Na_2S solution washing were both taken to investigate their effects on intensifying gold leaching. On this basis, the cyanide leaching of gold was also optimized primarily from mineral particle size, cyanide dosage and leaching time.

Experimental

Raw Materials and Reagents

The study was conducted on a refractory gold concentrate by floatation from a plant in Yunnan province of China. Representative samples were dry-milled in a planetary ball mill to the particle sizes of −0.045 mm over 85%, which is prepared for the composition and phase analyses and subsequent pretreatment and leaching tests.

As seen from Table 1, it is shown that the gold concentrate contained 1.62% As, 20.76% S, 6.95% C and 1.31% Sb, so it is a multi-refractory gold ore because of its complexity.

Mineralogical phase characterized by X-ray Diffraction (XRD) analysis (see Fig. 2) further indicates that the majority of gold concentrate was quartz and pyrite. In addition, there were relatively high contents of arsenopyrite and carbonaceous

Table 1 Chemical compositions of the gold concentrate sample

Element	Au[a]	Si	Ca	As	Fe	S	Sb	Al	C
Content (%)	18.05	20.08	5.3	1.62	19.6	20.6	1.31	3.07	6.95

[a]Unit g/t

Table 2 Phase analysis of Au occurrence in the concentrate sample

Phase type	Free and associated gold	Arsenopyrite	Pyrite	Iron oxide	Silicate	Total
Content (g/t)	0.31	12.07	3.69	1.87	0.11	18.05
Distribution (%)	1.72	66.87	20.44	10.36	0.61	100.00

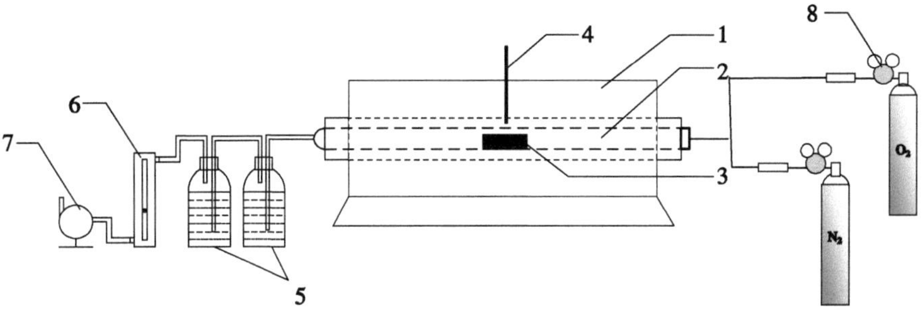

Fig. 1 Schematic diagram of the experimental facility for oxidative roasting, *1* horizontal tube furnace; *2* quartz tube; *3* raw material loader; *4* thermocouple; *5* exhaust gas absorption equipment; *6* gas flow meter; *7* air pump; *8* rotor flow meter

matter (i.e., elemental carbon). The phase analysis of Au occurrence (see Table 2) shows that only 1.72% of Au occurred in the formation of free and associated gold whereas 66.87% of Au was enwrapped in arsenopyrite and 20.44% in pyrite. Besides, a small quantity of Au (10.36%) was encapsulated in iron oxides. Carbonaceous matters are also important gold-carrier minerals [1, 10], but this situation was not found in the gold concentrate, namely, there was mainly the preg-robbing of gold caused by elemental carbon. Sodium hydroxide, sodium sulfide nonahydrate and sodium cyanide were all in analytically pure grade. The purity of nitrogen and oxygen used were both beyond 99%. De-ionized water was used throughout all experiments.

Experiment Methods

Oxidative roasting pretreatment was carried out to remove S and As to expose the gold inclusions in associated sulfide minerals and to remove C aiming to eliminate the preg-robbing effect of gold. The schematic diagram of the roasting facility is presented in Fig. 1. The effects of single-stage and two-stage oxidative roasting on the recovery of gold were both investigated. Considering that the gold concentrate contained a relatively high content of elemental C and its effective removal requires higher roasting temperatures and longer roasting time than sulfide gold ores [1], the single-stage oxidative roasting was conducted at the condition of temperature 700 °C, time 2.5 h and air flow (79%N_2 + 21%O_2 by volume) 3 L/min. For the two-stage oxidative roasting, the reaction conditions were I stage of temperature 450 °C or

500 °C, time 1 h and air flow 1 L/min and II stage of temperature 650–750 °C, time 1–3 h and air flow 3 L/min.

The pretreatment of alkaline sulfide washing was also taken to remove Sb and As in gold ores. The alkaline sulfide solution used was a solution of mixed NaOH and Na_2S. It was conducted in the mixed solution of NaOH 20.00 g/L and Na_2S 0–15 g/L under the conditions of liquid-solid ratio 3:1, agitation speed 300 rpm, leaching time 1 h, temperature 25 °C or 80 °C adjusted by a constant temperature bath. Finally, unless specified otherwise, the conditions of cyanide leaching of gold from gold ores were liquid-solid ratio 2.5:1, pH 11(±0.3), stirring speed 600 rpm, particle size −0.045 mm beyond 85%, ω(NaCN) 0.2%, leaching time 36 h.

Analytical Methods

A high frequency IR carbon and sulfur analyzer (HW2000B, Wuxi Yingzhicheng) was used to determine the contents of S and C in the gold concentrate and calcines. The other elements were all analyzed through acid digestion and an atomic absorption spectrometer (AA-6800, Shimadzu). Mineralogical compositions of the gold concentrate and calcines were obtained by an X-ray diffractometer (D/Max 2500, Rigaku). The size distributions of calcines were detected by the laser particle size analyzer (Mastersize 2000, Malvern).

Results and Discussion

Oxidative Roasting-Cyanidation Process

It has been known that stibnite (Sb_2S_3) is oxidized to volatile antimony(III) oxide (Sb_2O_3) using furnaces operating at approximately 850–1000 °C which is far beyond the temperature range of oxidative roasting for refractory gold ores. The reaction is shown as the following Eq. (1):

$$2Sb_2S_3 + 9O_2(g) = 2Sb_2O_3(g) + 6SO_2(g) \quad (1)$$

So during oxidative roasting, antimony in the gold concentrate is hardly removed in theory and the content of residual antimony in the calcine was not assayed in this section. The results of gold leaching rate from the gold concentrate by cyanidation are indicated in Table 3. It is shown that without oxidative roasting the gold leaching rate from the concentrate by conventional cyanidation was just as low as 1.33%. By the single-stage oxidative roasting, most of sulfur (93.09%) and carbon (98.57%) was removed but the removal rate of arsenic was only 58.84%. As a result, around 63.56% of gold could be leached from the calcine, which is unsatisfactory possibly due to the adverse effect of arsenic. Then, two-stage roasting

Table 3 Results of gold leaching rate under various oxidative roasting conditions

Roasting stages	T (°C)		t (h)		Removal rate (%)			Gold leaching rate (%)
	I	II	I	II	As	S	C	
No roasting	–		–		0	0	0	1.33
Single-stage	700		2.5		58.84	93.09	98.57	63.56
Two-stage	450	650	1	1	55.92	75.01	79.71	13.74
Two-stage	450	650	1	2	66.29	89.15	91.39	34.53
Two-stage	450	650	1	3	65.90	93.11	98.40	62.76
Two-stage	500	650	1	2	71.48	93.59	93.49	63.89
Two-stage	500	650	1	3	71.60	93.89	98.36	68.49
Two-stage	500	700	1	2	71.40	93.84	98.27	71.33

was carried out to minimize the influence of arsenic. After I stage roasting of temperature 450 °C, time 1 h and II stage roasting of temperature 650 °C, time increasing from 1 to 3 h, we found that the removal rate of sulfur and carbon increased to 93.11% (from 75.01%) and 98.40% (from 79.71%) respectively, but the removal rate of arsenic was still as low as 56–66%. Consequently, the optimum gold leaching rate reached only 62.76% which is undesirable. When the temperature of I stage roasting was increased to 500 °C and roasting time was still 1 h, and under II stage roasting of temperature 650 °C, time increasing from 2 to 3 h, it was shown that the removal rate of arsenic increased to around 68% accompanying with satisfactory removal rates of sulfur (93–94%) and carbon (96–98%), but the gold leaching rates were still not high (64–71%). As I stage roasting condition was unchanged but the temperature of II stage roasting was raised to 700 °C with roasting time 2 h, nearly 71% of arsenic, 94% of sulfur and 98% of carbon was removed accordingly and the gold leaching rate could be increased to 71.33% which is still not high enough.

Increasing the II stage roasting temperature to 750 °C, however, we found that the phenomenon of consolidation or severe secondary encapsulation occurred. Finally, we adopted the two-stage roasting of I stage roasting of temperature 500 °C, time 1 h and II stage roasting of temperature 700 °C and time 2 h with other conditions unchanged as mentioned previously.

Oxidative Roasting-Alkaline Na_2S Solution Washing-Cyanidation Process

After the optimal two-stage roasting, the main chemical compositions of gold concentrate and calcine are listed in Table 4. It is known that the content of arsenic in calcine was 0.42% (removal rate 71.40%) whilst that of antimony was still as

Table 4 Main chemical compositions of the gold concentrate and calcine (%)

Gold ore type	Au[a]	SiO_2	CaO	MgO	As	Fe	S	Sb	Al_2O_3	C	K_2O
Concentrate	18.05	27.68	4.26	2.00	1.62	19.6	20.76	1.31	6.19	6.95	2.10
Calcine	22.50	34.61	4.58	1.96	0.42	25.44	1.83	1.24	7.87	0.13	2.24

[a]Unit g/t

high as 1.24%. The residual arsenic and antimony in calcine could possibly impede the cyanide leaching of gold. Our previous research [11] had indicated that during the roasting process ferric arsenate was easily formed from arsenic and antimony (i.e., stibnite) was difficult to be removed.

Nevertheless, it had been shown that ferric arsenate and stibnite could be dissolved by alkaline Na_2S solutions as the following Eqs. (2)–(4) [8, 12]:

$$Sb_2S_3 + 3Na_2S(aq) = 2Na_3SbS_3(aq) \tag{2}$$

$$Sb_2S_3 + 6NaOH = Na_3SbS_3(aq) + Na_3SbO_3(aq) + 3H_2O \tag{3}$$

$$FeAsO_4 + 3NaOH + H_2O = NaFe(OH)_4(aq) + Na_2HAsO_4(aq) \tag{4}$$

Moreover, in alkaline Na_2S solutions, iron oxides on the surface of calcine particles would be sulfuretted, which can improve the diffusion rate of CN^- to Au^0 [13].

For these reasons, the calcine was further treated with the mixed solution of NaOH and Na_2S to extract antimony and arsenic before cyanide leaching. The mixed solution system presents a characteristic of stronger alkaline and reducing agent (rather negative redox potential) than that of cyanide leaching system, which can eliminate the disadvantageous effects of arsenic and antimony in significant measure [9, 13].

The results of gold leaching rate from the washed calcine are indicated in Table 5. It is shown that the gold leaching rate of calcine washed by the mixed solution of 20 g/L NaOH and 15 g/L Na_2S at 25 °C could go up significantly from 71.33 to 84.84%, suggesting the detrimental effects of arsenic and antimony on gold leaching from the calcine. It is also found that there was a remarkable effect of temperature on removing arsenic and antimony, especially for the latter. When the temperature was increased from 25 to 80 °C and after washing by the mixed solution of 20 g/L NaOH and 15 g/L Na_2S, the residual content of arsenic and antimony diminished from 0.28 to 0.21% and from 0.90 to 0.52%, respectively. Correspondingly, the gold leaching rate from the washed calcine ascended substantially from 84.84 to 93.52%. In addition, with the dosage of NaOH 20 g/L and at 80 °C, as the dosage of Na_2S rose from nil to 5 and 15 g/L the content of arsenic (0.3%) and antimony (0.58%) diminished gradually while the gold leaching rate could be improved from 87.07 to 93.52%.

Table 5 Results of gold leaching rate by pretreatment of optimal two-stage oxidative roasting and various alkaline Na_2S solution washing

Alkaline Na_2S solution		$T_{water\ bath}$ (°C)	Residual content of As and Sb (%)		Gold leaching rate (%)
NaOH (g/L)	Na_2S (g/L)		As	Sb	
0	0	25	0.43	1.15	71.33
20	15	25	0.28	0.90	84.84
20	0	80	0.30	0.58	87.07
20	5	80	0.27	0.55	90.09
20	15	80	0.21	0.52	93.52

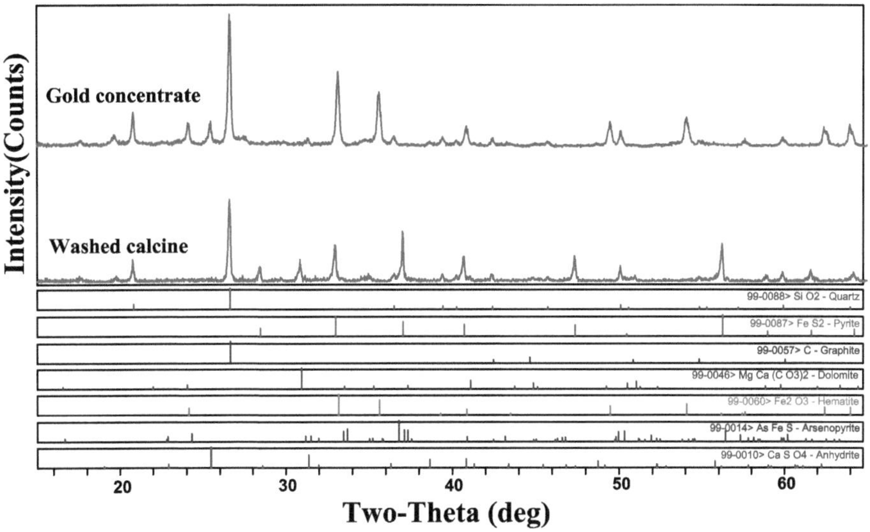

Fig. 2 XRD patterns of gold concentrate and washed calcine

Optimization of Gold Leaching from the Calcine by Cyanidation

Under the optimal pretreatment conditions of roasting and alkaline Na_2S solution washing, the mineralogical phase of gold concentrate characterized by XRD (see Fig. 2) indicates that pyrite, arsenopyrite and carbonaceous matter all disappeared whilst quartz and iron oxides with some gypsum were the main mineralogical phase.

Hence, based on the above analysis, the gold leaching from the calcine was further improved primarily from the aspects of particle size, NaCN concentration and leaching time. The results are presented in Fig. 3.

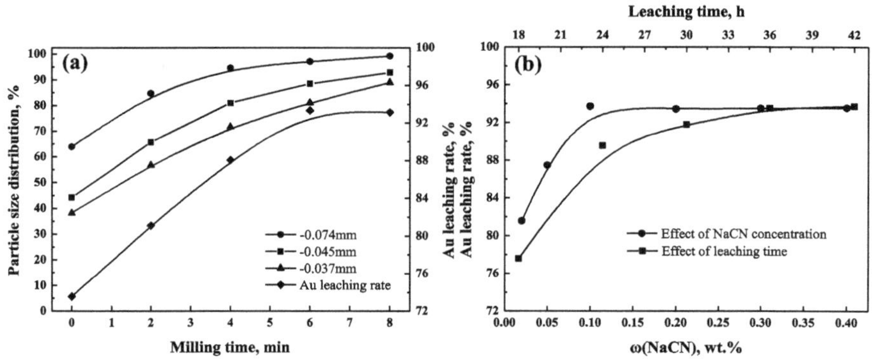

Fig. 3 Effects of **a** particle size distribution, **b** NaCN concentration and leaching time on cyanide leaching of gold

As shown from Fig. 3a, both the fineness of washed calcine particles and the gold leaching rate increased with the increase of milling time. When the milling time was 6 min, the particle size (−0.074, −0.045 and −0.037 mm) distribution of calcine increased from 63.92, 44.24 and 38.19% to 97.08, 88.53 and 81.11%, respectively. At the same time, the gold leaching rate rose from 73.55 to 93.32% markedly. However, with the milling time over 6 min, except the size of −0.037 mm, the fineness of calcine particles increased slightly and the gold leaching rate remained about the same.

Moreover, as we can see from Fig. 3b, the gold leaching rate rose with the increasing NaCN concentration and leaching time. When the concentration of NaCN and leaching time went up from 0.02 to 0.1% and from 18 to 36 h respectively, the gold leaching rate could increase from 81.56 to 93.73% and from 77.57 to 93.52% correspondingly. Then, the gold leaching rate remained stable as the NaCN concentration exceeded 0.1% and the leaching time was more than 36 h.

Therefore, the particle size of calcine was recommended as −0.045 mm more than about 85%, and the optimal NaCN concentration and leaching time was 0.1% and 36 h, respectively.

Conclusions

(1) It was difficult to leach gold from a multi-refractory gold concentrate which contained 1.62% As, 20.76% S, 6.95% C and 1.31% Sb by direct cyanidation. Undergoing the optimal oxidative roasting, i.e., two-stage roasting of I stage of 500 °C, 1 h and II stage of 700 °C, 2 h, most of sulfur (94%) and carbon (98%) was removed but the removal rate of arsenic was only 71% and most of antimony was residual in the calnie. Consequently, the gold leaching rate could only reach 71.33%.

(2) After washing the calcine with a solution of mixed 20 g/L NaOH and 15 g/L Na_2S at 80 °C, the residual arsenic and antimony could be removed effectively to much lower levels, i.e., the content of As and Sb decreasing from 0.43 to 0.21% and from 1.15 to 0.52% respectively. So the gold leaching rate could be increased from 71.33 to 93.52% substantially.
(3) By the optimal pretreatments of oxidative roasting and alkaline Na_2S washing, the XRD analysis showed that most of detrimental associated sulfide minerals and carbonaceous matter was removed. Additionally, the enhanced gold leaching from the calcine was primarily from particle size, NaCN concentration and leaching time whose optimized condition was indicated that the size distribution of −0.045 mm beyond around 85%, NaCN concentration 0.1% and leaching time 36 h, respectively.

Acknowledgements This work was financially supported by National Natural Science Foundation of China (Grant No. 51574284), National Science Foundation for Distinguished Young Scholars of China (Grant No. 51504293) and Science and Technology Program of Yunnan (No. 2013IB020).

References

1. G. Xue, W. Ren, Y. Xu, *Hydrometallurgy and Analysis Testing Process of Gold and Silver* (Science Press, Beijing, 2009)
2. G. Gao et al., Kinetics of high-sulphur and high-arsenic refractory gold concentrate oxidation by dilute nitric acid under mild conditions. Miner. Eng. **22**(22), 111–115 (2009)
3. Q. Li, D. Li, F. Qian, Pre-oxidation of high-sulfur and high-arsenic refractory gold concentrate by ozone and ferric ions in acidic media. Hydrometallurgy **97**(1), 61–66 (2009)
4. J.H. Kyle et al., Review of trace toxic elements (Pb, Cd, Hg, As, Sb, Bi, Se, Te) and their deportment in gold processing: part II: deportment in gold ore processing by cyanidation. Hydrometallurgy, s **111–112**(1), 10–21 (2012)
5. S. Ubaldini et al., Process flow-sheet for gold and antimony recovery from stibnite. Hydrometallurgy **57**(3), 187–199 (2000)
6. T. Jiang, *Chemistry of Extractive Metallurgy of Gold* (Hunan Science and Technology Press, Changsha, 1998)
7. G. Senanayake, A review of effects of silver, lead, sulfide and carbonaceous matter on gold cyanidation and mechanistic interpretation. Hydrometallurgy **90**(1), 46–73 (2008)
8. P. Raschman, E. Sminčáková, Kinetics of leaching of stibnite by mixed Na_2S and NaOH solutions. Hydrometallurgy, s **113–114**(3), 60–66 (2012)
9. C. Nunez, A. Roca, Recovery of gold and silver from a gossan type ore by a pre-sulphated process. Trans. Inst. Min. Metall. Sect. C **95**, 195–198 (1986)
10. M.D. Adams, A.M. Burger, Characterization and blinding of carbonaceous preg-robbers in gold ores. Miner. Eng. **11**(10), 919–927 (1998)
11. J. Hu, Research on the process of roasting pretreatment-cyanide extracting Au from sulfide-bearing refractory gold concentrate containing arsenic and high-carbon. Master thesis, Central South University, 2015, pp. 27–29
12. T. Fujita et al., Environmental leaching characteristics of scorodite synthesized with Fe(II) ions. Hydrometallurgy, s **111–112**(1), 87–102 (2012)
13. M. Stoychevski, Recovery of gold from As-bearing pyrite minerals whether or not roasted (1994)

Recovery of Cobalt from Copper Converter Slag by Reduction-Sulfurization Smelting at High Temperature

Shi Sun, Hongxu Li, Jiaqi Fan, Chao Li and Qi Liu

Abstract With the great demand of consumer electronics (CE) and electric vehicles (EV), cobalt has become one of the more scarce energy metals. The recovery of cobalt from a copper converter slag, which contained over 2.4% Co and was accumulated as industrial waste, by reduction-sulfurization smelting was explored. Quantitative analysis of treated slag showed that activated carbon and ferrous sulfide can be used to settle cobalt into matte as Co–Fe alloy. The experiment was carried out at 1673 K with a reducing atmosphere of burning electrode graphite crucible. It was found that the recovery of Co would increase with increased added activated carbon, was unaffected at higher carbon additions, and for low additions of carbon increased at higher FeS additions. Over 95% Co, 99% Cu and 62% Fe could be settled into matte. Preliminary separation of Co from converter slag by pyrometallurgical smelting is proposed.

Keywords Copper converter slag · Cobalt recovery · Reducing agent · Reduction-sulfurization smelting

Introduction

Cobalt is diversely used, with 22% used in batteries, 22% in superalloys manufacture, 11% in hard metals, 27% in a few other minor applications in 2005. In 2006, the increasing use of Co in batteries exceeded that in superalloys [1]. So far, about 13 million tons of cobalt resources have been identified and Co always occurs

S. Sun · H. Li · J. Fan · C. Li · Q. Liu
School of Metallurgical and Ecological Engineering, University of Science and Technology Beijing, 30# Xueyuan Road, Beijing 100083, China

S. Sun · H. Li (✉) · J. Fan · C. Li · Q. Liu
Beijing Key Laboratory of Recycling and Extraction of Metals (REM), University of Science and Technology Beijing, 30# Xueyuan Road, Beijing 100083, China
e-mail: lihongxu@ustb.edu.cn

J.-Y. Hwang et al. (eds.), *8th International Symposium on High-Temperature Metallurgical Processing*, The Minerals, Metals & Materials Series,
DOI 10.1007/978-3-319-51340-9_45

with Cu and Ni ores as companion metal [2]. The recovery efficiency of Co from the host ores is variable, and much of the Co easily deports to smelting slags [3].

During the pyrometallurgy process of copper smelting, Co is mainly oxidized and accumulated from Co–Cu ores into converter slag. Mineralogical studies showed that Co is present as CoO which is associated with the silicate/oxide phases together as magnetite and fayalite in converter slag [4]. It is difficult to directly recover Co from converter slag because of its much lower content than the main components such as Fe, Cu.

Reduction-sulfurization smelting has been used for the recovery of Co from copper smelting slags. CoO and CuO are reduced by Fe from the alloy to form metallic cobalt (and copper), resulting in the formation of FeO in the slag. The exchange of Co and Cu with Fe will take place only at the slag/metal interface [5]. A suitable addition of reducing agent (activated carbon, coke, or pyrite), slag modifier (TiO_2 or CaO), smelting temperature and smelting time were studied and the recovery rate of cobalt was in the range of 55–94% [6, 7]. Although activated carbon or coke-pyrite both work as reducing agents, the roles of C and FeS in the process of reduction-sulfurization smelting also need be studied further.

In this paper, correlations of Co and main metal elements in converter slag and the effects of reducing agent dosage (activated carbon and FeS) in reduction-sulfurization smelting have been investigated.

Materials and Experiments

Copper Converter Slag Sample

The converter slag from a Peirce-Smith converter blowing-refining process was ground, passed through a 200-mesh sieve, mounted in resin and analyzed by quantitative chemical analysis, X-ray fluorescence (XRF), scanning electron microscopy (SEM) and energy dispersive spectroscopy (EDS). The composition of the converter slag is shown in Table 1. The high content of Cu, Fe and Co was caused by over-oxidation operation to enrich Co into the slag. X-ray diffraction (XRD) analyses show that the main phases in the slag are magnetite (Fe_3O_4), fayalite (Fe_2SiO_4), copper (Cu) and cobalt iron oxide ($CoFe_2O_4$) and it means the metallization degree of Cu and the oxidation degree of Fe, Co are both high.

Table 1 Chemical composition of converter slag (mass fraction, %)

Fe	Si	Ca	Al	Cu	Co	S	Ni
44.17	14.55	1.03	0.87	14.83	2.42	1.22	0.26

Test Procedures

Dosages of activated carbon (AR) and FeS (AR) as reagents, were the controlled variables and dosages of simulated matte ($CuS{\cdot}H_2O$, AR) and slag modifier (CaO, AR) were always constant. In each test, one variable was changed; each test was repeated at least twice. As an experimental procedure, 9 g converter slag was mixed with the calculated amount of reagents in corundum crucibles (10 ml), which was put into an graphite crucible (300 ml). The electrode graphite crucible with 3 corundum crucibles was then put in the tube furnace.

The experiments of reduction-sulfurization were done under reducing atmosphere of burning electrode graphite crucible to consume oxygen in a vertical MoSi2 tube furnace at 1673 K. The temperature was controlled by computer program, which was firstly raised to the test temperature in 2 h, the electrode graphite crucible was then put into the furnace and held for 2 h. After the holding period, the graphite crucible was naturally cooled to room temperature outside the furnace with a graphite cover. In this process, the oxygen of air, which was around corundum crucibles, would always react with electrode graphite. This meant that samples were always in a reducing atmosphere.

The matte, which contains cobalt-bearing alloy, and depleted slag in the crucible were manually separated, weighed and ground. The composition of the depleted slag was analyzed by XRF. Based on the mass balance principle, metal recoveries were calculated. In addition, the depleted slag and matte were characterized by XRD. Phases in the solidified matte and slag were observed by SEM-EDS.

Results and Discussion

Correlation of Co–Fe–Cu in the Converter Slag

The distribution of main elements in the converter slag was analyzed at 2500× magnification by backscattered electron imaging and EDX elemental mapping. This showed that Co was distributed among phases of Fe–Si slag or Cu matte; see Fig. 1a, b.

38 data points of EDS were sorted according to the increasing of Fe and the mass fraction of Cu and Co distributed accordingly. The polyline trend of Co content (right y-axis) and column chart of Fe and Cu content (left y-axis) are shown in Fig. 2.

Two areas are evident in Fig. 2. The first area (points 1–7) has a higher Cu content, indicating the presence of Cu metal and with increasing Fe, the content of Co has an obvious increasing trend. The second area (points 8–37) has a much higher concentration of Fe as Fe_3O_4–Fe_2SiO_4–$CoFe_2O_4$ slag system with a higher average content of Co. It means that Co has an obvious affinity for Fe in both metalized Cu-matte, mainly as Co–Fe alloy, and Fe-containing converter slag,

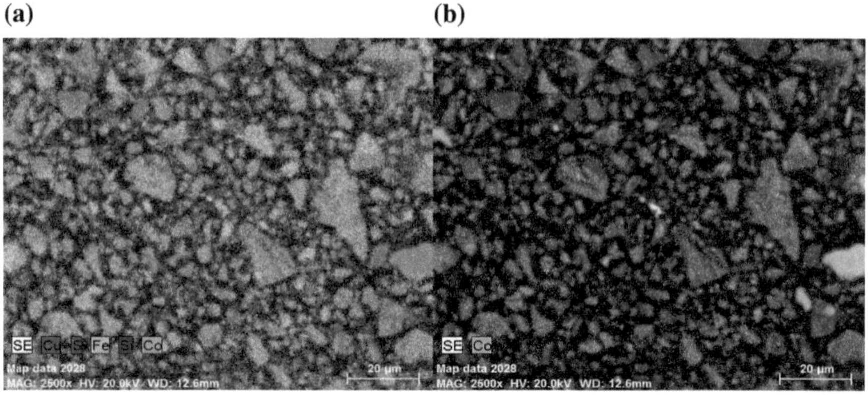

Fig. 1 **a** SEM image of main elements distribution. **b** SEM image of cobalt distribution

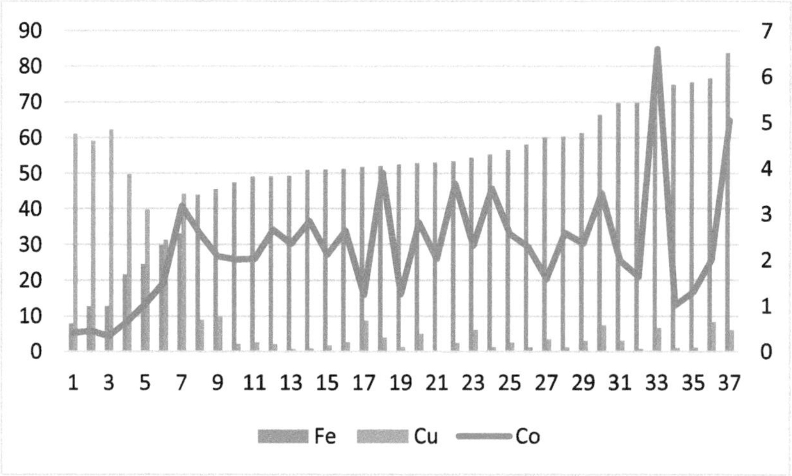

Fig. 2 EDS analyses of concentrations of Fe, Cu, Co in converter slag (mass fraction, %)

mainly as magnetite. This means that the recovery of cobalt is closely tied to recovery of iron. The key question concerns the roles of activated carbon and FeS in the reducing process of reduction-sulfurization.

Effect of Activated Carbon Dosage

Smelting experiments with activated carbon addition varying from 1 to 6% were conducted with 15% FeS, 20% $CuS{\cdot}H_2O$ and 5% CaO in a reducing atmosphere. The results are shown in Fig. 3.

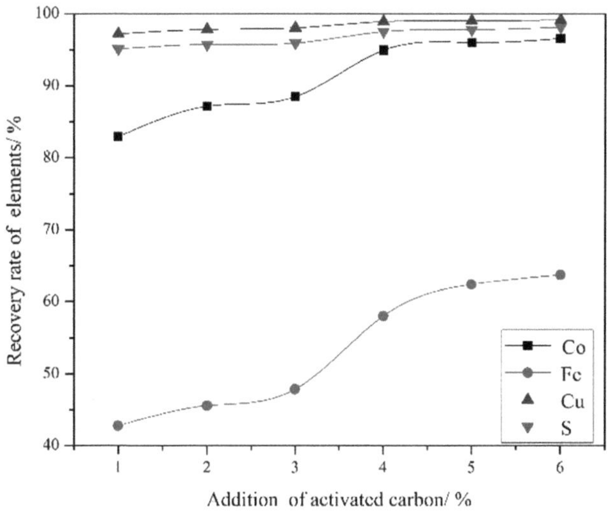

Fig. 3 The recovery rate curves of Co, Fe, Cu, S in activated carbon experiments

It was found that the recovery curve of Co was similar to that of Fe. The recovery rate of Co increased slightly from 1 to 3% additions of activated carbon, changed quickly from 3 to 5% and then leveled off up to 6%. The addition of activated carbon had an obvious influence on the recovery of Fe and Co but little influence on that of Cu and sulfides. This is due to the fact that Fe and were mainly recovered by reduction of oxides to metal, whereas Cu recovery mainly involved settling of metallic matte which also absorbed sulfides. The optimum activated carbon addition was 4–5% and the recoveries of Co, Cu and Fe reached 94.9–96.0%, 98.9–99.0%, 58–62.4%. The contents of Co and Cu in treated slag were 0.15–0.13% and 0.41–0.39%.

Effect of FeS (AR) Dosage

FeS Additions with Sufficient Activated Carbon

Smelting experiments with FeS addition varying as 5, 10, 15, 18, 20, 22% were conducted with 5% activated carbon, 20% $CuS{\cdot}H_2O$ and 5% CaO in a reducing atmosphere. The results are shown in Fig. 4.

It was found that the recovery of Co, Cu and sulfides was almost constant but excessive addition of FeS (more than 20%) would lower the recovery of Cu and sulfides. Recovery of Fe increased slightly from 5 to 22% of FeS addition. It showed that the addition of FeS, which promotes sulfurization, would not make any contribution of the recovery of valuable metals (Co, Cu etc.) under the condition of

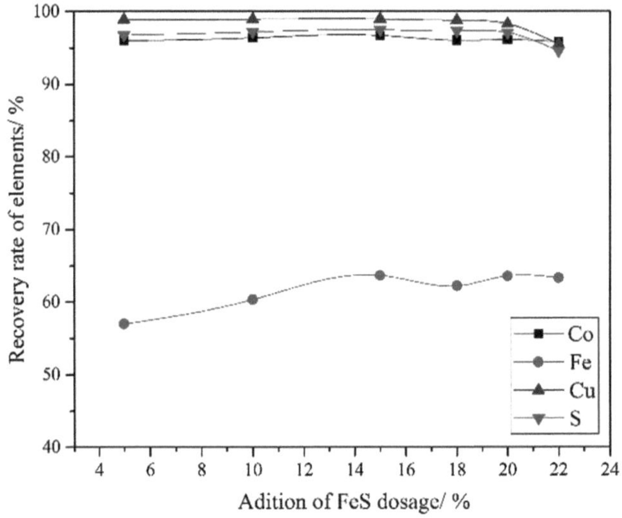

Fig. 4 The recovery rate curves of Co, Fe, Cu, S in FeS-5% C experiments

sufficient activated carbon addition. Cobalt was reduced as metal and settled into Fe-alloy.

FeS Additions with Inadequate Activated Carbon

Smelting experiments with FeS addition varying as 5, 10, 15, 18, 20, 22% were conducted with 1% activated carbon, 20% $CuS{\cdot}H_2O$ and 5% CaO in a reducing atmosphere. The results are shown in Fig. 5.

In Figs. 3 and 5, the average recovery rates of Co and Fe were respectively 82.9% (Co), 42.7% (Fe) and 83.9% (Co), 43.9% (Fe) with the dosage of 1% activated carbon and 15% FeS, which meant that the data of two different experiments was consistent.

In Fig. 5, it is shown that the recovery of Co and Fe increased from 5 to 18% of FeS dosage and then leveled off up to 22% of FeS dosage. It showed that FeS could contribute to the reducing process of Co if insufficient carbonaceous reductant is added. In the recovery of Co by reduction-sulfurization, both sulfurizing agent and carbonaceous reductant could be the main reducing agent under specific reaction conditions.

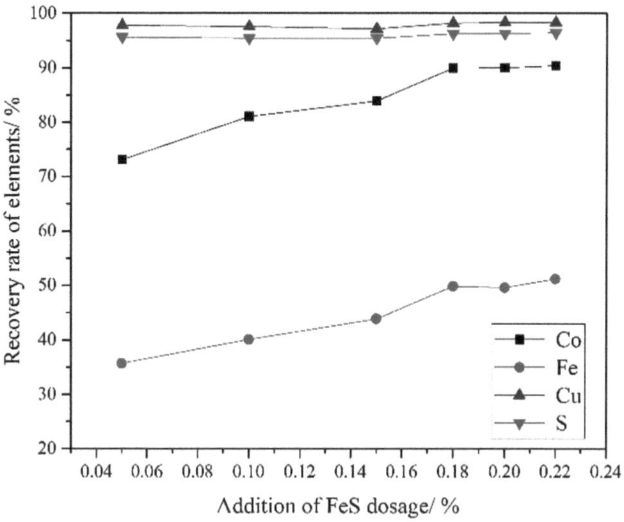

Fig. 5 The recovery rate curves of Co, Fe, Cu, S in FeS-1% C experiments

Characterization of Reduction-Sulfurization Smelting Products

The XRD patterns of original and depleted slags are shown in Fig. 6. It was found that magnetite (Fe_3O_4 and $CoFe_2O_4$) were mostly reduced and Cu settled from original slag. Hercynite ($FeAlO_2$) was formed by the reaction between molten slag and the corundum crucible. The formation of kilchoanite ($Ca_6Si_4O_{14}$) was caused by the addition of slag modifier (CaO). Likely reactions of magnetite in the process of reduction-sulfurization smelting are as follows.

$$\begin{aligned} FeS + 3Fe_3O_4 &= 10FeO + SO_2 \\ FeS + 3(CoO \cdot Fe_2O_3) &= 3CoO + 7FeO + SO_2 \end{aligned} \quad (1)$$

$$\begin{aligned} C + Fe_3O_4 &= 3FeO + CO \\ C + (CoO \cdot Fe_2O_3) &= 2FeO + CoO + CO \end{aligned} \quad (2)$$

The XRD analysis of representative matte-alloys from all experiments are shown in Fig. 7. It was found that Co was reduced to Fe–Co alloy, while the extra FeS would react with CuS (which was added to simulate the phase of Cu-matte and was found to have no influence on the recovery of elements in experiments with varying $CuS{\cdot}H_2O$ dosage) and formed bornite (Cu_5FeS_4), and much CoS was formed under the condition of inadequate activated carbon.

In the reduction-sulfurization smelting, likely reducing reactions of C would be reduction of FeO to Fe and then reducing Co from CoO, as follows.

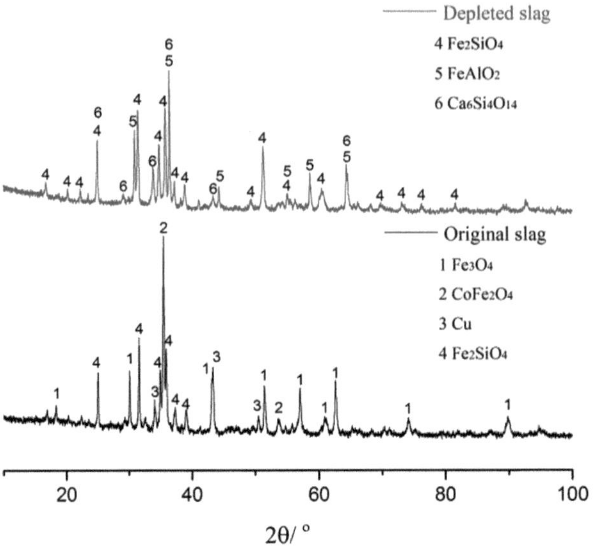

Fig. 6 XRD patterns of converter slag and depleted slag

Fig. 7 XRD spectrums of recovered matte

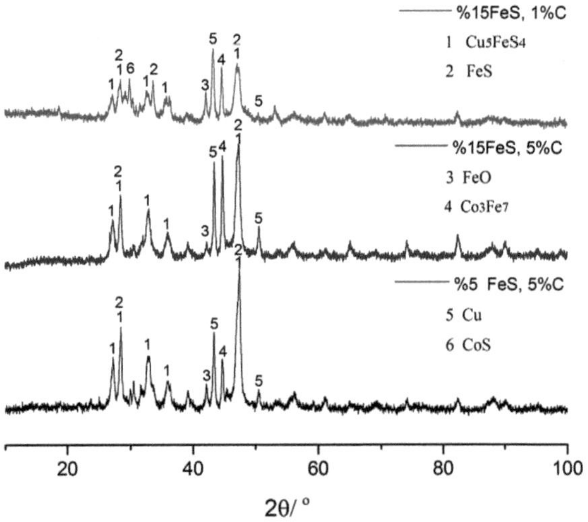

$$\begin{aligned} FeO + C &= Fe + CO \\ Fe + CoO &= Co + FeO \end{aligned} \tag{3}$$

FeS would reduce some CoO to CoS under the condition of inadequate carbonaceous reductant as follows.

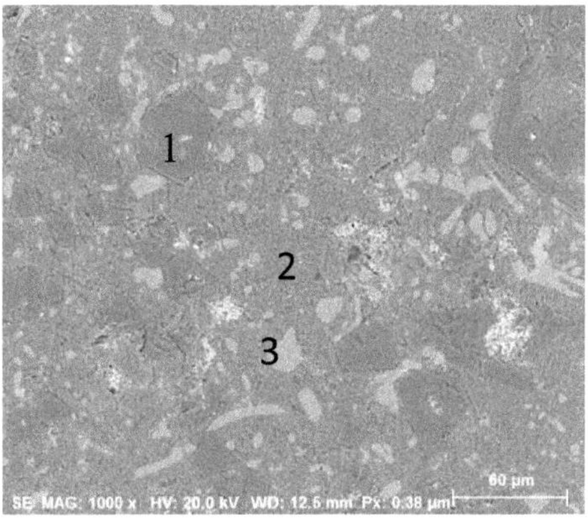

Fig. 8 SEM image of depleted slag

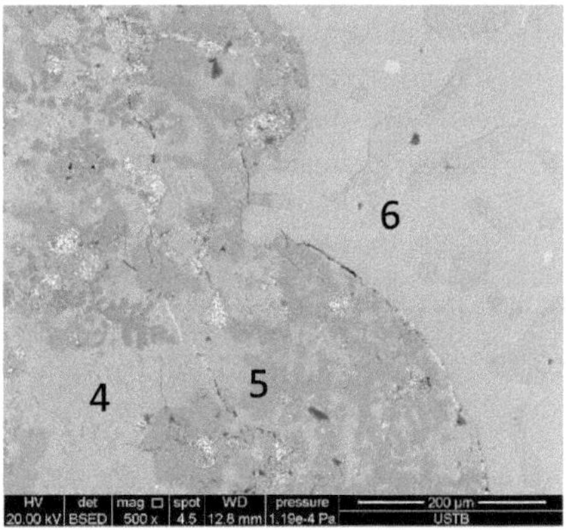

Fig. 9 SEM image of recovered matte

$$FeS + CoO = CoS + FeO \quad (4)$$

An SEM image of depleted slag and matte-alloy interface are shown in Figs. 8 and 9. The corresponding EDS results are shown in Table 2.

It was found that different phases of depleted slag still contained some Co which had obvious affinity for Fe and was closely associated with fayalite, hercynite and excess iron oxide as CoO. Under the condition of adequate activated carbon, CoO associating with iron oxide could be reduced thoroughly as alloy. However, CoO would also react with FeS to form CoS and be partly settled into bornite. In addition to the importance of reduction-sulfurization of Co, the physical mixing of slag and sulfide matte would retain some Co, which means that extending settling time and slow cooling might be worth considering.

Table 2 EDS composition and phases of marked points in SEM images

Point	w (Fe)%	w (Co)%	w (Cu)%	w (Al)%	w (Ca)%	w (Si)%	w (O)%	w (S)%	Phases
1	34.32	0.47	0.27	21.51	–	0.40	18.53	0.28	Hercynite
2	33.13	0.27	0.10	3.40	5.76	10.34	16.95	0.89	Fayalite, Kilchoanite
3	65.24	0.45	0.31	0.71	0.13	0.32	12.53	0.32	FeO
4	10.41	0.11	57.54	–	–	–	0.54	17.66	Bornite
5	62.30	0.84	0.54	–	–	0.38	14.28	0.41	FeS
6	59.27	16.4	8.32	–	–	–	0.16	0.30	Alloy

Conclusions

(1) Both activated carbon and FeS could be reducing agents to recover Co from the converter slag in the process of reduction-sulfurization smelting. CoO would be mainly reduced to Co–Fe–Cu alloy with adequate activated carbon and be reduced to Co–Fe–Cu alloy and CoS (associated with bornite) under the condition of insufficient activated carbon.

(2) Under the optimum conditions (addition of 5% activated carbon, 15% FeS, 5% CaO, 20% $CuS{\cdot}H_2O$, smelting temperature of 1673 K and smelting time of 2 h) 95% Co from converter slag, 99% Cu and 62% Fe of all reactants were recovered by the process of reduction-sulfurization smelting.

Acknowledgements The authors gratefully acknowledge the financial support of the National Science Foundation of PRC for the Research Project (2012-51234008).

References

1. E.M. Harper, G. Kavlak, T.E. Graedel, Tracking the metal of the Goblins: cobalt's cycle of use. Environ. Sci. Technol. **46**, 1079–1086 (2012)
2. X. Zhai, *Metallurgy on Heavy Metals* (Metallurgical Industry Press, Beijing, B, 2011), pp. 320–330 (in Chinese)
3. G.M. Mudd et al., Quantifying the recoverable resources of by-product metals: the case of cobalt. Ore Geol. Rev. **55**, 87–98 (2012)
4. T. Deng, Y. Ling, Process mineralogy of cobalt-bearing copper converter slag. Chin. J. Nonferrous Metal. **11**(5), 881–885 (2001) (in Chinese)
5. R.T. Jones, A.C. Deneys, Using a direct-current arc furnace to recover cobalt from slags. J. Mineral. Met. Mater. Soc. **50**(10), 53–56 (1998)
6. X. Zhai et al., Recovery of cobalt from converter slag of Chambishi copper smelter using reduction smelting process. Trans. Nonferrous Met. Soc. China **21**(9), 2117–2121 (2011)
7. C. Tang et al., Recovery of copper and cobalt from cobalt-bearing copper sulphide ore smelting slag by reduction matte smelting. Nonferrous Met. Extr. Metall. **1**, 1–5 (2015) (in Chinese)

Roasting of Celestite in Laboratory Scale Rotary Furnace

Selim Ertürk, Raşit Sezer, İbrahim Göksel Hizli, Ayşegül Bilen and Cüneyt Arslan

Abstract Turkish celestite ore was roasted in a lab scale rotary kiln furnace with metallurgical coke. Coke and celestite were mixed; the product contained 32% coke which is more than the stoichiometric requirement. 100 g of the mixed product was fed to a rotary furnace at 35 g/min. Temperature (1000 and 1100 °C), rotation speed (2.5, 5 and 10 rpm) and the slope of the furnace (2, 3 and 5°) were among the investigated parameters. Phase and chemical analyses were carried out by XRD and XRF techniques, respectively. The maximum solubility of the roasted product in water was 48% under the roasting conditions of 1100 °C, 2.5 rpm and 2° slope.

Keywords Turkish celestite · Roasting

Introduction

Strontium carbonate is the most important raw material to produce metallic strontium and other strontium chemicals (SrO, $SrCl_2$, $Sr(NO_3)_2$, Sr–Al master alloys, etc.) [1–4]. There are two commercial methods to obtain strontium carbonate from celestite. One method is hydrometallurgical route and the other is pyrometallurgical and hydrometallurgical. The hydrometallurgical route consists of the precipitation of strontium carbonate with a suitable carbonate agent (sodium carbonate, ammonium carbonate, ammonium bicarbonate, CO_2, etc.) [5–8].

S. Ertürk (✉) · R. Sezer · A. Bilen · C. Arslan
Istanbul Technical University, Istanbul, Turkey
e-mail: erturks@itu.edu.tr; erturks@gmail.com

R. Sezer
Karadeniz Technical University, Trabzon, Turkey

İ.G. Hizli
Istanbul University, Istanbul, Turkey

J.-Y. Hwang et al. (eds.), *8th International Symposium on High-Temperature Metallurgical Processing*, The Minerals, Metals & Materials Series,
DOI 10.1007/978-3-319-51340-9_46

The Pyro-Hydro route comprises the roasting of celestite with carbon to obtain soluble strontium sulfide and precipitating it with carbonating agents as in the hydrometallurgical route.

In the Pyro-Hydro route, which is called as "black ash" method, celestite ore is reduced to sulfide form (SrS) by carbothermic reduction at 900–1200 °C [8]. Then, soluble SrS is dissolved and filtered. $SrCO_3$ can be precipitated either by sodium carbonate (Na_2CO_3), ammonium carbonate ($NH_4(CO_3)_2$), or CO_2 from solution [6, 7].

Reactions in the high temperature roasting chamber are given below.

$$SrSO_4 + 4\,CO = SrS + 4\,CO_2$$

$$2C + O_2 = 2\,CO$$

Generally, the following reaction is used for the stoichiometric calculation for the reduction of celestite;

$$SrSO_4 + 2\,C = SrS + 2\,CO_2$$

In the present study, Turkish celestite ore was roasted with coke to obtain soluble SrS as a function of temperature, rotation speed and the slope of the lab-scale rotary furnace and efficiency of the roasting was calculated.

Experimental

Turkish celestite ore was obtained from Barit Maden Türk Ltd. with −2 mm in size. The chemical composition of the ore is given in Table 1. Metallurgical coke was obtained from Eti Elektrometalurji A.S. with 88% carbon content and chemical composition of which is given in Table 2.

From stochiometry, 194 g celestite ore would react with 27 g coke (183.68 g $SrSO_4$ requires 24 g carbon, stoichiometrically). In our experiments (all were repeated at least twice) we used 2.7 times stoichiometry which means that 170 g celestite ore was mixed with 80 g metallurgical coke and mechanically activated for

Table 1 Chemical composition of the celestite ore

Component	% wt.
$SrSO_4$	95.5
$CaSO_4$	3
$BaSO_4$	0.5
SiO_2	0.5
Fe_2O_3	0.5

Table 2 Chemical composition of metallurgical coke

Component	% wt.
Fixed carbon	88
Ash	10
Volatile components	1
Sulphur	1

2 min in a ring grinder. In all of the experiments, 100 g mechanically activated mixture was fed to the rotary furnace at a rate of 35 g/min.

A picture of the laboratory furnace was given in Fig. 1. Length of the alumina tube in the laboratory furnace is 1200 mm with a hot zone of 400 mm. Temperature profile along the length of the furnace was given in Fig. 2. Inner diameter of the tube is 90 mm with a thickness of 10 mm.

Phase analysis of the roasted product were carried out by X'pert Pro X-Ray Diffractometer. The conversion efficiency was assessed by dissolving the roasted

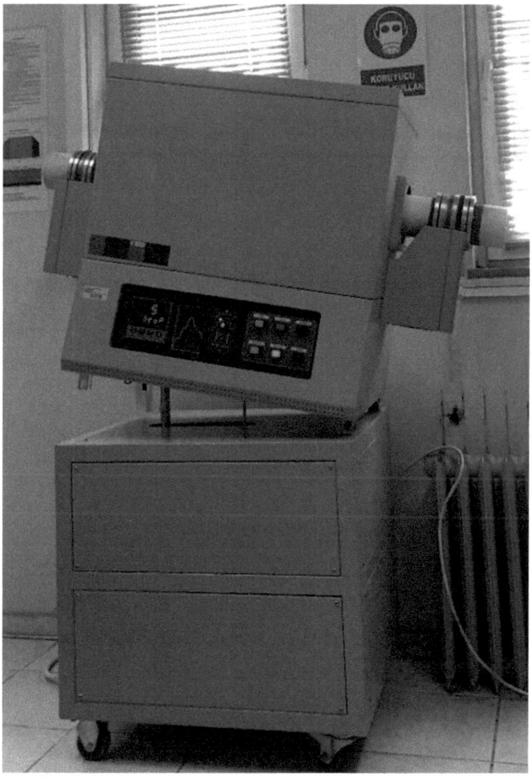

Fig. 1 Lab scale rotary furnace

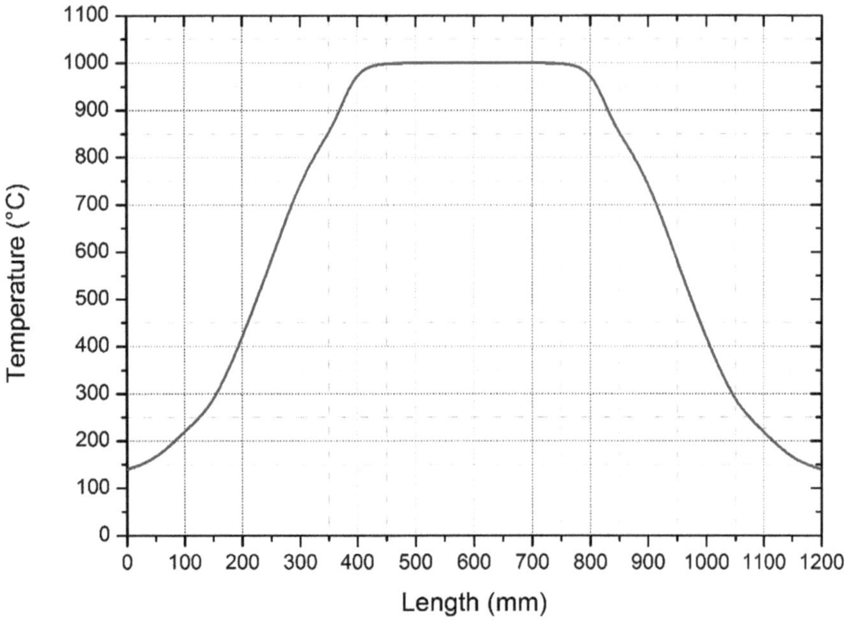

Fig. 2 Temperature profile of rotary furnace when the stable hot zone was set to 1000 °C

product in hot water. 10 g of the SrS product was reacted with 250 ml distilled water for 2 h at 97 °C on a magnetic stirrer. The residue (containing unreacted $SrSO_4$) was separated and dried in an oven for 24 h.

Results

The dwell time in the furnace was recorded for different slopes and rotation speeds at room temperature and test temperatures. The mean dwell time is shown in Fig. 3. for all of the conditions. The dwelling time increases if slope and rotation speed decreases. The highest time (70 min) was obtained at 2° and 2.5 rpm while the lowest (5 min) is obtained at 5° and 10 rpm.

XRD patterns of the roasted products are shown in Fig. 4. SrS peaks can be easily seen at 2°–2.5 rpm, 2°–5 rpm and 3°–2.5 rpm at 1000 °C. However, for material roasted at 1100 °C, except for the 5°–5 rpm and 5°–10 rpm conditions, the SrS peaks are strong and celestite peaks are very low.

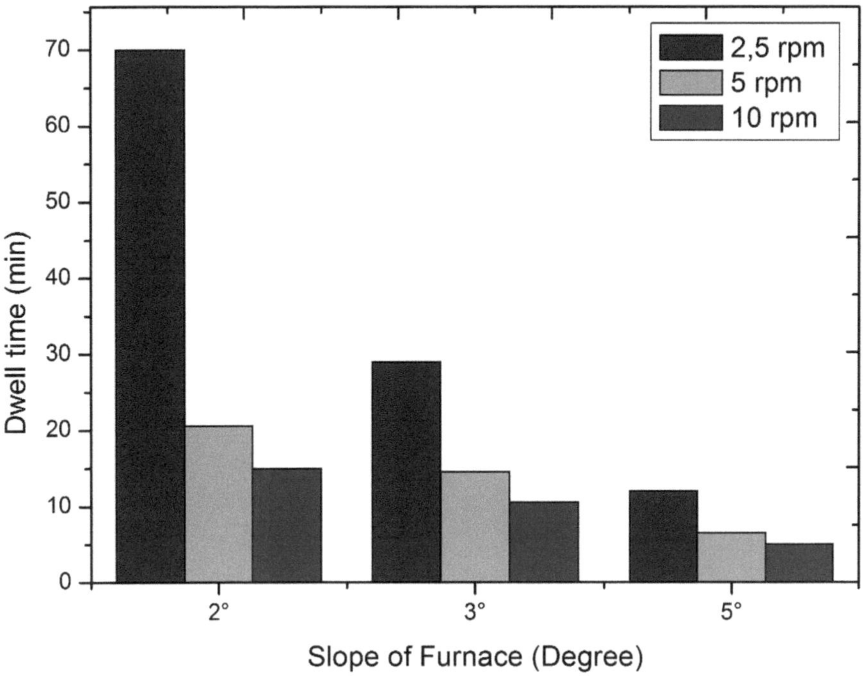

Fig. 3 Change in dwell time in the rotary furnace according to slope and rotation speed

SrS dissolves in water whereas celestite has no solubility in water. To assess the efficiency of the roasting process, 10 g of the roasted material was dissolved in 250 ml distilled water for 2 h at 97 °C by 400 rpm magnetic stirring. All of the SrS dissolved while unreacted celestite did not. After solid liquid separation, the residue (unroasted celestite) is dried in an oven for 24 h at 105 °C. According to the weight difference, the efficiency was calculated and given in Table 3.

Conclusion

Reduction of celestite was successful with the addition of excess of metallurgical coke in a laboratory scale rotary furnace. With increasing furnace rotation rate and slope, the residence time of the particles in the furnace decreased. The best furnace condition for both temperatures (1000 and 1100 °C) was 2° and 2.5 rpm with 70 min dwell time. The highest dissolution efficiency was obtained as 48% by 2°, 2.5 rpm and 1100 °C reduction conditions.

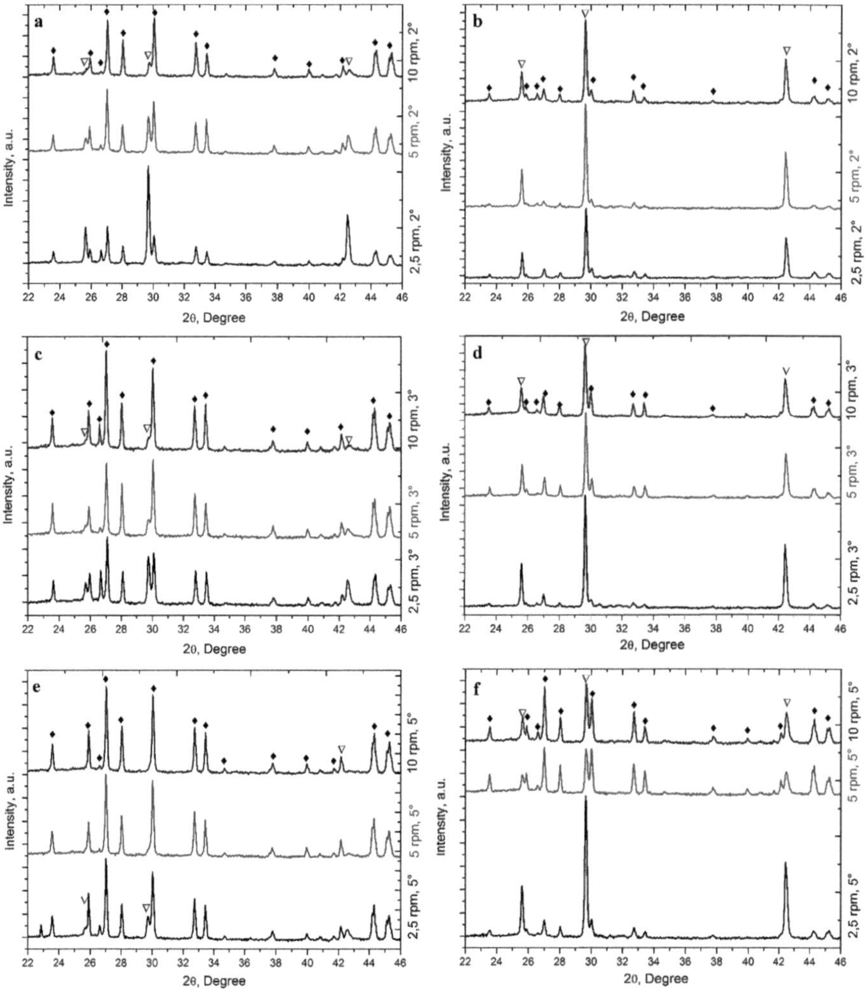

Fig. 4 XRD patterns of roasted products **a**, **c**, **e** 1000 °C, **b**, **d**, **f** 1100°C. *Black diamond* $SrSO_4$, *Nabla* SrS

Table 3 Efficiency of the dissolution experiments

Temperature (°C)	Rotation (rpm)	Slope (°)	Efficiency (%)	Temperature (°C)	Rotation (rpm)	Slope (°)	Efficiency (%)
1000	2.5	2	32	1100	2.5	2	**48**
1000	5	2	15	1100	5	2	42
1000	10	2	7	1100	10	2	31
1000	2.5	3	22	1100	2.5	3	47
1000	5	3	8	1100	5	3	38
1000	10	3	8	1100	10	3	34
1000	2.5	5	9	1100	2.5	5	40
1000	5	5	7	1100	5	5	19
1000	10	5	7	1100	10	5	12

Acknowledgements The authors greatly acknowledge the financial support provided by Scientific and Research Council of Turkey (TUBITAK) under the contract no 115M631 and would like to thank to Barit Maden Türk Ltd. for technical and materials support.

References

1. M. Sivakumar et al., Nanophase formation of strontium hexaferrite fine powder by the sonochemical method using $Fe(CO)_5$. J. Magn. Magn. Mater. **268**(1–2), 95–104 (2004)
2. R. Zhang et al., Gel combustion synthesis and luminescence properties of nanoparticles of monoclinic $SrAl_2O_4$:Eu^{2+}, Dy^{3+}. Mater. Chem. Phys. **113**(1), 255–259 (2009)
3. A.H.E. Castillejos, F.P.B. de la Cruz del, A.S. Uribe, The direct conversion of celestite to strontium carbonate in sodium carbonate aqueous media. Hydrometallurgy **40**(1–2), 207–222 (1996)
4. F.R.P. Carrillo, A.S. Uribe, A.H.E. Castillejos, A laboratory study of the leaching of celestite in a Pachuca tank. Miner. Eng. **8**(4–5), 495–509 (1995)
5. A.K. Kobe, N.J. Deiglmeier, Conversion from strontium sulfate by metathesis with alkali carbonate solution. Ind. Eng. Chem. **35**, 323–325 (1943)
6. M. Erdemoğlu, M. Canbazoğlu, The leaching of SrS with water and the precipitation of $SrCO_3$ from leach solution by different carbonating agents. Hydrometallurgy **49**(1), 135–150 (1998)
7. G. Owusu, J.E. Litz, Water leaching of SrS and precipitation of $SrCO_3$ using carbon dioxide as the precipitating agent. Hydrometallurgy **57**(1), 23–29 (2000)
8. R. Sonawane et al., Effect of a catalyst on the kinetics of reduction of celestite ($SrSO_4$) by active charcoal. Metall. Mater. Trans. B **31**(1), 35–41 (2000)

The Experimental Study of $CaCO_3$ in the Vanadium Extraction Process

Shu-Chao Wang, Yu Wang, Wei-Tong Du and Peng-cheng Li

Abstract $CaCO_3$ thermal decomposition produces CO_2 and CaO, and CO_2 can be used as a weak oxidant in vanadium extraction. CaO will also affect the vanadium extraction process. This experiment, taking self-blowing and temperature acquisition experiment system, through adding different quality $CaCO_3$, studies the molten iron in the oxidation behavior of C, V and P and its impact, and analysis of $CaCO_3$ decomposition of the heating effect of the molten pool. The result indicates that adding $CaCO_3$ in the vanadium extraction process had no obvious effect on the oxidation of carbon, but partly increased vanadium oxidation which reached more than 90%. Increased $CaCO_3$ additions also increased cooling the molten pool, with a maximum decrease in temperature of 20 K. In addition, the oxidation of P also increased, reaching 40%. The research has proved that $CaCO_3$ can serve to oxidize V and P is effective in vanadium extraction.

Keywords Carbon dioxide · Limestone · Vanadium extraction

Introduction

Carbon dioxide (CO_2), a major component of the greenhouse gases, is also the main constituent of metallurgical waste gas. With increasing recognition by international communities of the severity of climate change, the pressure for reduction of CO_2 emissions has become more prominent [1, 2]. In the field of metallurgical industry with huge dioxide carbon emission, the recycling of the emission will make a significant contribution to environmental protection and energy saving [3, 4].

Vanadium calcification is a new type of cleaning process, in which $CaCO_3$ can be used as a source of CaO, and CO_2 which can be used as oxidizer.

S.-C. Wang (✉) · Y. Wang · W.-T. Du · P. Li
College of Materials Science and Engineering, Chongqing University, Chongqing 400044, China
e-mail: 445814524@qq.com

J.-Y. Hwang et al. (eds.), *8th International Symposium on High-Temperature Metallurgical Processing*, The Minerals, Metals & Materials Series,
DOI 10.1007/978-3-319-51340-9_47

The traditional technique to extract vanadium during converter steelmaking process is to blow pure O_2 to oxidize V into V_2O_3 [5]. However, this process has several problems including strong oxidizers causing overheating and energy consuming, according to the industry vanadium extraction process. Due to the reaction of O_2 with Si and Mn is a strong exothermic reaction, the temperature rises quickly and in this juncture some coolant should be put into the converter to control the temperature of the converter at a proper level. In the process of extracting vanadium, $CaCO_3$ is used as a coolant, which improves vanadium extraction [6, 7].

Therefore, research about the effect of $CaCO_3$ on vanadium extraction in the converter has important theoretical and practical significance for the establishment of a new process of vanadium extraction and the reduction of CO_2 emission as well as environmental protection.

Experiment

Experimental Materials and Equipment

The vanadium-bearing metal was provided by Panzhihua Iron and Steel Group Company. The composition was shown in Table 1.

The experimental equipment is shown in Fig. 1, which includes a $MoSi_2$ electric resistance furnace and the flow rate control system. The device using silicon molybdenum rod as the maximum temperature of the heating element is 1500 °C. The use of the flow rate control system is to decide which one or two gases can be blown into the furnace and control its flow rate.

Table 1 Composition of vanadium containing hot metal

Composition	C	V	Si	Mn	P	S
wt%	3.73	0.32	0.053	0.14	0.081	0.145

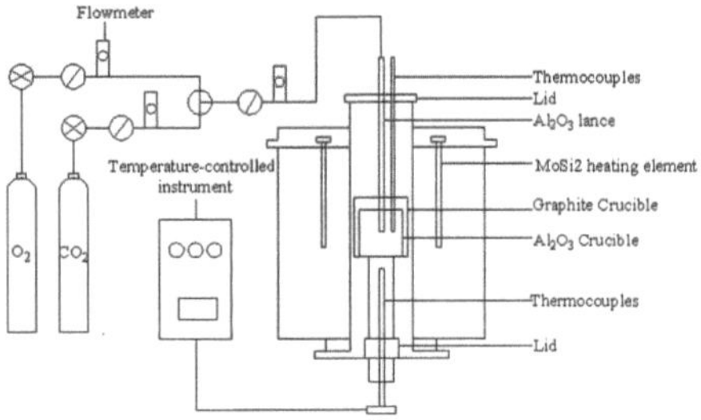

Fig. 1 Schematic drawing of experimental apparatus

Experimental Method

The 350 g vanadium-bearing metal was loaded in a corundum crucible (inner diameter, 46 mm; height, 120 mm) contained in a graphite crucible (inner diameter, 55 mm; height, 135 mm) heated in a $MoSi_2$ electric resistance furnace. Blowing began when the temperature of the melt reached the target temperature (1300 °C). Through the numerical control, the high temperature silicon molybdenum furnace was kept at constant temperature and lasted for 15 min. By measuring the temperature of the molten pool of tungsten-rhenium thermocouple in the process of experiments, the thermocouple was immersed to half the total bath depth. After confirming the complete melting and homogenization of the iron sample, the corundum lance (inner diameter, 12 mm; height, 800 mm) was immersed to half the total bath depth. The total flow of CO_2 and O_2 during mixed blowing was controlled at 0.6 L/min. with 10% CO_2 in the mixed gas, the rest of which was pure oxygen. According to the research of Li et al. [8], the Ca/V molar ratio of 1.1:1 in the calcified extraction of vanadium is beneficial to the extraction of vanadium. The different amounts of $CaCO_3$ to be added was calculated according to the CaO/V_2O_3 molar ratio in the slag, using molar ratios of 0, 2, 4, 6, with the addition of $CaCO_3$ made after blowing for 3 min later.

Results and Discussion

During the process of extracting vanadium, the $CaCO_3$ can decompose into CaO and CO_2, as the temperature is higher than the decomposition temperature of limestone. CaO does not react with the molten metal, but the CO_2 can react. CO_2 is a weak oxidizer and its reaction with [Fe] and [C] is endothermic reaction; compared with oxygen, the reaction of it with V and Si in molten iron is a slight exothermic reaction.

Chemical Analysis

The results show that the proportion of $n_{CaO}:n_{V2O3}$, does not affect oxidation of carbon (Tables 2, 3 and 4).

Table 2 Results of chemical analysis of iron samples (mass%)

$n_{CaO}:n_{V2O3}$	Initial [C]	Initial [V]	Initial [P]	Final [C]	Final [V]	Final [P]
2:1	3.73	0.325	0.081	2.65	0.028	0.073
4:1	3.73	0.325	0.081	2.65	0.025	0.071
6:1	3.73	0.325	0.081	2.65	0.026	0.066
8:1	3.73	0.325	0.081	2.68	0.023	0.048

Table 3 Change of the carbon concentration of the iron after blowing

$n_{CaO}:n_{V2O3}$	Initial [C]	Final [C]	Degree of oxidation of C (%)
2:1	3.73	2.65	28.95
4:1	3.73	2.65	28.95
6:1	3.73	2.65	28.95
8:1	3.73	2.68	28.15

Table 4 The vanadium in the molten iron change with blowing time

$n_{CaO}:n_{V2O3}$	Initial [V]	Final [V]	Degree of oxidation of V (%)
2:1	0.320	0.028	91.3
4:1	0.320	0.026	91.9
6:1	0.320	0.025	92.2
8:1	0.320	0.023	92.8

Table 5 The phosphorus in the molten iron change with blowing time

$n_{CaO}:n_{V2O3}$	Initial [P]	Final [P]	Degree of oxidation of P (%)
2:1	0.081	0.073	9.9
4:1	0.081	0.071	12.4
6:1	0.081	0.066	18.5
8:1	0.081	0.048	40.7

From the experimental results, it is found that the vanadium oxidation rate was slightly increased by adding limestone, with the extent of oxidation of vanadium exceeding 90%. Therefore, it can meet industry standard which is more than 90% vanadium oxidation (Table 5).

With increased additions of $CaCO_3$, the degree of oxidation of phosphorus also increased, to 40%.

$CaCO_3$ Cooling Effect

Figure 2 shows how the temperature of molten iron changed with the blowing time with different additions of $CaCO_3$. With the addition of $CaCO_3$, the temperature of the molten pool decreased obviously.

Figure 3 shows how the temperature of molten iron changed with no blowing gas with different additions of $CaCO_3$. After adding $CaCO_3$, the molten iron temperature decreased, confirming that the addition of $CaCO_3$ can reduce the temperature of the molten iron.

The results in Fig. 3 were recalculated by subtracting the underlying furnace heating rate, due to the numerical control, the high temperature silicon molybdenum

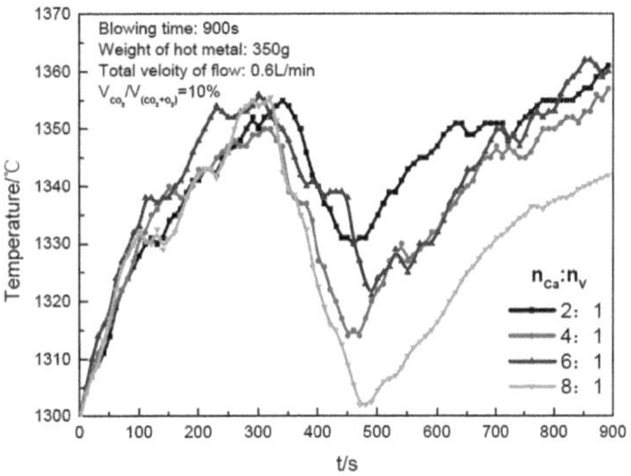

Fig. 2 Changes of the molten iron temperature of $CaCO_3$ under the conditions of blowing gas

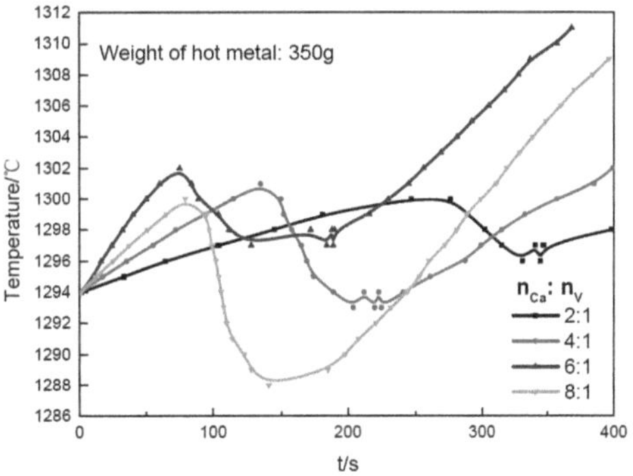

Fig. 3 Changes of the molten iron temperature of $CaCO_3$ under the conditions of no blowing gas

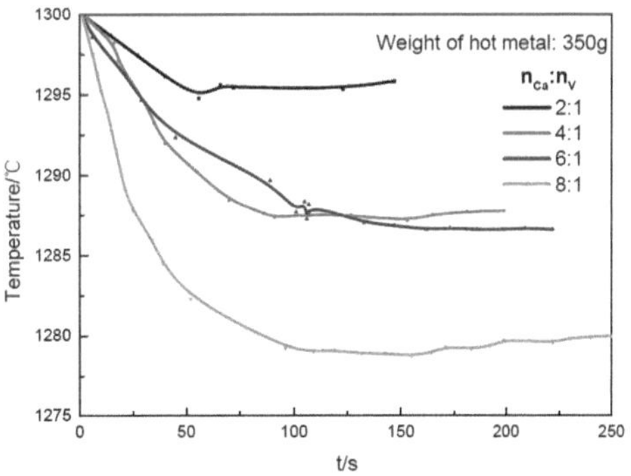

Fig. 4 Changes of the molten iron temperature of $CaCO_3$ under the conditions of no blowing gas

furnace was kept at constant temperature (the temperature rising rate of the furnace is removed, and the influence of the environment on the temperature of the molten pool is neglected) setting the zero time to the calcium carbonate addition time, with the results shown in Fig. 4. After the addition of $CaCO_3$, the molten iron temperature has a clear downward trend. With larger additions of $CaCO_3$, the temperature decrease is larger.

The temperature of the iron bath is influenced by $CaCO_3$, mainly because of the heat of decomposition of limestone.

$CaCO_3$ thermal decomposition producing CO_2 and CaO, the reaction is as following:

$$CaCO_3(s) = CaO(s) + CO_2(g) \quad \Delta G^\theta = -144.6T + 16{,}9120\ J/mol$$

Because $CaCO_3$ and CaO are solid, a_{CaO} and a_{CaCO3} is equal to 1, and the relationship between decomposition temperature and CO_2 partial pressure can be obtained:

$$T = \frac{20{,}339.15}{17.39 - \ln(P_{CO2}/P^\theta)}$$

For blowing under the condition of 10% CO_2,

$$\ln(P_{CO2}/P^\theta) = 0.1$$

$$T_{decomposition} = 1176\ K(903\ ^\circ C)$$

Conclusion

(1) Under the experimental conditions, the degree of oxidation of carbon was 27%, and was not affected by addition of limestone. The degree of oxidation of vanadium was up to 91% meeting industrial requirements. With increased additions of limestone, vanadium oxidation increased slightly. In addition, the degree of oxidation of P is also increased, reaching 40%.

(2) The bath temperature decreased with the increasing proportions of n_{CaO}/n_{V2O3}. When adding 8.96 g of $CaCO_3$, the molten pool temperature decreased by 20 K.

Acknowledgements This work was supported by National Natural Science Foundation of China (project No.51334001) and Sharing Fund of Large Scale Equipment, Chongqing University (project No. 201512150031).

References

1. J.G.I. Olivier, J.A.H.W. Peters, G. Janssens-Maenhout, in *Trends in global CO_2 emissions 2013 report*. PBL Netherlands Environmental Assessment Agency (2013)
2. K. Wang et al., Scenario analysis on CO_2 emissions reduction potential in China's iron and steel industry. Energy Policy **35**(4), 2320–2335 (2006)

3. J.X. Zhou, Summary of exploitation and utilization of carbon dioxide. Chem. Eng. Des. **14**(4), 7–8 (2004)
4. E. Worrell, L. Price, N. Martin, Energy efficiency and carbon dioxide emissions reduction opportunities in the US iron and steel sector. Energy **26**(5), 513–536 (2001)
5. R.R. Moskalyk, A.M. Alfantazi, Processing of vanadium: a review. Miner. Eng. **16**(9), 793–805 (2003)
6. H. Nomura, K. Mori, Kinetics of decarburization of liquid iron with high concentration of carbon. Tetsu-to-hagané **57**(9), 1468 (1971)
7. W.T. Du, Y. Wang, G. Wen, Thermodynamic study on vanadium extraction with CO_2 and O_2 mixed blowing, in *Proceedings of the EPD Congress 2015, Orlando, FL, United states, F 2015* (Wiley Online Library, 2015)
8. H.Y. Li, K. Wang, W.H. Hua et al., Selective leaching of vanadium in calcification-roasted vanadium slag by ammonium carbonate. Hydrometallurgy **160**, 18–25 (2015)

The Extraction of Zinc from Zinc Ferrite by Calcified-Roasting and Ammonia-Leaching Process

Zeqiang Xie, Yufeng Guo, Tao Jiang, Feng Chen and Lingzhi Yang

Abstract Zinc ferrite ($ZnFe_2O_4$), one of the major forms of zinc in Electric Arc Furnace (EAF) dust and zinc leaching residue, is very difficult to chemically dissolve in either acidic or alkaline solutions. In spite of the zinc resources in the Electric Arc Furnace (EAF) dust and zinc leaching residue are abundant, there is no mature utilization technology. This situation not only causes the waste of resources, but also produces the harm to the environment. In the paper, the basic principle of this process has been established by revealing the phase equilibria in the CaO–Fe_2O_3–ZnO system, where $Ca_2Fe_2O_5$ can be in equilibrium with ZnO rather than $ZnFe_2O_4$ at higher CaO region. It has also been demonstrated that the formed ZnO can be solved by ammonia-leaching and $Ca_2Fe_2O_5$ keep in solid residue, which can be used as a flux for steel refining.

Keywords Zinc ferrite · Calcified-roasting · Ammonia leaching · Extraction of zinc

Introduction

Recently, the China's iron and steel industry has entered a rapid development period, meanwhile, approximate 100 million ton Zinc-bearing dust produced in the process every year. Zinc-bearing dust contains a large number of valuable ingredients such as zinc, iron, which is an important secondary resource. In general, dust discharge of iron and steel enterprise production process account for steel output by 8–15% [1]. In 2015, China's crude steel production is close to 804 million ton, that is to say, only last year, China's iron and steel enterprises discharge of dust around

Z. Xie · Y. Guo (✉) · T. Jiang · F. Chen · L. Yang
School of Mineral Processing and Bioengineering, Central South University, Changsha 410083, Hunan, China
e-mail: guoyufengcsu@163.com

Z. Xie
e-mail: xiezeqiang@163.com

J.-Y. Hwang et al. (eds.), *8th International Symposium on High-Temperature Metallurgical Processing*, The Minerals, Metals & Materials Series,
DOI 10.1007/978-3-319-51340-9_48

80 million ton, of which zinc-containing 4–16 million ton, iron-containing about 24 million tons [2]. China's Zinc and iron resources are insufficient, therefore, adopting appropriate methods from the dust recovery of valuable metals such as zinc, iron is an important approach to solve the shortage of resources in China, there are both great economic benefits and remarkable environmental benefits.

Steel plant dust contains a lot of Fe and zinc element, in which the zinc mainly zinc oxide and zinc ferrite ($ZnFe_2O_4$), iron mainly exist as magnetite (Fe_3O_4), hematite (Fe_2O3), floating (FeO) and zinc ferrite austenite ($ZnFe_2O_4$) phase. However, zinc ferrite ($ZnFe_2O_4$) is a spinel structure, which has considerable stability. At room temperature and atmosphere pressure, Zinc ferrite is insoluble in water, dilute acid solution, alkali and ammonia solution. This led to the difficulty to process it with general method. Therefore developing on zinc ferrite ($ZnFe_2O_4$) comprehensive utilization of zinc dust research is particularly important.

In this paper, aiming at the intractable zinc ferrite ($ZnFe_2O_4$), a new fire metallurgy—hydrometallurgy process integration thinking is proposed, that is the extraction of Zinc from Zinc Ferrite by calcified roasting and ammonia leaching process. The first step: adding calcium oxide to make zinc ferrite phase reconstruction at high temperature. The chemical equation of the roasting reaction is:

$$ZnFe_2O_4 + 2CaO = ZnO + Ca_2Fe_2O_5 \quad (1)$$

The next step is using ammonia can form ammonia complexes with zinc oxide. Reaction of ammonia leaching process:

$$ZnO + nNH_3 \cdot H_2O = [Zn(NH_3)n]^{2+} + 2OH^- \quad (2)$$

The roasting reaction product ($Ca_2Fe_2O_5$) do not participate in the leaching reaction, eventually $Ca_2Fe_2O_5$ exist in ammonia leaching residue which can be used as sintering raw material and steelmaking solvent. After processing, zinc leach liquor can be used as raw material of zinc electrodeposition. In consequence, comprehensive utilization of iron and zinc in zinc dust can be realized. This method is studied only for pure Zinc ferrite ($ZnFe_2O_4$). But for the further research with high zinc ferrite phase of the comprehensive utilization of zinc dust, it provides a new way of thinking and guiding direction.

Experimental

Raw Materials

Because there is no pure zinc ferrite mineral in nature, sample is a synthetic zinc ferrite as raw materials. In order to simulate the consistent generating conditions of the smelting slag zinc ferrite, we adopt fire preparation of zinc ferrite under the condition of high temperature. The preparation process is as follows [2]: take a certain amount of zinc oxide and ferric oxide according to the theory of generated molar ratio as 1:1. In order to complete reaction, the reactant need to be mixed sufficiently. Then place the mixture in muffle furnace under 1000 °C roasting 4 h. Then take out the roasting samples to a mortar by repeated grinding to the particles less than 200 mesh. In order to remove unreacted ferric oxide and zinc oxide, put samples in 1 mol/L dilute hydrochloric by stir leaching (solution temperature: 40 ° C, leaching time: 1 h). After repeated washing three times again, then put the leaching residue to muffle furnace roasting 4 h under 1000 °C. Take out the sample after cooling, place samples in the dryer saved backup. The synthetic reaction equation is:

$$Fe_2O_3(s) + ZnO(s) \rightarrow ZnFe_2O_4(s) \quad (3)$$

The picture of Zinc Ferrite is shown in Fig. 1.

The XRD pattern of Zinc Ferrite sample is shown in Fig. 2.

The SEM of Zinc Ferrite sample is shown in Fig. 3.

The chemical composition of Zinc Ferrite sample is listed in Table 1.

Results in Fig. 1 and Table 1 show that Zinc Ferrite sample is pure after roasting transform. The zinc and Fe element content of Zinc Ferrite sample is 27.11 and 46.28 wt%, which is very close to theoretical content (27.12 and 46.33%) of Zinc Ferrite.

Fig. 1 Picture of zinc ferrite

Fig. 2 XRD of zinc ferrite

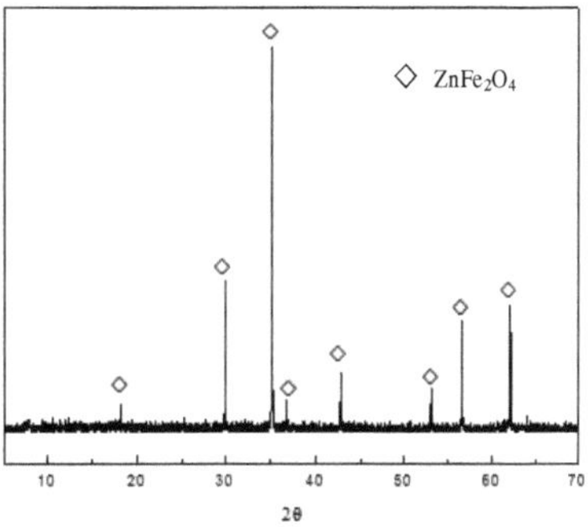

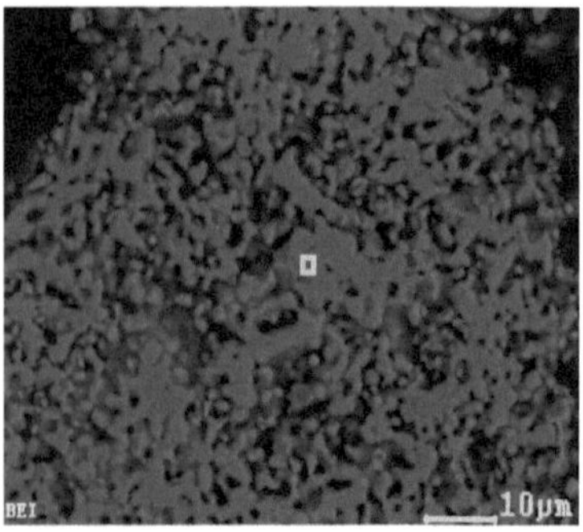

Fig. 3 SEM of zinc ferrite

Table 1 The chemical composition of zinc ferrite sample

Zn	Fe	O
27.11%	46.28%	26.61%

Experimental Method

(1) Roasting Experiment: Zinc Ferrite sample powder and analytical CaO powder are mixed by a certain percentage then the mixture is loaded in a corundum boat and roasted in a horizontal tube furnace at a target reaction temperature.
(2) Leaching Experiment: The roasted ore samples are ground into the particle with size <74 μm. Leaching experiments was conducted in a 250 mL beaker heated in a water bath at a target temperature. The leaching procedure initiated by pouring 100 mL of required concentration ammonia leaching agent into a 250 mL beaker with a plastic film covered. Then the leaching agent was heated to the experimental temperature with stirring. Time was counted when the Zinc Ferrite samples or calcified-roasting samples were added into the leaching beaker. When the required reaction time was up, the slurry was filtered, and then an appropriate dosage of dilution and leaching residue were taken for analysis. Each group of analysis is repeated for three times and the average value is taken as the final data.

Analysis Methods

Zinc content: EDTA Complexometric Titration (GB/T 8151.1-2012) [3].

Chemical phase analysis: X-ray Diffraction (Rigaku-TTRJII, Cu target, Kα, $\gamma = 0.154\ 06$ nm).

Results and Discussion

Phase Diagram Analysis

As seen from Fig. 4, the ZnO and $Ca_2Fe_2O_5$ can be stable coexistence without $ZnFe_2O_4$ at 1200 °C. To increase the ratio of CaO can increase dominant area of ZnO and $Ca_2Fe_2O_5$. The minimum theoretical CaO dosage is 2:1 to $ZnFe_2O_4$ and the appropriate range of CaO dosage is greater than 2:1.

The XRD of CaO-roasted sample is shown in Fig. 5. From the XRD, we can see that the mineral phase of actual experiment samples have been converted after calcified-roasting, that is Zinc Ferrite phase has disappeared.

The main affecting factors of the calcified-roasting process are the dosage of CaO additive, roasting temperature and roasting time were studied. Zinc Ferrite samples and calcified-roasting samples were leached in $NH_4Cl–NH_3·H_2O$ leaching system [4–8]. Leaching conditions: the total ammonia concentration was 6 mol/L, the mole ratio of NH_4Cl to $NH_3·H_2O$ was 1:1, solution temperature was 40 °C, the solid–liquid ratio was 20 g/L, the stirring speed was 350 r/min and the leaching time was 3 h.

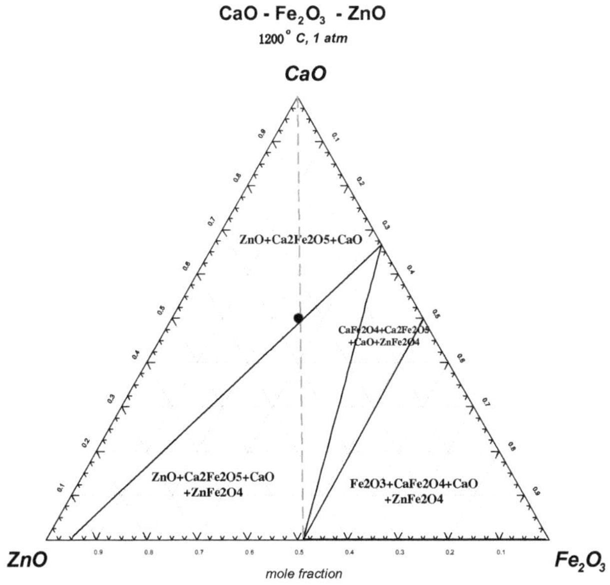

Fig. 4 Ternary-component phase diagram of CaO–ZnO–Fe_2O_3 (1200 °C)

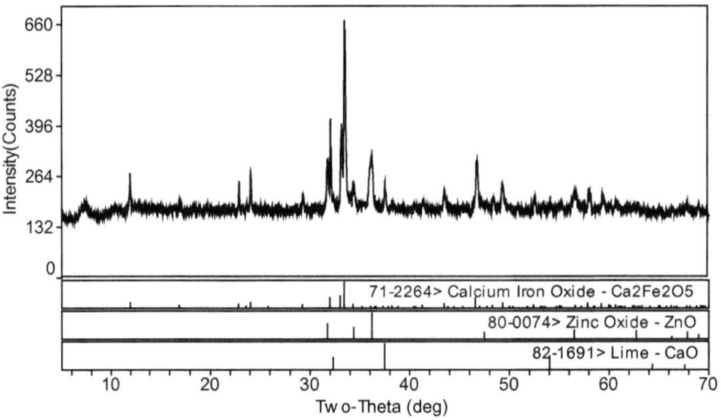

Fig. 5 XRD of CaO-roasted sample

The Influence of the Dosage of CaO Additives on Zinc Leaching Rate

The influence of the dosage of CaO additive on zinc leaching rate is very significant as shown in Fig. 6. When the dosage of CaO additives increases from 2:1 (theoretical mole ratio) to 2.6:1, the zinc leaching rate increases from 63.13 to 90.23%. This is because Zinc Ferrite samples cannot fully contact with CaO when CaO is added in the theoretical amount. So since CaO dosage increases, Zinc Ferrite could become better contact with CaO which promotes the reaction and results in the increase of zinc leaching rate. Zinc leaching rate increases significantly, almost shows a linear relationship when the dosage of CaO additive increases from 2:1 to 2.4:1 (mole ratio). After that, a further increase of CaO dosage would not increase zinc leaching rate obviously, which indicates the CaO dosage has reached a

Fig. 6 Influence of $CaO/ZnFe_2O_4$ (mole ratio) on zinc leaching rate

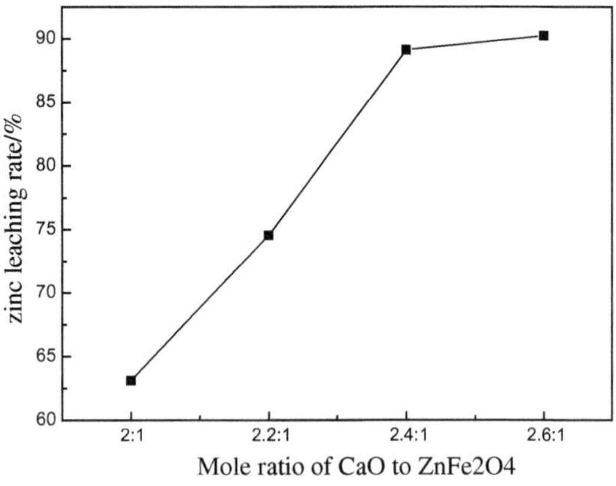

sufficient dosage and it would no longer be a restriction factor of Zinc Ferrite roasting-transformation and leaching.

The Influence of the Roasting Temperature on Zinc Leaching Rate

The fully mixed Zinc Ferrite and CaO samples ($CaO:ZnFe_2O_4$ = 2.4:1 mol ratio) were roasted at the temperature from 1180 to 1250 °C for 120 min. The influence of roasting time on zinc leaching rate is shown in Fig. 7. As seen from Fig. 7, the impact of temperature on zinc leaching rate is obvious. As the roasting temperature increases, zinc leaching rate increases as well. The zinc leaching rate continues to rise sharply from 60.13% (at 1200 °C) to 90.41% (at 1240 °C). The Zinc leaching rate decreased lightly at 1250 °C. The appropriate roasting temperature is 1220–1240 °C.

Influence of Roasting Time on Zinc Leaching Rate

The fully mixed Zinc Ferrite and CaO samples ($CaO:ZnFe_2O_4$ = 2.4:1 mol ratio, roasting temperature: 1220 °C) were roasted at the temperature from 30 to 120 min. The influence of roasting time on zinc leaching rate is shown in Fig. 8. As seen from Fig. 8, the impact of roasting time on zinc leaching rate is obvious. As the roasting time increases, zinc leaching rate increases as well. The zinc leaching rate continues to rise sharply from 40.17% (30 min) to 87.25% (120 min). The appropriate roasting time is 120 min.

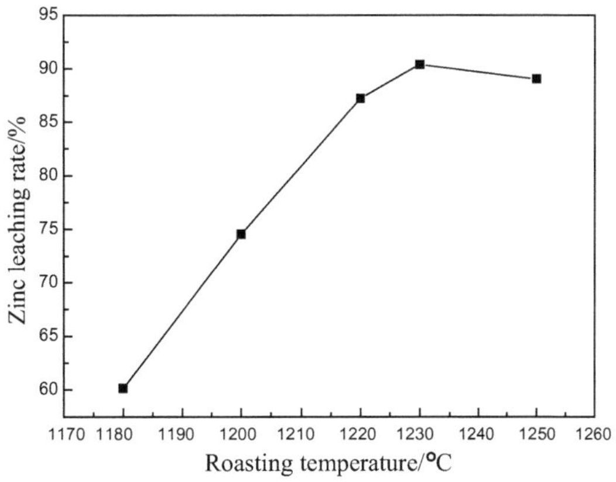

Fig. 7 Influence of roasting temperature on zinc leaching rate

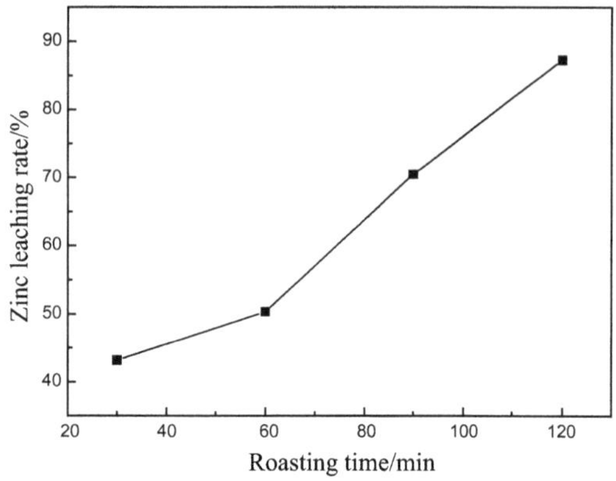

Fig. 8 Influence of roasting time on zinc leaching rate

Conclusions

(1) $ZnFe_2O_4$ can react with CaO and be transformed into ZnO and $Ca_2Fe_2O_5$ within the appropriate roasting conditions.
(2) Compared with $ZnFe_2O_4$ which can not leach in $NH_4Cl–NH_3·H_2O$ leaching system, the calcified-roasted samples can easier to leach in the leaching system.
(3) The zinc leaching rate of the roasted sample varied with the different roasting conditions. An appropriate increase of CaO dosage, roasting temperature or roasting time can promote the transformative reaction and lead to an obvious increase of zinc leaching rate.
(4) Under optimal roasting conditions, the calcified-roasting (transformative) reaction could react thoroughly, and most of $ZnFe_2O_4$ is transformed into ZnO, correspondingly, the zinc leaching rate is close to 90%.

Acknowledgements The authors are grateful to Major State Basic Research Development Program of China (973 Program Grant No. 2014CB643403).

References

1. X.F. She et al., The comparison of comprehensive utilization of zinc bearing dust process in steel works. Ironmaking **29**(4), 56–62 (2010)
2. M. Li, *Fundamental research on selective reduction of zinc calcine and separation of zinc and iron*, Ph.D. thesis, Central South University, 2013, p. 23
3. S.Y. Sun et al., *Ore and Non-ferrous Metals Analysis Manual* (Metallurgical Industry Press, BeiJing, 1990), p. 176
4. Z.W. Zhao, X.J. Jia, Leaching zinc oxide ore in alkaline solution. J. Central South Univ. Sci. Technol. **41**(1), 39–43 (2010)
5. Z.W. Zhao, S Long, Alkali leaching of refractory high silica zinc oxide ore by mechanical activation. J. Central South Univ. Sci. Technol. **41**(4), 1246–1250 (2010)

6. Z.Y. Liu, *Leaching mechanism of zinc oxide ores in ammonia-ammonium aqueous solution*, Ph.D. thesis, Central South University, 2012
7. A.L. Chen, Z.W. Zhao, Comprehensive utilization status and expectation of zinc oxide ores. Min. Metall. Eng. **28**(6), 62–66 (2008)
8. Y.M. Zhang, J. Li, Influence of ultrasonic irradiation on ammonia leaching of zinc from low-grade oxide zinc ore. Chin. J. Nonferrous Met. **19**(5), 961–966 (2009)

Part IX
Poster I

A New Method to Detect the High Temperature Distribution in the Ironmaking and Steelmaking Industry

Dongdong Zhou and Shusen Cheng

Abstract The high temperature distribution in ironmaking and steelmaking industry is very important to control the smelting process. Although equipment is available to measure the temperature in the smelting process, in general it could just measure one certain point or only at discrete times. The real time temperature distribution in ironmaking and steelmaking is important because it could reflect reaction information and help to control the operation parameters. In this paper, a new method to detect the temperature distribution in ironmaking and steelmaking through the application of digital imaging system based on the color CCD was studied. In addition, the two-color method and partial least square method were used in the calculation model to improve the accuracy. The results agreed well with the measured temperature.

Keywords High temperature measurement · New method · Ironmaking and steelmaking

Introduction

In the ironmaking and steelmaking industry, the temperature of the reactor is very important to estimate reaction information and control the smelting processes. For instance, the temperature of the raceway zone as viewed through the peephole of the tuyère in a blast furnace is essential to reveal coal combustion information and the thermal state in the lower part of blast furnace, while the temperature of hot metal in the trough could reflect the thermal state of hearth and indirectly the silicon content. Moreover, the temperature of the converter mouth flame is also very important to reflect the smelting process. For the continuous casting process, knowledge of the temperature distribution would also aid control of the solidification process.

D. Zhou (✉) · S. Cheng
School of Metallurgical and Ecological Engineering, University of Science and Technology Beijing, Beijing 100083, People's Republic of China
e-mail: dongdongzhou@live.com

J.-Y. Hwang et al. (eds.), *8th International Symposium on High-Temperature Metallurgical Processing*, The Minerals, Metals & Materials Series,
DOI 10.1007/978-3-319-51340-9_49

The induction melting furnace is used widely in ironmaking and steelmaking industry for smelting and refining the metal. It has advantages of high smelting efficiency, homogeneous metal product, little smelting loss and rapid temperature increase [1]. It is crucial to control the liquid steel temperature in the induction melting furnace. However, the existing real-time device could not measure the liquid steel temperature accurately in induction melting furnace, because the emissivity and over saturation problem were not addressed well in those systems.

The temperature measurement can be separated into contact and non-contact methods. Thermocouples are a typical contact method, but the diameter and volume of crucible used in the furnace are small, and the smelting temperature is also very high, so the contact temperature measurement method is not suitable for long and real-time temperature measurement in experimental induction melting furnaces. Of non-contact methods, digital imaging based on the color CCD (Charge Coupled Device) is one of the effective methods applied in temperature measurement. The characteristic parameters of the radiation could be obtained from images which are captured by a vision-based radiation monitoring system. Then the temperature distribution of radiation images could be calculated.

Many authors have measured the temperature of pulverized coal combustion flame, heavy oil-fired combustion test facility and gas fired combustion systems by using digital imaging systems [2–6]. Specifically, Wen measured the surface temperature of aluminum alloy [7]. Zauner used a CCD camera to detect the temperature changes of workpiece during heat treatment [8]. Zhou measured the three dimensional temperature distribution of coal combustion radiation [9]. In order to improve the accuracy of the radiation temperature measured by digital imaging system, Cencerrado using an advanced vision based system to obtain both luminous and spectral parameters from the grey values registered by each individual pixel to improve the accuracy of calculation model [10]. Fleury used images grabbed from CCD cameras to build a fuzzy classification set of rules to increase the accuracy of calculation model [11]. Jiang developed a new method to obtain the temperature from a color radiation image with the assumption of gray radiation for particle-laden combustion processes [12].

From the above descriptions, it is clear that digital imaging systems based on color CCD could detect the temperatures in the ironmaking and steelmaking industry in real time. In this paper, the temperature of liquid steel in an experimental induction melting furnace was detected through the application of digital imaging and image processing techniques. In addition, the two-color method and partial least square method were used in the calculation model to improve the accuracy. The results agreed well with the measured temperature. The study will contribute to controlling the melting process at high temperature and improving the automation level in ironmaking and steelmaking industry.

Theory

Liquid steel radiation is described by the Wien radiant law [13]

$$M_\lambda(\lambda, T) = C_1\varepsilon(\lambda, T)\lambda^{-5}e^{-\frac{C_2}{\lambda T}} \tag{1}$$

where M_λ (λ,T) is the spectral power, ε(λ,T) is the effective spectral emissivity of radiation. λ is wavelength, T is temperature, and C_1 and C_2 are the first and second radiation constants of Planck.

Physical Calculation Model

According to the physical model [5], the simplified image formation process of radiation is shown in Fig. 1. In the figure, S represents the radiation area, S′ is the area of camera target, O is the lens, d is the distance between radiation and lens, and l is the distance between lens and CCD target.

The illuminance of the lens E could be represented as follows

$$E = \frac{\tau}{4}\left(\frac{D}{f}\right)^2 \int_{-\infty}^{+\infty} \frac{\varepsilon(\lambda, T)C_1}{\lambda^5 e^{\frac{C_2}{\lambda T}}} d\lambda \tag{2}$$

where τ is the transmissivity of lens, A is the area of the lens. The CCD camera is an integral type device; the value of output electricity signal has linear relationship with illuminance of the lens and exposure time.

$$I = \mu Et \tag{3}$$

where I is the output electrical signal of CCD, μ is the photoelectric conversion coefficient, and t is exposure time. For the color CCD camera, the gray level in the radiation is given by

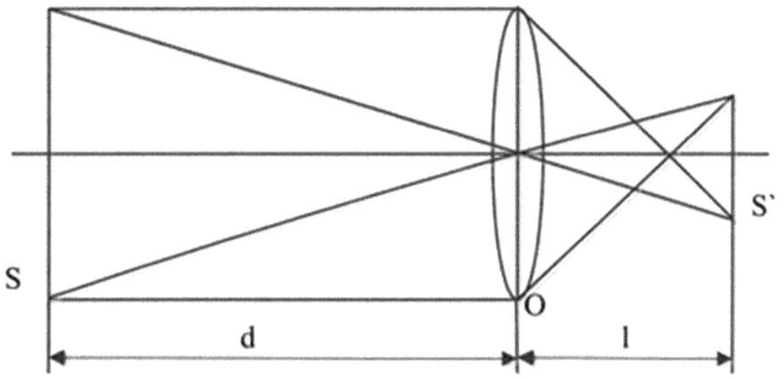

Fig. 1 Simplified image formation process

$$\begin{cases} H_R = \frac{\tau}{4} K_R \mu_R t_R \eta_R V(\lambda_R) \left(\frac{D}{f}\right)^2 \frac{\varepsilon(\lambda_R, T) C_1}{\lambda_R^5 e^{\frac{C_2}{\lambda_R T}}} \\ H_G = \frac{\tau}{4} K_G \mu_G t_G \eta_G V(\lambda_G) \left(\frac{D}{f}\right)^2 \frac{\varepsilon(\lambda_G, T) C_1}{\lambda_G^5 e^{\frac{C_2}{\lambda_G T}}} \\ H_B = \frac{\tau}{4} K_B \mu_B t_B \eta_B V(\lambda_B) \left(\frac{D}{f}\right)^2 \frac{\varepsilon(\lambda_B, T) C_1}{\lambda_B^5 e^{\frac{C_2}{\lambda_B T}}} \end{cases} \tag{4}$$

where H_R, H_G, H_B are the gray level of R, G, B. K_R, K_G, K_B are the transmittance of lens in the ranges wavelength of R, G, B. η_R, η_G, η_B are the conversion coefficient of gray level and output electricity signal of the CCD. $V(\lambda_R)$, $V(\lambda_G)$, $V(\lambda_B)$ are the spectral response function of R, G, B. λ_R, λ_G, λ_B are the wavelength of R, G, B. In the equation, the transmittance of lens, the wavelength, the spectral response function, the exposure time and the photoelectric conversion coefficient of R, G, B are determined by the CCD camera. The diameter and focal length of lens are also known.

The coefficient of the calculation model N_R, N_G, N_B can be written as

$$\begin{cases} N_R = \frac{\tau}{4} K_R \mu_R t_R \eta_R V(\lambda_R) \left(\frac{D}{f}\right)^2 \\ N_G = \frac{\tau}{4} K_G \mu_G t_G \eta_G V(\lambda_G) \left(\frac{D}{f}\right)^2 \\ N_B = \frac{\tau}{4} K_B \mu_B t_B \eta_B V(\lambda_B) \left(\frac{D}{f}\right)^2 \end{cases} \tag{5}$$

In the Eq. 4, the gray level can be calculated by the digital imaging processing software, the coefficient of the CCD camera is decided by CCD monitoring system, and emissivity of R, G, B can be calibrated by the blackbody furnace and the measured temperature of the radiation.

Calculation Method

In order to calculate the temperature of the radiation image, the two-color method was used. The expression for radiation temperature using the two-color method [13] can be written as

$$\frac{H_{\lambda_1}}{H_{\lambda_2}} = \frac{N_{\lambda_1}}{N_{\lambda_2}} \frac{\varepsilon(\lambda_1, T)}{\varepsilon(\lambda_2, T)} \frac{\lambda_1^{-5}}{\lambda_2^{-5}} \exp\left[\frac{-C_2}{T}\left(\frac{1}{\lambda_1} - \frac{1}{\lambda_2}\right)\right] \tag{6}$$

The radiation is assumed as the graybody

$$T = -C_2\left(\frac{1}{\lambda_1} - \frac{1}{\lambda_2}\right) / \left(In\frac{H_{\lambda_1}}{H_{\lambda_2}} - In\frac{N_{\lambda_1}}{N_{\lambda_2}} - In\frac{\varepsilon(\lambda_1, T)}{\varepsilon(\lambda_2, T)} + 5In\frac{\lambda_1}{\lambda_2}\right) \tag{7}$$

In Eq. 7, the ratio of two wavelengths emissivity is equal to 1. Consequently, the temperature of radiation can be expressed as

$$T = -C_2\left(\frac{1}{\lambda_1} - \frac{1}{\lambda_2}\right) / \left(In\frac{H_{\lambda_1}}{H_{\lambda_2}} - In\frac{N_{\lambda_1}}{N_{\lambda_2}} + 5In\frac{\lambda_1}{\lambda_2}\right) \quad (8)$$

In Eq. 8, the wavelengths of R, G and B are already known. The gray level of the two wavelengths can be obtained from the radiation images. The coefficients in the calculation model can be calibrated with a blackbody furnace.

Experiments and Results

Experiments

The simplified diagram of the digital imaging system based on the color CCD applied in the experiment induction melting furnace is shown in Fig. 2. The induction melting furnace consists of a furnace body, control cabinet, cooling water system and hydraulic system. In the furnace body, a crucible is used to contain the materials which should be melted to generate radiation. In this paper, solid steel was melted with slag. The medium frequency power is 60 kW, the rated temperature is 1780 °C, while the intermediate frequency is 4000 Hz. The inner diameter of the crucible was 100 mm, and the inner height was 180 mm. The digital imaging system used in this paper consisted of DAHENG CCD, lens, enclosure and computer. The parameters of the CCD are shown in Table 1, while the spectral response function of CCD camera is shown in Fig. 3. The focal length was 50 mm. The direct distance between the CCD and the surface of the crucible was about 550 mm, while the height is about 420 mm.

The radiation image saved in the computer was grey level, which is the total radiation intensity in Eq. 4. The relationship between the grey level and radiation temperature can be determined by the blackbody calibration method, as shown in

Fig. 2 Simplified diagram of digital imaging system in the experimental induction melting furnace

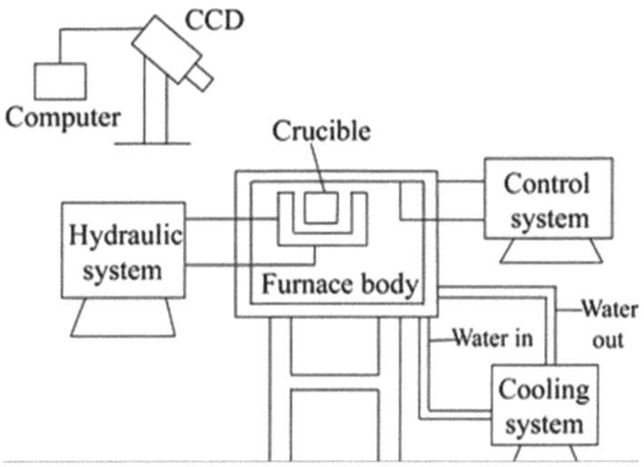

Table 1 Parameters of the CCD camera

Type	MER-125-30UM/UC
Image resolution	1292(H) × 964(V)
Frame rate	30 fps
Sensor	1/3′ Sony ICX445 CCD
Pixel size	3.75 μm × 3.75 μm
Spectrum	Color
Data interface	Mini USB2.0
Lens mount	C

Fig. 3 Spectral response function of CCD camera

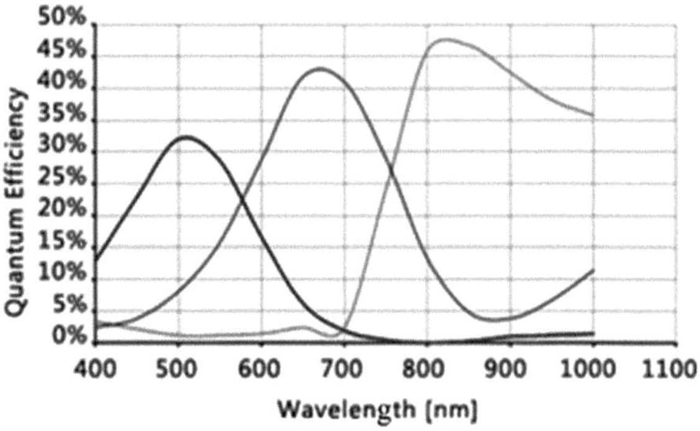

Fig. 4 Experimental blackbody calibration system, *a* temperature control modulation *b* blackbody furnace *c* CCD

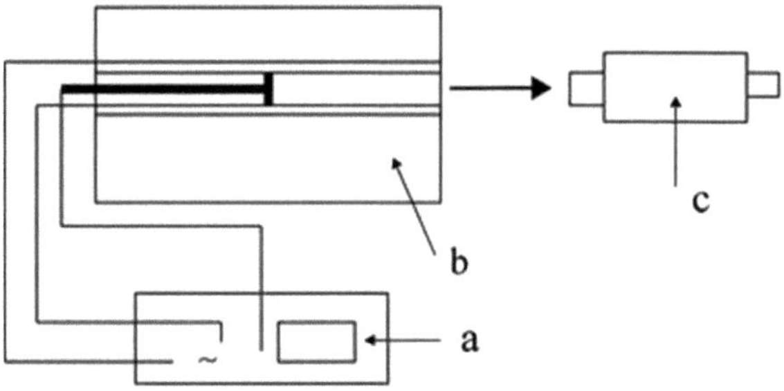

Fig. 4. A blackbody furnace was used to supply the radiation intensities at different temperatures, over the temperature range 1300–2000 °C, with steps of 50 °C. The CCD camera recorded the image brightness, and then the relationship between gray level and radiation temperature could be obtained.

Results

The temperature distribution of the liquid steel in the crucible of the experimental induction melting furnace is shown in Fig. 5a–c are radiation images in the early, medium and later stage of smelting process, while d–f are temperature distributions

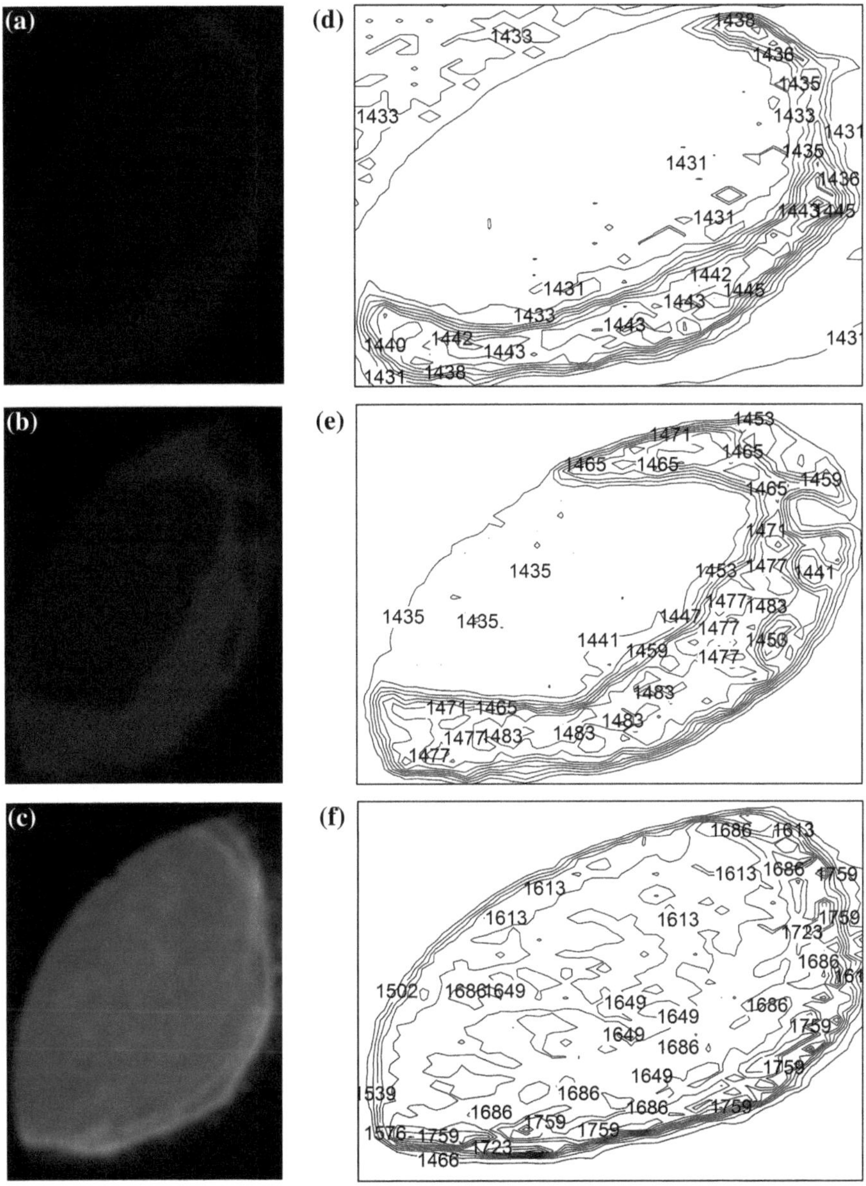

Fig. 5 Images and temperature distributions at different stages in the experimental induction melting furnace **a** image of case A **b** image of case B **c** image of case C **d** temperature distribution of case A **e** temperature distribution of case B **f** temperature distribution of case C

calculated by the model. The calculated temperatures were 1432.51, 1482.95 and 1651.01 °C, respectively. In the earlier stage, the maximum and minimum temperature are 1447.01 and 1432.46 °C, respectively. In the mid-term stage, the maximum and minimum temperature are 1489.08 and 1444.68 °C, respectively. In the later stage, the maximum and minimum temperature are 1759.65 and 1571.08 °C, respectively. In addition, the dark part in Fig. 5a, b is the slag on the surface of the liquid steel; the temperature in the area of that part were didn't calculated in the model.

In order to validate the calculated results, the temperature of the liquid steel in the crucible was measured with a thermocouple. The values were 1442.36, 1482.95 and 1651.01 °C in the earlier, mid-term and later stage, respectively. The relative errors were 0.68, 1.002 and 0.54%, respectively. The calculated temperatures agreed well with the measured ones. From the above analysis, it is clearly shown that the approach based on the digital imaging system could used to detect the high temperature distribution in the ironmaking and steelmaking industry.

Conclusions

The temperature in the ironmaking and steelmaking industry is important to provide melting information and to control the reaction process, but the traditional temperature measurement could not provide the real-time temperature monitoring process accurately. In this paper, a new method based on a digital imaging system is used to detect the liquid steel temperature in an experimental induction melting furnace. The calculated results agreed well with the measured ones. The study will help to control high-temperature melting processes and to improve automation in ironmaking and steelmaking.

Acknowledgements The authors would gratefully acknowledge the financial support from the National Natural Science Foundation of China (No. 61271303).

References

1. K. Kimura, M. Ishii, H. Yoshimura, An analysis of deformation, temperature, and microstructure for hot extruded titanium alloy. Metall. Trans. B **1**(24), 139–144 (1993)
2. G. Lu et al., Concurrent measurement of temperature and soot concentration of pulverized coal radiations. IEEE Trans. Instrum. Meas. **5**(51), 990–995 (2002)
3. C. Lou et al., Measurements of the radiation emissivity and radiation properties of particulate medium in pulverized-coal-fired boiler furnaces by image processing of visible radiation. Proc. Combust. Inst. **2**(31), 2771–2778 (2007)
4. F. Jiang et al., Visual radiation monitoring system based on two-color method. J. Therm. Sci. **3**(18), 284–288 (2009)
5. F. Wang, J. Yan, K. Cenl, Simultaneous measurements of two-dimensional temperature and particle concentration distribution from the image of the pulverized-coal radiation. Fuel **1**(89), 202–211 (2010)
6. T.S. Draper, D. Zeltner, D.R. Tree, Two-dimensional radiation temperature and emissivity measurements of pulverized oxy-coal radiations. Appl. Energy **95**, 38–44 (2012)

7. C. Wen et al., Emissivity characteristics of roughened aluminum alloy surfaces and assessment of multispectral radiation thermometry emissivity models. Int. J. Heat Mass Transf. **7**(48), 1316–1329 (2005)
8. G. Zauner, D. Heim, CCD cameras as thermal imaging devices in heat treatment processes. in *Proc Spie,* (2004), pp. 81–89
9. H. Zhou, S. Sheng, D.S. Han, Reconstruction of temperature distribution in a 2-D absorbing-emitting system from radiant energy images. JSME Int. J. **1**(43), 104–109 (2000)
10. A.G. Cencerrado, B. Peña, A. Gil, Coal radiation characterization by means of digital image processing in a semi-industrial scale PF swirl burner. Appl. Energy **6**(94), 375–384 (2012)
11. A.T. Fleury, F.C. Trigo, F.P.R. Martins, A new approach based on computer vision and non-linear Kalman filtering to monitor the nebulization quality of oil radiations. Expert Syst. Appl. **12**(40), 4760–4769 (2013)
12. Z.W. Jiang, Z.X. Luo, H.C. Zhou, A simple measurement method of temperature and emissivity of coal-fired radiations from visible radiation image and its application in a CFB boiler furnace. Fuel **6**(88), 980–987 (2009)
13. L. Xu et al., Comparative study of regression modeling methods for online coal calorific value prediction from radiation radiation features. Fuel **142**, 164–172 (2015)

A Study for Reconstructing the Three-Dimensional Temperature Field of a Blast Furnace Raceway Based on Monte Carlo Method

Yan Li, Shusen Cheng and Ruixuan Zhang

Abstract The measurement of raceway temperature in the blast furnace is a critical issue, which is very significant for judgment of, prediction and diagnosis of the raceway burning state. This paper, deals with mathematical models by Monte Carlo method to approximately simulate the radiation heat transfer in the race-way and establish matrix equations using CCD cameras, which are solved to obtain the three-dimensional temperature field. Due to the existence of fluctuation error, CCD camera measurement error and others, this is analyzed and random error is added in the measurement data to demonstrate the effectiveness and stability of the reconstructed temperature field. The results show that the reconstructed three-dimensional temperature field of the raceway is very close to the exact temperature.

Keywords Three-dimensional temperature · Raceway · Monte carlo method · CCD camera

Introduction

Blast furnace (BF) is the most widely used technology to produce iron, where raceway is an important region. The measurement of raceway temperature is a critical issue, which is significant for judgment, prediction and diagnosis of combustion in the raceway. The most common methods for temperature measurement are thermocouples and pyrometers. But these two methods cannot obtain and monitor the 2-D and 3-D temperature distributions in the high temperature raceway region of the blast furnaces for a long period. Thermal radiation is the dominant heat transfer mode in the high temperature devices such as combustion chambers and furnaces. Inverse radiation analysis is an effective method to obtain the temperature distributions in medium by the non-intrusive way and has been paid

Y. Li (✉) · S. Cheng · R. Zhang
School of Metallurgical and Ecological Engineering, University of Science and Technology Beijing, Beijing 100083, China
e-mail: s20150196@xs.ustb.edu.cn

J.-Y. Hwang et al. (eds.), *8th International Symposium on High-Temperature Metallurgical Processing*, The Minerals, Metals & Materials Series,
DOI 10.1007/978-3-319-51340-9_50

increasing attention to in recent years. Especially, Monte Carlo method [1] is popularly applied to radiative heat transfer and reconstructing temperature field. The comprehensive review of Monte Carlo method in radiative heat transfer has been given by Howell [2]. Zhou et al. [3, 4] and Lou [5] have numerically and experimentally reconstructed 2-D and 3-D temperature distribution of boiler furnaces using CCD cameras based on the Monte Carlo method. Wang [6] and Liu [7, 8] rebuilt the fired furnace temperature field based on the Monte Carlo method and the reconstruction matrix equations were solved by the least square QR decomposition(LSQR) method [9].

In this paper, Monte Carlo method was employed to describe the radiative energy propagation in the blast furnace raceway. Reconstructed 3-D temperature field of the blast furnace raceway using CCD cameras based on the forward and backward Monte Carlo methods. And the numerical reconstructions were utilized to validate the inverse method. It would be useful to get the temperature of raceway only by CCD cameras mounted at the blast furnace peephole and achieve real time monitoring.

Temperature Reconstruction Technique

In order to calculate conveniently, raceway region is assumed as three-dimensional rectangular enclosure containing inhomogeneous, absorbing, emitting and anisotropically scattering medium. Width is W, height is H, and depth is L. The grid system schematic diagram raceway shown in Fig. 1a. The three-dimensional raceway system is divided into volume elements and wall elements. Hypothetically, the rectangular enclosure is divided into 7 rows (x axis), 6 columns (y axis), 8 layers (z axis). In order to obtain the exit radiation information, one CCD camera is used and mounted at the blast furnace peephole, and blast furnace tuyere section is chosen as the CCD imagined object plane, as shown in Fig. 1b. The Monte Carlo method is employed to describe the radiative heat transfer in the blast furnace raceway.

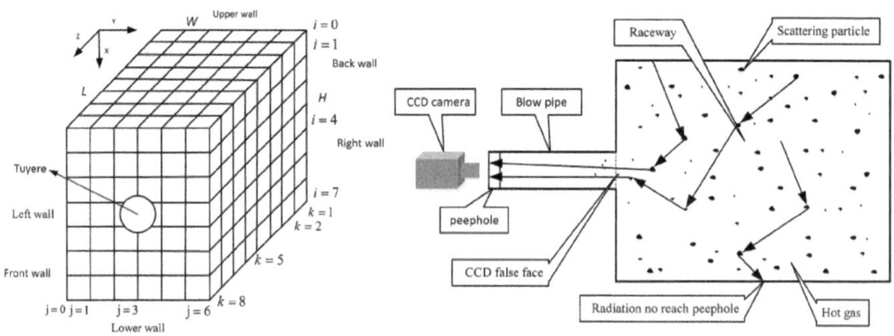

(a) three-dimensional grid system (b) raceway radiative heat transfer

Fig. 1 Schematic diagram raceway

In order to simplify radiative heat transfer in the raceway, wall is diffuse reflection gray body; radiative absorptivity of the wall is equal to its emissivity; absorption coefficient, scattering coefficient and the temperature of each volume element are equal; radiative absorptivity of each wall element is equal, hypothetically.

Reconstructing Model Based on the Forward Monte Carlo Method

Let N be the total number of elements of the raceway system containing n volume elements and m wall elements. The tuyere CCD camera target plane is be divided into M pixel elements. The forward Monte Carlo method is employed to describe the radiative heat transfer and trace bundles emitted by elements to get distribution of raceway radiative energy. Then, M equations can be obtained:

$$\begin{cases} \sum\limits_{i=1}^{n} Rd(i \to 1)4K_i\Delta V_i T_{Ei} + \sum\limits_{i=n+1}^{N} Rd(i \to 1)\varepsilon_{wi}\Delta S_{wi} T_{Ei} = P_1 \\ \vdots \\ \sum\limits_{i=1}^{n} Rd(i \to j)4K_i\Delta V_i T_{Ei} + \sum\limits_{i=n+1}^{N} Rd(i \to j)\varepsilon_{wi}\Delta S_{wi} T_{Ei} = P_j \\ \vdots \\ \sum\limits_{i=1}^{n} Rd(i \to j)4K_i\Delta V_i T_{Ei} + \sum\limits_{i=n+1}^{N} Rd(i \to j)\varepsilon_{wi}\Delta S_{wi} T_{Ei} = P_M \end{cases} \tag{1}$$

where K_i is the absorption coefficient of volume element i, ΔV_i is the volume of volume element i, ε_{wi} is the emissivity of wall element i, ΔS_{wi} is the area of wall element i, P_j is the radiative energy received by pixel element j. $Rd(i \to j)$ is the radiative energy fraction received by pixel j on the CCD target plane of total energy emitted by element i, calculated by forward Monte Carlo method. i = 1,2,…,N; j = 1,2,…,M.

Within the response wavelength range (380–780 nm) of CCD camera, T_E is blackbody visible light radiative energy and calculated by Eq. (2). The relationship between T_E and T is shown in Fig. 2.

$$T_E = \int_{0.38}^{0.78} \frac{c_1\lambda^{-5}}{\exp[c_2/\lambda T] - 1} d\lambda \tag{2}$$

where λ is the wavelength (μm), T is the temperature of elements, c_1 and c_2 are the first and second radiation constants, respectively. $c_1 = 3.743 \times 10^8 (\mathrm{W}\ \mu\mathrm{m}^4)/\mathrm{m}^2$, $c_2 = 1.4387 \times 104$ μm K.

Rewrite Eq. (2) in the matrix equation form,

$$\boldsymbol{A}\boldsymbol{T}_E = \boldsymbol{P} \quad \boldsymbol{A} \in \boldsymbol{R}^{M \times N}, \boldsymbol{T}_E \in \boldsymbol{R}^N, \boldsymbol{P} \in \boldsymbol{R}^M \tag{3}$$

where $\boldsymbol{A}$ is the coefficient matrix and can be obtained through forward Monte Carlo method, $\boldsymbol{T}_E$ is determined by Eq. (2), $\boldsymbol{P}$ is the radiative energy vector and can be obtained from pixels on the CCD target plane of the tuyere.

Solve Eq. (3) for the unknown $\boldsymbol{T}_E$ based on LSQR. Once $\boldsymbol{T}_E$ is obtained through inverse solution, three-dimensional temperature field of blast furnace raceway can be immediately obtained according to Fig. 2.

Reconstructing Model Based on the Backward Monte Carlo Method

In order to get meaning result using fewer bundles, Michael [10] put forward backward Monte Carlo method. Principles of backward Monte Carlo method are similar to forward Monte Carlo method, but emitting source of bundles is the CCD camera target instead of raceway system elements. The raceway system is divided into N volume elements and wall temperature is equal in this model. At the CCD camera imagined object plane, M pixel elements of emitted bundles were assumed with S bundles into each direction. Based on the backward Monte Carlo method, M equations can be obtained:

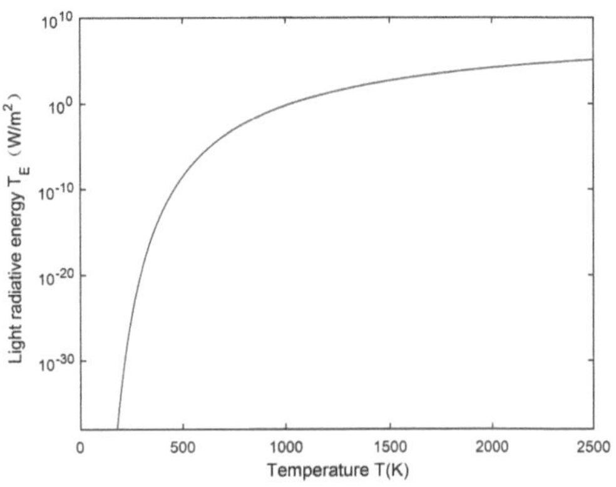

Fig. 2 Standard TE under different temperatures

$$\begin{cases} \frac{1}{S}\sum_{s=1}^{S}\left[\sum_{i=1}^{N}(K_i T_{Ei} l_{1si}) + w_{1s}\varepsilon_w T_{Ew}\right] = P_1 \\ \vdots \\ \frac{1}{S}\sum_{s=1}^{S}\left[\sum_{i=1}^{N}(K_i T_{Ei} l_{jsi}) + w_{js}\varepsilon_w T_{Ew}\right] = P_j \\ \vdots \\ \frac{1}{S}\sum_{s=1}^{S}\left[\sum_{i=1}^{N}(K_i T_{Ei} l_{Msi}) + w_{Ms}\varepsilon_w T_{Ew}\right] = P_M \end{cases} \tag{4}$$

where l_{jsi} is the path length in the volume element i for the bundle s of pixel element j, w_{js} denotes whether the bundle of number s of pixel element j is terminated by the enclosure wall (i.e., if the bundle is ended by the wall, its value is 1; otherwise, its value is 0). i = 1, 2,…,N; j = 1,2,…,M; s = 1,2,…,S.

Rewriting Eq. (4) in the matrix equation form,

$$\boldsymbol{A}\boldsymbol{T}_E = \boldsymbol{P} \quad \boldsymbol{A} \in \boldsymbol{R}^{M\times N}, \boldsymbol{T}_E \in \boldsymbol{R}^{N+1}, \boldsymbol{P} \in \boldsymbol{R}^{M} \tag{5}$$

where $\boldsymbol{A}$ is the coefficient matrix and can be obtained through backward Monte Carlo method.

The calculation procedures of Eq. (5) based on the backward Monte Carlo method are same as the forward Monte Carlo method model.

Results and Discussion

Errors could not only exist in radiative energy vector obtained by CCD cameras but also exist in the coefficient matrix. So random errors of normal distribution with zero average value and mean square deviation σ_1 and σ_2 were added to the energy vector $\boldsymbol{P}$ and the coefficient matrix $\boldsymbol{A}$ in Eqs. (6) and (7), respectively.

$$P_{\text{measure},j} = (\mu + \sigma_1\xi)P_j + P_j \tag{6}$$

$$A_{err,i} = (\mu + \sigma_2\xi)A_i + A_i \tag{7}$$

where $P_{measure,j}$ is the element of the energy vector $\boldsymbol{P}$ with errors and $A_{err,i}$ is the element of coefficient matrix $\boldsymbol{A}$ with errors. P_j is the element of the exact energy vector $\boldsymbol{P}$ and A_i is the element of exact coefficient matrix $\boldsymbol{A}$. Average value μ equals zero, ξ is a random variable of standard normal distribution and the probability lying in the range of $-2.576 < \xi < 2.576$ is 99%.

To examine the accuracy of the methods developed, three-dimensional temperature field was adopted as the exact raceway temperature in Eq. (8), and exact temperature of each element is shown in Fig. 3. In the backward Monte Carlo

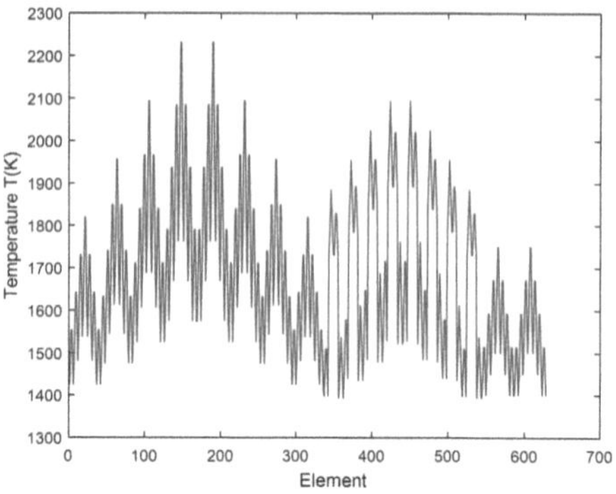

Fig. 3 Exact three-dimensional temperature field

method model, the exact temperature of all wall elements was assumed as 1500 K. Absorption coefficient equals to 0.3, scattering coefficient equals to 0.1 and emissivity equals to 0.5. H, W, L of blast furnace raceway equal to 2.1, 1.8 and 2.4 m, respectively. And tuyere diameter of blast furnace equals to 0.2 m. The CCD target plane of tuyere was divided into 26 × 26 pixel elements. The number of bundles is 2×10^6. Due to the use of random errors adding to the exact solution of direct problem, all the reconstruction results are the average results of 1000 samples.

$$T(x,y,z) = 1200\left[\left(1 - \frac{|x - H/2|}{H}\right)\left(1 - \frac{|y - W/2|}{W}\right)\left(1 - \frac{|z - L/2|}{L}\right) + 1\right] \quad (8)$$

The reconstruction error for the temperature field and the relative error for the temperature in each volume element are defined as

$$\delta_{re} = 100\sqrt{\frac{1}{N}\sum\nolimits_{i=1}^{N}(T_i^{re} - T_i^{ex})^2} \Big/ \left(\frac{1}{N}\sum\nolimits_{i=1}^{N} T_i^{ex}\right) \quad (9)$$

$$\delta_{re,i} = 100\frac{\left|T_i^{re} - T_i^{ex}\right|}{T_i^{ex}} \quad (10)$$

where T_i^{re} and T_i^{ex} represent reconstructed and exact temperature respectively. i = 1, 2,…,N.

Reconstruction Accuracy of Two Monte Carlo Methods

In this case, it is assumed that the coefficient matrix $\boldsymbol{A}$ had no fluctuating errors and only measured exit radiative energy vector $\boldsymbol{P}$ had errors. Temperature fields were reconstructed by forward and backward Monte Carlo method models. The reconstruction errors with various mean square deviations for models are shown in Fig. 4 and for comparison; σ = 0, 0.005, 0.01, 0.02, 0.03, 0.05, 0.07, 0.1. It can be seen

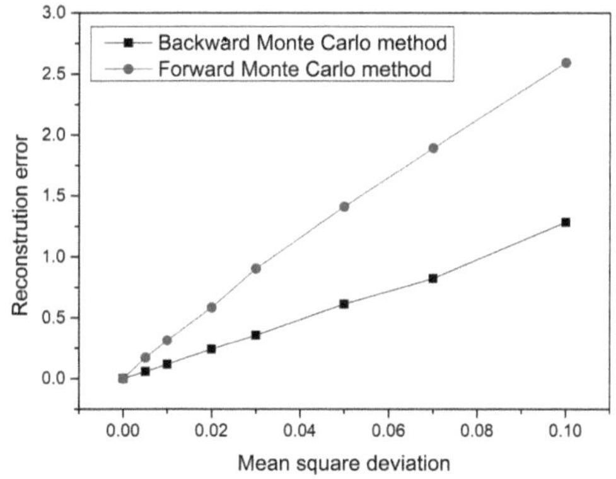

Fig. 4 Reconstruction errors for two Monte Carlo methods

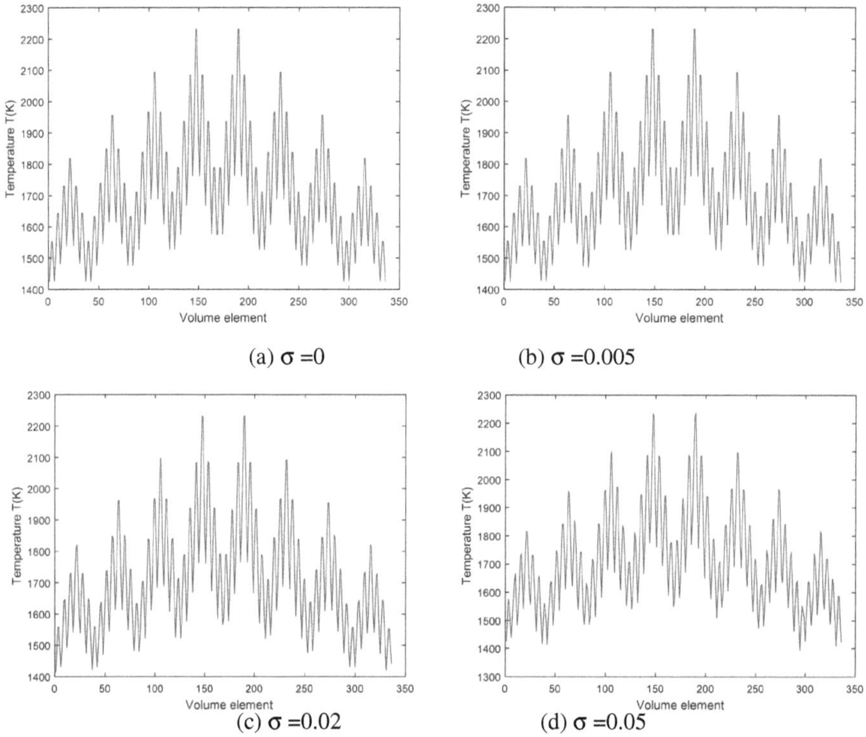

Fig. 5 Reconstructed temperature fields for different measurement errors

that reconstruction errors are low, and smaller than 3% when mean square deviation equals 0.1. So, the reconstructed three-dimensional temperature field of the raceway is very close to the exact temperature, based on the both Monte Carlo methods. What's more reconstruction errors of forward Monte Carlo method model are larger and increase more quickly than backward Monte Carlo method model with radiative energy vector errors. What's more, the forward Monte Carlo method computed time was little more than backward Monte Carlo method. Therefore, the reconstruction result based on backward Monte Carlo method is more accurate and get easy in the same conditions.

Reconstruction Results with Exact and Noisy Data

In this, it is assumed that only measured exit radiative energy vector $\boldsymbol{P}$ had errors in the backward Monte Carlo method model. The effects of measurement errors on the accuracy of estimation were examined.

Reconstructed temperature of each volume elements by backward Monte Carlo method is illustrated in Fig. 5 with no measurement errors and with mean square deviation of 0.005, 0.02, 0.05. Only few differences can be observed with mean square deviation. The relative errors for each volume element have also been calculated and shown in Fig. 6. It can be seen that the relative errors with no measurement errors are very small. Most relative errors increase with mean square

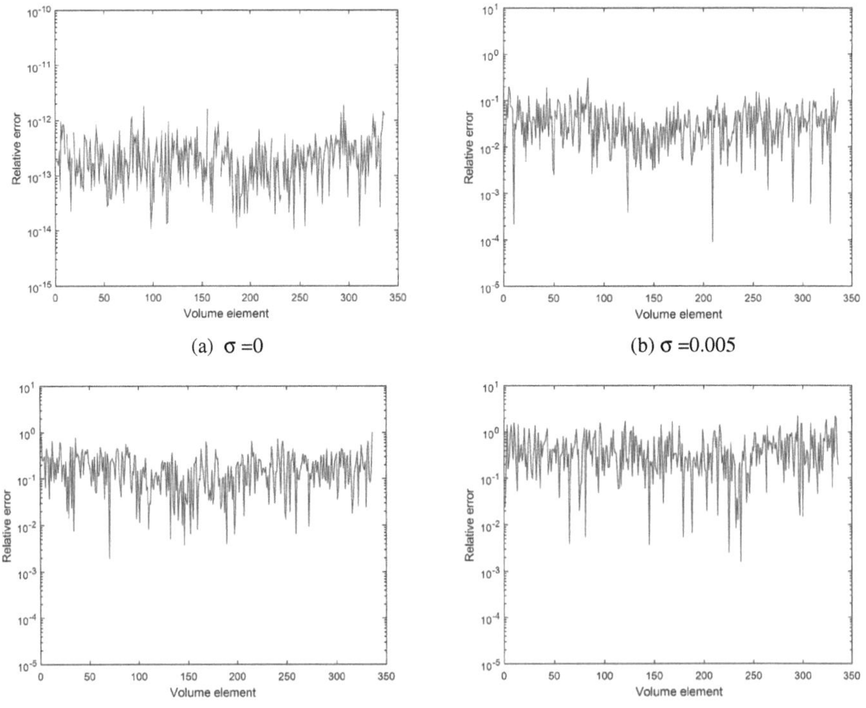

(a) σ =0 (b) σ =0.005

(c) σ =0.02 (d) σ =0.05

Fig. 6 Relative error for each volume element

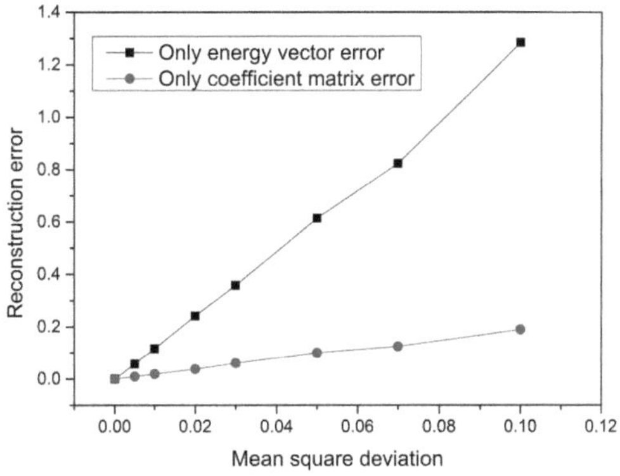

Fig. 7 Effects of two kinds of errors separately on the accuracy of estimation

deviation become larger but can be maintained at a very low level. Few points cannot be reconstructed well because the temperature values of these elements were lower and reconstruction for these elements can be influenced more easily by measurement errors. The overall reconstruction results indicate that the inverse backward Monte Carlo method can obtain accurate and reasonable temperature fields, even with noisy input data.

Effects of Coefficient Matrix and Radiative Energy Vector Errors on Reconstruction Accuracy

It is assumed that coefficient matrix $\boldsymbol{A}$ and radiative energy vector $\boldsymbol{P}$ had errors based on the backward Monte Carlo method. The reconstruction errors with various mean square deviations are shown in Fig. 7 and for comparison. As the increase of mean square deviation for coefficient matrix $\boldsymbol{A}$, the reconstruction errors for 3-D temperature fields are smaller and increase slower compared with that of radiative energy vector $\boldsymbol{P}$. This reveals that the effects of coefficient matrix $\boldsymbol{A}$ errors could be less compared with the effects of the measurement errors on the accuracy of estimation. Moreover, Fig. 7 also implies that, in order to improve the accuracy of the reconstruction of the temperature field, the measurement errors of the exit radiative energy of the CCD target plane should be confined within an appropriate range.

Conclusions

Two Monte Carlo methods are presented for determining the three-dimensional temperature field in blast furnace raceway of known radiative properties from the knowledge of the exit radiative energy received by CCD camera at peephole. In this study, the Monte Carlo method is employed to describe the radiative energy propagation and two reconstruction models of three-dimensional temperature field are built. The results show that the reconstructed three-dimensional temperature field of the raceway is very close to the exact temperature, with simulated random errors. And backward Monte Carlo method is preferential. Although the temperature is a function of space variables, only the measurements of exit radiative energy at the blast furnace peephole are required. Both measurement errors and coefficient fluctuating errors are considered. The results show that the present technique is robust and yields accurate three-dimensional temperature field of a blast furnace raceway for the exact and noisy input data.

References

1. W.J. Yang et al., Radiative heat transfer by the Monte Carlo method. Adv. Transf. **27**, 1–215 (1995)
2. J.R. Howell, The Monte Carlo method in radiative heat transfer. J. Heat Transf. **120**, 547–560 (1998)
3. H.C. Zhou, S.D. Han, et al, Visualization of three-dimensional temperature distributions in a large-scale furnace via regularized reconstruction from radiative energy images: numerical studies. **72**, 361–83 (2002)
4. H.C. Zhou, C. Lou et al., Experimental investigations on visualization of three-dimensional temperature distributions in a large-scale pulverized-coal-fired boiler furnace. Proc. Combust. Inst. **30**(1), 1699–1706 (2005)
5. C. Lou, H.C. Zhou, Deduction of the two-dimensional distribution of temperature in a cross section of a boiler furnace from images of flame radiation. Combust. Flame **143**(1–2), 97–105 (2005)
6. F. Wang, D. Liu et al., Efficient inverse radiation analysis of temperature distribution in participating medium based on backward Monte Carlo method. JQSRT **109**, 2171–2181 (2008)
7. D. Liu, F. Wang, J.H. Yan, Inverse radiation problem of temperature field in three-dimensional rectangular enclosure containing inhomogeneous, anisotropically scattering media. Int. J. Heat Mass Transf. **51**(13–14), 3434–3441 (2008)
8. D. Liu, J. Yan, F. Wang, Experimental reconstructions of flame temperature distributions in laboratory-scale and large-scale pulverized-coal fired furnaces by inverse radiation analysis. Fuel **93**, 397–403 (2012)
9. C.C. Paige, M.A. Saunders, LSQR: an algorithm for sparse linear equations and sparse least squares. AMC Trans. Math. **8**(1), 43–71 (1982)
10. F.M. Michael, Backward Monte Carlo simulations in radiative heat transfer. J. Heat Transf. **125**, 57–62 (2003)

Behaviour of Silicon in Nickel Laterite by Carbothermic Reduction in Vacuum

Lei Shi, Tao Qu, Dachun Liu, Yang Tian, Bin Yang and Yongnian Dai

Abstract The carbonthermal reduction of high Mg–Si nickel laterite was studied under vacuum. The effects of adding calcium oxide and the reducing agent on the reduction of silicon in residue were studied at pressures of 10–200 Pa and different temperatures, with varying proportions of carbon and CaO. The analysis of the residue was performed using X-ray diffraction and chemical composition analysis. The results show the following: At 1350 °C with a C/O ratio of 1.5:1, the recovery of Si reached a maximum 71.09%. Under these conditions, Si exists mainly in the forms of Fe_3Si, Fe_2Si, FeSi and SiC. The recovery of Si was above 80% when the temperature was 1450 °C, the C/O ratio 1:1, and 15% addition of CaO. When CaO addition was 22.61%, all of the solid phase was Ca_2SiO_4.

Keywords Nickel laterite · Vacuum metallurgy · Carbothermal reduction · Silicon behaviour

L. Shi · T. Qu (✉) · D. Liu · Y. Tian · B. Yang · Y. Dai
The National Engineering Laboratory for Vacuum Metallurgy, Kunming University of Science and Technology, Kunming 650093, People's Republic of China
e-mail: qutao_82@126.com

L. Shi
e-mail: 1421999318@qq.com

T. Qu · D. Liu · Y. Tian · B. Yang · Y. Dai
State Key Laboratory of Complex Nonferrous Metal Resources Clear Utilization, Kunming 650093, People's Republic of China

L. Shi · T. Qu · D. Liu · Y. Tian · B. Yang · Y. Dai
Key Laboratory for Nonferrous Vacuum Metallurgy of Yunnan Province, Kunming 650093, People's Republic of China

L. Shi
Faculty of Metallurgical and Energy Engineering of Kunming University of Science and Technology, Kunming 650093, People's Republic of China

J.-Y. Hwang et al. (eds.), *8th International Symposium on High-Temperature Metallurgical Processing*, The Minerals, Metals & Materials Series,
DOI 10.1007/978-3-319-51340-9_51

Introduction

Nickel plays an important role in a variety of industries, including the military, aerospace and machinery manufacturing [1]. The global consumption of nickel has increased, and large-scale exploitation of nickel resources has depleted nickel reserves, particularly sulphide ore resources. Hence in recent years, the focus is on the development of nickel laterite resources [2].

Current research into the development in the use of laterite ore resources can be divided into two processes: pyrometallurgical and leaching processes. The leaching processes include sulphuric acid pressure leaching [3, 4], atmospheric heap leaching [5], stir atmospheric leaching [6], bioleaching [7], sulphuric acid roasting-flooding [8], and reduction—ammonia leaching [9]. However, the current processes are primarily focused on the extraction of nickel. The recovery of other valuable metals with these processes has not been explored. Consequently, these processes do not effectively utilize nickel laterite resources [2].

Kunming University of Science and Technology studied the use of a reducing agent in thermal reductions in a vacuum [10, 11]. This study proposes a carbon thermal reduction method to treat Yuanjiang nickel laterite in a vacuum at high temperatures. This method can volatilize magnesium from the nickel laterite. Condensation is facilitated by controlling the collection of Ni, Fe, Si enrichments in the slag; the post-processing of slag allows for the comprehensive utilization of resources [12]. This paper studies the influence of reducing agents and additives on the behaviour of silicon in Yuanjiang nickel laterite during carbothermic reduction in a vacuum. During the experiment, the volatiles were difficult to collect. This paper analyses the behaviour of silicon in a reducing residue.

Experimental

Materials Used in the Experiments

The chemical composition of nickel laterite is shown in Table 1. The chemical composition of coal is shown in Table 2. The chemical composition of CaO is shown in Table 3.

Table 1 Chemical composition of nickel laterite (wt%)

Elements	Ni	MgO	Fe	SiO_2	Al_2O_3
Content	1.18	27.74	11.10	36.91	2.73

Table 2 Chemical composition of coal (wt%)

Elements	C	Volatile (V)	Ash (A)	Water (W)	Generate heat (cal/g)
Coal	47.14	21.00	30.65	1.21	5261.5

Table 3 Chemical composition of CaO (wt%)

CaO	Drying loss	Nitrogen compounds	Heavy metal	Fe	Silicate	Chloride	Si
≥98.5	≤0.4	≤0.005	≤0.003	≤0.003	≤0.05	≤0.01	≤0.01

Experimental Procedure

1. The main equipment used in the experiment was a vacuum furnace shown in Fig. 1.
2. Then, nickel laterite was placed in a resistance oven, and dried at 900 °C for 60 min, milled and sieved with −200 mesh, and bagged.
3. If CaO was not used, coke was added with the following molar ratios of C:Si: 1:1, 1.2:1, and 1.5:1. When CaO was used, the molar ratio of C and Si was 1:1. The mass ratio of nickel laterite: coal was 100:65.5. Different mass percentages of CaO used were as follows: 5, 10, 15, 20 and 22.64%. When the Si was present as fully formed Ca_2SiO_4, the mass percentage of CaO was 22.64%.
4. Approximately 10 g of dried, nickel laterite powder was mixed in accordance with the desired ratio. Then, agglomerates having a diameter of 20 mm were prepared under a pressure of 4 MPa.
5. A graphite crucible containing pellets was placed into the vacuum furnace. When the pressure of the furnace dropped to approximately 10 Pa, heating was started and controlled at a rate of 13 °C/min. When heated to 800 °C, the sample was incubated for 20 min and then heated to the set temperature (1350, 1400 and 1450 °C). Finally, the sample was incubated for 60 min.
6. After incubation at the set temperature, the sample was allowed to reach room temperature. Then, the cooling water was turned off, and the vacuum oven was opened. The original residue was weighed and tested.

The scheme of the experiments is shown in Fig. 2.

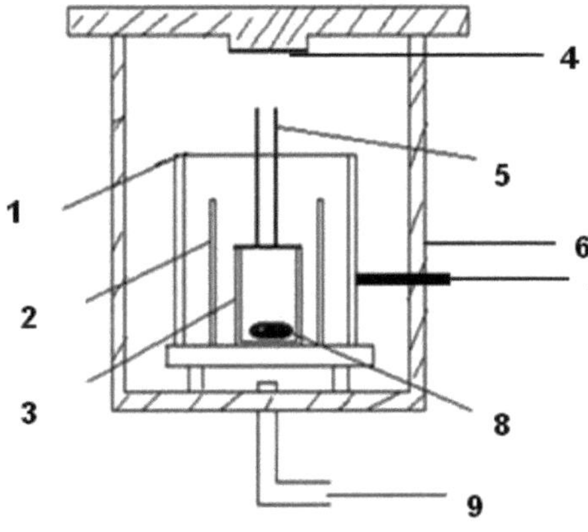

Fig. 1 Schematic model of the vacuum distillation furnace

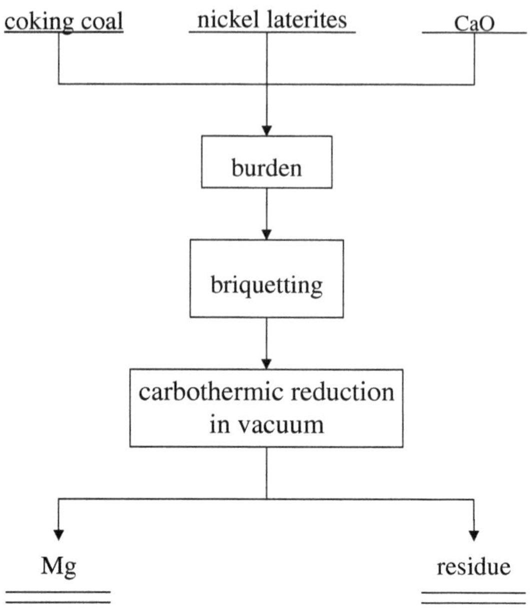

Fig. 2 The flow chart of the experiments on carbothermic reduction in vacuum

Analysis Methods

Using the X-ray diffraction (XRD) type of Rigaku D/max-3B, the phase of the reduction remainders were analysed, and a chemical composition analysis was performed. The straight yield of silicon in vacuum carbothermal reduction experiments was calculated according to the following formula.

$$\eta = \frac{M_1}{M_0} \times 100\%$$

where η represents the direct recovery of silicon, M_0 represents the quality of silicon in nickel laterite, and M_1 is the quality of silicon in the reducing residue.

The recoveries of Si for different C/O are shown in Table 4.

Table 4 Recoveries of Si for different C/O

T (°C)	C/O	M_0 (g)	M_1 (g)	η (%)
1350	1	12.81	7.26	56.68
	1.2	12.27	7.25	59.14
	1.5	12.77	9.08	71.09
1400	1	12.23	6.29	51.43
	1.2	12.56	7.32	58.13
	1.5	12.43	8.79	70.73
1450	1	12.49	5.67	45.43
	1.2	12.72	6.90	54.25
	1.5	12.52	8.62	68.86

Table 5 Recoveries of Si for different additions of CaO

T (°C)	CaO (%)	M_0 (g)	M_1 (g)	η (%)
1350	5	2.78	2.06	74.16
	10	2.62	2.02	76.91
	15	2.54	2.03	80.09
	20	2.79	2.26	81.02
	22.64	2.73	2.22	81.17
1400	5	2.90	2.14	73.93
	10	2.86	2.20	77.00
	15	2.61	2.10	80.30
	20	2.09	1.72	82.16
	22.64	2.42	1.99	82.31
1450	5	3.14	2.35	74.85
	10	2.71	2.11	77.77
	15	2.93	2.40	81.81
	20	2.42	2.02	83.45
	22.64	2.43	2.03	83.68

The recoveries of Si for different additions of CaO are shown in Table 5.

Results and Discussion

Influence of Reducing on the Behaviour of Silicon

When the system pressures were 10–200 Pa, the reaction temperatures were 1350, 1400 and 1450 °C. The carbon ratios were 1:1, 1.2:1, 1.5:1, and the reaction time was 60 min. The recovery of Si at different carbon ratios is shown in Fig. 3.

As observed from Fig. 3, when the temperature was constant, the carbon ratio rises from 1:1 to 1.5:1. The recovery of Si rises more quickly at 1350 °C, and the carbon ratio is 1.5:1. The recovery of Si reaches a maximum of 71.09%.

The XRD pattern of residues obtains with different carbon ratios at 1350 °C is shown in Fig. 4.

Figure 4 shows that, at 1350 °C and a carbon ratio of 1:1, Fe_3Si and Fe_2Si are prevalent in the residue. When the coal ratio was 1.2:1, Fe_3Si and Fe_2Si are prevalent in the residue, Si and SiC also present in the residue.

The possible chemical reactions of nickel laterite by carbothermic reduction under vacuum are reactions (1)–(7). The reactions' system pressure is 10–200 Pa. After a review of the Inorganic Thermodynamic Datasheets [13], we used the substances' Gibbs free energy function method [14] to obtain the relationship between the ΔG_T and T of reactions at the system pressures of 50 Pa as shown in Fig. 5.

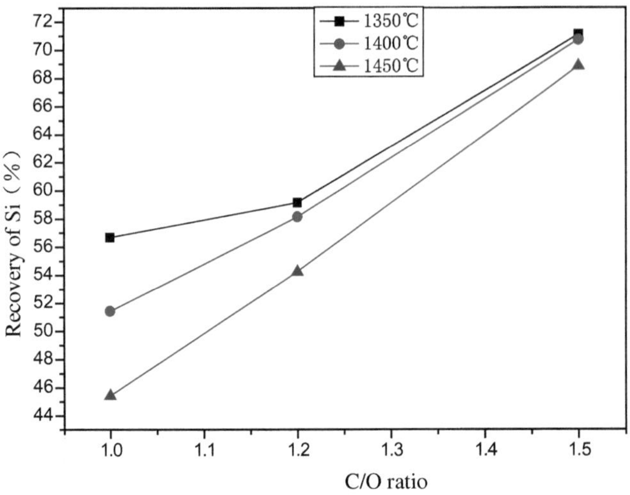

Fig. 3 Recoveries of Si for different C/O

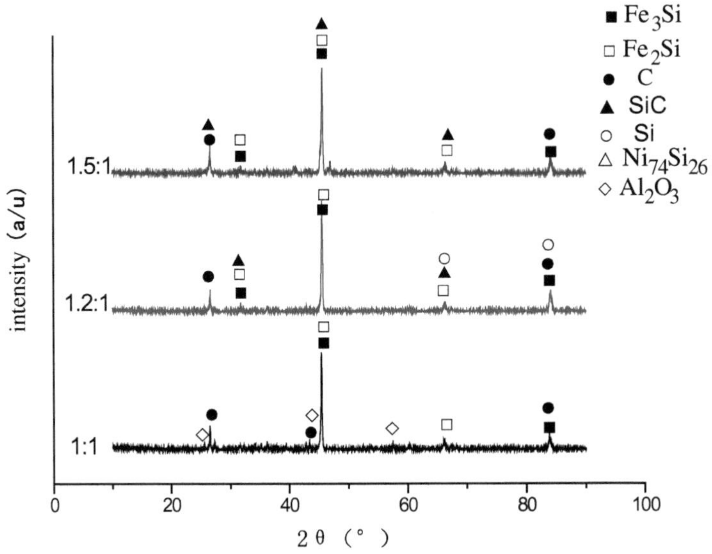

Fig. 4 XRD patterns of residue for different C/O at 1350 °C

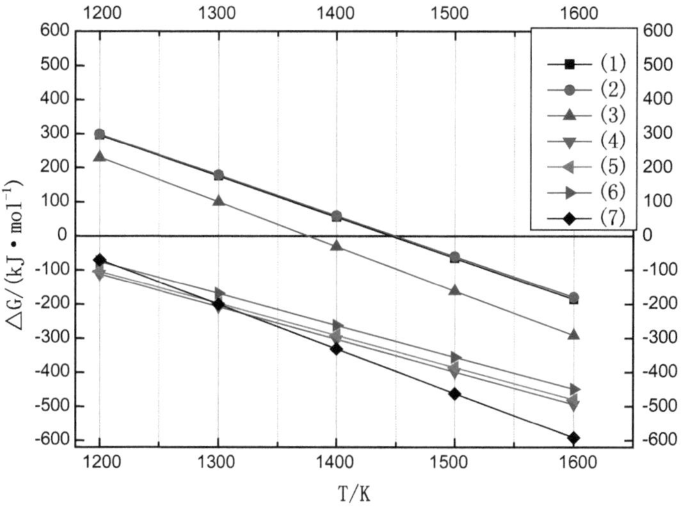

Fig. 5 The relationship between the ΔG_T and T of the reactions at the system

$$2MgO \cdot SiO_2(s) + 3C(s) + 4C(s) = 2Mg(g) + SiO(g) + 3CO(g) \quad (1)$$

$$2MgO \cdot SiO_2(s) + 4C(s) = 2Mg(g) + Si(s) + 4CO(g) \quad (2)$$

$$2MgO \cdot SiO_2(s) + 5C(s) = 2Mg(g) + SiC(s) + 4CO(g) \quad (3)$$

$$2FeO \cdot SiO_2(s) + 3C(s) = 2Fe(s) + SiO(g) + 3CO(g) \quad (4)$$

$$2FeO \cdot SiO_2(s) + 4C(s) = 2Fe(s) + Si(s) + 4CO(g) \quad (5)$$

$$2FeO \cdot SiO_2(s) + 5C(s) = 2Fe(s) + SiC(s) + 4CO(g) \quad (6)$$

$$Fe(s) + SiC(s) = FeSi(s) + C(s) \quad (7)$$

Based on the above chemical equations, reactions (1) and (4) have a low amount of C and generate more volatile SiO, resulting in a loss of Si. Reactions (2), (3), (5) and (6) have higher amounts of C and generate more stable Si and SiC. Therefore, in order to reduce the loss of Si in the residue, we should increase the amount of C.

Figure 5 shows that when the amount of C is low, SiO is generated in reactions (1) and (4). The onset temperature is lower than the experimental temperature. As the temperature increases, the Gibbs free energy becomes more negative, and the driving force of the reaction is greater. As the increase of carbon content, Si and SiC are generated in reactions (2), (3), (5) and (6). The Gibbs free energy is negative at the experimental temperature. The increase of carbon content promotes the generation of SiC and Si under the experimental conditions. Reaction (7) shows that when Fe and SiC coexist in the experimental temperature range, Fe and SiC are unstable, Si–Fe alloys are easy to be generated [15].

When the content of C is properly controlled, the reduction process is occurred in accordance with reactions (2) (3), (5), and (6). Whole Si generates in forms of Si and SiC. So it improves the recovery of Si, which is consistent with the XRD analysis.

Effect of Additives on the Behaviour of Silicon

The system pressures were 10–200 Pa. The reaction temperatures were 1350, 1400 and 1450 °C. The additions of CaO were 5–22.64%. The reaction time was 60 min. The recovery of Si at different additions of CaO is shown in Fig. 6.

Figure 6 shows that when the additions of CaO are 20 and 22.64%, the recovery of Si is more than 80%. In the case of different additions of CaO, Si recovery increases with increasing the reaction temperature. The recovery of Si reaches the maximum at 1450 °C.

The XRD patterns of residues with different additions for CaO at 1450 °C is shown in Fig. 7.

Figure 7 shows that with a temperature of 1450 °C and an addition of 5% CaO, Si primarily exists in the forms of Fe_3Si, $CaSiO_3$ and SiO_2. When the addition of CaO is 10%, Si primarily exists in the forms of Fe_3Si and SiO_2. When the addition of CaO is 15%, the Ca_2SiO_4 phase appears. When the addition of CaO is 15%, Si mainly exists in the forms of Al_2SiO_5 and Ca_2SiO_4, and the peak of Ca_2SiO_4 is more obvious. When CaO is 22.61%, all of the solid phase becomes Ca_2SiO_4.

When CaO was added, the possible chemical reactions of Nickel laterite by carbothermic reduction in vacuum are reactions (8)–(13) at the system pressure of

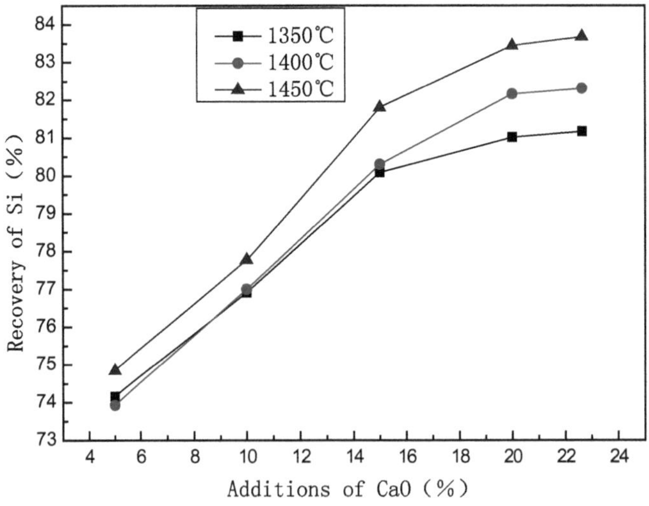

Fig. 6 Recoveries of Si for different additions of CaO

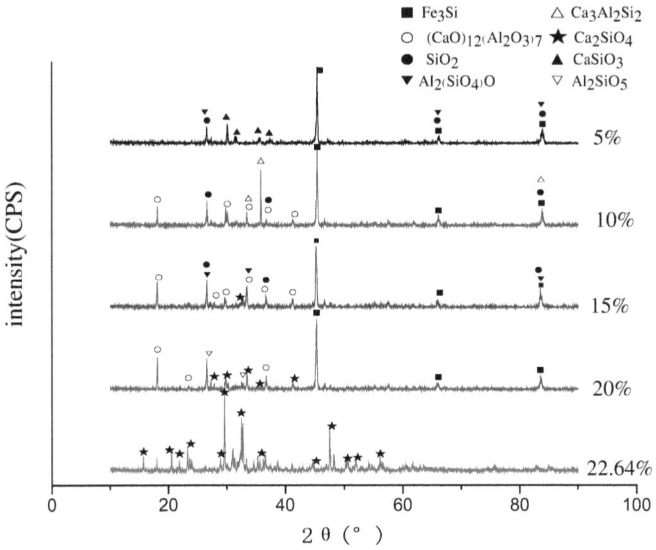

Fig. 7 XRD patterns of the residue with different additions of CaO at 1450 °C

10–200 Pa. The relationship between ΔG_T and T is determined for the reactions at the system pressures of 50 Pa (Fig. 8).

$$2MgO \cdot SiO_2(s) + 2CaO(s) + 2C(s) = 2Mg(g) + 2CaO \cdot SiO_2(s) + 2CO(g) \quad (8)$$

$$2[2MgO \cdot SiO_2](s) + 2CaO(s) + 6C(s) = 4Mg(g) + SiC(s) + 2CaO \cdot SiO_2(s) + 6CO(g) \quad (9)$$

$$2MgO \cdot SiO_2(s) + CaO(s) + 8C(s) = 2Mg(g) + SiC(s) + CaC_2(s) + 5CO(g) \quad (10)$$

$$2MgO \cdot SiO_2(s) + CaO(s) + 5C(s) = 2Mg(g) + Si(s) + Ca(g) + 5CO(g) \quad (11)$$

$$2CaO \cdot SiO_2(s) + 9C(s) = 2CaC_2(s) + SiC(s) + 4CO(g) \quad (12)$$

$$2MgO \cdot SiO_2(s) + 2CaO(s) = 2MgO(g) + 2CaO \cdot SiO_2(s) \quad (13)$$

Based upon the chemical equations, when the addition of CaO is low, SiC and $2CaO{\cdot}SiO_2$ are generated in reactions (9) and (10), or Si or SiC are generated in reactions (10) and (11). When the addition of CaO is high, all Si is generated in the form of $2CaO{\cdot}SiO_2$, in accordance with reaction (8). Therefore, in order to increase the generation of $2CaO{\cdot}SiO_2$, we should increase the addition of CaO.

Figure 8 shows that when the addition of CaO is low, Si or SiC is generated in reactions (9), (10) and (11), and the onset temperature is lower than the experimental temperature. When the addition of CaO is high, the onset temperature of

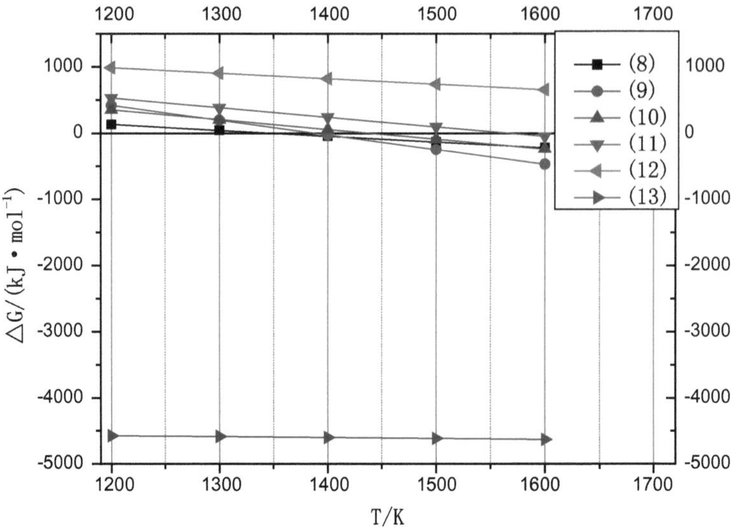

Fig. 8 The relationship between the ΔG_T and T of the reactions at the system pressure of 50 Pa

reaction (8) is the lowest; this reaction is the easiest to carry out, and all of the Si is generated in the form of $2CaO{\cdot}SiO_2$. The reaction of (12) shows that in the carbon thermal reduction reaction, Ca_2SiO_4 may be present and stable; the onset temperature is lower when all of the Si becomes $2CaO{\cdot}SiO_2$ than all of the Si is SiO in the reaction (1). Therefore, the reaction can effectively reduce the evaporation of Si. The reaction of (13) shows that the Gibbs free energy of the reaction is negative all the time, explaining that CaO can effectively destroy the structure of magnesium silicate.

Therefore, by properly controlling a higher addition of CaO, the reduction process is in accordance with reaction (8). All Si is in the form of Ca_2SiO_4 and the recovery of Si is improved, which is consistent with the XRD analysis.

Conclusions

(1) The recovery of Si is more than 50% via carbothermal reduction under vacuum. At 1400 °C, the carbon ratio is 1.5:1 and the recovery of Si is 71.09%. Si mainly exists as Fe–Si alloy and SiC, and the amount of SiC gradually increases as the carbon ratio increases.

(2) When CaO is added, the recovery of Si is more than 70% by carbothermic reduction under a vacuum. At 1400 °C, the carbon ratio is 1.5:1, the addition of CaO is more than 15%, and recoveries of Si are more than 80%. At 1450 °C, the carbon ratio is 1:1, the molar ratio of CaO:Si is 2:1, and all of Si exists in the form of $2CaO{\cdot}SiO_2$.

(3) CaO could effectively destroy the structure of magnesium silicate, strengthen the reduction of magnesium, and reduce the volatilization rate of Si.

Acknowledgements This work was supported financially by National Science Foundation of China (No. 51604133) and Academician Free Exploration Fund of Yunnan Province, China (No. 2015HA016 and 2016HA011).

References

1. F.G. Qiao, J.Y. Zhu, Y.L. Tian, Y. Liu, Nickel resources distribution in the world and nickel resources of Yunnan. Yunnan Geol. **24**, 395–401 (2005)
2. X.X. Fan, Y.H. Wang, H.G. Dong, Y.D. Wu, B.Y. Li, J.C. Zhao, B.J. Li, Study on a novel process of extracting nickel, cobalt and iron from laterites ore. Min. Metall. **21**, 39–43 (2012)
3. B.I. Whittington, R.G. McDonald, J.A. Johnson, D.M. Muir, Pressure acid leaching of arid-region nickel laterite ore: Part I: effect of water quality. Hydrometallurgy **70**, 31–46 (2003)
4. J.A. Johnson, R.G. Nald, D.M. Muir, J.P. Tranne, Pressure acid leaching of arid-region nickel laterite ore: Part IV: effect of acid loading and additives with nontronite ores. Hydrometallurgy **78**, 264–270 (2005)
5. S. Agatzini-Leonardou, I.G. Zafiratos, Beneficiation of a Greek serpentinic nickeliferous ore Part II. Sulphuric acid heap and agitation leaching. Hydrometallurgy **74**, 267–275 (2004)
6. S. Agatzini-Leonardou, I.G. Zafiratos, D. Spathis, Beneficiation of a Greek serpentinic nickeliferous ore—Part I. Mineral processing. Hydrometallurgy **3**, 259–265 (2004)
7. L. Le, J. Tang, D. Ryan, M. Valix, Bioleaching nickel laterite ores using multi-metal tolerant *Aspergillus foetidus* organism. Miner. Eng. **19**, 1259–1265 (2006)
8. B.B. Kar, Y.V. Swamy, B.V.R. Murthy, Design of experiments to study the extraction of nickel from lateritic ore by sulphatization using sulphuric acid. *Hydrometallurgy* **56**, 387–394 (2000)
9. Q.X. Zhou, Current situation and development direction of technology for oxidized nickel ore treatment. Yunnan Metall. **34**, 33–36 (2005)
10. Y. Dai, Bin Yang, *Vacuum Metallurgy of Non-ferrous Metal Materials* (Metallurgical Industry Press, Beijing, 2008)
11. Q.C. Yu, B. Yang, W.H. Ma, Z.H. Li, Y.N. Dai, Study of carbothermic reduction of magnesia in vacuum. Chin. J. Vac. Sci. Technol. **5**, 68–71 (2009)
12. B.Q. Xu, H.B. Pei, B. Yang, D.C. Liu, T. Qu, Y.N. Dai, The carbothermic reduction process for magnesium removal from nickel laterite in vacuum. Chin. J. Vac. Sci. Technol. **31**, 341–347 (2011)
13. D.L. Ye, et al. (eds.), *Handbook of thermodynamic data of inorganic compounds* (Metallurgical Industry Press, Beijing, 1981)
14. X.C. Fu et al., *Physical Chemistry* (Higher Education Press, Beijing, 2005)
15. Q. Luo, D.C. Liu, T. Qu, Y. Tian, B. Yang, Y.N. Dai, Volatile behavior of silicon by carbothermic reduction in vacuum. J. Central South Univ. Sci. Technol. **43**, 2900–2908 (2012)

Effect of CaO Addition on the Behavior of Vanadium and Phosphorus During Oxidation and Leaching Process

Tao Zhang, Zhou Wang, Dong-Wei Li and Diao Jiang

Abstract The feasibility of simultaneous vanadium extraction and phosphorus removal from vanadium-containing hot metal was demonstrated with laboratory experimental investigation. The effect of CaO addition on the distribution behavior of vanadium, phosphorus between basic vanadium slag and semi-steel as well as the influence of basicity on the leaching behavior of vanadium, phosphorus was particularly studied. Results show that vanadium oxidation ratio slightly decreases with the increase of basicity, but phosphorus oxidation ratio rapidly increases with the increase of basicity. The leaching of roasted vanadium slag with various CaO contents indicated that the optimal slag basicity for extracting vanadium was 1.5. The maximum leaching efficiency of vanadium could reach 84.02% and the sequence of vanadium leaching efficiency by using different leaching reagents was: 160 g/L Na_2CO_3 > 100 g/L $NaHCO_3$ > 400 g/L $(NH_4)_2CO_3$ > 5% H_2SO_4.

Keywords V-containing hot metal · Dephosphorization · Vanadium slag · Leaching

Introduction

Vanadium is an important metal to produce V_2O_5, VN, ferrovanadium (FeV) and Ti–Al–V alloy, which are widely used in steel industry and aerospace industry [1]. Vanadium–titanium magnetite ore is one of the important sources of vanadium productions. In China, vanadium slag is mostly produced from V-containing hot

T. Zhang (✉) · D.-W. Li
Department of Science and Research,
Chongqing University of Education, Chongqing 400065, China
e-mail: zhangtao@cque.edu.cn

Z. Wang · D. Jiang
College of Materials Science and Engineering,
Chongqing University, Chongqing 400044, China
e-mail: 45197669@qq.com

J.-Y. Hwang et al. (eds.), *8th International Symposium on High-Temperature Metallurgical Processing*, The Minerals, Metals & Materials Series,
DOI 10.1007/978-3-319-51340-9_52

metal during vanadium extraction process in LD converter [1, 2]. Considering the next roasting and leaching process of vanadium extraction, there is no lime added into the LD converter. So the conventional vanadium slag has no ability of dephosphorization, it is increasing the burden of dephosphorization in the semi-steel steelmaking process.

In order to dephosphorizing and vanadium extracting from the hot metal simultaneously, a novel vanadium extraction process was proposed to add lime into the converter during the vanadium slag production process. So part of phosphorus can be removed from the hot metal that dephosphorization burden was relieved in the semi-steel steelmaking process. As lime was added into the converter during the formation of vanadium slag, the salts will no longer required in the vanadium slag roasting process. The method that direct roasting vanadium slag with high calcium is similar to calcium roasting technique, which avoiding the emission of pollutant gases such as Cl_2, HCl and SO_2 in the sodium salts roasting process [3].

Several researches [4–13] were carried out on the distribution of V and P (or P_2O_5) between hot metal and slag. But there are little literatures about adding CaO into LD converter in the process of vanadium extraction. This study aims to clarify the effect of CaO addition on the distribution behavior of vanadium, phosphorus between basic slag and semi-steel as well as the influence of basicity on the leaching behavior of vanadium and phosphorus.

Experimental Procedure

Simultaneous Vanadium Extraction and Phosphorus Removal

To make the master alloy of hot metal containing vanadium and phosphorus, we melted pure iron, silicon iron, electrolytic Manganese, ferrophosphorus, ferrovanadium, and ferrotitanium in graphite crucible, in a high frequency induction furnace. The initial composition of the hot metal is shown in Table 1. Reagent-grade SiO_2, Fe_2O_3, V_2O_3, CaO, TiO_2, $MnCO_3$, MgO were used to produce the original slag. Firstly, 150 g pre-melted iron sample was charged into an alumina crucible (inner diameter, 46 mm; height, 120 mm) heated to 1380 °C under nitrogen atmosphere with a flow rate of 1 L/min, and maintained for 20 min to ensure pre-melted iron melt homogeneity in the 10 kW box-type electric resistance furnace with a proportional-integral-differential (PID) controller (±5 K). The prepared original slag dividually charged into the alumina crucible 5 times at 2 min

Table 1 Initial composition of hot metal (mass%)

	C	Si	Mn	P	V	Ti
Hot metal	4.77	0.29	0.16	0.088	0.26	0.14

intervals, then stirred after every time charged slag with preheating steel bar each time, and kept at the same temperature for a certain time. Then, the semi-steel and slag were taken out of the furnace and cooled down to room temperature. The composition of vanadium in semi-steel was analyzed by inductive coupled plasma-emission spectrometry (OPTIMA4300DV), while the carbon content was analyzed by infrared carbon/sulfur analyzer (CS-444LS).

The distribution ratio of vanadium and phosphorus between slag and semi-steel was calculated follows:

$$L_p = (\%P)/[\%P] \tag{1}$$

$$L_v = (\%V)/[\%V] \tag{2}$$

Where (%P) and (%V) indicate the mass percent of phosphorus and vanadium in the vanadium slag produced by simultaneous dephosphorization and vanadium extraction from the hot metal, respectively. The [%P] and [%V] indicate the phosphorus and vanadium content in the semi-steel, respectively.

Slag Sample Preparation and Leaching

To preparation the vanadium slag produced by dephosphorizing the V-containing hot metal, reagent-grade SiO_2, $FeC_2O_4{\cdot}H_2O$, V_2O_3, CaO, TiO_2, $MnCO_3$, MgO, P_2O_5 were mixed as Table 2. The chemical composition of slag was come from the modified Pan Steel industrial vanadium slag. Considering corrosion of alumina crucible the analytical reagent of Al_2O_3 was not added, which was proved to be about 2–3 mass% Al_2O_3 in synthesized slags by XRF. The index of w_{CaO}/w_{SiO2} was also listed in Table 2 to show basicity in the slag. 100 g of the mixture melted in Al_2O_3 crucible at 1723 K holding for 30 min to ensure homogeneity in the argon atmosphere with a flow rate of 1 L/min in the 10 kW box-type electric resistance furnace with a proportional-integral-differential (PID) controller (±5 K). Then the sample was dropped down to 1673 K holding for 1 h and then cooled down to about 1173 K in the same gaseous atmosphere and finally taken out for cooling in the air [3]. Part of the synthesized slag samples was roasted at 1173 K for 2 h to produce the roasted slag in the muffle furnace and taken out for cooling in the air.

Table 2 Chemical Compositions of Slag Samples (mass%)

S.No.	FeO	SiO_2	Al_2O_3	CaO	V_2O_3	TiO_2	MgO	MnO	P_2O_5	Basicity
Ca1	30.00	20.00	2.00	12.00	13.00	10.00	3.00	8.00	2.00	0.6
Ca2	30.00	14.55	2.00	17.45	13.00	10.00	3.00	8.00	2.00	1.2
Ca3	30.00	12.80	2.00	19.20	13.00	10.00	3.00	8.00	2.00	1.5
Ca4	30.00	11.43	2.00	20.57	13.00	10.00	3.00	8.00	2.00	1.8
Ca5	30.00	10.32	2.00	21.68	13.00	10.00	3.00	8.00	2.00	2.1

Slag samples of 4 g were heated in the electric resistance furnace to 900 °C and held for 2 h under the air atmosphere. After the roasting process was completed, the samples were taken out and cooled down to the room temperature. Then the slag sample were crushed and ground to −0.074 mm. After that, the samples leached with various leaching reagents of different temperature for 1 h stirred by magnetic force to extract soluble vanadium in liquid to solid ratio of 10:1 (ml:g). The slurry was separated via vacuum filtration, and the residue was washed several times by distilled water.

Vanadium concentration in the leaching solution was analyzed by the ammonium ferrous sulfate titration method. Phosphorus in the leaching solution has a negative effect on the quality of vanadium pentoxide which was the production of V-containing solution. So phosphorus in the leaching solution was determined by inductive coupled plasma-emission spectrometry (OPTIMA4300DV).

Results and Discussion

Distribution Behavior of Vanadium and Phosphorus Between Slag and Semi-steel

The result of final slag is shown in Table 3. The relationship between the basicity and the mass percent of V, P in the process of vanadium extraction is illustrated in Fig. 1. According to Fig. 1, the content of phosphorus in the semi-steel decreased with the increased of basicity. Evidently from this figure that the basicity of the slag is an important parameter to be considered for the evaluation of the efficient removal of phosphorus from V-containing hot metal in the process of vanadium extraction. The concentration of V in the semi-steel increased from 0.04 to 0.054%, with the basicity increased from 0.6 to 2.1.

Inoue et al. [14] pointed out that the influence of the basicity on the equilibrium partition ratio is no stronger for V than it is for P. This conclusion is consistent with other researches [4, 5]. In the present work, adding CaO in the vanadium extraction process decreased the concentration of FeO, so vanadium content increased with the increased of basicity. The increments of V lower than decrements of P in

Table 3 Chemical compositions of final slag samples (mass%)

S.No.	FeO	SiO_2	Al_2O_3	CaO	V_2O_3	TiO_2	MgO	MnO	P_2O_5	Basicity
Ca1	30.74	20.22	2.02	12.13	12.63	10.11	3.03	8.09	1.04	0.60
Ca2	23.17	16.14	2.22	19.37	13.54	11.10	3.33	8.88	2.26	1.20
Ca3	39.51	11.06	1.73	16.59	10.87	8.64	2.59	6.91	2.10	1.50
Ca4	22.89	12.57	2.20	22.62	13.76	11.00	3.30	8.80	2.87	1.80
Ca5	27.00	10.77	2.09	22.61	12.92	10.43	3.13	8.35	2.71	2.10

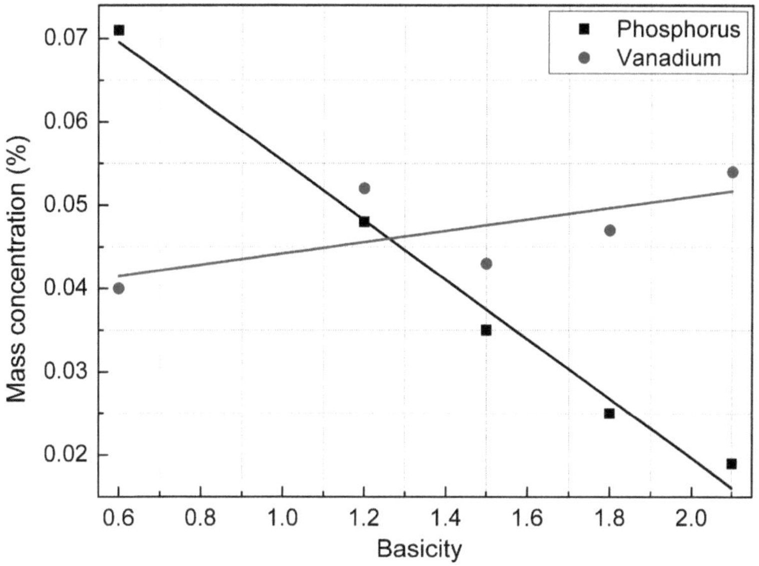

Fig. 1 Change of the mass percent P and V in relation to the slag basicity

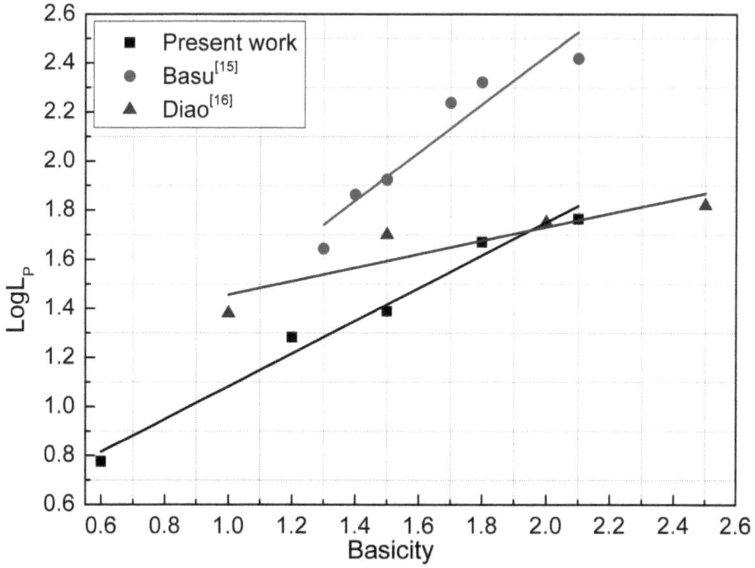

Fig. 2 Relationship between measured Lp and basicity

V-containing hot metal dephosphorization process, which indicate the dependence of vanadium oxidation on basicity lower than that of phosphorus.

The relationship between measured L_p and basicity is illustrated in Fig. 2 together with data previously reported by Basu et al. [15] and Diao [16]. The logarithm of phosphorus distribution ratio increased from 0.78 to 1.76, with the basicity increased from 0.6 to 2.1. Obviously, the L_p of present work has deviation with the data of Basu et al. [15] and Diao [16]. Considering the CaO content in the vanadium slag and vanadium extraction temperature, the dephosphorization ability of vanadium slag lower than that of Basu et al. [15] and Diao [16]. The relationship between measured L_v and basicity was shown in Fig. 3 together with data reported by Zhen et al. [11]. The logarithm of vanadium distribution ratio in present work slightly decreased from 2.33 to 2.21, with the basicity increased from 0.6 to 2.1. The data of Zhen et al. [11] close to the results of present work. That is, the addition of CaO in the vanadium extraction process has little effect on vanadium distribution ratio between vanadium slag and hot metal. The dependence of the vanadium distribution ratio on the slag basicity was smaller than that of phosphorus. In order to taking full advantage of vanadium, the basicity must be definite in dephosphorization V-containing hot metal process.

Leaching Behavior of Vanadium and Phosphorus in Various Reagengts

The vanadium leaching ratio in various leaching reagents was shown in Fig. 4. Sulfuric acid, soda, baking soda and ammonium carbonate are the common leaching reagents in the vanadium extraction process [12, 13, 17, 18]. From Fig. 4, the leaching ratio of vanadium used the acidic leaching reagent lower than that of

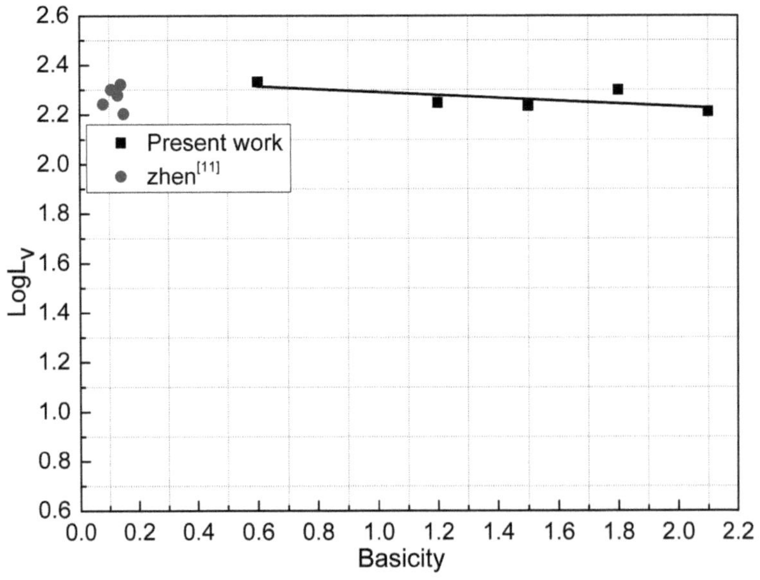

Fig. 3 Relationship between measured Lv and basicity

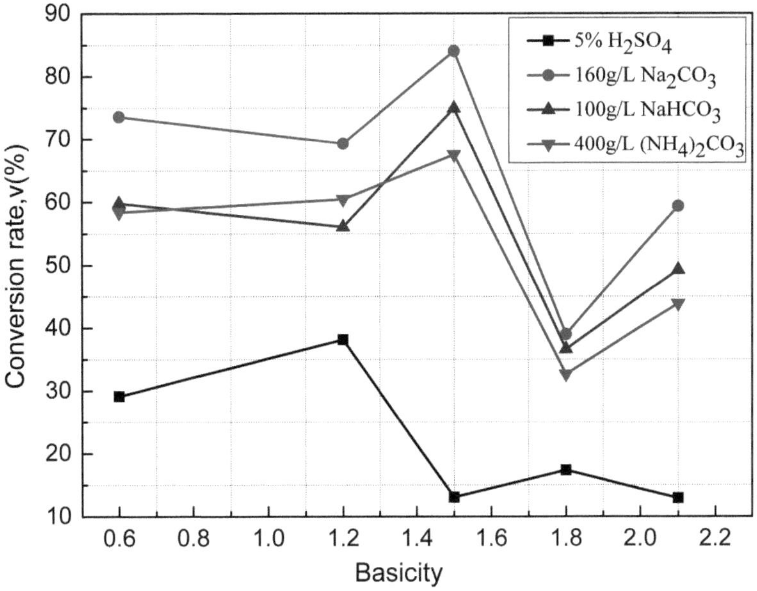

Fig. 4 The vanadium leaching ratio in various leaching reagents

used alkaline reagents. Maybe the concentration of acidic reagent has not higher enough [19], considering that the concentration of sulfuric acid is 5%. May be some insoluble materials generated at basicity of 1.8, due to the vanadium extraction ratio lowest in various leaching reagents. The extraction ratio of vanadium in the S.no. Ca3 were 74.93 and 67.53% used 100 g/L $NaHCO_3$ and 400 g/L $(NH_4)_2CO_3$ leaching, respectively. The highest leaching ratio is 84.02% in all of slag samples, appeared in S.no.Ca3 used the 160 g/L Na_2CO_3 leaching. It is that the solubility of baking soda in the water was lower than that of soda. In addition, the ammonium carbonate will decomposed and evaporated in the process of leaching at the temperature of 60 °C. So in the process of leaching have no enough baking soda and ammonium carbonate reacted with roasted samples that have disadvantages to the leaching ratio of vanadium.

The concentration of phosphorus in various leaching solution of S.no.Ca3 is shown in Fig. 5. The lowest concentration of phosphorus in solution is 0.0118 g/L used 400 g/L $(NH_4)_2CO_3$ as leaching reagent, the highest concentration of phosphorus is 0.0245 g/L in 5%H_2SO_4 leaching solution. Compared with the vanadium leaching ratio in various leaching reagents and the content of phosphorus in the leaching solution, the optimum leaching reagent is 160 g/L Na_2CO_3 with the leaching ratio of vanadium and the phosphorus content in solution are 84.02%, 0.0147 g/L, respectively.

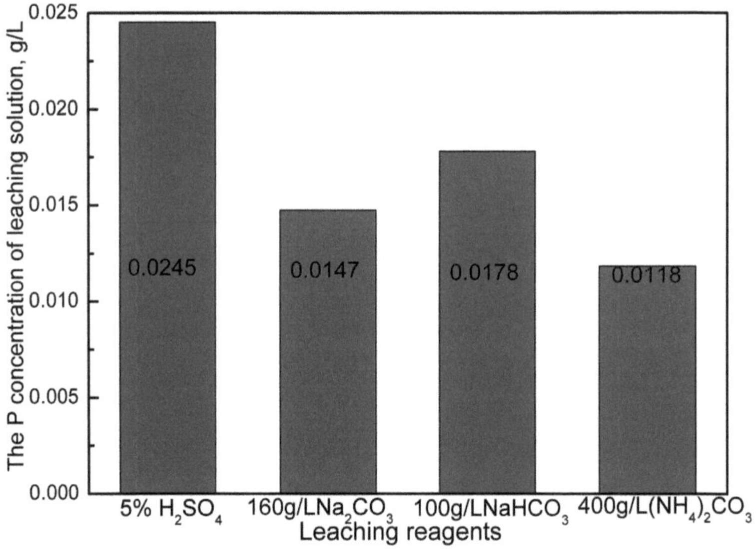

Fig. 5 The concentration of phosphorus in various leaching solution of S.no.Ca3

Conclusions

(1) Adding CaO into the conventional vanadium extraction process, the dephosphorization rate rapidly increasing with the increase of slag basicity. Conversely, the oxidation ratio of V slightly decreasing with the increase of slag basicity.
(2) The leaching of roasted vanadium slag with various CaO contents indicated that the optimal slag basicity for extracting vanadium was 1.5. The maximum leaching efficiency of vanadium could reach 84.02%. The content of P in all leaching solution lower than 0.0245 g/L, which fitting the requirements in vanadium precipitation process.
(3) The optimum slag basicity is 1.5. The V, P oxidation ratios are 84.6, 60.2%, respectively. The maximum V conversion ratio reached 84.02% and P content in the leaching solution is 0.0147 g/L.

Acknowledgements This work was financially supported by the National Basic Research and Development Program of China (No. 2013CB632604), Scientific and Technological Research Program of Chongqing Municipal Education Commission (Grant No. KJ1601403, KJ1601412), and Scientific Research Program of Chongqing University of Education (Grant No. KY201551C).

References

1. R.R. Moskalyk, A.M. Alfantazi, Processing of vanadium: a review. Miner. Eng. **16**, 793 (2003)
2. D.X. Huang, *Vanadium Extraction and Steelmaking* (Metallurgical Industry Press, Beijing, China, 2000), p. 35

3. D.S. Bradbury, The production of vanadium pentoxide, in *The 41st Annual Conference of Metallurgists of CIM, Vanadium-Geology, Processing and Applications*, Montreal, Quebec, Canada (2002), p. 115
4. Z.H. Tian, B.H. Li, X.M. Zhang, Z.H. Jiang, Double slag operation dephosphorization in BOF for producing low phosphorus steel. J. Iron. Steel Res. Int. **16**(3), 6 (2009)
5. X. Yang, F.M. Sun, J.L. Yang, F. Liu, K.S. Cheng, J.H. Wang, Optimization of low phosphorus steel production with double slag process in BOF. J. Iron. Steel Res. Int. **20**(8), 41 (2013)
6. Y.L. Zhang, F. Zhao, Y.G. Wang, Effects of influencing factors on distribution behaviors of vanadium between hot metal and FeO–SiO_2–MnO(–TiO_2) slag system. Steel Res. Int. **82**(8), 940 (2011)
7. S.Y. Gao, Z.H. Tian, L.X. Zhu, C.L. Zhao, Z.H. Yan, B.G. Luo, G. Wei, Study on dephosphoriztion of hot metal in coverter, J. Univ. Sci. Technol. Beijing, **33**(suppl.1), 63 (2011)
8. D.Y. Shin, C.H. Wee, M.S. Kim et al., Distribution behavior of vanadium and phosphorus between slag and molten steel. Met. Mater. Int. **13**(2), 171 (2007)
9. F. Pahlevani, H. Shibata, N. Maruoka et al., Behavior of vanadium and niobium during hot metal dephosphorization by CaO–SiO_2–Fe_tO slag. ISIJ Int. **51**(10), 1624 (2011)
10. X. Qiu, B. Xie, L. Jiang et al., Experimental study of phosphorus distribution between slag and metal during duplex dephosphorus converter processing, in *4th International Symposiumon High-Temperature Metallurgical Processing-Held during the TMS 2013 Annual Meeting and Exhibition*, San Antonio, Texas, USA (2013), p. 199
11. X.P. Zhen, B. Xie, C.Y. Zhao et al., Vanadium distribution behavior at the end point of vanadium extraction byconverter process. J. Univ. Sci. Technol. Beijing **34**(12), 1371 (1013)
12. H.Y. Li, N. Wang, B. Xie, Vanadium extraction from high calcium-content vanadium slag by calcification roasting, in *Chen Honorary Symposium on Hydrometallurgy, Electrometallurgy and Materials Characterization-Held during the TMS 2012 Annual Meeting and Exhibition.* Oakland, California, USA (2012), p. 621
13. X.S. Li, B. Xie, Extraction of vanadium from high calcium vanadium slag using direct roasting and soda leaching. Int. J. Miner. Metall. Mater. **19**(7), 595 (2012)
14. R. Inoue, H. Suito, Distribution of vanadium between liquid iron and MgO saturated slag of the system CaO–MgO–FeO_x–SiO_2. Trans. Iron Steel Inst. Jpn. **22**(9), 705 (1982)
15. S. Basu, A.K. Lahiri, S. Seetharaman, Phosphorus partition between liquid steel and CaO–SiO_2–FeO_x–P_2O_5–MgO slag containing 15 to 25 Pct FeO. Metall. Mater. Trans. B **38B**(4), 623 (2007)
16. J. Diao, *Applied fundamental research on medium and high phosphorus hot metal refining by duplex process in converter* [Dissertation] (in Chinese), ChongQing university, ChongQing, China (2010) p. 41
17. Q.M. Feng, Y.H. Shao, L.M. Ou et al., Vanadium recovery by hydrometallurgy from V-containing deposition residue of sodium aluminate solution. Nonferrous Met. **62**(4), 57 (2010)
18. H.Y. Li, N. Wang, B. Xie, Y. Wang, C. Li, H.X. Fang, *A Method for Extracting Vanadium from Vanadium Bearing Roasted Materials* (in Chinese), Chinese Patent, Appl. 201210066735.5 (2012)
19. Z.J. Ju, C.Y. Wang, F. Yin, Y.Q. Yang, D.F. Li, Process of vanadium extraction from stone coal vanadium ore byactivated sulfuric acid leaching at atmospheric pressure. Chin. J. Nonferrous Met. **22**(7), 2061 (2012)

Effect of Inner Shape on Blast Furnace Performance for Iron Making

Xun Zhang and Huiqing Tang

Abstract The inner shape of a blast furnace is of fundamental importance for blast furnace ironmaking to achieve goals of high yield, high quality, low fuel rate and long campaign life. Therefore, it is worthwhile to design better furnace inner shapes. This paper conducted numerical simulations of performance for the blast furnaces with conventional and modified inner shapes. By comparing the calculation results of the two blast furnaces with the same operation conditions, it can be found that there are some certain differences. It is necessary to optimize the inner shape of blast furnace considering the results of simulations. This paper provides references for future blast furnace inner shape design and related technology developments.

Keywords Blast furnace · Conventional inner shape · Modified inner shape · Numerical simulation

Introduction

The blast furnace inner-shape is important to achieve the goals of high-yield, high-quality, low consumption and long life for blast furnace ironmaking. Practice has proved that the appropriate design of the furnace inner shape can improve the blast furnace smelting index and extend furnace campaigns. Therefore, it is worthwhile to improve the designed furnace inner-shape. The complexity of the blast furnace interior follows from the blast furnace ironmaking process, which involves multiple simultaneous physical changes, chemical reactions and a variety of heterogeneous materials coexisting and interacting. There has been rapid progress in the development and application of the blast furnace mathematical

X. Zhang (✉) · H. Tang
State Key Laboratory of Advance Metallurgy, University of Science and Technology Beijing, 30 Xueyuan Rd, Beijing 100083, China
e-mail: 15650786169@163.com

J.-Y. Hwang et al. (eds.), *8th International Symposium on High-Temperature Metallurgical Processing*, The Minerals, Metals & Materials Series,
DOI 10.1007/978-3-319-51340-9_53

models in order to understand and improve the process of blast furnace ironmaking better. This paper presents a numerical study of the blast furnace interior conditions, comparing a blast furnace with a conventional inner shape and one with a modified shape by using a multi-fluid mathematical model. By comparing the calculation results of multiphase flow, heat and mass transfer in two kinds of furnaces with the same original fuel conditions and technical-economic indexes, it was found that it is necessary to optimize the inner shape of blast furnaces.

Parameters of Two Different Blast Furnaces and Operational Conditions

Figure 1 shows the internal dimensions of the conventional (Furnace A) and modified (Furnace B) blast furnaces. Furnace A will represent the conventional blast furnace and Furnace B will represent the modified blast furnace in each section of the following.

The operating conditions of both two furnaces are shown in Table 1.

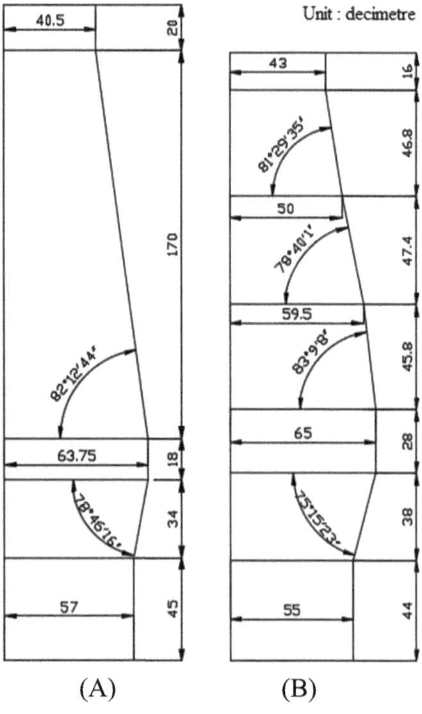

Fig. 1 The internal dimensions of Furnace A and B (unit: decimeter)

Table 1 Operating parameters of both two furnaces

Operating parameters	Amount
Volume of blast	4800 $Nm^3\ min^{-1}$
Hot blast temperature	1250 °C
Oxygen rate	4%
Volume of top gas	428,473 $Nm^3\ h^{-1}$
Temperature of top gas	150 °C
Ore ratio	1.68 t tHM^{-1}
Coke ratio	335 kg tHM^{-1}
Coal ratio	180 kg tHM^{-1}
Burden structure	60% high basicity sinter + 40% low basicity sinter
Ore grade	55.5–56.5% Fe
Charging temperature of fuel	25 °C
Hot metal produced per day	6250 t/d

Mathematical Model

To correctly simulate the internal state of blast furnace using the blast furnace mathematical model, the internal phases, characteristics, reactions and interactions must be taken into consideration. The multi-fluid mathematical model of the blast furnace [1] can predict the internal state of the blast furnace and provide valuable guidance for blast furnace operation by taking into account the above factors. The multi-fluid mathematical model of the blast furnace used in this study was based on previous work [2–4] with some specific assumptions. These assumptions are as follows.

(1) Only gas phase, solid phase and liquid phase were considered. The powder phase and other phases with special properties are not considered or included in the three phases for calculations.
(2) It was assumed that the pulverized coal instantaneously reacted with oxygen in the blast after being blown into blast furnace and then generated CO, CO_2, H_2, H_2O with the corresponding heat.
(3) Minor materials and reactions were ignored.
(4) The top pressure of furnace was set as 2.8×10^5 Pa.
(5) The temperature range of the cohesive zone is set between 1373 and 1673 K.
(6) The state distribution in blast furnace is axisymmetric.

Results and Discussion

Model Validation

Figure 2a, b show the radial distributions of CO, CO_2 in the top gas in the two furnaces. Figure 3 below shows the radial distributions of top gas temperature in the

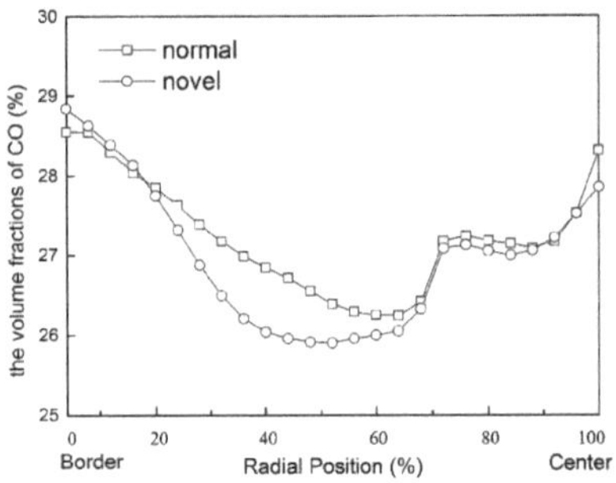

(a)

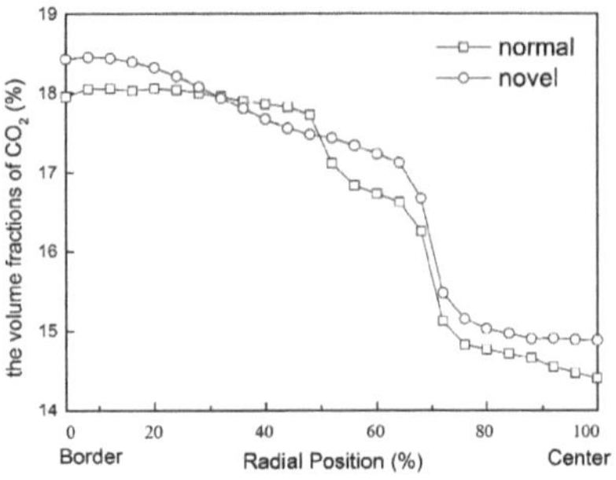

(b)

Fig. 2 The radial distributions of top gas components in two kinds of furnaces: **a** the radial distributions of CO in top gas and **b** the radial distributions of CO_2 in top gas

Fig. 3 The radial distributions of top gas temperature in two kinds of furnaces

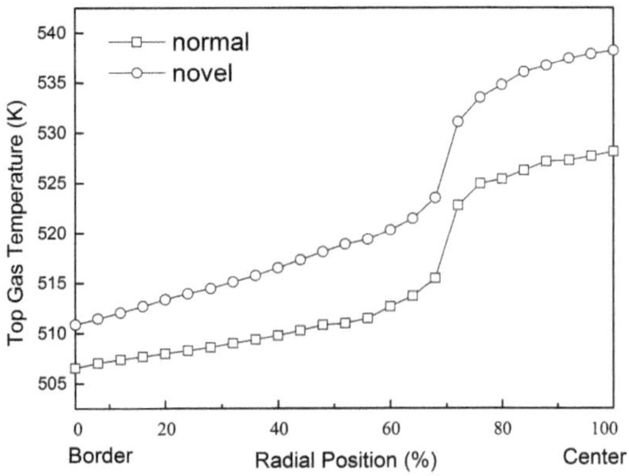

furnaces. According to the calculation, the average top gas temperature in both two furnaces is close to 510 K and the average volume fractions of CO and CO_2 in top gas are respectively close to 17 and 27%. Thus it can be seen that the simulation results of this mathematical model are consistent with the actual production process and the traditional theoretical analysis.

Comparison of Performances

Gas Behavior

Gas Temperature Distribution

Figure 4 shows that the gas temperature is higher in the furnace center and lower near the edge. The temperature difference from the border to the center is larger in

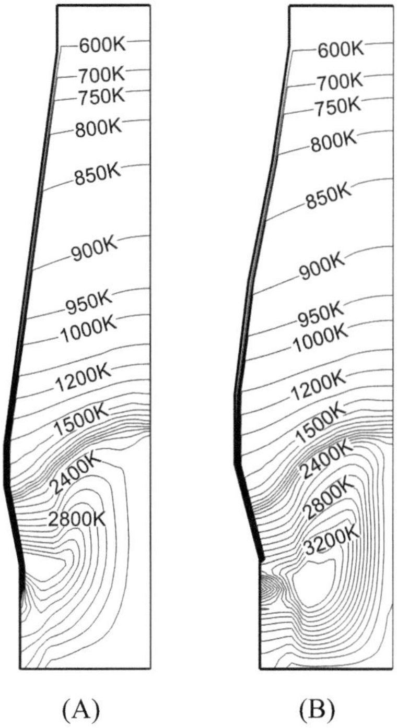

Fig. 4 The gas temperature distributions in Furnace A and Furnace B

Furnace B than that in Furnace A. The tuyere raceway zone in Furnace B is larger than that of Furnace A, which promotes extension of gas flow toward the center and keeps a higher center temperature. Moreover, if the center gas flow can be developed properly, it is conducive to lowering the root of the cohesive zone and decreasing the height of dropping zone. Also it can be seen from the figure that the gas temperature in Furnace A is significantly higher than that in Furnace B at the same elevation.

Distribution of Gas Component

The Fig. 5a shows that the volume fraction of CO in gas decreases first and then increases from the border to the center after gas passing through the cohesive zone in both two furnaces. However, the phenomenon mentioned above in Furnace B is more obvious, indicating the center gas flow is better developed in Furnace B. It would be beneficial to decrease the pressure of the top gas in Furnace B, to ensure the improvement of the smelting intensity and smooth running for burden. It can also be seen from the figure that the gas generated in the hearth is relatively concentrated in the zone of stack lower, belly and bosh upper after flowing through the softening zone.

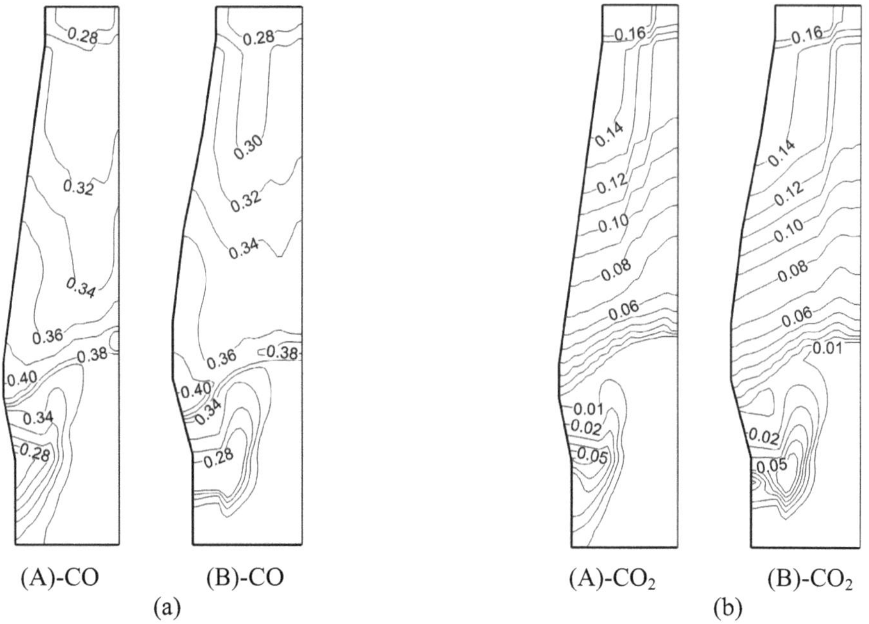

Fig. 5 The volume fraction distributions of gas components in Furnace A and Furnace B: **a** the volume fraction distributions of CO in gas and **b** the volume fraction distributions of CO_2 in gas

Figure 5b shows that the volume fraction of CO_2 in the upper center of the softening zone is obviously lower and the volume fraction of CO_2 near the lining is obviously higher in both two furnaces, which illustrates good gas utilization efficiency near the furnace lining. The reason for this phenomenon is that the center gas flow is developed well and the speed of center gas flow is high in both two furnaces. However, the volume fraction of CO_2 at the same height of the stack in Furnace B is significantly higher than that of Furnace A, indicating that the gas utilization efficiency in Furnace B is better. Moreover, the position where the volume fraction of CO_2 starts to increase in Furnace B is lower than that in Furnace A because of the influence of the softening zone position. This also shows that the indirect reduction region is relatively larger in Furnace B.

Solid Behavior

Distribution of Solid Component

According to Fig. 6a, b, the mass fractions of FeO and Fe in the solids are higher closer to the center in both two furnaces. This phenomenon is very similar to the gas temperature distribution, so we can see that the mass fraction distribution of solid components is controlled by the heat exchange between burden and high

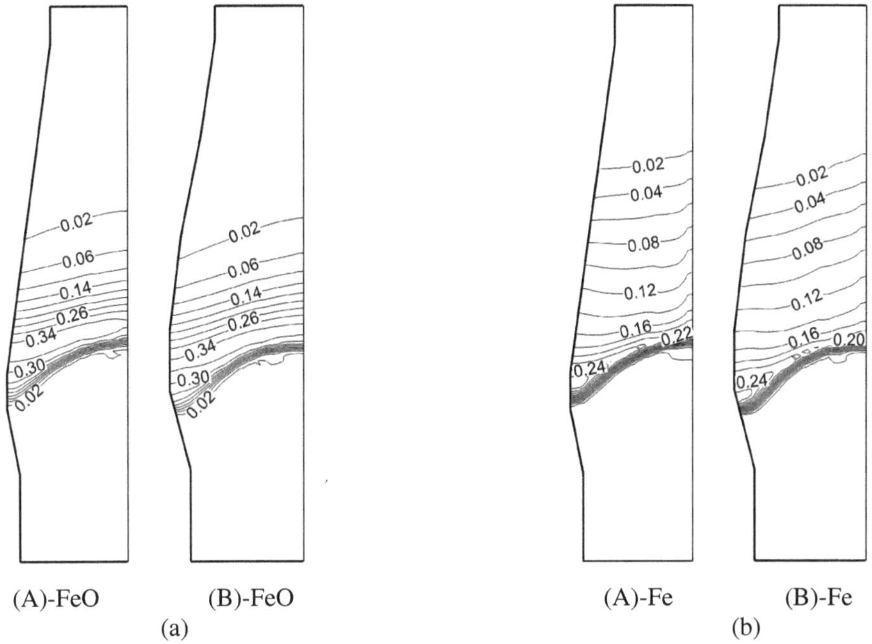

Fig. 6 The mass fraction distributions of solid components in Furnace A and Furnace B: **a** the mass fraction distributions of FeO and **b** the mass fraction distributions of Fe

temperature gas flow. It indicates that the reduction of iron ore is better closer to the center in furnace and this phenomenon is more obvious in Furnace B. Also the region containing FeO and Fe moves down overall in Furnace B because of the influence of softening zone position. Figure 6a shows that the FeO-containing region is thicker in Furnace A, indicating slower reduction of FeO. This may lead to poor permeability and increased energy consumption.

Cohesive Zone

As shown in Fig. 7, the shapes of the cohesive zone in two kinds of furnaces are the same inverted "V". However, the cohesive zone in Furnace B is at a lower elevation than in Furnace A, promoting indirect reduction by the relative increase in the size of the indirect reduction zone in Furnace B. Lowering the position of the cohesive zone is equivalent to reducing the height of the dropping zone in Furnace B. It may reduce the chance of the [C] in hot metal contacting with SiO that is expected to reduce the amount of silicon absorbed by hot metal, which is an essential condition for the smelting of low silicon pig iron. The center gas flow is not fully developed and the edge gas flow would be more easily overdeveloped in Furnace A because of the position of the cohesive zone compared with the Furnace B. It can also make the

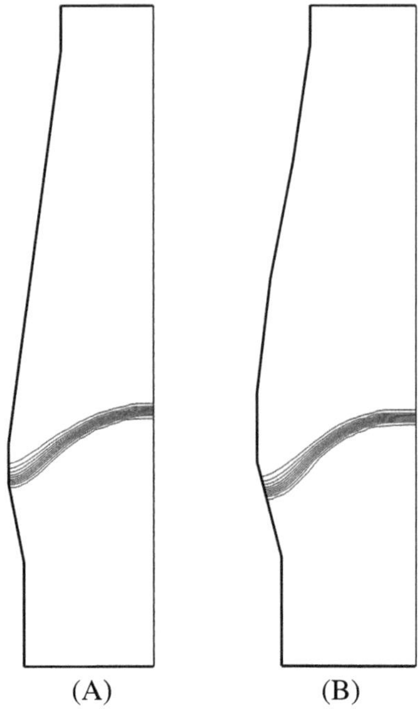

Fig. 7 The cohesive zone location in Furnace A and Furnace B

speed of bosh gas flow be high so that the root of cohesive zone would be unstable and the root may move upward in Furnace A.

Conclusions

By comparing the calculation results of two kinds of furnaces with the same the original fuel and technical-economic indexes, the following conclusions could be drawn:

(1) The gas temperature in the conventional blast furnace is significantly higher than that in the modified blast furnace, at the same elevation. The difference between the edge and center temperatures is smaller and the edge gas flow is more prominent in the conventional blast furnace.
(2) As is apparent from the CO and CO_2 distribution, center gas flow is developed well in the modified blast furnace, with improved gas utilization.
(3) From the mass fraction distributions of solid components, the region containing FeO and Fe is at a higher elevation in the conventional blast furnace because of the influence of softening zone position. The FeO-containing region is thicker in the conventional blast furnace, which can lead to poor permeability and increased energy consumption.

(4) The position of cohesive zone in the conventional blast furnace is higher compared with the modified blast furnace, which may cause difficulties in obtaining the hot metal of low silicon content.

Acknowledgement The authors are gratefully acknowledge the support of the State (No. 511).

References

1. J. Yagi, Mathematical modeling of the flow of four fluids in a packed bed. ISIJ Int. **33**(6), 619–639 (1993)
2. X. Dong, A. Yu, J. Yagi et al., Modelling of multiphase flow in a blast furnace: recent developments and future work. ISIJ Int. **47**(11), 1553–1570 (2007)
3. H. Nogami, M. Chu, J. Yagi, Multi-dimensional transient mathematical simulator of blast furnace process based on multi-fluid and kinetic theories. Comput. Chem. Eng. **29**(11), 2438–2448 (2005)
4. X. Shang, *The Blast Furnace Multiphase Flow Modeling Based on Computational Fluid Dynamics* (Master's thesis), Beike Zhang, guider. (Beijing University of Chemical Technology, Beijing, 2013)

Effect of Lance Configurations on Coal Flow and Combustion Characteristics

Hailong Huo, Zhenfeng Zhou, Jingsong Wang, Qingguo Xue, Yuanyuan Zhang and Yingli Liu

Abstract Coal burnout strongly depends on the residence time and dispersion of coal particles in the raceway region of a blast furnace. The residence time and dispersion of coal particles are directly related to the lance arrangement. In this work, a three dimensional model was developed to simulate the lance-blowpipe-tuyère-raceway region of a blast furnace. Double lance and eccentric configurations were considered. For the double lance, the double-lance and eccentric double-lance were considered. For the eccentric double lance, the effect of intersection angle of the two lances was investigated.

Keywords Lance configuration · Burnout · Raceway

Introduction

The blast furnace is the main manufacturing device to produce the crude materials of high quality steel. In the blast furnace, hot blast air of 1100–1200 °C is blown into furnace through the tuyères, and reacts with coke to produce heat and reduction

---

H. Huo (✉) · Z. Zhou · J. Wang · Q. Xue · Y. Zhang · Y. Liu
State Key Laboratory of Advanced Metallurgy, University of Science and Technology Beijing, Beijing 100083, People's Republic of China
e-mail: huohailong111@163.com

Z. Zhou
e-mail: zhouzhenfeng1989@sina.com

J. Wang
e-mail: wangjingsong@ustb.edu.cn

Q. Xue
e-mail: xueqingguo@ustb.edu.cn

Y. Zhang
e-mail: 15841693946@163.com

Y. Liu
e-mail: liuyingli112@126.com

J.-Y. Hwang et al. (eds.), *8th International Symposium on High-Temperature Metallurgical Processing*, The Minerals, Metals & Materials Series,
DOI 10.1007/978-3-319-51340-9_54

gases [1]. Pulverized coal has been injected (PCI) into the blast furnace to reduce the consumption of coke, since the 2nd oil crisis [2–5]. The implementation of PCI is increasing gradually on account of the advantage of coal, of low price and widespread distribution. However, unburnt char increases at increased pulverized coal injection rates. This can worsen the gas permeability and stability of the blast furnace. In brief, the effective combustion of coal in the raceway is main factor to improve the burnout of coal. Practically, the coal burnout can be increased to a certain degree by adjusting some operational parameters, for instance, blast temperature, oxygen enrichment, coal blending, and lance configurations [6, 7].

It is difficult to obtain detailed information of blast furnace, such as the gas distribution and temperature field and coal combustion behavior. At present, numerical simulation is used to guide furnace design and development [8, 9]. Computational fluid dynamics (CFD) is a common method for comprehending the characteristics and performance of coal in the blast furnace. Some CFD models have been developed to evaluate PCI operation. Majeski et al. [10] researched pulverized coal and natural gas injection into blast furnaces by means of CFD; the interaction of the fuel flow and blast air was investigated, and some effective ways to improve coal dispersion were found. Shen et al. [11] developed a simplified model of the raceway using CFD, and matched the calculation results against measurements obtaining by a test rig.

In this paper, a three-dimensional model was developed to simulate the lance-blowpipe-tuyère-raceway region of a blast furnace. The effects of the arrangement of a double lance and eccentric double lance were investigated.

Model Description

The gas-particle flow and coal combustion in the tuyère and raceway region were calculated based on the framework of the software package ANSYS–FLUENT. The mathematical formulation is described below.

Basic Equations

In this model, the gas phase is treated in an Eulerian framework and described by the steady-state Reynolds-averaged Navier–Stokes equations with a k–ε turbulence model [12]. The governing equations for the gas phases are summarized in Table 1.

Particles of pulverized coal were treated as a discrete phase, modeled using the Lagrangian method, where the trajectories of the discrete particles were determined by integrating Newton's second law of motion. The drag force and turbulence dispersion were included. Full coupling of mass, momentum, and energy of particles with the gas phase was implemented. The change of particle temperature was governed by three physical processes: convective heat transfer, latent heat transfer

Table 1 Governing equations for the gas phase

Mass	$\nabla \cdot (\rho U) = \sum_{n_p} \dot{m}$
Momentum	$\nabla \cdot (\rho UU) - \nabla \cdot \left[(\mu + \mu_t)\left(\nabla U + (\nabla U)^T\right)\right] = -\nabla\left(P + \frac{2}{3}\rho k\right) + \sum_{n_p} f_D$
Energy	$\nabla \cdot \left[\rho UH - \left(\frac{\lambda}{C_P} + \frac{\mu_t}{\sigma_H}\right)\nabla H\right] = \sum_{np} q$
Gas species	$\nabla \cdot \left[\rho UY_i - \left(\Gamma_i + \frac{\mu_t}{\sigma_{Y_i}}\right)\nabla Y_i\right] = W_i$
Turbulent kinetic energy	$\nabla \cdot \left[\rho Uk - \left(\mu + \frac{\mu_t}{\sigma_k}\right)\nabla k\right] = P_k - \rho\varepsilon$
Turbulent dissipation rate	$\nabla \cdot \left[\rho U\varepsilon - \left(\mu + \frac{\mu_t}{\sigma_\varepsilon}\right)\nabla \varepsilon\right] = \frac{\varepsilon}{k}(C_1 P_k - C_2\rho\varepsilon)$

associated with mass transfer, and radiative heat transfer. The governing equations for the particle phase are summarized in Table 2.

Where $\mu_t = C_\mu \rho \frac{k^2}{\varepsilon}$; $P_k = (\mu + \mu_t)\nabla U \cdot \left(\nabla U + (\nabla U)^T\right)$; i = O_2, CO_2, CO, VM, H_2, H_2O; $C_D = \max\left(24\left(1 + 0.15\mathrm{Re}^{0.687}\right), 0.44\right)$.

Combustion Process

Coal combustion is regarded as a multistage overlapping process: (i) preheating, (ii) devolatilization of raw coal, (iii) volatiles combustion, (iv) oxidation/gasification of the residual char.

(1) Release of volatiles

The devolatilization process releases volatiles ($C_\alpha H_\beta O_\gamma N_\delta$) and char (C(s)). The release of volatiles is simulated using the so-called two-competing-reactions model [13]. A pair of reactions with different rates (k_1, k_2) and volatile yields (α_1, α_2) compete to pyrolyse the raw coal.

Table 2 Governing equations for the particle phase

Mass	$\frac{dm_p}{dt} = -\dot{m}$
Momentum	$m_p \frac{dU_p}{dt} = -f_D$ $-f_D = \frac{1}{8}\pi d_p^2 \rho C_D \left\|U - U_p\right\|(U - U_p)$
Energy	$m_p C_p \frac{dT_p}{dt} = -q$ $\pi d_P \lambda N_\mu (T_g - T_p) + \sum \frac{dm_p}{dt} H_{reac} + A_p \varepsilon_p \left(\pi I - \sigma_B T_p^4\right)$

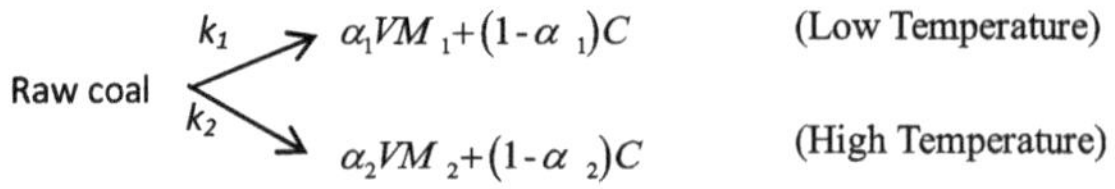

The rate constants k_1 and k_2 are in the Arrhenius format [14]:

$$k = A\ \exp(-E/T) \tag{1}$$

where A represents the Arrhenius rate constant, E represents the activation energy. A and E are $3.7 \times 10^5\ s^{-1}$ and 18,000 K at low temperature, and $1.46 \times 10^{13}\ s^{-1}$ and 30,189 K at high temperature.

The rate of volatiles production is given by:

$$\frac{dw(VM)}{dt} = (\alpha_1 k_1 + \alpha_2 k_2)C_0 \tag{2}$$

(2) Gas combustion

The volatiles are treated as one fuel gas ($C_\alpha H_\beta O_\gamma N_\delta$), and its combustion is simplistically represented by two overall reactions as follows:

$$\left(C_\alpha H_\beta O_\gamma N_\delta\right) + aO_2 \rightarrow bCO + cH_2O + dN_2 \quad R_1$$

$$CO + 0.5O_2 \rightarrow CO_2 \quad R_2$$

For the gas combustion, in addition to the combustion of volatiles, the combustion of hydrogen is included:

$$H_2 + 0.5O_2 \rightarrow H_2O \quad R_3$$

(3) Oxidation/gasification of the residual char

For the char reactions, the heterogeneous surface reaction model is used. The solid reactions include the following processes.

$$C(s) + 0.5O_2 \rightarrow CO \quad R_4$$

$$C(s) + CO_2 \rightarrow 2CO \quad R_5$$

$$C(s) + H_2O \rightarrow CO + H_2 \quad R_6$$

Table 3 Main parameters of the blast furnace

Blast volume (Nm^3/h)	Blast temperature (K)	Volume (m^3)	Coal ratio (kg/t)	Blast composition (wt%)	
				O_2	N_2
19,767	1473	120	150	21	79

Table 4 Properties of the coal

Proximate analysis (wt%)		Ultimate analysis	
Moisture	1	C	88.78
Volatiles	20.02	H	4.54
Ash	8.32	O	4.68
Fixed carbon	70.66	N	2

Geometry and Operating Condition

An oxygen blast furnace of 120 m^3 was built at a steel corporation in China. Therefore, some investigations were carried out based on the commercial blast furnace of 120 m^3 with 8 tuyères. The main operational parameters of the blast furnace are summarized in Table 3. The properties of coal used in this model are shown in Table 4.

The model simulated the lance-blowpipe-tuyère-raceway region of the practical blast furnace of 120 m^3. The blast furnace was dissected in 2007. The results of blast furnace dissection showed that the depth of the raceway was about 700 mm. The detailed parameters of this model are shown in Fig. 1. The base case is used for the investigation of the effect of conventional oxygen enrichment on coal combustion. The double coal lance and eccentric double coal lances are used for the

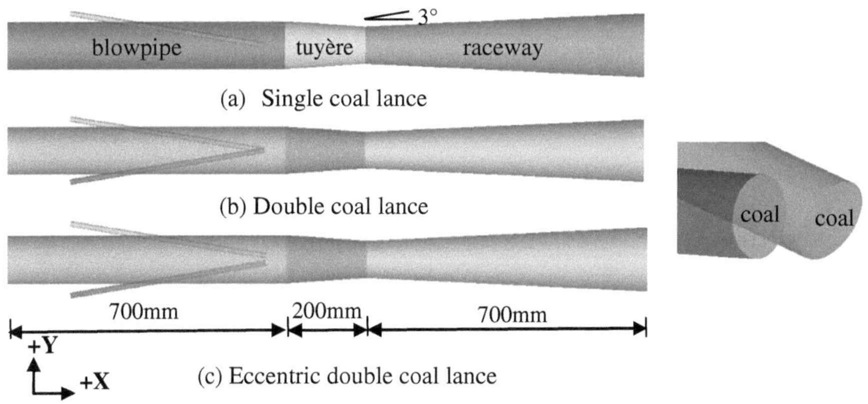

Fig. 1 Geometry and detailed information of models

investigation of the effect of the arrangement of lances on coal combustion. The main focus was the flow and combustion behavior of pulverized coal in the coal plume, which plays a significant role in determining the permeability of the bird nest. The raceway was designed as a tube of 700 mm long with a divergence angle of 3° based on previous work [15]. Because this geometric setting avoided flow recirculation, this simulation mainly focused on coal flow and combustion behavior of pulverized coal in the coal plume.

Results and Discussion

Effect of Double Coal Lance on Coal Combustion

Figure 2 shows the burnout with single and double lances. The burnout is 73.16% with the single coal lance and 72.69% with the double coal lance. The likely reason is that interaction between the two coal plumes limits the dispersion of coal particles, to the detriment to coal combustion, so the increase in the burnout is not obvious.

Figure 3 shows the coal burnout for different distances between coal lance tips in the Y direction. When the distance is 20 mm, the coal burnout is 72.69%, slightly decreasing to 71.67% at a distance of 40 mm, and 68.09% at 60 mm. It is expected that the coal particles would be more dispersed with increased distance between the coal lance tips. However, with double coal lances, the two coal plumes would contact at some position. This position moves downward with increased distance between the lance tips, and the coal particles would be less dispersed at the end of the raceway.

Figure 4 shows the coal burnout at different positions along the tuyère centerline for different coal lance tip distances. For distances less than 0.9 m, little coal burnout has occurred. This is because coal particles mainly absorb heat at the initial stage. At distances between 0.95 and 1.5 m, the burnout increases evidently. For lance tip separations of 20 and 40 mm, the burnout profiles are similar. When the distance between coal lance tips is 0.06 m, the burnout increases in the upstream part of coal plume.

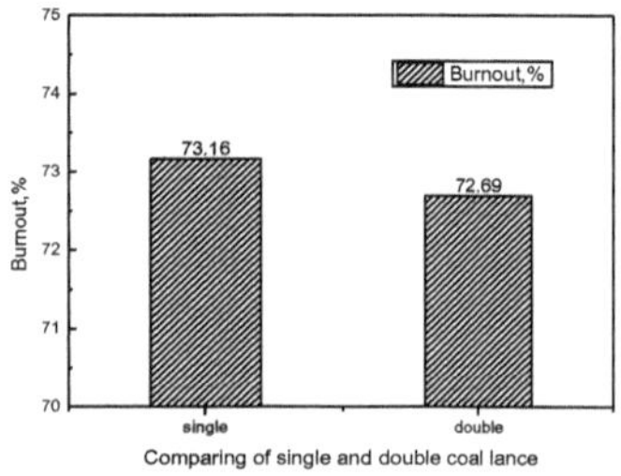

Fig. 2 The burnout of single and double lances

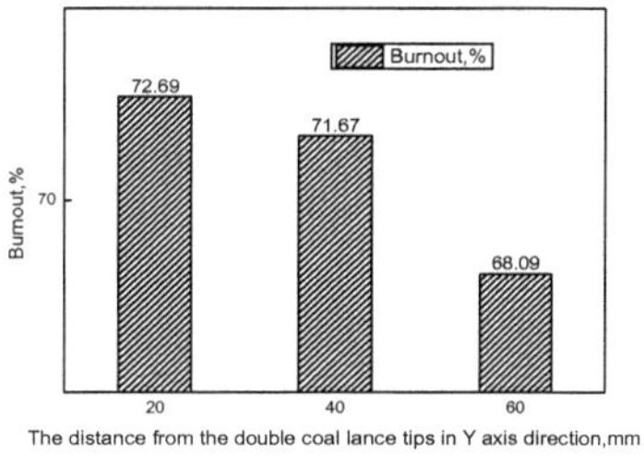

Fig. 3 The coal burnout for different distances between coal lance tips in the Y direction

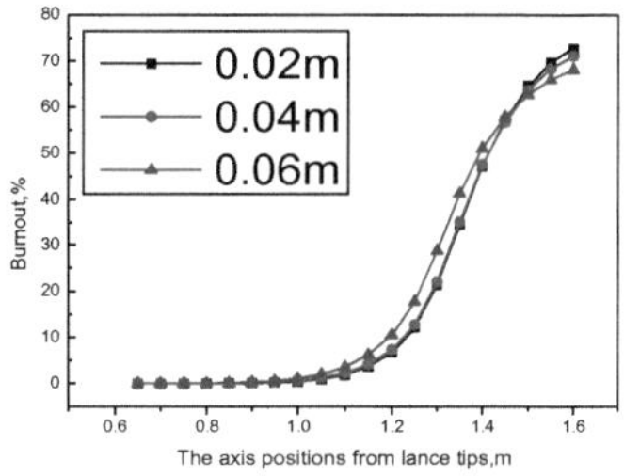

Fig. 4 The coal burnout at different positions along the tuyère centerline under different coal lance tip distances

Effect of Eccentric Double Coal Lance on Combustion

Figure 5 shows the burnout at the different intersection angles of the eccentric double coal lance. When the intersection angle is 20°, the burnout is 75.82%, increasing to 78.32% at 30° and 80.98% at 40°. In general, the burnout increases

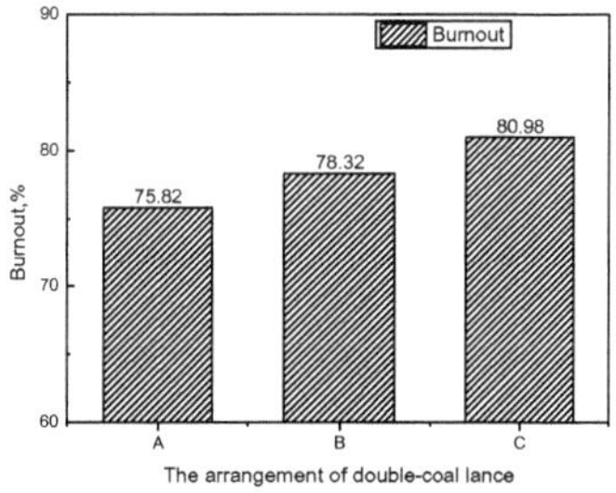

Fig. 5 The burnout at the different intersection angles for the eccentric double coal lance

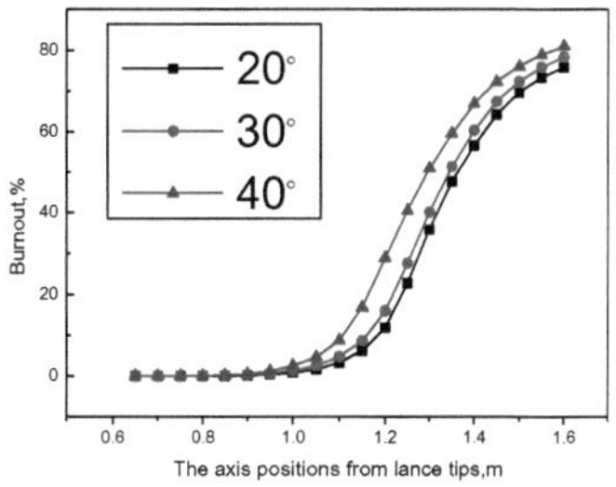

Fig. 6 The coal burnout at different positions along the tuyère centerline for different intersection angles

with increased intersection angle. The coal particles are more dispersed with the increase of intersection angle, improving coal combustion.

Figure 6 shows the coal burnout at different positions along the tuyère centerline for different angles. At all positions burnout increases with the increase of intersection angle. The main reason is that the coal particles are more dispersed with increased intersection angle. More oxygen is available around the particles.

Conclusions

The coal combustion behavior under different lance configurations was investigated. The main conclusions are as follows:

1. For a double coal lance, compared with a single lance, the increase in the burnout is not obvious. This is because the interaction between the two coal plumes limits coal combustion.
2. For eccentric double coal lances, burnout increases with the increase of intersection angle. The coal particles are more dispersed with increased of intersection angles; this is improves coal combustion.

References

1. S.W. Du, C.P. Yeh, W.H. Chen, C.H. Tsai, J.A. Lucas, Burning characteristics of pulverized coal within blast furnace raceway at various injection operations and ways of oxygen enrichment. Fuel **143**, 98–106 (2015)
2. T. Ariyama, M. Sato, Y.I. Yamakawa, Y. Yamada, M. Suzuki, Combustion behavior of pulverized coal in tuyere zone of blast furnace and influence of injection lance arrangement on combustibility. ISIJ Int. **6**, 476–483 (1994)
3. Q.F. Hou, M. Samman, J. Li, A.B. Yu, Modeling the gas-solid flow in the reduction shaft of COREX. ISIJ Int. **8**, 1772–1780 (2014)
4. K. Ishii, *Advanced Pulverized Coal Injection Technology and Blast Furnace Operation* (Elsevier, 2000)
5. R. Murai, Y. Kashihara, A. Murao, M. Sato, Convergent-divergent injection lance for the enhancement of combustion efficiency of pulverized coal at blast furnace. ISIJ Int. **5**, 770–776 (2016)
6. N. Vorobiev, M. Geier, M. Schiemann, V. Scherer, Experimentation for char combustion kinetics measurements: bias from char preparation. Fuel Process. Technol. **151**, 155–165 (2016)
7. S.G. Sahu, A. Mukherjee, M. Kumar, A.K. Adak, P. Sarkar, S. Biswas, P.K. Banerjee, Evaluation of combustion behaviour of coal blends for use in pulverized coal injection (PCI). Appl. Therm. Eng. **73**, 1014–1021 (2014)
8. M. Vascellari, S. Schulze, P. Nikrityuk, D. Safronov, C. Hasse, Numerical simulation of pulverized coal MILD combustion using a new heterogeneous combustion submodel. Flow Turbul. Combust. **92**, 319–345 (2014)
9. C.W. Chen, Numerical analysis for the multi-phase flow of pulverized coal injection inside blast furnace tuyere. Appl. Math. Model. **9**, 871–884 (2005)

10. A. Majeski, A. Runstedtler, J. D'alessio, N. Macfadyen, Injection of pulverized coal and natural gas into blast furnaces for iron-making: lance positioning and design, ISIJ Int. **7**, 1377–1383 (2015)
11. Y. Shen, A. Yu, P. Zulli, CFD modelling and analysis of pulverized coal injection in blast furnace: an overview. Steel Res. Int. **5**, 532–542 (2011)
12. B.E. Launder, D.B. Spalding, The numerical computation of turbulent flows. Comput. Methods Appl. Mech. Eng. **2**, 269–289 (1974)
13. S.K. Ubhayakar, D.B. Stickler, C.W. Von Rosenberg, R.E. Gannon, Rapid devolatilization of pulverized coal in hot combustion gases. Paper presented at the 16th Symposium (International) on Combustion (Pittsburgh, 1977), p. 427
14. Y. Shen, B. Guo, A. Yu, D. Maldonado, P. Austin, P. Zulli, Three-dimensional modelling of coal combustion in blast furnace. ISIJ Int. **48**, 777–786 (2008)
15. Y.S. Shen, D. Maldonado, B.Y. Guo, A.B. Yu, P. Austin, P. Zulli, Computational fluid dynamics study of pulverized coal combustion in blast furnace raceway. Ind. Eng. Chem. Res. **23**, 10314–10323 (2009)

Effect of Silicon on Removal of Phosphorus from High Phosphorus Si–Mn Alloy by CaO-Based Slag

Zhiqiang Zhou, Zizong Zhu, Yuchuan Ding and Shengnan Zhou

Abstract The effect of silicon on removal of phosphorus from high phosphorus Si–Mn alloy was investigated at 1673 K by CaO-based slag. It was found that the dephosphorization efficiency increases significantly as increasing the silicon content, which is due to the reaction between silicon and CaO at slag/alloy interface. The relation between silicon content with oxygen activity and phosphorus activity coefficient of Si–Mn melt was determined respectively as follows:

$$\lg a_O = -1.55x_{Si} - 1.67;\ \ln r_P = 9.26e^{\frac{x_{Si}}{0.0433}} + 11.0839$$

The carbon content in Si–Mn melt decreases continuously with the increasing silicon content, meanwhile, the decreasing rate of carbon content becomes rapid when the mole fraction of silicon is about 0.32. Si addition is beneficial for both dephosphorization process and low carbon Si–Mn alloy production.

Keywords Dephosphorization · Si–Mn alloy · Oxygen activity · Phosphorus activity coefficient

Introduction

Si–Mn alloy is widely used as deoxidizer and alloy additive in the metallurgical industry. Due to the requirement of light weight automobile for energy saving, the application of high strength and high ductility steel containing manganese for use in the automobile has become popular. Silicon and manganese are the main alloying elements used in the production of high manganese TRIP steel. The purity requirements of the alloy additives are also getting more and more rigorous. Phosphorus is one of the harmful microelements for most steel products, which can

Z. Zhou · Z. Zhu (✉) · Y. Ding · S. Zhou
College of Materials Science and Engineering, Chongqing University,
Chongqing 400044, People's Republic of China
e-mail: zhuzizong@163.com

J.-Y. Hwang et al. (eds.), *8th International Symposium on High-Temperature Metallurgical Processing*, The Minerals, Metals & Materials Series,
DOI 10.1007/978-3-319-51340-9_55

get into steel from alloy additives and influence adversely the quality of steel [1, 2]. Manganese ore resources in China are poor in quality and almost all of the phosphorus gets into the Si–Mn products during the production. Accordingly, dephosphorization (De-P) has been one of the most significant issues in the production of Si–Mn alloy.

Phosphorus exists mainly in the form of Mn_3P, MnP and SiP [3] and it is often distributed in the grain boundaries [4]. Mn_3P, MnP and SiP are more stable than Fe_3P from the point of view of standard Gibbs free energy values, which makes it more difficult to remove phosphorus from Si–Mn alloy. It is well known that silicon or manganese will be oxidized before phosphorus under oxidizing conditions. Therefore, the study on reducing Phosphorous of Si–Mn alloy by dephosphorization is important.

Great efforts have been made on thermodynamic studies of reducing De-P of Si–Mn alloy using metallic calcium [5–7] or CaO-based slags [8, 9]. Phosphorus in Si–Mn alloy is expected to be removed to the slag as phosphide by the following mechanism:

$$\frac{3}{2}[\mathrm{Ca}] + [\mathrm{P}] = (\mathrm{Ca}_{1.5}\mathrm{P})(\mathrm{s}) \tag{1}$$

$$[\mathrm{P}] + \frac{3}{2}(\mathrm{O}^{2-}) = (\mathrm{P}^{3-}) + \frac{3}{4}\mathrm{O}_2 \tag{2}$$

It is generally accepted that the De-P rate increases with the increasing silicon content in Si–Mn alloy [10–12]. However, there is little specific and systematic research on De-P mechanism in Si–Mn melt. Hence, the current study focuses on the thermodynamic behavior of silicon in Si–Mn alloy melt in order to clarify its role in the De-P of Si–Mn alloy.

Experimental

The experimental apparatus is composed of a molybdenum silicide ($MoSi_2$) furnace, a proportional integral differential control cabinet and an Argon gas cylinders. The temperature was controlled within ±2 K using a B-type (Pt30Rh-Pt6Rh) thermocouple and a DWT-702-type PID controller. The Si–Mn alloy mixed with varying amounts of Si–Fe alloy. The chemical composition is shown in Table 1.

Table 1 Chemical composition of Si–Mn alloy and Si–Fe alloy wt%

Name	Mn	Si	C	P	S	Fe
Si–Mn	67.49	17.58	1.50	0.95	0.04	12.14
Si–Fe	0.41	75.05	0.09	0.031	0.01	24.18

They were supplied from industry refining process and slag samples were prepared with reagent grade CaF_2 and CaO calcined from $CaCO_3$ at 1273 K for 8 h. The experimental set up is given in Fig. 1.

The Si–Mn alloy (100 g) mixed with varying Si–Fe alloy and CaO–CaF_2 (2:3) slag (20 g) were held in the graphite crucible and heated to 1673 K under Argon atmosphere at a flow rate of 200 ml/min. High purity argon contains approximately 10 ppm Oxygen. In order to remove Oxygen, Argon gas was purified using silica gel in addition to Mg turnings at 723 K. The reaction time was determined as 30 min on the basis of preliminary experiments. The change of phosphorus content showed a gradual declining trend with time and the De-P process was almost finished after reacting for 30 min when the slag and alloys were in the molten condition. Then the graphite crucible was taken out from the furnace quickly and quenched by dipping half of the crucible into brine. The slags and alloy samples were then crushed for chemical analyses. The Si, Mn, P and Ca of the alloys were determined by the inductively coupled plasma-atomic emission spectroscopy (ICP-OES). The C and S content of alloys were determined by the infrared carbon-sulfur analyzer (CS-444LS). The crystalline phases of powdered slags were identified by X-ray diffraction analysis (D/Max2500pc). Each component in final slags were measured by X-ray fluorescence spectroscopy (XRF; Bruker-S4 Explorer). The final compositions of the alloys and slags are listed in Table 2.

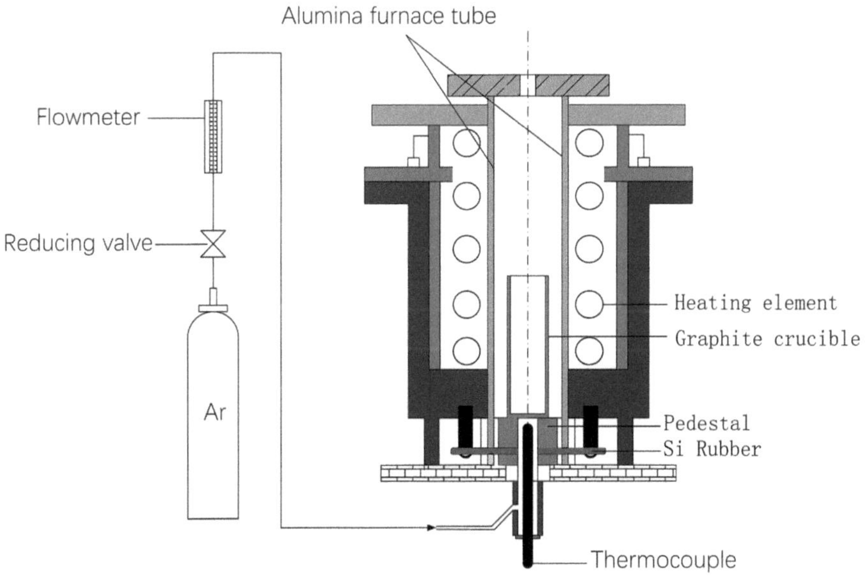

Fig. 1 Schematic of the experimental apparatus

Table 2 Chemical composition of Si–Mn alloy and slag samples after De-P

Metal (wt%/x_i^*)					Slag (wt%)			
Mn	Si	Fe	C	P	CaO	SiO_2	$Ca_{1.5}P$	SiC
66.19/0.5187	19.28/0.310	12.07/0.097	1.68/0.063	0.78/0.0113	40.89	4.52	3.62	1.18
65.04/0.5193	20.63/0.326	12.28/0.096	1.30/0.048	0.75/0.0107	37.21	4.97	3.96	1.84
65.32/0.5321	21.08/0.335	12.21/0.097	0.70/0.026	0.69/0.0099	35.26	6.32	4.88	2.03
64.38/0.5187	22.47/0.357	11.96/0.095	0.54/0.020	0.65/0.0093	33.19	8.13	6.75	4.17
64.43/0.5242	22.96/0.361	11.58/0.091	0.41/0.015	0.62/0.0088	29.57	9.54	7.13	4.88
63.93/0.5108	23.81/0.378	11.34/0.090	0.35/0.013	0.57/0.0082	27.05	11.31	7.84	5.13
63.60/0.5092	24.35/0.385	11.26/0.089	0.24/0.009	0.55/0.0078	26.81	12.25	8.16	5.84

x_i^* mean, mole fraction of element "i" in Si–Mn alloy

Results and Discussion

Influence of Silicon Content on De-P Efficiency

In the present study, the De-P efficiency (η_{De-P}, pct) is determined from the following equation:

$$\eta_{De-P}\,(\text{pct}) = \frac{w[\text{P}]^i - w[\text{P}]^f}{w[\text{P}]^i} \times 100\,\text{pct} \tag{3}$$

where the superscripts "i" and "f" represent the content of phosphorus before and after the De-P process, respectively.

Figure 2 illustrates that the De-P efficiency increases significantly as increasing the Si content, which is consistent with reports by other researchers [12]. The reaction between silicon and CaO at slag/alloy interface provides the required Ca to the De-P process. The corresponding reaction can be expressed as follows:

$$[\text{Si}] + 2(\text{CaO}) = (\text{SiO}_2) + 2[\text{Ca}] \tag{4}$$

$$\Delta G^{\theta} = 750.18 - 0.111T\ (\text{kJ/mol})$$

$$\Delta G = \Delta G^{\theta} + RT\ \ln J = 750.18 - 0.111T + RT\ \ln\frac{a_{\text{Ca}}^2 \cdot a_{\text{SiO}_2}}{a_{\text{CaO}}^2 \cdot a_{[\text{Si}]}} < 0 \tag{5}$$

where a_{Ca} a_{Si} a_{CaO} and a_{SiO_2} are the activities of Ca, Si, CaO and SiO_2, respectively. It is obvious that the reaction (4) cannot happen in the standard state. However, the CaO and $2CaO \cdot SiO_2$ in CaO–CaF_2-based slags are both saturated, the activities of these components can be considered as unity. Besides, the Ca content in Si–Mn melt is considerably lower than the Si content and hence $a_{Ca} \ll a_{Si}$. Therefore, the reaction (4) can happen under this experimental condition according to Eq. (5). Furthermore, as shown in Fig. 3, the crystalline phases of powdered slags identified

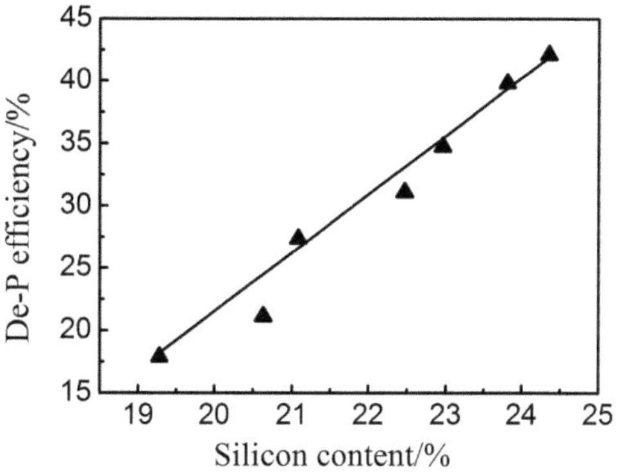

Fig. 2 Effect of silicon content on De-P efficiency

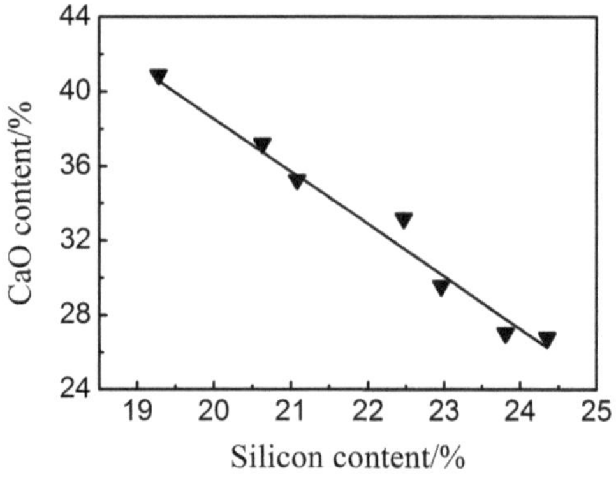

Fig. 3 Relationship between CaO content and Si content

by X-ray diffraction analysis illustrate that the CaO content reduces gradually as increasing the silicon content, which confirms that the reducing mechanism by silicon is due to the transfer of Ca from CaO-based slags to the Si–Mn melt according to the reaction (4).

Influence of Silicon Content on Oxygen Activity in Si–Mn Melt

The De-P product exists as calcium phosphide (Ca_3P_2) in the De-P refining process of Si–Mn alloy. The De-P efficiency would be significantly reduced if the oxygen activity in Si–Mn melt is high and part of required Ca will be oxidized. Thus, the oxygen activity should be maintained low enough in order to improve De-P efficiency. The oxygen activity in Si–Mn melt can be obtained from the Si–O equilibrium at slag/alloy interface. The corresponding reaction can be expressed as follows:

$$[\mathrm{Si}] + 2[\mathrm{O}] = (\mathrm{SiO_2}) \tag{6}$$

$$\Delta G^{\theta} = -594{,}285 + 229.76T = -RT \cdot \ln K \ (\mathrm{kJ/mol}) \tag{7}$$

$$K = \frac{a_{\mathrm{SiO_2}}}{a_{\mathrm{Si}} \cdot a_{\mathrm{O}}^2} \tag{8}$$

where a_{Si} and a_{O} are the activity of Si and O in Si–Mn melt, respectively. $a_{\mathrm{SiO_2}}$ is unity. The equilibrium constants of Eqs. (7) and (8) can be rewritten as follows:

$$\lg K = \frac{31{,}038.02}{T} - 12.0 \tag{9}$$

$$\lg K = -\lg a_{\mathrm{Si}} - 2\lg a_{\mathrm{O}} = -\lg r_{\mathrm{Si}} \cdot x_{\mathrm{Si}} - 2\lg a_{\mathrm{O}} \tag{10}$$

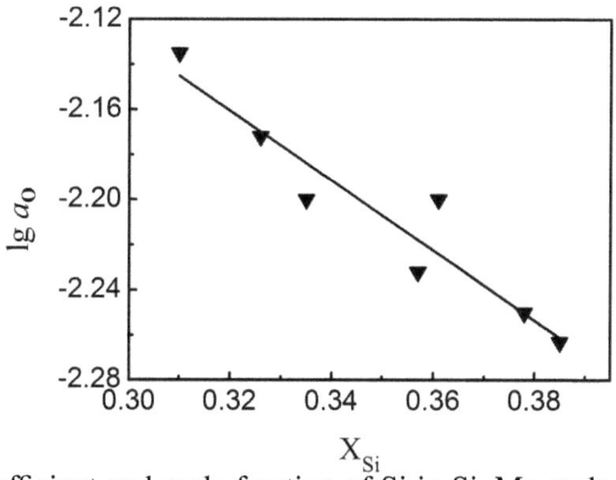

Fig. 4 Effect of silicon content on oxygen activity

where r_{Si} and x_{Si} are the activity coefficient and mole fraction of Si in Si–Mn melt, respectively. By coupling Eqs. (9) and (10), The oxygen activity in Si–Mn melt can be expressed as function of T and r_{Si} as follows:

$$\lg a_O = -\frac{15{,}519.01}{T} + 6 - \frac{\lg r_{Si} + \lg x_{Si}}{2} \tag{11}$$

Much work has been done to determine the activity coefficient of Si in Si–Mn melt [13–15]. The r_{Si} used in this paper in carbon-saturated Si–Mn melt has been determined over the range of compositions and temperatures similar to the present work [16]:

$$\begin{aligned} \ln r_{Si} &= 8.2237 - 13.4747x_{Si} - 123.5x_{Fe} - 28.5236x_{Mn}^2 \\ &\quad + 199.7x_{Mn}x_{Fe} + 79.5894x_{Si}x_{Fe} \end{aligned} \tag{12}$$

Thus, as shown in Fig. 4, the relationship between oxygen activity and Si content in Si–Mn melt is obtained by combining Eqs. (11) and (12). It is well known that reducing oxygen activity is beneficial for the De-P process. The oxygen activity in Si–Mn melt gradually reduces with the increase of Si content as expected. The oxygen activity in Si–Mn melt as a function of the molar content of Si is determined by linear fitting as follows:

$$\lg a_{[O]} = -1.55x_{Si} - 1.67 \tag{13}$$

Influence of Silicon Content on Activity Coefficient of Phosphorus in Si–Mn Melt

It is obvious that each element in Si–Mn melt has a certain effect on the activity coefficient of phosphorus. As shown in Fig. 2, the De-P efficiency increases significantly with increasing the silicon content. Hence, it is necessary to study the

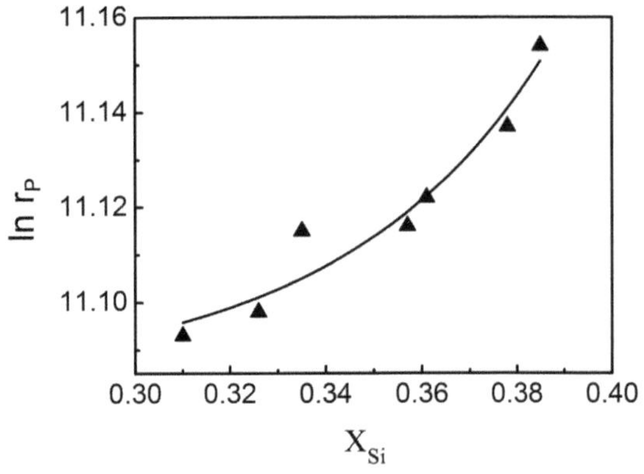

Fig. 5 Effect of activity coefficient of phosphorus

interaction relation between the activity coefficient of phosphorus and Si content in Si–Mn melt. In the Si–Mn–Fe–C–P system, the activity coefficient of phosphorus is determined as follows:

$$\ln r_{\mathrm{P}} = \varepsilon_{\mathrm{P}}^{\mathrm{Mn}} X_{\mathrm{Mn}} + \varepsilon_{\mathrm{P}}^{\mathrm{Si}} X_{\mathrm{Si}} + \varepsilon_{\mathrm{P}}^{\mathrm{Fe}} X_{\mathrm{Fe}} + \varepsilon_{\mathrm{P}}^{\mathrm{C}} X_{\mathrm{C}} + \varepsilon_{\mathrm{P}}^{p} X_{p} \quad (14)$$

where the standard state for Si, Mn, Fe and C is the hypothetical pure liquid at 1673 K, $\varepsilon_{\mathrm{P}}^{\mathrm{Mn}} = 12.0$ [16], $\varepsilon_{\mathrm{P}}^{\mathrm{Si}} = 10.7$ [12], $e_{\mathrm{P}}^{\mathrm{C}} = 16.6$ [12], $\varepsilon_{\mathrm{P}}^{\mathrm{Fe}} = 7.43$ [16], the fifth item in the right side of Eq. (14) can be neglected for the molar content of phosphorus in Si–Mn melt is minimum. Figure 5 shows the relationship between activity coefficient of phosphorus and Si content in Si–Mn melt of the present work. The activity coefficient of phosphorus in Si–Mn melt increases with the increase of Si content, which is beneficial for the removal of phosphorus in form of Ca_3P_2 and promote the De-P efficiency. The activity coefficient of phosphorus in Si–Mn melt as a function of the molar content of Si is determined by linear fitting as follows:

$$\ln r_{\mathrm{P}} = 9.26\mathrm{e}^{\frac{x_{\mathrm{Si}}}{0.0433}} + 11.0839 \quad (15)$$

Influence of Silicon Content on Carbon Solubility

Both Si and C have significant effect on De-P process of Si–Mn alloy. As shown in Fig. 6, the De-P efficiency linearly decreases with the increase of C content in Si–Mn melt. Therefore, it is necessary to reduce the C content in Si–Mn melt. In this study, high pure graphite crucible was used. Small amounts of graphite from the crucible walls can get into the melt inevitably, which may slightly reduce De-P efficiency. However, it is believed that the overall tendency of influence of silicon content on carbon solubility will not be affected. Many authors [17–20] have studied experimentally the solubility of C in Mn–Si–C melt. In order to clarify its

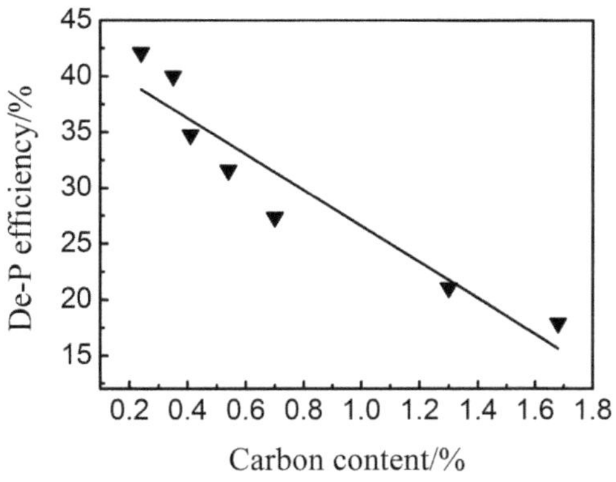

Fig. 6 Effect of carbon content on De-P efficiency

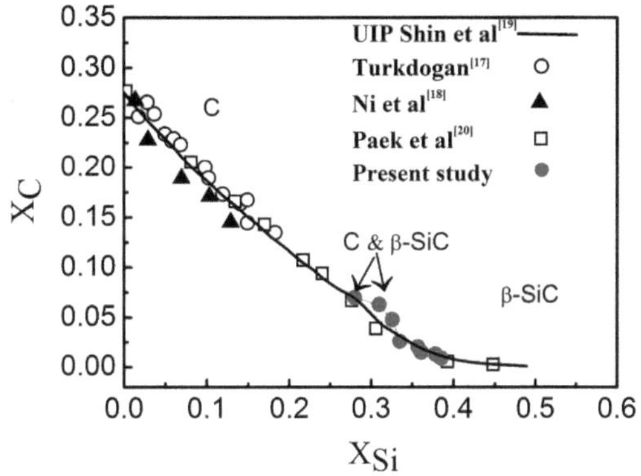

Fig. 7 Dependence of carbon solubility on Si content in Mn–Si–C melt

characteristic behavior, the obtained experimental data is integrated based on the dependence of C solubility on Si content in Mn–Si–C melt and it is reproduced in Fig. 7. Obviously, the C solubility in Si–Mn melt decreases continuously with increasing Si content. The decrease rate becomes rapid when the mole fraction of silicon is about 0.32, where the Mn–Si–C melt is in saturation with both SiC and C. It is considered that graphite is the stable phase in the lower Si range, whereas SiC becomes the stable phase under higher Si content [21]. Therefore, addition of Si decreases the solubility of C in Si–Mn melt, resulting in the nucleation and growth of SiC particles. Thereafter, carbide particles are believed to float up and dissolve into the slags at metal/slag interface [21]. The corresponding reaction can be expressed as follows:

$$Si(l) + C = SiC(\beta) \tag{16}$$

$$\Delta G^{\theta} = -99{,}098 + 29.798T \ (\text{J/mol})$$

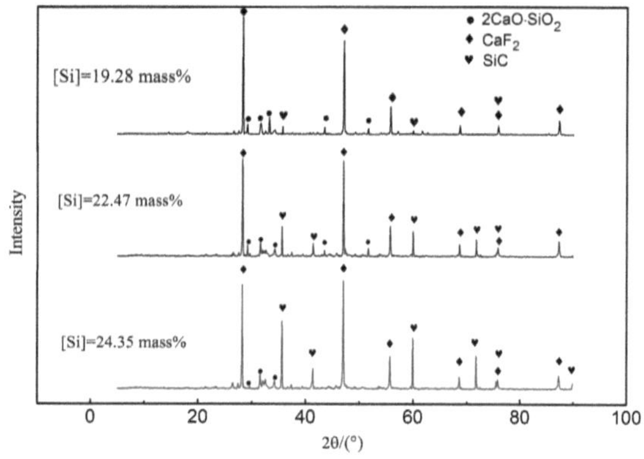

Fig. 8 XRD patterns of De-P slags with different silicon content

As shown in Fig. 8, SiC is first found in the De-P slag by XRD analyses, which is in agreement with the discussion presented above [22]. Hence, it is considered that the Si addition in Si–Mn alloy is beneficial for both the removal of carbon and phosphorus.

Conclusions

In this work, the influence of silicon on De-P efficiency, oxygen activity, phosphorus activity coefficient and carbon solubility was investigated by CaO-based slag at 1673 K. Based on results obtained in this paper, it can be concluded that:

(1) The De-P efficiency increases significantly as increasing the silicon content, which is mainly due to the reaction between silicon and CaO at slag/alloy interface.

(2) With the increase of silicon content, the oxygen activity gradually reduces and phosphorus activity coefficient increases, both of which are beneficial for the removal of phosphorus in the form of Ca_3P_2 and promote the De-P efficiency. The relation between silicon content with oxygen activity and phosphorus activity coefficient of Si–Mn melt was determined respectively as follows:

$$\lg a_O = -1.55x_{Si} - 1.67;\ \ln r_P = 9.26e^{\frac{x_{Si}}{0.0433}} + 11.0839$$

(3) The De-P efficiency linearly decreases with the increase of C content in Si–Mn melt. However, the C solubility in Si–Mn melt decreases continuously with increasing Si content. The rate of decrease becomes rapid when the mole fraction of silicon is about 0.32, where the Mn–Si–C melt is in saturation with both SiC and C. Hence, it is considered that Si addition in Si–Mn alloy is beneficial for both the removal of carbon and phosphorus.

References

1. L.J. Zhu, W.U. Di, X.M. Zhao, Effect of silicon content on thermodynamics of austenite decomposition in C–Si–Mn TRIP Steels. J. Iron. Steel Res. Int. **13**(3), 57–60 (2006)
2. S.I. Rudyuk et al., Effect of sulfur and phosphorus on the properties of steel 18B. Met. Sci. Heat Treat. **16**(12), 1056–1059 (1974)
3. X.S. Wen, H.Y. Li, Y.D. Sun, Application of reduction dephosphorization on low or extra low carbon ferromanganese-silicon. Ferroalloys **42**(2), 1–3 (2011)
4. S.P. Liu, P. Tang, Review of dephosphorization processes for ferromanganese. Ferroalloys **34**(4), 5–6 (2003)
5. D.M. Ni, W. Cheng, S.K. Wei, Study on dephosphorization of silicomanganese ferroalloy under reducing atmosphere with metallic calcium. Iron Steel **24**(9), 49–53 (1989)
6. T. Ogasawara et al., The activity of calcium in calcium-calcium halide fluxes. ISIJ Int. **36** (Suppl), S30–S33 (1996)
7. N. Masumitsu, K. Ito, R.J. Fruehan, Thermodynamics of Ca–CaF_2 and Ca–$CaCI_2$ systems for the dephosphorization of steel. Metall. Trans. B **19**(4), 643–648 (1988)
8. S. Tabuchi, N. Sano, Thermodynamics of phosphate and phosphide in CaO–CaF_2 melts. Metall. Mater. Trans. B **15**(2), 351–356 (1984)
9. J.H. Shin, J.H. Park, Thermodynamics of reducing refining of phosphorus from Si–Mn alloy using CaO–CaF_2 slag. Metall. Mater. Trans. B **43**(6), 1–4 (2012)
10. S.L. Zeng, J. Liu, H. Li, Experiments on outside furnace dephosphorization for ferromanganese-silicon. Ferroalloys **35**(1), 1–5 (2004)
11. S.L. Zeng et al., Optimization of technics on precipitation and dephosphorization for ferromanganese-silicon. Ferroalloys **40**(6), 1–5 (2009)
12. E. Aida, D. Min, N. Sano, Phosphorus distribution between Mn–Si melts and CaO–SiO_2–MnO–CaF_2 Slags, Tetsu-to-Hagane(J. Iron Steel Inst. Jpn.) **74**(10), 1931–1938 (1988)
13. K.Y. Hong et al., Reassessment of solution properties of liquid Mn–Si alloy. ISIJ Int. **53**(3), 544–546 (2013)
14. R. Gee, T. Rosenqvist, Scand. J. Metall. **5**, 57–62 (1976)
15. D.R. Swinbourne, W.J. Rankin, R.H. Eric, The effect of alumina in slag on manganese and silicon distributions in silicomanganese smelting. Metal. Mater. Trans. B **26**(1), 59–65 (1995)
16. S. Ueda, K. Morita, N. Sano, Thermodynamics of phosphorus in molten Si–Fe and Si–Mn alloys. Metall. Mater. Trans. B **28**(6), 1151–1155 (1997)
17. E.T. Turkdogan, R.A. Hancock, Trans. Inst. Min. Metall. **67**, 573–600 (1957)
18. R. Ni, Z. Ma, S. Wei, Thermodynamics of Mn–Fe–C and Mn–Si–C System. Steel Res. **61**, 113–116 (1990)
19. J.P. Shin, Y.E. Lee, Assessment of Mn–Fe–Si–C melt in unified interaction parameter formalism. Metall. Mater. Trans. B **47**(1), 216–227 (2016)
20. M.K. Paek et al., Thermodynamic interactions among carbon, silicon and iron in carbon saturated manganese melts. Korean J. Met. Mater. **50**(1), 45–51 (2012)
21. K. Tang, S.E. Olsen, Computer simulation of equilibrium relations in manganese ferroalloy production. Metall. Mater. Trans. B **7**(37), 599–606 (2006)
22. S.H. Ahn et al., Thermodynamic assessment of liquid Fe–Si–C system by unified interaction parameter model. ISIJ Int. **54**(4), 750–755 (2014)

Effect of Super Gravity on the Solidification Structure and C Segregation of High-Carbon Steel

Yuhou Yang, Bo Song, Gaoyang Song and Zeyun Cai

Abstract In this research, super gravity field was introduced in the solidification process of high carbon steel. The experimental results show that the solidification structure of Fe–1 wt% C steel can be largely refined by applying super gravity at the cooling rate of 10 °C/min. Both the primary and secondary dendrite length was markedly reduced. In addition, the eutectic structure in the inter-dendrite was also refined with the obvious reduce of eutectic space. However, the macro-segregation of C along the direction of super gravity was caused with higher weight percent at the upper area than that at the bottom of the sample. The mechanism of solidification structure refinement and the C macro-segregation of Fe–1 wt% C steel in super gravity field were also discussed in this paper.

Keywords Super gravity · High-carbon steel · Solidification structure · C segregation

Y. Yang · B. Song (✉) · G. Song · Z. Cai
School of Metallurgical and Ecological Engineering, University of Science and Technology Beijing, Beijing 100083, China
e-mail: songbo@metall.ustb.edu.cn

Y. Yang
e-mail: yyhyangyuhou@163.com

G. Song
e-mail: sgy_beike@163.com

Z. Cai
e-mail: zexiaoyun@126.com

Y. Yang · B. Song · G. Song · Z. Cai
State Key Laboratory of Advanced Metallurgy, University of Science and Technology Beijing, Beijing 100083, China

J.-Y. Hwang et al. (eds.), *8th International Symposium on High-Temperature Metallurgical Processing*, The Minerals, Metals & Materials Series,
DOI 10.1007/978-3-319-51340-9_56

Introduction

External fields exhibit great advantages in the solidification process of metals. Many researches have conducted with respect to electric current, magnetic and the ultrasonic fields [1–6]. However, rarely reports were related to the effect of super gravity field on the solidification structure of metals. The existing researches associated with super gravity field are mainly focus on the separation of impurities or valuable components [7, 8] and the centrifugal filling [9–12]. Previous studies about the effect of super gravity on the solidification structure of metals were mainly conducted on the low-melting temperature metals [13, 14], and few attempts had been made on the high-melting temperature metals due to the strict requirement for the apparatus. As a result, uncertainties still remain for the influence of super gravity field on the solidification structure of high-melting metals.

In addition, the segregation behavior of solute elements in super gravity field is also attractive, which markedly influence the mechanical properties of the castings. Zhao et al. [15] found that the impurity elements of Fe and Si in Al segregated to the bottom of the sample along the direction of super gravity. Tabushi et al. [16] found that the copper content at the outer surface of the casting is higher than that in the inner in their study about the density and hardness gradients of Al–Al_2Cu alloy by centrifugal method. While Watanabe et al. [17] found the opposite experimental results that the copper content at the outer surface is less than that at the inner surface of the ring when they studied the functionally graded materials of Al–3 wt% Cu alloy by a centrifugal in situ method. However in another research [18] on the formation mechanism of graded composition in Al–20 wt% Cu materials by centrifugal method, they found that the copper content increased along the direction of centrifugal force. Therefore, the results concerning the segregation behavior of solute elements in super gravity field between researchers varied much.

In this paper, the solidification structure evolution and the C segregation behavior of high carbon steel in super gravity field were investigated.

Experimental

In this research, super gravity field was generated by the autorotation of the heating chamber. The sample was heated by microwave controlled by a temperature controller within the observed precision range of ±5 °C by an R type thermocouple. The super gravity field is evaluated by gravity coefficient which is defined as the ratio of centrifugal acceleration to the normal acceleration via Eq. (1).

$$G = \sqrt{g^2 + (\omega^2 r)^2}/g = \sqrt{g^2 + \left(\frac{N^2\pi^2 r}{900}\right)^2}/g \tag{1}$$

where G is the gravity coefficient; ω is the rotational angular velocity, rad s^{-1}; N is rotating speed, r min^{-1}; R is the distance between the rotation center and the sample, 0.1 m; g is the normal gravity acceleration, 9.8 m s^{-2}.

The high carbon steel containing about 1 wt% C (Fe–1 wt% C) for super gravity experiments is manufactured by technically pure Fe and pig iron (wt% C = 4.8). Pure Fe was melted at 1600 °C with argon gas protection in molybdenum wire furnace, then specific mass of pig iron was added into the melt. The molten alloy was then water quenched after heat preservation for 10 min at 1600 °C. The basic steel for super gravity experiments was obtained in this process.

In the super gravity experiments, the basic steel (140 g) were put into an alumina crucible (inner diameter 24 mm) to be heated to 1600 °C in the heating furnace with argon gas protection. After holding at 1600 °C for 10 min, the sample was cooled down at a cooling rate of 10 °C/min and the centrifuge apparatus was started and adjusted to the target angular velocity to apply super gravity field when temperature fell to the liquidus temperature. The apparatus was not shut off until temperature fell to 50 °C lower than the solidus temperature to ensure complete solidification. Then the sample was furnace cooled to room temperature. Simultaneously, a parallel sample was obtained in normal gravity field under the same cooling conditions. Every sample was sectioned longitudinally into two halves along the central plane, one of which was burnished, polished and etched with 50% hydrochloric acid solution for metallographic examination. The macrostructure and the microstructure were examined by a digital camera and an optical microscope (9XB-PC type), respectively. The carbon content was detected by Carbon-Sulfur Analyser (EMIA-820V) at different positions of the samples in normal gravity field and super gravity field.

Results and Discussion

Macrostructures of Fe–1 wt% C Steel in Normal Gravity Field and Super Gravity Field

Figure 1 shows the macrostructures of Fe–1 wt% C steel obtained in normal gravity field and super gravity field at the cooling rate of 10 °C/min. When in normal gravity field, as exhibited in Fig. 1a, the solidification structure is extremely developed dendrite. Both the primary and the secondary dendrite are coarse, which is somewhat different with the conventional casting structure with the three distinct zones: the chilling zone close to the mold surface, the columnar zone and the equiaxed zone in the center of the casting. That is because a large temperature gradient in the melt will be generated when the molten metal is poured into a mold conventionally. While in the present experiments, the sample was cooled down in the furnace at the controlled cooling rate of 10 °C/min without the chilling effect, in which condition, temperature gradient was not that large. While in super gravity

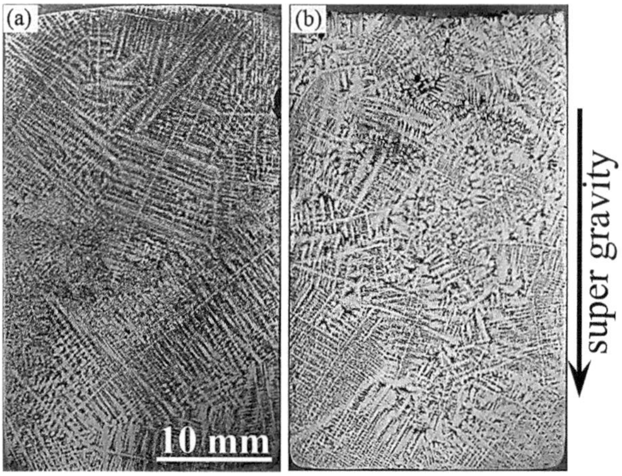

Fig. 1 Macrostructures of Fe–1 wt% C steel in normal gravity field and in super gravity field at the cooling rate of 10 °C/min. **a** G = 1 and **b** G = 80

field of $G = 80$, significant solidification structure refinement is observed as the number of the dendrites increased obviously in the whole section and both the primary and the secondary dendrite length was markedly reduced as shown in Fig. 1b.

In general, the solid phase will float upward or settle down in the liquid if the density difference is present between the solid and liquid theoretically. However, considering the large viscosity of the molten metal and the small size of the crystal nuclei, the crystal nuclei can hardly move in the melt during solidification process. Therefore, in general obvious floatation or sedimentation of crystals can not occur naturally. However, obvious crystals floatation or sedimentation can be obtained in super gravity field for the fact that the motion of a spherical solid particle in the viscous liquid can be expressed as follows according to Stokes' law [19, 20].

$$V_r = \frac{d^2(\rho_l - \rho_p)Gg}{18\eta} \tag{2}$$

where V_r is the particle's moving velocity in the viscous liquid, m s^{-1}; d is the diameter of the particle, m; ρ_l and ρ_p are the densities of the liquid and solid particle, kg m^{-3}; η is the dynamic viscosity of the liquid, Pa s.

The particle's moving velocity in equilibrium is in direct proportion to the square of its diameter and gravity coefficient according to Eq. (2). So, with respect to the present research, when the primary γ-Fe nuclei formed in the melt, they began to settle down under the effect of super gravity since the density of γ-Fe is larger than that of the remaining liquid shown in Fig. 2, calculated via Jmatpro 7.0 software. As a result, large amount of crystals sedimentation happened, causing the crystal

Fig. 2 Density change of primary γ-Fe and the remaining liquid during solidification process

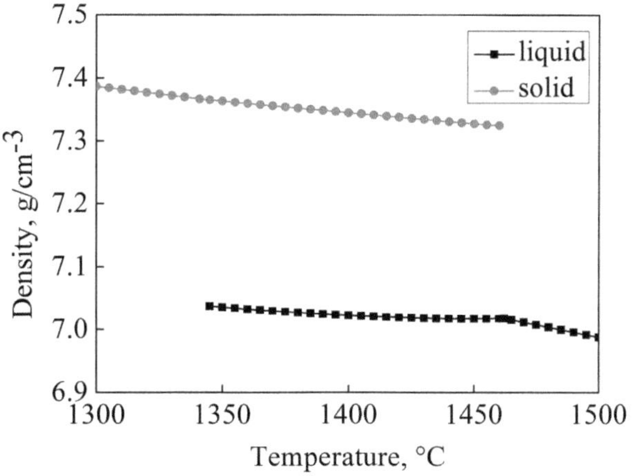

rain and leading to the multiplication of crystal nuclei, which was responsible for the solidification structure refinement. In fact, the crystal rain mechanism is also reported in the study of electric current pulse and pulse magneto-oscillation on the grain refinement of pure Al [21, 22].

Microstructure of the Eutectic Phase in Normal Gravity Field and Super Gravity Field

In order to investigate the effect of super gravity on the eutectic structure, both the eutectic phase of the samples obtained in normal gravity field and super gravity field were observed at the center region of the casting. The results are shown in Fig. 3. It can be seen that in normal gravity field, the eutectic cementite is coarse and the eutectic space is very large. While in super gravity field, the cementite phase is markedly refined and the eutectic space is significantly reduced compared with that in normal gravity field. This may be because in super gravity field, the turbulent flow is very violent [23], leading to a faster heat transfer, which will make the eutectic structure more refined.

In fact, concerning the evolution of the eutectic structure in super gravity field, the mechanism is not very clear. In convectional casting, the turbulent convection is generated at the interface of the solid and the liquid because of the abrupt mold rotation after the molten metal was poured into the mold. While this is not the case like that in convectional casting for the present research since the metal was melted and solidified in the furnace. As a result, problems remain that how the turbulent convection is generated by super gravity and how the turbulent convection affect the eutectic structure. Therefore, more investigations are needed.

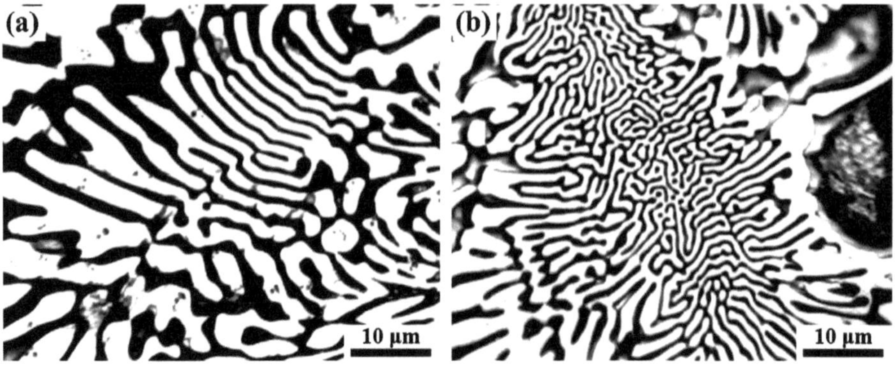

Fig. 3 Microstructures of eutectic phase in normal gravity field and super gravity field. **a** G = 1 and **b** G = 80

Macro-segregation of C in Super Gravity Field

In order to investigate the segregation behavior of element of carbon in super gravity field, the carbon content at different positions of the samples in normal gravity field and super gravity field, which is depicted in Fig. 4. We can know that in normal gravity field, no obvious segregation appears, but in super gravity field, the carbon content at the top of the sample is much higher than the average content (wt% C = 1) and it gradually decreases with the position approaching the bottom of the sample. As a result, the macro-segregation of carbon was generated by super gravity.

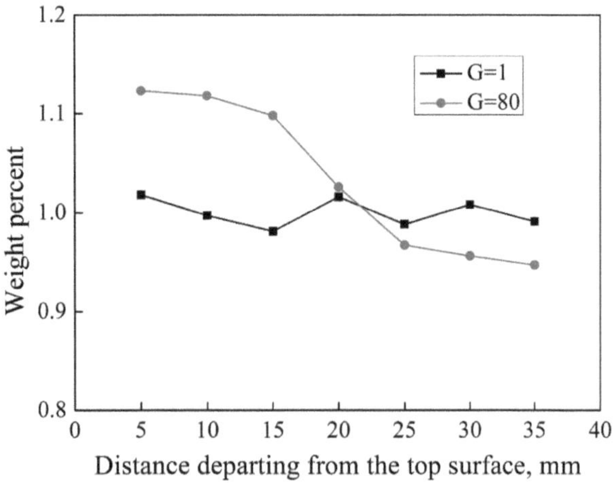

Fig. 4 The content of C at different positions of samples in normal gravity and super gravity field

It is known that one of the most important characteristics of super gravity field is the great intensification of segregation behavior as long as the density differences exist. Fu et al. [24] found that the element whose density is greater than molten steel will segregated along the direction of centrifugal force, while the element whose density is lower than the molten metal will segregate along the opposite direction of the centrifugal force in the study on the segregation mechanism of high speed steel rolls in centrifugal cast. They believed that the mechanism of the segregation is that the atom clusters are formed in the melt and such clusters have different densities with the melt, leading to the relative movement under centrifugal force. While Zhao et al. [15] believed that the element partial concentration of the liquid ahead of the freezing interface would be greater than that in the main liquid, leading to the density difference. When applying super gravity, the local liquid enriching more elements migrated along the direction of super gravity field.

In the present research, the carbon is more enriched in the remaining liquid phase along with the proceeding solidification process shown in Fig. 5, since the partition coefficient of carbon is much less than 1 [25]. Then it is also demonstrated in Fig. 6 that the density of liquid steel gradually decreases with increasing carbon content. Therefore, the local liquid enriched of carbon, which has a lower density than the main melt will move along the opposite direction of super gravity. As a result, the carbon content at the top of the sample is higher than that at the bottom part, which is consistent with the results by Hanguang Fu.

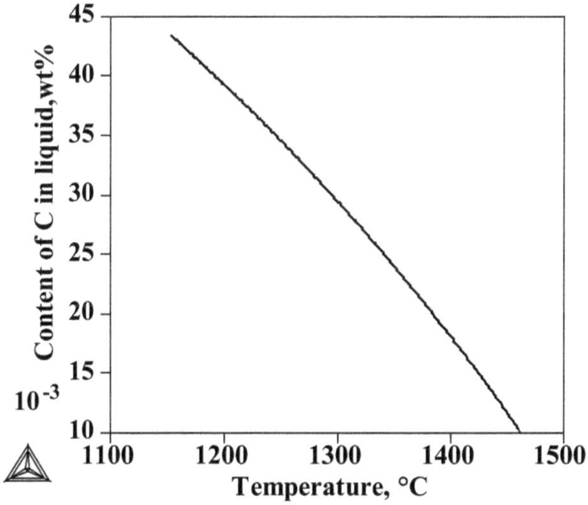

Fig. 5 The content of carbon in liquid during solidification process calculated via Thermal-Calc software

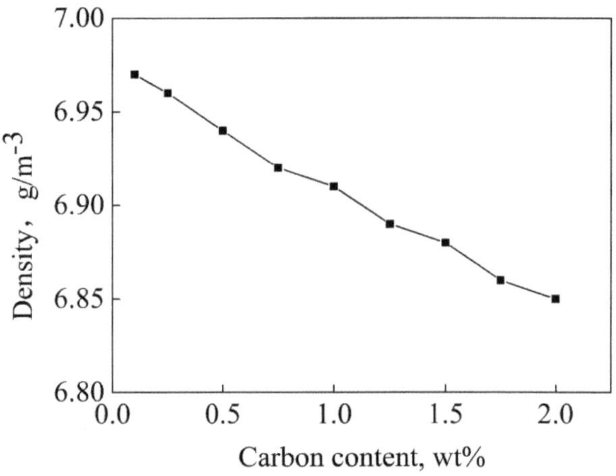

Fig. 6 Density change of liquid Fe–C alloy with carbon content at 1600 °C calculated via Jmatpro7.0 software

Conclusions

(1) The solidification structure of Fe–1 wt% C steel can be obviously refined in super gravity field with the great reduce of both the primary dendrite length and the secondary dendrite size. The mechanism is the extremely intensification of crystal settling by super gravity, leading to the crystal multiplication.
(2) The eutectic structure is markedly refined by super gravity with the large reduce of eutectic space.
(3) The macro-segregation of carbon is generated in super gravity with the higher content at the top of the sample and the lower content at the bottom. The largely intensified movement of the carbon-enriched liquid, which has a lower density than the bulk melt contributes the main reason.

Acknowledgements This work was supported by the National Natural Science Foundation of China (No. 51234001). The authors would like to sincerely thank the State Key Laboratory of Advanced Metallurgy, University of Science and Technology Beijing for using the super gravity apparatus.

References

1. J. Li et al., Research on solidification structure refinement of pure aluminum by electric current pulse with parallel electrodes. Mat. Sci. Eng. A **490**(1), 452–456 (2008)
2. J. Ma et al., Grain refinement of pure Al with different electric current pulse modes. Mater. Lett. **63**(1), 142–144 (2009)
3. Y.Y. Gong et al., Structure refinement of pure aluminum by pulse magneto-oscillation. Mat. Sci. Eng. A **497**(1), 147–152 (2008)
4. X. Jian et al., Effect of power ultrasound on solidification of aluminum A356 alloy. Mater. Lett. **59**(2), 190–193 (2005)

5. Y. Cui et al., Effect of ultrasonic vibration on unmixed zone formation. Scripta Mater. **55**(11), 975–978 (2006)
6. X. Jian et al., Refinement of eutectic silicon phase of aluminum A356 alloy using high-intensity ultrasonic vibration. Scripta Mater. **54**(5), 893–896 (2006)
7. G.Y. Song et al., Separating behavior of nonmetallic inclusions in molten aluminum under super-gravity field. Metall. Mater. Trans. B **46**(5), 2190–2197 (2015)
8. J.C. Li et al., Innovative methodology to enrich britholite (Ca_3Ce_2 [(Si, P)O_4] 3F) phase from rare-earth-rich slag by super gravity. Metall. Mater. Trans. B **45**(4), 1272–1280 (2014)
9. K. Watanabe et al., Casting behavior of titanium alloys in a centrifugal casting machine. Biomaterials **24**(10), 1737–1743 (2003)
10. G. Chirita et al., Advantages of the centrifugal CASTING technique for the production of structural components with Al–Si Alloys. Mater Des. **29**(1), 20–27 (2008)
11. K. Liu et al., Single step centrifugal casting TiAl automotive valves. Intermetallics **13**(9), 925–928 (2005)
12. J. Bohacek et al., Simulation of horizontal centrifugal casting: mold filling and solidification. ISIJ Int. **54**(2), 266–274 (2014)
13. L.X. Zhao et al., Influences of super-gravity field on aluminum grain refining. Metall. Mater. Trans. A **41**(3), 670–675 (2010)
14. C.C. Battalle et al., The effect of enhanced gravity levels on microstructural development in Pb-50 wt pct Sn alloys during controlled directional solidification. Metall. Mater. Trans. A **25** (4), 865–870 (1994)
15. L.X. Zhao et al., Removal of low-content impurities from Al by super-gravity. Metall. Mater. Trans. B **41**(3), 505–508 (2010)
16. K. Tabushi, H. Sato, Y. Watanabe, Effect of casting condition on density and hardness gradients of Al–Al_2Cu alloy FGM fabricated by centrifugal in situ method. Mater. Sci. Forum **631**, 449–454 (2010)
17. Y. Watanabe et al., Density and hardness gradients of functionally graded material ring fabricated from Al-3 mass% Cu alloy by a centrifugal in-situ method. Mater. Trans. **48**(11), 2945–2952 (2007)
18. Y. Watanabe et al., Formation mechanism of graded composition in Al–Al_2Cu functionally graded materials fabricated by a centrifugal in situ method. Acta Mater. **53**(6), 1631–1641 (2005)
19. Y. Watanabe et al., A novel fabrication method for functionally graded materials under centrifugal force: the centrifugal mixed-powder method. Materials **2**(4), 2510–2525 (2009)
20. Y. Watanabe et al., Particle size distributions in functionally graded materials fabricated by the centrifugal solid-particle method. Compos. Sci. Technol. **62**(6), 881–888 (2002)
21. X.L. Liao et al., Refining mechanism of the electric current pulse on the solidification structure of pure aluminum. Acta Mater. **55**(9), 3103–3109 (2007)
22. Y.Y. Gong et al., Structure refinement of pure aluminum by pulse magneto-oscillation. Mater. Sci. Eng., A **497**(1), 147–152 (2008)
23. C.N. Wei et al., A study of IN-713LC superalloy grain refinement effects on microstructure and tensile properties. Mater. Chem. Phys. **80**(1), 89–93 (2003)
24. H.G. Fu et al., A study of segregation mechanism in centrifugal cast high speed steel rolls. Mater. Sci. Eng., A **479**(1), 253–260 (2008)
25. M.C. Schneider et al., Formation of macrosegregation by multicomponent thermosolutal convection during the solidification of steel. Metall. Mater. Trans. A **26**(9), 2373–2388 (1995)

High Temperature Distribution Measurement of the Blast Furnace Raceway Through Imaging Techniques and Optimization Algorithms

Ruixuan Zhang, Shusen Cheng and Yan Li

Abstract The high temperature distribution of the blast furnace raceway plays an important role in continuous and steady production. There are a number of techniques which have been developed for flame temperature measurements, but most of them have considerable errors and instability. To obtain more accurate results, an efficient optimization algorithm is necessary. This paper presents a novel measurement approach based on the two-color method and an optimization algorithm using images of raceway flame captured by the RGB CCD (Charged-couple Device) camera to calculate the projection temperature distribution. Furthermore, both the back-propagation neural network and the principal component analysis are applied to calibrate the correlation between blackbody temperature and the gray-level intensity. Experimental results obtained from the blast furnace raceway demonstrate that the proposed optimization approach performs well in the measurement of flame temperature compared with the measured temperature.

Keywords Flame temperature · RGB CCD camera · Blast furnace raceway · Back-propagation neural network · Principal component analysis

Introduction

Combustion temperature of the blast furnace raceway is a significant characteristic parameter that is important in the blast furnace operation. The flame temperature can reach up to more than 2000 °C. It's difficult to use the thermocouple and other traditional measurement methods to obtain accurate results. As a result, it's necessary to find a kind of simplified and efficient method to measure the high temperature of the blast furnace raceway. With the development of the CCD

R. Zhang (✉) · S. Cheng · Y. Li
School of Metallurgical and Ecological Engineering, University of Science and Technology Beijing, Beijing 100083, China
e-mail: 373277846@qq.com

J.-Y. Hwang et al. (eds.), *8th International Symposium on High-Temperature Metallurgical Processing*, The Minerals, Metals & Materials Series,
DOI 10.1007/978-3-319-51340-9_57

(Charged-couple Device) camera which has a stable and reliable performance, high image resolution and resistance at high temperature, the temperature measurement method based on the digital imaging technique has become popular.

In recent years, substantial efforts have been made in developing temperature measurement methods using digital images. Huang et al. demonstrated a CCD camera system to measure transient two-dimensional temperature [1, 2]. Lu et al. proposed an imaging-based multicolor pyrometric system for the monitoring of the temperature distribution in a coal-fired flame. The emissivity and the calibration coefficient were also reported in detail [3, 4]. Lou et al. reported experimental investigations on the estimation of radiative properties and temperature distributions in a coal-fired boiler by an imaging processing system where RGB CCD cameras and a radiative imaging model were used [5].

Although extensive research have been conducted on the temperature measurement methods based on the digital imaging technique. There are few studies contraposing to the combustion temperature of the blast furnace raceway. This paper focuses on the measurement of high temperature distributions of the blast furnace raceway based on the digital imaging technique and the two-color method which need calibration on the blackbody furnace to obtain the calibration coefficient. However, the calibration coefficient obtained from the experiment generally has weak universality for the temperature limit and the large temperature interval of the blackbody furnace. The high temperature calculated by this coefficient may be quite inaccurate. Consequently, the back-propagation neural network and the principal component analysis are applied to obtain accurate estimation of the calibration coefficient of high temperature.

Methodology

Two-Color Method

Temperature of the blast furnace raceway is under 2500 °C. Wavelengths of Visible light range from 380 to 780 nm. As a result, Planck's radiation law can be replaced by Wien's radiation law in this measurement approach, i.e.,

$$E(\lambda, T) = \varepsilon(\lambda, T)\frac{C_1}{\lambda^5}\exp\left(-\frac{C_2}{\lambda T}\right) \tag{1}$$

where $E(\lambda, T)$ is the monochromatic emissive power (W/m^3), λ is the wavelength (nm), $\varepsilon(\lambda, T)$ is the monochromatic emissivity, T is absolute temperature (K), C_1 and C_2 are the first and second Planck's constants. Then temperature can be given by the ratio between $E(\lambda_1, T)$ and $E(\lambda_2, T)$ at wavelengths λ_1 and λ_2, i.e.,

$$T = \frac{C_2\left(\frac{1}{\lambda_2} - \frac{1}{\lambda_1}\right)}{\ln\frac{E(\lambda_1, T)}{E(\lambda_2, T)} - \ln\frac{\varepsilon(\lambda_1, T)}{\varepsilon(\lambda_2, T)} + 5\ln\frac{\lambda_1}{\lambda_2}} \tag{2}$$

The flame emissivity has little change when two different wavelengths are close to each other. Consequently, the radiating body can be approximately assumed as a graybody which means:

$$\frac{\varepsilon(\lambda_1, T)}{\varepsilon(\lambda_2, T)} = 1 \tag{3}$$

Equation (3) can be written as,

$$T = \frac{C_2\left(\frac{1}{\lambda_2} - \frac{1}{\lambda_1}\right)}{\ln\frac{E(\lambda_1, T)}{E(\lambda_2, T)} + 5\ln\frac{\lambda_1}{\lambda_2}} \tag{4}$$

Measurement Principle

The gray-level intensities of each pixel in images grabbed by the RGB CCD camera is composed of red (700 nm), green (546.1 nm) and blue (435.8 nm) three colors. Through the establishment of the relationship between the gray-level intensities of three colors and the monochromatic emissive power, temperature of each pixel can be obtained using the two-color method [6]. The gray-level intensities of three colors of the RGB CCD camera can be expressed by,

$$\begin{cases} R = A_R \int_{380}^{780} E(\lambda, T) T_R(\lambda)\, d\lambda \\ G = A_G \int_{380}^{780} E(\lambda, T) T_G(\lambda)\, d\lambda \\ B = A_B \int_{380}^{780} E(\lambda, T) T_B(\lambda)\, d\lambda \end{cases} \tag{5}$$

where R, G and B are the gray-level intensities of red (R), green (G) and blue (B) colors which represent response values of three channels of the RGB CCD camera. $T_R(\lambda)$, $T_G(\lambda)$ and $T_B(\lambda)$ are the spectral response characteristics of the three channels, respectively. A_R, A_G and A_B are the product of the gain and the photoelectric conversion coefficient of three channels. $T_R(\lambda)$, $T_G(\lambda)$ and $T_B(\lambda)$ can be seen as the delta function ideally which may lead to some errors. Consequently, Eq. (5) can be simplified as,

$$\begin{cases} R = K_R E(\lambda_R, T) \\ G = K_G E(\lambda_G, T) \\ B = K_B E(\lambda_B, T) \end{cases} \tag{6}$$

Considering that the CCD sensor gives higher spectral sensitivities for the R and G wavelength bands than that for the blue band, these two bands are selected for the temperature calculation and thus a better signal-to-noise ratio can be achieved. The temperature expression is as following:

$$T_{R\&G} = \frac{C_2\left(\frac{1}{\lambda_G} - \frac{1}{\lambda_R}\right)}{\ln\frac{R}{G} + \ln\frac{K_G}{K_R} + 5\ln\frac{\lambda_R}{\lambda_G}} \tag{7}$$

where λ_R and λ_G are the central wavelengths of R channel and G channel of the RGB CCD camera, K_R and K_G are the conversion coefficient between radiant energy and gray-level intensities of images, $\ln(K_G/K_R)$ is the calibration coefficient represented by C.

The calibration coefficient (C) can be determined by the calibration experiment using a blackbody furnace which has an emissivity of approximately 0.99. This property of the blackbody furnace which meets the graybody hypothesis can contribute to reducing the errors of the measurement model. Under a certain distance and the constant lighting conditions, the calibration coefficient is only related to the temperature and the gray-level ratio (R/G) using the same temperature measurement system. Once the relationship between the calibration coefficient and the gray-level ratio is known, the temperature can be obtained easily. However, the blackbody furnace generally gives a large temperature interval in terms of controlling the temperature. And it is difficult for the blackbody furnace to reach the combustion temperature of the blast furnace raceway. In addition, the relationship between the calibration coefficient and the gray-level ratio is nonlinear that cannot be obtained directly according to the experience of other research. But this relationship is necessary for the temperature measurement of the blast furnace raceway. Consequently, the next section focus on using the back-propagation neural network and the principal component analysis to obtain more accurate relationship.

Optimization Algorithm

Back-Propagation Neural Network

The back-propagation neural network is a feedforward network using the learning theory of counterpropagation. It can train the network by calculating the errors between the actual output of samples and the output of the neural network to adjust the weight and threshold between each neuron backwardly from the input layer to the hidden layer, then the output layer step by step according to these errors with the purpose of making the final output of the neural network to approximate the actual output more closely.

As the most widely used network, the back-propagation neural network does not need to establish the accurate mathematical equations of the calibration coefficient. But it still can extract implicit useful information from a large number of data. In addition, the back-propagation neural network with the adaptive and learning function can approximate any nonlinear functional relationship with any degree of accuracy. As a result, it can effectively estimate the relationship between the calibration coefficient and the gray-level ratio and obtain the temperature distribution of the flame [7].

10 sets of data of the measurements of the temperature and the calibration coefficient and gray-level ratio were extracted from images of a radiating body whose temperature was known captured by the RGB CCD camera in the calibration experiment using a blackbody furnace. One set of the data is shown in Table 1.

In this model, a three-layer back-propagation neural network consisting of an input layer, a hidden layer, and an output layer has been developed. The number of neurons in the input and the output layer are both one in terms of the question being investigated. The gray-level ratio is the input and the calibration coefficient is the output. All the input must be normalized; while the output also needs to be reversely normalized. The Levenberg–Marquardt optimization algorithm has been used to train the network [8]. A log-sigmoid transfer function is used in the hidden layer and a purelin transfer function is used in the output layer. These 9 sets of data have been input to train the network. The last 1 set of data is used to test the accuracy of the network and show the results.

According to the empirical formula, the number of neurons in the hidden layer should range from 3 to 11. Different number of neurons may lead to a different convergence rate. Comparing with the convergence rate at different number of neurons during the training process, the number was finally determined as 5.

Table 1 Measurements of the temperature and the calibration coefficient and gray-level ratio

Blackbody temperature (°C)	Gray-level ratio (R/G)	Calibration coefficient (C)
1000	2.75	3.52
1050	2.28	3.43
1100	1.68	3.49
1150	1.49	3.38
1200	1.35	3.27
1250	1.24	3.16
1300	0.92	3.28
1350	0.83	3.22
1400	0.85	3.05
1450	0.76	3.02

Figure 1 shows the structure of the back-propagation neural network. Figure 2 shows the variation of the residual error during a training process. Termination of the training means that the mean squared error doesn't decrease in 6 consecutive epochs.

It can be seen that the mean squared error has dropped down to 1×10^{-4} from Fig. 2. The total epochs show a convergence rate which is not very fast. In general, the training process presents a good trend. Although the network has been tested during the training process, using another sample can provide more details about the accuracy of the network.

The last 1 set of data was input into the trained network. Comparison between the measured and the estimated results for calibration coefficient is summarized in Table 2. The maximum and minimum relative errors are 4.28 and 0.31%, respectively. The estimation accuracy was verified for the mean relative errors less than 5%.

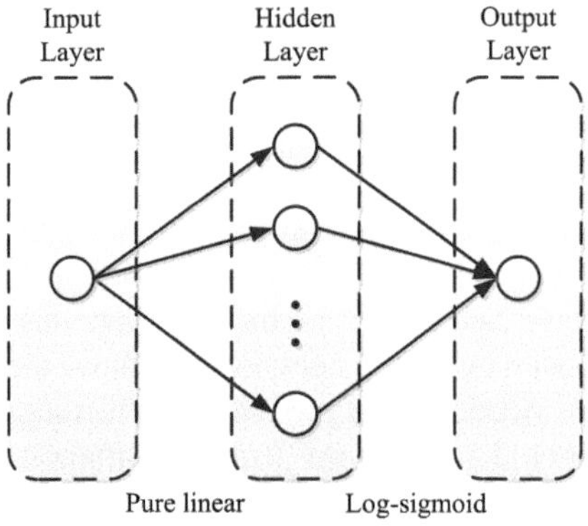

Fig. 1 Structure of the back-propagation neural network

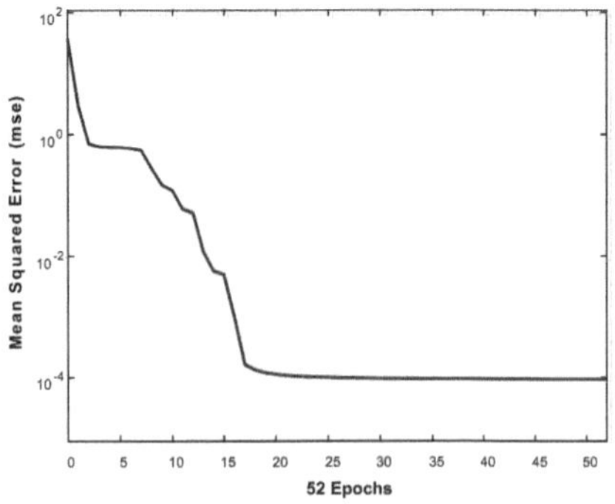

Fig. 2 Variation of the residual error during the training process

Table 2 Comparison between the measured coefficient and the estimated coefficient

Blackbody temperature (°C)	Gray-level ratio (R/G)	Calibration coefficient (C)		Relative errors (%)
		Measured value	Estimated value	
1000	2.77	3.51	3.47	1.14
1050	2.25	3.46	3.44	0.58
1100	1.66	3.49	3.35	4.01
1150	1.47	3.39	3.32	2.06
1200	1.35	3.27	3.28	0.31
1250	1.25	3.15	3.22	2.22
1300	0.93	3.27	3.13	4.28
1350	0.84	3.21	3.11	3.12
1400	0.83	3.07	3.10	0.98
1450	0.76	3.02	3.08	1.99

Principal Component Analysis (PCA)

Principal Component Analysis is a kind of statistical method which can compress multiple related variables into a few orthogonal variables. It can be regarded as a kind of orthogonalization that transforms a vector of variables from the original–dimensional space to a lower space comprised by orthogonal principal components.

The training result of the back-propagation neural network depends on the choice of the sample to a large degree. Too few samples can lead to underfitting and larger errors. The predictive ability of a network trained by these samples is usually limited. As a result, 9 sets of data were used to train the neural network. In another hand, excessive samples may include too much superfluous information that can result in overfitting in which all the training points are well fit but the fitting curve takes wild oscillations between these points. But noise and superfluous information that can also lead to a slow convergence rate and the lower accuracy inevitably exist in these 9 sets of data. As a result, it's necessary to compress the samples.

Principal Component Analysis can extract the main information from the samples using the criterion for minimum residual sums of squares function. Less input data generally require less training time. Another obvious advantage of using PCA should be that the orthogonalization of input data can further speed up the training of the neural network. In addition, part of the noise can be removed so that the accuracy of the neural network will improve then [8]. Through PCA, the 9 sets of data were then transformed into the principal components. Figure 3 shows the relative significance represented by each principal component. It can be seen that the first principal component possesses 99.09% of the information of the original data. The remaining 8 principal components are hence insignificant.

Through compressing the 9 sets of data, only the first principal component were selected to train a new neural network to approximate the relationship between the calibration coefficient and the gray-level ratio. Figure 4 shows the variation of the

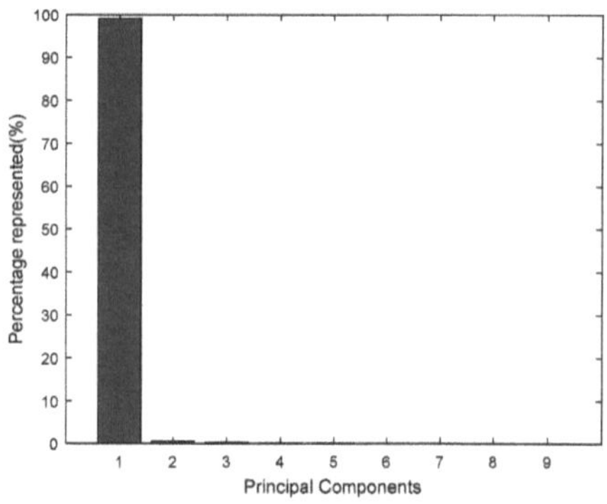

Fig. 3 Significance represented by each principal component in percentage

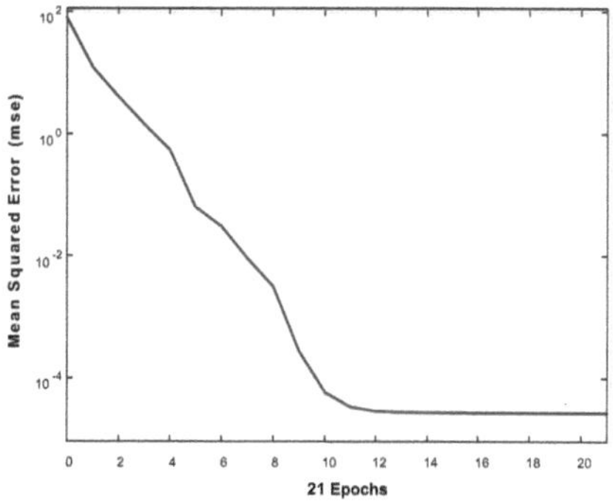

Fig. 4 Variation of the residual error during the training process using PCA

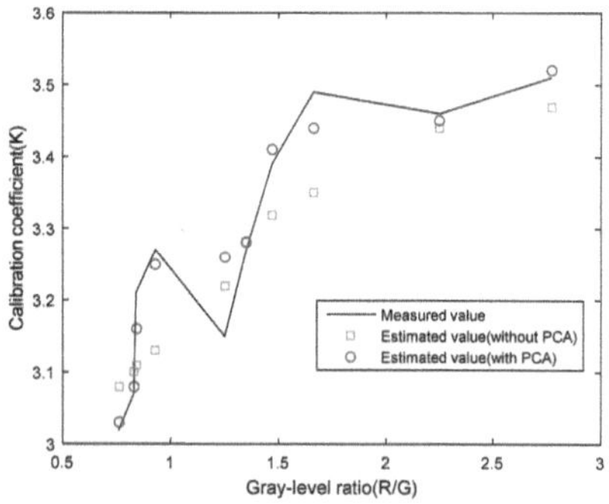

Fig. 5 Comparison between the estimated coefficient with PCA and that without PCA

residual error during a new training process using PCA. In this training process, the new neural network achieves a smaller error using only 21 epochs compared with the training process shown in Fig. 2. This result completely reflects the positive influence benefitting from PCA on speeding up the training of the neural network. The new estimated values obtained from the test data compared with the measured values and estimated values without PCA are shown in Fig. 5.

It can be seen that the data points estimated by the neural network incorporating PCA shown in Fig. 5 are quite close to the measured values. A direct comparison between the two sets of coefficient indicates that using PCA can effectively reduce the error and improve the accuracy of the neural network because a lot of noise was removed and more useful data was input. Through this method, the accuracy of the calibration coefficient used in the two-color model for the blast furnace raceway can be guaranteed.

Results and Discussion

Experiments were conducted at a platform of the blast furnace raceway. A large number of flame images of the raceway were taken by the RGB CCD camera. The high temperature distribution of the raceway were calculated using the two-color method incorporating the calibration coefficient estimated from the back-propagation neural network using PCA.

Figure 6 illustrates a flame image of the blast furnace raceway and its temperature distribution in two forms. It can been obviously found that temperature distributions calculated are quite consistent with gray-level intensities of the flame image. The central temperature is around 1950 °C. Although it is difficult to find accurate comparative temperature of the blast furnace raceway, it still has a good agreement with the results obtained from other research [9]. In addition, the adiabatic combustion temperature of the blast furnace raceway can also be a reference to verify the effectiveness of the proposed optimization algorithm. Comparison between the calculated results of this paper and results obtained from other research on the same blast furnace raceway is summarized in Table 3.

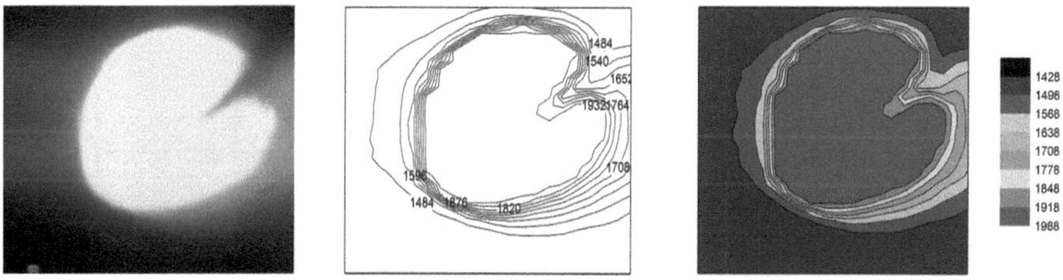

Fig. 6 Flame image of the blast furnace raceway and corresponding temperature distribution in two forms

Table 3 Comparison between the calculated results of this paper and results obtained from other research

Average central temperature calculated in this paper (°C)	Average raceway temperature of other research (°C)	Relative errors (%)
1951	1980	1.49

Conclusions

A measurement approach based on the two-color method and the optimization algorithm has been applied to calculate the high temperature distribution of the blast furnace raceway. Optimization algorithms incorporating the Back-propagation neural network and the principal component analysis have successfully obtained the relationship between the calibration coefficient and the gray-level ratio. The calibration coefficient estimated by the algorithms has a good agreement with the results measured in the calibration experiment. The flame temperature distribution has been calculated by the estimated calibration coefficient and gray-level intensities of the red and green colors. The central temperature of the blast furnace raceway has been verified according to the experimental temperature and the adiabatic combustion temperature. The use of the back-propagation neural network and the principal component analysis have given rise to the higher accuracy of the measurement approach contraposing to the temperature of the blast furnace raceway. In general, the novel measurement approach can be considered effective and feasible.

References

1. Y. Huang, Y. Yan, Transient two-dimensional temperature measurement of open flames by dual-spectral image analysis. Trans. Inst. Meas. Control **5**(22), 371–384 (2000)
2. Y. Huang et al., Vision-based measurement of temperature distribution in a 500-kW model furnace using the two-colour method. Measurement **3**(28), 175–183 (2000)
3. G. Lu, Y. Yan, Temperature profiling of pulverized coal flames using multicolor pyrometric and digital imaging techniques. IEEE Trans. Instrum. Meas. **4**(55), 1303–1308 (2006)
4. G. Lu et al., Concurrent measurement of temperature and soot concentration of pulverized coal flames. IEEE Trans. Instrum. Meas. **5**(51), 990–995 (2002)
5. W.H. Li et al., Estimation of radiative properties and temperature distributions in coal-fired boiler furnaces by a portable image processing system. Exp. Thermal Fluid Sci. **2**(35), 416–421 (2011)
6. Z.W. Jiang et al., A simple measurement method of temperature and emissivity of coal-fired flames from visible radiation image and its application in a CFB boiler furnace. Fuel **6**(88), 980–987 (2009)
7. V. Tipsuwanporn et al., Fire detection using neural network, SICE-ICASE Int. Joint Conf. 5474–5477 (2006)
8. L.J. Xu et al., Online fuel tracking by combining principal component analysis and neural network techniques. IEEE Trans. Instrum. Meas. **4**(54), 1640–1645 (2005)
9. M. Lundgren et al., The evolution of structural order as a measure of thermal history of coke in the blast furnace. Metall. Mater. Trans. B **2**(45), 603–616 (2014)

Kinetics and Reduction Behavior of Self-reducing Briquettes Containing Blast Furnace Dust

Shengli Wu, Feng Chang, Jianliang Zhang and Hua Lu

Abstract Kinetic studies of the blast furnace dust (BFD) reduction process have been carried out under non-isothermal conditions. BFD pellets were made with bentonite as the binder. A Comprehensive Thermal Analyzer (TG-DSC) was used to determine the mass loss of pellets, which were heated at three heating rates of 10, 20, 30 K/min up to 1493 K. Furthermore, the metallization rate and mineral composition of BFD pellets heated to several different temperatures were analyzed by chemical analysis and X-ray diffraction (XRD). It was found that, with its inherent carbon resources as the reducing agent, the metallization rate of BFD pellets could reach 88% at most. The reduction and weight loss process of iron oxides mainly happens above 1273 K. In addition, from application of the Coats-Redfern equation the apparent activation energy of the reduction from wustite to Fe was 443 kJ/mol and the mechanism was interface reaction R_2.

Keywords Blast furnace dust · Briquette · Kinetics · Reduction behavior

Introduction

In integrated steel mills large blast furnaces are the most important element for the production of hot metal. Preheated air is blown into the lower part of the blast-furnace from the tuyères. After closely contacting with the descending iron-bearing burden, a dusty gas leaves the blast-furnace at the top. Every one ton of hot metal will produce 1500–2000 m^3 of top gas. The gaseous phase of this top gas contains about 26% combustible gases such as carbon monoxide and hydrogen. The calorific value of the gas is 3000–4000 kJ/m^3 [1, 2]. For trouble-free operation a low dust concentration is required. Typically, dust separation is performed in two stages. First, the coarse dust is collected in a dry dust separator (dust catcher). In a

S. Wu · F. Chang (✉) · J. Zhang · H. Lu
School of Metallurgical and Ecological Engineering, University of Science and Technology Beijing, Beijing, China
e-mail: bkchangfeng@163.com

J.-Y. Hwang et al. (eds.), *8th International Symposium on High-Temperature Metallurgical Processing*, The Minerals, Metals & Materials Series,
DOI 10.1007/978-3-319-51340-9_58

second separation step the remaining fine dust is collected, resulting in a dust concentration of less than 5 mg/Nm^3 in the clean top gas. The second dust separation step is usually performed by a scrubber system, where most of the scrubber water is recycled after separation of the collected dust particles from the scrubber water by means of sedimentation. Recently, totally dry cleaning using fabric filters has become an increasingly popular technology for top gas cleaning in newly built blast furnaces. In such systems a fabric filter replaces the wet scrubber of the second dedusting stage. Besides other advantages, one major advantage of the totally dry gas cleaning is that the collected dust is obtained as dry powder. Thus, storage and handling of the dust material is quite simple and there is no water in the residue, which would usually have to be removed prior to recycling.

Guo et al. [3] proposed a novel method which can automatically recognize the blast furnace dust composition by multi-feature analysis method based on comprehensive image-processing techniques. Lanzerstorfer attempted to reduce the Zn content of the recycled blast furnace dust (to increase the recycling rates) through air classification [4]. By separating a small fraction of fines from the BF dust, the reduction of the mass of Zn in the coarse dust recycled to the sinter plant only reach 40–60%. Some researchers used the blast furnace sludge as an absorbent to remove Pb^{2+} and Ni^{2+} from aqueous solutions [5, 6]. Herck et al. [7] investigated the removal of zinc and lead from blast furnace sludge with a hydrometallurgical process. However, the leaching rate of zinc and lead is relatively low in the hydrometallurgical process. In addition, it has such many problems as a great amount consumption of leaching agent, serious equipment corrosion and so on. Moreover, it will cause new environmental pollution with respect to sulfur and chlorine. Sakamoto recovered iron and zinc from zinc bearing dusts by using the sensible heat of hot metal [8]. Nowadays, the pyrometallurgical process is increasingly adopted in treating iron-bearing dust. However, there is no information available regarding the kinetics and reduction behavior of self-reducing briquettes only containing blast furnace dust. The aim of this paper is to study the kinetics and reduction behavior of self-reducing briquettes only containing blast furnace dust, using thermal analysis and X-ray diffraction.

Experimental

Raw Material

The blast furnace dust used in this study was kindly supplied by Baosteel in China. The main chemical composition of the initial sample is given in Table 1. The total

Table 1 The main chemical composition of the blast furnace dust (wt%)

TFe	FeO	CaO	SiO_2	Al_2O_3	MgO	K	Na	Pb	Cl	C
44.28	2.87	2.90	8.15	4.26	0.53	0.06	0.08	0.23	0.05	19.59

iron and carbon contents in blast furnace dust are about 44 and 20%, respectively. However, it is worth mentioning that the zinc concentration is below the detection limit by inductively coupled plasma atomic emission spectroscopy (ICP-AES). The particle size of the blast furnace dust is between 1 and 253 μm, and based on the size distribution its surface area is 6.14 m^2/cm^3. According to laser diffraction analysis, the mean particle diameter was estimated as 51 μm. The XRD pattern of the dust sample showed mainly hematite with small magnetite and quartz reflection peaks. The BET specific surface area of the blast furnace dust sample was 12.13 m^2/g. The true density of the blast furnace dust sample was 3359 kg/m^3. In terms of particle morphology, some carbon particles were present as isolated particles; some were partly covered by iron oxides; some carbon particles were almost entirely covered by iron oxides [9] (Figs. 1, 2 and 3).

Test and Analysis Method

TG/DSC analysis was performed with an integrated thermal analyzer (STA409C, Netzsch, Germany). About 5 g in mass of the samples were homogenized and successive quartering provided, following by a step of grinding and blending in the mortar to make that the samples are representative. And the samples in the integrated thermal analyzer were 20–30 mg in mass. Two alumina crucibles, one

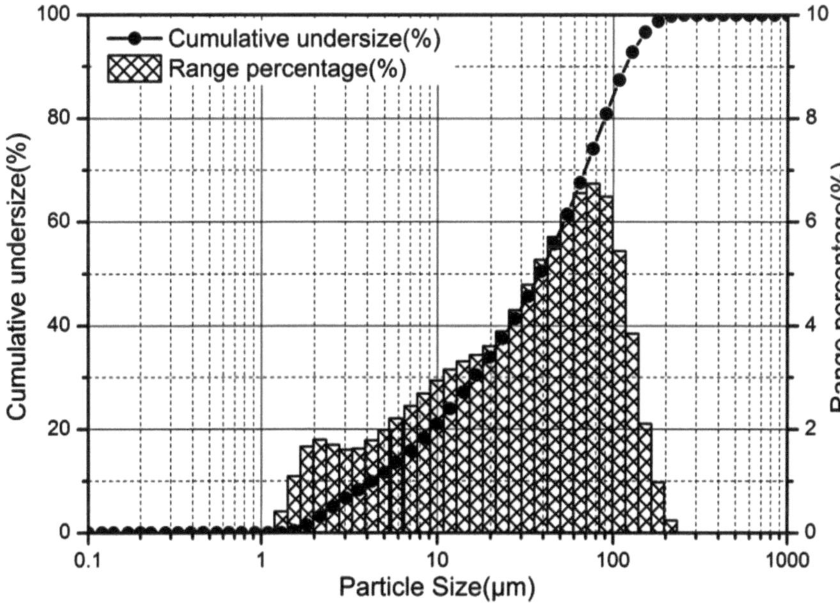

Fig. 1 The particle size distribution of blast furnace dust

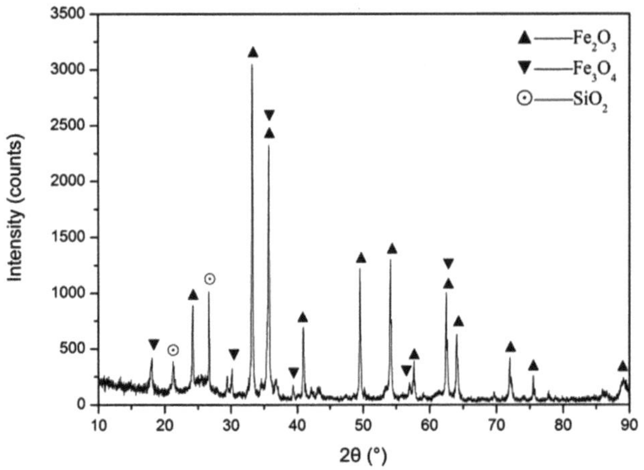

Fig. 2 The XRD patterns of the blast furnace dust

Fig. 3 The particles morphology of the blast furnace dust

containing alumina powder as a reference and the other a dust sample, were heated from room temperature to 1493 K at a heating rate of 10 K/min in argon with a constant flow rate about 30 ml/min. To estimate the apparent activation energy of the thermal decomposition process, heating rates of 20 and 30 K/min were also used.

The fundamental rate equation used in all kinetic studies is generally described as

$$d\alpha/\mathrm{d}t = kf(\alpha) \tag{1}$$

where k is the rate constant and $f(\alpha)$ is the reaction model, a function depending on the actual reaction mechanism. Equation (1) expresses the rate of conversion, $\mathrm{d}\alpha/\mathrm{d}t$, at a constant temperature as a function of the reactant concentration loss and rate constant. In this study, the conversion rate α is defined as:

$$\alpha = (m_0 - m_t)/(m_0 - m_f) \tag{2}$$

where m_t, m_0, and m_f are time t, initial and final weights of the sample, respectively. The rate constant k is generally given by the Arrhenius equation:

$$k = A\ \exp(-E/RT) \tag{3}$$

where E is the apparent activation energy (kJ/mol), R is the gas constant (8.314 J/K mol), A is the pre-exponential factor (s^{-1}), T is the absolute temperature (K). The combination of Eqs. (1) and (3) gives the following relationship:

$$d\alpha/\mathrm{d}t = A\ \exp(-E/RT)f(\alpha) \tag{4}$$

For a dynamic TGA process, introducing the heating rate, $\beta = \mathrm{d}T/\mathrm{d}t$, into Eqs. (4), (5) is obtained as:

$$d\alpha/\mathrm{d}T = (A/\beta)\ \exp(-E/RT)f(\alpha) \tag{5}$$

Equations (4) and (5) are the fundamental expressions of analytical methods to calculate kinetic parameters on the basis of TGA data.

Through variable separation and integration, Eq. (5) leads to:

$$g(\alpha) = \int_0^{\alpha} d\alpha/f(\alpha) = (A/\beta)\int_0^{T} \exp(-E/RT)dT \tag{6}$$

where $g(\alpha)$ is the integral conversion function.

If $E/(RT)$ is replaced by x and integration limits transformed, Eq. (6) becomes:

$$g(\alpha) = \frac{AE}{\beta R}\int_x^{\infty} \frac{\exp(-x)}{x^2}dx \tag{7}$$

Equation (9) can be written as

$$g(\alpha) = \frac{AE}{\beta R}p(x) \tag{8}$$

$p(x)$ has no analytical solution but has many approximations [10–12], with one of the most popular being the Coats-Redfern method [13]. This method utilizes the asymptotic series expansion for approximating the exponential integral in Eq. (8), giving

Table 2 Solid state rate equations

Reaction Model		$f(\alpha)$	$g(\alpha)$
Nucleation and growth	A_2	$2(1-\alpha)/[-\ln(1-\alpha)]^{1/2}$	$[-\ln(1-\alpha)]^{1/2}$
	A_3	$3(1-\alpha)/[-\ln(1-\alpha)]^{2/3}$	$[-\ln(1-\alpha)]^{1/3}$
	A_4	$4(1-\alpha)/[-\ln(1-\alpha)]^{3/4}$	$[-\ln(1-\alpha)]^{1/4}$
Interface reaction	R_1	1	α
	R_2	$2(1-\alpha)^{1/2}$	$1-(1-\alpha)^{1/2}$
	R_3	$3(1-\alpha)^{2/3}$	$1-(1-\alpha)^{1/3}$
Diffuse	D_1	$1/(2\alpha)$	α^2
	D_2	$-1/\ln(1-\alpha)$	$\alpha+(1-\alpha)\ln(1-\alpha)$
	D_3	$3(1-\alpha)^{2/3}/\{2[1-(1-\alpha)^{1/3}]\}$	$[1-(1-\alpha)^{1/3}]^2$
	D_4	$3/\{2[(1-\alpha)^{-1/3}-1]\}$	$1-2\alpha/3-(1-\alpha)^{2/3}$
Chemical reaction	F_1	$1-\alpha$	$-\ln(1-\alpha)$
	F_2	$(1-\alpha)^2$	$(1-\alpha)^{-1}$
	F_3	$(1-\alpha)^3/2$	$(1-\alpha)^{-2}$

$$\ln\frac{g(\alpha)}{T^2}=\ln[\frac{AR}{\beta E}(1-\frac{2RT}{E})]-\frac{E}{RT} \tag{9}$$

Since in general $2RT/E \ll 1$ and it exhibits a small variation with T, for practical considerations it is assumed that the term $(1-2RT/E)$ is approximately constant and equal to unity. Thus, for a given form of $g(\alpha)$, the plot of $\ln[g(\alpha)/T^2]$ versus $1/T$ gives a straight line whose slope and intercept allow an estimation of the values of the activation energy and pre-exponential factor, respectively. In order to identify the most probable mechanism function, $g(\alpha)$, Coats-Redfern proposed the choice of that one which ensures the best linearity, corresponding to the maximum value of the correlation coefficient. The commonly seen solid state thermal decomposition reaction mechanism functions are listed in Table 2 [14]. The Coats-Redfern method is a single heating rate application of the Coats-Redfern equation.

Results and Discussion

Non-isothermal Reduction Kinetics

The TG results of blast furnace dust briquettes at different heating rates are given in Fig. 4. Samples have wt. losses only between 27.30 and 31.67%, indicating that some unreacted carbon remained in the dust. Figure 5 shows TG, DTG results of blast furnace dust briquettes at 20 K/min. As we can see from the figures, the reduction process of blast furnace dust can be roughly divided into three stages: (1) 25–380 °C, the volatilization of free moisture and chemically bonded water; (2) 380–990 °C, the slow reduction of Fe_2O_3 to Fe_3O_4 and Fe_3O_4 to FeO; (3) 990–1220 °C, extensive carbon gasification and rapid reduction of wustite to iron.

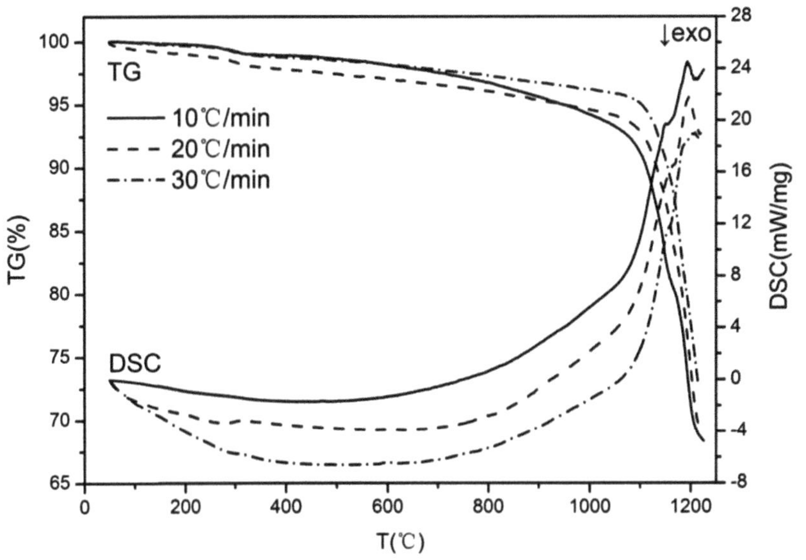

Fig. 4 TG-DSC results of blast furnace dust briquettes

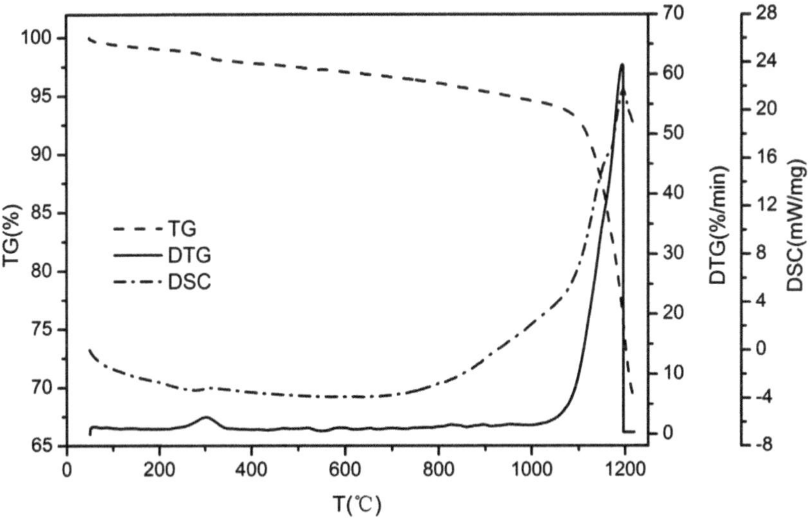

Fig. 5 TG, DTG, DSC results of blast furnace dust briquettes at 20 °C/min

Figure 6 shows the plots of $\ln[g(\alpha)/T^2]$ versus $1/T$ using values of α extracted from Fig. 5 at the heating rate of 20 °C/min. According to Coats-Redfern equation, if a correct model is selected for the reaction, the plot of $\ln[g(\alpha)/T^2]$ versus $1/T$ will be linear with high-correlation coefficient. Table 3 lists the calculated kinetic

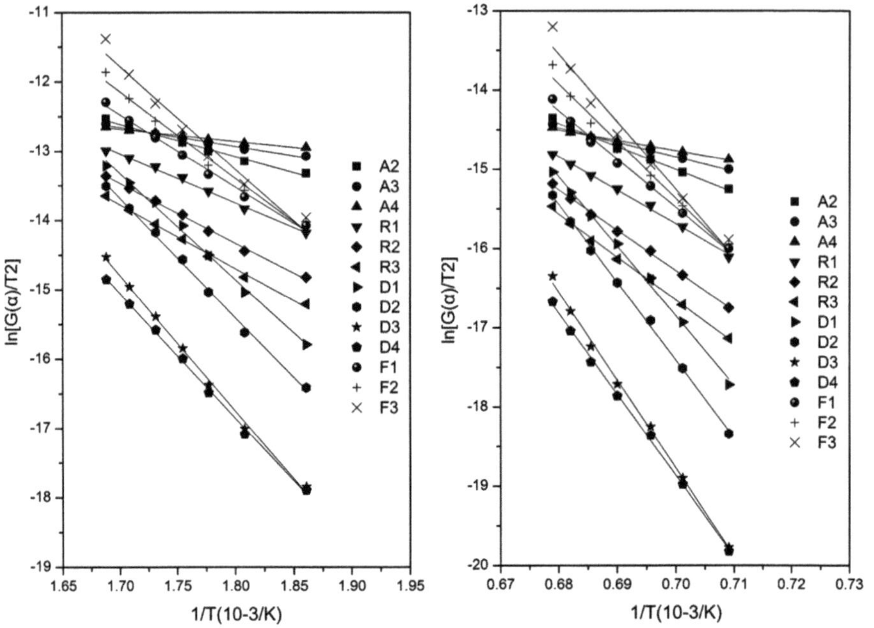

(a) The decomposition of chemically bonded water

(b) Carbon gasification and rapid reduction of wustite to iron

Fig. 6 The plots of $\ln[g(\alpha)/T^2]$ versus $1/T$ using values of α at the heating rate of 20 °C/min

Table 3 Metallization of blast furnace dust briquettes heated under different conditions

Reduction temperature (°C)	Reduction time (min)	Metallization rate (%)
950	10	2.8
1050	10	45.0
1150	10	80.6
1250	10	87.9
1150	5	15.4
1150	15	88.1

parameters for different models with α from TG simulated curves by the Coats-Redfern method during the decomposition of chemically bonded water. The calculated results indicate that the decomposition of chemically bonded water obeys chemical reaction F_1, therefore, the reaction function is $f(\alpha) = 1 - \alpha$. The effect of heating rate on the apparent activation energy is insignificant. The mean apparent activation energy is 82 kJ/mol. Table 3 lists the calculated kinetic parameters for different models with α from TG simulated curves by the Coats-Redfern method during carbon gasification and rapid reduction of wustite to iron. The calculated results indicate that this stage obeys interface reaction mechanism R_2 with the reaction function $f(\alpha) = 2(1 - \alpha)^{1/2}$. The effect of heating rate on the apparent activation energy is insignificant. The mean apparent activation energy is 443 kJ/mol. The kinetic equations of these two stages are:

(a) The decomposition of chemically bonded water

$$\frac{d\alpha}{dt} = 10^{5.09} \cdot (1 - \alpha) \cdot e^{-9863/T} \tag{11}$$

(b) Carbon gasification and rapid reduction of wustite to iron (Fig. 6)

$$\frac{d\alpha}{dt} = 10^{13.36} \cdot (1 - \alpha)^{1/2} \cdot e^{-53284/T} \tag{12}$$

Direct Reduction Behavior

Table 3 shows the metallization rate results of blast furnace dust briquettes heated under specific conditions. Figure 7 shows XRD analysis results of blast furnace dust briquette samples heated at different holding temperatures. The samples were promptly sent inside the chamber of Muffle furnace at the corresponding temperature. The heating rate is 10 °C/min until the temperature hits the goal and then maintains 10 min at this level. As we can see, reduction temperature influences the metallization greatly. Metallization increases with increased temperature, from 45.0 to 87.9% for a temperature from 1050 to 1250 °C. After heating to 950 °C, the Fe_2O_3 peaks disappeared while only peaks of Fe_3O_4 were detected.

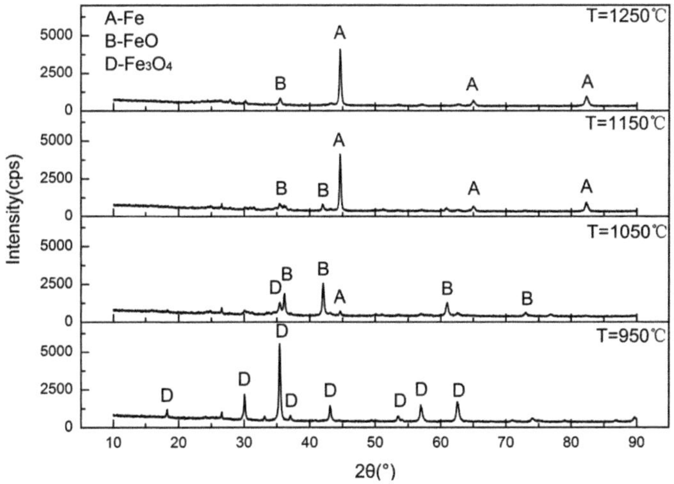

Fig. 7 XRD analysis of blast furnace dust briquette heated at different temperatures

The corresponding metallization is merely 2.8%. At 1050 °C, a large proportion of Fe_3O_4 transformed into FeO and even Fe in the sample. When the temperature was 1150 °C or even higher, the reduction of iron oxide was almost complete after 15 min.

Conclusions

(1) Using only the inherent carbon as reducing agent, the metallization rate of BFD pellets reached 88% at most.
(2) The reduction and weight loss of iron oxides mainly occur above 1273 K. The apparent activation energy of the rapid reduction stage is about 443 kJ/mol and the mechanism function is interface reaction mechanism R_2.

Acknowledgements This work is supported by National Natural Science Foundation of China (U1260202).

References

1. T. Mansfeldt et al., Chemical and mineralogical characterization of blast-furnace sludge from an abandoned landfill. Environ. Sci. Technol. **38**(22), 5977–5984 (2004)
2. X.L. Wang, *Ferrous Metallurgy (Iron-Making Part)*, 3rd edn. (Metallurgical Industry Press, Beijing, 2013), p. 22
3. H. Guo et al., Novel recognition method of blast furnace dust composition by multifeature analysis based on comprehensive image-processing techniques. JOM **66**(11), 2377–2389 (2014)
4. C. Lanzerstorfer, Air classification of blast furnace dust catcher dust for zinc load reduction at the sinter plant. Int. J. Environ. Sci. Technol. **13**(2), 755–760 (2016)
5. A. Bhatnagar et al., Removal of lead ions from aqueous solutions by different types of industrial waste materials equilibrium and kinetic studies. Sep. Sci. Technol. **41**(9), 1881–1892 (2006)
6. A. Radjenovic et al., Removal of Ni^{2+} from aqueous solution by blast furnace sludge as an adsorbent. Desalination Water Treat. **21**(1–3), 286–294 (2010)
7. P.V. Herck et al., Zinc and lead removal from blast furnace sludge with a hydrometallurgical process. Environ. Sci. Technol. **34**(17), 3802–3808 (2000)
8. N. Sakamoto et al., Zinc recovery from zinc bearing dusts by use of sensible heat of hot metal. ISIJ Int. **35**(11), 1323–1330 (1995)
9. F. Chang et al., in *Characterization of Sintering Dust, Blast Furnace Dust and Carbon Steel Electric Arc Furnace Dust. Characterization of Minerals, Metals, and Materials 2015*, TMS2015 144th Annual Meeting, Orlando, Florida, USA, 2015
10. C.D. Doyle, Kinetic analysis of thermogravimetric data. J. Appl. Polym. Sci. **5**(15), 285–292 (1961)
11. R.K. Agrawal et al., Integral approximations for nonisothermal kinetics. AIChE J. **33**(7), 1212–1214 (1987)
12. H.J. Chen et al., Methods for determining the kinetic parameters from nonisothermal thermogravimetry: a comparison of reliability. J. Chem. Eng. Jpn. **37**(9), 1172–1178 (2004)

13. A.W. Coats et al., Kinetic parameters from thermogravimetric data. Nature **201**(4914), 68–69 (1964)
14. R.Z. Hu et al., *Thermal Analysis Kinetics*, 2nd edn. (Science Press, Beijing, 2004), pp. 151–155

Model Analysis of the Phenomena of Pulverized Coal Injection in Blast Furnace

Tian Chen, Shusen Cheng and Wenxuan Xu

Abstract Pulverized coal injection (PCI) can generate heat and reducing gas in the raceway and replace some expensive metallurgical coke in the blast furnace iron-making process. In this paper, a three dimensional mathematical model of the raceway including the coke bed has been developed to simulate the complicated phenomena of pulverized coal injection. The effects of oxygen enrichment, particle diameter in the raceway are investigated in this study. The establishment of the 3D model is expected to analyze the raceway working state accurately and optimising the PCI operation of blast furnaces.

Keywords Pulverized coal injection · CFD · Raceway

Introduction

The ironmaking blast furnace plays a dominant role in an integrated steelwork in terms of energy consumption. For reducing the use of coke, pulverized coal injection (PCI) technology is widely used in this process [1]. However, with the increasing PCI rate, the injected coal powder cannot burn completely inside the raceway and the unburnt coal powder may accumulate in the boundary of raceway, affecting the coke bed permeability and thereby the furnace stability. Under high PCI rate operation, the ways to improve the combustion of coal powder has become one of topics for the ironmaking researchers. But it is difficult to physically measure the combustion behavior of pulverized coal in actual blast furnace directly, because of the hazardous and high temperature in furnace environment. The development of computational fluid dynamics (CFD) provides another way for the study of the pulverized coal injection in furnace. Many researchers [2–5] have tried to use CFD method to investigate PCI. However, few researches have considered the influence

T. Chen (✉) · S. Cheng · W. Xu
School of Metallurgical and Ecological Engineering, University of Science and Technology Beijing, Beijing 100083, China
e-mail: 1126626806@qq.com

J.-Y. Hwang et al. (eds.), *8th International Symposium on High-Temperature Metallurgical Processing*, The Minerals, Metals & Materials Series,
DOI 10.1007/978-3-319-51340-9_59

of coke bed, which surrounds the raceway. In this paper, a mathematical model of the gas-solid flow, heat exchange and combustion reaction has been developed. With the aid of CFD, the behavior of pulverized coal combustion is systematically studied.

Model Description

Basic Equation

According to the characteristics of gas-solid turbulent flow in the blast furnace, the transfer process of the pulverized coal combustion should satisfy the control equations of the gas-solid two phases, such as the continuous, momentum, energy and component transport equations. The basic equations are summarized in Table 1.

Turbulence Model

The standard $k-\varepsilon$ model has good robustness and convergence, so it is widely used in the numerical simulation of the combustion process of pulverized coal. In the standard $k-\varepsilon$ model, the corresponding transmission differential equations are as follows:

$$\frac{\partial(\rho k)}{\partial t}+\frac{\partial(\rho\mu_j k)}{\partial x_j}=\frac{\partial}{\partial x_j}\left(\left(\mu+\frac{\mu_t}{\sigma_k}\right)\cdot\frac{\partial k}{\partial x_j}\right)+G_k+G_b-\rho\varepsilon \tag{1}$$

$$\frac{\partial(\rho\varepsilon)}{\partial t}+\frac{\partial(\rho\mu_j\varepsilon)}{\partial x_j}=\frac{\partial}{\partial x_j}\left(\left(\mu+\frac{\mu_t}{\sigma_\varepsilon}\right)\cdot\frac{\partial\varepsilon}{\partial x_j}\right)+C_{\varepsilon_1}\frac{\varepsilon}{k}(G_k+C_{\varepsilon_3}G_b)-C_{\varepsilon_2}\rho\frac{\varepsilon^2}{k} \tag{2}$$

In the formula, G_k, G_b expressions are $G_k=\mu_t\left(\frac{\partial\mu_j}{\partial x_i}+\frac{\partial\mu_i}{\partial x_j}\right)\frac{\partial\mu_j}{\partial x_j}$, $G_b=\beta g_i\frac{\mu_t}{\mathrm{Pr}_t}\frac{\partial T}{\partial x_i}$. β is the coefficient of thermal expansion and Pr_t is prandtl number. C_{ε_1}, C_{ε_2}, C_μ, σ_k, σ_ε are empirical constant. For compressible fluid flow, when the main direction is

Table 1 Basic equations

Continuity equation	$\frac{\partial\rho}{\partial t}+\frac{\partial}{x_i}(\rho u_i)=0$
Momentum equation	$\frac{\partial}{\partial t}(\rho u_i)+\frac{\partial}{\partial x_j}(\rho u_j u_i)=\frac{\partial}{\partial x_j}\tau_{ji}+f_i$
Energy equation	$\frac{\partial}{\partial t}(\rho E)+\frac{\partial}{\partial x_j}(\rho\mu_j E)=\frac{\partial}{\partial x_j}\left(\Gamma_e\frac{\partial}{\partial x_j}E\right)+\frac{\partial p}{\partial t}+S_e$
Component transport equation	$\frac{\partial\rho Y_i}{\partial t}+\frac{\partial}{\partial x_j}(\rho\mu_i Y_i)=\frac{\partial}{\partial x_j}\left(\Gamma_i\frac{\partial}{\partial x_j}Y_i\right)+R_i$

parallel to the gravity direction, $C_{\varepsilon_3} = 1$, and when the main direction is perpendicular to the direction of gravity, $C_{\varepsilon_3} = 0$; C_b is the turbulent kinetic energy caused by the action of buoyancy. According to Lauder recommendation and related experimental verification: $C_{\varepsilon_1} = 1.44$, $C_{\varepsilon_2} = 1.92$, $C_\mu = 0.09$, $\sigma_k = 1.0$, $\sigma_\varepsilon = 1.3$.

Radiation Model

Optical depth αL is an important index to select the radiation model, α is the absorption coefficient of the medium, L is the characteristic length. The P1 model is suitable for the optical depth is greater than 1 and less than 3 and the optical depth is about 1.4 in blast furnace, so the P1 model is used to calculate the radiation of pulverized coal combustion. The radiation heat flux equation q_r is

$$q_r = \frac{1}{3(\alpha + \sigma_s) - C\sigma_s} \nabla G \tag{3}$$

G is the incident radiation heat flux, and its expression is

$$-\nabla q_r = \alpha G - 4\alpha\sigma T^4 \tag{4}$$

Discrete Phase Model

There is a large amount of gas phase flow in the blast furnace. Pulverized coal by pulverized coal injection gun is injected into blast furnace, and the coal concentration in blowpipe and raceway is very low, due to the volume of hot air is much larger than the carrier gas and the amount of pulverized coal. Therefore, it is not appropriate to take the pulverized coal as a continuous phase. According to the BF multiphase flow characteristics, it is reasonable to describe by Lagrange method.

In Lagrange coordinates, assuming the pulverized coal for spherical particles, the pulverized coal particles are affected by the traction resistance of airflow, which plays a role of accelerating the motion of particles. The combined forces of the discrete particles of the unit mass equal to its acceleration, the equation for the particle motion is

$$\frac{du_p}{dt} = F_D(u - u_p) + \frac{g_i(\rho_p - \rho)}{\rho_p} + F_i \tag{5}$$

The expression of F_D is

$$F_D = \frac{18\mu}{\rho_P d_p^2}\frac{C_D\mathrm{Re}}{24} \tag{6}$$

The temperature of pulverized coal is mainly influenced by convection heat transfer, chemical reaction and heat radiation, and its expression is

$$m_p C_p \frac{dT_P}{dt} = \pi d_p \lambda N\mu\left(T_g - T_p\right) + \sum \frac{dm_p}{dt} H_{reac} + A_P \varepsilon_P \left(\pi I - \sigma T_p^4\right) \tag{7}$$

Combustion Reaction Model

Blast furnace coal injection requires consideration of many complex process. In addition to the gas-solid two-phase turbulence flow, the pyrolysis and combustion of pulverized coal, residual carbon combustion, radiation heat transfer and other reaction processes should also be considered.

1. Gas phase chemical reaction model

The C_mH_n is used to represent the volatile component of coal, and the gas phase combustion chemical reaction is

$$C_mH_n + (m + n/4)O_2 \rightarrow mCO_2 + nH_2O/2 \tag{8}$$

The rate of the reaction is determined by the minimum value of the "dynamics" and "diffusion" factor, so the gas phase reaction rate is described by the EBU-Arrhenius model [6], and the formula is

$$\begin{aligned} w_s &= \min(w_E, w_A) \\ w_E &= A_E \rho \frac{\varepsilon}{k} \min(Y_s, Y_{ox}/\beta) \\ w_A &= B_s \rho^2 Y_s Y_{ox} \exp(-E/RT) \end{aligned} \tag{9}$$

2. Pulverized coal combustion reaction model

The evaporation process of moisture in pulverized coal is described by a diffusion model, which is similar to the droplet evaporation [7].

$$\dot{m}_w = \pi d_p N\mu D\rho \ln\left(1 + \frac{Y_{w,s} - Y_{w,g}}{1 - Y_{w,s}}\right) \tag{10}$$

$$Y_{w,s} = B_w \exp\left(-\frac{E_w}{RT_P}\right) \tag{11}$$

The devolatilization of raw coal uses two-competing model. In this model, the pyrolysis of coal is controlled by two first order reactions with different rate and volatile yield.

$$\begin{cases} \alpha_1 VM_1 + (1 - \alpha_1)\mathrm{Char}_1 & \text{(Low temperature)} \\ \alpha_2 VM_2 + (1 - \alpha_2)\mathrm{Char}_2 & \text{(High temperature)} \end{cases} \tag{12}$$

After the pyrolysis of pulverized coal particles, the residual carbon is formed, and the particle surface combustion process using Kinetics-diffusion combustion model to solve.

$$\dot{m}_h = \sum \pi d_p^2 \rho Y_{ss} B_s \exp(-\frac{E}{RT_P}) \tag{13}$$

3. Coke combustion reaction model

Assuming that coke as elemental carbon, and reaction is first order reaction, then there are three surface chemical reactions.

$$\begin{aligned} &\mathrm{C} + \mathrm{O_2} \rightarrow \mathrm{CO_2} \\ &2\mathrm{C} + \mathrm{O_2} \rightarrow 2\mathrm{CO} \\ &\mathrm{C} + \mathrm{H_2O} \rightarrow \mathrm{CO} + \mathrm{H_2} \end{aligned} \tag{14}$$

4. Operating conditions

The model geometry is based on a 3200 m^3 blast furnace in China. The main dimensions of the model geometry are shown in Table 2. In this CFD model, one computational domain covering blowpipe, raceway and coke bed is developed, which is shown in Fig. 1, so that the in-furnace phenomena of PCI operation can be studied under various conditions. The raceway region is simplified to a cylinder, and the coke bed is treated as a porous media. The design parameters and operating parameters of blast furnace are summarized in Table 3. Some coal properties are listed in Table 4. Parameters, which have been published in page [8], in Tables 3 and 4 are also derived from the 3200 m^3 blast furnace in China.

5. Boundary conditions

In this paper, the CFX software is used to solve the problem. The blast and carrier gas inlet boundaries use velocity inlet, and the values are 250 and 10 m/s,

Table 2 Geometry dimension

Coke bed/mm	Raceway/mm	Blowpipe/mm
Depth 7000	Radius 1410	Radius 80
Height 6000	Width 2570	Length 600
Width 3600		

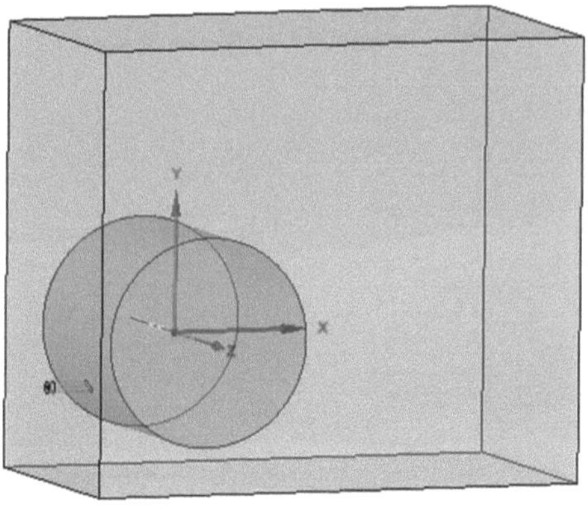

Fig. 1 Model geometry

Table 3 Parameters of blast furnace

Parameter	Pressure/KPa	Blast flow/$Nm^3\ h^{-1}$	Coal flow/kg h^{-1}	Lance diameter/mm	Lance angle/°
Value	404	11,188	1562.5	20	0

Table 4 Parameters of coal

Proximate analysis	Moisture, %	3.4	Ultimate analysis	C,%	84.7
	Volatile matter, %	32.6		H,%	5.6
	Ash, %	9.3		O,%	7.0
	Fixed carbon, %	54.7		N,%	2.7
Specific heat	1000 J $kg^{-1}\ k^{-1}$		Gross specific energy	31.15 MJ kg^{-1}	
Carrier gas flow	41.67 $Nm^3\ h^{-1}$		Density	1300 kg m^{-3}	
Coal temperature	360 K		Carrier gas temperature	360 K	

respectively. The coal inlet boundary uses mass flow rate, and the value is 0.43 kg/s. The outlet boundary uses pressure outlet, and the pressure is 0.4 MPa. The boundary condition of walls in this paper all uses the no-slip wall. The grid is constructed by the tetrahedral mesh.

Results and Discussion

According to the actual situation, this paper establishes a raceway model and the control equations of the gas-solid flow. Firstly, this paper studies the temperature field, flow field and gas composition change and discusses these phenomena in detail. Secondly, this paper studies the effects of oxygen enrichment on coal burnout and gas composition. Finally, this paper studies the influences of particle size on coal burnout and temperature field.

Model Validation

Gas composition is an important concern in BF practice, so the model is validated in terms of gas composition. It is generally known that the situation of the raceway is extremely complex, and many conditions are not clear in the raceway, so the direct measurement is very difficult. This is why many scholars use CFD to simulate and analyze the possible situations in the raceway. It is also very difficult for us to measure the blast furnace that we simulated directly, so we have to use the method adopted by authors Shen et al. [9] to verify the model. The method is that they verify model by simulating the actual situation, which has been measured by others. This paper verifies the model with the same measured data from the author Shen's article [9]. Figure 2 shows the measured and calculated gas mass fractions of O_2, CO_2, and CO along the tuyere axis. It can be seen that calculated gas composition agrees with the measured data broadly. It is proved that the model is relatively reliable and accurate.

Typical In-Furnace Phenomena

Figure 3a shows the temperature contour of the raceway and surrounding coke bed. Along the main coal plume, the temperature is much higher than other regions and can reach about 3000 K, because the reaction mainly occurs here. Due to the gas and coal particles flow in the raceway and its subsequent heat releases from chemical reactions, the temperature is reduced to about 2200 K in the recirculation region. The temperature of the coke bed, which is far from the raceway, is further reduced to about 1700 K, because of the long distance of heat transfer and porous medium.

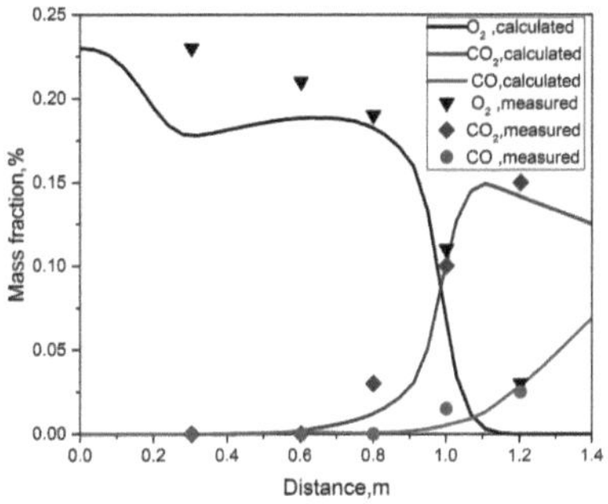

Fig. 2 Gas compositions against measurements in a real BF

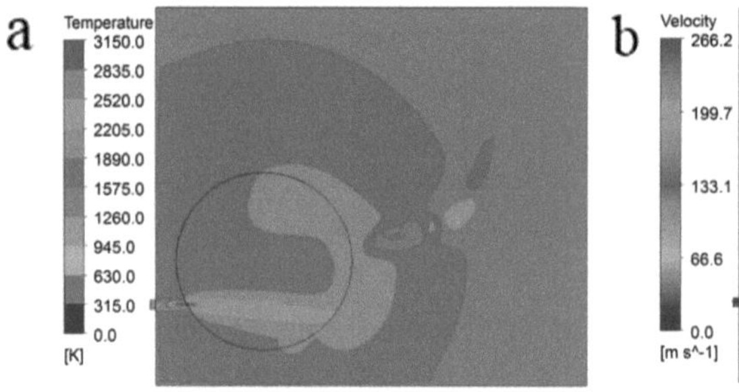

Fig. 3 **a** Temperature field; **b** velocity field

Fig. 4 Distributions of gas species along the tuyere axis

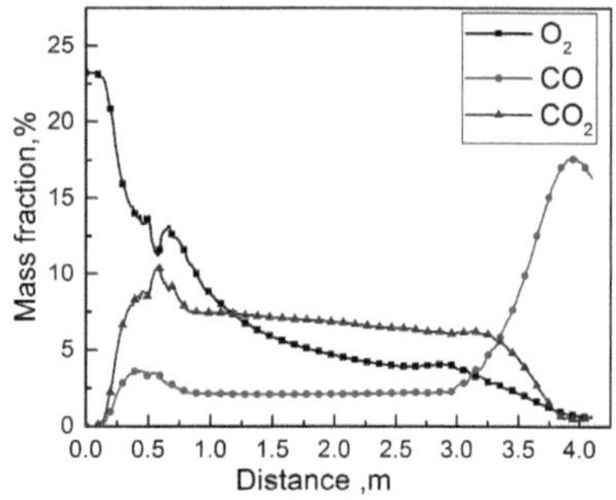

Figure 3b shows the velocity vectors in the raceway. It is shown that the blast, carrier gas and coal flow into the blowpipe at the same time and gas is accelerated through the tuyere. As a result, a high-speed jet of up to 266.2 m/s forms along the tuyere axis. Subsequently, the gas reaches the boundary of the raceway, due to the influence of the coke bed, and the gas velocity decreases rapidly to 40–60 m/s, which is affected by the coke bed. At the same time, the gas flow starts a large-scale recirculation above the main gas flow jet in the raceway. That is, inside the raceway, the flow pattern can be divided into two parts: a high speed jet and a large-scale recirculation.

Figure 4 shows the change in the content of the gas composition along the tuyere axis. It can be concluded from the figure that the initial O_2 is sufficient, but the O_2 is consumed continuously along with the occurrence of the chemical reaction. Subsequently, coal reaction with oxygen generates a large amount of CO_2 and a small amount of CO, so that carbon dioxide levels increase rapidly. Finally, the oxygen is almost depleted at the boundary of raceway, and CO_2 is completely converted to CO in the coke bed due to the strong coke reactions outside the raceway cavity.

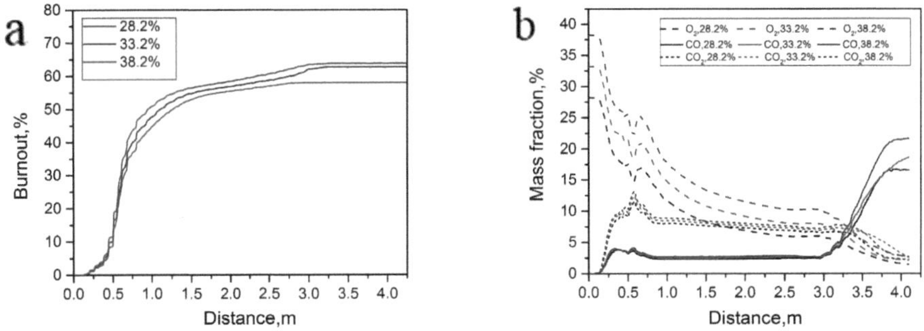

Fig. 5 **a** Burnout; **b** distributions of gas species along the tuyere axis

Effects of Oxygen Enrichment

In this work, the coal burnout is defined according to the ash balance,

$$\text{Burnout} = \left(1 - \frac{m_a^0}{m_a}\right) / \left(1 - m_a^0\right) \tag{15}$$

It represents the total weight loss of the coal due to devolatilization and char reactions. The effect of oxygen enrichment on burnout of coal along the tuyere axis is shown in Fig. 5a. This paper studied the change of coal burnout under different oxygen enrichment. It can be seen that the increase of oxygen enrichment can improve the burnout of coal. However, the effect is less and less with the increase of oxygen enrichment.

Figure 5b shows the effect of oxygen enrichment on gas composition along the tuyere axis. The oxygen content in the raceway increases obviously by increasing the oxygen content of blast, and the content of carbon dioxide and carbon monoxide can also be affected. The higher oxygen content can accelerate the reaction rate of oxygen and coke, so the content of carbon dioxide is increase. Then the higher carbon dioxide content speeds up the reaction rate of carbon dioxide and coke when the gas reaches the boundary of raceway, which generates a large amount of carbon monoxide. It is indicated the gas composition in both raceway and coke bed is sensitive to the oxygen enrichment.

Effects of Pulverized Coal Size

Figure 6a shows the effect of particle size on burnout of coal. It can be seen from the figure that the change of coal burnout has two main features: (i) The coal burnout increases rapidly with the devolatilization of coal. The residual carbon begins to burn, after burnout reaches a certain value, which causes the coal burnout

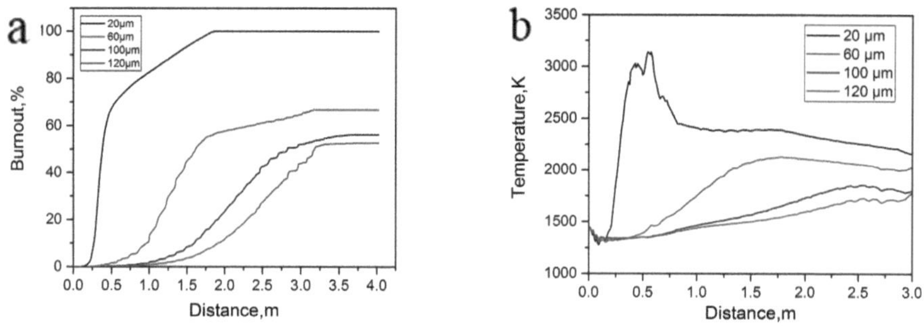

Fig. 6 **a** Burnout; **b** temperature along the tuyere axis

rises slowly. It is indicated that the volatile matter of the pulverized coal particle influences the coal combustion process very much. (ii) The small particle size of coal could give a slightly earlier and faster burnout than other cases, and the burnout becomes higher with the decrease of particle size of coal. The main reason for these features is that the temperature of small particles increases fastly, so the thermal decomposition of coal is earlier. As a result, there is enough time for residual carbon to react and burn.

Figure 6b shows the effect of particle size on the temperature field. Fine particles increase their temperature faster and always have a higher particle temperature, so the devolatilization and combustion of coal becomes more intense, which causes the raceway temperature increases rapidly. Because of the short residence time of the pulverized coal particles and the temperature of larger particle size rise slowly, and therefore a lot of pulverized coal can't fully burn, resulting in a lower temperature of raceway. And this trend becomes more and more obvious with the increase of the size of coal particle. In short, the particle size of coal has a great influence on the temperature field of raceway.

Conclusions

1. The raceway is divided into two regions: main coal plume and recirculation region. And these two regions show different patterns. Compared to main coal plume, the temperature and gas velocity is relatively lower. And more burning reactions take place in main coal region.
2. Oxygen enrichment can improve the burnout of coal, but the effect is less and less with the increase of oxygen enrichment. Gas composition distribution is sensitive to oxygen enrichment, and the content of carbon dioxide and carbon monoxide increased with the increase of oxygen content in blast.
3. The burnout and temperature become higher with the decrease of particle size of coal, and coal devolatilization is the main contributor to raise the final burnout level in the raceway.

List of Symbols

ρ	Density, $kg\,m^{-3}$
σ	Stefan-Boltzmann constant
μ_i	Fluid velocity vector, m/s
d_p	Particle diameter, m
τ_{ji}	Generalized Newtonian viscous stress, N
C_D	Drag coefficient
μ	Viscosity coefficient
Re	Reynolds number
f_i	Volume force, N
$N\mu$	Nusselt number
E	Energy per unit mass, J/kg
λ	Thermal conductivity
Γ_i	Mass transfer coefficient
ε_p	Blackness coefficient
P	Pressure, Pa
I	Radiation intensity
S_e	Internal heat source or radiation source term
H_{reac}	Reaction heat, $J\,kg^{-1}$
Y_i	Mass fraction of component i
w	Reaction rate, $kg\,m^{-3}\,s^{-1}$
R_i	Generation rate of component i
β	Equivalent proportion
k	Turbulent kinetic energy, J
B	Pre-exponential factor
ε	Dissipation rate of turbulent kinetic energy
$\dot{m}$	Quality change rate, kg/s
μ_t	Turbulent viscosity
VM	Volatile matter
α	Absorption coefficient
m_a	Ash mass fraction
σ_s	Scattering coefficient
m_a^0	Original ash mass fraction
C	Phase function coefficient

References

1. F. Shen et al., Industrial practice of BiPCI process of pulverized coal injection for blast furnace ironmaking at SSAB. Steel Res. Int. **1**(79), 11–16 (2008)
2. Y.S. Shen et al., Computational fluid gynamics study of pulverized coal combustion in blast furnace raceway. Ind. Eng. Chem. Res. **23**(48), 10314–10323 (2009)

3. M. Gu et al., Numerical analysis of pulverized coal combustion inside tuyere and raceway. Steel Res. Int. **1**(79), 17–24 (2008)
4. H. Nogami, H. Yamaoka, K. Takatani, Raceway design for the innovative blast furnace. ISIJ Int. **12**(44), 2150–2158 (2004)
5. B.Y. Guo et al., Three-dimensional simulation of flow and combustion for pulverised coal injection. ISIJ Int. **9**(45), 1272–1281 (2005)
6. L.X. Zhou, S.J. Zhang, J.H. Chen, Numerical simulations of three-dimensional gas-particle turbulent flow and pulverized coal combustion in blast furnace. J. Nanjing Univ. Sci. Technol. **2**(31), 163–167 (2007)
7. L.X. Zhou, L. Li, R.X. Li, Simulation of 3-D gas particle flows and coal combustion in a tangentially fired furnace using a two fluid trajectory model. Powder Technol. **2**(125), 226–233 (2002)
8. S.Y. Zhang et al., Numerical simulation of pulverized coal combustion in blowpipe. Iron Steel **4**(49), 6–12 (2014)
9. Y.S. Shen, et al., Three-dimensional modelling of in-furnace coal/coke combustion in a blast furnace. Fuel **90**, 728–738 (2011)

Sintering Performance of Blends Containing High Proportion of Limonite

Feng Zhang, De-qing Zhu and Jian Pan

Abstract Sintering performance of a series of sinter blends with high ratio of limonite was investigated. The metallurgical performance and mineralogy of product sinter were also examined. The results show that good sintering performance could be obtained when the ratio of limonite was below 43% in the blends of iron ore fines, while good microstructure and metallurgical properties were achieved under optimum sintering conditions. In contrast, sintering performance will rapidly deteriorate when the ratio of limonite is elevated up to 63% in the blends.

Keywords Sintering performance · Iron ore fines · Limonite

Introduction

Sintering is one of the most important agglomeration processes used to convert iron ore fines into a suitable lumpy product for blast furnace [1]. At present, 75–85% of feed burden charged into the blast furnace worldwide is in the form of iron ore sinter [2]. However, due to the depletion of traditional high-grade iron ore resources, there have been considerable changes in iron ore resources available for sintering process throughout the world, for example, the Fe grade has decreased while the gangues (such as Al_2O_3, SiO_2, loss on ignition (LOI), etc.) have increased [3, 4]. In recent years, the production of domestic magnetite concentrate has been declining and the ratio of imported iron ore fines in sintering blends continuously has been increasing because of the high production costs of mines and the lower economy growth rate in China. Therefore, the degree of external dependence of iron ores for China reaches 84% in 2015. As a result, the usage of Australian iron ores has been increasing steadily over this period to reduce costs, which means that the ore resources changed to goethitic type ores such as pisolitic or Marra Mamba ores.

F. Zhang (✉) · D. Zhu · J. Pan
School of Minerals Processing and Bioengineering,
Central South University, Changsha 410083, China
e-mail: zhangfengcsu@126.com

J.-Y. Hwang et al. (eds.), *8th International Symposium on High-Temperature Metallurgical Processing*, The Minerals, Metals & Materials Series,
DOI 10.1007/978-3-319-51340-9_60

Thus, how to effectively make sinter from blends containing high proportion of limonite becomes a very important subject for Chinese steel industry.

It is well known that the assimilation rate of the limonitic ores with fluxes is very high because the limonitic ores are porous when the dehydration of the combined water occurs during sintering. In addition, an increase in the amount of limonite in the raw materials for sinter blends has been found to lower the permeability of the sintering bed and the sinter yield and strength with higher coke consumption, and therefore these raw materials have been regarded as having inferior quality for sintering [5]. Since around 1990, extensive works have been carried out, primarily in Japan and Australia, on sintering behaviour of the significantly goethite ores in iron ore blends. Kasai et al. [6] studied the melt properties of pisolitic goethite ore. Otomo et al. [7] studied suppression of the formation of large pores in the assimilated parts of sinter produced using high ratio pisolitic ores, the result revealed that by increasing CaO concentration of the shell layer from 10 to 26 mass%, the suppression of the large pores formation was achieved in terms of the decrease in both of the diameter and amount of the pores. Yang et al. [8–10] studied sintering of blends containing magnetite concentrate and hematite or/and goethite ores and the behavior of ore blends containing hematite, magnetite concentrate or/and goethite ores, the results have demonstrated that the goethite ore-Yandi ore-can be successfully blended into a ore blend at up to 40% for sintering. Loo et al. [11] summarised the differences in sintering behaviour between hematite ores and goethitic ores based on past research programs carried out at BHP Billiton and recommended steps towards overcoming losses in productivity. Zhu et al. [12] researched the effect of the separation of large limonite ore particles on the granulation process of sinter raw materials, the result indicated that the technique could be useful in improving sintering behavior of limonite and obtaining better sintering indexes. Zhu et al. [13] researched the Granulation behaviour of limonite fines and specularite fines, found that the granulation behaviour of specularite fines can be improved by blending with limonite fines. At present, the amount of impurities in iron ore has been slowly increasing, and the iron ore fines tend to be more goethitic [3]. Therefore, the chemical and physical characteristics of iron ore fines have been changed. In addition, the proportion of limonite in sintering blends has an increasing trend in China. Thus, it is necessary to study the sintering performance of blends with high proportion of limonite.

In the paper, efforts have been made to study the effects of replacing the imported hematite ore or mixed-type iron ores with limonite ores on sintering performance of the blends by conducting sintering pot tests while determining the metallurgical performance and mineralogy of product sinter.

Experimental

Raw Materials

The raw materials used in the research included iron ore fines, burnt lime, limestone, dolomite and coke breeze. Ore A and Ore B were hematite type from Brazil, Ore C and Ore D were mixed-type iron ore fines from Australia, Ore E and Ore F were limonite type from Australia. The chemistry and particle size distributions of all the materials used for sintering were given in Tables 1 and 2, respectively.

As can be seen in Table 1, Iron ore fines A, B, C and D bear high iron grade above 60%, Iron ore E and F possess low iron ore grade and high LOI value. All of iron ore fines have wide size distributions ranging from 0.25 to 8 mm. Iron ore fine B and D possess high ratio of fine size fraction below 0.25 mm, as shown in Table 2.

Table 1 Chemical compositions of raw materials (mass/%)

Materials	TFe	FeO	SiO_2	CaO	Al_2O_3	MgO	P	S	LOI
Ore A	65.28	0.00	2.71	0.05	1.08	0.08	0.02	0.02	2.36
Ore B	61.15	0.72	6.62	0.06	1.58	0.06	0.04	0.05	3.02
Ore C	62.28	0.18	4.02	0.12	2.07	0.13	0.07	0.03	3.84
Ore D	60.36	0.00	4.11	0.02	2.16	0.05	0.06	0.05	6.41
Ore E	57.04	0.00	5.64	0.06	1.38	0.08	0.04	0.02	10.09
Ore F	58.50	0.00	5.30	0.20	2.38	0.13	0.07	0.03	7.20
Burnt lime	0.27	0.00	4.51	82.28	0.26	1.65	0.01	0.08	10.08
Limestone	0.78	0.00	2.52	54.39	0.51	0.64	0.02	0.01	39.88
Dolomite	0.56	0.00	0.75	33.50	0.23	22.29	0.01	0.01	44.28
Coke	1.65	0.00	6.02	0.35	4.55	0.08	0.01	0.40	86.87

Table 2 Size distributions of raw materials (mass/%)

Size (mm)	+8	6.3–8	4–6.3	2–4	1–2	0.5–1	0.25–0.5	−0.25	Total
Iron ore A	24.94	4.79	10.65	14.74	16.49	7.84	7.69	12.86	100
Iron ore B	18.45	7.57	16.43	10.32	10.57	3.79	4.26	28.62	100
Iron ore C	6.54	10.93	23.84	18.03	16.12	5.12	6.02	13.40	100
Iron ore D	7.04	6.66	10.36	13.10	16.25	7.94	11.16	27.49	100
Iron ore E	30.01	5.67	8.94	11.90	20.70	8.35	7.26	7.17	100
Iron ore F	9.87	7.42	13.79	16.38	23.37	8.05	8.41	12.71	100
Burnt lime	0.00	0.00	1.48	9.00	28.34	11.17	11.14	38.87	100
Limestone	0.00	0.00	0.88	10.06	11.03	31.57	11.37	35.09	100
Dolomite	0.00	0.00	9.23	20.28	14.25	22.83	5.93	27.48	100
Coke	0.39	0.00	9.12	18.26	14.57	34.59	8.43	14.64	100

Burnt lime and limestone were used to control the sinter basicity and dolomite was used to adjust the MgO content of sinter. The target binary basicity of sinter (CaO/SiO_2) product was 1.8 and MgO was 1.65 mass%, respectively.

Methods

The ore blends given in Table 3 were used to investigate the sintering behaviour with various ratio of limonite ore (Ore E, F). The chosen blends were designed to simulate part of industrial sinter blends in Chinese sintering plants, which generally use various levels of imported hematite, mixed-type and limonite ores in the ore blends.

Pilot-scale sinter pot tests of the above ore blends were carried out in a sinter pot with a diameter of 200 mm and height of 700 mm. The sinter mix was homogenized for 2 min and then granulated for 3 min in a drum mixer of Φ600 mm × 1400 mm at 20 r/min. The granulated mix was charged into the sinter pot. The sintering conditions are as follows: ignition at 1100 °C for 1.5 min and with a suction of 5 kPa, sintering at 10 kPa, and cooling at 5 kPa for 5 min. After the sintering was completed, sinter cake was freely dropped from 2 m height for 3 times, and finally was screened into 6 size fractions: +40, 40–25, 25–16, 16–10, 10–5 and −5 mm. The sinter performance was evaluated by measuring tumble index, productivity and solid fuel rate. RDI (low temperature reduction degradation) and RI (reducibility index) were determined according to ISO standards to predict metallurgical behavior of sinter product in BF.

Sinter mineralogy was observed by setting samples in resin followed by sectioning of the sample and polishing to obtain a smooth surface for optical microscopy.

Table 3 Composition of the ore blends for sintering (mass/%)

Ores	Base blend	Blend 2	Blend 3	Blend 4
Hematite ore A	18	18	18	18
Hematite ore B	25	25	25	5
Mixed-type ore C	20	10	0	10
Mixed-type ore D	4	4	4	4
Limonite ore E	33	23	13	23
Limonite ore F	0	20	40	40
Total	100	100	100	100

Results and Discussion

Pilot-Scale Pot Tests Results

Sintering Performance of Ore Blends

Sinter pot tests are carried out for the 3 potential ore blends and the base case are listed in Table 3. Ore F (Limonite ore) was introduced to the Base blend step by step to replace Ore C (Mixed-type), Ore E (limonite) and Ore B (hematite) forming Blends 2 (43% limonite ore), Blend 3 (53% limonite ore) and Blend 4 (63% limonite ore).

Table 4 shows the particle size of ore blends. As shown in Table 4, the proportion of coarse particle (+3 mm) obviously increases with the rise of the ratio of limonite ores in the blends. E specially the ratio of unclei particles size (5–3 mm) increase sharply from 26.64% (Base) to 51.45% (Blend 4), result in the increase of average size from 3.77 to 4.11 mm. Therefore, blending more coarse limonite ores probably improve the granulation of sinter mixtures.

The sintering results are summarised in Table 5. For the Base Blend at 7.5% mix moisture and 5.35% coke, the productivity is 1.51 t/(m^2 h), the sinter tumble index is 61.8% and coke rate is 81.44 kg/t.

Table 4 Particle size distributions of ore blends (mass/%)

Size (mm)	+5	5–3	3–1	1–0.5	−0.5	Total	Average (mm)
Base blend	44.28	26.64	21.42	7.48	0.18	100	3.77
Blend 2	40.47	36.85	18.68	3.94	0.06	100	3.90
Blend 3	42.68	48.50	8.06	0.75	0.02	100	4.24
Blend 4	37.61	51.41	7.15	3.79	0.04	100	4.11

Table 5 Summary of the sintering results

Blends	Base blend	Blend 2	Blend 3	Blend 4
Moisture (mass%)	7.50	8.50	8.00	8.00
Dosage of coke (mass%)	5.35	5.35	5.20	5.35
Permeability index (J.P.U.)	29.35	34.72	36.56	36.15
Sintering time (min)	28.50	23.67	24.92	23.83
Productivity (t/(m^2 h))	1.51	1.86	1.72	1.73
Yield (mass%)	76.60	77.04	76.85	75.67
Solid fuel (kg/t)	81.44	80.80	78.84	82.77
Tumble index (+6.3 mm%)	61.80	60.53	60.13	58.53

For sintering of Blend 2 at 8.5% moisture and 5.35% coke, the productivity is increased to 1.86 t/(m^2 h) from 1.5 t/(m^2 h), and the sinter tumble index is 60.53%, slightly lower than the Base Blend. The solid fuel is 80.8 kg/t, which is lower than the Base Blend. At mix moisture of 8.0%, the productivity, solid fuel and sinter strength for Blend 3 are lower than Blend 2, while Blend 3 having higher productivity than the Base Blend. For Blend 4, the productivity is comparable to Blend 3, however, the sinter strength is the lowest and the solid fuel is the highest.

Effect of High Proportion of Limonite Ores in Blends on Sintering Performance

Figure 1 shows the effect of high proportion of limonite ores on sintering performance. With an increase in the amount of limonite ores in blends from 33 to 63%, the productivity rises first then falls, while tumble index show a declining trend. The solid fuel shows a trend of decrease with the content of limonite ores from 33% (Base Blend) up to 53% (Blend 3). However, the solid fuel sharp rises when the content of limonite ores is up to 63% (Blend 4). The highest productivity can be

Fig. 1 Effect of high proportion of limonite ores on sintering performance

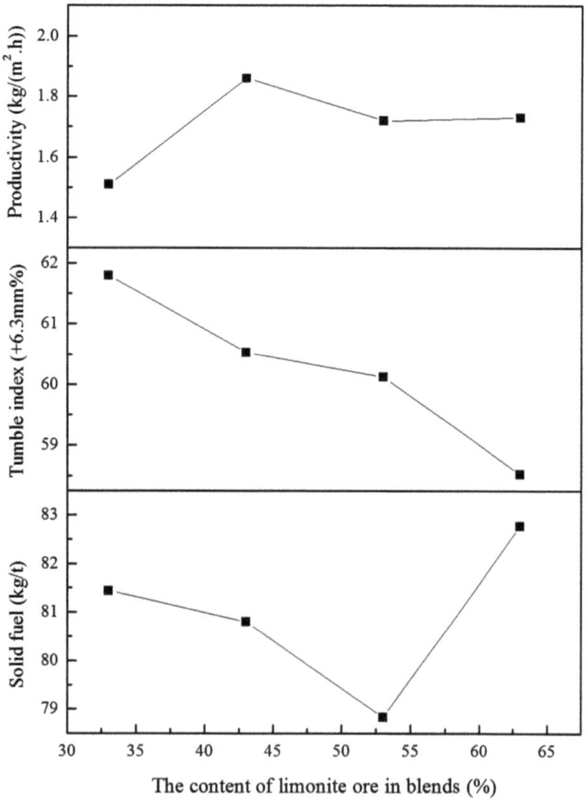

obtained from the blends containing 43% limonite ores (Blend 2) at increased mix moisture while sinter strength and solid fuel are slightly decreased compared with the Base Blend.

Generally, sinter blends comprising high amount of limonite ores possess lower tumble strength and higher solid fuel consumption [14]. In addition, it was widely observed that introducing pisolite ore into blends composed predominantly of Australian and Brazilian hematite ores caused sinter plant productivity to drop [11]. In this work, the trend of tumble index accords with the previous study, while the trend of productivity and solid fuel is different with earlier works. This is because the permeability of blends is improved with the increase of proportion of limonite ores in blends due to the highly proportion of coarse particle (+3 mm), as shown in Tables 4 and 5. Moreover, for the same bed permeability, the bed containing limonite ores have a faster flame front speed than the bed with denser ores [11]. Hence, the improvement of productivity and the reducing of solid fuel can be attained in high proportion limonite ore blends. As shown in Table 5, high productivity of Blends 2, 3 and 4 and low solid fuel of Blends 2 and 3 were attributed to the lesser sintering time than the Base Blend. It should be noticed that the sintering performance rapidly deteriorate when the proportion of limonite ores reaches 63%. It is the main reason that excessive limonite ores reduce the heat available during the melting period and results in the formation of a more viscous melt thus the deterioration of sinter strength because of higher alumina in limonite ores.

Characterisation of Product Sinter

Chemical Compositions

Chemical compositions of the product sinter are given in Table 6. There are similar chemical compositions for the four groups of ore blends under corresponding optimal sintering conditions as shown in Table 5. As the level of limonite ore increases gradually from 33% in the Base Blend to 63% in Blend 4, the content of Al_2O_3 increases from 2.15 to 2.57% due to the addition of high Al_2O_3 limonite ore F, which is harmful for the strength and metallurgical properties of sinter.

Table 6 Chemical compositions of product sinter (mass%)

Blends	TFe	FeO	SiO_2	CaO	MgO	Al_2O_3	P	S
Base blend	56.67	8.69	5.45	9.98	1.35	2.15	0.04	0.03
Blend 2	56.59	9.39	5.67	9.99	1.78	2.35	0.06	0.03
Blend 3	56.61	8.68	5.75	10.17	1.42	2.51	0.06	0.01
Blend 4	56.61	8.89	5.61	10.15	1.65	2.57	0.06	0.01

Table 7 RDI and RI of the sinters (%)

Blends	RI	$RDI_{+6.3mm}$	$RDI_{+3.15mm}$	$RDI_{-0.5mm}$
Base blend	81.65	47.21	73.98	6.40
Blend 2	84.13	47.61	78.03	5.10
Blend 3	83.30	46.10	77.24	5.07
Blend 4	78.23	45.77	76.79	5.46

Metallurgical Properties

The RDI and RI of product sinter were listed in Table 7. All of sinter product from four groups of ore blends has excellent metallurgical properties, which can be attributed to the high porosity of the sinter product, as shown in Fig. 2. With the increase of limonite ores in ore blends, the $RDI_{+3.15mm}$ decrease after sharply increase, and the RI rises first then fall. The sinters from Blend 2 have the best metallurgical properties. However, the RI of sinter sharply decreased when the proportion of limonite fines from 53 up to 63% (Blend 4) due to the content of SFCA decreases and the content of silicate increases, as shown in Table 8.

Mineralogy Examination

Mineral compositions of the product sinter were measured by using Image-Pro plus 6.0 image analysis software and under microscopy. The results are given in Table 8. The product sinter of Blend 4 contains lower SFCA and higher silicate than other sinter products.

Figure 2 shows the distributions and shape of pores in product sinter produced from four groups of ore blends. As shown in Fig. 2, all of sinter contains a lot of pores due to the high proportion of limonite in ore blends. Porous structure of sinter is contributed to improving the reducibility of sinter, but it has a bad effect on tumble index of sinter. Product sinter of Base Blend contained numerous small circular pores, and the pores evenly distributed in sinter. In addition, product sinter of Blend 2 contains many middle thick wall pores. On the contrary, product sinter of Blend 3 and Blend 4 contain many large irregular pores and the sinter is porous with large thin wall holes. So, the product sinter of Base Blend and Blend 2 have higher tumble index than Blend 3 and Blend 4.

Figure 3 shows the morphology of SFCA of product sinter. The product sinter of Base Blend, Blend 2 and Blend 3 contain abundant SFCA with the melting corrosion structure while that of Blend 4 contain more columnar SFCA than others.

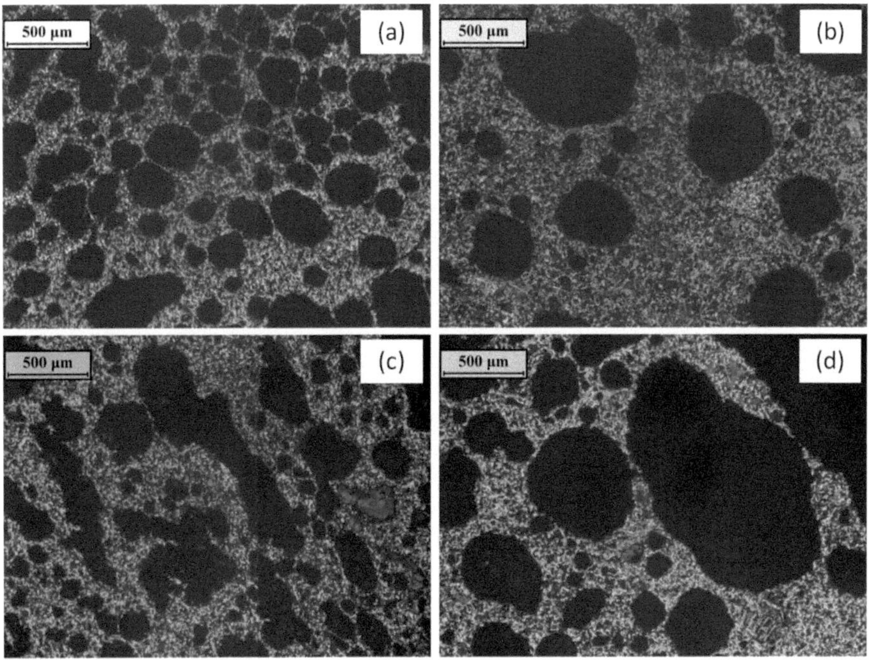

Fig. 2 Optical micrographs of sinter showing the distribution of pore produced from four kinds of blends (**a**-base blend, **b**-blend 2, **c**-blend 3, **d**-blend 4)

Table 8 Mineral phase compositions of product sinters (areas %)

Blends	Magnetite	Hematite	SFCA	Silicate	Magnesium ferrite	Glass and others
Base blend	37.16	25.20	25.16	7.93	1.89	2.66
Blend 2	37.68	24.53	25.51	7.21	2.08	2.99
Blend 3	38.76	26.37	24.65	6.87	1.72	1.63
Blend 4	36.49	25.98	23.54	9.14	2.15	2.70

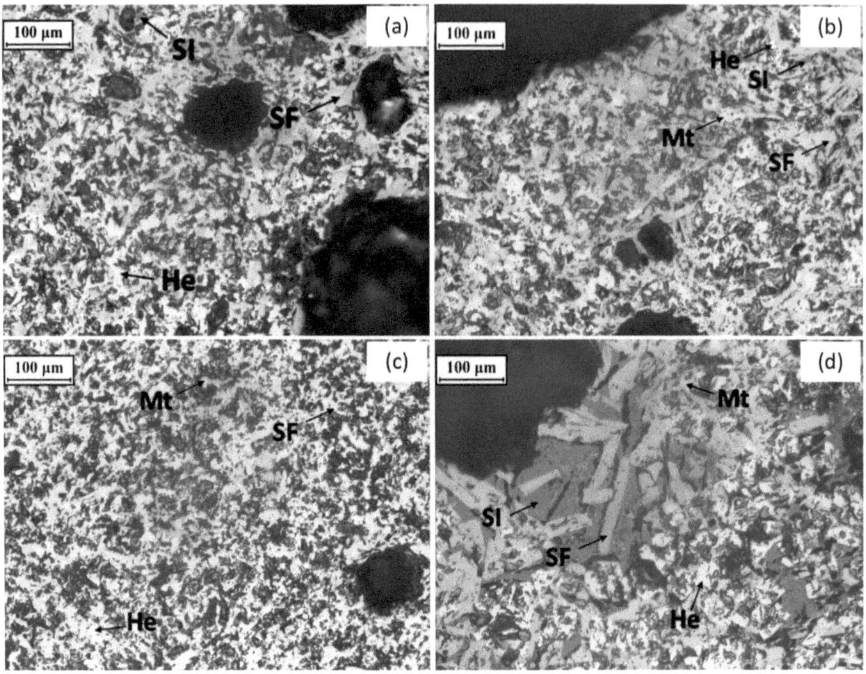

Fig. 3 Optical micrographs of sinter showing SFCA produced from four kinds of blends (**a**-base blend; **b**-blend 2; **c**-blend 3; **d**-blend 4; *SF*-SFCA; *He*-hematite; *Mt*-magnetite; *SI*-silicate)

Conclusions

(1) With an increase in the ratio of limonite ores in ore blends from 33 to 63%, the productivity rise first then falls, while tumble index gradually declines. The optimum ore blend containing 43% of limonite ores was recommended, good sintering indexes are attained as follows: the productivity of 1.86 t/(m^2 h), tumble index of 60.53%, and solid fuel consumptions of 80.8 kg/t. The sintering performance rapidly deteriorates when the proportion of limonite ores is up to 63%.

(2) With an increase in the ratio of limonite in ore blends, the $RDI_{+3.15mm}$ and the RI rises first then falls. The sinter from Blend 2 has the best metallurgical properties: $RDI_{+3.15mm}$ was 78.03%, and RI was 84.13% because the product sinter of Blend 2 possesses optimum microstructure and reasonable mineral compositions.

References

1. C.E. Loo, W. Leung, Factors influencing the bonding phase structure of iron ore sinters. ISIJ Int. **43**, 1393–1402 (2003)
2. S.L. Wu, H.F. Xu, Y.Q. Tian, Evaluation of lump ores in modern blast furnaces as part of mixed burden practice. Ironmaking Steelmaking **36**, 19–23 (2009)
3. L. Lu, O. Ishiyama, Recent advance in iron ore sintering. Iron ore conference 2015, AUSIMM, PERTH, (2015), p. 61–67
4. J. Pan, B. Shi, D. Zhu, Y. Mo, Improving sintering performance of specularite concentrates by pre-briquetting process. ISIJ Int. **56**, 777–785 (2016)
5. T. Kawaguchi, T. Usui, Summarized achievements of the porous meso-mosaic texture sinter research project. ISIJ Int. **45**, 414–426 (2005)
6. E. Kasai, Y. Sakano, T. Kawaguchi, T. Nakamura, Influence of properties of fluxing materials on the flow of melt formed in the sintering process. ISIJ Int. **40**, 857–862 (2000)
7. T. Otomo, N. Taguchi, E. Kasai, Suppression of the formation parts of sinter produced using of large pisolitic pores in the assimilated ores. ISIJ Int. **36**, 1338–1343 (1996)
8. L.X. Yang, D. Witchar, Z.N. Yu, Introduction of pisolitic goethite ore into a Chinese ore blend. ISIJ Int. **40**, 647–653 (2000)
9. L.X. Yang, C.E. Loo, Structure of sinters formed from complex Chinese iron ores. ISIJ Int. **37**, 449–457 (1997)
10. L.X. Yang, D. Witchar, Sintering of blends containing magnetite concentrate and hematite or/and goethite ores. ISIJ Int. **38**, 1069–1076 (1998)
11. C.E. Loo, Perspective of goethitic ore sintering fundamentals. ISIJ Int. **45**, 436–448 (2005)
12. J. Zhu, S.L. Wu, J. Fan, G. Zhang, Z. Que, Effect of the separation of large limonite ore particles in the granulation process of sinter raw materials. ISIJ Int. **53**, 1529–1537 (2013)
13. D.Q. Zhu, X.L. Zhou, J. Pan, B.J. Shi, Granulation behaviour of specularite fines in ferrous sinter mixtures. Mineral Process. Extr. Metall. **125**, 172–177 (2016)
14. C.E. Loo, D.J. Wong, Fundamental insights into the sintering behaviour of goethitic ore blends. ISIJ Int. **45**, 459–468 (2005)

Thermodynamics Study on Phosphorus Distribution Between $2CaO{\cdot}SiO_2$-$3CaO{\cdot}P_2O_5$ Solid Solution and Liquid Slag

Chao Jiang, Ming-Mei Zhu, Rui-Rui Zhao and Zhang-Guang Gao

Abstract This paper focuses on the CaO-SiO_2-FeO-MgO-MnO-P_2O_5 slag system that represents the major components of dephosphorization slag in BOF. Phosphorus distribution between $2CaO{\cdot}SiO_2$-$3CaO{\cdot}P_2O_5$ Solid Solution and Liquid Slag were analyzed by FactSage. The effects of slag content, basicity on the distribution ratio of P between the solid solution phase and the liquid phase (L_P) and the percentage content of P in solid solution phase and liquid phase have been discussed. The results show that, with the increase of basicity, L_P (wt%) decreases until the basicity reaches 1.9, then it turns to increase. L_P (wt%) increases gradually with the increasing amount of FeO or P_2O_5.

Keywords Thermodynamics study · Phosphorus distribution · $nCao{\cdot}SiO_2$-$3CaO{\cdot}P_2O_5$

Introduction

There is abundance of iron ore resources with low grade in China, especially the existence of phosphorus and sulfur aggravate the burden of steel manufacture. At present, dephosphorization (de-P) in Basic Oxygen Furnace (BOF) has been widely considered as an effective way to utilize the high phosphorus iron ores [1–3]. The commonly used de-P slag is CaO-SiO_2-Fe_tO-P_2O_5 system, the contents of which are usually saturated with dicalcium silicate (C_2S), resulting the coexisting of C_2S solid phase and liquid phase. It is considered that the solid solution (SS) phase can enrich P and thus decreases the P concentration of liquid phase of the slag.

In the 1970s, Healy [4] obtained the relationship between phosphorus distribution ratio and content of slag and temperature by the improved Flood-Grjotheim

C. Jiang · M.-M. Zhu (✉) · R.-R. Zhao · Z.-G. Gao
College of Materials Science and Engineering,
Chongqing University, Chongqing 400044, China
e-mail: menglaiji@qq.com

J.-Y. Hwang et al. (eds.), *8th International Symposium on High-Temperature Metallurgical Processing*, The Minerals, Metals & Materials Series,
DOI 10.1007/978-3-319-51340-9_61

ionic model on the basis of a large number of experimental data. Gaskell [5] formulated equilibrium distribution ratio (2) by optical basicity.

$$\log\frac{(\%P)}{[\%P]} = \frac{22350}{T} + 0.08(\%\text{CaO}) + 2.5\log(\text{T}\cdot\text{Fe}) - 16 \pm 0.4 \quad (1)$$

$$\log\frac{(\%\text{P}_2\text{O}_5)}{[\%P][\%O]^{5/2}} = \{-558.874 + \frac{2175100}{T} - \frac{1930041500}{T^2}\}\Lambda - 24.33 \quad (2)$$

$$L_P = (\%P_2O_5)_{SS}/(\%P_2O_5)_L \quad (3)$$

$$L_P = (P_2O_5)_{SS}/(P_2O_5)_L \quad (4)$$

where (%X) and [%X] are X contents in slag and hot metal in mass% respectively, T is the temperature of slag, T.Fe is the total iron, Λ is optical basicity, L_P is phosphorus distribution ratio and the subscripts SS and L denote solid solution and liquid phase, respectively.

The module Equilib in FactSage 6.3 was used for theoretical calculation, and the temperature of calculation is 1350 °C, the compositions of the slag are shown in Table 1 and the compositions of the slag in the position of CaO-FeO-SiO_2 pseudo-ternary phase diagram is shown in Fig. 1, the all slag is basically in liquidus of 1350 °C, and the composition of the slag is basically in primary crystals area of C_2S.

Calculation of Phosphorus Distribution Ratio

The distribution ratio of P_2O_5 between $2CaO{\cdot}SiO_2$-$3CaO{\cdot}P_2O_5$(nC_2S-C_3P) solid solution phase and liquid slag is $L_P(\text{wt.\%}) = (\%P_2O_5)_{SS}/(\%P_2O_5)_L$ where the subscripts SS and L denote solid solution and liquid phase, respectively.

Table 1 Compositions of the slag

No.	R	CaO	SiO_2	FeO	P_2O_5	MgO	MnO
1	1.3	33.9	26.1	23	4	7	6
2	1.6	36.9	23.1	23	4	7	6
3	1.9	39.3	20.7	23	4	7	6
4	2.2	41.3	18.8	23	4	7	6
5	2.5	42.9	17.1	23	4	7	6
6	1.9	33.4	17.6	32	4	7	6
7	1.9	35.4	18.6	29	4	7	6
8	1.9	37.3	19.7	26	4	7	6
9	1.9	41.3	21.7	20	4	7	6
10	1.9	38.7	20.3	26	2	7	6
11	1.9	36.0	19.0	26	6	7	6
12	1.9	34.7	18.3	26	8	7	6
13	1.9	33.4	17.6	26	10	7	6

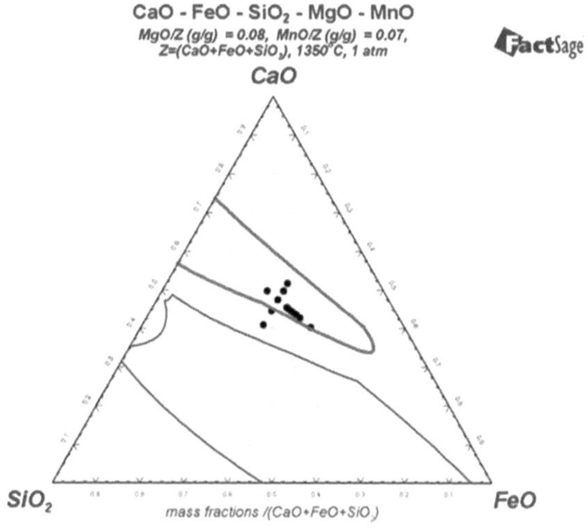

Fig. 1 Slag composition in the pseudo-ternary phase diagram of the CaO-FeO-SiO_2 system

Activity coefficient of P_2O_5 in CaO-SiO_2-FeO-MnO-MgO-P_2O_5 slag are calculated by regular solution model [6], using Eqs. (5) and (6).

$$\begin{aligned} RT\ \ln \gamma PO_{2.5}(R.S) = {} & 31380X_{\text{FeO}}^2 - 251040X_{\text{CaO}}^2 - 37660X_{\text{MgO}}^2 \\ & + 83680X_{\text{SiO}_2}^2 - 251040X_{\text{FeO}} \cdot X_{\text{CaO}} - 102510X_{\text{FeO}} \cdot X_{\text{MgO}} \\ & + 94140X_{\text{FeO}} \cdot X_{\text{SiO}_2} - 188280X_{\text{CaO}} \cdot X_{\text{MgO}} - 33470X_{\text{CaO}} \cdot X_{\text{SiO}_2} \\ & + 112960X_{\text{MgO}} \cdot X_{\text{SiO}_2} \end{aligned} \tag{5}$$

$$RT\ \ln a_{\text{P}_2\text{O}_5(l)} = 2RT\ \ln a_{PO_{2.5}(R.S.)} + 52720 - 230.706(\text{J}) \tag{6}$$

When liquid phase and solid solution reached equilibrium, activity coefficient of P_2O_5 in this two phases are about the same, so

$$a_{\text{P}_2\text{O}_5(SS)} = a_{\text{P}_2\text{O}_5(L)} \tag{7}$$

$$L_P = \frac{(\%\text{P}_2\text{O}_5)_{SS}}{(\%\text{P}_2\text{O}_5)_L} = k\frac{a_{\text{P}_2\text{O}_5(SS)} \times \gamma_{\text{P}_2\text{O}_5(L)}}{a_{\text{P}_2\text{O}_5(L)} \times \gamma_{\text{P}_2\text{O}_5(SS)}} = k\frac{\gamma_{\text{P}_2\text{O}_5(L)}}{\gamma_{\text{P}_2\text{O}_5(SS)}} \tag{8}$$

$$\gamma_{\text{P}_2\text{O}_5(SS)} = k\frac{\gamma_{\text{P}_2\text{O}_5(L)}}{L_P} \tag{9}$$

where $k = \frac{(\%\text{P}_2\text{O}_5)_{SS} \times (x_{\text{P}_2\text{O}_5})_L}{(\%\text{P}_2\text{O}_5)_L \times (x_{\text{P}_2\text{O}_5})_{SS}}$ is the conversion coefficient of P_2O_5 in liquid phase and solid solution between in mole fraction and mass fraction.

Results of Calculations

The calculation results are shown in Tables 2, 3, and 4. The activity coefficients of P_2O_5 in solid solution phase and liquid slag by regular solution model and L_P [calculated by (9)] are shown in Table 5.

Activity coefficient of P_2O_5 in CaO-SiO_2-FeO-MnO-MgO-P_2O_5 slag is calculated by regular solution model.

Table 2 Compositions of phases in slag/wt%

No.	Liquid slag	Solid solution		Other phases	
		Ca_2SiO_4	$Ca_3(PO_4)_2$	Olivine	Monoxide
1	43.76	17.34	4.65	24.07	10.18
2	34.13	35.09	4.88	6.35	19.55
3	21.26	44.70	5.11	3.44	25.50
4	26.47	44.23	4.73	1.19	23.48
5	36.07	37.29	5.04	0.60	21.00
6	24.90	35.79	4.45	3.04	31.81
7	23.60	38.92	4.63	3.14	29.72
8	22.54	41.65	4.87	3.37	27.56
9	19.98	47.70	5.37	3.50	23.45
10	12.52	52.09	1.90	2.56	30.92
11	34.08	30.85	7.70	3.30	24.07
12	45.92	20.09	10.46	2.64	20.89
13	57.11	9.88	13.20	1.56	18.25

Table 3 Compositions of solid solution phase and liquid slag/wt%

No.	Liquid slag					Solid solution		
	MgO	FeO	SiO_2	CaO	P_2O_5	SiO_2	CaO	P_2O_5
1	6.25	31.02	26.85	31.62	4.28	27.51	62.82	9.67
2	6.94	29.09	25.60	33.21	5.17	30.62	63.78	5.59
3	3.65	36.27	18.35	33.90	7.82	31.31	64.01	4.69
4	1.54	45.64	11.06	34.82	6.94	31.51	64.07	4.42
5	1.18	43.04	10.71	40.39	4.69	30.73	63.82	5.45
6	3.28	42.80	16.31	29.73	7.88	31.03	63.92	5.06
7	3.33	41.18	16.66	30.85	7.96	31.17	63.96	4.87
8	3.57	38.48	17.72	32.39	7.85	31.22	63.97	4.79
9	3.75	33.73	19.15	35.66	7.73	31.35	64.02	4.63
10	1.98	53.37	9.69	25.94	9.03	33.66	64.74	1.61
11	4.74	33.16	20.84	33.99	7.26	27.92	62.95	9.14
12	5.61	30.27	22.65	34.49	7.00	22.93	61.39	15.67
13	6.24	28.42	23.87	34.53	6.94	14.93	58.90	26.18

Table 4 Mole fraction of components in solid solution phase and liquid slag/wt%

No.	Liquid slag				Solid solution			
	MgO	FeO	SiO_2	CaO	P_2O_5	SiO_2	CaO	P_2O_5
1	0.095	0.265	0.274	0.346	0.019	0.278	0.680	0.041
2	0.106	0.248	0.261	0.363	0.022	0.302	0.674	0.023
3	0.058	0.324	0.196	0.387	0.035	0.307	0.673	0.019
4	0.025	0.416	0.121	0.406	0.032	0.309	0.673	0.018
5	0.019	0.384	0.114	0.461	0.021	0.303	0.674	0.023
6	0.053	0.388	0.177	0.345	0.036	0.305	0.674	0.021
7	0.054	0.373	0.180	0.357	0.036	0.306	0.673	0.020
8	0.057	0.345	0.190	0.372	0.036	0.307	0.673	0.020
9	0.059	0.299	0.203	0.404	0.035	0.308	0.673	0.019
10	0.033	0.502	0.109	0.312	0.043	0.325	0.669	0.007
11	0.074	0.292	0.219	0.383	0.032	0.281	0.680	0.039
12	0.087	0.263	0.235	0.384	0.031	0.241	0.690	0.070
13	0.096	0.246	0.246	0.382	0.030	0.168	0.708	0.124

Table 5 Activity coefficient and L_P in solid solution phase and liquid slag

No.	R	$\gamma P_2O_5(L)$	$\gamma P_2O_5(SS)$	L_P (wt%)
1	1.3	4.06E−14	1.82E−14	2.26
2	1.6	1.90E−14	1.82E−14	1.08
3	1.9	3.25E−15	5.89E−15	0.60
4	2.2	5.02E−16	8.76E−16	0.64
5	2.5	5.26E−17	4.90E−17	1.16
6	1.9	1.03E−14	1.76E−14	0.64
7	1.9	7.15E−15	1.29E−14	0.61
8	1.9	4.97E−15	8.89E−15	0.61
9	1.9	2.05E−15	3.70E−15	0.60
10	1.9	1.41E−14	9.24E−14	0.18
11	1.9	5.36E−15	4.44E−15	1.26
12	1.9	6.70E−15	2.96E−15	2.24
13	1.9	8.66E−15	2.11E−15	3.77

Discussion

1. Effect of basicity on phosphorus distribution ratio

Figure 2a shows the influence of basicity of slag to distribution ratio of P between in solid solution phase and liquid slag phase. L_P (wt%) decreased from 2.26 to 0.6 with the basicity increased from 1.3 to 1.9, after basicity is more than 1.9, L_P (wt%) increases with basicity increased.

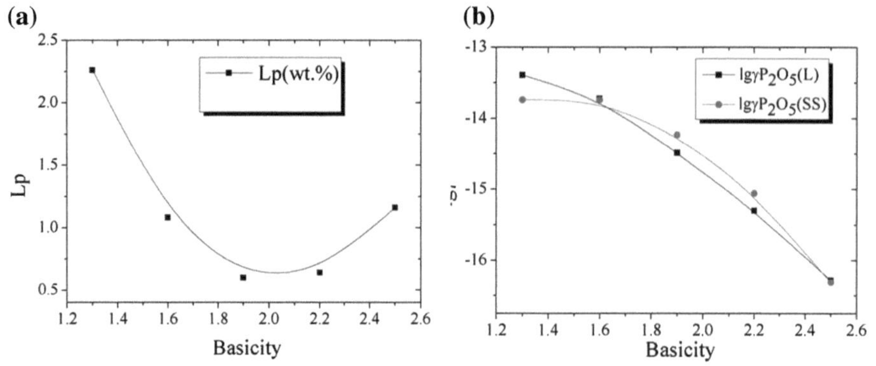

Fig. 2 Effect of basicity on phosphorus distribution ratio and activity coefficient of P_2O_5. **a** Effect of basicity on L_P. **b** Effect of basicity on activity coefficient of P_2O_5

Fig. 3 Effect of basicity on phases content in slag

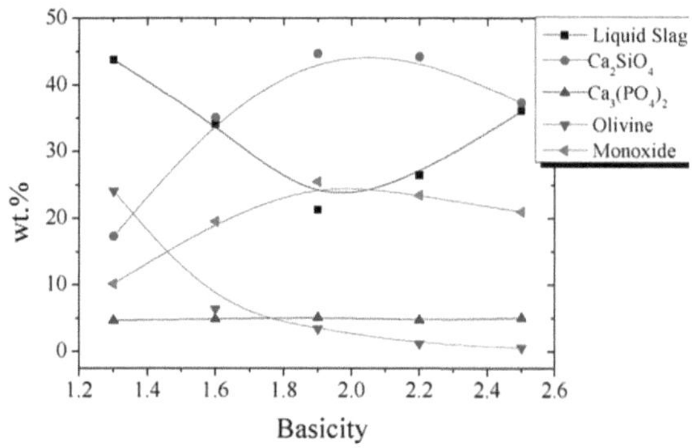

Figure 2b shows the influence of basicity of slag on the activity coefficients of the liquid slag phase and the SS phase. The activity coefficients of the liquid slag phase and the SS phase decreases with increasing basicity. When the bas basicity increases from 1.9 to 2.5, the activity coefficient of P in solid solution is more than that in liquid phase.

Figure 3 shows that at the process of basicity increases from 1.3 to 1.9, Ca_2SiO_4 content in SS phase increased significantly while $Ca_3(PO_4)_2$ content has no significant change. The precipitation of C_2S has diluted the P concentration in SS phase, which leads to leads decrease of P_2O_5 content in SS phase; the liquid phase decreased at the process of basicity increases from 1.3 to 1.9. This all can cause L_P (wt%) to decrease with the increases of basicity. Similarly, L_P (wt%) increased with the increases of basicity at the process of basicity increases from 1.9 to 2.5.

2. Effect of FeO content on phosphorus distribution ratio

Figure 4a shows the influence of the amount of FeO to the distribution ratio of P. At the process of the content of FeO increases from 20 to 32%, L_P (wt%)

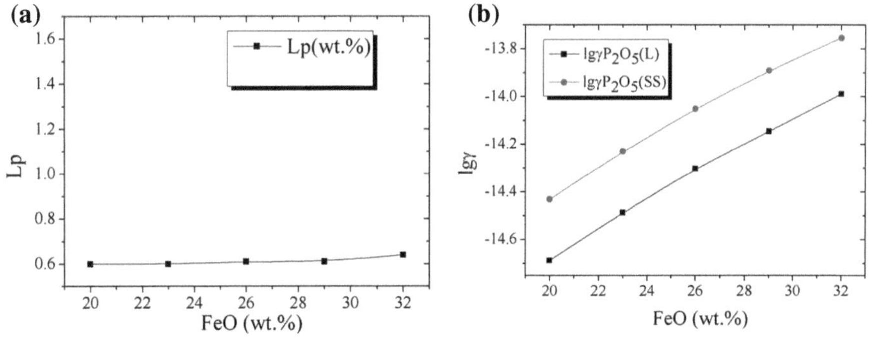

Fig. 4 Effect of FeO on phosphorus distribution ratio and activity coefficient of P_2O_5. **a** Effect of FeO on phosphorus distribution ratio. **b** Effect of FeO on activity coefficient of P_2O_5

Fig. 5 Effect of FeO on phases content in slag

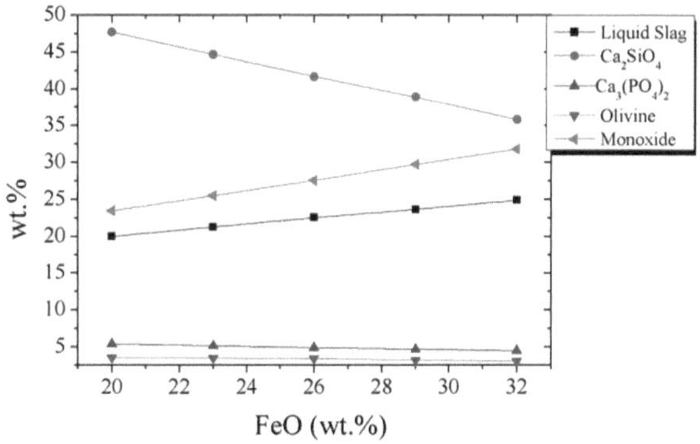

increases activity coefficient of P_2O_5 in solid solution phase and liquid phase increases with the increase of the content of FeO, and the activity coefficient of P_2O_5 in SS phase is always more than which in liquid phase. Figure 5 combined with Fig. 6 show that the liquid phase content and the T.Fe content in liquid phase increased with the increases of the content of FeO, and the increases of the content of T.Fe can help to reduce the melting point and viscosity of the slag, and can help phosphorus in liquid slag to diffuse into the surface of the solid solution phase, and then solid solution reaction happened. So C_2S and C_3P content in SS phase decreased with the increases of FeO. L_P (wt%) increased with the increases of the content of FeO. Ito [7] results show that, L_P increases with T.Fe content increased, which is in agreement with this result of calculation.

3. Effect of P_2O_5 content on phosphorus distribution ratio

Figure 7a shows the influence of the content of P_2O_5 to the distribution ratio of P. L_P (wt%) increases with the increase of the content of P_2O_5 in slag, and the rate of increase is growing. This result is in agreement with experiment which was done by Dou Xiaofei [8]. Figure 7b shows the influence of the content of P_2O_5 on the

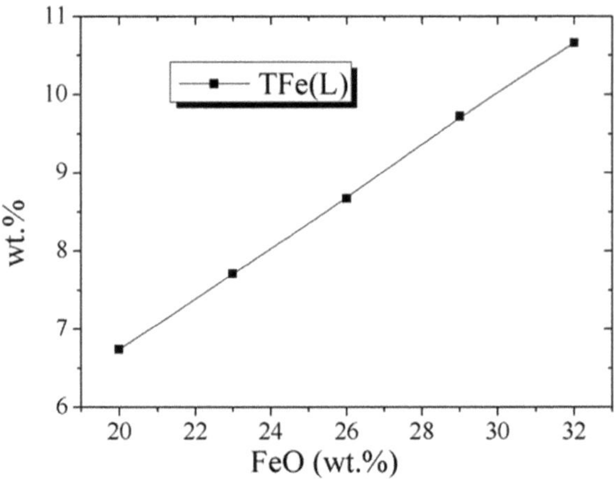

Fig. 6 Effect of FeO on T.Fe content

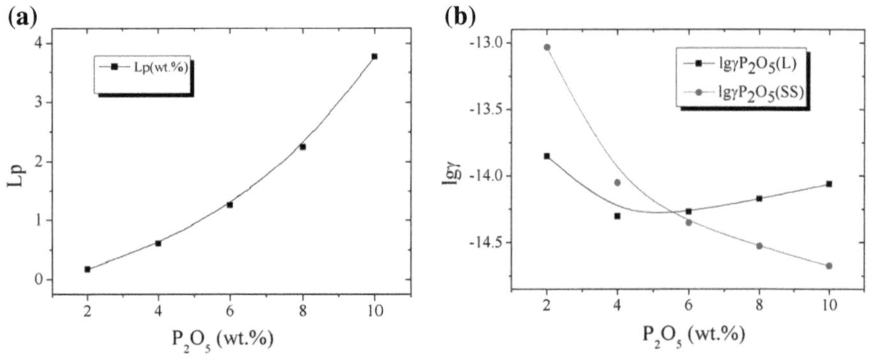

Fig. 7 Effect of P_2O_5 on phosphorus distribution ratio and activity coefficient. **a** Effect of P_2O_5 on phosphorus distribution ratio. **b** Effect of P_2O_5 on activity coefficient

activity coefficient of P_2O_5 in SS phase and liquid phase. With the content of P in slag increased the activity coefficient of P_2O_5 in SS phase decreases while that in liquid phase increases after the first decrease, the activity coefficient is minimal when the content of P_2O_5 is 4%.

Figure 8 shows that the liquid phase and C_3P content increased with the increases of P_2O_5 in slag while the C_2S content and the oxide in slag decreased. With the continuous increases of P content in slag, the related SS phase has decreased continuously with the decreases of C_2S and increases of C_3P, there are more P exist in C_3P phase which is more stable, so the activity coefficient of P_2O_5 in SS would be decreased. Though the continuous increases of P content in slag can cause P_2O_5 content in slag to increase, it can also conduct more liquid phase, which can dilute P content in liquid phase. Therefore, the activity coefficient of P_2O_5 in liquid phase decreased at first and then increased.

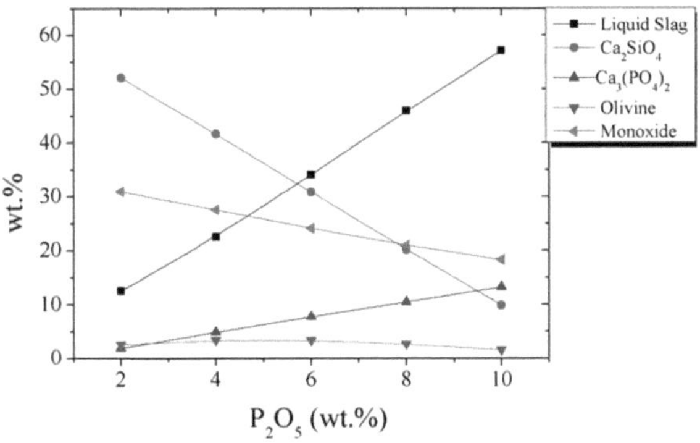

Fig. 8 Effect of P_2O_5 on phases content in slag

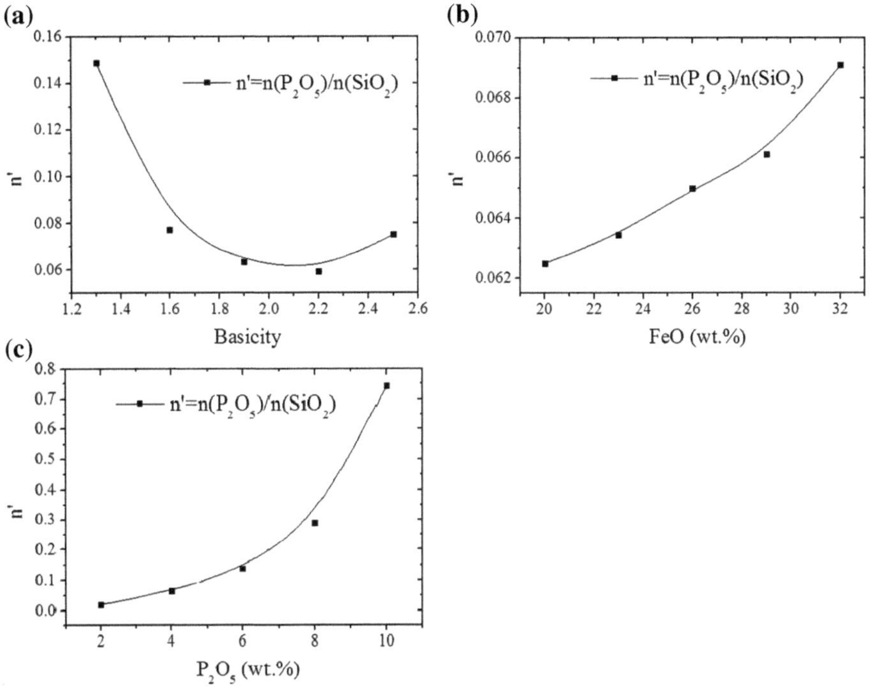

Fig. 9 Effect of the content of slag on phosphorus contents. **a** Effect of basicity on phosphorus content. **b** Effect of FeO on phosphorus content. **c** Effect of P_2O_5 on phosphorus content

4. Effect of the content of slag on phosphorus distribution ratio

Figure 9 shows the influence of the content of the slag to the content of P in solid solution. Figure 9a shows that at the process of basicity increases from 1.3 to 2.5, the content of P in solid solution increases after the first decrease with the increases of basicity. The reason is that basicity was low at first, CaO combined with SiO_2

preferentially. After basicity increased to between 1.9 and 2.2, the content of C_2S in slag is high and P_2O_5 dissolved into C_2S then generated C_2S-C_3P. With the increases of basicity, the content of P dissolved in solid solution has increased gradually. Figure 9b shows that with the increases of FeO content in slag, the content of P in solid solution has increased gradually, so the increases of FeO content has increased the driving force of P diffused into solid solution phase. So with the increases of P_2O_5 content in slag, the content of P diffused in solid solution has increased gradually, which was showed in Fig. 9c. Zhou et al. [9] did the experiment shows that P_2O_5 content in SS increases with P_2O_5 content in slag increased.

Conclusions

(1) L_P (wt%) decreased from 2.26 to 0.6 with the basicity increased from 1.3 to 1.9, after basicity is more than 1.9, L_P (wt%) increases with basicity increased.
(2) At the process of the content of FeO increases from 20 to 32%, L_P (wt%) increases with the content of FeO increased, but the rate of increase is not obvious. The activity coefficient of P_2O_5 in solid solution phase and liquid phase increases with the increase of the content of FeO.
(3) L_P (wt%) increases with the increase of the content of P_2O_5 in slag, and the rate of increase is growing. With the content of P in slag increased, the activity coefficient of P_2O_5 in SS phase decreases while which in liquid phase increases after the first decrease, the activity coefficient is minimal at the content of P_2O_5 is 4%.
(4) The content of P in solid solution increases after the first decrease with the increases of basicity. With the increases of FeO content in slag, the content of P in solid solution has increased gradually, so the increases of FeO content has increased the driving force of P diffused into solid solution phase. With the increases of P_2O_5 content in slag, the content of P diffused in solid solution has increased gradually.

References

1. G. Thornton, D. Anderson, Low phosphorus basic oxygen steelmaking practices in British Steel. Ironmaking Steelmaking **21**(3), 247–251 (1994)
2. P.K. Tripathy, A. Banerjee, B. Singh et al., Approaches for conversion of high phosphorus hot metal to steel for flat products. ISIJ Int. **48**(5), 578–583 (2008)
3. B.J. Monaghan, R.J. Pomfret, K.S. Coley, The kinetics of dephosphorization of carbon-saturated iron using an oxidizing slag. Metall. Trans. B **29B**(1), 111–118 (1998)
4. G.W. Healy, A new look at phosphorus distribution. Iron Steel Institute **208**(7), 664–668 (1970)

5. D.R. Gaskell, On the correlation between the distribution of phosphorus between slag and metal and the theoretical optical basicity of the slag. Trans. ISIJ **22**(12), 997–1000 (1982)
6. S. Ban-ya, Mathematical expression of slag-metal reactions in steelmaking process by quadratic formalism based on the regular solution model. ISIJ **33**, 2–11 (1993)
7. K. Ito, M. Yanagisawa, N. Sano, Tetsu-to-Hagané **68**, 342–346 (1982)
8. X. Dou, Fundamental research on phosphorus distribution between $2CaO{\cdot}SiO_2$-$3CaO{\cdot}P_2O_5$ solid solution and liquid slag phase in CaO-SiO_2-Fe_tO-P_2O_5 slags. Chongqing University, 37–48 (2015)
9. Zhou Hanmei, Bao Yanping, Lu Lin, Effect of P_2O_5 on rich phosphorus phase of CaO-SiO_2-Fe_tO-P_2O_5 slag. China Metallurgy **1**(23), 45–49 (2013)

Two-Step Copper Smelting Process at Dongying Fangyuan

Zhi Wang, Zhixiang Cui, Chuanbing Wei and Haibin Wang

Abstract Two-step copper smelting process (hereinafter Fangyuan Process) has been recently developed and patented by Dongying Fangyuan Company. This is the first and only process of this kind to produce anode copper in the world. And it can be summarized as below. The raw materials including concentrates and flux are simply mixed without treatment and fed into Submerged Lance Smelting Furance (SLS). High grade matte (>73% Cu) produced by SLS is sent to Submerged Lance Converting Refining Furnace (SLCR) via matte conduit. Two SLCRs work in turn to maintain continuous operation. Anode copper from SLCR is sent to the casting wheel to get anode copper plates. The new technology started operation in October 2015 with a capacity to treat 1.5 Mt/a copper concentrate proved with various advantages. This paper will describe its development process at Dongying Fangyuan, with detailed discussion about the operation conditions and parameters.

Keywords Two-step · Fangyuan · SLS · SLCR · Development

Introduction

With the continuous improvements of copper smelting technique as well as increasingly stringent requirements of environmental protection, the experts and researchers of the nonferrous metal metallurgy industry strive to find a new smelting process to relief the environmental protection issues caused by matte transportation, etc. In addition, continuous converting has been one of the research topics for many years. At present, the continuous converting processes adopted in industrial production mainly include the Mitsubishi Process [1] researched and developed by Japan, double flash continuous converting [2] invented by Finland and the Noranda continuous converting process and so on.

Z. Wang (✉) · Z. Cui · C. Wei · H. Wang
Dongying Fangyuan Nonferrous Metals Co., Ltd, Dongying, Shandong, People's Republic of China
e-mail: wangzhi@sdfygroup.com

J.-Y. Hwang et al. (eds.), *8th International Symposium on High-Temperature Metallurgical Processing*, The Minerals, Metals & Materials Series,
DOI 10.1007/978-3-319-51340-9_62

Compared with the above mentioned processes, Fangyuan Process (patent number: 201410237182.4) independently innovated by Fangyuan can not only solve issues such as ladle transportation of copper matte, low altitude pollution and limited capacity, but also shorten the procedures from three steps to two steps, which will further enhance the overall level of the copper smelting industry from the aspects of energy conservation and environment protection, CAPEX, OPEX, personnel allocation and labor intensity and so forth.

The phase II project of Dongying Fangyuan, which has applied the new process successfully, has been officially commenced on October 23rd, 2015. The production line is composed of "1 SLS + 2 SLCRs". The copper content in the product produced by the SLCR in the second step can be over 99.2%, thus such product can be directly casted into anode plates and further sent to electrolytic refining section.

Fangyuan process is the result of Fangyuan Company through several years of bold tests and innovations based on both its Phase I project "production line of processing 500,000 t/a polymetallic concentrate" (namely "bottom blowing smelting—PS converting—pyro-refining") [3] and "Fangyuan's production line for scrap copper direct to anode copper". The process is independently researched and developed by Fangyuan Company. It is a Chinese technology and Chinese process, which is also the first one around the globe.

Brief Introduction of Fangyuan Process Flow

Compared with conventional copper smelting processes, Fangyuan Process can produce qualified anode plates in only two steps, with its process sketch as Fig. 1.

All the minerals of Fangyuan Company are imported. All kinds of copper concentrates with various compositions are blended in the feeding preparation plant as per relevant feeding requirements. Then such concentrates and materials such as quartz, slag concentrates, smoke dust, etc. are respectively stored in corresponding bins for blending according to charge sheet prior to be transported to three charging bins above the SLS via belt conveyors for storage. After measurement, the materials are charged as per production requirements into the SLS through feeding ports. In addition, in order to avoid damage to belt of the long-distance transportation of cold feeding, a separate set of lifting device for cold feeding is equipped for SLS; therefore, such cold dope can be directly charged nearby.

Chemical reactions on blended materials take place in SLS. Pure oxygen and air are injected from the bottom and side of SLS. Thus chemical reactions will finish quickly producing copper matte (%Cu $\geq$ 73% or so) to be continuously discharged through siphon hole and further directed into the SLCR via matte conduit. The slag produced is discharged into the slag ladle via the slag taphole, and transported by

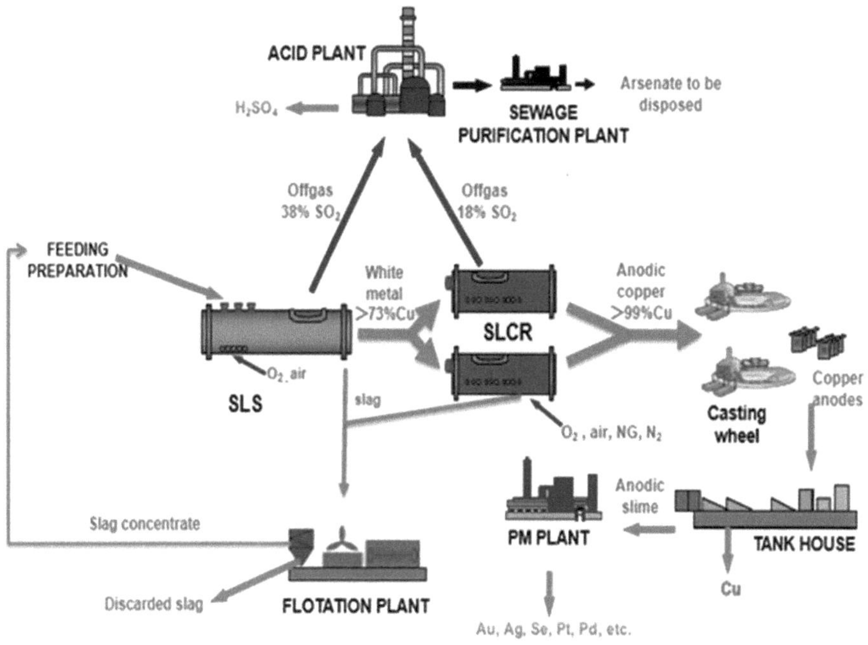

Fig. 1 Process sketch of Fangyuan process

slag hauler to a yard for slow cooling. Another product, the smoke dust, goes through uptake to the waste heat boiler, electrostatic precipitator successively, after which it will be sent into the acid plant via high temperature draught fan for acid-making. The high-grade matte produced from SLS is directed constantly into the SLCR via matte conduit. Two SLCRs work alternatively to guarantee the continuous discharge of matte from the SLS. We choose the injection of the four kinds of gases including nitrogen, natural gas, oxygen and air through lances equipped at the bottom SLCR with absolute possibility to change the availability among them. When the feedstock in SLCRs meets relevant air supply requirements, the furnace will be rotated to position for blasting and production. Slag is discharged through the slag taphole and transported via slag ladle. When it approaches to the end of anode-making phase, starts the operation to switch nitrogen/natural gas by accurate control of flow in order to keep 99.21% Cu of liquid anode so as to meet the casting requirements of anode plates. Then it is directly sent into the constant disc casting machine via the taphole for plate casting.

The schematic diagram of the process arrangement is shown in Figs. 2 and 3.

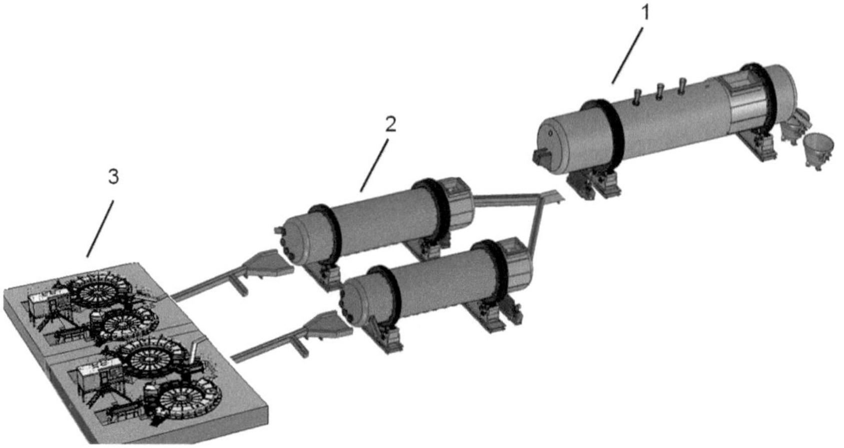

1 SLS; 2 SLCR; 3 Disc casting machine

Fig. 2 Connection diagram of main devices of Fangyuan process

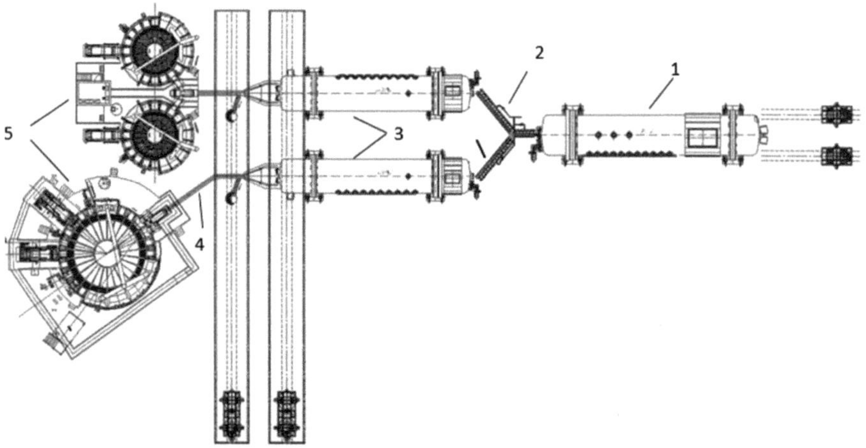

1 SLS; 2 Launder; 3 SLCR; 4 Launder; 5 Disc casting machine

Fig. 3 Schematic diagram of Fangyuan process arrangement

Main Process Configuration and Technical Indicators

Smelting Section

The smelting section of Fangyuan Process is one SLS with size of $\Phi 5.5 \times 28.8$ m equipped with 23 oxygen lances arranged in two rows. Also there are lances arranged on the side wall of furnace near slag discharging end. Such oxygen lances is designed with special structure, that is it is shrouded tube in which the inner layer is for running pure oxygen and the outer for air; and the air in the outer layer also plays the role to protect the lance. Two slag tapholes are located on the end wall of the SLS; the number of slag tapholes decided is to match the slag tapping and transportation of slag ladles. Such slag is transported by 12 m^3 slag ladles and after slow cooling transported to flotation plant for slag cleaning.

There are two offgas outlets. They are located on the upper parts of slag discharging end and copper discharging end, respectively. Siphon hole is chosen for matte discharge, through which copper matte is discharged continuously and directed into SLCR by matte conduit. SLS and SLCRs are connected by matte conduit made of special inadhesive material with good thermal insulation to avoid low altitude pollution caused by offgas.

The schematic diagram of the SLS construction is shown in Fig. 4.

After months of production practices, the technological process is smooth; all the operating parameters of SLS are basically stable and all indicators reach the designed values. At present, the maximum feeding amount of SLS has reached 207

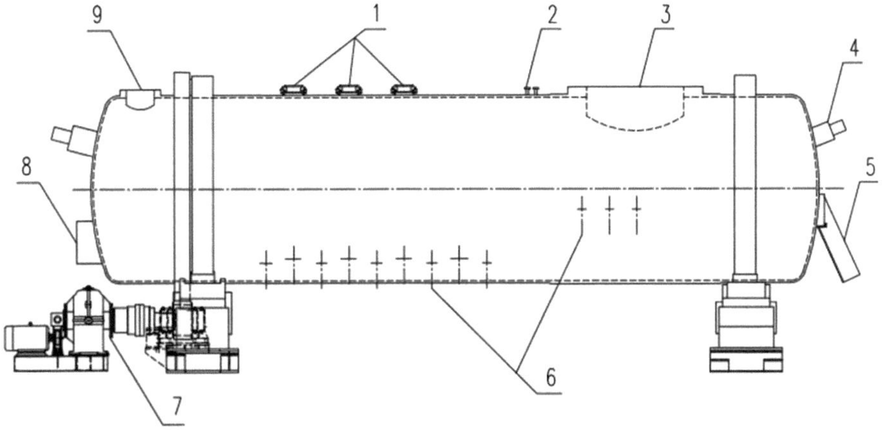

Fig. 4 Schematic diagram of SLS construction *1* feeding port; *2* thermometer hole and liquid level measuring hole; *3* flue outlet; *4* combustor; *5* slag taphole; *6* oxygen lance; *7* actuator; *8* copper discharge port; *9* second flue

Table 1 Main economic and technical indicator

Name	Unit	Designed value/desired value	Actual value
Total quantity of feedstock	t/h	260	180
Conc. throughput	t/h	220	158–170
Coal blending ratio	%	0	0
Ratio of iron to silicon in slag	%	1.8–2.0	1.8–2.0
Copper content in slag	%	≤ 3	1.9–3.1
Furnace temperature	°C	1180	1180
Dust yield	%	<2	1.7–2.0
Oxygen concentration	%	70–75	73
Matte grade	%	>73	73–75
S capture	%	>96.64	>99.5

t/h, which is somewhat limited by the restricted oxygen capacity of the current oxygen plant with mere 25,668 m^3/h and the air 13,706 m^3/h. The main indicators for current SLS production and operation are shown in Table 1.

Converting and Refining Section

The copper matte with a grade of about 73% produced from SLS is continuously sent into SLCR via the matte conduit. The process needs two SLCRs. The SLCRs are $\Phi 4.8 \times 23$ m horizontal vessel. Two SLCRs and one SLS form a simple triangle. The two SLCRs work alternatively. Blasting is injected from bottom and side into the SLCR through 17 lances arranged in two rows. Such oxygen lances have a special structural design, and they can provide and switch among four kinds of gases as per process requirements, in which the outer layer is for nitrogen and natural gas while the inter one for oxygen and air. The flows and ratios of such gases can be accurately controlled through a computer.

The liquid feeding port of SLCR is located in the center of the end wall. In addition, there is a cold feeding port at the top. A copper taphole is on the other end wall. The flue is located at the upper part close to end with the liquid feeding hole. Slag taphole is located at one side of the furnace barrel being close to the copper discharging end. According to the production cycle, slag tapping operation can be realized by rotating furnace.

The matte in SLS is discharged continuously. Therefore, two SLCRs work alternatively to realize the continuity of the overall smelting process. The working cycle of two SLCRs is shown in Fig. 5.

Take the #1 furnace as an example. Charge of copper matte is started upon the end of the last casting. At this moment, the furnace is rotated out. When the melt level inside the furnace meets the blasting requirements, rotate the furnace to production position and starts blasting simultaneously. Feeding is stopped when the

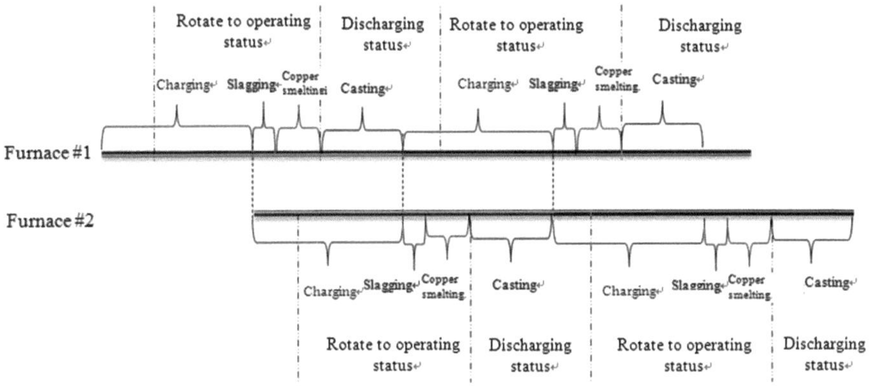

Fig. 5 Schematic diagram of the cycle for SLCR

Table 2 Anode copper components

	Components % of anode copper								
Cu	Fe	S	O	As	Zn	Sb	Pb	Ni	Bi
99.32	0.002	0.006	0.18	0.006	0.011	0.009	0.08	0.041	<0.001
99.24	0.004	0.017	0.13	0.002	0.013	0.008	0.11	0.084	0.001
99.21	0.003	0.021	0.15	0.005	0.008	0.007	0.06	0.058	<0.001
99.07	0.003	0.014	0.14	0.05	0.002	0.012	0.09	0.047	<0.001
99.16	0.002	0.009	0.17	0.05	0.003	0.015	0.05	0.054	<0.001

melt level inside the furnace is enough for slag skimming (at this moment, launder of copper matte from SLS is changed to enter the # 2 furnace). Converting of # 1 furnace continues and it will be rotated for slag skimming when slagging finishes. When approaching the end of copper smelting stage, samples are taken for laboratory tests to determine the conditions of the copper liquid. Moreover, the feed volumes of four kinds of gases, namely oxygen, air, nitrogen and natural gas are appropriately adjusted to make the copper liquid reach the component requirements of anode copper. Then the furnace is rotated out, drilled the taphole and casting is started. Natural gas can be used for thermal insulation during casting process. Next cycle is started upon the completion of the casting.

As per production practices, the grade of the copper matte produced from the SLS is above 73% or even higher, which makes few slags inside the SLCR. Slag can be discharged after accumulated for few cycles to be certain amount. So far, the pyro-refining copper grade can generally reach about 99.21% and can be directly sent into a constant casting machine via a copper taphole for casting. The anode copper components meet the electrolytic refinery requirements. The analysis of anode copper components is shown in Table 2.

Table 3 Main economic and technical indicators

Name	Unit	Value
Charging amount (excluding cold dope)	t/h	53–57
Oxygen concentration of gas containing oxygen	%	21–35
Copper content in slag	%	8–10
S capture	%	>99.5

Currently, the main indicators of the production and operation of SLCRs are shown in Table 3.

Advantages, Disadvantages and Prospects

Advantages

- Short process: the production process from copper concentrate to anode copper is shortened from previous three steps to current two steps. Anode plates are directly casted; equipment configuration is compact, which makes the process simple and efficient.
- Friendly to environment: the oxygen concentration of the SLCRs is high and the sealing is favorable; SLS and SLCRs are connected via a sealing launder so as to avoid the low altitude dissipation of SO_2 offgas.
- Energy saving and efficient: the production of anode plates is shortened from previous three steps to current two steps eliminating pyro-refining section, which lowers energy consumption.
- Less converting slag. Less slag is produced from SLCR due to high grade matte produced from SLS, which relieves the pressure of treating converting slag.
- Less investment: the investment on equipment and supporting devices such as refining furnaces, overhead cranes and ladles is reduced, which saves 10–15% of investment compared to conventional processes.
- Low operating costs, low comprehensive energy consumption and few personnel allocation, which saves 20% compared to the advanced values of the industry.

Improvement to Be Done

- The processing capacity of the SLS shall be further improved with improvement and optimization of the matte-inducing launder, slag-inducing launder, the second flue duct and both structures and arrangements of oxygen lances.

- The harmonious matching between the capacity of the SLCRs and that of the SLS shall be realized. In addition, the copper discharge ports of the two SLCRs and the arrangement of oxygen lances shall be optimized.
- The slag treatment process shall be improved and slag cleaning furnace shall be installed to replace slag flotation process.

As a new copper smelting process, Fangyuan Process has achieved favorable effects even at the beginning of the production. We do believe that the all indicators of the process will be definitely improved with further explorations and continuous improvement, coordination of all working procedures and the accumulation of workers' experiences. By then, Fangyuan Process will definitely change the pattern of worldwide copper smelting industry.

References

1. Y. Zhao, Mitsubishi process applied at the Naoshima smelter. *Non-ferrous Mining and Metallurgy* (1996)
2. S. Yao, Optimization and improvement of double flash copper smelting process in China. *Nonferrous Metal (Smelting Section)* (2008)
3. Z. Cui, D. Shen, Z. Wang, Theory and practice of copper smelting with oxygen enriched bottom blowing. China Nonferrous Metallurgy (Book A, Production Practice, Heavy Metal), June 2010

Part X
Poster II

Comparison of the Ringing Characteristics Between Acid and Alkaline Iron Ore Pellets Powder in Kiln

Xin Min, Yongbin Yang, Qian Li, Bin Xu, Tao Jiang, Xiaoliang Liu and Yan Zhang

Abstract The ringing characteristics of acid and alkaline iron ore pellets powder in kiln were studied by the thermodynamic calculation using software FactSage7.0, roasting experiments using briquetting method and analysis of SEM-EDS. The thermodynamic calculation demonstrated that liquid phase could be generated from alkaline pellets powder at a lower temperature and the amount of formed liquid phase was much more than that from acid pellets powder at the same temperature. Roasting experiments indicated that the compression strength of alkaline pellets powder briquettes were also higher than that of acid pellets powder briquettes. Additionally, the analysis of SEM-EDS suggested that the alkaline pellets powder briquette presented more liquid phase and a more compact structure comparing with the acid pellets powder briquette, which was consistent with the results of thermodynamic calculation and roasting experiments. Hence, in comparison with acid pellets powder, alkaline pellets powder is much easier to ring in the kiln.

Keywords Ringing · Pellets kiln · Alkaline pellets

Introduction

As a kind of excellent raw materials of blast furnace, pellets have attracted a large amount of attention owing to their characteristics of high iron grade, wonderful compression strength and fine metallurgical properties. At present, there are mainly three kinds of processes manufacturing pellets around the world, including shaft furnace process, travelling grate process and grate-kiln process. Grate-kiln process, especially, due to its superior quality of products, low cost, huge production capacities as well as flexible raw materials and fuels, has been developing

---

X. Min · Y. Yang (✉) · Q. Li · B. Xu · T. Jiang · X. Liu · Y. Zhang
School of Minerals Processing and Bioengineering,
Central South University, Changsha 410083, People's Republic of China
e-mail: ybyangcsu@126.com

J.-Y. Hwang et al. (eds.), *8th International Symposium on High-Temperature Metallurgical Processing*, The Minerals, Metals & Materials Series,
DOI 10.1007/978-3-319-51340-9_63

rapidly in China [1–4]. However, ringing in kiln is the biggest problem puzzling relative producers, which wastes raw materials, squanders fuels and even stops production [5].

In the recent decades, the researches of ringing had been almost focused on acid pellets kiln. Fu [6] found that too much powder in kiln, inappropriate temperature control and superabundant low melting point substances were the main reasons why the kiln easily got ringing. In addition, abundant powder in kiln was the common reason. A further research of Fan [7] considered that too much powder was the main cause of ringing in the head and tail part of kiln, but the primary reason of ringing in the middle part of kiln is the highest temperature of roasting and the quantity of low melting point substances and superfine powder. Si [8, 9] researched the effects of coal on ringing characteristics in kiln, which suggested that abnormal temperature distribution resulted by coal's incomplete combustion, gathering of coal ash and unburned coal were the dominating causes. Furthermore, there were some researches [10, 11] indicating that Na, K and F could aggravate the ringing in kiln.

As for the alkaline pellets kiln, Zhanjiang pelletizing plant in China created the first production line of alkaline pellets in 2009 and now 5 million tons pellets can be produced every year. Comparing with acid pellets, alkaline pellets have better qualities of green balls, roasted pellets and reduced pellets. Few attempts, however, have been concentrated on the ringing of alkaline iron ore pellets powder in kiln, and the ringing characteristics in alkaline pellets kiln are still unclear.

Based on the above, it is important to study the ringing characteristics of alkaline pellets powder in kiln. In this paper, the ringing characteristics of alkaline pellets powder in kiln were mainly studied by the thermodynamic calculation using software FactSage7.0, roasting experiments using briquetting method and analysis of SEM-EDS. More importantly, the ringing characteristics between acid and alkaline pellets powder were compared.

Experimental

Raw Materials

Iron ore concentrates (named $1^{\#}$, $2^{\#}$ and $3^{\#}$), bentonite and limestone samples were obtained from Zhanjiang pelletizing plant, Zhanjiang, China. The chemical compositions of raw materials were detected by chemical titration and listed in Tables 1

Table 1 Chemical composition of iron ore concentrates (mass fraction, %)

Iron ore concentrates	TFe	FeO	Fe_2O_3	SiO_2	Al_2O_3	CaO	MgO	S	LOI
$1^{\#}$	63.39	–	90.49	2.84	2.54	0.14	0.11	0.013	3.13
$2^{\#}$	67.15	–	96.00	2.01	0.43	0.30	0.74	0.014	0.10
$3^{\#}$	68.65	29.21	65.69	1.88	0.62	0.10	0.018	0.024	0.67

Table 2 Chemical composition of bentonite and limestone (mass fraction, %)

Materials	TFe	SiO_2	Al_2O_3	CaO	MgO	S
Bentonite	12.72	40.43	15.26	1.84	2.10	0.040
Limestone	0.10	0.66	0.05	51.72	3.19	0.66

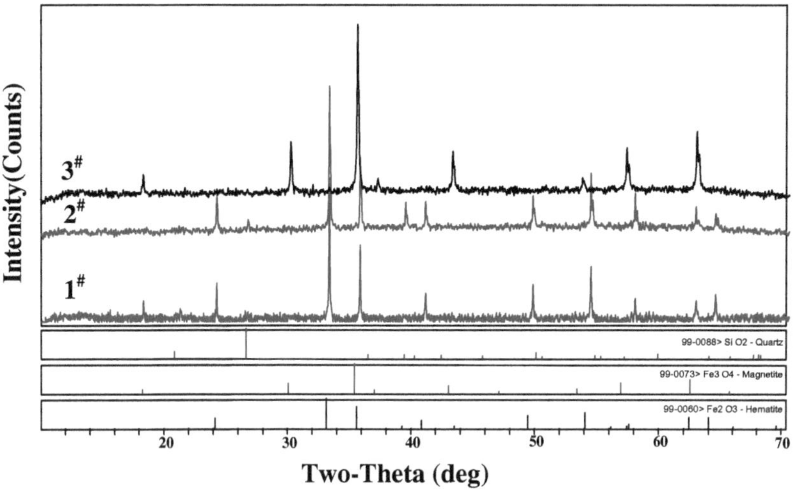

Fig. 1 XRD of iron ore concentrates

and 2, respectively. X-ray Diffraction (XRD) was also used to determine the mineralogy of iron ore concentrates, and the results were shown in Fig. 1.

As shown in Table 1, Iron ore concentrate $1^{\#}$ and $2^{\#}$ have similar chemical compositions, and the contents of Fe_2O_3 amounted to as high as 90.49 and 96.00%, respectively. Meanwhile, neither of them had FeO. Comparatively, the iron ore concentrate $3^{\#}$ not only had 65.69% Fe_2O_3, but also had 29.21% FeO. In association with the crystal phase of iron ore concentrates characterized by XRD, we could know that iron ore concentrate $1^{\#}$ and $2^{\#}$ are mostly hematite whilst $3^{\#}$ is basically magnetite. It is obviously found from Table 2 that the main compositions of bentonite and limestone both fulfil the experiments' quality requirements.

Experiment Methods

Briquetting-roasting Experiment Briquetting-roasting experiment was firstly conducted to study ringing characteristics between acid and alkaline iron ore pellets powder. According to Zhanjiang pelletizing plant, acid ore pellets powder was consisted of hematite, magnetite and bentonite (the proportion was 7:3:0.07), and

Table 3 Chemical components of acid and alkaline ore pellets powder (mass fraction, %)

Powder type	FeO	Fe_2O_3	SiO_2	Al_2O_3	CaO	MgO
Acid	8.80	82.60	2.82	2.06	0.14	0.10
Alkaline	8.50	84.45	2.20	0.58	1.43	0.60

alkaline ore pellets powder was composed of hematite, magnetite, bentonite and limestone (the proportion was 7:3:0.07:0.23).

At the beginning, iron ore concentrates pretreated by High Pressure Grinding Rolls (HPGR) and additives (bentonite or limestone) were sufficiently mixed in a particular proportion. Muffle furnace then was used to heat mixed powder to simulate the preheating process of pellets production in grate, and the whole simulative preheating process lasted 9 min at the temperature of 1060 °C. After the preheated powder was cooled down naturally, 8% (mass fraction) water was added to the mixed it, then 2.6 g sample was put in a Φ 10 × 10 mm cylindrical briquette mold and subsequently compressed into a Φ 10 × 10 mm briquette at the pressure of 250 N/cm^2. It is worth mentioning that in order to simulate the ringing behavior of powder in kiln, we preheated the powder rather than briquette.

Afterwards, all briquettes were dried on an oven for 2 h at 100 °C. Roasting experiments were carried out in a horizontal furnace where briquettes were roasted. The ringing degree was characterized by the roasted briquette strength which was measured by the compression strength testing machine (ZQYC-10000NIII) invented by Central South University, and the higher the strength was, the more serious the ringing was.

Thermodynamic Calculation To distinguish the different ringing characteristics between acid and alkaline pellets kiln, thermodynamic calculation was carried out by software Factsage7.0 which is one of the largest fully integrated database computing systems in the field of chemical thermodynamics in the world. Equilib module of this software was used to quantitatively calculate the content of liquid phase under different conditions. The $CaO–SiO_2–Fe_2O_3–Fe_3O_4–MgO–Al_2O_3$ system was selected and the oxygen partial pressure was set as 2.1 × 104 Pa. The actual chemical components of acid and alkaline ore pellets powder which were used to conduct the thermodynamic calculation were listed in Table 3.

As shown in Table 3, acid and alkaline pellets powder have similar FeO, Fe_2O_3 and SiO_2 contents, while alkaline pellets powder has more alkaline substances than acid pellets powder, which means alkaline pellets powder is easier to form low melting point substances (like $CaO_x·FeO_{2-x}·SiO_2$ and $CaO·Fe_2O_3$, etc.) by solid state reactions. And low melting point substances may form liquid phase at a higher temperature.

Analytical Methods

A part of roasted briquettes were polished in the semi-automatic polishing machine and then coated carbon on their surface to acquire electrical conductivity. After this pretreatment, SEM/EDS (Quanta 250 FEG scanning electron microscope and Genesis XM2 scanning electron microscopy spectrometer) was used to determine their internal morphology and composition.

Results and Discussion

Briquetting-Roasting Experiment

To ensure the reliability of this experiment, the compression strength of 10 briquettes were tested and the average value was calculated as the final result. The compression strength of acid and alkaline roasted briquettes under the conditions of roasting temperature 1220–1300 °C and roasting time 3–6 min was listed in Table 4.

As indicated in Table 4, the compression strength of alkaline roasted briquettes was much higher (about 3.5–8.7 times) than that of alkaline roasted briquettes under the same roasting temperature and time. With the increase of temperature between 1220–1300 °C and time between 3 and 9 min, respectively, compression strength of acid and alkaline roasted briquettes both increased continuously. Furthermore, when the roasting time was 6 min, the strength of both acid and alkaline roasted briquettes rapidly increased from 1220 to 1260 °C, but slowly rising from 1260 to 1300 °C. However, at the roasting temperature of 1260 °C and with the increase of time between 3 and 9 min, there was a significant difference between acid and alkaline roasted briquettes—the strength of acid roasted briquettes dramatically increased from 491 to 1548 N/Briquette (over 3 times) but alkaline roasted briquettes slightly increased from 4257 to 5473 N/Briquette (just around 1.3 times).

There are three important details at least we can discover from the above analyses

(1) Alkaline pellets kiln is much easier to ring than acid pellets kiln.
(2) Higher roasting temperature and longer roasting time can increase the degree of ringing in both acid and alkaline pellets kilns.

Table 4 Compression strength of acid and alkaline roasted briquettes

Roasting temperature/°C	Roasting time/min	Compression strength (N/Briquette)	
		Acid	Alkaline
1220	6	660	2943
1260	3	491	4257
1260	6	857	4869
1260	9	1548	5473
1300	6	943	5124

(3) Time is a more important factor than temperature to form ring in acid pellets kiln while the result is just the opposite in alkaline pellets kiln.

Thermodynamic Calculation

The simulative contents of liquid phase of acid and alkaline pellets powder were shown in Fig. 2. With the increase of temperature between 1220 and 1320 °C, the simulative contents of liquid phase of acid and alkaline pellets powder both increase continuously. Furthermore, the contents of acid and alkaline powder both increase rapidly between 1220 and 1260 °C (increasing from 0 to 1.1% and from 4.0 to 5.6%, respectively), but slowly between 1260 and 1320 °C. It is easily to find that the effect of temperature on simulative content of liquid phase is very similar with it on compression strength of roasted briquettes. So liquid phase possibly play a significant role on the ringing in kiln, especially in alkaline kiln.

SEM and EDS Analysis

SEM images (Fig. 3) suggest that the acid pellets powder mainly consists of Fe_2O_3 as well as some independent SiO_2. The integral connection is relatively loose while the connection of Fe_2O_3 is rather compact, and the diameter of crystalline grain is about 5–10 μm. The solid phase consolidation is dominated in acid pellets powder.

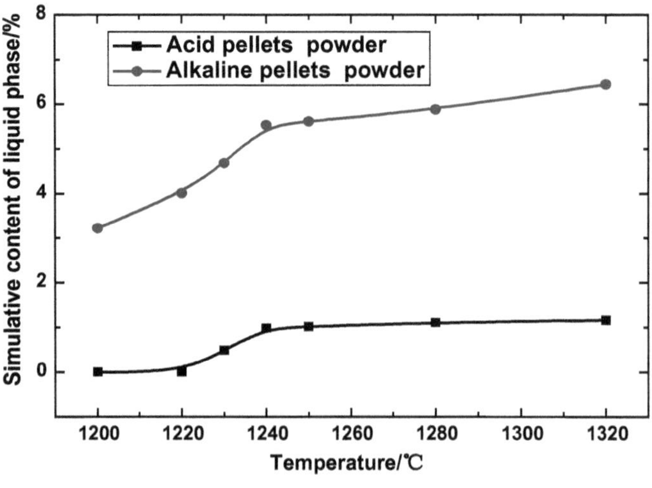

Fig. 2 Simulative content of liquid phase of acid and alkaline pellets powder (mass fraction, %)

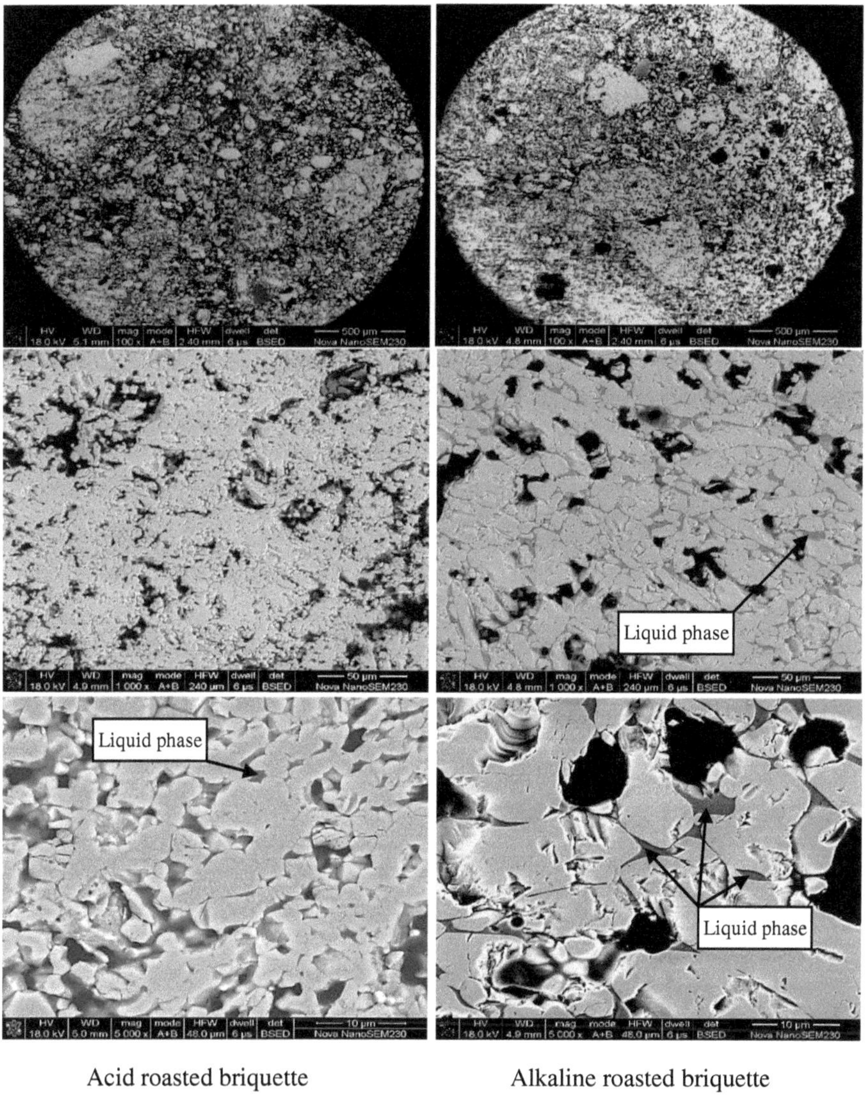

Fig. 3 SEM images of acid and alkaline roasted briquette

Unlike acid pellets powder, the integral connection of alkaline pellets powder is much closer; furthermore, besides solid phase consolidation, liquid phase consolidation acts an important role in the crystallizing process.

Furthermore, EDS analysis (Fig. 4) illustrates that there is a big difference in the composition of liquid phase between acid and alkaline roasted briquette. The elements of liquid phase of acid roasted briquette are mainly Si, O and Fe, while alkaline roasted briquette are largely Si, O and Ca. The results indicate that the

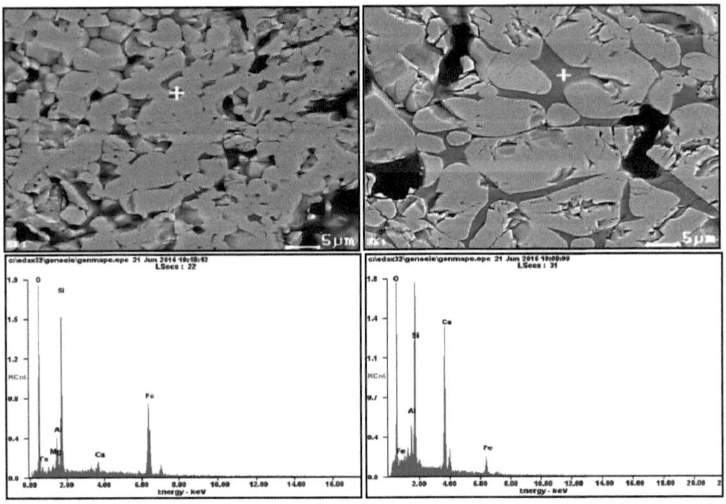

Acid roasted briquette | Alkaline roasted briquette

Fig. 4 EDS analysis of acid and alkaline roasted briquette

liquid phase of acid roasted briquette is mostly fayalite and alkaline roasted briquette is calcium silicate. Calcium ferrite is easy to form when alkaline substance exists at low temperatures. If the temperature is over 1250 °C, calcium ferrite will transfer to calcium silicate, as shown in Eq. (1).

$$CaO \cdot Fe_2O_3 + SiO_2 \rightarrow CaSiO_3 + Fe_2O_3 \quad (1)$$

This is a recrystallization reaction of Fe_2O_3, which can significantly increase the strength of alkaline roasted briquettes.

Conclusions

(1) Alkaline pellets kiln is much easier to form ring than acid pellets kiln. Higher roasting temperatures and longer roasting time can increase the degree of ringing in both acid and alkaline pellets kilns. Time is a more important factor than temperature to form ring in acid pellets kiln while the result is just the opposite in alkaline pellets kiln.

(2) Liquid phase play a significant role on the ringing in kiln, especially in alkaline kiln, because liquid phase can not only increase the cohesiveness of the powder but also promote the recrystallization reaction of Fe_2O_3.

References

1. K.W. Ye, The present state and the forecasting of our country pelletizing industry. Sinter. Pelletizing **28**(1), 1–4 (2003)
2. M.X. Xu, in *The Present State and The Forecasting of Pelletizing Industry in China [C]*. The national conference corpus of iron making production technology in 2012 (2012, Part 1), pp. 385–389
3. K.W. Ye, Review of three kinds of pelletizing process [J]. Sinter. Pelletizing **27**(1), 4–7 (2002)
4. Y.J. Zhao, Review of roasting process of pelletizing [J]. Technol. Wind **13**(1), 113 (2012)
5. A. Chatterjee, P.K. Mukhopadhyay, Flow of materials in rotary kilns used for sponge iron manufacture: part III. Effect of ring formation within the kiln [J]. Metall. Mater. Trans. B **14** (3), 393–399 (1983)
6. J.Y. Fu, The study of an steel's ring forming [J]. Sinter Palletizing **3**(3), 2–6 (2004)
7. X.H. Fan, Study of mechanism on ring formation in grate-kiln of acid pellet. Iron and Steel **43** (3), 15–20 (2008)
8. J.F. Si, Structure of rings and ring-forming mechanism of rotary pelletizing kiln [J]. Iron Steel **49**(1), 17–21 (2014)
9. J.F. Si, Characteristics of microstructure for ring-forming material of oxidized pellet rotary kiln [J]. Res Iron Steel **42**(1), 5–8 (2014)
10. G.P. Luo, Influence of F, K, Na on the ring formation properties of oxidized pellet rotary kiln [J]. Sinter. Pelletizing **3**(6), 29–32 (2013)
11. J.T. Liu, Study on ring-forming cause of baosteel oxidized pellet rotary kiln [J]. Sinter. Pelletizing **38**(5), 31–35 (2013)

Comprehensive Utilization of Ludwigite Iron Concentrate by Gas-based Direct Reduction

Luxing Feng, Jun Deng, Xun Xue, Zhicheng Cao, Jingkun Tang and Daohong Wu

Abstract The effect of reducing temperature, H_2/CO mole ratio, and reduction gas flow rate on the metallization rate of oxide pellets were investigated at laboratory scale, with ludwigite iron concentrate used as raw materials. The reducing temperature varied from 850 to 950 °C, the H_2/CO mole ratio changed from 1.5 to 4.5, the range of reduction gas flow rate was from 2 to 5 L/min. The results indicate that the metallization degree of pellets increase with increase of the temperature, H_2/CO mole ratio, and gas flow rate. However, after temperature rises to 850 °C, metallization degree increases slowly. The effect of H_2/CO on the metallization degree is not very obviously with the ratio of H_2/CO increasing from 1.5 to 4.5. When the gas flow rate was increased from 2 to 3 L/min, the metallization degree increased greatly from 72.2 to 92.1%. In the flow range from 3 to 5 L/min, the metallization degree was increased slowly. The results obtained could be useful for comprehensive utilization of ludwigite iron concentrate by gas-based direct reduction method.

Keywords Ludwigite iron concentrate · Gas-based direct reduction · Metallization degree

Introduction

Boron and its compounds are commercially important and have more than 300 commercial end uses. Boron ores mainly consist of ascharite and ludwigite in China. Recently, the ascharite is nearly exhausted; the Ludwigite is quite abundant in Liaoning province.The total reserve of ludwigite ore is about 280 Mt, of which about 21.84 Mt is B_2O_3, accounting for 58% boron reserves in China [1]. It is the main alternate boron resource of the future. However, ludwigite cannot be directly

L. Feng (✉) · J. Deng · X. Xue · Z. Cao · J. Tang · D. Wu
Beijing Shenwu Environment & Energy Technology Corp.,
No. 18, Shenniu Road, Machikou Town, Changping District,
102200 Beijing, China
e-mail: chpdiz@163.com

J.-Y. Hwang et al. (eds.), *8th International Symposium on High-Temperature Metallurgical Processing*, The Minerals, Metals & Materials Series,
DOI 10.1007/978-3-319-51340-9_64

utilized as a single resource by traditional processing methods because of its low grades of total iron (TFe) and B_2O_3. The average composition of the ore is approximately 6–8 wt% B_2O_3, 27–30 wt% TFe [2, 3]. In China, the comprehensive utilization of ludwigite is becoming increasingly urgent because of the increased demand for boron compounds and the depletion of the high-grade ascharite resource.

The key problems for the comprehensive utilization of ludwigite are the separation degree of boron and iron and their qualities for further production. Many methods and technologies have been developed to utilize ludwigite, which include magnetic separation [4], selective reduction and magnetic separation [5, 6], hydrometallurgy [7], blast-furnace reduction-melting separation [8], and iron-nugget process [9–12]. Due to the complicated mineral phase composition, it is difficult to get single boron concentrate and iron concentrate through magnetic separation or magnetic separation combined with selective reduction. Although the hydrometallurgical method can separate boron and iron perfectly, the excessive acid consumption causes very high cost, equipment erosion and environmental pollution. The traditional pyrometallurgical method can also achieve a high degree of separation of boron and iron. However, the separation slags face the same problem that a low grade of boron content and low efficiency of extraction of boron (EEB).

In order to efficiently utilize ludwigite ore and overcome the existing problems (i.e. coke or coal consumption, low grade of boron content and low EEB of the slag and soon), a new pyrometallurgical process has been proposed by the present authors for ludwigite iron concentrate oxide pellet direct reduction via gas-based shaft furnace, which can avoid the pollution of reduction product from reducing agent and additive, reduce the amount of slag generated during smelting separation process, greatly increase the grade and activity of boron. The experiments of gas-based reduction of ludwigite iron concentrate oxide pellets were performed at laboratory scale to investigate the isothermal reduction behavior.

Experimental

Raw Materials

The ludwigite iron concentrate used in this study was obtained from Liaoning Province, China. The chemical composition of the complex ore sample were tested by ICP-AES (OPTIMA 7000 DV, Perkinelmer, America) and XRF (XRF-1800X, Shimadzu, Japan), which is shown in Table 1. Mineralogical analysis of the concentrate was performed using X-ray diffraction (XRD; MAX-RB, 12 kW). The

Table 1 Chemical composition of ludwigite iron concentrate wt%

B_2O_3	TFe	FeO	MgO	SiO_2	Al_2O_3	CaO
4.74	54.66	22.98	12.19	3.62	0.40	0.59

results indicate that the main crystalline phases are magnetite (Fe_3O_4), szaibelyite ($Mg_2(OH)(B_2O_4(OH))$), and chrysotile ($Mg_3(Si_2O_5)(OH)_4$) 95% of the iron ore concentrate particles were smaller than 0.074 mm.

Experimental Procedure

The ludwigite iron concentrate and water were fully mixed. The moisture content of the mixture was controlled at 9 wt%. The resulting mixture was pelletized in a disc pelletizer. The diameter of the pellet was 10–14 mm. The wet green pellet was dried at 105 °C for 4 h, and then oxidized roasting in a muffle furnace at 1200 °C for 20 min. The chemical composition of oxide pellets were also tested by ICP-AES (OPTIMA 7000 DV, Perkinelmer, America) and XRF (XRF-1800X, Shimadzu, Japan), which is shown in Table 2. The reduction experiments were performed in a shaft furnace to simulate the gas-based direct reduction process of oxide pellets. A schematic of the experimental apparatus is shown in Fig. 1.

In each run, the oxide pellet was placed into a reduction tube and then heated in the furnace. A perforated stainless steel plate and certain amount of porcelain balls put into the reduction tube before the pellets were added into it, which could make sure the pellets touched with reducing gas equally. The pellets were heated to the

Table 2 Chemical composition of oxide pellets wt%

B_2O_3	TFe	FeO	MgO	SiO_2	Al_2O_3	CaO
4.52	53.86	0.62	12.12	6.38	0.53	<0.5

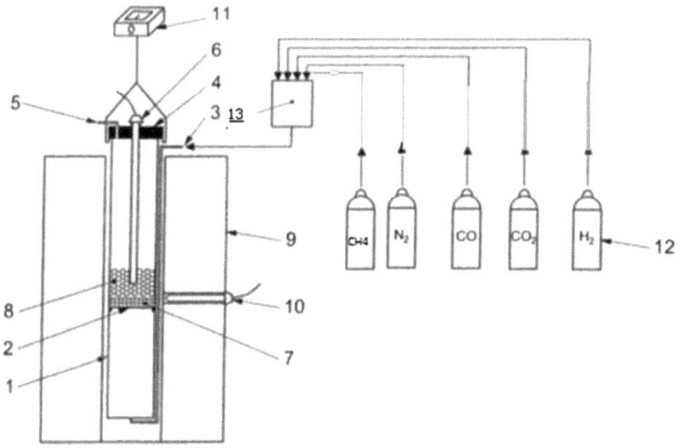

Fig. 1 Schematic of the experimental apparatus: *1* reduction tube; *2* perforated stainless steel plate; *3* inlet; *4* reduction tube cover; *5* outlet; *6* thermocouple; *7* porcelain balls; *8* pellets; *9* electronic furnace; *10* thermocouple; *11* electronic balance; *12* gas cylinder; *13* gas distribution system

target temperature before reducing gas was fed into the reduction tube. The reduced sample was removed from the furnace with reduction tube at set time intervals and cooled to ambient temperature under the protection of nitrogen.

The pellets, which became into metalized pellets and didn't melt, were tested by the index of metallization degree (η). The η was calculated by the following formula:

$$\eta = \frac{\text{MFe}}{\text{TFe}} \times 100\% \tag{1}$$

where MFe is the content of metallic iron after reduction, TFe is the content of total iron after reduction. Both of them were analyzed by chemical method.

Analysis and Characterization

The chemical analyses of metalized pellets were conducted by Chemical Analysis Center of Beijing Shenwu Environment and Technology Corp using ICP-AES (OPTIMA 7000 DV, Perkinelmer, America) and XRF (XRF-1800X, Shimadzu, Japan). The XRD patterns of the samples were detected with a Rigaku diffractometer (DMAX-RB 12 kW) using Cu Kα radiation; the scanning angles were in the range from 10 to 100° (2θ) at a speed of 10°/min.

Results and Discussion

Effect of Temperature on the Metallization Degree

To optimize the reduction temperature, a series of reduction tests were performed at various reduction temperatures, including 800, 850, 900 and 950 °C, when other parameters remain constant as reduction time of 210 min with certain H_2/CO mole ratio, gas flow rate and effective gas content.

Effect of temperature on the metallization degree of the reduced pellets of ludwigite iron concentrate is shown in Fig. 2. It can be observed that the degree of metallization is greatly affected by temperature. The metallization degrees of the pellets are 81.5, 90.8, 92.7 and 94.2% when reduced at 800, 850, 900 and 950 °C for 210 min, respectively. It shows that the metallization degree increases with the increasing of temperature. Synthetically considering the heating capacity of industrial reduction equipment and the metallization degree of reduced production, the reduction temperature should be above 850 °C for the subsequent experiments.

The main reactions in the pellets are just direct reductions, which could be shown as:

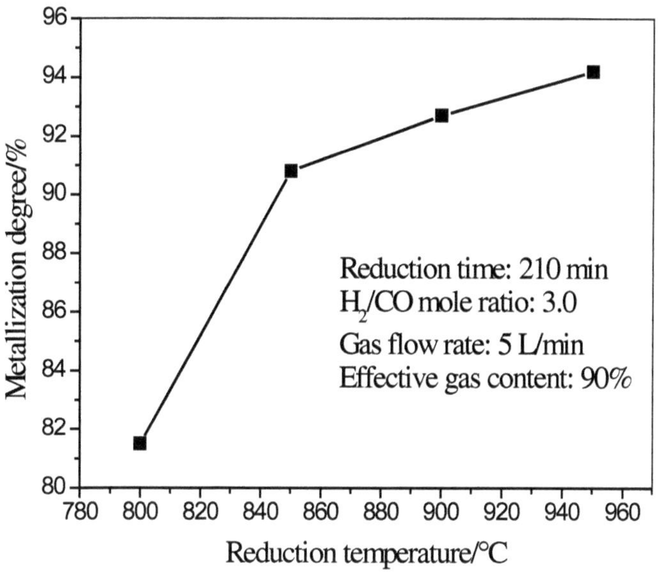

Fig. 2 Effect of reduction temperature on metallization rate of reduced pellets

$$3Fe_2O_3 + CO = 2Fe_3O_4 + CO_2 \tag{2}$$

$$Fe_3O_4 + CO = 3FeO + CO_2 \tag{3}$$

$$FeO + CO = Fe + CO_2 \tag{4}$$

and

$$3Fe_2O_3 + H_2 = 2Fe_3O_4 + H_2O \tag{5}$$

$$Fe_3O_4 + H_2 = 3FeO + H_2O \tag{6}$$

$$FeO + H_2 = Fe + H_2O \tag{7}$$

The main reaction process included the gas diffusion and the interface reduction reaction of iron oxide. It was found that for more than 90% of reduction process, interfacial chemical reaction was the rate controlling step under all experimental conditions for gas content [13]. Occupying almost the whole stage, the reduction is dominated by the interfacial chemical reaction, which increased with increasing of temperature.

Effect of H_2/CO (Mole Ratio) on the Metallization Degree

To optimize the H_2/CO (mole ratio), a series of reduction tests were performed at various H_2/CO, including 1.5, 2.0, 3.0 and 4.5, when other parameters remain constant as reduction time of 210 min with certain reduction temperature, gas flow rate and effective gas content.

Effect of H_2/CO on the metallization degree of the reduced pellets of ludwigite iron concentrate is shown in Fig. 3. Results show that, with increasing the H_2/CO from 1.5 to 4.5, the metallization degree of reduction products increases from 92.7 to 94.8%. The metallization degrees of pellets were 92.7, 93.2, 94.1 and 94.8% with the H_2/CO mole ratio 1.5, 2.5, 3.0, and 4.5, respectively, when the pellets were heated at 900 °C for 210 min. It means that the metallization degree increases with the increasing of H_2/CO to some extent, but the effect of H_2/CO on the metallization degree is not very obviously. In other words, the oxide pellets of ludwigite iron concentrate has a wide range of adaptability to reducing gas composition.

In general, the metallization degree increases with the increasing of H_2/CO. It proved that H_2 can improve the dynamic condition, and the interface reaction rate constant increases with the proportion raise of H_2 in reducing gas [13, 14]. That means that raising the ratio of H_2/CO can improve the reduction process. In additional, H_2 caused the decrease of the diffusion resistance due to the smaller molecule diameter. As a result, the diffusion control step was significantly shortened with the increase H_2/CO mole ratio.

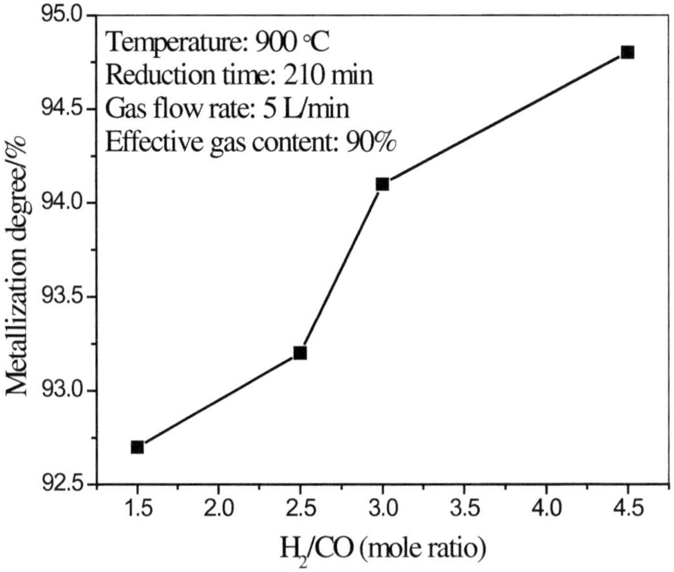

Fig. 3 Effect of H_2/CO (mole ratio) on metallization degree of reduced pellets

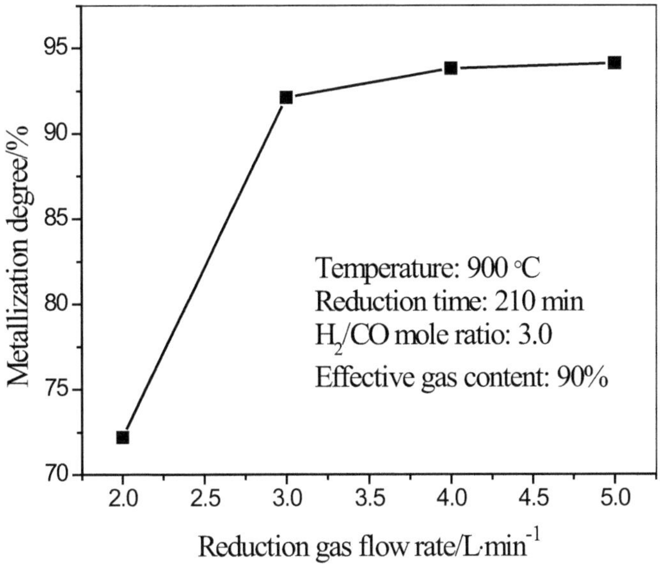

Fig. 4 Effect of reduction gas flow rate on metallization degree of reduced pellets

Effect of Reduction Gas Flow Rate on the Metallization Degree

To optimize the reduction gas flow rate, a series of reduction tests were performed at various reduction temperatures, including 2, 3, 4 and 5 L/min, when other parameters remain constant as reduction time of 210 min with certain reduction temperature, H_2/CO (mole ratio) and effective gas content.

Effect of reduction gas flow rate on the metallization degree of the reduced pellets of ludwigite iron concentrate is shown in Fig. 4. As evident in the figure, the metallization degree gradually increased with increasing reduction gas flow rate. However, the metallization degree increased greatly when the gas flow rate was increased from 2 to 3 L/min and then increased slowly during the flow of reduction gas was increased from 3 to 5 L/min. The metallization degree of pellets were 72.2, 92.1, 93.8 and 94.1% with the gas flow rate 2, 3, 4, and 5 L/min, respectively, when the pellets were heated at 900 °C for 210 min. These experimental results indicate that reduction gas rapidly improves the reduction rate when its flow is small and remains slowly increasing when its flow rate is high.

Conclusion

1. Generally, the increase of temperature leads to an obvious increase of metallization degree. However, after temperature rises to 850 °C, metallization

degree increases slowly. So, the optimum reduction temperature is about 850 °C.

2. The metallization degree increases with the increasing of H_2/CO mole ratio to some extent. However, the affection of H_2/CO on the metallization degree is not very obviously with the ratio of H_2/CO increasing from 1.5 to 4.5. In other words, the oxide pellets of ludwigite iron concentrate has a wide range of adaptability to reducing gas composition.
3. Increasing reduction gas flow rate can gradually increase the metallization degree. When the gas flow rate of reduction gas was increased from 2 to 3 L/min, the metallization degree increased greatly from 72.2 to 92.1%. In the flow range from to 5 L/min, the metallization degree was increased slowly.

References

1. S.L. Liu, C.M. Cui, X.P. Zhang, Pyrometallurgical separation of boron from iron in ludwigite ore. ISIJ Int. **38**(10), 1077–1079 (1998)
2. X.P. Zhang, J.F. Lang, C.M. Cui, S.L. Liu, Comprehensive utilization of low grade ludwigite ore with blast furnace smelting. Iron Steel **30**(12), 9–11 (1995)
3. Z.T. Sui, P.X. Zhang, C. Yamauchi, Precipitation selectivity of boron compounds from slags. Acta Mater. **47**(4), 1337–1344 (1999)
4. C.R. Wang, X.Q. Lian, On Some Problems of the Beneficiation of Boron Iron Ores. Met. Mine **9**, 33–35 (1995)
5. J.L. Zhang, H.T. Cai, Enrichment of boron in low-grade paigeite. J. Univ. Sci. Technol. Beijing **31**(1), 36–40 (2009)
6. X.J. Fu et al., Comprehensive utilization of ludwigite ore based on metallizing reduction and magnetic separation. J. Iron. Steel Res. Int. **22**(8), 672–680 (2015)
7. R. Liu et al., Comprehensive utilization of ludwigite and its prospect. Multipurpose Util. Miner. Res. **2**, 33–37 (2006)
8. Q.J. Zhao, Separating iron from boric iron ore by selective reduction. J. Northeast. Univ. Technol. **11**(2), 122–126 (1990)
9. G. Wang et al., New separation method of boron and iron from ludwigite based on carbon bearing pellet reduction and melting technology. ISIJ Int. **52**(1), 45–51 (2012)
10. G. Wang et al., Effect of carbon species on the reduction and melting behavior of boron-bearing iron concentrate/carbon composite pellets. Int. J. Miner. Metall. Mater. **20**(6), 522–528 (2013)
11. G. Wang et al., Reduction-melting behaviors of boron-bearing iron concentrate/carbon composite pellets with addition of CaO. Int. J. Miner. Metall. Mater. **22**(9), 926–932 (2015)
12. G. Wang, Q. Xue, X. She et al., Carbothermal reduction of boron-bearing iron concentrate and melting separation of the reduced pellet. ISIJ Int. **55**(4), 751–757 (2015)
13. L.Y. Yi et al., Action rules of H_2 and CO in gas-based direct reduction of iron ore pellets. J. Northeast. Univ. Technol. **19**(8), 2291–2296 (2012)
14. Y.M. Wang, Z.F. Yuan, H. Matsuura, Reduction extraction kinetics of titania and iron from an ilmenite by H_2–Ar gas mixtures. ISIJ Int. **49**(2), 164–170 (2009)

Decarburization of Spent Petrochemical Catalysts via Microwave Oxidation Roasting

Libo Zhang, Peng Liu, Haigang Dong, Bingguo Liu and Jinhui Peng

Abstract Parameters of technique to decarburize from spent petrochemical catalysts via microwave oxidation roasting were investigated. Decarburization process was performed on a microwave heating system. The results showed that the decarburization percentage of spent petrochemical catalysts was significantly affected by roasting temperature and sample thickness in the range studied. The decarburization percentage was 99.5% in microwave fields under the optimized conditions, roasting temperature 600 °C, roasting time 40 min and sample thickness 2 cm respectively. The residuum after microwave decarburization was characterized by field emission scanning electron microscopy (FE-SEM) which showed that the deposit carbon had been removed effectively.

Keywords Decarburization · Spent petrochemical catalysts · Oxidation roasting · Microwave

L. Zhang · P. Liu · B. Liu · J. Peng
Faculty of Metallurgical and Energy Engineering, Kunming University of Science and Technology, 650093 Kunming, Yunnan, China

L. Zhang · P. Liu · B. Liu · J. Peng
Yunnan Provincial Key Laboratory of Intensification Metallurgy, 650093 Kunming, Yunnan, China

L. Zhang · P. Liu · B. Liu · J. Peng
State Key Laboratory of Complex Nonferrous Metal Resources Clean Utilization, Kunming University of Science and Technology, Kunming, Yunnan, China

L. Zhang · P. Liu · B. Liu (✉) · J. Peng
National Local Joint Laboratory of Engineering Application of Microwave Energy and Equipment Technology, 650093 Kunming, Yunnan, China
e-mail: bingoliu@126.com

H. Dong
State Key Laboratory of Advanced Technology for Comprehensive Utilization of Platinum Metals, 650093 Kunming, Yunnan, China

J.-Y. Hwang et al. (eds.), *8th International Symposium on High-Temperature Metallurgical Processing*, The Minerals, Metals & Materials Series,
DOI 10.1007/978-3-319-51340-9_65

Introduction

Catalysts with platinum group metals as the active ingredient are extensively used in the petroleum refining and petrochemical industries [1]. The catalysts used in the petrochemical processes loss their effectiveness over time. Petrochemical catalysts are deactivated during use due to fouling (carbon deposit), poisoning or thermal degradation/sintering while platinum group metals are still in the spent catalysts. In China, large amount of spent catalysts containing platinum group metals are accumulated [2]. Therefore, it is necessary to recycle platinum group metals from spent catalysts.

Recycling of platinum group metals from the spent catalysts has been investigated by different authors [3–6]. Pt, Pd, Rh and other associated metals were recycled using smelting, chemical leaching in UK. At the Umicore precious metals refining plant, Belgium, spent catalysts are smelted in a furnace to produce a concentrate containing Cu, Pt and other metals, which is further processed by hydrometallurgical route to recover the metals. However, the applicability of high rates of recovery in conventional processing is limited because of deposit carbon with strong stability and absorption ability. So, It is necessary to remove the deposit carbon by thermal pretreatment and to improve leaching efficiency. It is currently manufactured commercially using conventional electric furnace under air, due to the process simplicity. However, the major drawbacks of conventional thermal pretreatment process are high energy consumption and long duration of the process. The microwave technology has been effectively utilized for material processing due to its faster heating rate, which possibly reduces the processing time and minimizes the energy cost.

Microwaves are a form of electromagnetic energy with associated electric and magnetic fields. Microwave energy is extremely efficient in the selective heating of materials, as energy is not wasted in "bulk heating". Microwave heating has a clear advantage as compared to conventional heating methods, such as: rapid and efficient heating of materials, improvement in product quality and cleaner process with better environmental acceptability. In addition microwave heating has better control over the heating process and has potential to heat the materials directly, at higher heating rates than conventional heating. However, a detailed investigation on the thermal pre-treatment of the spent catalysts has not been reported in the literatures. Therefore a humble effort is needed to be made in that direction.

The aim of this study is to investigate decarburization process of the spent catalysts by microwave heating and to provide a new method to decarburize from the spent catalysts. The decarburization effects between conventional heating and microwave heating were also compared.

Experimental

Materials

The spent catalysts used in the study were obtained from Kunming Institute of Precious Metals. Chemical analysis of the spent catalyst is presented in Table 1.

Equipment for Microwave Heating and Method

Microwave reactor employed in the present study is made by Key Laboratory of Unconventional Metallurgy, Ministry of Education, Kunming University of Science and Technology, which has the ability to alter the power intensity from 0 to 3000 W and to operate at a frequency of 2.40 GHz. The multimode cavity is shown in Fig. 1.

The crucible that hold the sample material is made of silicon carbide material (diameter was 6 cm), while the temperature is measured using a type K thermocouple placed inside. The decarburization percentage of the spent catalyst at different reaction times was calculated.

Table 1 Composition of the spent catalysts

Components	Al_2O_3	SiO_2	MgO	C	H_2O
Content (wt%)	31.641	46.612	12.507	7.24	2

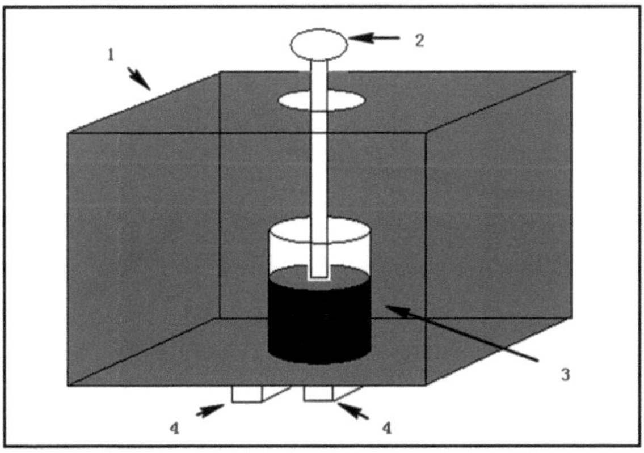

Fig. 1 Multimode cavity of microwave heating (*1* multimode cavity; *2* thermal couple; *3* crucible; *4* magnetrons with waveguides)

Calculation of the Decarburization Rate

The decarburization percentage β was calculated by the following formula:

$$\beta = \frac{(0.98\mathrm{m}_0 - \mathrm{m}_i)}{\mathrm{m}_0 \times \mathrm{C}_\mathrm{T}} \times \%$$

where,

- i is the reaction time of the decarburization under a reaction temperature;
- m_0 is the initial mass of the spent catalyst sample ($0.98\mathrm{m}_0$ is the sample mass without water);
- m_i is the sample mass at reaction time i;
- C_T is the total carbon content of the spent catalyst, $C_\mathrm{T} = 0.0724$ (7.24%);
- β is the decarburization percentage at reaction time i

Results and Discussion

Analysis of TGA and DTG Curves

The thermogravimetric investigations were carried out on a thermal analyze (NETZSCH STA 409 PC/PG) under air atmosphere. The sample weight was 30 mg, heating rate was 10 K/min, and air flow rate was 30 ml/min. The TGA results of the spent catalyst are shown in Fig. 2.

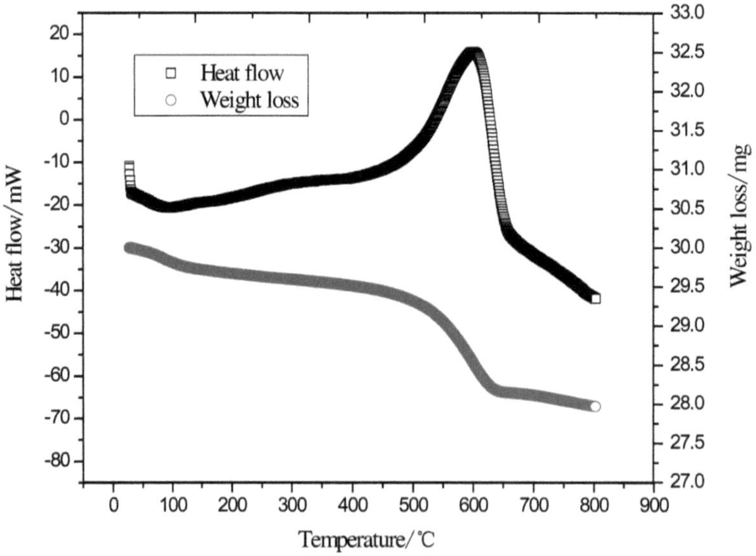

Fig. 2 Thermogravimetric analysis of spent catalysts

The results show that there are two weight loss stages in the TGA curves. The stage I begins at 23 °C and stops at 500 °C, accompanied by 1.7% weight loss. It is attributed to the dehydration of the spent catalyst. Stage II begins at 500 °C and stops at 650 °C, The observed weight loss by 5.1% is attributed to the combustion of the deposit carbon in the spent catalyst.

Conventional Heating Decarburization Process

The combustion reaction of the deposit carbon in the spent catalyst mainly occurred in the temperature range between 500 and 650 °C as shown in Fig. 2. The decarburization percentage under different reaction times and sample thicknesses at 600 °C in the muffle furnace are presented in Fig. 3.

We can see that the decarburization percentage increased with the increasing heating time in Fig. 3. While the thicker the spent catalyst, the lower the decarburization percentage. It is attributed to the low combustion reaction rate of the deposit carbon in the spent catalyst and the slow the diffusion rate of the air and the reaction product CO_2. Heat flowed from the outside to the inside of the sample in conventional heating. The higher sample thickness resulted in longer conduction time. For conventional heating, the decarburization percentage was only 99.5% under 600 °C, 90 min and 1.5 cm sample thickness.

The decarburization percentage was presented in Fig. 4 at different reaction times and temperatures when the sample thickness was 2 cm. It is obvious that decarburization percentage increased with the increasing temperature and time.

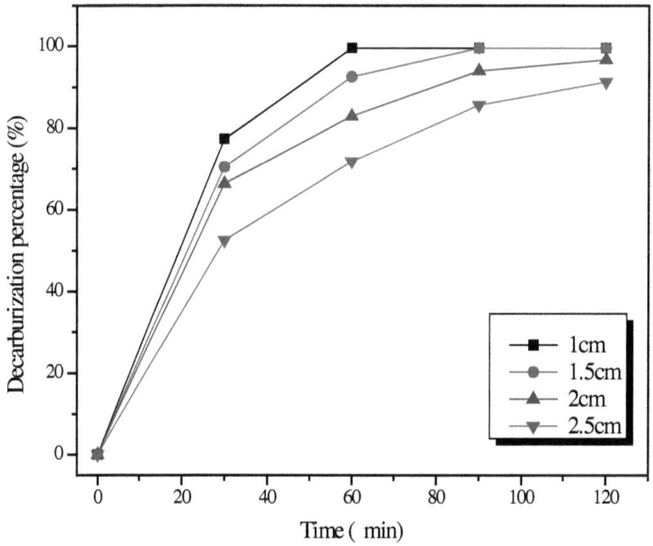

Fig. 3 Decarburization percentage of spent catalysts under different thicknesses by conventional heating

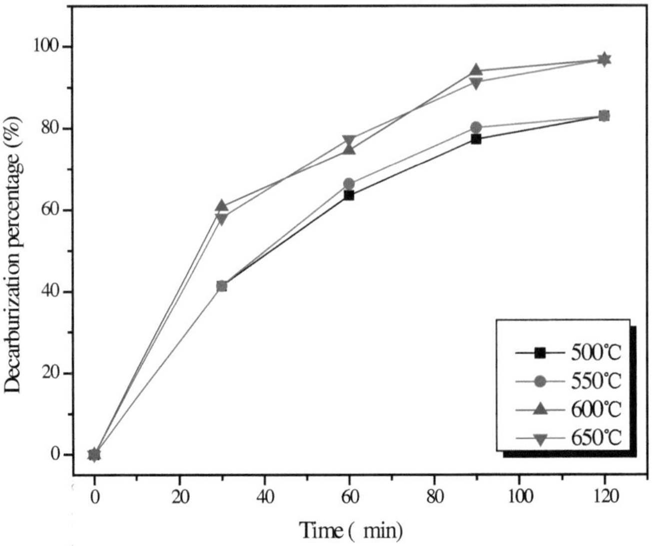

Fig. 4 Decarburization percentage of spent catalysts under different reaction temperatures by conventional heating

Microwave Heating Decarburization Process

The decarburization percentages are presented in Figs. 5 and 6 at different reaction times and temperatures in microwave fields.

The decarburization percentage increased with the increasing temperature and heating time and decreased with the increasing sample thickness within the experimental range. Compared with conventional heating, within 40 min, at 600 °C a complete decarburization (99.5%) could be attained by microwave heating when the sample thicknesses were 1, 1.5 and 2 cm. It is known that microwave absorption ability can be described by the dielectric properties [7–9]. Since the microwave absorption abilities of the components in a sample are different (Table 2), dielectric constant (ε_r') and loss (ε_r'') of carbon are bigger than any of other components, microwave had a preferred heating interaction on deposit carbon. Therefore decarburization reaction time could be shortened from 90 to 40 min by microwave heating.

For conventional heating, the heat is generated outside of the sample, the surface of the sample is heated first, and then the heat flows to the inside by conduction and the temperature distribution of the sample decreased from the surface to the interior, therefore it took longer heat conduction time to attain a complete decarburization. On the other hand, microwaves reach the inside of the sample and interact with the materials making microwave energy transformed into heat directly, and the heat flows back to the surface. Therefore the temperature distribution of microwave heating and conventional heating are different, the temperature decreases from the interior to the surface for microwave heating and from the surface to the interior for

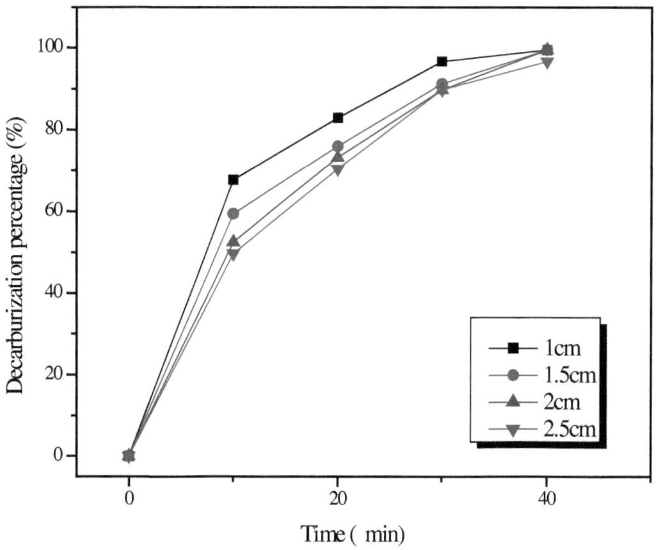

Fig. 5 Decarburization percentage of spent catalysts under different thicknesses by microwave heating

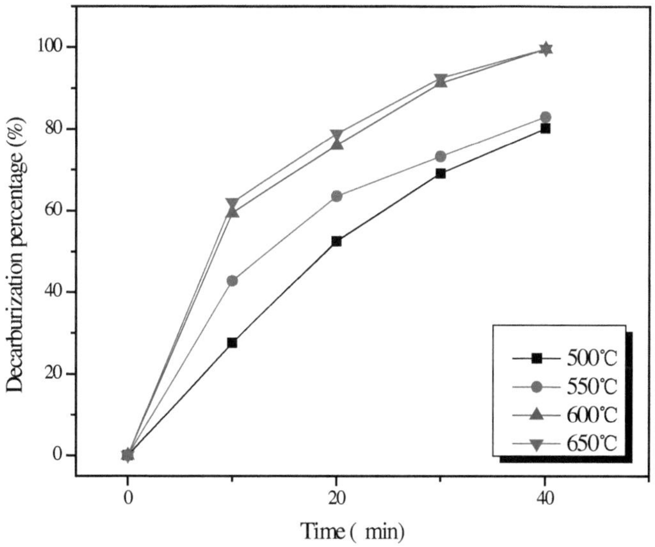

Fig. 6 Decarburization percentage of spent catalysts under different reaction temperatures by microwave heating

Table 2 Dielectric properties of the components in the spent catalyst

Components	Al_2O_3 (Powder)	SiO_2 (Powder)	MgO (Powder)	C (Particle)
Dielectric constant ε_r'	2.465	2.435	1.67	10.772
Dielectric loss ε_r''	8.61×10^{-3}	9.46×10^{-3}	2.08×10^{-2}	2.86×10^{-1}
Loss tangent tan δ	3.49×10^{-3}	3.89×10^{-3}	1.25×10^{-2}	2.66×10^{-2}

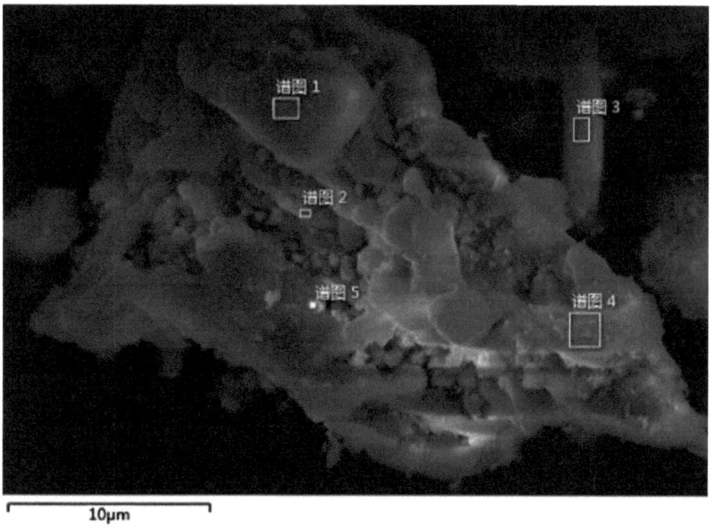

Fig. 7 FE-SEM picture of the spent catalysts before decarburization

conventional heating [10, 11]. Compared with the conventional heating, the heat could generate quickly and uniformly in the spent catalyst for microwave heating. Microwave heating shortened the reaction time and increased the sample thickness in the decarburization process. The optimized conditions for microwave heating were as follows: roasting temperature 600 °C, roasting time 40 min and sample thickness 2 cm, respectively.

Energy Spectrum Analysis

The field emission scanning electron microscopy (FE-SEM) of the spent catalyst before and after microwave heating decarburization under the optimum conditions were presented in Figs. 7 and 8. And the corresponding energy spectrum analysis results are presented in Tables 3 and 4.

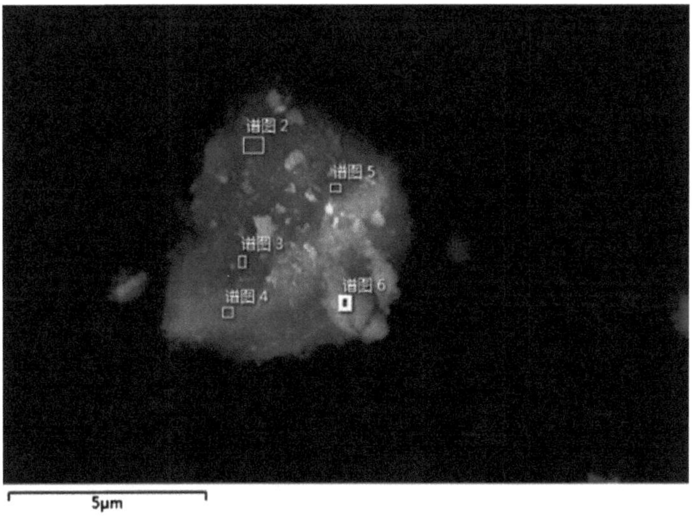

Fig. 8 FE-SEM picture of the spent catalysts after decarburization

Table 3 FE-SEM energy spectrum analysis of the spent catalysts before decarburization

Components	Al	Si	Mg	O	C
Energy spectrum1 (%)	14.7	23.7	8.2	45.5	7.9
Energy spectrum2 (%)	18.5	20.3	6.3	45.9	9
Energy spectrum3 (%)	19.2	19.6	9	39.1	13.1
Energy spectrum4 (%)	14.6	18.7	5.8	53.9	7
Energy spectrum5 (%)	14.2	23.3	7.1	49.4	6

Table 4 FE-SEM energy spectrum analysis of the spent catalysts after decarburization

Components	Al	Si	Mg	O	C
Energy spectrum2 (%)	20.9	20.5	8.1	50.3	0.2
Energy spectrum3 (%)	21.8	23.2	8.9	45.6	0.3
Energy spectrum4 (%)	19.5	25.7	6.4	48.3	0.1
Energy spectrum5 (%)	20.5	22.5	7.8	49.1	0.1
Energy spectrum6 (%)	24.4	21.6	7.3	46.6	0.1

The results show that the deposit carbon of the spent catalysts was less than 0.3% after microwave heating decarburization under the optimum condition. Deposit carbon was removed effectively. Therefore it is feasible to decarburize from the spent catalysts by microwave heating.

Conclusions

Parameters of technique to decarburize from spent petrochemical catalysts via microwave oxidation roasting were investigated. The results showed that the decarburization percentage of spent petrochemical catalysts was significantly affected by roasting temperature and sample thickness in the range studied. The decarburization percentage was 99.5% in microwave fields under the optimized conditions, roasting temperature 600 °C, roasting time 40 min and sample thickness 2 cm respectively. The residuum after microwave decarburization was characterized by field emission scanning electron microscopy (FE-SEM) which showed that the deposit carbon had been removed effectively. Therefore it is feasible to decarburize from the spent catalysts by microwave heating.

Acknowledgements Financial support for this work from the National Natural Science Foundation of China (No. 51564033), Yunnan Applied Basic Research Project (No. 2016FA023), the State Key Laboratory of Advanced Technologies for Comprehensive Utilization of Platinum Metals (No. SKL-SPM-201514), State Key Laboratory of Complex Nonferrous Metal Resources Clean Utilization, Kunming University of Science and Technology (No. CNMRCUXT1403).

References

1. H. Dong, J. Zhao, J. Chen, W. Yuedong, B. Li, Recovery of platinum group metals from spent catalysts: a review. Int. J. Miner. Process. **145**, 108–113 (2015)
2. S.R. Izatt, N.E. Izatt, R.L. Bruening, Metal separations of interest to the Chinese metallurgical industry. J. Rare Earths **28**, 22 (2010)
3. M.A. Barakat, M.H.H. Mahmoud, Recovery of platinum from spent catalyst. Hydrometallurgy **72**, 179–184 (2014)
4. M.A. Barakat, M.H.H. Mahmoud, Y.S. Mahrous, Recovery and separation of palladium from spent catalyst. Appl. Catal. A **301**, 182–186 (2006)
5. Z. Tavassolirizi, K. Shams, M.R. Omidkhah, Platinum recovery from model media and a Pt–Sn/alumina spent catalyst extract using corn husk-based adsorbent. J. Ind. Eng. Chem. **23**, 119–127 (2015)
6. C.N. Mpinga, J.J. Eksteen, C. Aldrich, L. Dyer, Direct leach approaches to platinum group metal (PGM) ores and concentrates: a review. Miner. Eng. **78**, 93–113 (2015)
7. X.F. Zhang, X.L. Dong, H. Huang, Y.Y. Liu, W.N. Wang, X.G. Zhu, B. Lv, J.P. Lei, C.G. Lee, Microwave absorption properties of the carbon-coated nickel nanocapsules. Appl. Phys. Lett. **89** (2006)
8. D. Beneroso, A. Albero-Ortiz, J. Monzó-Cabrera, A. Díaz-Morcillo, A. Arenillas, J.A. Menéndez, Dielectric characterization of biodegradable wastes during pyrolysis. Fuel **172**, 146–152 (2016)
9. Zhiwei Peng, Jiann-Yang Hwang, Byoung-Gon Kim, Joe Mouris, Ron Hutcheon, Microwave absorption capability of high volatile bituminous coal during pyrolysis. Ener. Fuels **26**, 5146–5151 (2012)
10. M.E.C. Oliveira, A.S. Franca, Microwave heating of foodstuffs. J. Food Eng. **53**(4), 347–359 (2012)
11. H.W. Yang, S. Gunasekaran, Comparison of temperature distribution in model food cylinders based on Maxwell's equations and Lambert's law during pulsed microwave heating. J. Food Eng. **64**(4), 445–453 (2014)

Effect of FeO Content in Laterite Nickel Slag on the Corrosion Behaviour of Refractory Materials

Shuming Huang, Jilai Xue and Zengie Wang

Abstract The corrosion behaviors of chrome corundum/magnesia in laterite nickel slag were investigated by static crucible method. The results show that the molten slag can penetrate into chrome corundum or magnesia which are used as lining materials in smelter furnace, and the corrosion can occur at the slag-refractory interface and the inner area of the refractories as well. Both the penetrated amount of slag and the penetration depth of Fe, Si increase with increasing FeO content (10–25 wt%) in slag. Such penetration in magnesia is stronger than that in chrome corundum under the same corrosion conditions, as more pores in magnesia may provide channels for slag penetration and corrosion product transportation. A damage of the refractory linings and short service life of smelter furnace may happen during metallurgical processing slag with high FeO content.

Keywords Laterite nickel ore · Slag corrosion · Refractory material

Introduction

Most of nickel exists as sulfides and oxides on the earth [1, 2]. With the depletion of sulfide nickel ores, the utilization of laterite-nickel ores has increased in recently years for making Ni-Fe alloys in electric furnaces [3, 4]. This process usually generates a large volume of acidic slag that can corrode the refractory materials and shorten the service life of the furnace [5–8].

S. Huang · J. Xue
School of Metallurgical and Ecological Engineering,
University of Science and Technology Beijing,
Xueyuan Road 30, 100083 Beijing, China
e-mail: 17888816295@163.com

Z. Wang (✉)
College of Material Science and Engineering,
Beijing University of Technology, Pingleyuan 100, 100124 Beijing, China
e-mail: wangzj@bjut.edu.cn

J.-Y. Hwang et al. (eds.), *8th International Symposium on High-Temperature Metallurgical Processing*, The Minerals, Metals & Materials Series,
DOI 10.1007/978-3-319-51340-9_66

Since the chemistry of laterite nickel ores and the composition of target alloys may vary, the metallurgical slag and the corrosion behaviors of the lining refractories can be quite different from smelter to smelter. As for today, chrome corundum and magnesia are the major refractory materials for nickel smelting furnace. In practice, an adjustment in the reduction process has to be made based on the variation of nickel contents in the ores and the target alloy. For instance, a partial reduction for the metals could be performed in order to produce higher Ni content in Ni-Fe alloys. It is, therefore, very common that such operation may result in change of FeO content ranging from 10 to 25 wt% in the slag. Although the chemical reactions and the characterization of corrosion zone at the slag-refractory interface have been reported in the references above, the information on the penetration of the laterite nickel slag into the chrome corundum and magnesia refractories are still rare in the literature. The corrosion of the refractories and the slag penetration can also be associated with the changing FeO content in slag.

This work is aimed to have better understanding of the corrosion behaviors of chrome corundum and magnesia in contact with molten laterite nickel slag containing varying FeO. The investigation has been carried out by static crucible method at 1550 °C that is around the common operating temperatures of industrial furnace. The corrosion of the refractory materials and penetration of the molten slag were characterized using SEM-EDS techniques. The results obtained will be useful in selecting refractory materials and improving the resistance performance of the furnace linings during ferronickel production.

Materials and Methods

Materials

Table 1 shows the chemical compositions of the chrome corundum brick and the magnesia brick (commercial products for industrial application) used in this work.

Table 1 Composition of the refractory materials determined by XRF

Composition	(wt%)	
	Chrome corundum brick	Magnesia brick
Cr_2O_3	60.75	/
Al_2O_3	18.48	5.36
SiO_2	8.46	4.30
ZrO_2	5.22	/
P_2O_5	1.78	/
Fe_2O_3	0.27	0.67
CaO	0.12	2.05
MgO	/	87.01
TiO_2	/	0.25

Table 2 Slag compositions (wt%) with varying FeO content used in corrosion tests

Composition (%)	SiO_2	MgO	FeO	Al_2O_3	CaO
FeO = 10	61.68	18.82	10.00	5.00	4.50
FeO = 15	58.00	17.50	15.00	5.00	4.50
FeO = 20	54.32	16.18	20.00	5.00	4.50
FeO = 25	50.63	14.87	25.00	5.00	4.50

Refractory samples were taken from these bricks and were machined into a crucible for corrosion test. Each crucible was made in size of 60 mm × 60 mm × 70 mm, with a central groove in dimension of ϕ40 mm × 35 mm. In order to be more closer to the experimental composition, the final slag was taken from a Ni-Fe alloy factory as well as added SiO_2, Fe_2O_3, Al_2O_3, CaO (trace of alkaline oxide converted into MgO and the acidic oxide into SiO_2). The normalized slag composition for corrosion tests is listed in Table 2.

Corrosion Test The static corrosion test method was applied, where each crucible made of the refractory material was filled with 40 g slag with varying composition (see Table 2). The crucible containing slag was heated in a furnace under Ar atmosphere up to 1550 °C and then maintained for 2 h. After the corrosion test, the crucible sample was cut through vertically at its middle, and the cross-section of the sample was used for observation and examination.

The corrosion zone at the slag-refractory interface was characterized using SEM (JSM-6480LV) and EDS (Noran System six) techniques. As the separation of the slag and the refractory parts after corrosion was difficult, the amount of the slag penetration into the refractory material was estimated through image analysis to obtain the cross-section area of the remaining slag on the crucible bottom (more details will appear in later section).

Results and Discussion

Slag Penetration in Chrome Corundum

Figure 1 shows the cross-section of chrome corundum crucibles containing laterite nickel slag with different FeO contents after corrosion at 1500 °C for 2 h. Uneven contours can be observed at the bottom of each crucible showing that the laterite nickel slag has already corroded into chrome corundum.

In order to evaluate the degree of the corrosion for chrome corundum material, the area of slag remained outside the corroded refractory region in the crucible sample can be used to represent the amount of unpenetrated slag, which can be estimated through the area obtained by image analysis method. This means that the more the amount of unpenetrated slag remains, the less the penetrated slag goes into the refractory. The data on the slag penetration estimated in such a way are listed in

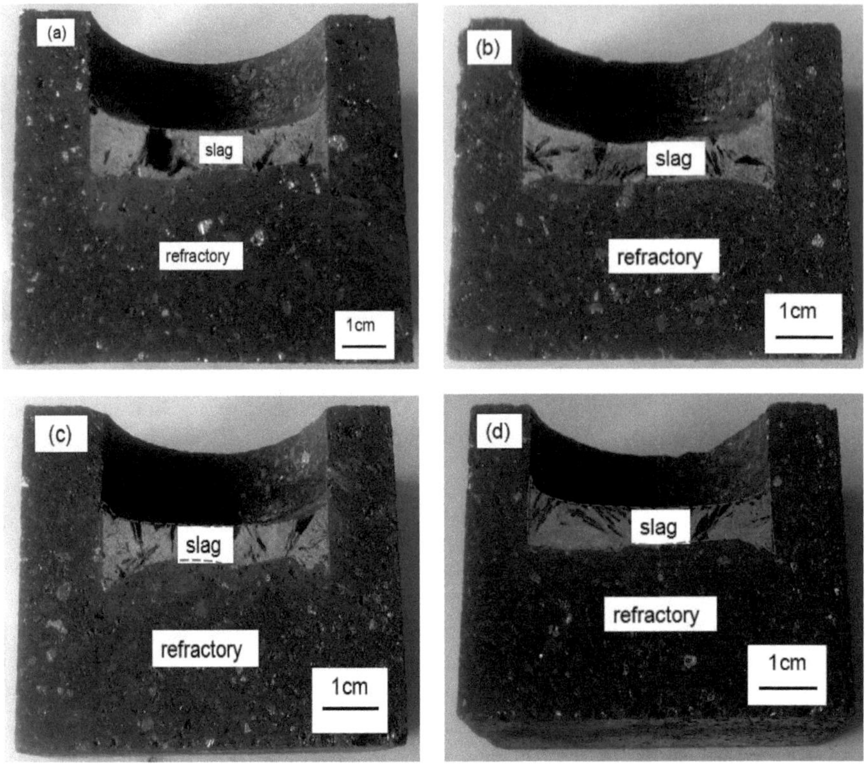

Fig. 1 Cross section of chrome corundum crucibles containing slag with varying FeO content **a** FeO = 10 wt%, **b** FeO = 15 wt%, **c** FeO = 20 wt% and **d** FeO = 25 wt%

Table 3, which were based on the areas shown on the cross-section (see its half part in Fig. 1a–d). It can be seen that with increasing FeO content from 10 to 25%, the average remained slag reduced, as demonstrated by the unpenetrated area decreasing from 351.84 to 255.66 mm^2, This means that the slag penetration into the refractory material increases with increased FeO contents in slag.

Slag Penetration in Magnesia

Figure 2 shows the cross-section of magnesia crucibles containing laterite nickel slag after corrosion at 1500 °C for 2 h. Compared to Fig. 1 with the same FeO contents, more slag penetrated into magnesia than that in chrome corundum. In Fig. 2d, almost no slag left at the bottom of the Magnesia crucible when FeO content in slag is about 25%. In Table 4, the average slag remaining area decreases

Table 3 Slag remaining areas on the cross-sections of the chrome corundum crucibles after corrosion tests

FeO in slag (%)		Slag remaining area on cross-section (mm^2)			Average (mm^2)
FeO = 10	Left-half	343.25	357.79	343.39	351.84
	Right-half	355.89	356.50	354.23	
FeO = 15	Left-half	318.62	319.83	324.27	310.93
	Right-half	300.93	299.38	302.56	
FeO = 20	Left-half	300.28	302.52	300.87	305.51
	Right-half	310.95	308.66	309.76	
FeO = 25	Left-half	241.98	244.32	245.87	255.66
	Right-half	266.55	264.77	270.46	

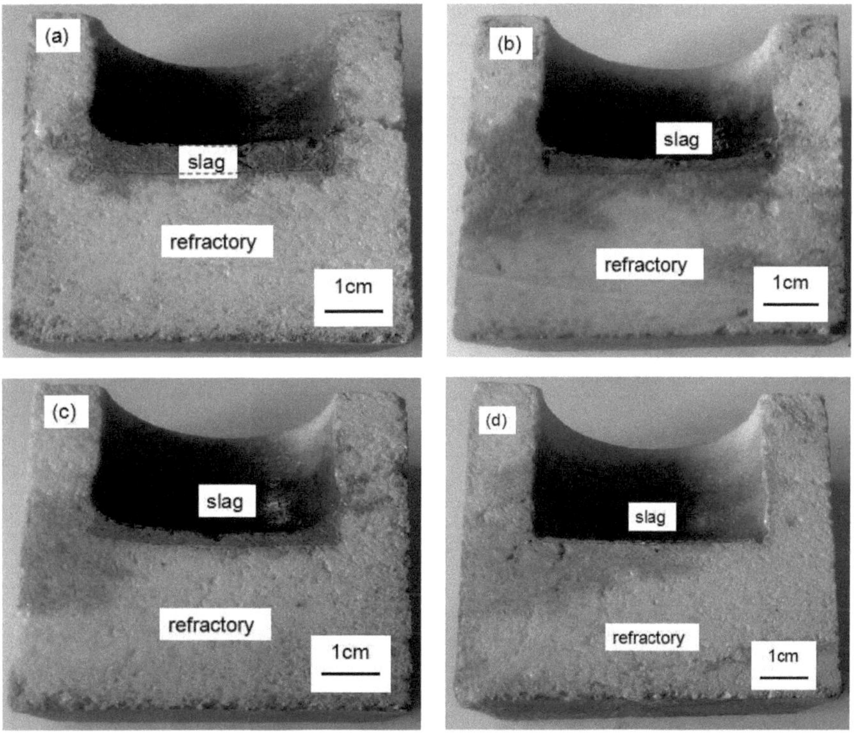

Fig. 2 Cross section of magnesia crucibles containing slag with varying FeO content **a** FeO = 10 wt%, **b** FeO = 15 wt%, **c** FeO = 20 wt% and **d** FeO = 25 wt%

from 219 to 44.7 mm^2 as FeO content in slag increases from 10 to 25%. It means that the penetrated slag into magnesia is more than that in chrome corundum and the chrome corundum has better resistance to the penetration of laterite nickel slag.

Corrosion at the Slag Refractory Interface

Figure 3 shows the cross-section of chrome corundum and magnesia before corrosion, where more pores can be seen on the magnesia than on the chrome corundum. In Fig. 4, a larger amount of penetrated slag exists on the porous magnesia than on the dense chrome corundum. This is due to the pores in magnesia may provide the channel to the penetration of molten slag [9]. At the same time, the lower solubility of chrome oxides in laterite nickel slag [10] may give a better resistance to the slag corrosion.

Figure 5 presents the profiles of Fe and Si contents against the penetration depth of the slag in chrome corundum, in which the penetration depth accounted from the slag-refractory interface toward the inner area of the testing sample. Fe and Si contents were obtained through SEM-EDS scanning analysis of elements in the selected corrosion areas along the penetration depth. The results show a general trend of penetrated Fe and Si contents decreased with increased penetration depth within the corroded refractory material. At the same time, a higher Fe content in the chrome corundum can be found when the original FeO content in slag is higher. This means that more penetration or corrosion can occur within the inner area of the

Table 4 Slag remaining areas on the cross-sections of the magnesia crucibles after corrosion tests

FeO in slag (%)		Slag remaining area on cross-section (mm^2)			Average (mm^2)
FeO = 10	Left-half	203.85	212.65	209.34	219.46
	Right-half	239.64	223.94	227.31	
FeO = 15	Left-half	115.09	120.39	117.55	122.42
	Right-half	127.55	129.28	124.64	
FeO = 20	Left-half	103.66	105.91	109.86	109.48
	Right-half	113.68	116.52	107.24	
FeO = 25	Left-half	41.35	43.29	39.84	44.70
	Right-half	47.96	46.22	49.53	

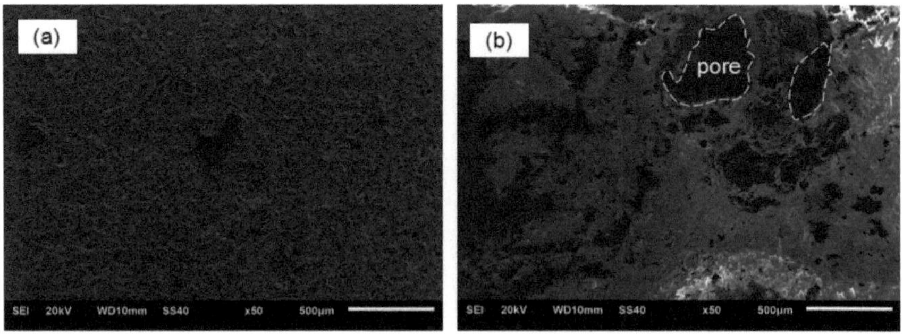

Fig. 3 SEM images showing the cross-section of refractory material before corrosion: **a** chrome corundum and **b** magnesia

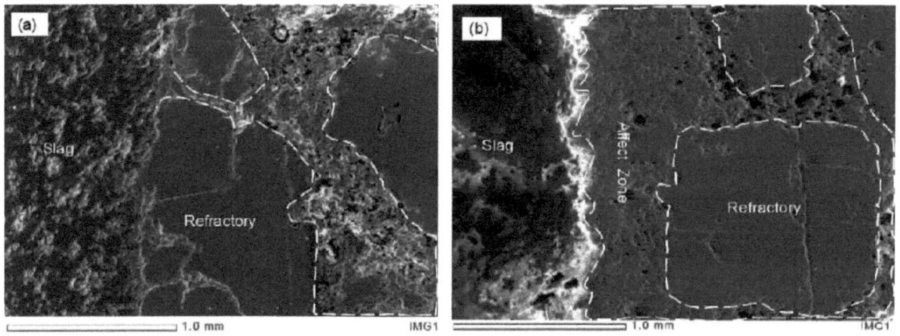

Fig. 4 SEM images of cross-section of refractory materials after corrosion in slag with 10 wt% FeO: **a** chrome corundum and **b** magnesia

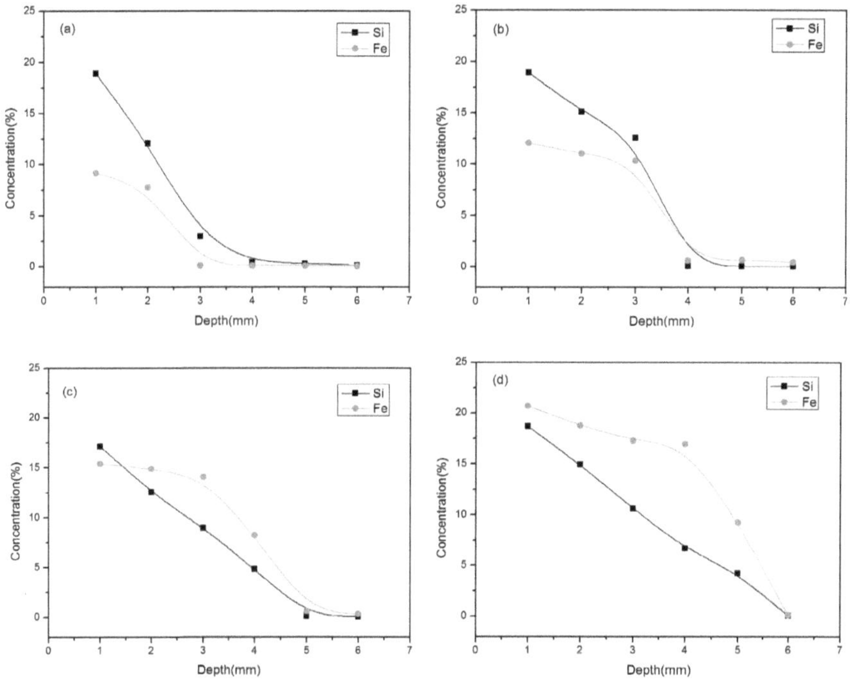

Fig. 5 Fe and Si contents in chrome corundum versus penetration depth of the slag containing **a** FeO = 10 wt%, **b** FeO = 15 wt%, **c** FeO = 20 wt% and **d** FeO = 25 wt%

chrome corundum lining when the smelter furnace operates with the metallurgical slag contains more FeO. In practice, this result suggests that a certain damage of the refractory lining or a shorter service life of smelter furnace should be expected when a higher FeO content slag appears in metallurgical operation using the laterite-nickel ores.

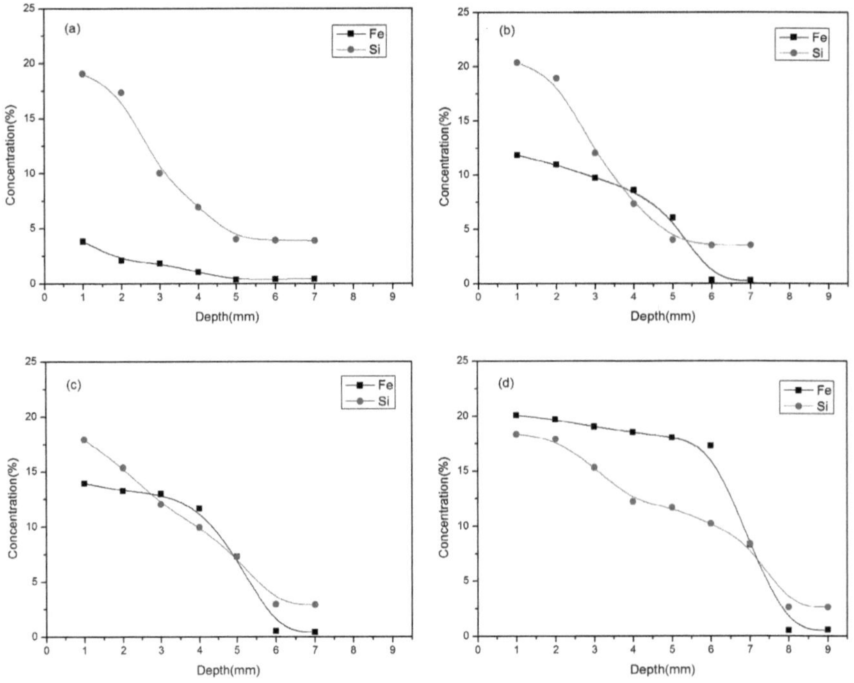

Fig. 6 Fe and Si contents in magnesia versus penetration depth of the slag containing **a** FeO = 10 wt%, **b** FeO = 15 wt%, **c** FeO = 20 wt% and **d** FeO = 25 wt%

According to the compositions of slag and refractory materials, the major corrosion reactions can be as follows Eqs. (1)–(7).

$$Cr_2O_3 + FeO = Cr_2FeO_4 \quad (1)$$

$$3CaO + 2SiO_2 = 3CaO \cdot 2SiO_2 \quad (2)$$

$$2CaO + SiO_2 = 2CaO \cdot SiO_2 \quad (3)$$

$$3CaO + SiO_2 = 3CaO \cdot SiO_2 \quad (4)$$

$$SiO_2 + Al_2O_3 = Al_2SiO_5 \quad (5)$$

$$Fe_2O_3 + 2CaO = 2CaO \cdot Fe_2O_3 \quad (6)$$

$$2MgO + SiO_2 = Mg_2SiO_4 \quad (7)$$

$$Fe_2O_3 + CaO = CaO \cdot Fe_2O_3 \quad (8)$$

The corrosion reaction involving Cr_2O_3 is quite limited thermodynamically [10], while the other reactions in the slag with lower FeO content, Eqs. (1)–(4), may occur, and the resulting products have high melting point, for instances, 1737 K for $3CaO \cdot 2SiO_2$, 2427 K for $2CaO \cdot SiO_2$, 2072 K for $3CaO \cdot SiO_2$. Thus the corresponding viscosity of such slag is higher, making more difficult for the slag penetrating in refractory materials [11]. When FeO content gets higher, not only Eqs. (1)–(4) but also Eqs. (5) and (7) can appear, which result in $CaO \cdot Fe_2O_3$ with lower melting point of 1513 K and lower viscosity as well as relative easier slag penetration.

With the same SEM-EDS analysis method applied to magnesia refractory, the obtained results are presented in Fig. 6. It can be seen that as FeO content in slag gets higher, the penetration of Si and Fe becomes stronger. Comparing to Fig. 5a–d, the values of ending penetration depth for Fe or Si in magnesia are larger than those in chrome corundum under the same corrosion conditions. This can be mainly attributed to the more porous microstructure of magnesia than that of chrome corundum. Besides, the reaction of Eq. (6) will take place for magnesia and its resulting product has a lower melting point and viscosity which favor the slag penetration and corrosion.

Conclusions

1. Molten slag from laterite nickel ores can penetrate into chrome corundum and magnesia which are used as lining materials in smelter furnace, and the corrosion can occur at the slag-refractory interface and the inner area of the refractory material as well.
2. Both the penetrated amount of the molten slag and the penetration depth of Fe, Si increase with increasing FeO content (10–25 wt%) in slag.
3. The penetrated amount of the slag and the penetration depth of Fe, Si in magnesia are more than those in chrome corundum under the same corrosion conditions, as more pores in magnesia may provide channels for penetration of molten slag and transportation of the corrosion product.
4. A damage of the lining materials and short service life of smelter furnace may happen with metallurgical operation using laterite nickel slag with higher FeO contents.

Acknowledgements Financial support from National Science and Technology Support Program of China (2015BAB03B00) is gratefully acknowledged.

References

1. A.D. Dalvi, W.G. Bacon, R.C. Osborne, in *The Past and the Future of Nickel Laterites*. 2004 International Convention Canada (2004), pp. 1–4
2. H.H. He, Q.F. Cai, *Nickel and Cobalt Metallury of China* (Metallurgical Industry Press, 2000)
3. L.G. Chen, A. Malflie, P.T. Jones, Comparison of the chemical corrosion resistance of magnesia-based refractory by stainless steelmaking slags under vacuum conditions. Ceram. Int. **42**, 743–751 (2016)

4. J. Poirier, F. Qafassaoui, J.P. Ildefonse, Analysis and interpretation of refractory microstructures in studies of corrosion mechanisms by liquid oxides. J. Eur. Ceram. Soc. **28**, 1557–1568 (2008)
5. P. Gehre, C.G. Aneziris, Investigation of slag containing refractory materials for gasification processes. J. Eur. Ceram. Soc. **32**, 4051–4062 (2012)
6. P. Gehre, C. Wenzel, C.G. Aneziris, Investigation of shape alumina based refractories used in slagging gasifiers. Ceram. Int. **37**, 1701–1704 (2011)
7. K. Tanaka, T. Nishimura, J. Moda, Al_2O_3–Cr_2O_3 castables with excellent spalling resistance. Taikabutsu Overseas **33**, 270–279 (2013)
8. A.P. Luz, F.C. Leite, Slag conditioning effects on MgO–C refractory corrosion performance. Ceram. Int. **39**, 7507–7515 (2013)
9. F.M. Stabile, M. Piccico, Viscosity and thermal evolution of density and wetting angle of a commercial glaze by means of hot stage microscopy. Procedia Mater. Sci. **9**, 563–570 (2015)
10. M. Reinmöller, M. Klinger, E. Thieme, Analysis and prediction of slag-induced corrosion of chrome oxide-free refractory materials during fusion of coal and biomass ash under simulated gasification conditions. Fuel Process. Technol. **149**, 218–230 (2016)
11. L.P. Fu, H.Z. Gu, A. Huang, Slag resistance mechanism of lightweight microporous corundum aggregate. Am. Ceram. Soc. **98**, 1658–1663 (2015)

Effects of Blowing Conditions on the Dispersion States of Materials Charged into Bottom Blown Oxygen Smelting Furnace

Dongxing Wang, Yan Liu, Zhang Ting-an and Xiaolong Li

Abstract Recent years, bottom blown oxygen copper smelting process has been widely used in China. Besides many advantages such as autothermal operation, this technology is still in developing stage. This topic aims at promoting the dispersion of materials charged into the bottom blown smelting furnace used in the process. Based on the similarity theory, a 1/5 scaled physical model of smelting furnace was constructed. Polypropene particles were used as charged materials. High-speed camera was used to record the phenomenon of the particles dispersion under different gas-flow rates, nozzle diameters and installation angles of nozzles. Image processing software was used for obtaining the particles concentration under different conditions. Results show that when the gas flow rate increases, charged materials disperse better. With the same gas flow, the increasing of nozzle diameter promotes the dispersion effect of materials. The effects of installation angles are opposite for nozzles with diameter of 2.50 and 5.92 mm.

Keywords Physical simulation · Copper smelting · Bottom blown oxygen · Dispersion of materials · Copper converter · Bottom blown · Matte-smelting · Water model

Introduction

Bottom blown oxygen copper smelting process (SKS process) is a new bath smelting technology in copper making industry. Be different from traditional top blown process, such as Ausmelt and side blown process, such as Noranda and

D. Wang · Y. Liu · Z. Ting-an (✉) · X. Li
Key Laboratory of Ecological Metallurgy of Multi-metal Intergrown Ores of Education Ministry, Special Metallurgy and Process Engineering Institute, School of Metallurgy of Northeastern University, Liaoning, 110819 Shenyang, China
e-mail: zta2000@163.net

D. Wang
e-mail: wangdongxing@foxmail.com

J.-Y. Hwang et al. (eds.), *8th International Symposium on High-Temperature Metallurgical Processing*, The Minerals, Metals & Materials Series,
DOI 10.1007/978-3-319-51340-9_67

Teniente process, oxygen-enriched air is blown into the liquid matte from the nozzles installed at the bottom of the bottom blown oxygen smelting furnace [1, 2]. In recent years, SKS process (invented by Shui Kou Shan Mining Bureau of China) has been widely applied in China. Since the first SKS smelting furnace was put into production in 2008, it is about to be more than 8 bottom blown furnaces in China by 2016 [3–5].

The SKS smelting furnace is a horizontal cylindrical furnace which is similar to the Noranda furnace [3]. Raw materials are poured into the furnace from the feed inlets on the top of the furnace continuously. High grade copper matte produced in the furnace is tapped regularly through the matte taphole. Through the slag taphole at one end of the furnace, slag will be tapped regularly and then transported to slow cooling process [3, 4].

SKS process has many merits, for instance, its high oxygen enrichment and thermal efficiency. Autothermal operation can be achieved in the SKS smelting furnace without additional fuel. It has a good adaptability for raw materials such as the wet material and block material, so the preparation of raw material is very simple [3, 4].

The application of high speed submerged jetting technology (without blockage of nozzles) is another significant feature of SKS smelting furnace. When the air is injected into the bath from submerged nozzles at a high velocity, approaching sonic velocity, the air flow in the molten pool will be in a stable state of continuous air column and at the end of the air column, air will burst into many little bubbles. High speed jetting provides power for the flowing of fluid in the molten pool and is directly related to the energy and mass transfer and the reaction rate. The little bubbles can promote the utilization efficiency of oxygen and the rate of chemical reactions [4–6].

Besides the advantages, there are still some problems need to be solved urgently in the SKS smelting furnace. Some of the cold raw materials charged into the SKS smelting furnace float on the surface of the pool, mixing with slag and copper matte. The viscous mixture can be splashed to the lower edge of feed inlet and adhere onto the surface of it. As time passes, it would block the inlet. To knock through the blockage, the feed inlets cannot be completely sealed, leading to the leakage of exhaust gas and deterioration of the working environment. In addition, the blockage of inlets increases the labor intensity of workers [4–6].

In order to prevent the blockage of inlets, this topic aims at promoting the dispersion of materials charged into the bottom blown smelting furnace used in the SKS process. Water model experiment is an important method to study the flow phenomena in metallurgical reactors [7–11]. Based on the similarity theory, a scaled-down water model of SKS smelting furnace was built. The effects of different factors such as gas-flow rates, nozzle diameters and installation angles of nozzles on the dispersion states of raw materials were systematically studied.

Experimental

Principles

The water model experiments are on the basis of similarity criterion, the flowing regularity in the model can be generalized to the prototype.

According to geometric similarity, a 1:5 scaled-down water model was built. Dynamic similarity achieved through that the modified Froude number Fr' of the model is equal to that of the prototype.

$$Fr' = \frac{\rho_g \cdot u^2}{\rho_l \cdot g \cdot H} \tag{1}$$

Where ρ stands for the density; u is the characteristic velocity; H is the characteristic length; g stands for the gravitational acceleration; the subscripts l and g stand for the liquid and gas respectively.

Characteristic velocity is given in Eq. (2).

$$u = \frac{4Q \cdot \rho_g}{\pi \cdot d^2} \tag{2}$$

Where ρ_g stands for the gas density; Q is the gas flow rate; d is the diameter of nozzle.

Rely on the fact that $Fr'_{\mathrm{model}} = Fr'_{prototype}$, considering Eqs. (1)–(3) was derived. Using Eq. (3), the standard gas flow rate used in the model can be calculated.

$$Q_m = \left(\frac{\rho_{gp}}{\rho_{gm}} \cdot \frac{\rho_{rm}}{\rho_{rp}} \cdot \frac{d_m^4}{d_p^4} \cdot \frac{H_m}{H_p}\right)^{1/2} \cdot Q_P \tag{3}$$

Where ρ stands for the density; d stands for the diameter of nozzle; H is the depth of the pool; Q is gas flow rate; the subscripts p, m, l and g stand for the prototype, model, liquid and gas respectively.

Apparatus and Materials

The water model is a slice model simplified from a company's SKS smelting furnace as shown in Fig. 1. For easy to observe the phenomenon, the model contains only one nozzle in axis. The parameters of model and prototype are shown in Table 1. The material of the model is acrylic. Figure 2 shows the sketch of experimental set-up. On the bottom of the model, there are different positions for nozzles to be installed.

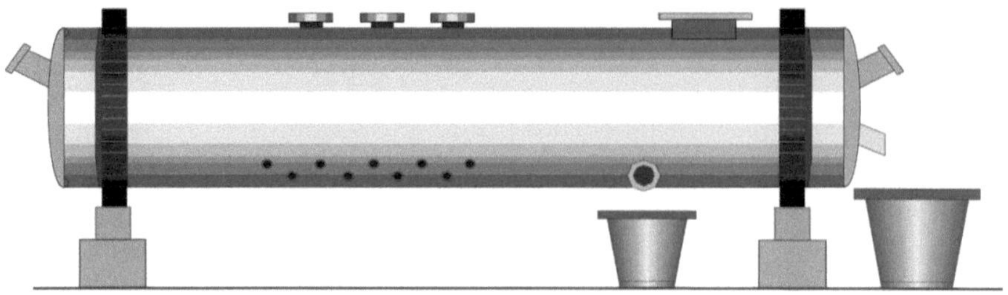

Fig. 1 Bottom blown oxygen copper smelting furnace

Table 1 Parameters of model and prototype

System	Radius of the pool (mm)	Width of the pool (mm)	Height of liquid level (mm)	Density of liquid (kg m^{-3})	Density of gas (kg m^{-3})
Water model	349	360	270	1000	1.25
Prototype	1710	15,000	1320	4600	1.37

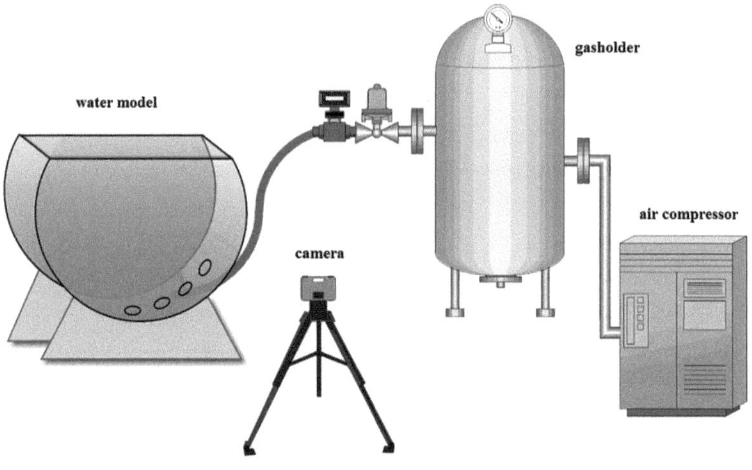

Fig. 2 Sketch of experimental set-up

An air compressor and a gasholder were linked as air source. 60 g polypropene particles were charged into the water model as the raw materials. High-speed camera was used to record the dispersion states of raw materials in the furnace. Then clearer pictures were selected. Image processing software was used to count the particles concentration under different conditions.

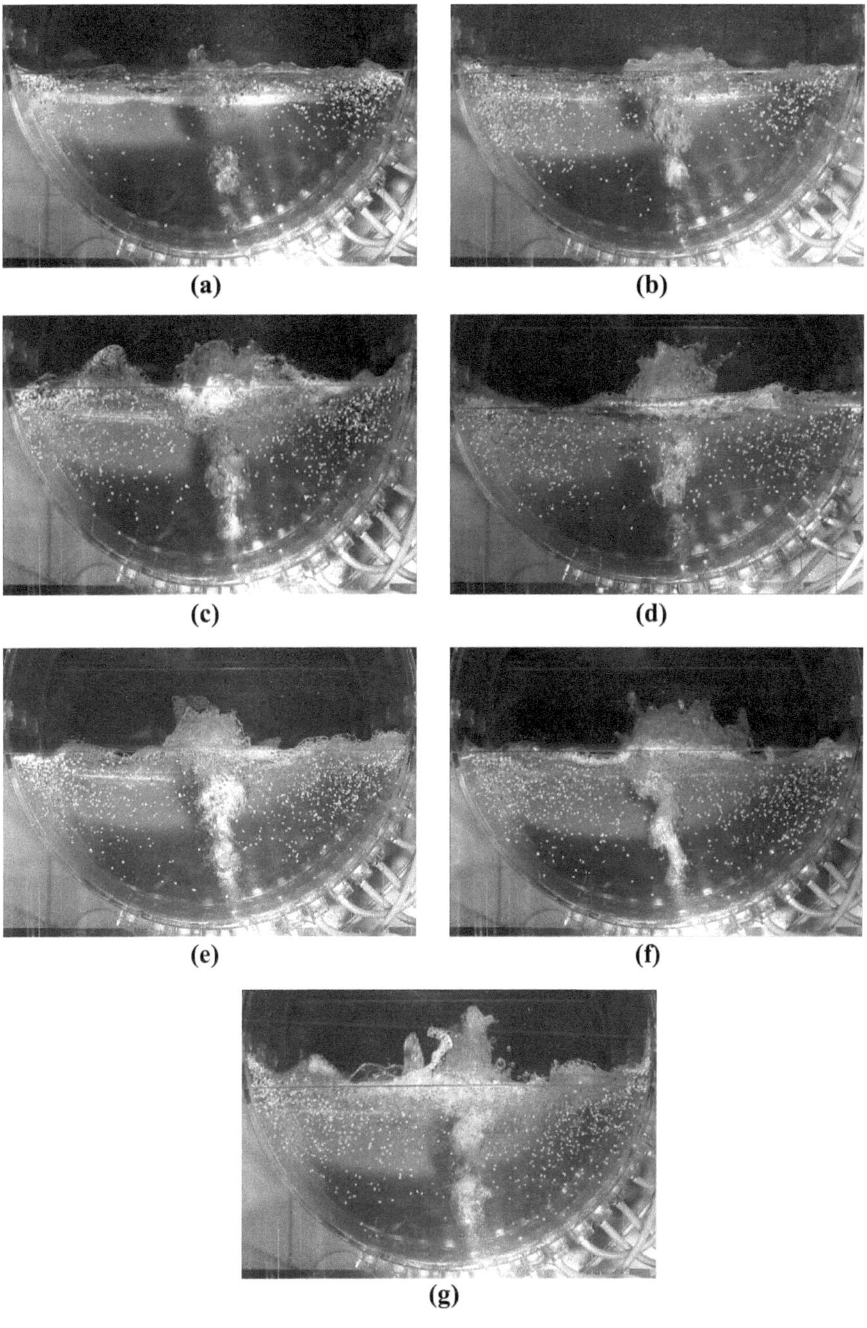

Fig. 3 Dispersion states of materials under different gas-flow rates **a** Q = 2.5 Nm^3/h **b** Q = 4.8 Nm^3/h **c** Q = 8.0 Nm^3/h **d** Q = 9.5 Nm^3/h **e** Q = 11.5 Nm^3/h **f** Q = 13.5 Nm^3/h **g** Q = 15.1 Nm^3/h

Results and Discussion

Effects of Gas-Flow Rates

The diameter of nozzle was 5.92 mm. Installation angle of nozzle was 7°. Gas flow rate was 2.5, 4.8, 8.0, 9.5, 11.5, 13.5 and 15.1 Nm^3/h respectively. The experiment phenomena are shown in Fig. 3.

Image processing software was used to calculate the particles involved into the pool and the results are shown in Fig. 4.

Figure 4 illustrates that the amount of particles involved into the pool has an increasing tendency as the gas-flow rate increases. When the gas flow rate is 2.5 Nm^3/h, the amount of particles is 248. When the gas flow rate increases to 15.1 Nm^3/h, there are 396 particles involved into the pool.

It can be seen from the experimental phenomena that with the increasing of the gas-flow rate, better decentralized result of charged materials is obtained. Because when the gas flow rate increases, more gas was injected in the same amount of time. With the same nozzle's diameter, the gas velocity increases, so the gas can provide more kinetic energy to the pool. Hence the more particles are involved into the pool.

Effects of Nozzle Diameters

The gas flow rate was 9.5 Nm^3/h. Installation angle of nozzle was 7°, diameter of nozzle was 2.50, 3.00, 3.50, 4.00, 4.50, 5.00 and 5.92 mm respectively. The experiment phenomena are shown in Fig. 5.

The amount of particles involved into the pool is shown in Fig. 6. As shown in Fig. 6, when the diameter of nozzle increases, the amount of particles involved in

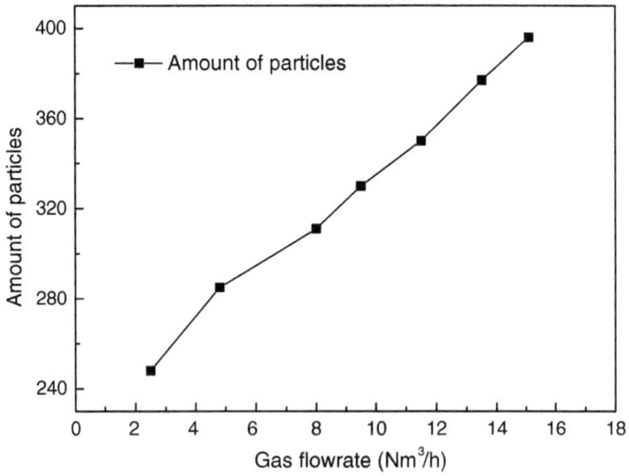

Fig. 4 Amount of particles under different gas-flow rates

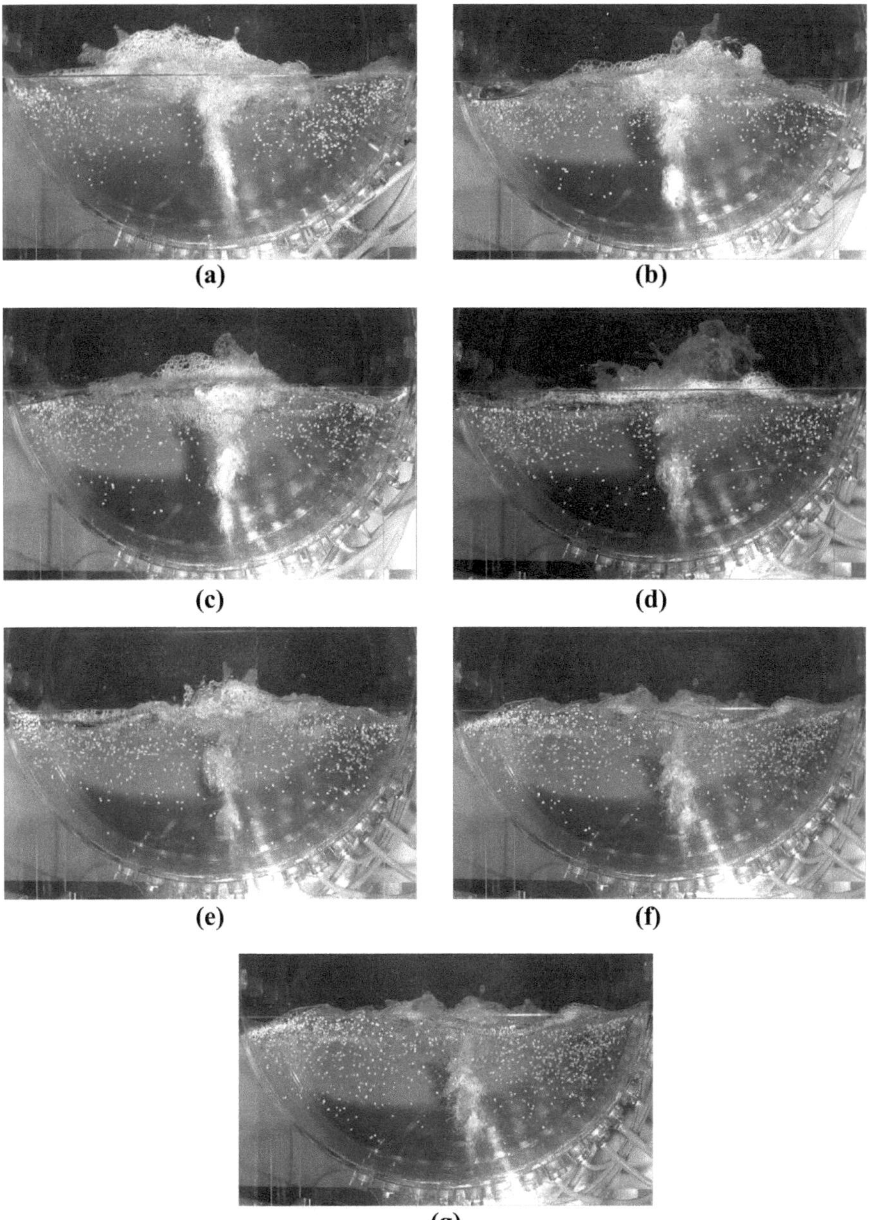

Fig. 5 Dispersion states of materials under different nozzle diameters a D = 2.50 mm, b D = 3.00 mm, c D = 3.50 mm, d D = 4.00 mm, e D = 4.50 mm, f D = 5.00 mm, g D = 5.92 mm

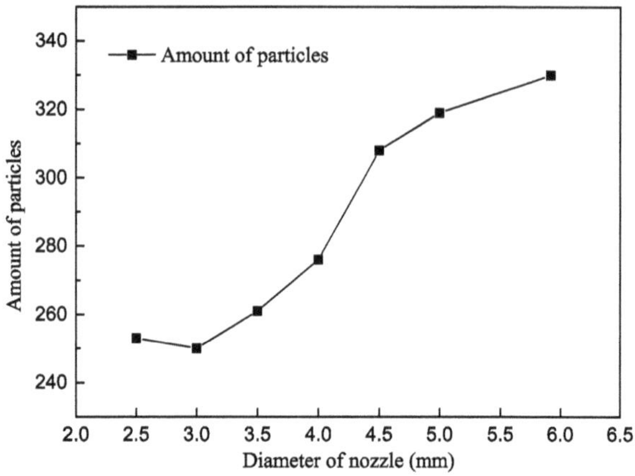

Fig. 6 Amount of particles under different nozzle diameters

the pool shows an increasing tendency on the whole. When the diameter is 2.50 mm, the amount of particles is 253, almost the same as the result when the diameter is 3.00 mm. When the diameter increases to 5.92 mm, 330 particles are involved into the pool. This is because with the smaller diameter, the gas velocity is higher and the gas column is thinner, so that the effective action area between gas and liquid is smaller. The work done by the buoyancy which drives the liquid is much less and the action time is shorter. As a result, the injected gas escapes from the surface of liquid before the gas kinetic energy fully transfers to the liquid.

Effects of Installation Angles of Nozzles

Diameter of Nozzle is 2.50 mm

The gas flow rate was 9.5 Nm^3/h. Diameter of nozzle was 2.50 mm. Installation angle of nozzle was 0°, 7°, 14°, 22°, 29°, 36°, 43°, 50°, 57° and 64° respectively. The experiment phenomena are shown in Fig. 7.

The amount of particles involved in the pool is shown in Fig. 8. It can be seen from Fig. 8 that with the increasing of installation angles, the amount of particles decreases firstly and then increases. The growth trend increases rapidly especially when the angles become larger than 36°. As a whole, the charged materials disperse better and the splashing becomes weak as the installation angles increase.

Diameter of Nozzle is 5.92 mm

The gas flow rate was 9.5 Nm^3/h. Diameter of nozzle was 5.92 mm. Installation angle of nozzle was 0°, 7°, 14°, 22°, 29°, 36°, 43°, 50°, 57° and 64° respectively. The experiment phenomena are shown in Fig. 9.

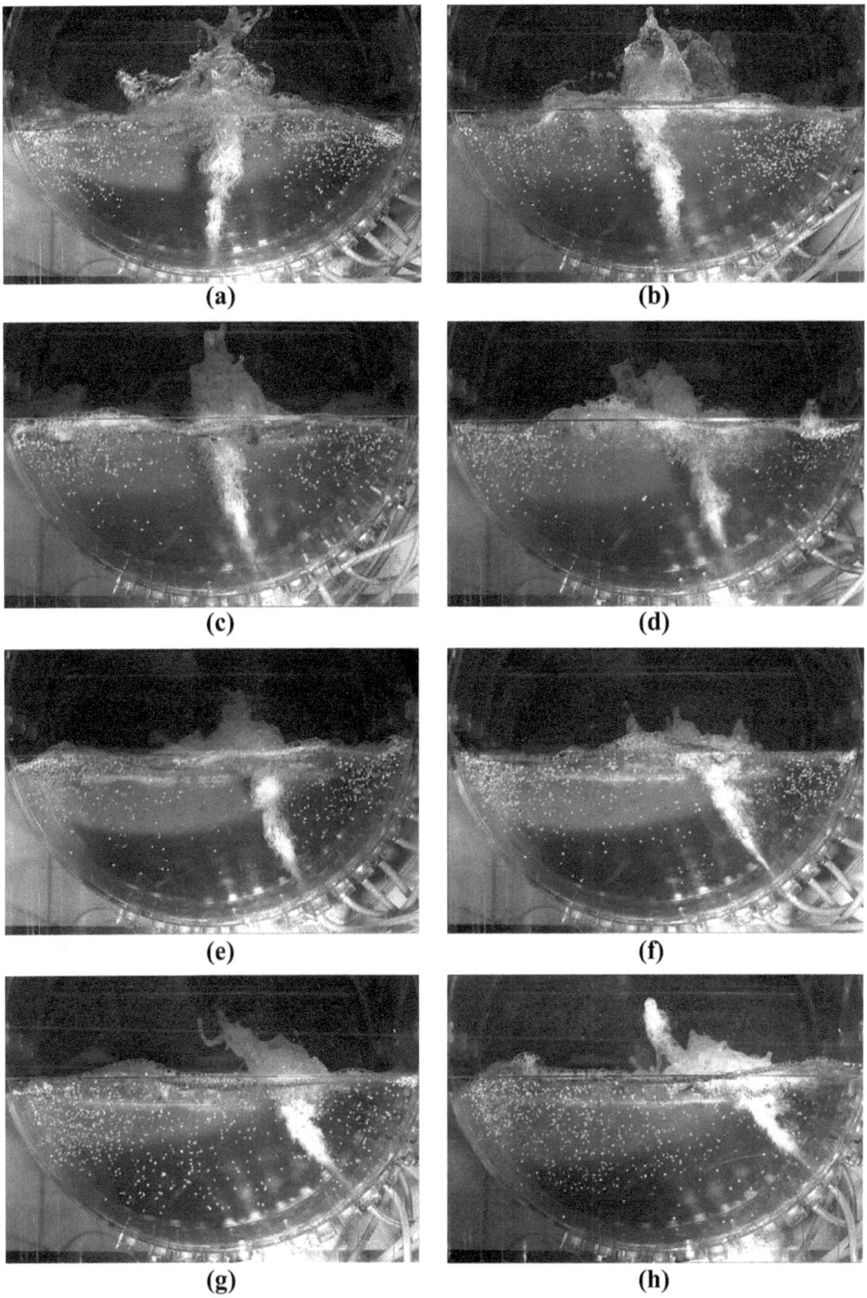

Fig. 7 Dispersion states of materials under different installation angles of nozzles (diameter of nozzle is 2.50 mm) **a** 0°, **b** 7°, **c** 14°, **d** 22°, **e** 29°, **f** 36°, **g** 43°, **h** 50°, **i** 57°, **j** 64°

(i)

(j)

Fig. 7 (continued)

Fig. 8 Amount of particles under different installation angles of nozzles (diameter of nozzle is 2.50 mm)

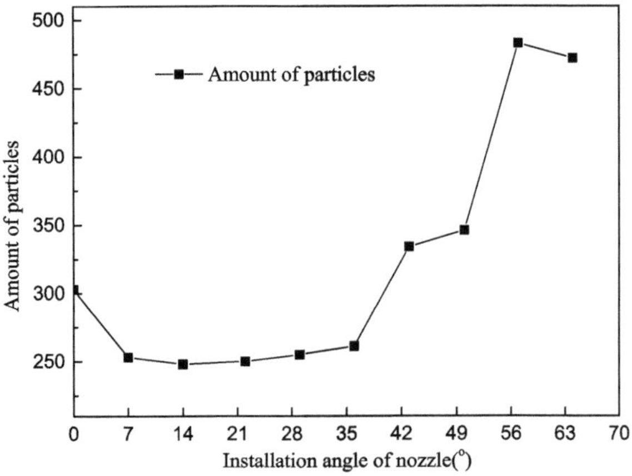

The amount of particles involved in the pool is shown in Fig. 10. Figure 10 shows a contrary tendency against Fig. 8. When the installation angles increase, the amount of particles decreases gradually. Because when the angle is 0°, the route of gas in the liquid and the action time between them are the longest. The work done by the buoyancy is largest. The pool obtains the most kinetic energy from the gas buoyancy and the stirring effect is the best. Hence, the particles disperse better when the angle is smaller. Compared with 2.50 mm nozzle, the gas velocity of 5.92 mm nozzle is far lower. So the kinetics of liquid is most from the buoyancy, not from the kinetics of gas. The splashing of 5.92 mm nozzle is less than that of the 2.50 mm nozzle.

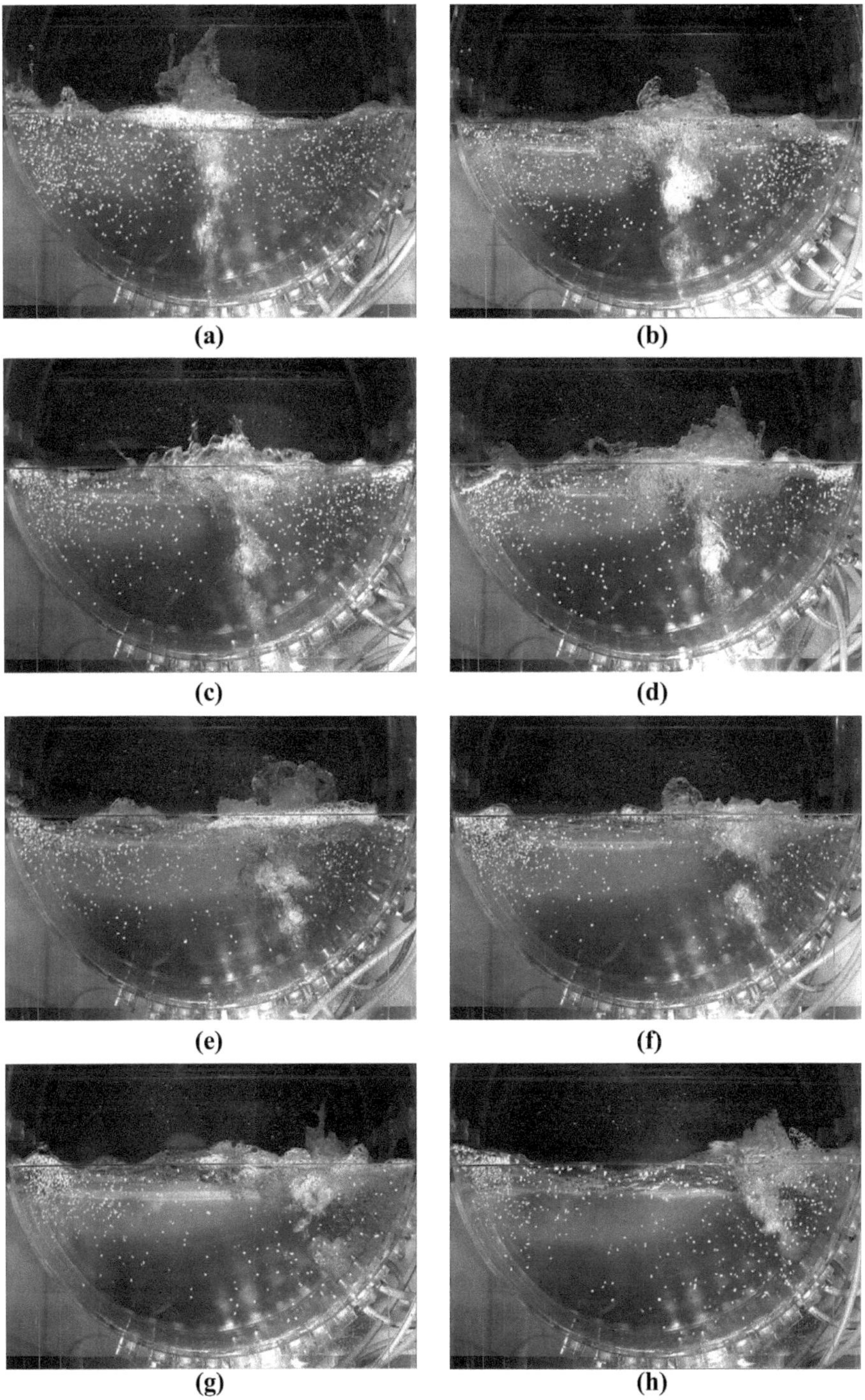

Fig. 9 Dispersion states of materials under different installation angles of nozzles (diameter of nozzle is 5.92 mm) **a** 0°, **b** 7°, **c** 14°, **d** 22°, **e** 29°, **f** 36°, **g** 43°, **h** 50°, **i** 57°, **j** 64°

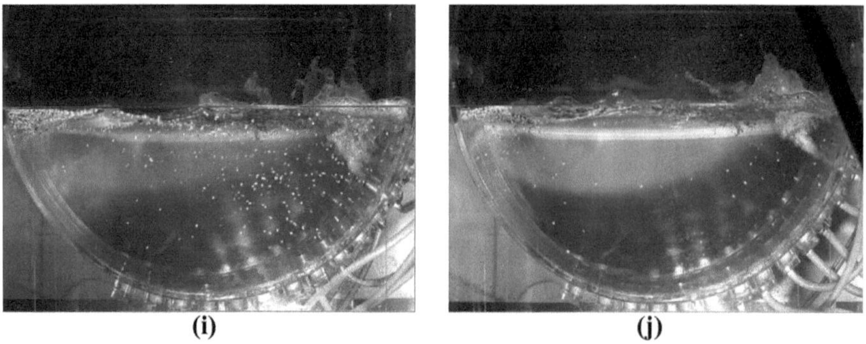

Fig. 9 (continued)

Fig. 10 Amount of particles under different installation angles of nozzles (diameter of nozzle is 5.92 mm)

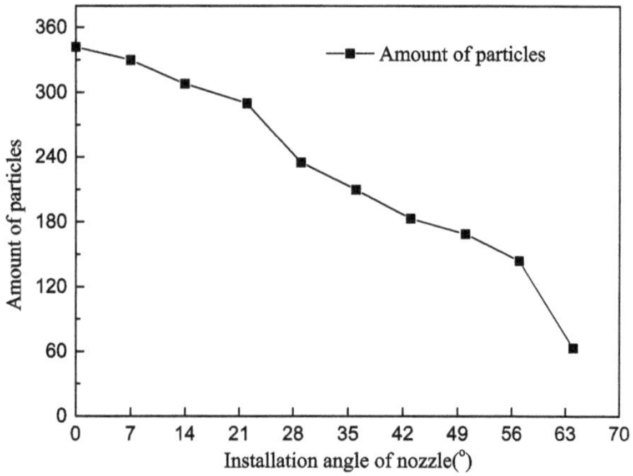

Conclusion

The dispersion states of charged raw materials in SKS smelting furnace under different gas-flow rates, nozzle diameters and installation angles of nozzles were researched. The experimental results showed that:

(1) With the increase of gas flow rate, the charged materials disperse better and the splashing of liquid increases.
(2) With the same gas flow, the increasing of nozzle diameter promotes the dispersion effect of materials and weakens the splashing.
(3) The effects of installation angles are opposite for nozzles with diameter of 2.50 and 5.92 mm. With the same gas flow, when the diameter of nozzle equals to

2.50 mm, the charged materials disperse better and the splashing becomes weak as the installation angles increase.

(4) With the same gas flow, when the diameter of nozzle equals to 5.92 mm, the charged materials disperse worse as the installation angles increase. Compared with 2.50 mm nozzle, the splashing is general weak.

Acknowledgements This work was supported by National Science & Technology Support Program (No. 2012BAE01B02), National High-Tech R&D Program of China (863 Program No. 2012AA062303) and Joint Funds of the National Natural Science Foundation of China No. U1402271.

References

1. G. King, M. Schlesinger, A.K. Biswas, in *Extractive Metallurgy of Copper*, 4th edn. (Pergamon Press, Elsevier Science Ltd., Kidlington, UK, 2002)
2. L. Hu, D. Li, Development and application of oxygen bottom blowing melting furnace. Non-Ferrous Metall. Equip. **1**, 34–37, 53 (2011)
3. B. Zhao, Z. Cui, Z. Wang, in *A New Copper Smelting Technology-Bottom Blown Oxygen Furnace Developed at Dongying Fangyuan Nonferrous Metals*. TMS Annual Meeting 2013, 4th International Symposium on High-Temperature Metallurgical Processing, pp 3–10
4. Z.X. Cui, D.B. Shen, Z. Wang, A new process of copper smelting with oxygen enriched bottom blowing technology. World Metall. Erzmetall. **64**(5), 254–259 (2011)
5. Jiang Jimu, The new technology and device of continuous copper smelting using oxygen bottom blown furnace. China Metal. Bull. **17**, 29–31 (2008)
6. Shandong Fang-yuan Nonferrous Metals Group and China ENFI Engineering Corporation, The industrial practice of oxygen bottom blown smelting technology for polymetallic resources. Resour. Recycl. **11**, 46–49 (2009)
7. L. Shui, Z.X. Cui, X.D. Ma, M.A. Rhamdhani, A. Nguyen, B.J. Zhao, Mixing phenomena in a bottom blown copper smelter: a water model study. Metall. Mater. Trans. B **46**(3), 1218–1225 (2015)
8. D.X. Wang, Y. Liu, Z.M. Zhang, P. Shao, T.A. Zhang, in *Experimental Study of Bottom Blown Oxygen Copper Smelting Process for Water Model*. 7th International Conference on Micromechanics of Granular Media (Powders and Grains, Sydney, NSW, Australia, July 8–12, 2013, 1542), pp. 1304–1307
9. M.A. Cooksey, W. Yang, in PIV measurement on physical models of aluminium reduction cells, in *Light Metals* 2006. Proceedings of the Technical Sessions Presented by the TMS Aluminum Committee at the 135th TMS Annual Meeting, Warrendale, USA, March 12–16 (2006), pp. 359–365
10. Y. Liu, Z.M. Zhang, J.N. Liu, J.H. Zhang, M. Sano, J. Zhang, Cold model study on mg desulfurization of hot metal under mechanical stirring. J. Iron. Steel Res. Int. **20**(12), 1–6 (2013)
11. Y. Liu, Y.D. Li, T.A. Zhang, N.X. Feng, Research on the penetration depth in aluminum reduction cell with new type of anode and cathode structures. JOM **66**(7), 1202–1209 (2014)

Effects of Pre-oxidation and Additive on Carbothermic Reduction of Ilmenite Concentrate

Wei Lv, Yingyi Zhang, Gangqiang Fan, Kai Tang and Xuewei Lv

Abstract The EAF smelting process of ilmenite concentrate consumes large amounts of energy. Semi-melting reduction can lower the reaction temperature and separate the iron and slag more thoroughly. However, this needs additives and pre-oxidation to promote the reaction. In this paper, the pre-oxidized ilmenite powders were reduced isothermally with graphite and Na_2SO_4. The influence of reduction conditions on the metallization ratio of samples, such as reaction temperature, and sodium sulfate additive were investigated. The results show that additive such as Na_2SO_4 can significantly decrease the melting point of pre-oxidized ilmenite concentrate.

Keywords Pre-oxidation · Ilmenite concentrate · Carbothermic reduction · Sodium sulfate

Introduction

Titanium is the ninth-most abundant element in the Earth's crust [1]. The world production of titanium mineral concentrates was about 6 million metric tons in 2015, and ilmenite accounts for about 92% of the world's consumption of titanium minerals [2]. The Panzhihua ilmenite concentrate in Sichuan, southwest of China is one of the largest reserves in the world with about 870 million tons, which accounts for more than 90% of the total resource of China and over 35% of the world [3, 4].

Ilmenite concentrate is the major feedstock to produce titanium pigment, paper, plastics and titanium alloys especially ferrotitanium and the aluminum-alloy [5, 6].

W. Lv · Y. Zhang · G. Fan · K. Tang · X. Lv (✉)
College of Materials Science and Engineering, Chongqing University,
400044 Chongqing, China
e-mail: lvxuewei@163.com

W. Lv
e-mail: yjlvwei@163.com

J.-Y. Hwang et al. (eds.), *8th International Symposium on High-Temperature Metallurgical Processing*, The Minerals, Metals & Materials Series,
DOI 10.1007/978-3-319-51340-9_68

However, the Panzhihua ilmenite is a complex polymetallic rock ore, with a high ratio of FeO/Fe_2O_3 and a high grade of MgO, CaO and SiO_2 [4]. Therefore it is difficult to separate by conventional beneficiation process [7]. The smelting process must keep certain FeO in order to improve its fluidity. The semi-melting reduction can lower the reaction temperature and separate the iron and slag more thoroughly; additives or catalyst could promote the reaction.

Over the past decades, strengthen solid-state reduction of ilmenite concentrate has been studied by many researchers. Gou et al. [8] found that the oxidation process can increase the specific surface area of the ilmenite concentrate, and reduce the activation energy of carbothermic reduction. Chen et al. [9] found that pre-oxidation destroyed the crystal lattice of raw titanomagnetite concentrate, formed plenty of pores in the particles, which can accelerate the reduction process. Wu et al. [10] found that alkali chloride can significantly accelerate the carbothermal reduction reaction of pre-oxidized ilmenite concentrate. Bao et al. [11] found that the addition of foreign ions can significantly promote the reduction reactivity of ilmenite. Guo et al. [12] found that preoxidation ilmenite with $Na_2B_4O_7$ additive significantly advance the reduction of iron. In order to efficiently utilize the ilmenite concentrate and reduce consumption of refining energy, the pre-oxidation process and the additive process were combined together to study the effects of pre-oxidation and additives on carbothermic reduction of ilmenite concentrate.

The isothermal reduction behavior of pre-oxidation ilmenite concentrate with graphite and sodium sulphate additive was investigated in this work.

Experimental

Materials

The ilmenite concentrate used in this investigation was supplied by Panzhihua Iron and Steel Group. The pre-oxidized ilmenite concentrate used in this study was oxidized from ilmenite concentrate in muffle furnace at 1100 °C for 2 h in air. The size of the particles for pre-oxidation was 150–200 mesh. The ilmenite and pre-oxidized ilmenite concentrates were also examined by the X-ray diffraction (XRD) which were shown in Fig. 1. As can be seen from Fig. 1, the $FeTiO_3$ and Fe_3O_4 were converted to Fe_2O_3, TiO_2 and Fe_2TiO_5 by oxidation. The chemical analysis of both concentrates were listed in Table 1. The main impurity in ilmenite concentrate are MgO, CaO, SiO_2 and Al_2O_3.

Apparatus and Procedure

The pre-oxidized ilmenite concentrate was mixed homogenously with the graphite and the sodium sulfate powder. The molar ratio of C to O (Iron oxide) in mixtures

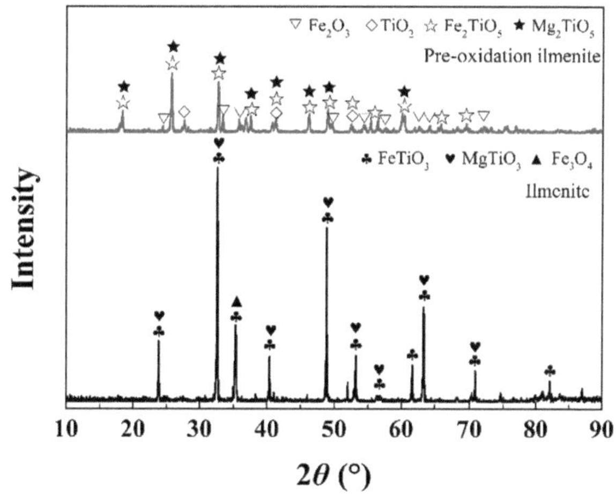

Fig. 1 XRD pattern of Ilmenite and pre-oxidation ilmenite concentrate powder

was set to 1.2:1.0, the sodium sulfate (4%, wt) was used as additive. The mixtures of 10 g were filled into the mold (with the diameter of 16 mm, pressure 8 MPa) and pressed to cylindrical briquettes. The carbothermic reduction was investigated by a vertical tube globar furnace under Argon atmosphere (0.6 L/min). The schematic diagram of the experimental apparatus was shown in Fig. 2. The temperatures for the isothermal carbothermic reduction were 1350, 1400 and 1450 °C, respectively. The experimental procedure was as follows: Firstly, when the temperature reached setting numerical value, samples were filled into furnace for 30 min, then taken out of the furnace, and quenched in the water to room temperature. Finally, the experimental samples were dried at 80 °C for 5 min.

Reaction Mechanism

According to the Fe–S–C phase diagram calculated by FactSage 6.2 (Fig. 3), it can be indicated that the iron is possibly not melted without sufficient carburization and sulfurization. Thus, the carburization and sulfurization can reduce the melting point of metal to achieve the effect of semi-melting reduction.

Results and Discussion

Phase Transitions During Reduction

The XRD pattern of samples, which were reduced by graphite with 4% Na_2SO_4 additive at different temperatures were shown in Fig. 4, The main phases in the reduced sample are Fe, TiO_2, Ti_3O_5, Ti_2O_3, and solid solution. The temperature changes have

Table 1 Experiment of ilmenite and pre-oxidation ilmenite sample chemical composition analysis (wt%)

	TiO_2	FeO	Fe_2O_3	MgO	SiO_2	Al_2O_3	CaO	MnO	V_2O_5	Na_2O	S	P
Ilmenite	46.82	34.52	6.10	5.66	3.23	1.00	0.842	0.635	0.058	<0.005	0.14	<0.005
Pre-oxidized ilmenite	45.17	0.772	42.76	4.78	2.95	1.25	0.761	0.745	0.063	<0.005	0.09	<0.005

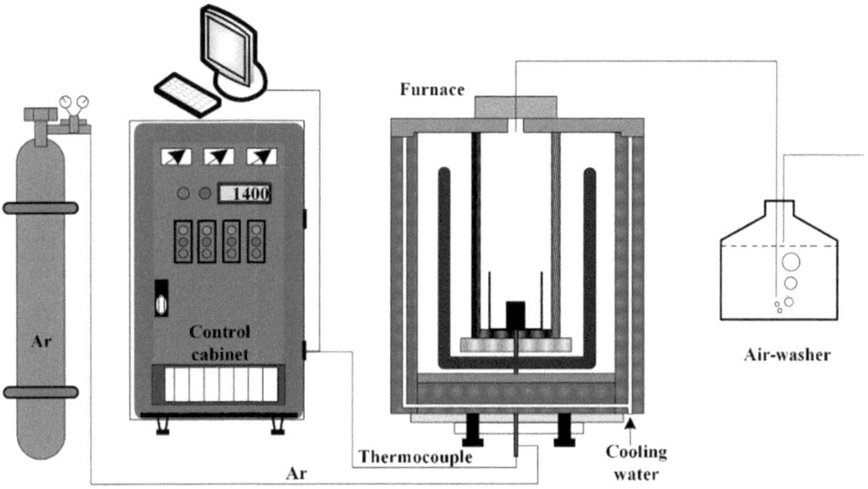

Fig. 2 Experimental apparatus of $MoSi_2$ furnace

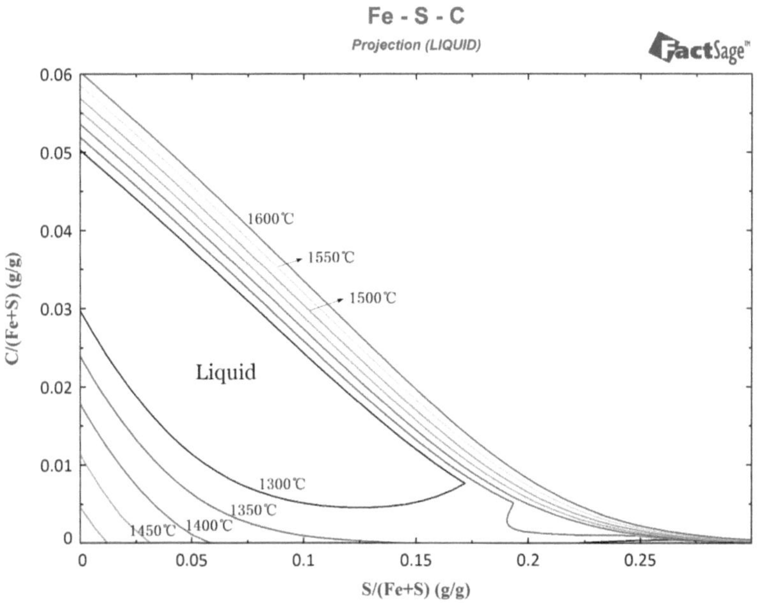

Fig. 3 Phase diagram of Fe–S–C

little effect on the phase composition. The XRD pattern of the reduced samples with different content of Na_2SO_4 additives is shown in Fig. 5. By comparison of Na_2SO_4 addition (without and with 4% of Na_2SO_4), the low valent titanium was increased significantly, which indicates that the Na_2SO_4 additives promote the reduction.

Fig. 4 XRD pattern of the reduced samples at different temperatures

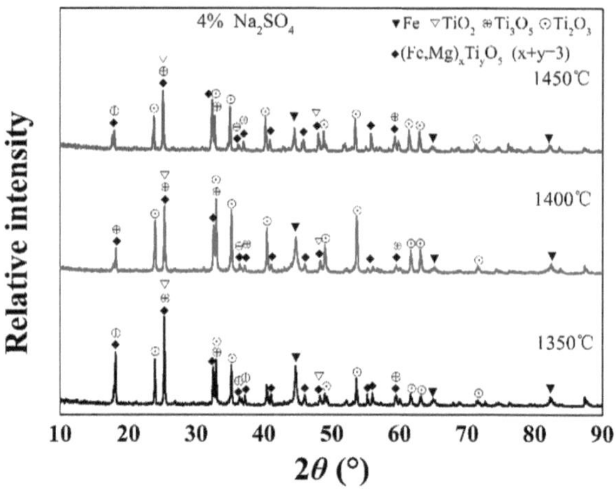

The Metallization Ratio of Iron on the Reduced Samples

The metallization ratio of the reduction was defined as:

$$\text{Metallization ratio} = \frac{MFe}{TFe} \times 100\% \tag{1}$$

where *MFe* and *TFe* are the mass fraction of the metallic iron and total iron in the reduced samples, respectively.

Figure 6 shows the trend of metallization ratio changing with temperature. It can be seen that the metallization ratio of the reduced sample with 4% Na_2SO_4 additive increased with the increasing temperature. The highest metallization ratio was reached 94%. However, the metallization ratio of the reduced sample with 0% Na_2SO_4 additives didn't change with the temperature variation, and maintained at around 89%.

The titanium grade of reduced sample as a function of the temperature is shown in Fig. 7. Comparing with the sample without Na_2SO_4, the addition of Na_2SO_4 significantly increased the content of low valent titanium oxide in the reduced samples. Thus, Na_2SO_4 increases reduction, which is consistent with the XRD pattern in Fig. 5.

SEM Studies for Morphological Analysis

Morphology and elements distribution images of the reduced sample with different content of sodium sulfate additive were shown in Figs. 8 and 9. It can be seen from Fig. 8 that the metallic iron grain gradually grew up, forming spherical structure.

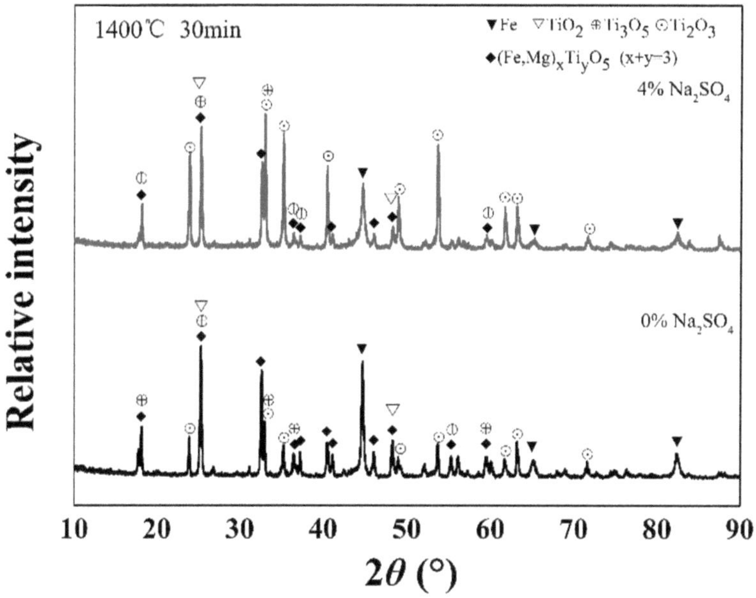

Fig. 5 XRD pattern of the reduced samples with different content of Na_2SO_4 additives

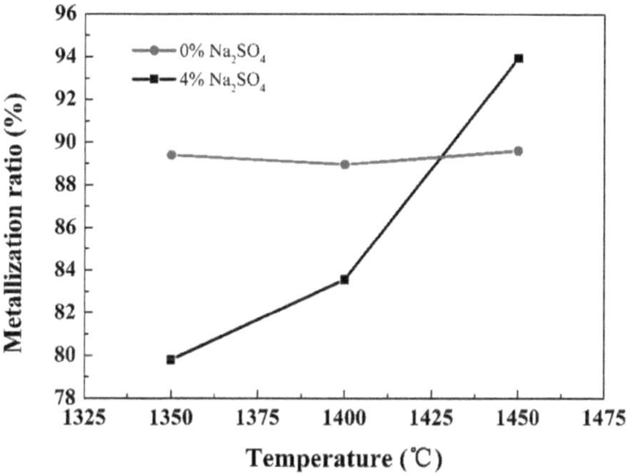

Fig. 6 The trend of metallization ratio as a promotion of temperature variation

Figure 9 shows that metallic iron grain have not yet assembled grow and the shape of the metallic iron grain is irregular. It can be concluded that the additive of sodium sulfate can reduce the melting point, which is consistent with the theoretical calculation results from Fig. 3, thus significantly promoting both the aggregation and growth of metallic iron grain particles and separation of slag and iron.

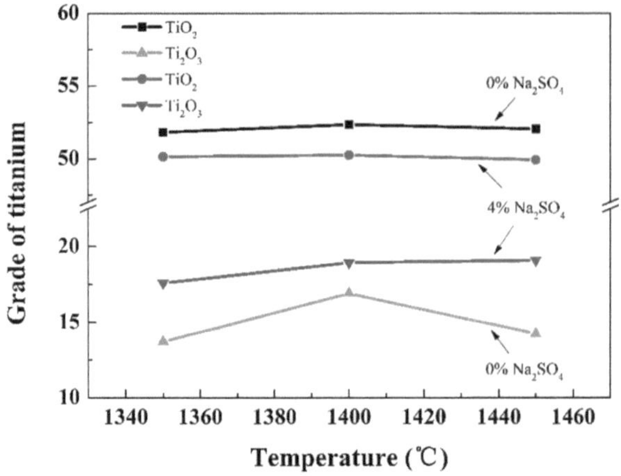

Fig. 7 Titanium grade of reduced sample as a function of temperature variation

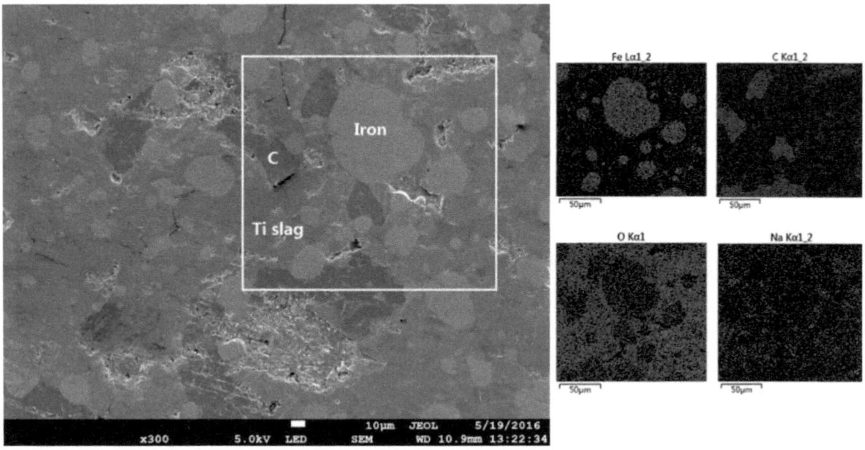

Fig. 8 Morphology and elements distribution images of the sample reduced (1400 °C, Na_2SO_4 4%, C/O 1.2, 30 min)

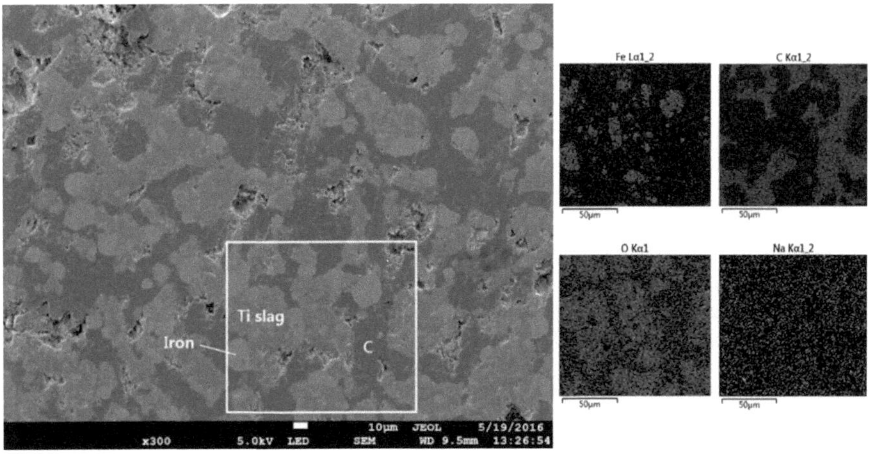

Fig. 9 Morphology and elements distribution images of the reduced sample (1400 °C, Na_2SO_4 0%, C/O 1.2, 30 min)

Conclusions

The reduction behavior of pre-oxidized ilmenite concentrate was experimentally demonstrated. Following conclusions were obtained:

(1) The metallization ratio changes with temperature, Na_2SO_4 additives and C/O molar ratio. By adding Na_2SO_4, the highest metallization ratio reached 94%.
(2) The sodium sulfate significantly promote the aggregation and growth of metallic iron grain particles, which is beneficial for the magnetic separation process.

Acknowledgements This work was supported by the National Natural Science Foundation of China (Grant No. 51374262).

References

1. G. Mazzocchitti, I. Giannopoulou, D. Panias, Silicon and aluminum removal from ilmenite concentrates by alkaline leaching. Hydrometallurgy **96**, 327–332 (2009)
2. By, U.S., Geological survey, mineral commodity summaries (2012)
3. F. Wu et al., Hydrogen peroxide leaching of hydrolyzed titania residue prepared from mechanically activated Panzhihua ilmenite leached by hydrochloric acid. Int. J. Miner. Process. **98**, 106–112 (2011)

4. L. Zhang et al., Hydrochloric acid leaching behavior of different treated Panxi ilmenite concentrations. Hydrometallurgy **107**, 40–47 (2011)
5. M.R. Lanyon, T. Lwin, R.R. Merritt, The dissolution of iron in the hydrochloric acid leach of an ilmenite concentrate. Hydrometallurgy **51**, 299–323 (1999)
6. R. Huang et al., Enhancement reduction of Panzhihua ilmenite concentrate with coke and conglomeration of metal with ferrosilicon. Steel Res. Int. **84**, 892–899 (2013)
7. T. Ping, H.U. Hui-Ping, Z. Li, Effects of mechanical activation and oxidation-reduction on hydrochloric acid leaching of Panxi ilmenite concentration. Trans. Nonferrous Met. Soc. China **21**, 1414–1421 (2011)
8. H.P. Gou, G.H. Zhang, K.C. Chou, Influence of pre-oxidation on carbothermic reduction process of ilmenite concentrate. ISIJ Int. **55**, 928–933 (2015)
9. D.S. Chen et al., Solid state reduction of Panzhihua titanomagnetite concentrates with pulverized coal. Miner. Eng. **24**, 864–869 (2011)
10. W.U. Jian-Hui et al., The co-operative catalytic effect of alkali chlorides on the carbothermic reduction of pre-oxidized ilmenite. J. Guangdong Non-ferrous Met. **15**, 865–878 (2000)
11. J. Bao, Z. Li, N. Cai, Promoting the reduction reactivity of ilmenite by introducing foreign ions in chemical looping combustion. Ind. Eng. Chem. Res **52**, 6119–6128 (2013)
12. Y.F. Guo, et al., in *Preparation of titanium-rich material ilmenite concentrate by reduction-rustion process*. xxiv international mineral processing congress (2008)

Influence of Converter Slag on Decomposition Behavior of Limestone During BOF Steelmaking Process

Hua Lu, Chenxiao Li, Wenwen Mao and Hong Li

Abstract Directly charging limestone into slag in BOF is a newly emerging technique which broke through the traditional BOF slagging principle. It is necessary to study the decomposition behavior of limestone in converter slag during BOF steelmaking. FeO_x, SiO_2, MnO_x and pre-melting slags were mixed with limestone according to different quality proportion to simulate the environment of limestone in converter slag. TG-DTA results indicated that slag with individual component, including FeO_x, SiO_2 and MnO_x, all had obvious effects on reducing the decomposition temperature of limestone. Pre-melting slags with multiple components had further reducing effects than individual component. XRD results showed converter slag component would react with decomposition product of $CaCO_3$ and generate solid solution or oxide, reducing the activity of CaO and the decomposition temperature of $CaCO_3$.

Keywords Limestone · Converter slag · TG-DTA · Decomposition temperature · Activity

Introduction

In conventional basic oxygen furnace process, lime is used as primary flux. The purpose of lime addition is to promote the removing of phosphorus from hot metal by enhancing slag basicity [1]. To reduce the burden of converter, many great efforts have been made to improve the activity of lime. It is considered that it is unreasonable to obtain lime by calcining limestone separately and then charge lime into BOF for slagging. On the basis of long-term laboratory studies, a process of

H. Lu (✉) · C. Li · W. Mao · H. Li
School of Metallurgical and Ecological Engineering, University of Science and Technology Beijing, 30 Xueyuan Road, Haidian District, Beijing 100083, People's Republic of China
e-mail: luhuayejin@163.com

J.-Y. Hwang et al. (eds.), *8th International Symposium on High-Temperature Metallurgical Processing*, The Minerals, Metals & Materials Series,
DOI 10.1007/978-3-319-51340-9_69

charging limestone rather than lime into slag in BOF was proposed which broke through the traditional BOF slagging principle [2]. Theoretical research and operation process have been gradually mature which aroused wide concern with its remarkable effects of energy saving and emission reduction [3, 4]. Deng [5] claimed that the dissolution rate of limestone in converter slag was in general slower than the burned lime and the addition of limestone instead of lime would not be beneficial in the converter process. However, Maruoka [6] found that CO_2 content in quicklime could promote the dissolution rate of quicklime into molten slag. Previous studies showed that with oxygen blowing time being unchanged, slag forming rate increased significantly and final steel compositions and temperature reached the required standard during the steelmaking process with limestone used [7] and the heat is sufficient for limestone to completely replace lime on the condition that the ratio of hot metal to ferrous scrap is adjusted properly [3]. When limestone is charged directly into BOF, lime can be produced to slag by limestone calcination in a very short time, combining with CO_2 release during the BOF process [8]. CO_2 can react with elements in hot metal, producing CO and reducing O_2 consumption [9]. Slag with high oxidation property, high basicity and relatively low temperature can be formed quickly by using limestone, which is advantageous for dephosphorization. Industrial tests showed the endpoint Lp rises with the proportion of lime replaced by limestone increasing [10].

Limestone is surrounded by plenty of converter slag which mostly consists of Fe_xO, SiO_2 and MnO_x except CaO. Therefore it is necessary to study the effects of converter slag components on limestone decomposition behavior. It has been reported in many literature that limestone could promote the slag-making process and improve the oxidation property of converter slag[11], but few paper about influence of converter slag components on decomposition behavior of limestone has been published.

In this work, mixture systems were constituted by Fe_xO, SiO_2, MnO_x or pre-melting slags mixed with limestone to simulate the environment of limestone in converter slag. Mixture systems were studied through thermogravimetry and differential thermal analysis (TG-DTA). Thermal decomposition parameters were calculated according to the TG-DTA test results for the sake of investigating the influence of converter slag components on decomposition behavior of limestone. The compositions of the limestone and mixture systems are listed in Tables 1 and 2.

Table 1 Compositions (mass%) of the limestone in current study

CaO	H_2O	MgO	S	SiO_2	CO_2
54.16	0.09	0.48	0.011	0.7	43

Table 2 Compositions (mass%) of the mixture systems in current study

Number	Mixture system	Number	Mixture system
C	limestone	S6	40% limestone + 60% SiO_2
F2	80% limestone + 20% Fe_2O_3	S8	20% limestone + 80% SiO_2
F4	60% limestone + 40% Fe_2O_3	M2	80% limestone + 20% $MnCO_3$
F5	50% limestone + 50% Fe_2O_3	M4	60% limestone + 40% $MnCO_3$
F6	40% limestone + 60% Fe_2O_3	M6	40% limestone + 60% $MnCO_3$
F8	20% limestone + 80% Fe_2O_3	M8	20% limestone + 80% $MnCO_3$
S2	80% limestone + 20% SiO_2	Sg1	50% limestone + 50% slag1
S4	60% limestone + 40% SiO_2	Sg2	50% limestone + 50% slag2

Materials and Methods

Natural limestone and analytically pure chemical reagents including Fe_2O_3, SiO_2 and $MnCO_3$ are utilized in this study. Fe_2O_3 is used for simulating FeO_x in converter slag. $MnCO_3$ is applied as MnO which is easily oxidized under the air atmosphere. Pre-melting slags consisted of Fe_2O_3 and SiO_2 with or without $MnCO_3$ were also prepared for simulating slag systems of multiple components. The Pre-melting slags used in current experiments are named as slag1 and slag2 respectively. Slag1 is FeO_x–SiO_2–MnO ternary system with Fe_2O_3:SiO_2:$MnCO_3$ of 5:1:1 in mass ratio. Slag2 is FeO_x–SiO_2 binary system with the mass ratio of Fe_2O_3:SiO_2 equal to 5:1.

Two kinds of pre-melting slags were prepared by blending the pure oxide compounds in a graphite crucible. Before mixing, limestone and all the oxides were milled to less than 200 mesh and dried in a muffle furnace at 300 °C for 24 h. After obtaining a homogeneous mixture, the crucible was placed in a carbon tube furnace under purified argon atmosphere and held for at least 2 h at a temperature of 1600 °C in order to get the fully melted slag. As soon as the pre-melting finished, the slags were quenched in the water and crushed into powder.

Thermal analysis of mixture systems was investigated by thermogravimetry and differential thermal analysis (TG-DTA). About 20 mg of sample was taken in a platinum crucible and heated from 300 to 1273 K under air atmosphere at heating rate of 10 K/min. For mixture systems containing $MnCO_3$ (M2–M8), they were heated in nitrogen environment (with N_2 of 99.99% purity). Samples F4, S4 and M4 were investigated by X-ray diffraction analysis to find out the main compounds and present phases after being roasted in a tubular resistance furnace with the same heating system as the TG-DTA test mentioned above.

TG and DTA Results of Different Mixture Systems

Extrapolated onset temperature (T_e), peak temperature (T_m) and extrapolated end temperature (T_c) are used to describe the thermal decomposition process according to the regulation of the International Confederation for Thermal Analysis and Calorimetry (ICTAC). TG-DTA curves of limestone are shown in Fig. 1a. It can be seen that T_e, T_m and T_c of decomposition process of limestone are 748.0, 819.7 and 834.0 °C respectively. TG-DTA curves of pure Fe_2O_3 are shown in Fig. 1b, in which the TG curve has a slow weightlessness because small amount of Fe_2O_3 generated Fe_3O_4 gradually in the heating process. TG-DTA curves of pure SiO_2 are shown in Fig. 1c. SiO_2 has two kinds of crystal forms and changes from β-crystal form to α-crystal form when heated to 573 °C. As is shown in Fig. 1c, the endothermic peak at 578 °C should be the crystal transformation of SiO_2. The phase transition temperature of the SiO_2 is outside the scope of $CaCO_3$ decomposition temperature. Hence it will not affect the results of $CaCO_3$.

TG and DTA curves of F2, F4, F6 and F8 are shown in Fig. 2a–d. The shapes of the DTA curves almost have no change after adding Fe_2O_3 into limestone. T_e, T_m and T_c determined by DTA extrapolation method were confirmed in DTA curves. As is shown in Fig. 2, T_e, T_m and T_c will be decreased with the increase of Fe_2O_3 in Limestone-Fe_2O_3 mixture systems.

TG and DTA curves of various systems with different Limestone-SiO_2 mixtures are shown in Fig. 3a–d. T_e, T_m and T_c determined by DTA extrapolation method were confirmed in differential thermal curves. As is shown in Fig. 4, T_e, T_m and T_c

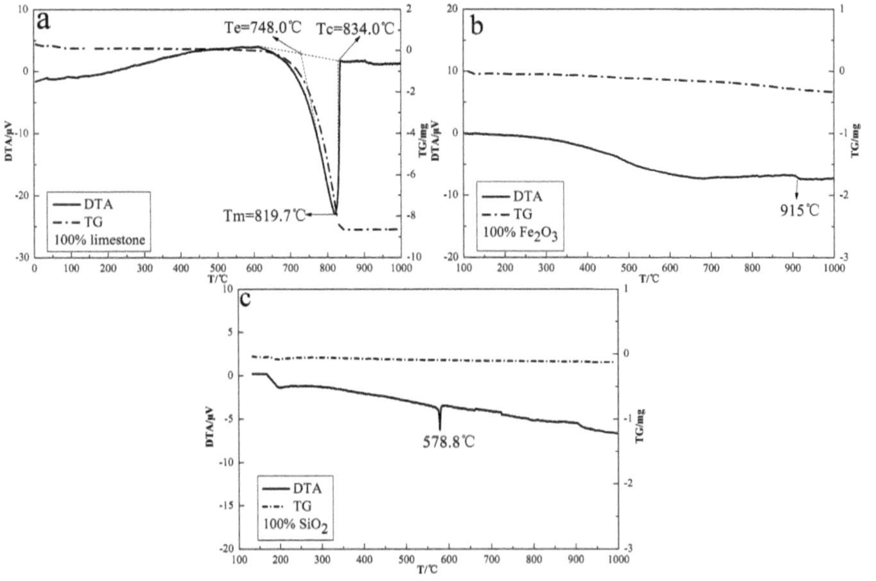

Fig. 1 TG and DTA curves of samples: **a** Pure limestone **b** Pure Fe_2O_3 **c** Pure SiO_2

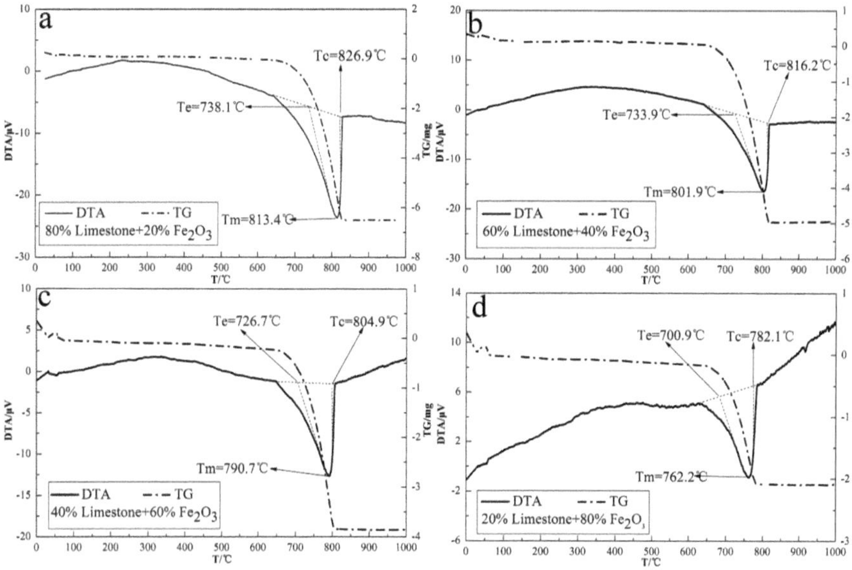

Fig. 2 TG and DTA curves of various systems with different Limestone-Fe_2O_3 mixtures **a** F2 **b** F4 **c** F6 **d** F8

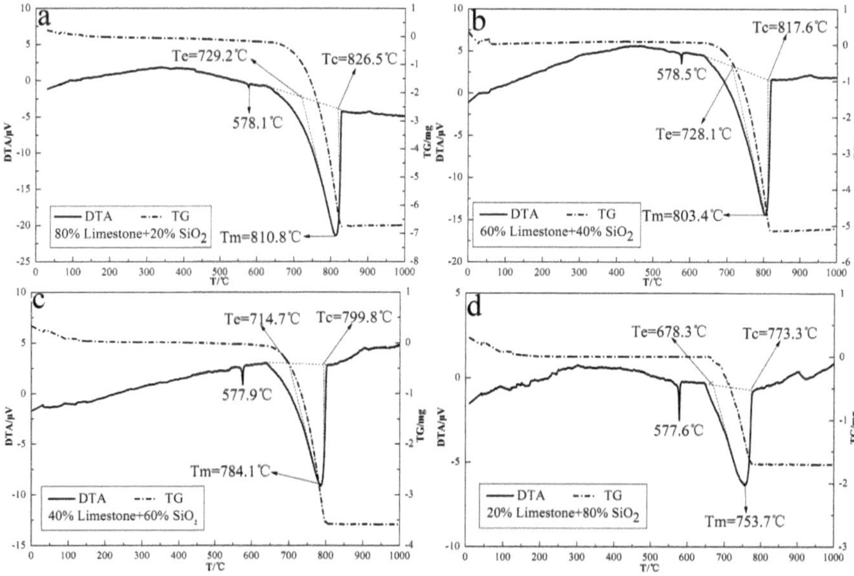

Fig. 3 TG and DTA curves of various systems with different Limestone-SiO_2 mixtures: **a** S2 **b** S4 **c** S6 **d** S8

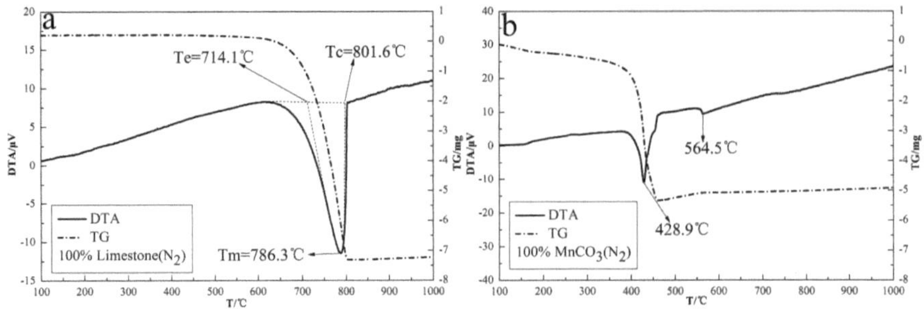

Fig. 4 TG and DTA curves under nitrogen atmosphere: **a** Limestone **b** Pure $MnCO_3$

will be greatly decreased with the increase of SiO_2 in systems with different Limestone-SiO_2 mixtures. TG-DTA curves of pure limestone and $MnCO_3$ under nitrogen atmosphere are shown in Fig. 4a–b. T_e, T_m and T_c in Fig. 4a are much lower than those in Fig. 1a. For example, T_m reduces from 819.7 to 786.3 °C. TG-DTA curves in Fig. 4b present a weightlessness and endothermic peak before 500 °C which arose from thermal decomposition of $MnCO_3$. Then TG curves slightly rebounded, which suggest that MnO produced by decomposition of $MnCO_3$ is slightly oxidized.

As all the reactions represented in the TG-DTA curves of $MnCO_3$ are outside the scope of $CaCO_3$ decomposition temperature, it will not affect the results of limestone. TG and DTA curves of various systems with different Limestone-$MnCO_3$ mixtures are shown in Fig. 5a–d and T_e, T_m and T_c determined by DTA extrapolation method are confirmed in Differential Thermal Analysis curves. To compare the different influences of slag with multiple components and slag with individual component on limestone decomposition temperature, mixture systems of Sg1, Sg2 and F5 were investigated by TG-DTA test method. Results are shown in Fig. 6a–c. It can be seen that shapes of TG-DTA curves are very similar except the location of the endothermic peak.

Influence of Different Slag Components on Limestone Decomposition

Influence of Slag with Individual Component on Limestone Decomposition

All the temperature data of mixture systems of Limestone-Fe_2O_3 (SiO_2, $MnCO_3$ and pre-melting slags) are listed in Table 3. T_e, T_m and T_c of pure limestone are compared with various mixture systems according to the data in Table 3 and results are shown in Fig. 7a–d. It can be seen that both Fe_2O_3 and SiO_2 can reduce the decomposition temperature of $CaCO_3$ and with the increase of Fe_2O_3 or SiO_2 in

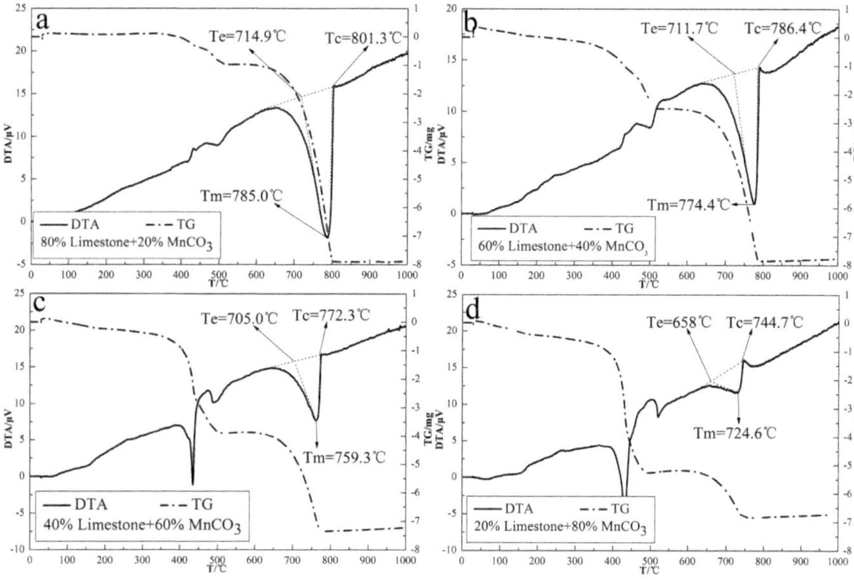

Fig. 5 TG and DTA curves of various systems with different Limestone-$MnCO_3$ mixtures: **a** M2 **b** M4 **c** M6 **d** M8

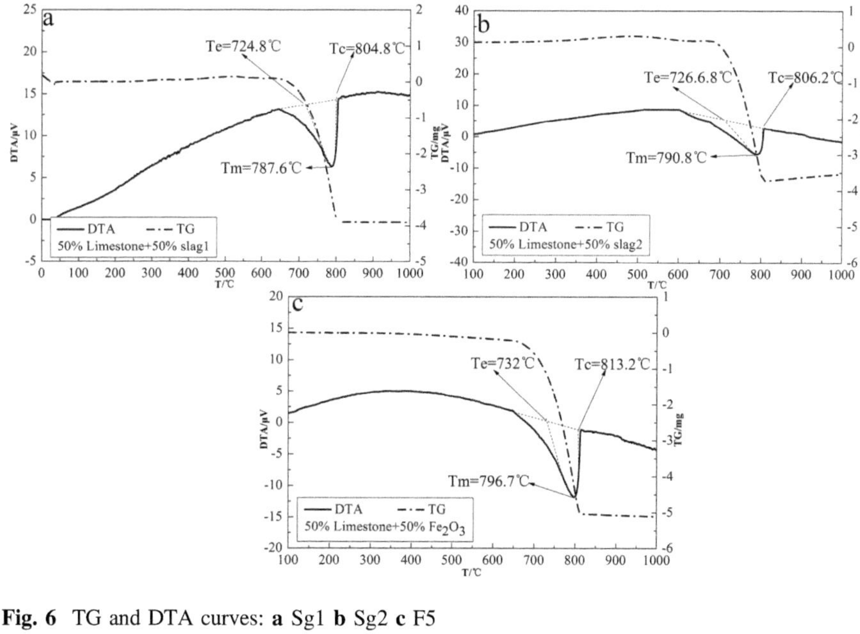

Fig. 6 TG and DTA curves: **a** Sg1 **b** Sg2 **c** F5

Table 3 T_e, T_m and T_c of different samples acquired by DTA extrapolation method

Number	Sample	T_e/°C	T_m/°C	T_c/°C
F2	80% limestone + 20% Fe_2O_3	738.1	813.4	826.9
F4	60% limestone + 40% Fe_2O_3	733.9	801.8	816.2
F5	50% limestone + 50% Fe_2O_3	731.1	796.7	812.3
F6	40% limestone + 60% Fe_2O_3	726.7	790.7	804.9
F8	20% limestone + 80% Fe_2O_3	700.9	762.2	782.1
S2	80% limestone + 20% SiO_2	729.2	810.8	826.5
S4	60% limestone + 40% SiO_2	728.1	803.4	817.6
S6	40% limestone + 60% SiO_2	714.7	784.0	799.8
S8	20% limestone + 80% SiO_2	678.3	753.7	773.3
M2	80% limestone + 20% $MnCO_3(N_2)$	714.9	785.0	801.3
M4	60% limestone + 40% $MnCO_3(N_2)$	711.7	774.4	786.4
M6	40% limestone + 60% $MnCO_3(N_2)$	705.0	759.3	772.3
M8	20% limestone + 80% $MnCO_3(N_2)$	658.8	724.6	744.7
Sg1	50% limestone + 50% slag1	724.8	787.6	804.8
Sg2	50% limestone + 50% slag2	726.8	790.8	806.2

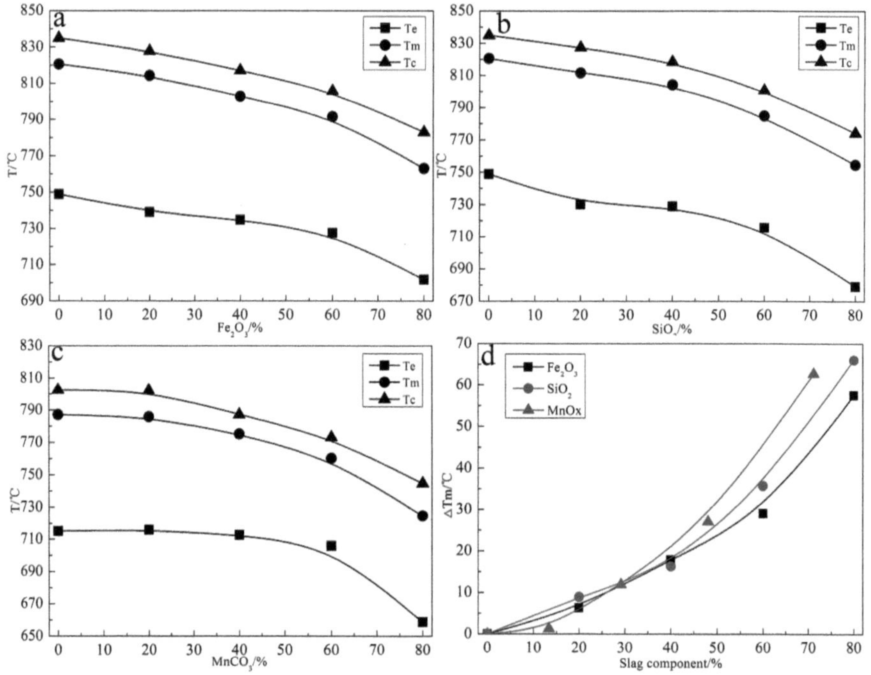

Fig. 7 Influence of slag with individual component on decomposition temperature of $CaCO_3$: **a** Fe_2O_3 **b** SiO_2 **c** $MnCO_3$ **d** individual slag component

mixture systems of Limestone-Fe_2O_3 or Limestone-SiO_2, decomposition temperature of $CaCO_3$ will become lower.

It can be seen from Fig. 7c that decomposition temperature of $CaCO_3$ will also continuously decrease with the increase of $MnCO_3$ in mixture systems of Limestone-$MnCO_3$. T_e, T_m and T_c of limestone in nitrogen atmosphere are 714.1, 786.3 and 801.6 °C respectively. As to M8, T_e, T_m and T_c have decreased to 658, 724.6 and 744.7 °C. It can be concluded that three slags with individual component, all can reduce the decomposition temperature of $CaCO_3$ and the decomposition of $CaCO_3$ will take place more easily when limestone is surrounded by large amount of converter slag which mostly consists of Fe_xO, SiO_2 and MnO_x except CaO.

$MnCO_3$ will decompose into MnO_x in the heating process. In order to compare the influence of different slag components on the decomposition temperature of limestone, $MnCO_3$ is substituted by MnO_x according to the weight loss ratio of $MnCO_3$ in the heating process and the results are shown in Fig. 7d. It can be seen that the more slag components in mixture systems, the bigger ΔTm. However there are also significant differences for Fe_2O_3, SiO_2 and MnO_x. Generally speaking, SiO_2 has a better effect on reducing the decomposition temperature of $CaCO_3$ than Fe_2O_3. MnO_x has the weakest effect on the decomposition temperature of $CaCO_3$ when the addition amount is less than 30% in mixture system, but it changes into the strongest one with the addition amount exceeding 30%.

Influence of Slag with Multiple Components on Limestone Decomposition

Influence of slag with multiple components on decomposition temperature of limestone is presented in Fig. 8. The result suggests that both slag1 and slag2 have

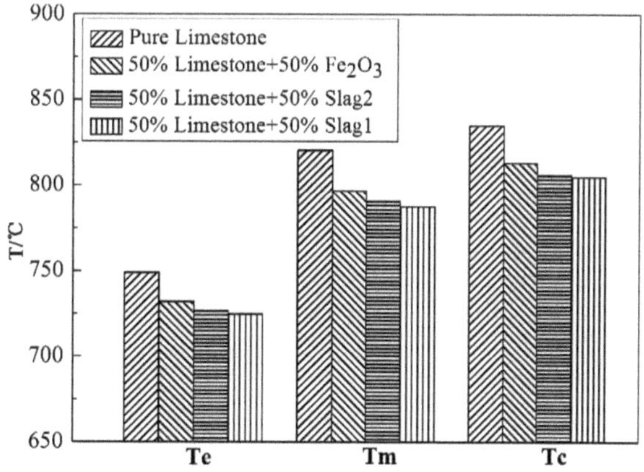

Fig. 8 Influence of slag with multiple components on decomposition temperature of limestone

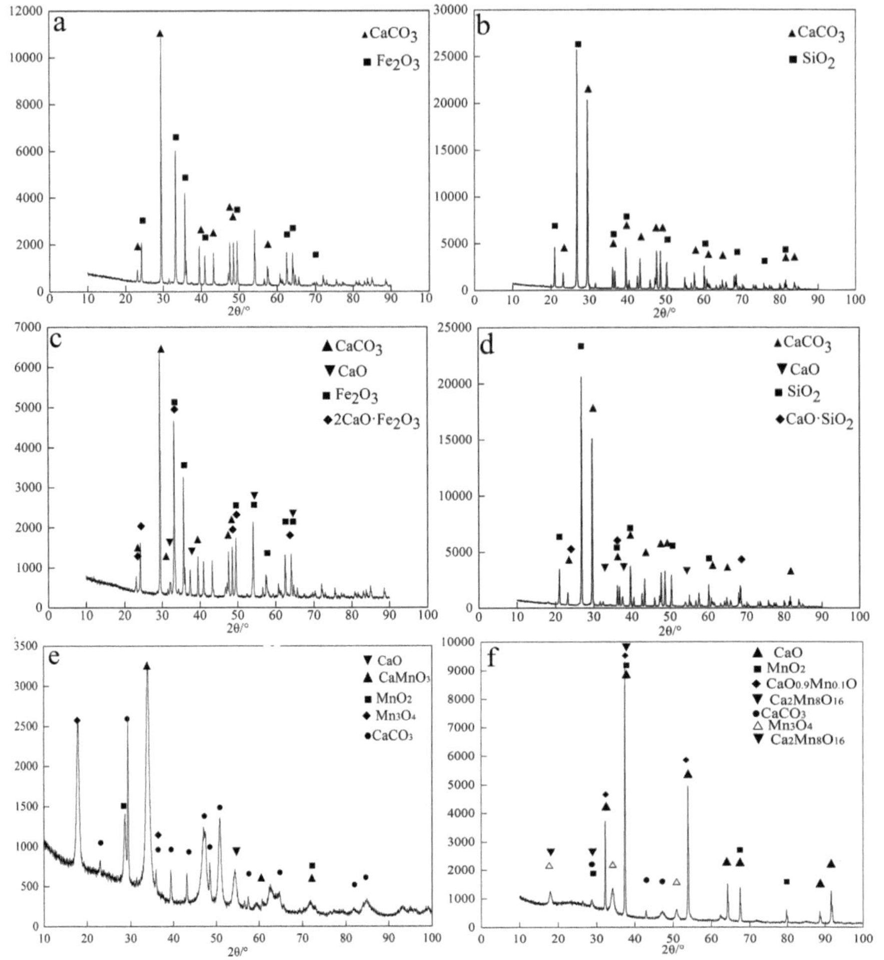

Fig. 9 XRD results of roasted samples: **a** F4 heated to 720 °C **b** S4 heated to 720 °C **c** F4 heated to 760 °C **d** S4 heated to 760 °C **e** M4 heated to 720 °C **f** M4 heated to 760 °C

a greater ability of reducing the decomposition temperature of $CaCO_3$ than Fe_2O_3. As is shown in Fig. 7d, small amount of MnO_x has no obvious influence on reducing the decomposition temperature of $CaCO_3$. However, slag1, which contains small amount MnO_x has a stronger influence than slag2. This indicates that slag with multiple components have further reducing effects than slag with individual component. Therefore, limestone can decompose much more quickly in converter slag environment.

Results of Roasting Mixture Systems

As is shown in Table 2, T_e of sample F4, S4 and M4 are 733.9, 728.1 and 711.7 °C respectively. For sample F4 and S4, they are just beginning to decompose when heated to 720 °C. Figure 9a, b show X-ray diffraction patterns of F4 and S4 after heated to 720 °C. XRD results do not show characteristic diffraction peaks of CaO, which indicates that $CaCO_3$ decomposes rarely. Figure 9c, d show X-ray diffraction patterns of sample F4 and S4 after heated to 760 °C. It can be seen that characteristic diffraction peaks of $2CaO \cdot Fe_2O_3$ and $CaO \cdot SiO_2$ are observed respectively which suggest that part of $CaCO_3$ have decomposed into CaO in 760 °C and CaO have reacted with Fe_2O_3 or SiO_2 to generate solid solution or oxide simultaneously.

X-ray diffraction patterns of M4 heated to 720 °C is presented in Fig. 9e. No characteristic diffraction peaks of CaO are found, but there are very obvious characteristic diffraction peaks of $CaMnO_3$ whose intensity is the strongest in all diffraction peaks. These suggest that part of $CaCO_3$ have decomposed into CaO, but all CaO will react with MnO_x to form $CaMnO_3$ simultaneously. X-ray diffraction patterns of M4 after heated to 760 °C is shown in Fig. 9f. With the heating temperature increasing to 760 °C, $CaCO_3$ continue to break down and characteristic diffraction peaks of CaO are found in Fig. 9f, but most of the characteristic diffraction peaks are compounds of $CaMnO_x$. XRD results of three mixture systems indicate that Fe_2O_3, SiO_2 and MnO_x can react with CaO which got from $CaCO_3$ decomposition in high-temperature environment.

Conclusions

Converter slags that mostly consist of Fe_xO, SiO_2 and MnO_x except CaO at the early stage of converter blowing can promote the decomposition of limestone. All individual slag components including Fe_xO, SiO_2 and MnO_x have the effect of reducing the decomposition temperature of limestone and multiple slag components acting together have more significant effect than individual ones. The mechanism of converter slag reducing limestone decomposition temperatures can be attributed to the reactions between slag components and CaO got from $CaCO_3$ decomposition, which reduces the activity of decomposition product.

Acknowledgements The financial support of the National Key Technology Support Program (2012BAC27B02) is gratefully acknowledged.

References

1. C.J. Lewis, Acceleration of the Dissolution of Lime in the BOF Progress. American Patent, 3,574,575, 13 Mar 1971
2. H. Li, Y. Qu, A Method of Using Limestone Instead of Lime for Slagging in BOF Steelmaking. CN Patent, CN101,525,678, 20 Oct 2009
3. H. Li, Y. Qu, Discussion on limestone addition instead of lime for energy saving and emission reduction in BOF steelmaking. China Metall. **20**, 59–63 (2010)
4. D.X. Dong, J. Feng, W. Nian, Experimental study on using limestone instead of lime for slagging in 60t converter. China Metall. **23**, 58–62 (2013)
5. T.F. Deng, P. Nortier, M. Ek, S.C. Du, Limestone dissolution in converter slag at 1873 K. Metall. Mater. Trans. B. **44**, 98–105 (2013)
6. N. Maruoka, A. Ishikawa, H. Shibata et al., Dissolution rate of various limes into steelmaking slag. High Temp. Mater. Process. (London) **32**, 15–24 (2013)
7. H. Li, L.F. Guo, Y.Q. Li et al., Industrial experiments of using limestone instead of lime for slagging during LD-steelmaking Process. Adv. Mater. Res. **233–235**, 2644–2647 (2011)
8. W.C. Song, H. Li, L.F. Guo et al., Analysis on energy-saving and CO_2 emissions reduction in BOF steelmaking by substituting limestone for lime to slag. Int. Conf. Mater. Renewable Energy Environ. **1**, 991–994 (2011)
9. H. Li, J. Feng, Y.Q. Li et al., Thermodynamic analysis of limestone decomposition and CO_2 oxidation effect in the early stage of BOF steelmaking. J. Uni. Sci. Technol. Beijing, S **1**, 83–87 (2011)
10. J. Feng, H. Li, C.X. Li et al., Research on using limestone for slagging during BOF steelmaking process. Adv. Mater. Res. **690–693**, 237–245 (2013)
11. Z.Q. Li, H. Li, L.F. Guo et al., Preliminary research on the behavior of limestone charged into converter in the process of making slag. Steelmaking **27**, 33–36 (2011)

Influence of Hot Charge on Blast Furance Performance for Iron Making

Huiqing Tang, Xun Zhang, Lifeng Zhang and Zhilong Zhao

Abstract A two-dimensional gas-solid model has been developed for examination of the blast furnace (BF) performance under hot charge operation. Simulation conditions were obtained from the actual operation conditions of the BF with inner volume of 2500 m^3 and molten iron output of 6250 ton hot metal (THM) per day. The simulation results indicate that hot charge operation could change the thermal state of the BF in the region above cohesive zone (CZ) and has little influence on the thermal state of CZ zone and the region below CZ; it could intensify both the indirect reduction of the ore and the gasification of the coke in the region above the CZ. The optimum temperature for hot charge operation is 750 K, under which, a coke rate reduction of 25 kg/THM could be obtained.

Keywords Hot charge · Blast furnace · Iron making

Introduction

The blast furnace (BF) is the chemical reactor to produce molten iron from iron ore. Although various alternative non-blast iron making technologies have been developed, BF ironmaking currently is the leading ironmaking process with many features as high productivity and high thermal efficiency. It is noted that more than 90% of pig iron is produced by BF. As BF iron making is an energy intensive

H. Tang (✉) · X. Zhang
State Key Laboratory of Advance Metallurgy, University of Science and Technology Beijing, 30 Xueyuan Rd., Beijing 100083, China
e-mail: hqtang@ustb.edu.cn

L. Zhang
School of Metallurgical and Ecological Engineering, University of Science and Technology Beijing, 30 Xueyuan Rd., Beijing 100083, China

Z. Zhao
MCC Capital Engineering and Research Incorporation Limited, Beijing Economic-Technological Development Area, Beijing 100176, China

J.-Y. Hwang et al. (eds.), *8th International Symposium on High-Temperature Metallurgical Processing*, The Minerals, Metals & Materials Series,
DOI 10.1007/978-3-319-51340-9_70

process, successive efforts have been devoted by researchers to utilize the waste-heat of the solid burden including the sinter and the coke. One proposed method is hot charge operation, in which, it is envisaged that the sensible heat of hot sinter and hot coke could be utilized to improve the BF operation for coke rate reduction. In the present research, the influence of hot charge operation on BF performance for iron making was studied by numerical simulation.

Description of the Modeled Blast Furnace

The model geometry in this study was based on the industrial BF of 2500 m^3 with an output of 6250 t hot metal per day (THM/day). Its operation condition is shown in Table 1. The properties of the lump ore were: a total iron content of 55 mass%, an average size of 2.0 mm, an average porosity of 0.34, and bulk density of 1750 kg/m^3. The properties of the lump coke were: a fixed carbon of 90 mass%, an average size of 5.4 mm, an average porosity of 0.54 and bulk density of 500 kg/m^3.

Mathematical Model

Theory

Four phases exist in the BF iron making process and they are solid (lumpy ore and coke), gas (CO, CO_2, O_2 and N_2), liquid (molten iron and molten slag) and powder (PC and dust). In the present research, the complicated multi-phase and chemically

Table 1 Operation conditions for the BF

Parameter	Value
Productivity (THM/day)	6250
Blast temperature (K)	1250
Blast rate (m^3/THM)	4800
Oxygen enrichment (volume %)	4.0
Top absolute pressure (Pa)	2.8×10^5
[a]PC injection rate (kg/THM)	180
Ore rate (kg/THM)	1680
Coke rate (kg/THM)	335
Batch weight of ore (ton)	76
Batch weight of coke (ton)	15

[a]*PC* pulverized coal

reacting flow in the BF was treated as a gas-solid counter flow and the liquid phase and the powder phase were treated in a simple method.

The momentum transfer between gas and solid was described using Ergun Equation, The heat transfer was described using modified Ranz Equation, and the involved reactions were given in Table 2.

The liquid phase was generated in the CZ, which is region where solid temperature is from 1373 to 1673 K. After generation, the liquid droplets undergo a temperature increase during they percolate the coke bed to the hearth. The heat is mainly supplied from the gas phase; therefore, this heat is treated as an energy source of gas phase. The PC is as assumed to be completely gasified in the raceway. The gaseous product and the left blast form the inlet conditions for the gas phase.

Governing Equations for Gas and Solid Phases

The general governing equation for the gas phase is Eq. (1).

$$div(\rho_g \vec{U}_g \Phi_g) = div(\Gamma_g grad\Phi_g) + S_g \quad (1)$$

Where, ρ_g is gas density, kg/m^3; and $\vec{U}_g$ is superficial gas velocity vector, m/s; Φ_g is the general gas phase variable which is 1.0 for gas mass continuity equation, the component velocities for the gas momentum equations, the enthalpy of the gas energy equation and mass fractions of CO_2, CO and O_2 for the gas species equations. Γ_g is the diffusion coefficient, which represents effective dynamic viscosity in

Table 2 Chemical reactions involved in the simulation

No.	Reactions	Reaction rate kmol/(m^3 s)	Description
1	$3Fe_2O_3(s) + CO(g) = 2Fe_3O_4(s) + CO_2(g)$	R1	Indirect reduction, three interface shrinking core model [1, 2]
2	$Fe_3O_4(s) + CO(g) = 3FeO(s) + CO_2(g)$	R2	
3	$FeO(s) + CO(g) = Fe(s) + CO_2(g)$	R3	
4	$C(s) + CO_2(g) = 2CO(g)$	R4	Carbon solution loss reaction, field model [3]
5	$C(s) + 1/2O_2(g) = CO(g)$	R5	Carbon combustion, field model [3]
6	Fe(s) = Fe(l)	R6	Solid Components Melting, enthalpy method [4, 5]
7	FeO(s) = FeO(l)	R7	
8	Gangue(s) = Slag(l)	R8	
9	FeO(l) + C(s) = Fe(l) + CO	R9	Direct reduction, R9 = R6

the momentum equations, effective thermal conductivity in the energy equations and effective diffusion coefficient in the species equations. S_g is the source term and is determined by Ergun equation, Ranz equation and reaction rates of the reactions in Table 2.

The general governing equation for the solid phase is Eq. (2).

$$div(\rho_s \vec{V}_s \Phi_s) = div(\Gamma_s grad \Phi_s) + S_s \quad (2)$$

Where, ρ_s is solid bulk density, kg/m^3; and $\vec{V}_s$ is physical solid velocity vector, m/s; Φ_s is the general solid phase variable which is 1.0 for solid mass continuity equation, the component velocities for the solid momentum equations, the enthalpy for the solid energy equation and mass fractions of Fe_2O_3, Fe_3O_4, FeO, Fe and gangue for the solid species equations. Γ_s is diffusion coefficient which represents effective dynamic viscosity in the momentum equations, effective thermal conductivity in the energy equation and 0.0 in the species equations. S_s is the source term and is determined by solid gravity force, Ergun equation, modified Ranz equation and reaction rates of the reactions in Table 2.

Solution Method

The above two sets of equations form the basis for the present simulation. The CFD package FLUENT (V12.0) was employed. The solution method is described in Ref. [6].

Results and Discussion

In the present research, only the hot charge temperature was investigated. Seven cases were studied corresponding to the normal operation case under 300 K and six hot charge operation cases under charge temperature from 450 to 1200 K with an interval of 150 K.

Influence on Solid Temperature Distribution

Figure 1a–g show the solid temperature distributions for the simulated cases. In these figures, the CZ is indicated by dash lines.

In Fig. 1a, it could be seen that the solid burden gets heated as it descends. The solid temperatures were generally higher near the furnace center because more coke particles were charged at the furnace center, allowing more gas flow of higher

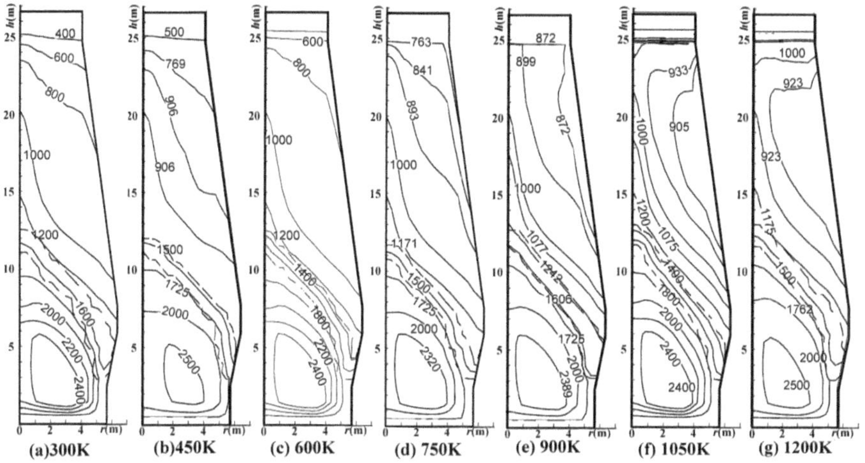

Fig. 1 Distributions of solid temperature for the studied cases. (unit: K)

temperature and thus higher solid temperature in the center. In Fig. 1b–e, the solid temperature profile was similar to that of conventional operation. The difference was that the contour line of a specific temperature, e.g. line of 800 K moved to a higher location. However, in Fig. 1f–g, the solid burden was cooled first. This phenomenon is more obvious near the wall than near the center line. Accordingly the heat transfer from gas to solid was reversed.

From Fig. 1a–g, it could be observed that the CZ positions and shapes were nearly the same for all cases; moreover, no significant change of solid temperature distributions below CZ could be observed in Fig. 1a–g. It was considered that hot charge operation has little influence on the thermal state in the CZ region and below.

Figure 2 shows the effect of hot charge temperature on BF top gas temperature. It could be observed that the top gas temperature linearly increases with a linear increase of hot charge temperature. Combining the analysis of Figs. 1 and 2, it was considered that a large portion of the sensible heat of the solid burden charge was used for increasing the sensible heat of BF top gas. Therefore the influence of hot charge operation on BF in-furnace thermal state was significantly weakened.

These findings indicate that the heat of the burden is prohibited to effectively transfer to the BF lower part by the ascending BF gas.

Influence on Solid Reduction Behavior

Figure 3a–g show ore reduction degree distributions for the studied cases. Figure 3a shows the distribution of the ore reduction degree from the BF top to bottom under BF normal operation. It was observed that the reduction degree was

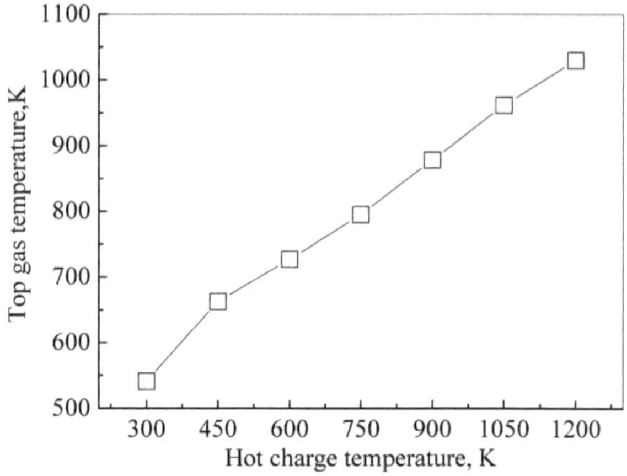

Fig. 2 Variation of top gas temperature with the increase of hot charge temperature

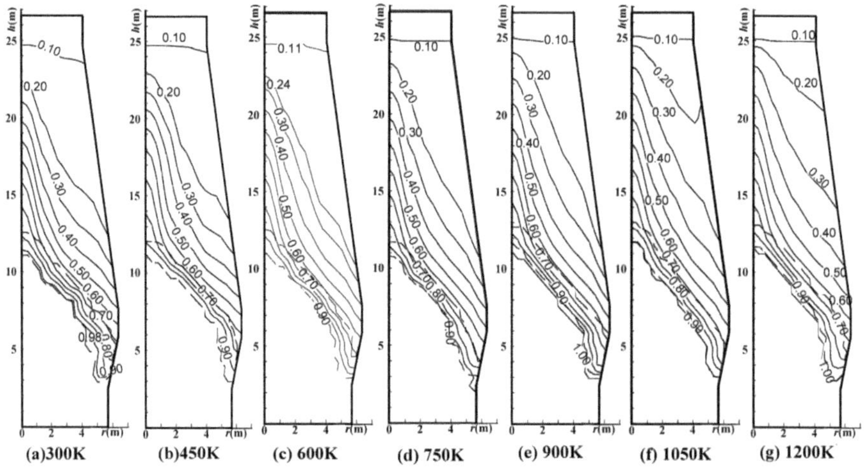

Fig. 3 Distributions of ore reduction degree for the studied cases (unit: -)

generally high at the furnace center due to the high availability of reducing gas in this region. In the lumpy zone, the reduction degree was gradually increased from 0.0 to around 0.6 upon reaching the CZ. In the CZ, the reduction degree was increased rapidly, where an early increase was observed near the furnace center due to the preferential gas flow and high coke/ore ratio. After passing the CZ, the ore was fully reduced to liquid iron, meaning 100 percent reduction degree. By increasing hot charge temperature, the locations of contour lines with reduction degree of 0.2, 0.3 and 0.4 moved to a higher BF position (Fig. 3b–g). The reduction

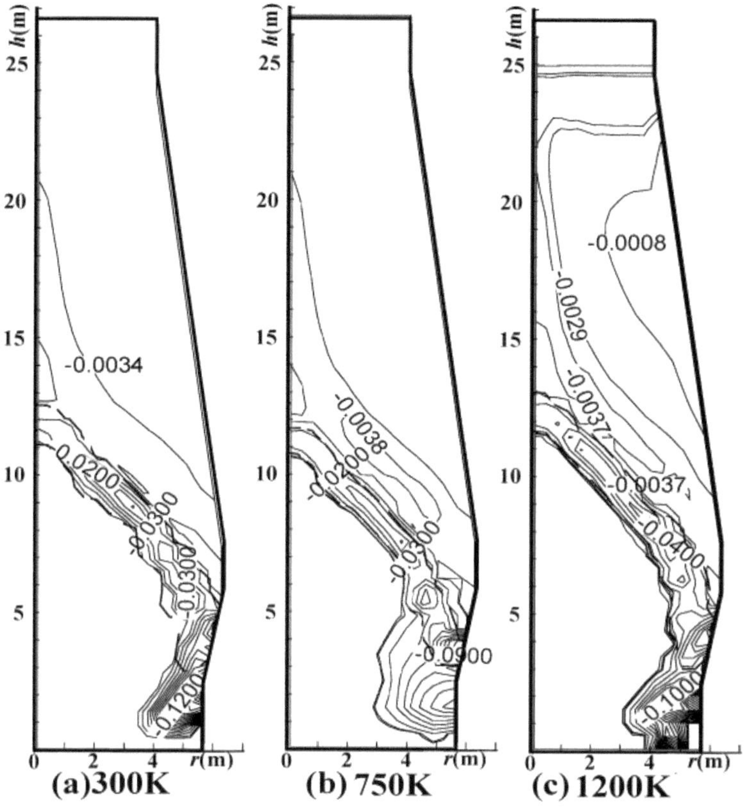

Fig. 4 Distributions of coke conversion rate for the selected cases (unit: kg/(m^3 s))

degree distribution could reflect the reduction rates of reactions (R1-R3); therefore, it is indicated that the hot charge operation could improve the indirect reduction of the ore in the BF upper part.

An examination of coke gasification behavior of the selected cases was shown in Fig. 4a–c, comparing the coke conversion rate distributions in Fig. 4a, b indicated that the coke gasification did not show significant changes when hot charge temperature increased from 300 to 750 K. Comparing the coke conversion rate distributions in Fig. 4a, c indicated that coke gasification was significantly increased in the BF upper part when hot charge temperature increased from 300 to 1200 K. These results show that the hot charge operation could simultaneously intensify the ore reduction and the coke gasification. This becomes more obvious as hot charge temperature reaches more than 750 K.

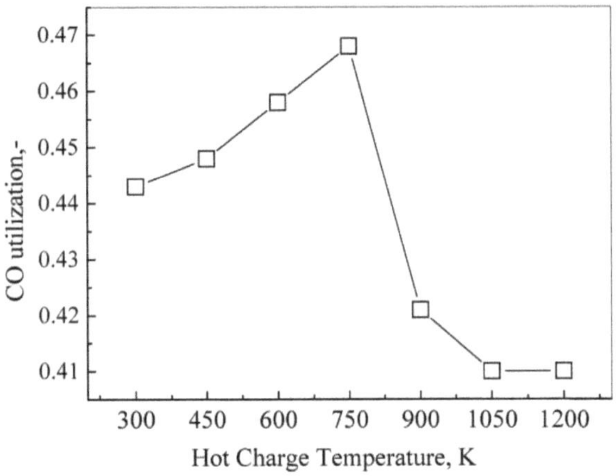

Fig. 5 Variation of CO utilization of BF top gas with the increase of hot charge temperature

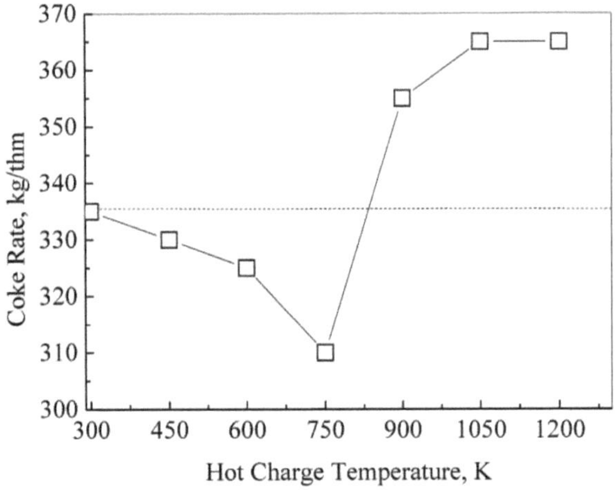

Fig. 6 Variation of coke rate with the increase of hot charge temperature

Potential of Coke Rate Reduction

CO utilization of BF top gas is an important performance indicator, and could provide guidance for process optimization. Effect of hot charge temperature on CO utilization was shown in Fig. 5. From full-scale mass balances of carbon and oxygen and together with CO utilization of BF top gas, the corresponding coke rate could be calculated and the results were presented in Fig. 6. It can be seen that the optimum temperature was 750 K. Under 750 K, a coke rate reduction of 25 kg/THM could be reached.

Conclusions

Influence of hot charge operation on BF performance was investigated in the present research. The following conclusions could be drawn:

(1) Hot charge operation could change the thermal state of the BF in the region above CZ. It has little influence on the thermal state of CZ zone and the region below CZ.
(2) Hot charge operation could simultaneously intensify both the indirect reduction of the ore and the solution loss reaction of the coke in the region above CZ.
(3) The optimum temperature for hot charge operation is 750 K, under which a coke rate reduction of 25 kg/THM could be obtained.

Acknowledgements The authors wish to thank State Key Laboratory of Advanced Metallurgy of University of Science and Technology Beijing (USTB) for its financial supports.

References

1. H. Tang, Z. Guo, K. Kitagawa, Simulation study on performance of z-path moving-fluidized bed for gaseous reduction of iron ore fines. ISIJ Int. **52**(7), 1241–1249 (2012)
2. S. Natsui, T. Kikuchi, R.O. Suzuki, Numerical analysis of carbon monoxide-hydrogen gas reduction of iron ore in a packed bed by an euler-lagrange approach. Metall. Mater. Trans. **45B**, 2396–2431 (2014)
3. P.R. Austin, H. Nogami, J. Yagi, A mathematical model for blast furnace reaction analysis based four fluid model. ISIJ Int. **37**(8), 748–755 (1997)
4. Y.S. Shen et al., Three-dimensional modelling of in-furnace coal/coke combustion in a blast furnace. Fuel **90**, 728–738 (2011)
5. Y. Dutil et al., A review on phase-change materials: mathematical modeling and simulations. Renew. Sustain. Energy Rev. **15**, 112–130 (2011)
6. H. Hu, S.A. Argyropoulos, Mathematical modelling of solidification and melting: a review. Model. Simulation Mater. Sci. Eng. **4**, 371–396 (1996)

Investigation and Application of Evaluation System of Stock Surface Gas Flow Distribution in Blast Furnace

Wenxuan Xu, Shusen Cheng and Guolei Zhao

Abstract According to cross beam temperature measuring data of stock surface gas flow in blast furnace production process, using zoning method to deal with stock surface gas flow cross beam temperature measuring data, the evaluation system of stock surface gas flow distribution in radial direction and circumferential direction was established. Central index, platform index and edge index were proposed to evaluate stock surface gas flow distribution in radial direction, global uniform index and local uniform index were proposed to evaluate stock surface gas flow distribution in circumferential direction. Combined with actual production of 5500 m^3 blast furnace and typical morphology of stock surface gas flow distribution, the strength of stock surface gas flow distribution was divided into five grades, to provide guidance for the optimization of blast furnace operation.

Keywords Cross beam temperature measuring · Gas flow distribution · Evaluation system

Introduction

With the increase of blast furnace volume, the stability and uniformity of the operation of large blast furnace is crucial, controlling the reasonable distribution of gas flow is the basis of stable and high production of blast furnace. In the operation of large blast furnace, it is particularly concerned with the coordinated development of the center and the edge of the two streams. The dissection, simulation and experiment of blast furnace show that, the formation of gas flow in the stock surface of blast furnace mainly consist of three steps: the first step is gas flow produce in the raceway, then spread upward; the second step is gas flow spread in the coke slit in

W. Xu (✉) · S. Cheng · G. Zhao
School of Metallurgical and Ecological Engineering,
University of Science and Technology Beijing, Beijing 100083, China
e-mail: xuwenxuan_ustb@126.com

J.-Y. Hwang et al. (eds.), *8th International Symposium on High-Temperature Metallurgical Processing*, The Minerals, Metals & Materials Series,
DOI 10.1007/978-3-319-51340-9_71

the cohesive zone; and finally the gas flow redistribute in the lumpy zone. The distribution of gas flow in stock surface reflects the reduction rate of iron ore in the blast furnace, and is one of the most important parameters of the status of blast furnace, which can judge the distribution of gas flow in stock surface accurately and provide guidance to the burden distribution in a blast furnace.

In the traditional blast furnace operation, the gas flow distribution of the blast furnace is judged mainly by experience. In recent years, many scholars have proposed many methods to estimate the gas flow detection model, but most of them are based on the single factor, such as cross beam temperature measuring in throat, the composition of gas flow, infrared image, burden distribution model to estimate the distribution pattern of gas flow [1–4]. Although the cross beam temperature measuring can detect the temperature of stock surface, but in some cases, cross beam temperature measuring cannot reflect the stock surface temperature and the gas flow distribution. The composition of gas flow can only reflect the comprehensive parameters of shaft, only can be used as a reference in the gas flow distribution. Infrared image can reflect the distribution of gas flow, but the image has many interference factors and infrared thermometer measurement range of the problem, the detection of gas flow distribution is not very reliable. Therefore, the limitations of a single discriminant factor, the estimate of the gas flow will have a large deviation. On the other hand, some researchers have modeled the blast furnace gas flow with fluid dynamics method, but the modeling is very complicated [5–9].

In this study, based on the production practice of 5500 m^3 blast furnace, combined the cross beam temperature measuring data, blast furnace production index and gas flow distribution, the evaluation system of gas flow distribution in radial and circumferential direction of blast furnace throat was established, and the strength of gas flow was divided into different grades, to provide the guidance for the burden distribution and furnace condition judgement in blast furnace.

Mathematical Model

Mathematical Model of Gas Flow Distribution in Radial Direction

According to the cross beam temperature measuring data and the installation position of temperature measuring point in 5500 m^3 blast furnace, as shown in Fig. 1. The four cross beam were installed in four directions of the northeast, the southeast, the southwest and the northwest, respectively, the total number of thermocouples in the four cross beam is 28, and the distance of temperature measuring point to the furnace wall in each direction is shown in Table 1, the seventh thermocouple in southeast direction is located in the center of the blast furnace throat. According to the location of temperature measuring point in each direction, take the middle of two adjacent thermocouple locations as the boundary,

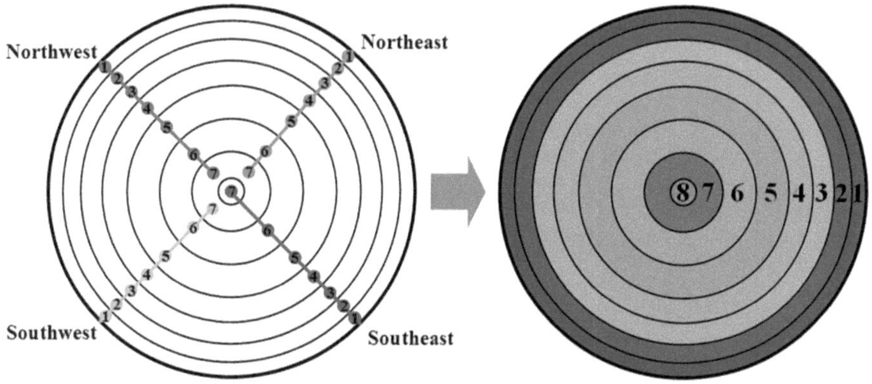

Fig. 1 Schematical drawing of the temperature measure point and radial zone division

Table 1 The distance between thermocouple to furnace wall of the cross thermocouple in each direction

Thermocouple number	NE 1	NE 2	NE 3	NE 4	NE 5	NE 6	NE 7	SE 7
Distance to furnace wall/m	0.2	0.7	1.3	2.0	2.8	4.0	4.8	5.6

Notes NE represents northeast, *SE* represents southeast

furnace throat was divided into seven ring regions and a circular area, it is assumed that the eight regions from the center to the wall of the furnace followed by radial region 1 (RR1) to radial region 8 (RR8), as shown in Fig. 1.

The characteristic temperature (CT) of each radial region is the average temperature of all temperature measuring points in each radial region, according to the CT of the eight radial regions, the radial gas index was proposed, calculation equation is shown as follows:

$$\mathrm{RGI}_i = \mathrm{CT}_i / \sum_{i=1}^{8} \mathrm{CT} \tag{1}$$

where, the CT_i is the characteristic temperature of radial region i, calculation equation is shown as follow:

$$\mathrm{CT}_i = \sum_{i=1}^{N} \mathrm{T}_i / N \tag{2}$$

where, the T_i is the temperature of temperature measuring points i, N is the number of temperature measuring points locate in the radial region i.

Due to there are many radial gas indexes, which are difficult to comprehensively judge the distribution of gas flow in the stock level of blast furnace throat, so according to the stock surface profile and the actual production of blast furnace, the blast furnace throat was divided into edge region, platform region and center region, accordingly, center index (CI), platform index (PI), edge index (EI) were proposed, the calculation formula are shown as follows:

$$CI = RGI_6 + RGI_7 + RGI_8 \tag{3}$$

$$PI = RGI_2 + RGI_3 + RGI_4 + RGI_5 \tag{4}$$

$$EI = RGI_1 \tag{5}$$

Mathematical Model of Gas Flow Distribution in Circumferential Direction

Due to the 5500 m^3 blast furnace top equipped with parallel hopper bell-less top system, burden materials discharged from two hoppers, then entered to the center throat, but burden materials not move along the center line of center throat, leading to the moving distance of burden materials in rotate chute is different, then burden materials fell on the stock surface is not circle in circumferential direction, which have influence on the uniform distribution of gas flow in circumferential direction. Therefore, it is necessary to investigate the uniform distribution of gas flow in circumferential direction. According to the installation position of the four cross beam temperature measuring in blast furnace throat, whose four cross beams are located in the direction of northeast, southeast, southwest, northwest, respectively, so the blast furnace throat is divided into four fan-shape regions, as shown in Fig. 2. The four cross beams in each direction are located in the middle of fan-shape region, it is assumed that the northeast fan-shape region, southeast fan-shape region, southwest fan-shape region and northwest fan-shape region are redefined by

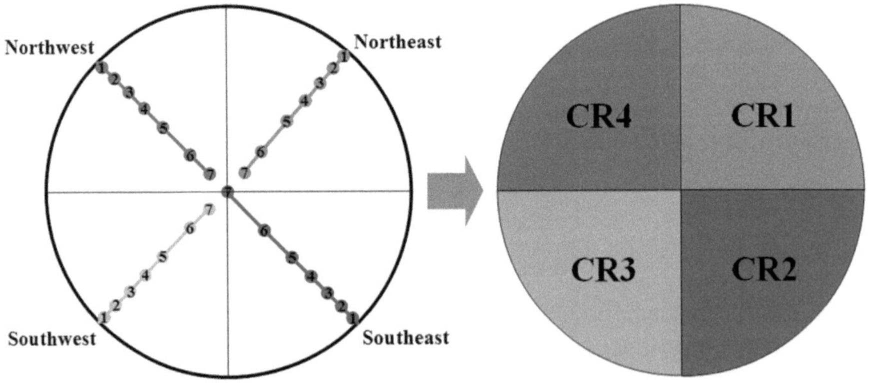

Fig. 2 Schematical drawing of the temperature measure point and circumferential zone division

circumferential region 1 (CR1), circumferential region 2 (CR2), circumferential region 3 (CR3) and circumferential region 4 (CR4), as show in Fig. 2.

The characteristic temperature of four circumferential region is the average temperature of temperature measuring points in one of the four circumferential regions, the calculation formula is shown as follows:

$$\mathrm{CT}_i = \frac{1}{6}\sum_{j=1}^{6} \mathrm{T}_{ij} \tag{6}$$

where, T_{ij} is the temperature of temperature measuring point j of circumferential region i, according to the characteristic temperature (CT) of circumferential region i, global uniform index (GUI) and local uniform index (LUI) were proposed, GUI is the global uniform distribution of gas flow in blast furnace throat, LUI is the local uniform distribution of gas flow in special circumferential zone in blast furnace throat, the GUI and LUI calculation formula are shown as follows:

$$\mathrm{GUI} = 1/\sqrt{\frac{1}{4}\sum_{i=1}^{4}\left(\frac{\mathrm{CT}_i - \overline{\mathrm{CT}_i}}{\overline{\mathrm{CT}_i}}\right)^2} \tag{7}$$

$$\mathrm{LUI}_i = 1/\left(\frac{\mathrm{CT}_i}{\sum_{i=1}^{4}\mathrm{CT}_i} - \frac{1}{4}\right) \tag{8}$$

where, CT_i is the characteristic temperature of circumferential zone i, $\overline{\mathrm{CT}_i}$ is the average characteristic temperature of four circumferential zone, it is could be concluded that with the increase of GUI or LUI, the distribution of gas flow in circumferential direction is more uniform.

Model Application

Evaluation of Gas Flow Distribution in Radial Direction

Figure 3 shows the comparison of the variation of gas temperature and radial gas index in radial region, in the RR1–RR6, the change trend of the gas temperature and radial gas index is consistent, however, for RR7 and RR8. The change trend of the gas temperature and radial gas index is contrary. Therefore, in the actual production of blast furnace, not just through the gas temperature changes to evaluate the strength of gas flow, it is necessary to evaluate the strength of gas flow in different radial region by radial gas index.

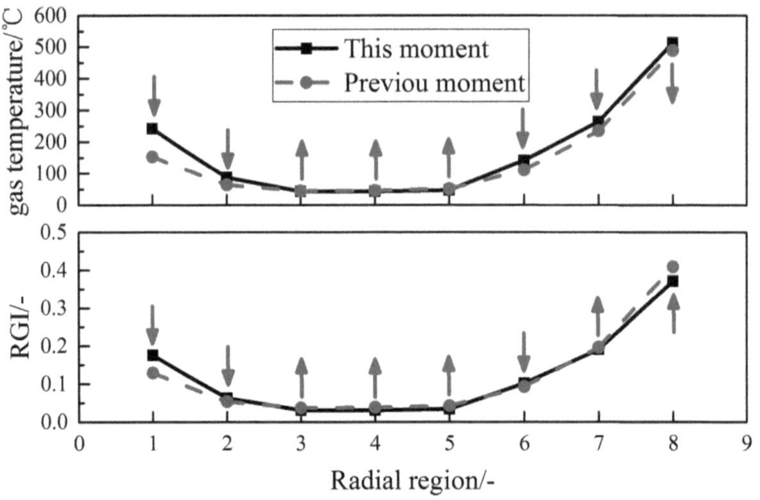

Fig. 3 Comparison of the variation of gas temperature and radial gas index in radial region

Figure 4 shows the variation of radial gas index in different region of a blast furnace during 2009–2015 years, the red solid line represents the average radial gas index (R_O) in this period, based on the average radial gas index (R_O), the radial gas index was divided into five grades, which is extremely strong state, strong state, normal state, weak state and extremely weak state, it is assumed that $R_1 = 1.3R_O$, $R_2 = 1.1R_O, R_3 = 0.9R_O, R_4 = 0.7R_O$. When the radial gas index is bigger than R_1 and smaller than 1, it is indicated that the strength of gas flow in radial region in extremely strong state. When the radial gas index is bigger than R_2 and smaller than R_1, it is indicated that the strength of gas flow in radial region in strong state. When the radial gas index is bigger than R_3 and smaller than R_2, it is indicated that the strength of gas flow in radial region in normal state. When the radial gas index is bigger than R_4 and smaller than R_3, it is indicated that the strength of gas flow in radial region in weak state. When the radial gas index is bigger than 0 and smaller than R_4, it is indicated that the strength of gas flow in radial region in extremely weak state.

According to the R_O, R_1, R_2, R_3 and R_4, the proportion of different strength of gas flow in different region during 2009–2015 years was shown in Fig. 5. In the RR1, the proportion of weak state is highest, the proportion of strong state is lowest. In the RR2, the proportion of extremely strong state is highest, the proportion of extremely weak state is lowest. In the RR3, the proportion of extremely strong state is highest, the proportion of extremely weak state is lowest. In the RR4, the proportion of weak state is highest, the proportion of strong state is lowest. In the RR5, the proportion of extremely weak state is highest, the proportion of strong state is lowest. In the RR6, the proportion of extremely weak state is highest, the proportion of extremely strong state is lowest. In the RR7, the proportion of extremely strong state is highest, the proportion of extremely weak state is lowest. In the RR8, the proportion of weak state is highest, the proportion of extremely strong state is lowest.

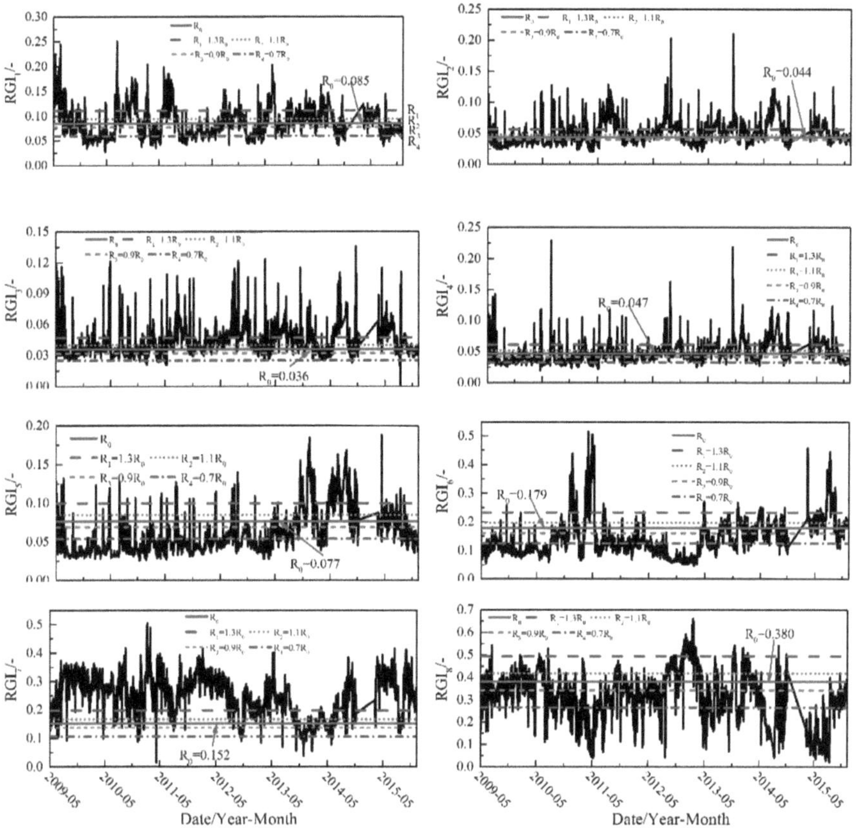

Fig. 4 The variation of radial gas index in different radial region during 2009–2015 years

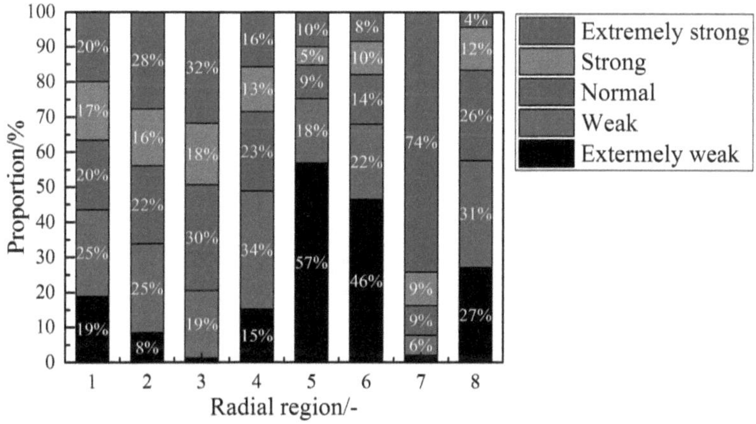

Fig. 5 The proportion of different strength of gas flow in different area during 2009–2015 years

Figure 6 is the morphology of radial gas index in blast furnace throat, RR1 is the blast furnace wall, and RR8 is the center of blast furnace throat, it is could be obtained that the maximum radial gas index is located in the center of throat. However, owing to charging coke into the center of blast furnace, therefore, radial gas index in RR6 is bigger than RR7. In addition, the radial gas index in RR1 is bigger than RR2, which reflects the idea of the operation of large blast furnace: Development Center, control edge and suppression platform.

Figure 7 shows the variation of center index, platform index and edge index during 2009–2015 years, center index has slight decline since 2009, but the change trend of platform index is contrary to center index, edge index almost keep constant

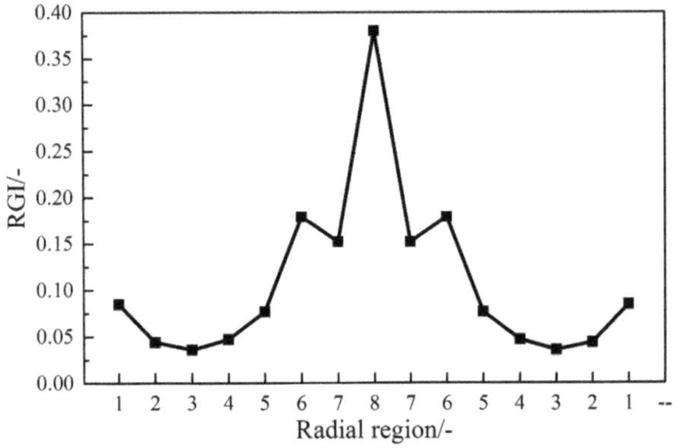

Fig. 6 Morphology of radial gas index in blast furnace throat

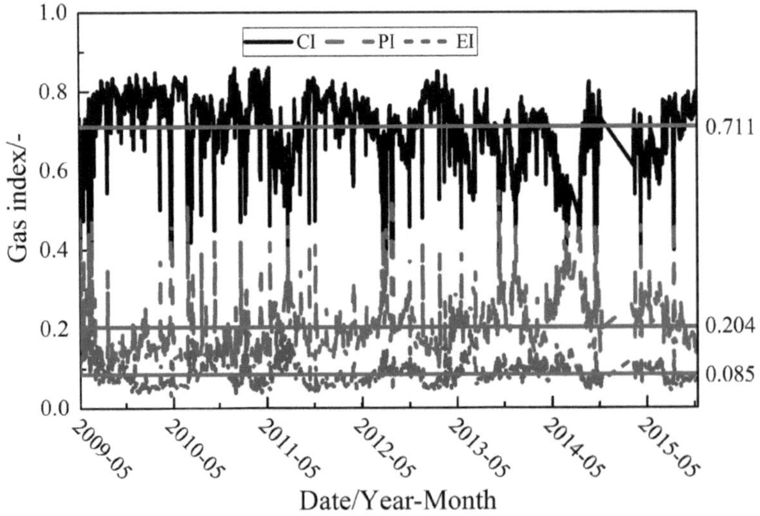

Fig. 7 Variation of center index, platform index and edge index during 2009–2015 years

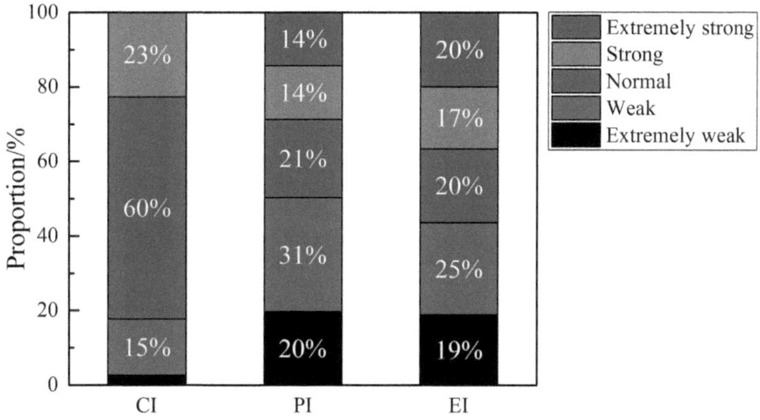

Fig. 8 The proportion of gas intensity in different area during 2009–2015 years

fluctuation, in addition, the sum of center index, platform index and edge index is 1. The proportion of center index, platform index and edge index in different region during 2009–2015 years were statistically analyzed, as shown in Fig. 8, for the central index, the proportion of normal level is highest, reaching 60%; for the platform index and edge index, the proportion of weak level is highest, reaching 31 and 25%, respectively.

Evaluation of Gas Flow Distribution in Circumferential Direction

Figure 9 shows the variation of global uniform index in 2014 year, from January to February, global uniform index remained stable. During March–April, GUI began to increase gradually, on behalf of the gas flow the circumferential distribution becomes more uniform. After 4 months, gas flow in the circumferential direction, the uniform level remained stable; the average uniform index is 755 in 2014 year. It is assumed that when GUI >10, the distribution of gas flow in circumferential direction is uniform; when 0< GUI <10, the distribution of gas flow in circumferential direction is not uniform, the distribution of gas flow in circumferential direction reached to uniform level accounted for 16.72% in 2014 year.

Figure 10 shows the variation of local uniform index in 2014 year, it is assumed that when LUI is bigger than 40 or smaller than −40, the distribution of gas flow in circumferential region *i* is uniform. When LUI is bigger than −40 and smaller than 40, the distribution of gas flow in circumferential region *i* is not uniform; the average local uniform index in CR1 is 4.31, the proportion of uniform state of CR1 in 2014 is 60.18%. The average local uniform index in CR2 is −1351.70, the proportion of uniform state of CR2 in 2014 is 36.78%. The average local uniform

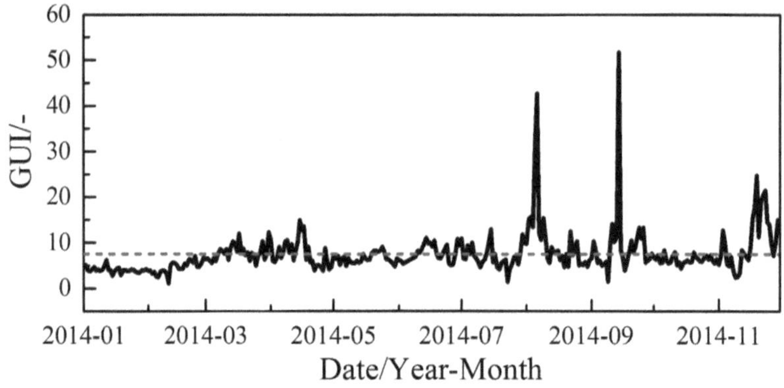

Fig. 9 Variation of global uniform index in 2014 year

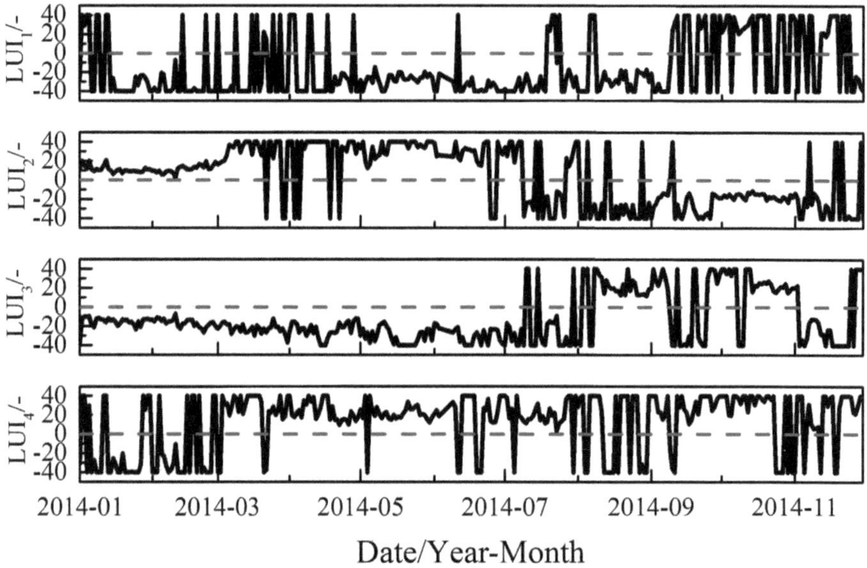

Fig. 10 Variation of local uniform index in 2014 year

index in CR3 is −16.03, the proportion of uniform state of CR3 in 2014 is 26.14%. The average local uniform index in CR4 is −10.16, the proportion of uniform state of CR4 in 2014 is 49.54%. Therefore, the uniform distribution of gas flow in CR1 is controlled better.

Conclusion

Combined with the actual production of 5500 m^3 blast furnace, the evaluation system of gas flow distribution base on cross beam temperature measuring data was established. In the radial direction, the center index, platform index and edge index were proposed, the distribution of gas flow in the radial direction of blast furnace throat was systematically evaluated. In the circumferential direction, the global uniform index and local uniform index were proposed. Combined with expert opinion and production data, the different strength grade of the gas flow distribution is given. Finally, the model was applied to 5500 m^3 blast furnace in China, and provide guidance for optimizing burden distribution and furnace condition judgement of blast furnace.

References

1. Y.H. Xu et al., Measurement of the temperature profile based on infrared image processing and its application in blast furnace. Control Eng. China **6**(34), 354–356 (2005)
2. X.F. Dong et al., Gas-powder flow in blast furnace with different shapes of cohesive zone. Appl. Math. Model. **11**(30), 1293–1309 (2006)
3. B. Abhay, H. Saxen, Classification of blast furnace probe temperatures using neural networks. Steel Res. **6**(66), 231–236 (1995)
4. M. Nikus, H. Saxén, Prediction of a blast furnace burden distribution variable. ISIJ Int. **9**(36), 1142–1150 (1996)
5. Z.S. Alex et al., Sophisticated multi-phase multi-flow modeling of the blast furnace. ISIJ Int. **40**(4), 322–331 (2000)
6. G.Y. Pan et al., Analysis gas and stock temperature field in blast furnace by upwind finite element method. Iron Steel **1**(36), 14–18 (2001)
7. R. Timmer et al., Radial gas distribution in the blast furnace top. Steel Res. **2**(68), 47–53 (1997)
8. X.F. Dong et al., Modelling of multiphase flow in ironmaking blast furnace. Ind. Eng. Chem. Res. **1**(48), 214–226 (2009)
9. N.K. Nath, Simulation of gas flow in blast furnace for different burden distribution and cohesive zone shape. Mater. Manuf. Processes **5**(17), 671–681 (2002)

Investigation of the Carbothermic Reduction of Chromium-Containing Vanadium Extraction Residue

Pan Gu, Jiang Diao, Wen-feng Tan, Bing Xie, Wang Zhou and Zhen Zhang

Abstract In order to recover the chromium resource in chromium-containing vanadium extraction residue, a promising method is to prepare ferrochromium with the residue. The synthetic residue has been reduced under 1300, 1350 and 1400 °C at the reduction time of 1, 2 and 3 h respectively. The reduced samples have been smelted and separated by an induction furnace. The contents of chromium and iron in metallic phase have been determined by an inductively-coupled plasma spectrometer. The results indicate that chromium iron carbide $(Cr, Fe)_3C_7$ is the intermediate phase and titanium oxides have been reduced during the reduction process. While the sample has been reduced under 1400 °C for 3 h, metallic phase becomes the dominate phase and reduction reactions almost complete. Moreover, the chromium content in the metal phase reaches 10.19%.

Keywords Vanadium-Chromium slag · Residue · Carbothermic reduction

Introduction

Vanadium titano-magnetite ore is a resource which riches in Fe, Ti and contains the accompanying element of V, Cr, Ga, Sc, Co, Ni and Pt et al. It is a significant resource to produce V-containing products. In order to take full advantage of the rare elements in vanadium titano-magnetite ore, a lot of approaches had been proposed by previous investigations. It is a most common method to extract vanadium from V-containing slag obtained by oxidizing the V-containing hot metal. Vanadium slag usually contains 5–20% V_2O_3 and is a direct resource for vanadium extraction. For the sake of extracting the vanadium in vanadium slag,

P. Gu · J. Diao (✉) · W. Tan · B. Xie · W. Zhou · Z. Zhang
College of Materials Science and Engineering, Chongqing University,
Chongqing 400044, China
e-mail: diaojiang@163.com

J.-Y. Hwang et al. (eds.), *8th International Symposium on High-Temperature Metallurgical Processing*, The Minerals, Metals & Materials Series,
DOI 10.1007/978-3-319-51340-9_72

several methods had been put forward. Usually, the vanadium slag is pretreated with sodium or calcium salts by roasting in oxidizing atmosphere before leaching. The additives can be sodium salts [1, 2] or calcium oxide [3, 4], and the leaching agents can be water, acid solution or aqueous alkali. It brings about sodium roasting water leaching method, calcium roasting acid leaching method and roasting alkali leaching method. In addition, NaOH molten salt roasting has been reported for vanadium extraction [5]. Nevertheless, there exists some technical problems to be addressed in utilizing the Hongge ore which located in Panzhihua-Xichang region. The chromium content in Hongge ore is higher than Vanadium content [6, 7]. If this ore was utilized in vanadium extraction and steelmaking, vanadium slag with high chromium content (V-Cr slag) would be produced due to similar characters of vanadium and chromium elements. For extracting vanadium and chromium in V-Cr slag, lots of techniques have been proposed. For example, the method of submolten slats which can accomplish the extraction rate exceeding 88% of vanadium and chromium by NaOH–$NaNO_3$ binary melt medium or NaOH/KOH sub-molten salt medium [6, 8]. The subsequent work is to separate vanadium and chromium in the solution containing the both element. Therefore, extraction method, chemical precipitation method, ionic exchange method, crystallization process and electrochemical process have been proposed to achieve it [9].

A lot of investigations based on Hongge ore have been done recently. Zhang [10] and Diao et al. [11] investigated the existing forms of vanadium and chromium, and found that both the vanadium and chromium concentrate in spinel. Moreover, they studied the distribution of vanadium and chromium in V-Cr spinel and concluded that chromium concentrated in the center of spinels and vanadium primarily distributed in the outer shell of spinels. Zhao et al. [7] found that chromium spinels are more thermodynamically stable than vanadium spinels. Furthermore, when V-Cr slags are roasted with sodium salts at 700–850 °C, the sodium vanadates previously produced will surround the Cr(III) ions. So the oxidation of Cr will be hindered as well. Based on the previous investigation, Li et al. [12]. proposed asynchronous extraction of vanadium and chromium by stepwise sodium roasting-water leaching to accomplish the separation of vanadium and chromium in V-Cr slag. The overall extractions of vanadium and chromium are 98.9 and 96.6%, respectively. However, as reported by Li et al. after the first stage of asynchronous extraction, 87.9% of vanadium have been extracted and the vanadium content in leaching residue is rather low. So the carbothermic reduction of these residue has been proposed in the present work to avoid the multifarious operation in the process of dealing with leaching liquor containing chromium obtained from second stage of asynchronous extraction. In the present work, a novel technique of the preparation of ferrochrome have been proposed based on the chromium-containing vanadium extraction residue.

Experimental Procedure

The composition of synthetic slag refers to the composition of first stage leaching residue of Li et al. and the objective composition is shown in Table 1. Analytical regent (AR) pure CaO, Fe, SiO_2, Fe_2O_3, $MnCO_3$, Cr_2O_3, Al_2O_3, MgO, TiO_2 and Na_2CO_3 were used as raw materials to prepare the slags. The addition of MnO and Na_2O were substituted by $MnCO_3$ and Na_2CO_3, respectively. FeO was added by Fe and Fe_2O_3 on the mole ratio of 1:1. All oxides had been dried with drum wind drying oven for 24 h before weighting except CaO which had been roasted at 1000 °C for 12 h before weighting. All the agents were weighted and mixed together in mortar.

A $MoSi_2$ furnace (Fig. 1) with six U-shape heating elements was employed for heating and melting. Apt-10 pct Rh/Pt thermocouple protected by alumina tube was

Table 1 Objective composition of synthetic slag, wt%

Cr_2O_3	SiO_2	TiO_2	Al_2O_3	MgO	CaO	Fe_2O_3	Na_2O	FeO	MnO
16.80	15.10	3.74	1.40	3.51	1.65	38.22	1.76	15.92	1.91

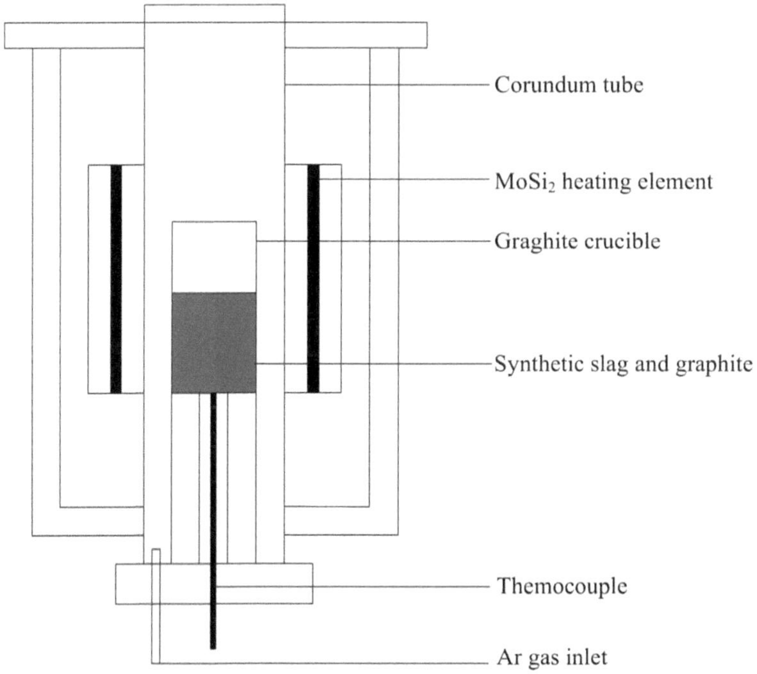

Fig. 1 Experimental furnace

placed mid-beneath the bottom of crucible and was held constant within ±5 K (±5 °C) to monitor the experimental temperature. The corundum crucible was used to prepare the slag. In a general run, the furnace was heated to 1400 °C, and then placed the crucible along with 150 g slags in the uniform-temperature zone of the furnace. Argon atmosphere (1.0 L/min) was used for avoiding the oxidation of Fe^{2+}. The slags were heated for 1 h at 1400 °C since the temperature back to 1400 °C. Subsequently, the crucible was taken out and naturally cooled. Fraction synthetic slags were crushed to 320 meshes for X Ray Diffraction (XRD), 200 meshes for X-ray fluorescence (XRF) and the following carbothermic reduction in the same furnace respectively.

Graphite (≤ 30 μm) was used as reductant for its high purity. 45 g synthetic slags mixed with 8 g graphite were reduced under 1300, 1350 and 1400 °C for 1, 2 and 3 h respectively. The crucibles used were made of graphite and argon atmosphere (1.0 L/min) was used during the whole reduction process. Partially reduced sample was crushed to 200 meshes and subjected to XRD. In order to obtain ferrochrome, the fraction reduced samples were smelting separately by an induction furnace for 20 min in corundum crucible, and the peak power used was 8 kW. For the sake of chromium detection, the metallic phase obtained was crushed and an inductively-coupled plasma spectrometer (iCPA 6300 Duo, Thermo Fisher) was employed. The solutions for Inductively Coupled Plasma (ICP) were prepared according to the following process. 0.100 g samples were weighted by electronic scales, and then were made into solutions of 1000 ml by alkali fusion.

Results and Discussion

Synthetic Slag Composition and Phase

Figure 2 and Table 2 show the XRD result and the composition of the synthetic slag. It concludes that the phase of synthetic slag is primarily spinel phase. Compared with Table 1, other elements have less discrepancy with addition except chromium. The burning loss of chromium is about 50%. It attributes to the evaporation of CrO_3 forming for the oxidation of Cr_2O_3. The same phenomenon had been perceived by Jadid [13] and Warshaw et al. [14].

Phase Transformation During Reduction Process

Figure 3 shows the morphology of reduction products in the different reduction process. Visible metallic phase can be seen from it.

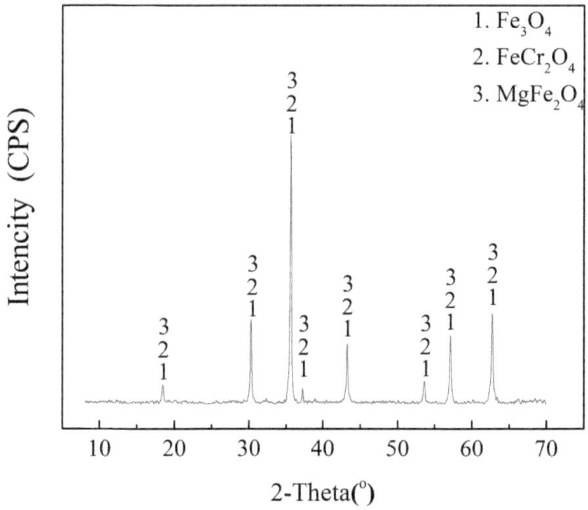

Fig. 2 XRD results of synthetic slag

Table 2 XRF results of the synthetic slag, wt%

Fe	O	Na	Si	Cr	Ti	Mg	Al	Mn	Ca
39.65	27.78	10.98	7.34	6.29	2.62	2.13	1.75	1.46	1.38

Figure 4 shows the XRD patterns under different reduction time and reduction temperature. It is clear that the peak intensities of graphite decreases with the increasing reduction time and reduction temperature, which indicates the graphite have been consumed as reductant. With reduction processing, the intensities of reduction products (Cr, Fe, FeC_3, Cr_3C_7) increase. Tomoyuki [15] had investigated the reduction mechanism of chromium oxide and concluded that Cr_3C_2, Cr_7C_3, $Cr_{23}C_6$ and Cr were formed in turn during the carbothemic reduction of Cr_2O_3 and the carbide plays a role of intermediate products. In the present work, the peak of reduction products above overlaps and appears the strongest peak in the XRD patterns. The intensities of FeC_3 and Cr_3C_7 have deceasing tendency with reduction time increasing at 42.9° (2θ). It indicates that the carbide is the immediate phase during reduction process and it is in accordance with Tomoyuki. However, the peaks of Cr_3C_2 and $Cr_{23}C_6$ have not been detected out. In addition to Fe and Cr, the reduction of titanium oxides has been detected from Fig. 4. Ti_nO_{2n-1} and TiO have been found in the Fig. 4. Moreover, the intensities of Ti_nO_{2n-1} have a decreasing tendency and the intensities of TiO has an increasing tendency at 1350 °C. It indicates that Ti_nO_{2n-1} and TiO formed in turn in the reduction process which is in agreement with Kucukkaragoz [16].

Fig. 3 Morphology of reduction products in the different reduction process

The Chromium Content Variation Under Different Reduction Condition

The iron content and chromium content of separated metallic phase have been detected by ICP. One melt separated sample have been displayed in Fig. 5. The bright region is metal and the other is slag.

Figure 6 shows the iron content and chromium content under different reduction condition. The iron content of metallic phase has no obvious changes under different reduction conditions. While the reduction time is 1, 2 and 3 h respectively, the chromium contents are 5.15, 5.94 and 6.38 ppm at 1300 °C, 2.17, 3.37 and 7.53 ppm at 1350 °C, 6.52 pm, 6.85 pm and 7.38 pm at 1400 °C respectively. The chromium content displays an obvious increment with increasing reduction time on the same reduction temperature. However, the chromium content at 1350 °C have an anomaly variation. Possible reason for this is that metallic phase has been oxidized in the process of grinding for preparing the ICP solution. Nevertheless, the tendency of chromium content various can be obtained through Fig. 6. It is a

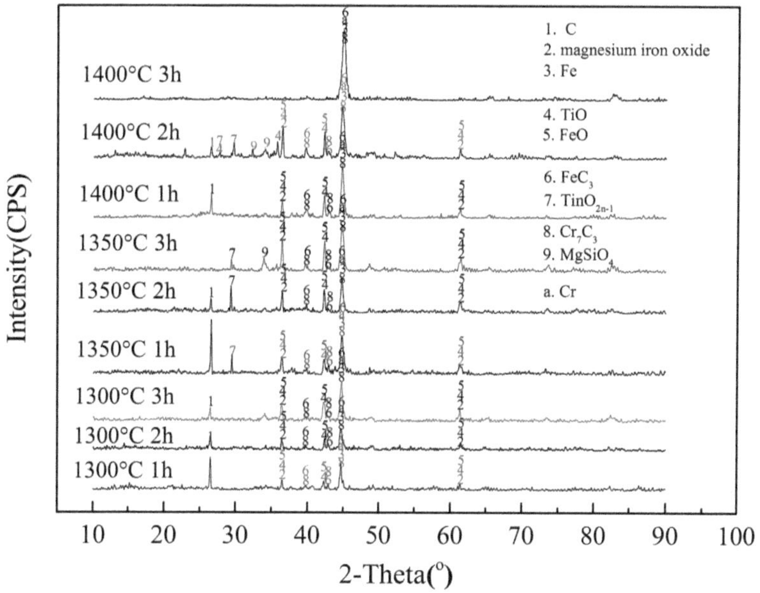

Fig. 4 XRD results under different reduction conditions

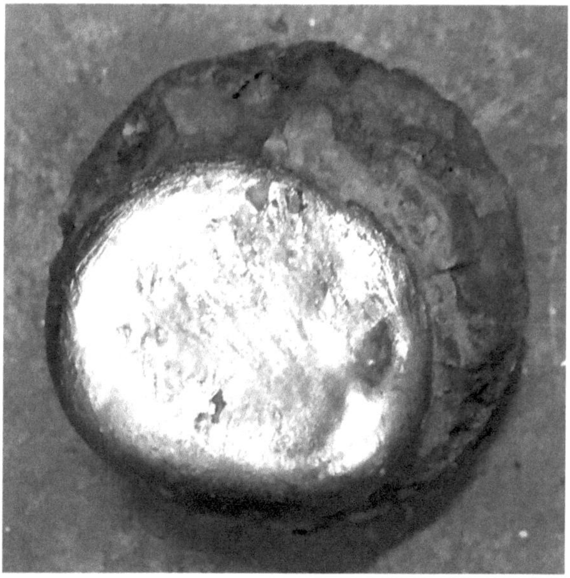

Fig. 5 Melt separation sample

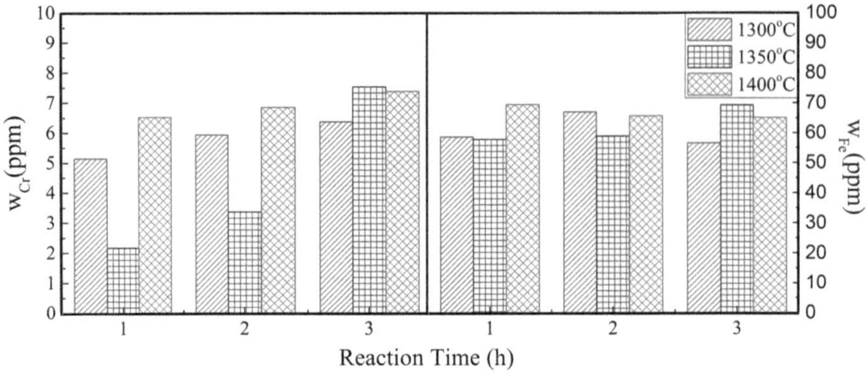

Fig. 6 Iron content and chromium content in melt separation samples

noticeable phenomenon that the chromium content is low. This is due to the chromium content in the synthetic slag is low because of the evaporation of chromium during preparation process. As previously mentioned, the existence of C comes from carbide which is reduction products during reduction process, and reason for the existence of O is oxidization during grinding. The content of these element will be detected in the future work. If only considering Fe and Cr in metallic phase, the highest chromium content reaches 10.19%.

Conclusions

(1) The chromium containing slag can be reduced by carbothemic reduction. FeC_3 and Cr_7C_3 are generating during reduction process. These carbides play the role of immediate products. However, there are no obvious peaks of Cr_3C_2 and $Cr_{23}C_6$ in the XRD patterns.
(2) Except for Fe and Cr, titanium oxides have been reduced during reduction process. Ti_nO_{2n-1} and TiO come into being in turn with increasing reduction time.
(3) The chromium content in the metallic phase increases with increasing reduction time at the same reduction temperature. If only considering Fe and Cr in metallic phase, the highest chromium content reaches 10.19%. This is due to the chromium content in the synthetic slag is low for the evaporation of chromium during preparation process.

Acknowledgements This work was financially supported by National Natural Science Foundation of China (grant no. 51604047, no. 51474041 and no. 21105134), National Basic Research Program of China (no. 2013CB632604) and National Key Technology R&D Program (no. 2015BAB19B02).

References

1. M.R. Tavakoli, D.B. Dreisinger, Separation of vanadium from iron by solvent extraction using acidic and neutral organophosporus extractants. Hydrometallurgy **141**, 17–23 (2014)
2. C.P.J. Van Vuuren, P.P. Stander, The oxidation of FeV_2O_4 by oxygen in a sodium carbonate mixture. Miner. Eng. **14**, 803–808 (2001)
3. H.Y. Li, N. Wang B. Xie. *Vanadium Extraction from High Calcium-Content Vanadium Slag by Calcification Roasting*, (Wiley, 2012), pp. 621–627
4. Z. Yang, H.Y. Li, X.C. Yin et al., Leaching kinetics of calcification roasted vanadium slag with high CaO content by sulfuric acid. Int. J. Miner. Process. **133**, 105–111 (2014)
5. D. Chen, L. Zhao, Y. Liu et al., A novel process for recovery of iron, titanium, and vanadium from titanomagnetite concentrates: NaOH molten salt roasting and water leaching processes. J. Hazard. Mater. **244**, 588–595 (2013)
6. H. Liu, D.U. Hao, D. Wang et al., Kinetics analysis of decomposition of vanadium slag by KOH sub-molten salt method. Trans. Nonferrous Metals Soc. China **23**, 1489–1500 (2013)
7. L. Zhao, L. Wang, T. Qi et al., A novel method to extract iron, titanium, vanadium, and chromium from high-chromium vanadium-bearing titanomagnetite concentrates. Hydrometallurgy **149**, 106–109 (2014)
8. Z. Wang, S. Zheng, S. Wang et al., Research and prospect on extraction of vanadium from vanadium slag by liquid oxidation technologies. Trans. Nonferrous Metals Soc. China **24**, 1273–1288 (2014)
9. A.G. Chmielewski, T.S. Urbański, W. Migdał, Separation technologies for metals recovery from industrial wastes. Hydrometallurgy **45**, 333–344 (1997)
10. X. Zhang, B. Xie, J. Diao et al., Nucleation and growth kinetics of spinel crystals in vanadium slag. Ironmaking Steelmaking **39**, 147–154 (2012)
11. J. Diao, B. Xie, Y. Wang et al., Mineralogical characterisation of vanadium slag under different treatment conditions. Ironmaking Steelmaking **36**, 476–480 (2009)
12. H.Y. Li, H.X. Fang, K. Wang et al., Asynchronous extraction of vanadium and chromium from vanadium slag by stepwise sodium roasting–water leaching. Hydrometallurgy **156**, 124–135 (2015)
13. A.M. Mirzayousef-Jadid, K. Schwerdtfeger, Redox equilibria of chromium in calcium silicate base melts. Metall. Mater. Trans. B **40**, 533–543 (2009)
14. I. Warshaw, M.L. Keith, Solid Solution and Chromium Oxide Loss in Part of the System MgO-Al2O3-Cr2O3-SiO2. J. Am. Ceram. Soc. **37**, 161–168 (1954)
15. T. Mori, J. Yang, M. Kuwabara, Mechanism of carbothermic reduction of chromium oxide. ISIJ Int. **47**, 1387–1393 (2007)
16. C.S. Kucukkaragoz, R.H. Eric, Solid state reduction of a natural ilmenite. Miner. Eng. **19**, 334–337 (2006)

Molecular Dynamics Study of the Structural Properties with Varying B_2O_3/SiO_2 Ratios in the System $CaO–SiO_2–B_2O_3$

Xiao-Ping Liang, Wei-Tong Du and Yu Wang

Abstract The structural properties for ternary molten $CaO–SiO_2–B_2O_3$ system with varying B_2O_3/SiO_2 ratios at a fixed CaO content were calculated by molecular dynamics (MD) simulation. The simulation results revealed that with the increasing of B_2O_3/SiO_2 mol ratio, the bond length of B-O was approximately constant at 1.35 Å, and the coordination number for B ($CN_{B\text{-}O}$) was observed to vary from 3.40 to 3.51, and the average angle of O-B-O increases from 105° to nearly 120°, implying that the geometric stability of structure is reducing sharply. Also Q^3 and Q^4 tetrahedral species (tetrahedron with three and four bridging oxygens respectively) decrease with the increasing of B_2O_3/SiO_2 mol ratio, and the proportion of (Q^3 + Q^4) decrease linearly from 0.6 to nearly 0, suggesting that B_2O_3 may behave as a network-breaking basic oxide in the $CaO–SiO_2–B_2O_3$ system.

Keywords $CaO–SiO_2–B_2O_3$ · Molecular dynamics · B_2O_3/sio$_2$ ratio · Structure

Introduction

Mold fluxes are critical in continuous casting process and play an important role to improve final products quality. Since the fluorides in mold flux can cause pollution to environment, more and more researchers realized the importance of studying the fluorine free mold fluxes during recent years [1, 2]. With regard to the substitutes for F^- in mold fluxes, an increasing attention has been paid for B_2O_3 to use it in continuous casting process since it is an effective fluxing agent to lower the melting point of mold fluxes [3, 4]. Consequently, it is of practical importance to acquire the structure's atomic-scale information in the $CaO–SiO_2–B_2O_3$ system, and to investigate how the structure changes with varying B_2O_3/SiO_2 ratio.

X.-P. Liang (✉) · W.-T. Du · Y. Wang
College of Materials Science and Engineering, Chongqing University, Chongqing 400044, China
e-mail: xpliang@cqu.edu.cn

J.-Y. Hwang et al. (eds.), *8th International Symposium on High-Temperature Metallurgical Processing*, The Minerals, Metals & Materials Series,
DOI 10.1007/978-3-319-51340-9_73

Therefore, in present work, an attempt has been made to explore the structural properties for ternary molten $CaO–SiO_2–B_2O_3$ system, with varying B_2O_3/SiO_2 ratios at a fixed CaO content by molecular dynamics (MD) simulation. The aim of this study is to reveal the structural features changes with the increasing of B_2O_3/SiO_2 mol ratio, and to clarify the microscopic behavior of B_2O_3 in this system.

Simulation Details

In the molecular dynamic simulations of $CaO–SiO_2–B_2O_3$ system, selecting a fitted potential function with suitable parameters is the first step for a correct simulation. The generally used potential function, Born-Mayer-Higgins (BMH), which has been successfully used to describe the interactions between ions of ferrosilicate melts and glasses [4–7] was applied to simulate the $CaO–SiO_2–B_2O_3$ system. The general expression for potential between two atoms i-j is obtained as follows:

$$U_{ij} = \frac{q_i q_j}{r_{ij}} + A_{ij}\exp(-B_{ij} \cdot r_{ij}) \tag{1}$$

where U_{ij} is the pair potential energy, q_i and q_j are the effective charges and r_{ij} is the distance between the two atoms i–j. A and B are the potential parameters and the determined values are listed in Table 1. The ionic charges follow the setup: +2 for Ca, +4 for Si, +3 for B and −2 for O.

As mentioned, the present simulation chooses the samples of the ternary system $CaO–SiO_2–B_2O_3$ with varying B_2O_3/SiO_2 mol ratio at a fixed mole ratio of CaO. The samples were divided into 10 groups with B_2O_3/SiO_2 mol ratio from 0.17 to 1.92 as shown in Table 2, also demonstrating the atomic numbers, cell size and density values adopted.

The numbers of different atoms were then determined according to the mole fractions, containing approximately 4000 atoms in total, and periodic boundary conditions were applied. The density of every sample was calculated by the empirical formula.

Table 1 Potential parameters used in this study

Atom i	Atom j	A_{ij} (gÅ^2/fs^2)	B_{ij} (1/Å)
Ca	Ca	5.27E−21	6.25
Ca	Si	4.27E−22	6.25
Ca	B	6.89E−23	6.25
Ca	O	1.15E−20	6.06
Si	Si	3.46E−23	6.25
Si	B	5.58E−24	6.25
Si	O	1.01E−21	6.06
B	B	9.01E−25	6.02
B	O	1.71E−22	6.06
O	O	2.39E−20	2.88

Table 2 The composition of molten flux, atomic numbers, cell size and density

Sample No.	Composition (mol %)			B_2O_3/ SiO_2 Mole ratio	Atomic number					Cell size Å	ρ g·cm^{-3}
	CaO	SiO_2	B_2O_3		Ca	Si	B	O	Total		
1	30.0	60.0	10.0	0.17	517	1034	345	3103	4999	40.0502	2.666
2	30.0	56.0	14.0	0.25	503	940	470	3087	5000	39.7658	2.668
3	30.0	52.0	18.0	0.35	490	850	588	3072	5000	39.4900	2.670
4	30.0	48.0	22.0	0.46	478	764	701	3057	5000	39.2241	2.672
5	30.0	44.0	26.0	0.59	466	683	807	3043	4999	38.9659	2.674
6	30.0	40.0	30.0	0.75	455	606	909	3030	5000	38.7219	2.676
7	30.0	36.0	34.0	0.94	444	533	1006	3018	5001	38.4855	2.678
8	30.0	32.0	38.0	1.19	434	462	1098	3006	5000	38.2573	2.679
9	30.0	28.0	42.0	1.50	424	395	1186	2994	4999	38.0316	2.681
10	30.0	24.0	46.0	1.92	414	331	1271	2983	4999	37.8133	2.683

$$\rho = \sum_{i=1}^{N} \frac{X_i M_i}{V_i} \tag{2}$$

where X_i is the molar fraction of the oxide i and M_i is the molar mass (g/mol). V_i gives the molar volume fraction of the oxide i (m^3/mol), and N is the number of atoms in molten mold flux.

As for molten mold flux, the molar volume fraction of each oxide is decided by both temperature and pressure as follows:

$$V_i(T,P) = \bar{V}_i + \left[\frac{\partial \bar{V}_i}{\partial T}\right]_P (T - 1673) + \left[\frac{\partial \bar{V}_i}{\partial P}\right]_T P \tag{3}$$

$\bar{V}_i$ represents the partial molar volume of the oxide i at the temperature of 1673 K and pressure of 0.1 Mpa.

Due to the restriction of the simulated calculation, the number of atoms is always far smaller than what in actual conditions. Therefore, periodic boundary conditions are applied on all faces of the model box to create an infinite system with random initial configurations. The calculation was carried out at Ewald summation method with an accuracy of 10^{-5} and a cutoff of 10 Å. Canonical ensemble, NVT (i.e. the volume and temperature of the samples are controlled), were performed for all simulations in this study.

Before the simulation process started, atoms Ca, Si, B and O were placed into the model box and equilibrated at 5000 K for 8000 time steps, the time for each step was 1 fs, to make up the initial configuration. Subsequently, the system was cooled down from 5000 to 1673 K in steps of 2000. Then, the temperature was fixed at 1673 K for another 10,000 time steps to achieve the final configuration. The partial radial distribution functions (RDF) were collected to perform the microscopic structure analysis during this period.

Results and Discussion

Radial Distribution Functions (RDF) and Coordination Numbers (CN)

The RDFs indicate the average number of atoms j that are located at a distance r from a target atomic species i atoms. It is defined as follows:

$$g_{\mathrm{ij}}\mathrm{r} = \frac{V}{N_i N_j}\sum_{k}^{N_i}\frac{n_{kj}(r-\frac{\Delta r}{2}, r+\frac{\Delta r}{2})}{4\pi r^2 \Delta r} \tag{4}$$

where N_i (N_j) is the total number of atomic species i (j) atoms included in the system, n_{kj} is the number of atomic species k atoms included in a spherical shell of thickness Δr located at a distance r, and V is the volume of the system.

The RDF shows the basic structural information such as the average bond lengths and average coordination numbers. The running CN is calculated through the corresponding RDF and the chosen cutoff is the position of the first valley of RDF.

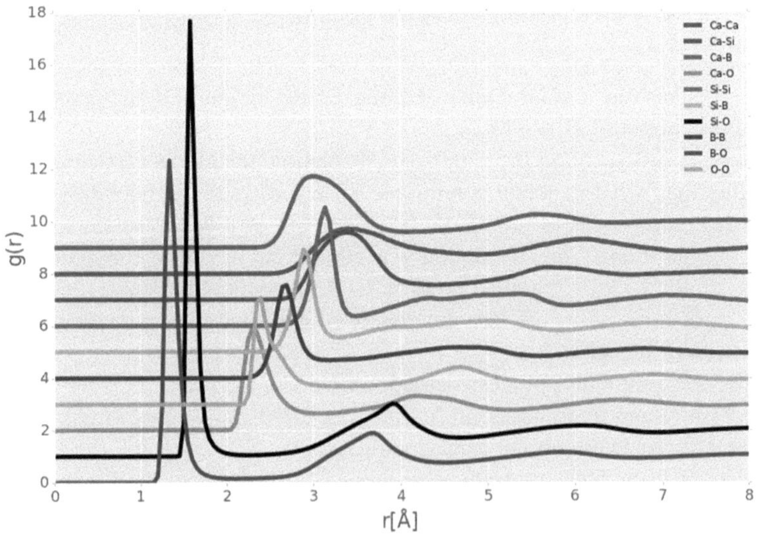

Fig. 1 Radial distribution functions (RDFs) of sample No. 8

$$N_{ij}(r) = \frac{4\pi N_j}{V} \int_0^r r^2 g_{ij}(r) dr \quad (5)$$

As an example, the curves of RDFs and running CNs for all atomic pairs of the sample No. 8 are presented in Fig. 1, and the content of this flux is close to the Boron containing fluorine-free mold flux. The first peaks of Si–O, B–O and Ca–O RDFs, which correspond to the most probable bond distances, are equal to 1.60, 1.35 and 2.325 Å, respectively. These are in agreement with the previous reports [3–6] that obtained through MD simulations using the SW potential function.

What is noteworthy is that as the most basic role in the mold flux, the first peak of Si-O shows a narrow and sharp shape, suggesting that the [SiO_4] tetrahedron is very stable and indicating that the overwhelming majority of Si atoms are 4-coordinated with oxygen. In turn, it can be concluded that the basic unit with tetrahedron geometry features has strong stability, i.e. [BO_4] tetrahedron also has strong stability in $CaO–SiO_2–B_2O_3$ mold flux system.

Coordination Numbers functions also help us to understand the nature of the structure of the melt. For the boron element, the vertical coordinates of the Coordination Numbers functions has a relatively flat platform at about 3.47 (presented in Fig. 2), indicating that the CN of No. 8 mold flux is between 3 and 4 on average. At the same time it shows that compared with Si-O, B-O is less stable than Si-O. In addition, except for the [BO_4] tetrahedron structure, there should still exist quite amount of other low coordination forms of B-O. For the Ca element, there is no obvious platform of the CN, and the approximate CN is 7.08.

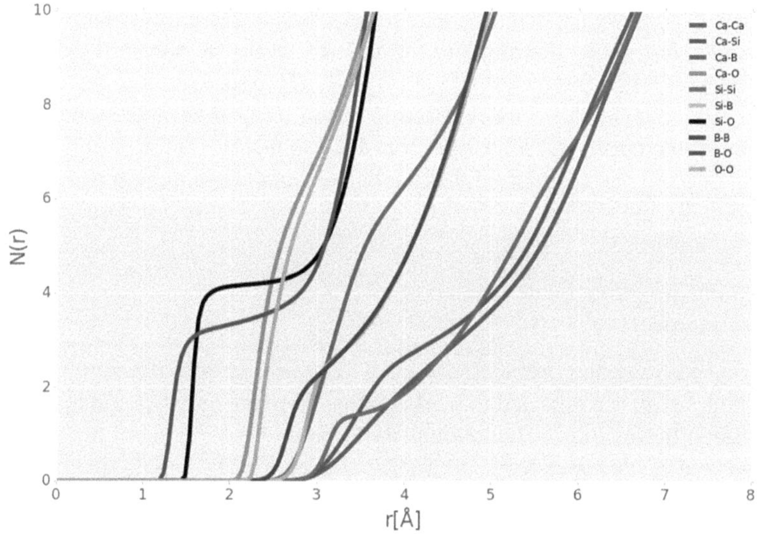

Fig. 2 The running CNs of sample No. 8

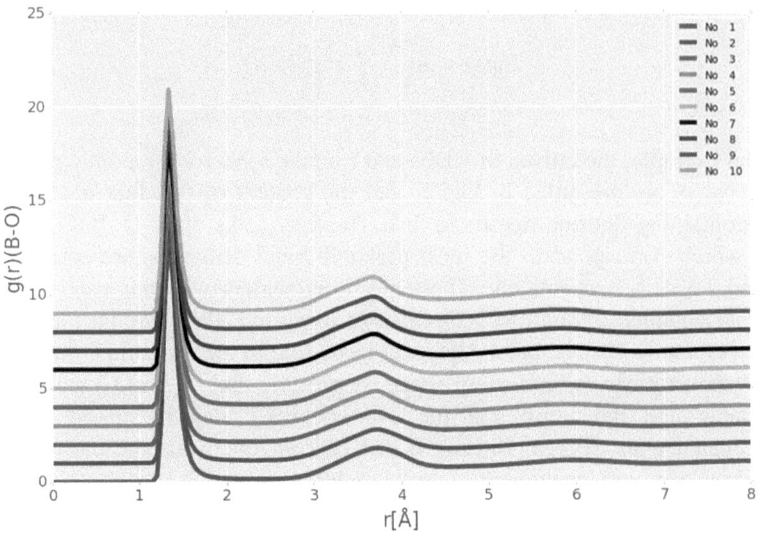

Fig. 3 RDFs for B-O pairs of all molten flux

The change of the mole ratio of the B_2O_3/SiO_2 can also impact the simulated microstructure of mold flux sample. The results of the proximity effect of B-O are shown in Fig. 3. The average bond length of B-O was not change with the mole ratio of the B_2O_3/SiO_2, illustrated that although the addition of B_2O_3 made the component a huge change, in statistical sense, it affected the specific of bond length changes for a large number of atoms.

Distributions of Bond Angles

Figure 4 shows the distributions of O-B-O angle in boron-oxygen triangle (a) and tetrahedral units (b) with varying B_2O_3/SiO_2 ratio. From Fig. 4a we can find that dramatic changes have taken place in boron-oxygen triangle, and the axis of symmetry of the curve was increased from 105° to nearly 120° with the B_2O_3/SiO_2 ratio increased. It can be inferred that the B_2O_3 could cause the viscosity deceased. As for Fig. 4b, the stability of O-B-O angle showed a little decreased with B_2O_3/SiO_2 ratio increased.

Distributions of 4-coordinated B and Q^n

Figure 5a illustrates the concentration of B coordinated with different numbers of O. The simulation is conducted to work out that there are only two kinds of forms of coordinated boron, i.e., 3-coordinated B and 4-coordinated B. Generally, 3-coordinated B increases with the mole ratio of the B_2O_3/SiO_2 increasing, and 4-coordinated B behaves just the opposite. It should be noted that the concentration of 3-coordinated B in No. 10 preponderates over others with a percentage more than 67%. However, when the content of B_2O_3 is between 10 and 26%, 3-coordinated B

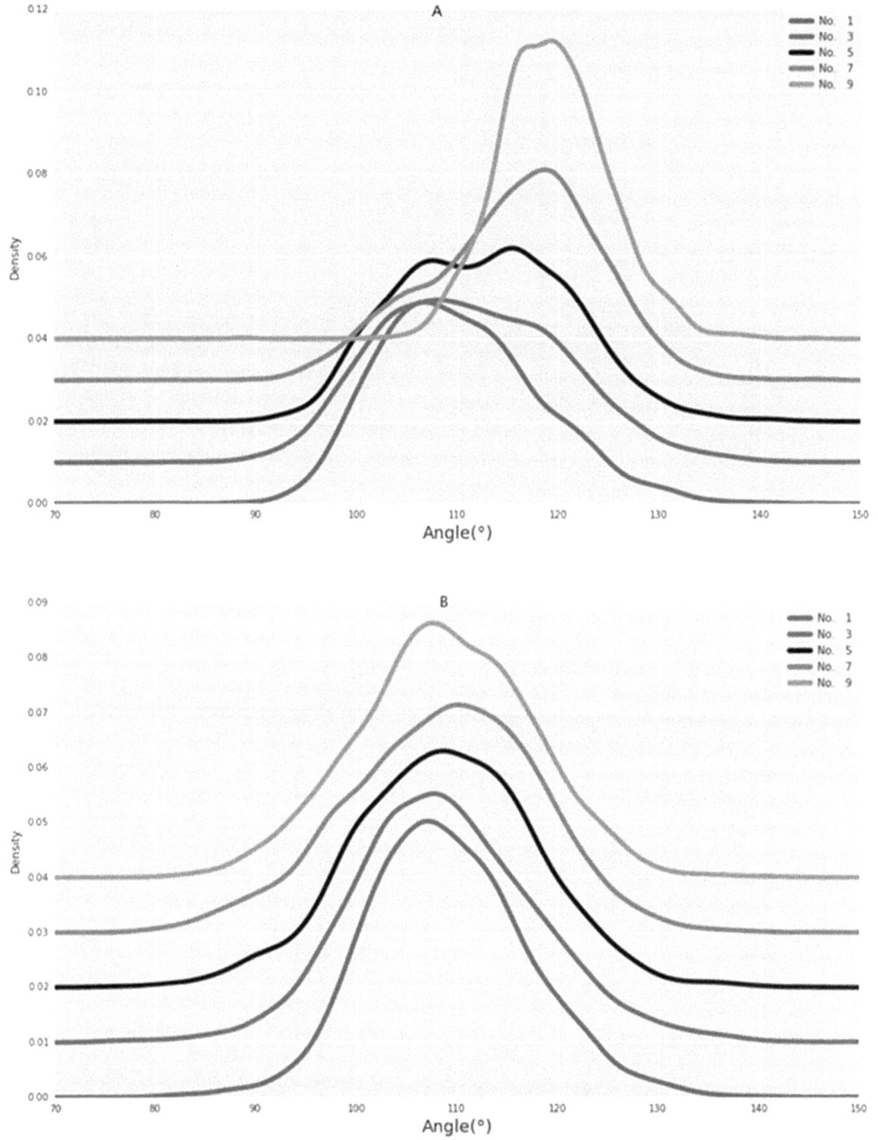

Fig. 4 Distributions of O-B-O angle in boron-oxygen triangle **a** and tetrahedral units **b** with varying B_2O_3/SiO_2 ratio

increases firstly then decreases. The results showed that the concentration of B coordinated with different numbers of O had non-linear change tendencies.

Figure 5b, c elucidate the proportion of Q^n shifts with B_2O_3/SiO_2 ratio. It can be seen that the concentration of Q^0 and Q^1 increases dramatically at the cost of a

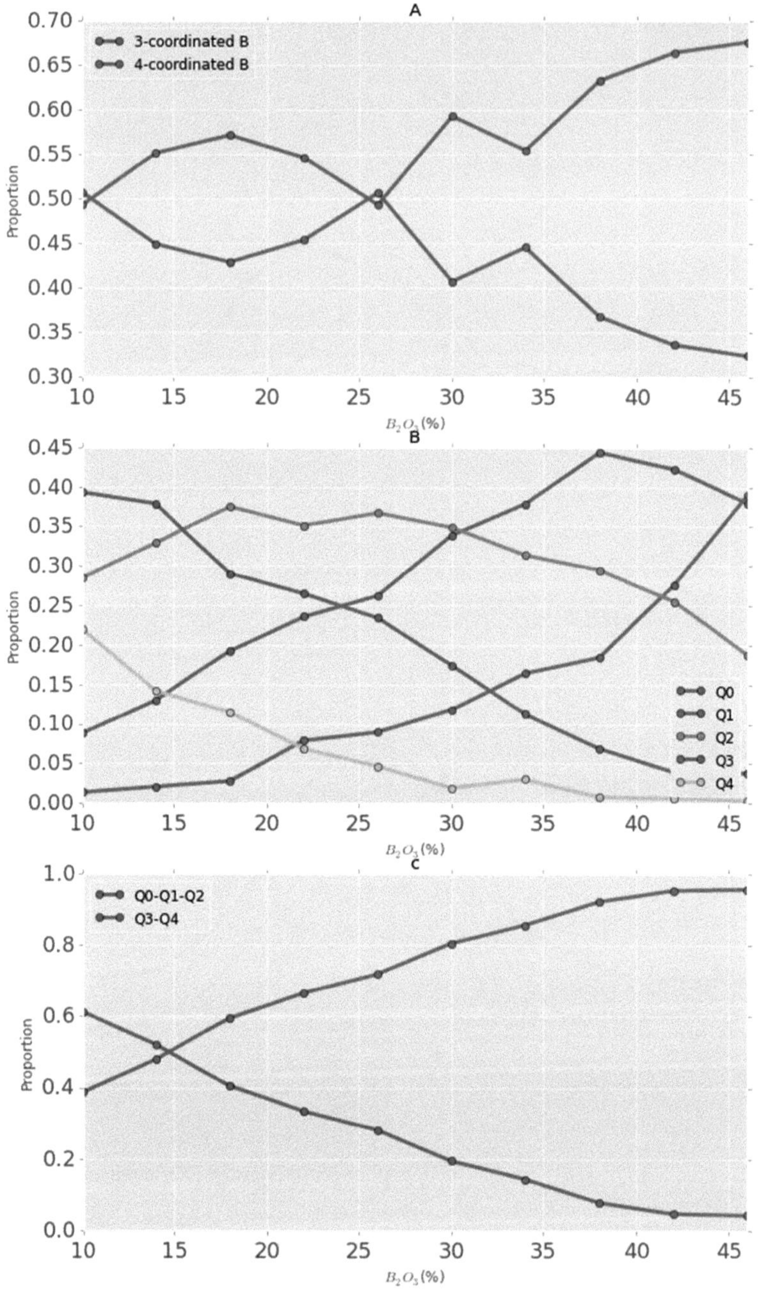

Fig. 5 The concentration of B coordinated with different numbers of O **a** and the proportion of Q^n (n is the number of BO and oxygen tricluster in one tetrahedron) shifting with B_2O_3/SiO_2 ratio **b**, **c**

decreasing in Q^3 and Q^4. The degree of polymerization (DOP) can be measured with the equation given below:

$$\mathrm{DOP} = \frac{Q^3 + Q^4}{Q^0 + Q^1 + Q^2} \tag{6}$$

The DOP decreased linearly with the increasing of B_2O_3/SiO_2 ratio. This indicates that the viscosity of mold flux may decrease as SiO_2 is replaced by B_2O_3, for a given CaO content.

Conclusions

The structural properties for ternary molten $CaO–SiO_2–B_2O_3$ system with varying B_2O_3/SiO_2 ratios at a fixed CaO content were calculated by molecular dynamics (MD) simulation.

1. With the increasing of B_2O_3/SiO_2 mol ratio, the bond length of B-O was approximately constant at 1.35 Å, and the $CN_{B\text{-}O}$ was observed to vary from 3.40 to 3.51.
2. The average angle of O-B-O increases from 105° to nearly 120°, implying that the geometric stability of structure is reducing sharply.
3. Q^3 and Q^4 tetrahedral species decrease with the increasing of B_2O_3/SiO_2 mol ratio, and the proportion of Q^3 + Q^4 decrease linearly from 0.6 to nearly 0.
4. B_2O_3 behave as a network-breaking basic oxide in the $CaO–SiO_2–B_2O_3$ system.

Acknowledgements This work was financially supported by National Natural Science Foundation of China (grant No. 51574051).

References

1. P. McMillan, Structural studies of silicate glasses and melts—applications and limitations of Raman spectroscopy. Am. Mineral. **69**(7–8), 622–644 (1984)
2. D. Massiot et al., Structure and dynamics of oxide melts and glasses: a view from multinuclear and high temperature NMR. J. Non-Cryst. Solids **354**(2–9), 249–254 (2008)
3. J.H. Park et al., FT-IR spectroscopic study on structure of $CaO\text{-}SiO_2$ and $CaO\text{-}SiO_2\text{-}CaF_2$ slags. ISIJ Int. **42**(4), 344–351 (2002)
4. Y. Waseda, J.M. Toguri, The structure of the molten $FeO\text{-}SiO_2$ system. Metall. Trans. B **9**, 595–601 (1978)
5. W.G. Seo, F. Tsukihashi, Thermodynamic and structural properties for the $FeO\text{-}SiO_2$ system by using molecular dynamics calculation. Mater. Trans. **46**(6), 1240–1247 (2005)
6. D.K. Belashchenko, O.I. Ostrovski, Computer Simulation of Noncrystalline Ionic-Covalent Oxides in the $SiO_2–CaO–FeO$ System. Inorg. Mater. **38**(8), 799–804 (2002)
7. S. Plimpton, Fast parallel algorithms for short-range molecular dynamics. J Comp Phys **117**, 1–19 (1995)

One-Step Extraction of Lead from Spent Lead-Acid Battery Paste via Reductive Sulfur-Fixing Smelting: Thermodynamic Analysis

Yun Li, Chaobo Tang, Yongming Chen, Shenghai Yang, Lulu Guo, Jing He and Motang Tang

Abstract A new innovative process for one-step and cleaner extraction of lead from spent lead-acid battery by reductive sulfur-fixing smelting was presented. This paper summarized and discussed several potential sulfur-fixing agents and molten salts which can be used in this new technique. Thermodynamic analysis involving reaction mechanism, ΔG_T and predominance-area diagrams of different smelting systems were performed, respectively. The results proved the technical feasibility and high-efficiency of this process. Copper oxide ash (CuO), zinc oxide ash (ZnO) and iron oxide bearing residues (Fe_xO_y) can be used as sulfur-fixing agent to extract lead through exchange reactions. Sulfur-fixed efficiency of CuO, ZnO and FeO decrease slightly in turn. Sulfur contained in $PbSO_4$ is fixed in form of Cu_2S, ZnS and FeS respectively which avoids SO_2 emission. The addition of molten salt can significantly decrease temperature of reductive sulfur-fixing reactions. Molten salt can be regenerated and recycled in smelting system after filtering solid residues.

Keywords Spent lead-acid battery · Reductive sulfur-fixing smelting · Sulfur-fixing agent · Thermodynamic analysis · Molten salt

Introduction

Secondary lead material produced by the recycling of lead-acid batteries [1] has become the significant source of lead in the world [2]. Recently in China, advances in auto, transportation and telecommunication industries are quickly increasing the

Y. Li · C. Tang · Y. Chen (✉) · S. Yang · L. Guo · J. He · M. Tang
School of Metallurgy and Environment, Central South University, 410083 Changsha, Hunan, People's Republic of China
e-mail: csuchenyongming@163.com

Y. Li
e-mail: AllenLeecsu@163.com

C. Tang
e-mail: chaobotang@163.com

J.-Y. Hwang et al. (eds.), *8th International Symposium on High-Temperature Metallurgical Processing*, The Minerals, Metals & Materials Series,
DOI 10.1007/978-3-319-51340-9_74

amount of the application of lead acid battery. The reported numbers of scrap-lead acid battery annually in China are more than 2.6 million tons [3]. Typically, the lead acid battery comprises 30–40% lead paste, 24–30% grid, 22–30% plastic shell and 11–30% H_2SO_4 electrolyte. Lead paste, a mixture containing 50–60% $PbSO_4$, 15–35% PbO_2, 5–10% $PbO/Pb(OH)_2$ and 2–5% elemental Pb, is a high quality lead-bearing materials. It accounted for 80–85% of the total lead second resource [3].

Typically, the recycling technologies for the recovery of lead from spent lead paste can be based on both hydrometallurgical and pyrometallurgical processes. Recycling through hydrometallurgy basically consists of the acid or base leaching of scrap to put the metals in a solution [2, 4]. Once in a solution, metals can be recovered by precipitation, altering the pH of the solution or adding some reaction agent, or by electrolysis [5]. Pyrometallurgy technique treats the battery paste by thermal means in furnace to recover the contained lead. Some smelteries also feed small amount of lead pastes together with galenite concentrate in traditional lead smelting furnace [2]. Presently, lead paste is mainly treated with a chemical desulfurization process to recover the metal [6]. The desulfurization process is employed in quite a number of flowsheet of the processes in Western countries [4].

Lead paste, however, belonged to manufactured mineral. Its components are totally different from the natural galenite minerals (PbS). As a result, numerous material differences exist in thermal decomposition, oxidation-reduction reaction and other relative thermodynamic behaviors between lead paste and galenite during smelting process. Therefore, the recovery of lead from lead acid battery sometimes cannot simply rely on and apply mechanically existing lead smelting techniques and equipments. It needs metallurgists to pay more attention on the basic theories' studies and investigations of the recovery of lead from lead acid battery and other lead-bearing secondary materials.

Based on existing metallurgical methods and processes and aim at the treatment of lead paste and other man-made secondary materials, new innovative lead metallurgy technologies should be explored and developed. Our research group have investigated lead metallurgy technology for years [7–9] and proposed a new innovated process to cleanly extract lead by one-step from lead-bearing secondary materials through reductive sulfur-fixing smelting [10]. The principle flow chart was shown in Fig. 1.

Compared with traditional lead metallurgical techniques, this new process can extract and recover lead in one step. Sulfur-fixing agent was a kind of metal oxide. The sulfur contained in lead paste will be fixed and exchanged into relative sulfide due to the presence of sulfur-fixing agent during smelting, which avoids the emission of low concentrate SO_2 gas. Meanwhile, the selective addition of molten salt can dramatically decrease the reaction temperature and extract valuable metals in low-temperature ($\leq$900 °C/1173 K). The sulfur-fixing agents used in this innovative process generally contain iron oxides, zinc oxides and copper oxides. In order to decrease the smelting cost and simultaneously treat and recover secondary resources, sulfur-fixing agent usually is chose from metallurgical secondary wastes, such as pyrite cinders, steelmaking flue dusts, magnetite, hematite, ZnO ashes, copper-bearing oxidized ores or copper ashes and so on. The molten salt added in

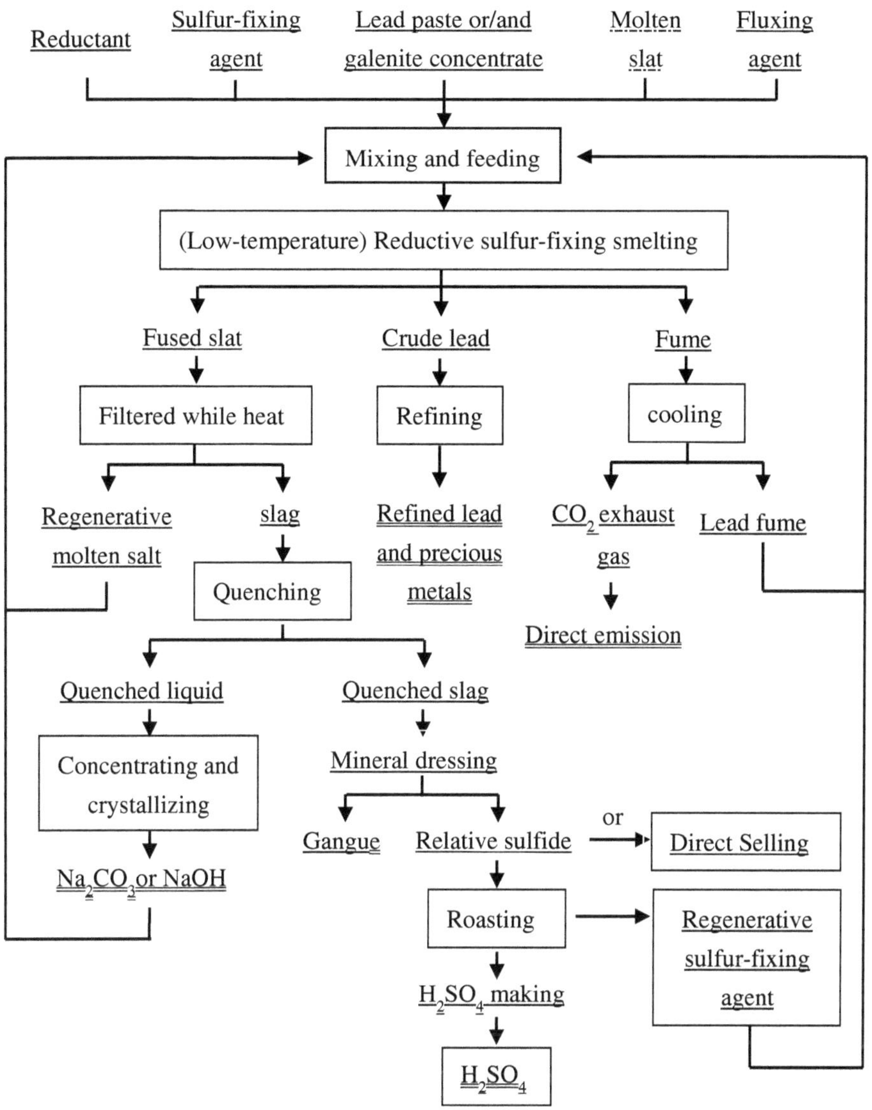

Fig. 1 Principle flow chart of innovate process for one-step extraction of lead from lead-bearing materials

the reductive sulfur-fixing smelting process contains Na_2CO_3, NaOH, KOH and NaCl etc.

According to the difference in sulfur-fixing agents, this paper summarized and discussed three different sulfur-fixing system, namely iron oxide (Fe_xO_y) sulfur-fixing system, zinc oxide (ZnO) sulfur-fixing system and copper oxide (CuO) sulfur-fixing system. Combining different molten salt compositions, every system can be an option to be used in real smelting process. In this paper, thermodynamic analysis involving reaction mechanism, ΔG_T and predominance-area diagrams of different smelting systems were performed, respectively. Finally, this paper tried to establish thermodynamic basic theory of this innovative smelting process and provided theoretical guidance for real industrial production.

Research Methods

Metallurgical process is a complex multi-dimensional system. It is difficult to observe directly and clearly the various physical and chemical reactions during smelting. However, metallurgical thermodynamics can calculate and investigate the feasibility, direction and limitation of reactions and the form of resultant products.

Gibbs free energy ΔG_T [11] can judge whether the reaction occurs favorably and which direction the reaction carries out to. Meanwhile, predominance-area diagram can tell us intuitively the smelting conditions about producing desired metallurgical productions and how to create these conditions. The major reactions that may occur in real reductive sulfur-fixing smelting systems briefly contain:

- Personal reduction and decomposition reactions of reactants;
- Reductive sulfur-fixing reactions with Fe_xO_y, ZnO and CuO;
- Reductive sulfur-fixing reactions with Na_2CO_3 and NaOH;
- Molten salt regeneration reactions.

Every possible reaction occurred in previous four sub-systems will be formulated and discussed in detailed from the perspective of thermodynamics in the following sections.

Results and Discussion

Gibbs Free Energy Calculation of Reactions

Personal Reduction and Decomposition Reactions of Reactants

$PbSO_4$ may be reduced and decomposed at smelting temperature. The specific reactions are shown as follows:

$$PbSO_4 = PbO + SO_{3(g)} \tag{1}$$

$$PbSO_4 + C = PbO + CO_{(g)} + SO_{2(g)} \tag{2}$$

$$PbSO_4 + CO_{(g)} = PbO + CO_{2(g)} + SO_{2(g)} \tag{3}$$

$$PbO + C = Pb + CO_{(g)} \tag{4}$$

$$PbO + CO_{(g)} = Pb + CO_{2(g)} \tag{5}$$

$$PbSO_4 + 4C = PbS + 4CO_{(g)} \tag{6}$$

$$PbSO_4 + 4CO_{(g)} = PbS + 4CO_{2(g)} \tag{7}$$

$$PbSO_4 + PbS = 2Pb + 2SO_{2(g)} \tag{8}$$

$$PbO + 0.5PbS = 1.5Pb + 0.5SO_{2(g)} \tag{9}$$

$$PbSO_4 + 2C = Pb + SO_{2(g)} + 2CO_{(g)} \tag{10}$$

$$PbSO_4 + 2CO_{(g)} = Pb + SO_{2(g)} + 2CO_{2(g)} \tag{11}$$

Figure 2 illustrated $\Delta G_T - T$ relational expression of reactions (1)–(11). It was observed that ΔG_T of $PbSO_4$ reduction and decomposition reactions decreased with temperature increasing. Therefore, the curve of ΔG_T would drop down and finally under the zero line at a certain temperature. That indicated the corresponding reactions became thermodynamically favorable. $\Delta G_T - T$ curves of reactions (1)–(5) showed that $PbSO_4$ was stable at non-reductive atmosphere and even the trend of decomposition at high temperature was small. However, $PbSO_4$ would decompose and generate PbO easily at reductive atmosphere (mole ratio of $PbSO_4$ and C is 1:1). PbO also continued to decompose into Pb to a certain extent. The $\Delta G_T - T$ curves of reactions (6)–(11) showed that $PbSO_4$ had a strong trend to decompose into PbS at reductive atmosphere (mole ratio of $PbSO_4$ and C is 4:1), in addition, a portion of $PbSO_4$ would be directly reduced to metallic Pb. In summary, the type of $PbSO_4$ decomposition products is related to reductive atmosphere and temperature.

Reductive Sulfur-Fixing Reactions with Fe_xO_y, ZnO and CuO

Typically, Fe_xO_y will decompose into FeO at reductive atmosphere. Therefore, this section directly investigates the reactions between FeO, ZnO, CuO and other reactants. The possible reactions were shown as follow:

$$PbSO_4 + FeO + 5C = Pb + FeS + 5CO_{(g)} \tag{12}$$

$$PbSO_4 + FeO + 5CO_{(g)} = Pb + FeS + 5CO_{2(g)} \tag{13}$$

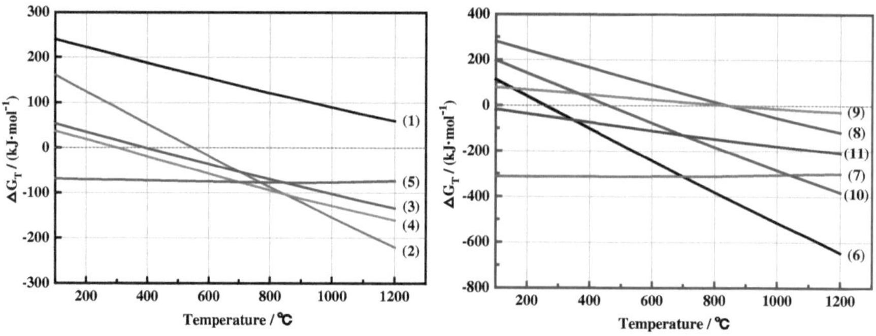

Fig. 2 Relationship between ΔG_T and T of reactions (1)–(11)

$$PbS + FeO + C = Pb + FeS + CO_{(g)} \tag{14}$$

$$PbS + FeO + CO_{(g)} = Pb + FeS + CO_{2(g)} \tag{15}$$

$$PbSO_4 + ZnO + 5C = Pb + ZnS + 5CO_{(g)} \tag{16}$$

$$PbSO_4 + ZnO + 5CO_{(g)} = Pb + ZnS + 5CO_{2(g)} \tag{17}$$

$$PbS + ZnO + C = Pb + ZnS + CO_{(g)} \tag{18}$$

$$PbS + ZnO + CO_{(g)} = Pb + ZnS + CO_{2(g)} \tag{19}$$

$$PbSO_4 + 2CuO + 6C = Pb + Cu_2S + 6CO_{(g)} \tag{20}$$

$$PbSO_4 + 2CuO + 6CO_{(g)} = Pb + Cu_2S + 6CO_{2(g)} \tag{21}$$

$$PbS + 2CuO + 2C = Pb + Cu_2S + 2CO_{(g)} \tag{22}$$

$$PbS + 2CuO + 2CO_{(g)} = Pb + Cu_2S + 2CO_{2(g)} \tag{23}$$

$\Delta G_T - T$ curves of reactions (12)–(23) were presented in Fig. 3. It was illustrated that FeO, ZnO and CuO all could react with $PbSO_4$ and PbS to produce metallic Pb at a certain temperature and reducing atmosphere. In addition, the reaction trend increased with temperature increasing. Comparison of three kinds of sulfur-fixing agents, the efficiency of sulfur-fixation was CuO, ZnO and FeO in turn. The sulfur-fixing reaction tendency of ZnO and FeO was roughly similar. However, the trend of reaction between PbS and three sulfur-fixing agents FeO/ZnO and CuO were much smaller than that of $PbSO_4$. The reductive sulfur-fixing reactions between PbS/FeO and ZnO only became favorable at high temperature. Therefore, smelting temperature was a key factor for efficiently recovering lead from $PbSO_4$ and PbS.

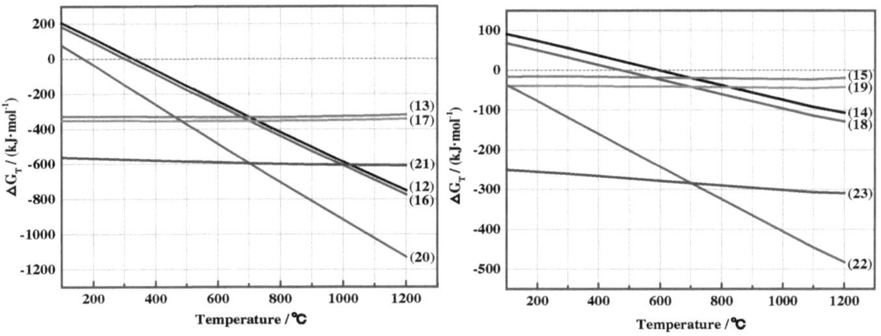

Fig. 3 Relationship between ΔG_T and T of reactions (12)–(23)

Reductive Sulfur-Fixing Reactions with Na_2CO_3 and NaOH

Generally, molten salt, such as Na_2CO_3, NaOH and NaCl, have a relative low melting point [12]. Therefore, the addition of molten salt could decrease the smelting temperature of the whole smelting system. The selective additive molten salt Na_2CO_3 and NaOH would also involve in reductive sulfur-fixing reaction. Nevertheless, molten salt NaCl was inert and didn't involve in any chemical reactions. Following reactions are shown and summarized below.

$$PbSO_4 + Na_2CO_3 + 6C = Pb + Na_2S + 7CO_{(g)} \tag{24}$$

$$PbSO_4 + Na_2CO_3 + 5CO_{(g)} = Pb + Na_2S + 6CO_{2(g)} \tag{25}$$

$$PbSO_4 + 2NaOH + 5C = Pb + Na_2S + H_2O_{(g)} + 5CO_{(g)} \tag{26}$$

$$PbSO_4 + 2NaOH + 5CO_{(g)} = Pb + Na_2S + H_2O_{(g)} + 5CO_{2(g)} \tag{27}$$

$$PbS + Na_2CO_3 + 2C = Pb + Na_2S + 3CO_{(g)} \tag{28}$$

$$PbS + Na_2CO_3 + CO_{(g)} = Pb + Na_2S + 2CO_{2(g)} \tag{29}$$

$$PbS + 2NaOH + C = Pb + Na_2S + H_2O_{(g)} + CO_{(g)} \tag{30}$$

$$PbS + 2NaOH + CO_{(g)} = Pb + Na_2S + H_2O_{(g)} + CO_{2(g)} \tag{31}$$

Molten Salt Regeneration Reactions

The additive molten salt can be regenerated and recycled through intermediate products (Na_2SO_4 and Na_2S) and sulfur-fixing agents during smelting. The specific reaction equations can be briefly represented as follows:

$$Na_2SO_4 + FeO + 3C = Na_2CO_3 + FeS + 2CO_{(g)} \quad (32)$$

$$Na_2SO_4 + FeO + 4CO_{(g)} = Na_2CO_3 + FeS + 3CO_{2(g)} \quad (33)$$

$$Na_2SO_4 + ZnO + 3C = Na_2CO_3 + ZnS + 2CO_{(g)} \quad (34)$$

$$Na_2SO_4 + ZnO + 4CO_{(g)} = Na_2CO_3 + ZnS + 3CO_{2(g)} \quad (35)$$

$$Na_2SO_4 + 2CuO + 4C = Na_2CO_3 + Cu_2S + 3CO_{(g)} \quad (36)$$

$$Na_2SO_4 + 2CuO + 5CO_{(g)} = Na_2CO_3 + Cu_2S + 4CO_{2(g)} \quad (37)$$

$$Na_2S + FeO + CO_2(g) = Na_2CO_3 + FeS \quad (38)$$

$$Na_2S + ZnO + CO_2(g) = Na_2CO_3 + ZnS \quad (39)$$

$$Na_2S + 2CuO + 0.5C + 0.5CO_2(g) = Na_2CO_3 + Cu_2S \quad (40)$$

$$NaOH + 0.5CO_{2(g)} = 0.5Na_2CO_3 + 0.5H_2O_{(g)} \quad (41)$$

$\Delta G_T - T$ curves of reactions (24)–(31) presented in Fig. 4 implied that increasing temperature facilitated the reactions between $PbSO_4$, PbS and Na_2CO_3, NaOH. Compared the reaction tendency of Na_2CO_3 with that of NaOH, reductive sulfur-fixing reactions involved NaOH was more favorable than that of Na_2CO_3 at relative low temperature, while, at high temperature (>1000 °C/1273 K), $PbSO_4$ was more easy to react with NaOH. Therefore, considering from the thermodynamics aspect, NaOH was a suitable molten salt to recover lead from spent lead paste in low temperature. On the contrary, Na_2CO_3 was suitable for relatively high temperature smelting technique.

$\Delta G_T - T$ curves of reactions (32)–(41) presented in Fig. 4 indicated that the intermediate product Na_2SO_4 can continue to react with sulfur-fixing agents FeO, ZnO and CuO to regenerate Na_2CO_3. The reaction of CuO and Na_2SO_4 was the most favorable. In addition, Na_2S also could react with sulfur-fixing agents to regenerate and recycle molten salt. The $\Delta G_T - T$ curves of reactions (33), (35), (37) and (38)–(41) indicated CO and CO_2 gas also could partly react and regenerate molten salt at low temperature and reductive atmosphere. That indicated the generation of molten salt could achieve by controlling cooling atmosphere and cooling method in real smelting process.

Predominance-Area Diagram

Predominance-area diagram can tell us intuitively the smelting conditions for producing desired metallurgical productions and how to create and optimize these conditions. Therefore, metallurgical investigators can avoid tedious thermodynamic

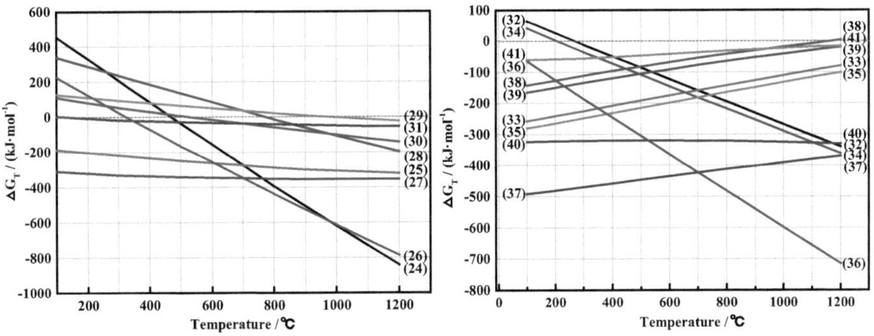

Fig. 4 Relationship between ΔG_T and T of reactions (24)–(41)

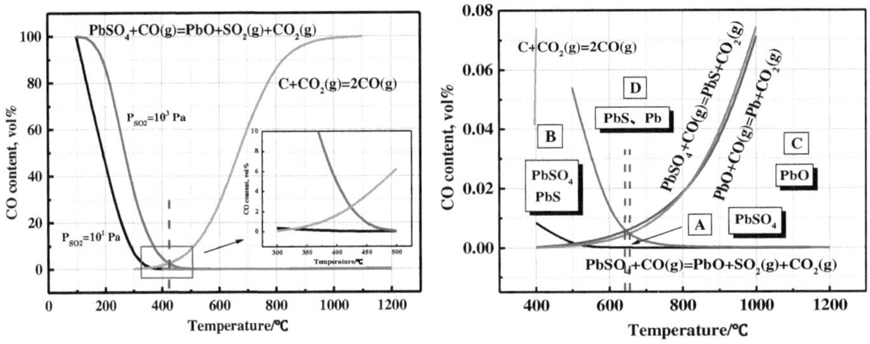

Fig. 5 Reductive equilibrium diagram of $PbSO_4$

calculation. $PbSO_4$ can occur a series of physical and chemical reactions at high temperature, reductive and sulfurized atmosphere. The reductive equilibrium diagrams of $PbSO_4$ were illustrated in Fig. 5.

It indicated that four primary stable areas existed at variable temperature and reductive atmosphere. A, B, C and D represented stable areas of $PbSO_4$, $PbSO_4$ and PbS, PbO, PbS and Pb respectively. It was observed from the curve of $PbSO_4 \rightarrow PbO$ reaction at different P_{SO_2} that decreasing sulfur potential and gas resultants partial pressure could efficiently reduce the initial reduction temperature and equilibrium-desired CO% of $PbSO_4 \rightarrow PbO$ reaction. Actually, when smelting temperature was above 400 °C (673 K), the equilibrium-desired CO% of $PbSO_4 \rightarrow PbO$ reaction and $PbO \rightarrow Pb$ reaction were very low. $PbSO_4$ was decomposed easily into PbO and further reduced into Pb from thermodynamic perspective. Figure 5 further illustrated that $PbSO_4$ was also very easy to react to generate PbS. However, PbS was very hard to continually reduce to produce metallic Pb. As a result, it was very difficult to achieve higher Pb recovery rate from lead paste only through strong reductive atmosphere and high temperature smelting technique. Some measures must be taken to decrease sulfur potential and extract Pb

Fig. 6 Reductive sulfur-fixing equilibrium diagram of $PbSO_4$ and PbS

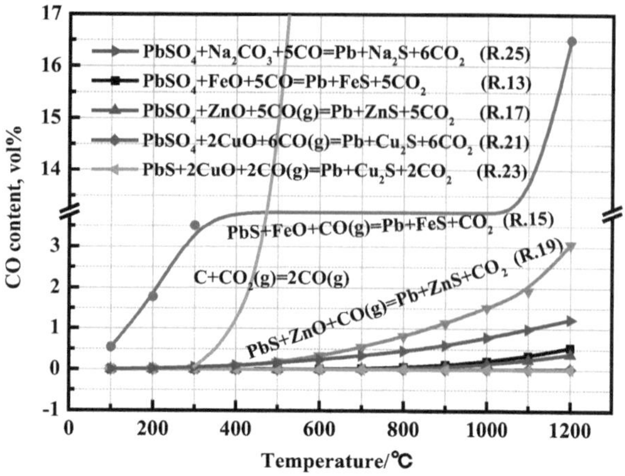

form PbS, or some efforts be made to find the directional smelting technique to inhibit $PbSO_4 \rightarrow PbS$ reaction and recover Pb through $PbSO_4 \rightarrow PbO \rightarrow Pb$ path.

Reductive sulfur-fixing equilibrium diagram of $PbSO_4$ and PbS shown in Fig. 6 implied that equilibrium-required CO% of reactions involved different sulfur-fixing agents were different. At the same smelting conditions, equilibrium-required CO% of FeO reacting with $PbSO_4$ and PbS respectively was the highest, next to ZnO and CuO in turn. However, the equilibrium-required CO% of all three sulfur-fixing agents increased as temperature increasing. The difference of equilibrium-required CO% of three reductive sulfur-fixing reactions involved $PbSO_4$ were not very obvious. All corresponding reactions could carry out smoothly at weak reductive atmosphere. Nevertheless, the difference of reactions involved PbS were obvious. Equilibrium-required CO% of PbS and CuO reaction was lowest. The reductive sulfur-fixing reaction of PbS and ZnO also could occur smoothly at less than 3 CO %. While, the reaction equilibrium-required CO% involved PbS and FeO varied greatly with smelting temperature increasing. It indicated high equilibrium-required CO% was needed to guarantee the reaction carried out smoothly when used FeO as sulfur-fixing agent.

Conclusions

The innovative cleaner metallurgical process for one-step extraction of lead from spent lead-acid battery paste via reductive sulfur-fixing smelting is technically feasible. This new technique is characterized by high comprehensive recovery of valuable metals, elimination of SO_2 emission, energy conservation and environment-friendly. It also can co-treat pyrite cinder and other industrial solid wastes. Two different paths, low-temperature smelting ($\leq$900 °C/1173 K) and high-temperature smelting, can be chosen to implement this innovative technique.

Thermodynamic analysis results show that smelting temperature and reductive atmosphere are key factor to ensure Pb recovery during smelting. Sulfur-fixed efficiency of CuO, ZnO and Fe_xO_y decrease slightly in turn. The trend of the reductive sulfur-fixing reaction between PbS and CuO, ZnO and Fe_xO_y were much smaller than that of $PbSO_4$. Sulfur contained in $PbSO_4$ and PbS were fixed in form of Cu_2S, ZnS and FeS respectively which avoids SO_2 emission. The addition of molten salt can significantly decrease temperature of reductive sulfur-fixing reactions. Molten salt can be regenerated and recycled in smelting system after filtering solid residues.

Acknowledgements The authors thank the Applied Science and Technology Specialized Research Project of Guangdong Province (No. 2016B020242001), the National Natural Science Foundation of China (No. 51234009), the Student Innovation and Entrepreneurship Organization of Mittal, Central South University [No. 15MX23] and the Fundamental Research Funds for the Central Universities of Central South University [No. 2016zzts039] for founding this research.

References

1. R.A. Huggins *Lead-Acid Batteries, Energy Storage*, (Switzerland, Springer International Publishing), pp. 309–323 (2016)
2. Q. Zhang, The current status on the recycling of lead-acid batteries in China. Int. J. Electrochem. Sci. **8**, 6457–6466 (2013)
3. Xi. Tian et al., The lead-acid battery industry in China: outlook for production and recycling. Waste. Manag. Res. **33**(11), 986–994 (2015)
4. V.A.S.I.L.E. Hotea, Clean technology of lead recovery from spent lead paste. Recent Res Appl. Econ. Manag. Econ. Aspects Environ. **2**, 263–270 (2013)
5. D.C.R. Espinosa et al., An overview on the current processes for the recycling of batteries. J. Power Sources **135**(1), 311–319 (2004)
6. J. Weijma et al., Biological conversion of anglesite ($PbSO_4$) and lead waste from spent car batteries to galena (PbS). Biotechnol. Prog. **18**(4), 770–775 (2002)
7. Hu. Yujie et al., Extraction of lead from secondary lead through a low-temperature alkaline and sulfur-fixing smelting process. Chin. J. Eng. **37**(5), 588–594 (2015)
8. Hu. Yujie et al., Reductive smelting of spent lead-acid battery colloid sludge in molten salt of sodium at low temperature. China Nonferrous Metall. **43**(1), 75–79 (2014)
9. Huang. Chao et al., Sulfur-fixing reduction smelting of spent lead-acid battery colloid sludge in fused salt at low temperature. Min. Metall. Eng. **32**(2), 84–87 (2012)
10. Li. Yun et al., One-step extraction of antimony in low temperature from stibnite concentrate using iron oxide as sulfur-fixing agent. Metals **6**(7), 153 (2016)
11. A Roine. HSC Chemistry for windows, version 6.0, chemical reaction and equilibrium software with extensive thermochemical database (2006)
12. Ye Longgang et al., The thermal physical properties and stability of the eutectic composition in a Na_2CO_3–NaCl binary system. Thermochim. Acta. **596**, 14–20 (2014)

Removal of Cd(II) Ion from Aqueous Solution by Adsorption on Wasted Low Grade Phosphorus-Containing Iron Ore

Xiaoli Yuan, Wentang Xia, Juan An, Xiaoyan Xiang, Xuejiao Zhou, Jianguo Yin and Wenqiang Yang

Abstract Low grade phosphorus-containing iron ore (LGPIO), a raw and wasted industrial solid material was used as an adsorbent to remove Cd(II) ion from wastewater. The effects of initial pH value, adsorptive time, initial Cd(II) ion concentration, adsorptive temperature and LGPIO dosage on the Cd(II) ion removal efficiency were studied. The results show that the Cd(II) ion removal efficiencies exceed 99.8% and Cd(II) ion concentrations are less than 0.10 mg/L under the pH value 6, temperature 20 °C, adsorptive time 90 min, initial Cd(II) ion concentration 50 mg/L, particle size <0.15 mm, adsorbent dosage 50 g/L and stirring speed 290 r/min. After Cd(II) ion removal reaction, the Cd(II) ion concentrations completely complies with the requirement of national discharge standard of water pollutants for cadmium and zinc industry (GB 20466-2010) (TCd(II) = 0.1 mg/L) in China. Therefore, it can be concluded that LGPIO is a new low-cost adsorbent which is suitable for the adsorption of Cd(II) ion from wastewater.

Keywords Low grade phosphorus-containing iron ore (LGPIO) · Cd(II) ion removal · Wastewater · Adsorbent

Introduction

As well known, cadmium is one of the most toxic pollutants released into the environment. If its concentration exceed the tolerance limit, cadmium can be accumulated throughout the food chain and can cause harmful biochemical and many health hazards effect on human beings, such as kidney damage, high blood pressure, red blood cells and destruction of testicular tissue [1, 2]. Especially the

---

X. Yuan · W. Xia (✉) · J. An · X. Xiang · X. Zhou · J. Yin · W. Yang
School of Metallurgy and Material Engineering, Chongqing University of Science and Technology, 401331 Chongqing, China
e-mail: wentangx@163.com

X. Yuan
e-mail: yuanxiaoli1981@126.com

J.-Y. Hwang et al. (eds.), *8th International Symposium on High-Temperature Metallurgical Processing*, The Minerals, Metals & Materials Series,
DOI 10.1007/978-3-319-51340-9_75

Cd(II) ion-containing wastewater discharged from industries has caused environmental pollution, which has become an serious environmental problem all over the world. Therefore, the removal of Cd(II) ion from wastewater has gained widespread attention. A number of technologies have been developed for Cd(II) ion removal from wastewaters at different concentrations, such as adsorption, flotation, coagulation, chemical precipitation, ion exchange, and electrochemical methods [3, 4]. Of all these methods, adsorption is an efficient and economic method to remove Cd(II) ion from wastewater. In recent years, more and more studies have been focused on seeking out economical and high-affinity adsorbents including activated carbon, grafted silica, olive stone, natural kaolinite clay, goethite, waste mineral containing Fe(III)/Cr(III) hydroxide, etc [4–10]. Among these, mineral waste has been recognized as one of the most competitive candidate adsorbents due to the favorable characteristics including environmental friendliness, easy accessibility, chemical stability (low solubility), and low cost.

In the present study, low grade phosphorus-containing iron ore (LGPIO) was used as the adsorbent to remove Cd(II) ion from wastewater. The LGPIO is recognized as a solid waste product due to too low grade iron to be extracted. However, LGPIO contains a large amount of hematite, silicate, dolomite and apatite, and all these phases act as the efficient adsorbents for the removal of Cd(II) ion from the wastewater [5, 6, 9].

The objective of this work was to explore the feasibility of LGPIO as an adsorbent for the removal of Cd(II) ion from wastewater. The influences of adsorption parameters including initial pH value, adsorptive temperature, initial Cd (II) ion concentration, adsorbent dosage, adsorptive time, and particle size on the Cd(II) ion removal efficiency of LGPIO were investigated.

Materials and Methods

Materials

The LGPIO used in this experimental research was obtained from Wushan, Chongqing, China. The chemical composition of LGPIO is listed in Table 1. It was crushed and screened to lower than 0.25 mm in particle size.

Experimental Methods

A defined volume of the Cd(II) ion stock solution with a concentration of 1 g Cd(II)/L was diluted to the initial concentration by adding deionized water. About

Table 1 Chemical composition of LGPIO

TFe	SiO_2	CaO	MgO	Al_2O_3	MnO	V_2O_5	K_2O	Na_2O	P	F	S
25.49	16.13	20.76	7.99	5.55	0.36	0.097	0.25	0.057	1.26	0.77	0.04

100 mL of Cd(II) ion solution was poured into the glass beaker of 200 mL and its pH value was adjusted to desired value with moderate doses of HCl or NaOH. Then the LGPIO was added into the wastewater. The experimental set up is listed in Table 2. After equilibrium was reached, the sample was filtrated and the pH of the wastewater was measured by pH meter (METTLER-TOLED). The Cd(II) ion concentrations of all samples including wastewater were analyzed by flame atomic absorption spectrophotometry method (TAS-986, Beijing Puxi Science and Technology Instrument Co Ltd China).

Results and Discussion

Effect of Initial PH Value

The plots of Cd(II) ion removal efficiency and Cd(II) ion residual concentration versus initial pH were presented in Fig. 1. It can be seen from Fig. 1 that the initial pH value is a very important parameter for the removal of Cd(II) ion from the wastewater. With the initial pH value increasing from 1.5 to 6, the Cd(II) ion removal efficiency increased rapidly from 77.0 to 95.6%, and the residual Cd(II) ion concentration decreased from 9.762 to 2.092 mg/L. Based on the previous work [7], the fact that the amount of Cd(II) ion removal at low pH (pH <6) is considerably lower may be attributed to the competition between Cd^{2+} and H^{+} ions sites on LGPIO surface. Similar results have been reported by other researchers [10]. However, when the initial pH value ranged from 6 to 12, the Cd(II) ion removal efficiency decreased. This can be usually explained due to formation of soluble hydroxyl complexes. Previous work indicates that OH^{-} ions in the alkaline pH value affects firstly hydrolysis products of $Cd(OH)^{+}$, then $Cd(OH)_2$ hydrolysis complexes, and finally decrease the adsorption [7]. Therefore, the maximum removal efficiency was achieved with 95.6% at pH 6.

Table 2 Experimental conditions of Cd(II) ion adsorption

No.	Initial pH	Temperature/°C	Stirring speed/ (r/min)	The range of LGPIO particle size/mm	Reaction time/min	LGPIO dosage/ (g/L)	Initial Cd(II) ion concentration/ (mg/L)
1	1–12	20	290	<0.25	60	50	50
2	6	20–60	290	<0.25	60	50	50
3	6	20	290	<0.25	60	30–70	50
4	6	20	290	<0.25	60	50	20–100
5	6	20	290	<0.25	30–150	50	50
6	6	20	290	0–0.25	60	50	50

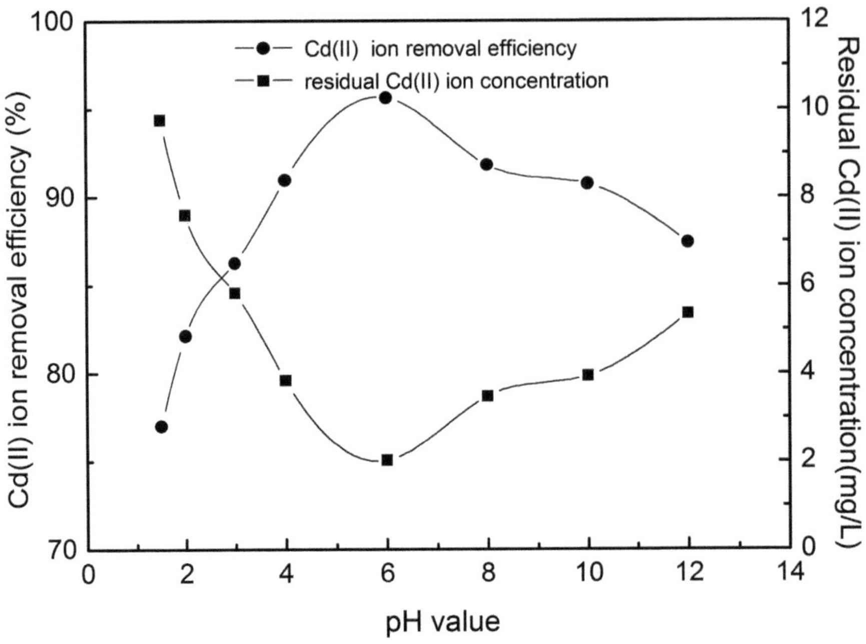

Fig. 1 Effect of initial pH on Cd(II) ion removal efficiency

Effect of Adsorptive Temperature

Figure 2 showed the Cd(II) ion removal efficiencies and residual Cd(II) ion concentrations at different adsorptive temperatures. It can be seen from Fig. 2 that when reaction temperature was increased from 20 to 60 °C, the Cd(II) removal efficiency and the residual Cd(II) ion concentration have not obviously changed. Thus as indicated in Fig. 2 that the adsorptive temperature had almost no effect on the Cd(II) ion removal efficiency. Therefore 20 °C would be appropriate temperature for the industrial application.

Effect of LGPIO Dosage

Figure 3 shows the relationship between LGPIO dosage and the Cd(II) ion removal efficiency. As can be seen form the figure, the LGPIO dosage notably influenced the process of Cd(II) ion removal from wastewater. When LGPIO dosage was varied from 30 to 50 g/L, the Cd(II) ion removal efficiency increased remarkably with increasing LGPIO dosage. The highest Cd(II) removal efficiency (95.6%) was reached when LGPIO dosage was 50 g/L. With further increase of LGPIO dosage,

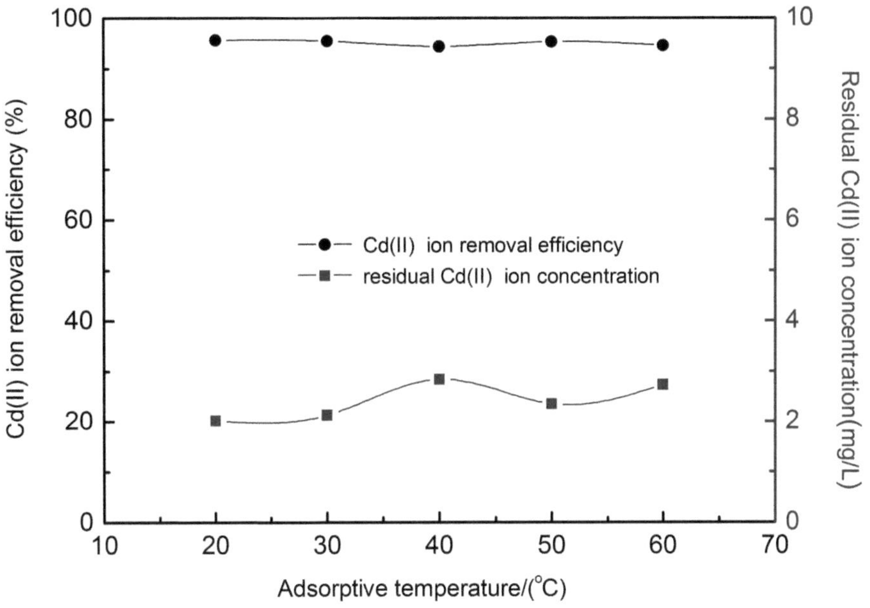

Fig. 2 Effect of adsorptive temperature on Cd(II) ion removal efficiency

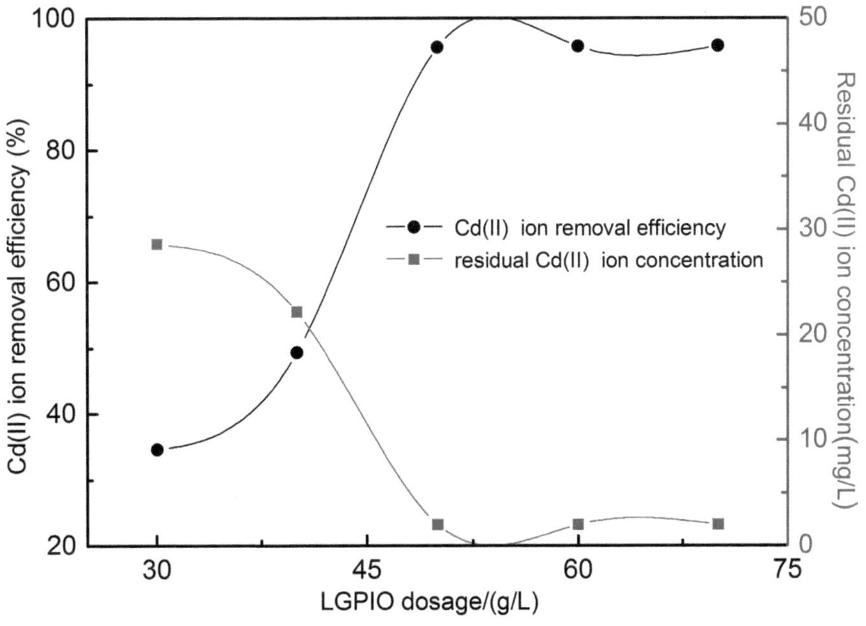

Fig. 3 Effect of LGPIO dosage on Cd(II) ion removal efficiency

the Cd(II) ion removal efficiency and residual Cd(II) concentration were nearly constant.

Effect of Initial Cd(II) Ion Concentration

The plots of Cd(II) ion removal efficiency and Cd(II) ion residual concentration versus initial Cd(II) ion concentration were presented in Fig. 4. It can be seen that with increased initial Cd(II) ion concentration from 20 to 50 mg/L, the Cd(II) ion removal efficiency increased rapidly from 67.0 to 95.6%, and the equilibrium decreased from 6.617 to 2.092 mg/L. With further initial Cd(II) ion concentration increase, the Cd(II) ion removal efficiency decreased and the residual Cd(II) ion concentration increased rapidly. Thus the appropriate initial Cd(II) ion concentration would be maintained at 50 mg/L.

Effect of Adsorptive Time

The effect of adsorptive time on the Cd(II) ion removal efficiency was shown in Fig. 5. With longer reaction time (e.g. from 30 to 90 min), the Cd(II) ion removal

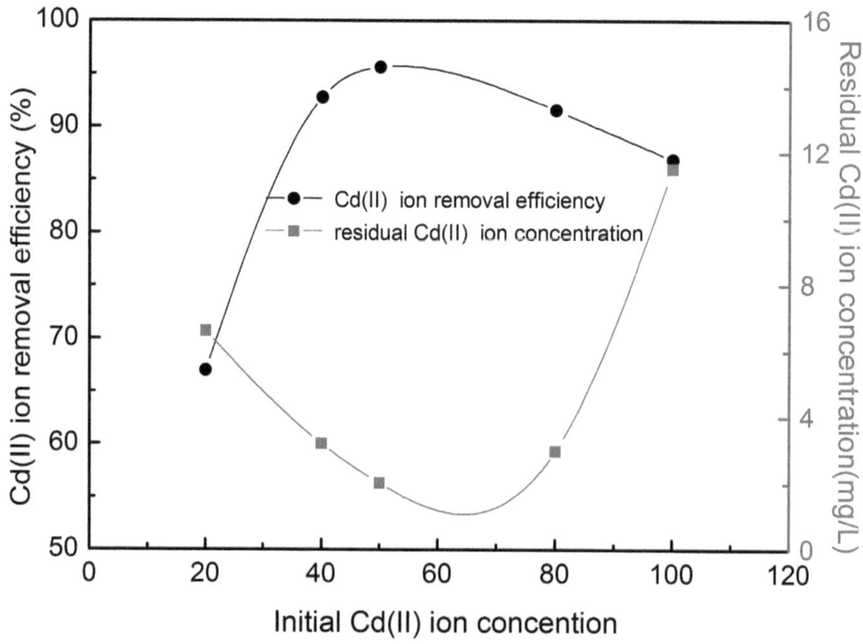

Fig. 4 Effect of initial Cd(II) ion concentration on Cd(II) ion removal efficiency

efficiency increased rapidly. The highest Cd(II) ion removal efficiency was 99.8% at 90 min. With further increase in the reaction time, the Cd(II) ion removal efficiency increased slowly.

Effect of Particle Size

Figure 6 showed the Cd(II) ion removal efficiencies and residual Cd(II) ion concentrations at the different range of LGPIO particle size. As can be seen from Fig. 6, the Cd(II) ion removal efficiency increased with decreased the particle size. The Cd(II) ion removal efficiency increased slightly when the particle size was in the range from 0.15 to 0.075 mm to <0.075 mm. Considering the factors including Cd(II) removal efficiency, processing cost and the cycle utilization of adsorbent, particle size range of <0.15 mm would be appropriate for the industrial application.

Optimal Experiment

To verify the optimal conditions, five further experiments were performed under experimental conditions showed in Table 3. The optimal reaction conditions were

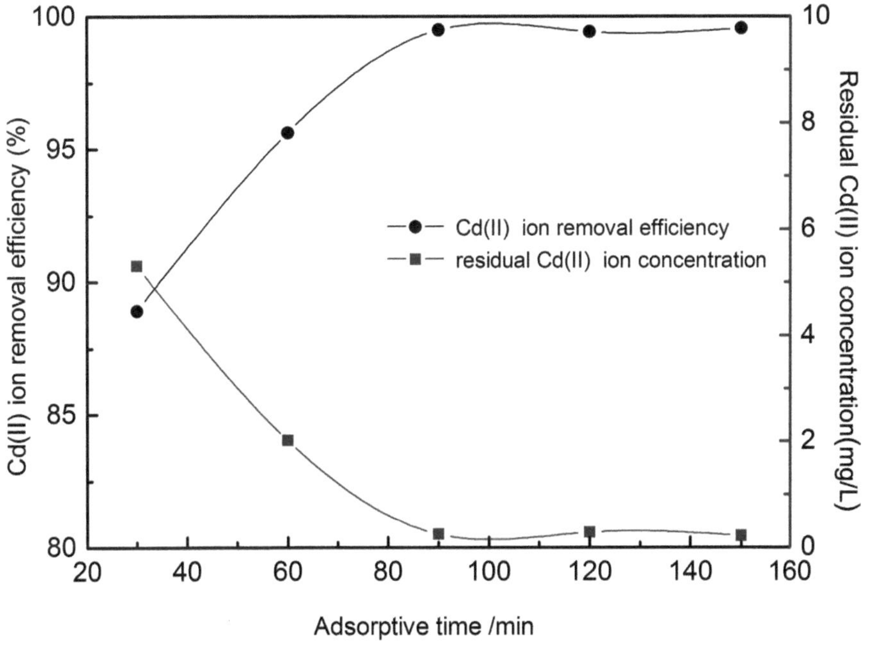

Fig. 5 Effect of reaction time on Cd(II) ion removal efficiency

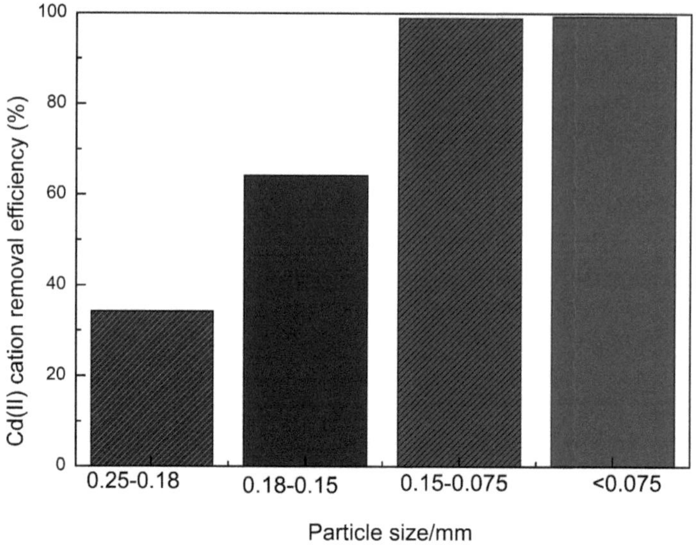

Fig. 6 Effect of particle size on Cd(II) ion removal efficiency

Table 3 Effect of wastewater volume on Cd(II) ion removal

Solution volume/mL	After Cd(II) ion removal concentration/ (mg/L)	Cd(II) ion removal efficiency/%
200	0.095	99.81
400	0.085	99.83
600	0.080	99.84
800	0.080	99.84
1000	0.070	99.86

determined as follows: pH 6.0, the range of LGIOWP particle size <0.15 mm, temperature 20 °C, initial Cd(II) ion concentration 50 mg/L, reaction time 90 min, adsorbent dosage 50 g/L and stirring speed 290 r/min.

As can be seen from the Table 3, the experimental results indicated that the optimal reaction conditions were reliable. The Cd(II) ion removal efficiencies exceeded 99.8% in the experimental work reported above, and the Cd(II) ion concentrations completely meet with the requirement of national discharge standard of water pollutants cadmium and zinc industry (GB 20466-2010) (TCd(II) = 0.1 mg/L) in China.

Conclusions

The Cd(II) ion removal adsorption experiments results indicate that the low grade phosphorus-containing iron ore (LGPIO) is a good adsorbent for Cd(II) ion removal from the wastewater. The optimal parameters were determined as follows: initial pH value 6, reaction time 90 min, the range of LGPIO particle size <0.15 mm, temperature 20 °C, initial Cd(II) concentration 50 mg/L, adsorbent dosage 50 g/L and stirring speed 290 r/min. Under the optimal experimental conditions, the Cd(II) ion removal efficiency exceeded 99%, and the Cd(II) concentrations from wastewater were lower than 0.1 mg/L. The results completely meet the requirement of national discharge standard of water pollutants for cadmium and zinc industry (GB 20466-2010) (TCd(II) = 0.1 mg/L) in China. From the above results show that LGPIO could be used as adsorbent for Cd(II) ion removal from the wastewater.

Acknowledgements This research was funded by Research Foundation of Chongqing University of Science & Technology (Grant No.CK2014Z23) and Project Supported by the Scientific and Technological Research Program of Chongqing Municipal Eduion Commission (Grant No. KJ1501306). This work was also financially supported by the National Natural Science Foundation of China (Grant No.51504051, 51174246 and 51574053) and Chongqing Research Program of Basic Research and Frontier Technology (Grant No. cstc2015jcyjA90001).

References

1. S.G. Susmita, G.B. Krishna, Immobilization of Pb(II), Cd(II) and Ni(II) ions on kaolinite and montmorillonite surfaces from aqueous medium. J. Environ. Manage. **87**, 46–58 (2008)
2. G.J. Copello, L.E. Diaz, V.C. Dall'Orto, Adsorption of Cd(II) and Pb(II) onto a one Step-Synthesized polyampholyte: Kinetics and equilibrium studies. J. Hazard. Mater. **217**, 374–381 (2012)
3. A.A. Ruhan, S. Ahmet, T. Mustafa, Equilibrium, thermodynamic and kinetic studies on biosorption of Pb(II) and Cd(II) from aqueous solution by macrofungus (Lactarius scrobiculatus) biomass. Chem. Eng. J. **151**, 255–261 (2009)
4. M.Q. Jiang et al., Adsorption of Pb(II), Cd(II), Ni(II) and Cu(II) onto natural kaolinite clay. Desalination **252**, 33–39 (2010)
5. C. Namasivayam, K. Ranganathan, Removal of Cd(II) from wastewater by adsorption on "waste" Fe(III)/Cr(III) hydroxide. Water Res. **29**, 1737–1744 (1995)
6. K. Kadirvelu, C. Namasivayam, Activated carbon from coconut coir pith as metal adsorbent: Adsorption of Cd(II) from aqueous solution. Adv. Environ. Res. **7**, 471–478 (2003)
7. K. Ibrahim et al., Adsorption of Cd(II) ions from aqueous solutions using activated carbon prepared from olive stone by $ZnCl_2$ activation. Bioresour. Technol. **99**, 492–501 (2008)
8. H.B. Michael, B. Bart, Modelling the sorption of Mn(II), Co(II), Ni(II), Zn(II), Cd(II), Eu(III), Am(III), Sn(IV), Th(IV), Np(V) and U(VI) on montmorillonite: Linear free energy relationships and estimates of surface binding constants for some selected heavy metals and actinides. Geochim. Cosmochim. Acta. **69**, 875–892 (2005)
9. C. Nathalie, G. Richard, D. Eric, Adsorption of Cu(II) and Pb(II) onto a grafted silica: Isotherms and kinetic models. Water Res. **37**, 3079–3086 (2003)
10. G. Blazquez et al., Removal of cadmium ions with olive stones. Process Biochem. **40**, 2649–2654 (2005)

Research on the Flow Behavior of Molten Slag Through Pore

Yingli Liu, Qingguo Xue, Jingsong Wang and Guang Wang

Abstract In the dropping zone of blast furnace (BF), primary slag absorbing ash of pulverized coal and coke become to be intermediate slag. Due to viscosity and surface tension of slag and the pore between cokes, some slag remained in coke layer and affected the permeability of blast furnace. The flow behavior of slag with different basicity (0.55–1.10) and Al/Si (0.35–0.50) were studied. Meanwhile, the viscosity and surface tension of slag were calculated. The results showed that, with the increasing basicity of slag, surface tension increased and viscosity decreased. With the Al/Si increase, surface tension and viscosity increased. The smallest pore that all slag can pass through was 4.5 mm. However, the height between slag droplets and the bottom of crucible indicated that the ability of slag to pass through pore were weakened with the increasing basicity and Al/Si. Thus, slag which absorbed ash from fuel become difficult to pass through coke layer.

Keywords Ash · Pore · Molten slag · Flow behavior

Introduction

In the dropping zone of the blast furnace, molten slag and iron pass through coke bed. Coke is the only solid in dropping zone. Between the lump cokes, the channel was formed and molten iron and slag can flow to furnace hearth [1]. Because of the

Y. Liu · Q. Xue · J. Wang (✉) · G. Wang
State Key Laboratory of Advanced Metallurgy,
University of Science and Technology Beijing, 100083 Beijing, PR, China
e-mail: wangjingsong@ustb.edu.cn

Y. Liu
e-mail: liuyingli112@126.com

Q. Xue
e-mail: xueqingguo@ustb.edu.cn

G. Wang
e-mail: wangguang@ustb.edu.cn

J.-Y. Hwang et al. (eds.), *8th International Symposium on High-Temperature Metallurgical Processing*, The Minerals, Metals & Materials Series,
DOI 10.1007/978-3-319-51340-9_76

surface tension of molten iron and slag, when the surface tension and gravity reached balance, the liquid slag cannot successfully drip in a small pore. The slag which remained in the pore affects the permeability of lower part of blast furnace in dropping zone [2, 3].

Currently, pulverized coal injection (PCI) is a normal production. After combustion of pulverized coal with hot air in cohesive zone, the upward pulverized coal ash was mixed with the downward liquid slag and iron. Primary slag absorbing ash of pulverized coal become to be intermediate slag. Morimasa ichida [4] founded that, the primary slag that directly contacted and mixed with the ash accounted for 15–20% of the total slag in dropping zone. Therefore, the influence of ash from coal and coke on the property, especially on the flow behavior, was significant. The main component of ash were Al_2O_3 and SiO_2. After mixing with ash, basicity and Al/Si were obviously changed. These leads to a series of changes such as viscosity, melting point, surface tension, etc.

Top gas recycling-blast furnace (TGR-BF) [5] is a new ironmaking process. It adopts pure oxygen at room temperature to replace the traditional hot blast operation and recycles heated recycle top gas after removal CO_2. Due to pure oxygen blast, combustion rate of pulverized coal increased significantly. Thus the PCI rate increased from 150 to 220 kg/tFe, which means that more ash will be mixed with primary slag. In the condition of BF and TGR-BF, the variation of slag property affects the permeability in lower part of the blast furnace. In this paper, the flow behavior of slag with different basicity and Al/Si passed through pore were studied.

Experimental

Raw Materials

The slag used in this experiment was composed of CaO, SiO_2, MgO and Al_2O_3 with high purity analytical powder. The compositions of slag with different basicity and Al/Si were showed in Table 1. Mixed slag was placed into high temperature furnace and pre-melted for 1 h at 1500 °C under Argon. After pre-melting, slag were cooled to room temperature and then grounded into powder for experimental work.

According to the mathematical model of multi-zone [6] in top gas recycling-oxygen blast furnace, coke ratio and coal ratio would be 210 kg/tFe and 200 kg/tFe, respectively. In traditional blast furnace, they are 352.5 kg/tFe and 150 kg/tFe, respectively. Based on the composition of coke and coal showed in Table 2, slag from fuel ash were 46.1 and 59.2 kg/tFe, which was 14 and 17% of the total amount of slag in TGR-BF and traditional blast furnace (TBF). When 20% of primary slag was mixed with ash of fuel, the basicity of slag was about 0.6 and Al/Si was about 0.4 in TGR-BF, respectively. In the condition of TBF, basicity and Al/Si were about 0.5. Therefore, basicity of slag varied from 0.55 to 1.10 and Al/Si was set from 0.35 to 0.50.

Table 1 Compositions of slag with different R and Al/Si (mass%)

	CaO	SiO_2	Al_2O_3	MgO	R	Al/Si
1	26.1	47.4	16.6	10.0	0.55	0.35
2	25.4	46.2	18.5	10.0	0.55	0.40
3	24.8	45.0	20.3	10.0	0.55	0.45
4	24.1	43.9	22.0	10.0	0.55	0.50
5	40.4	36.7	12.9	10.0	1.10	0.35
6	39.6	36.0	14.4	10.0	1.10	0.40
7	38.8	35.3	15.9	10.0	1.10	0.45
8	38.1	34.6	17.3	10.0	1.10	0.50
9	30.0	42.9	17.1	10.0	0.70	0.40
10	35.2	39.1	15.7	10.0	0.90	0.40

Table 2 Chemical Composition of coke and coal (mass%)

	C	Vol	Ash				
			SiO_2	Al_2O_3	CaO	MgO	FeO
Coke	84.96	2.38	7.26	5.05	0.10	0.15	0.10
Coal	78.67	11.33	6.90	2.31	0.10	0.10	0.32

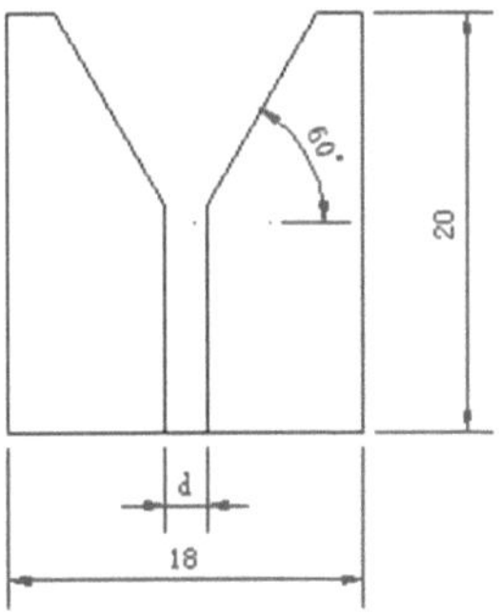

(a) Funnel graphite crucible

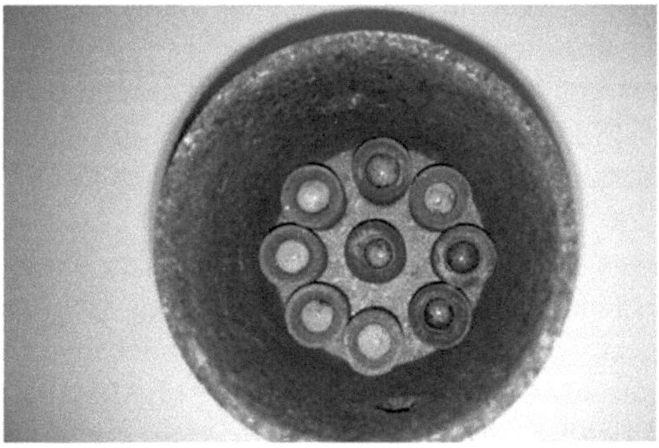

(b) Schematic of crucible placement

Fig. 1 Graphite crucible in the experiment

Experimental Apparatus and Conditions

The funnel graphite crucible of Φ18 mm × h 20 mm was used in this experiment, as showed in Fig. 1a. By changing the pore diameter of funnel graphite crucible, the void formed by coke particles in coke bed was simulated. During the test, 9 crucibles with different inner diameters were placed. The pore diameter started from 2 mm and was gradually increased to 6 mm (0.5 mm increments).

First of all, about 1.5 g slag was pressed into 9 mm diameter cylinder and was placed in the funnel graphite crucible. After each crucible was placed slag sample, they were installed into a big graphite crucible. Graphite crucible was showed in Fig. 1b. In a high-temperature tube furnace, this big crucible was heated to 1500 °C in a high purity (99.99%) argon atmosphere at a flow rate of 2.0 L/min. After thermostatic for 30 min, turn off the power. When the temperature dropped to <1300 °C, the crucible with solidification slag was took out for further investigation.

Results and Discussion

Influence of Al/Si on Physical Properties of Slag

In order to study the influence of Al/Si ratio and basicity on slag viscosity, models of Riboud [7, 8] and KTH [9] were used to calculate viscosity of quarternary system $CaO\text{-}SiO_2\text{-}Al_2O_3\text{-}MgO$, respectively. The surface tension of slag were obtained by weighted average of each pure reagent's surface tension.

The variation of viscosity and surface tension with Al/Si ratio in the basicity of 0.55 and 1.10 were showed in Fig. 2. At the same basicity, the viscosity and surface tension were increased with the rising of Al/Si. To understand the reason of this phenomenon, infrared spectroscopy was used. The results were presented in Fig. 3. It is known that the slag viscosity depend on the slag structure e.g. the complexity of $[SiO_4]^{4-}$ and $[AlO_4]^{5-}$. In the infrared spectra, it is symmetric stretching vibration of the silicon-oxygen bond of $[SiO_4]^{4-}$ tetrahedron between the Bands of 1200^{-1} and 750 cm^{-1}. Bands of 630^{-1} and 400 cm^{-1} corresponds to the Si-O-Al bond of glass phase structure. Between the bands of 750^{-1} and 630 cm^{-1}, it corresponds to the symmetric stretching vibration of $[AlO_4]^{5-}$ tetrahedron. Infrared spectra showed that, the width and intensity of the corresponding bands of $[AlO_4]^{5-}$ was enhanced with the increasing of Al/Si. This indicated that the relative amounts of $[AlO_4]^{5-}$ was increased. Due to the coupling of $[AlO_4]^{5-}$ and $[SiO_4]^{4-}$, degree of polymerization (DOP) of the slag structure was elevated. Therefore, these promoted the increase of slag viscosity.

Influence of Basicity on Physical Properties of Slag

The variation of viscosity and surface tension with basicity at the fixed Al/Si ratio of 0.4 was shown in Fig. 4. The results indicated that, when basicity increased from 0.55 to 1.10, viscosity decreased. This can be correlated with the theoretical basis [10] that the structure of $[SiO_4]^{4-}$ was simpler when the basicity increased. This phenomenon was confirmed in Fig. 5. Compared with the infrared spectra at the basicity of 0.55, the intensity of $[SiO_4]^{4-}$ was lower between the bands of 1200^{-1}

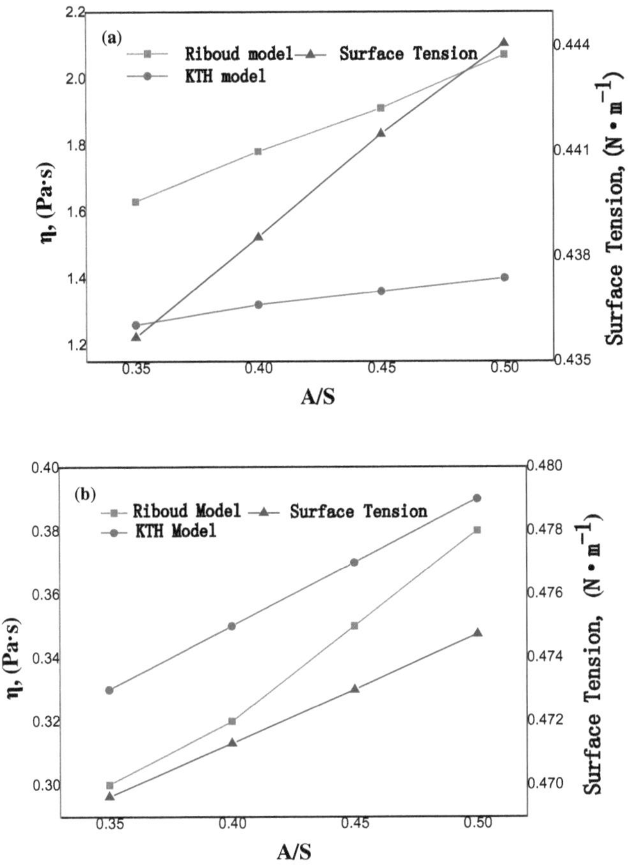

Fig. 2 Variation of viscosity and surface tension with Al/Si **a** R = 0.55; **b** R = 1.10

and 750 cm^{-1}. In Fig. 4, the value of surface tension showed a rising trend. This was determined by the surface tension of pure material e.g. SiO_2 was surfactant and CaO has a higher surface tension.

Influence of Al/Si on the Flow Behavior of Slag

In the melt, the phenomenon of slag through pore was observed. The results indicated that, slag with different Al/Si ratio passed through ≥ 4.5 mm pore. When the pore was less than 4.5 mm, slag could not pass through in 30 min. However, after funnel crucible was cut along the axis, the distance of slag drops to the

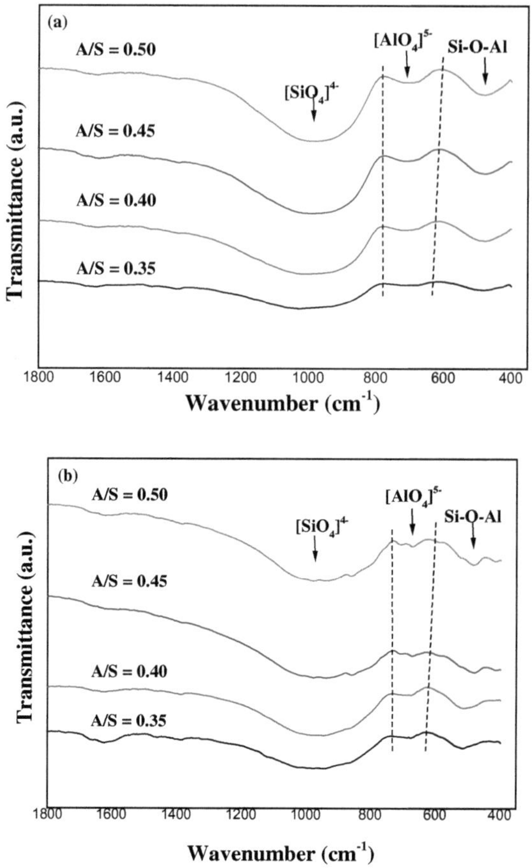

Fig. 3 Change of slag structure with Al/Si **a** R = 0.55; **b** R = 1.10

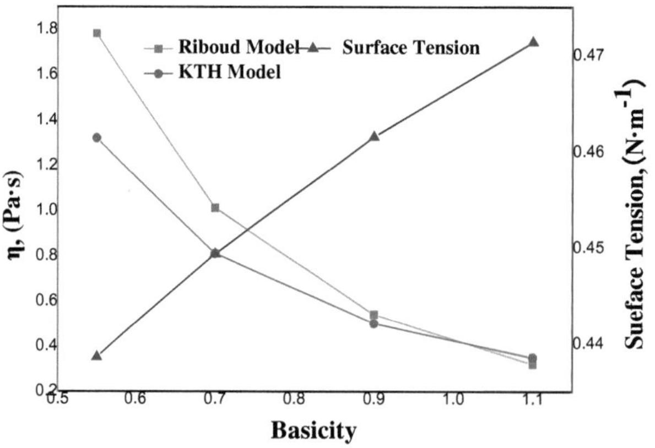

Fig. 4 Variation of viscosity and surface tension with basicity Al/Si = 0.4

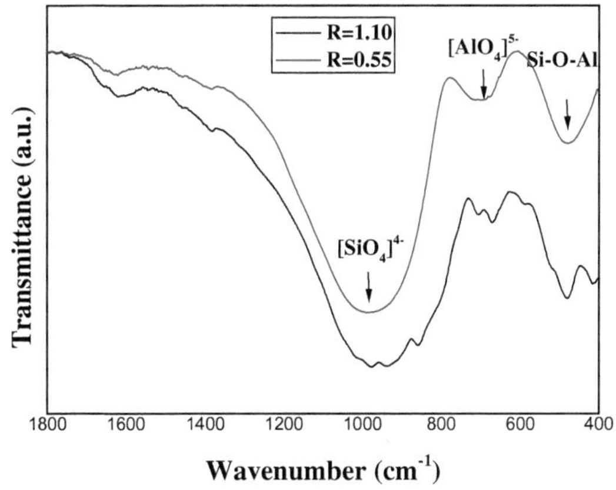

Fig. 5 Variation of viscosity and surface tension with basicity Al/Si = 0.4

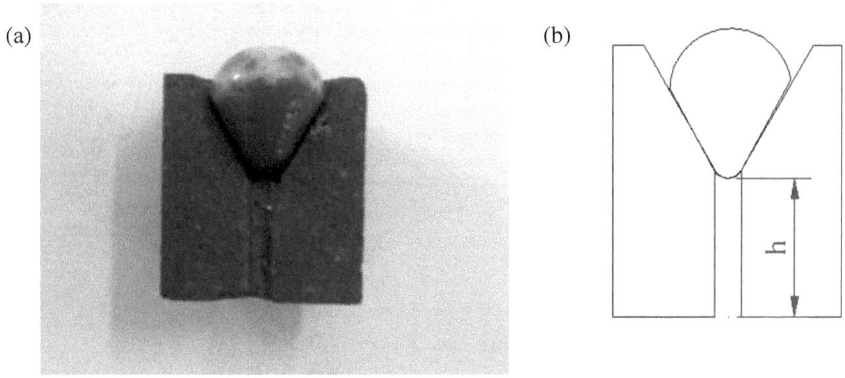

Fig. 6 Slag of static holdup in pore. **a** Actual experimental result. **b** Measurement schematic

horizontal surface (as presented in Fig. 6) still showed some differences. It also can reflect the influence of change of the slag composition on the capability of slag through pore.

Height change with the pore size at different Al/Si ratio were showed in Fig. 7. When the Al/Si ratio was varied from 0.35 to 0.50, the height increased. This indicated that with the increase of Al/Si ratio, the ability of slag through pore decreased. According to the previous data (in Fig. 2), the upward force of surface tension were increased and the downward force of slag droplets was reduced. Therefore, the flow ability of slag was weakened.

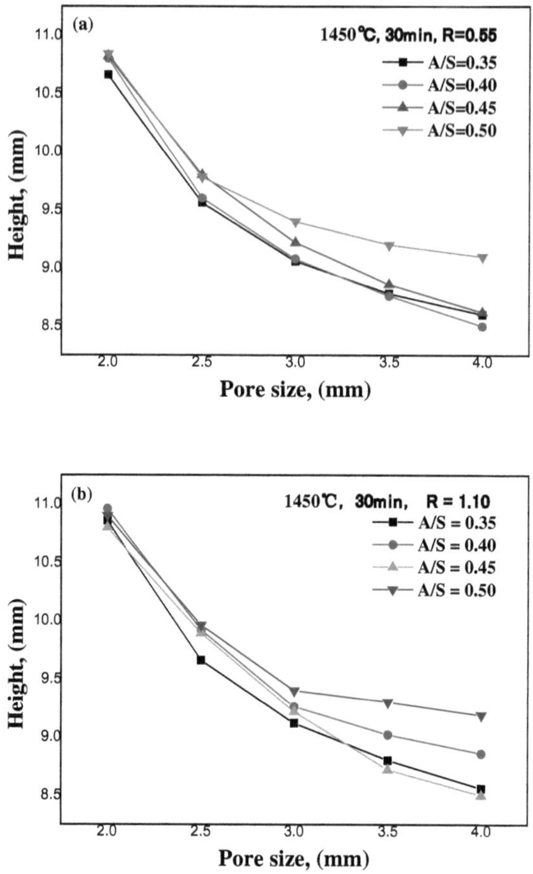

Fig. 7 Height change with the pore size **a** R = 0.55; **b** R = 1.10

Influence of Basicity on the Flow Behavior of Slag

Height change with the pore size in the basicity from 0.55 to 1.10 at fixed Al/Si ratio of 0.4 was showed in Fig. 8. The results indicated that the height increased with the increased basicity. These indicated that the ability of slag penetration through pore decreased. Combined with the results shown in Fig. 4, the viscosity decreased and surface tension increased with the increased basicity. In can be conducted that surface tension was the key factor affecting slag flow behavior.

Fig. 8 Height change with the pore size Al/Si = 0.4

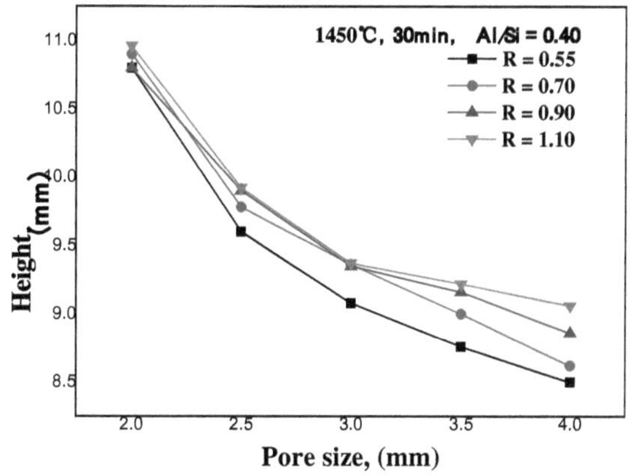

Conclusions

To understand the flow behavior of molten slag in dropping zone, the experiment of slag penetration through pore with different basicity and Al/Si ratio were carried out. The results showed that:

(1) With the increase of Al/Si ratio, surface tension and viscosity increased. With the increase of slag basicity, surface tension increased and viscosity decreased. Surface tension was the key factor affecting slag flow behavior.
(2) In this experiment, the smallest pore that all slag pass through was 4.5 mm.
(3) The ability of slag to pass through pore were weakened with the increasing basicity and Al/Si ratio. Namely, slag after absorbing ash from fuel become difficult to pass through coke layer.

Acknowledgements The authors gratefully acknowledge the National Key Research and Development Program (No. 2016YFB0601304).

References

1. T. Fukutake, V. Rajaykuma, Liquid holdup and abnormal flow phenomena in packed bed under conditions simulating the flow in the dripping zone of a blast furnace. ISIJ Int. **22**, 355–363 (1982)
2. I.H. Jeong, H.S. Kim, Y. Sasaki, Trickle flow behaviors of liquid iron and molten slag in the lower part of blast furnace. ISIJ Int. **53**(12), 2090–2098 (2013)
3. H.L. George, B.J. Monaghan, S.J. Chew, Flow of molten slag through coke channels. ISIJ Int. **53**(7), 1172–1179 (2013)
4. M. Ichida, T. Orimoto, T. Tanaka, Behavior of pulverized coal ash and physical property of dripping slag under high pulverized coal injection operation. ISIJ Int. **41**(4), 325–332 (2001)
5. M.S. Qin et al., Blast furnace operation with full oxygen blast. Iromaking steelmaking **6**(15), 287–292 (1988)

6. Han Yihua et al., Comprehensive mathematical model of top gas recycling-oxygen blast furnaces. J. Univ. Sci. Technol. B **33**(10), 1280–1286 (2011)
7. P.V. Riboud, Y. Roux, L.D. Lucas, Improvement of continuous casting powders. Fachberichte Hutten praxis Metallweiterverarbeitung **19**, 859–869 (1981)
8. K.C. Mills, L. Yuan, R.T. Jones, Estimating the physical properties of slags. J. S. Afr. Inst. Min. Metall. **111**, 649–658 (2011)
9. Du. Sichen, J. Bygden, S.A. Seetharaman Model for estimation of viscosities of complex metallic and ionic melts. Met. Mat. Trans. B **25B**, 1–9 (1994)
10. X. Wang, Iron and steel metallurgy (Ironmaking part). (Metallurgical Industry, Beijing, Press, 2008). 105–118 (1991)

Study on the Influence of Materials on Heat Transfer Characteristics of Blast Furnace Cooling Staves

Feng-guang Li, Ju Zhang and Rui Guo

Abstract Cooling stave is one of the most important equipment of blast furnace for iron making. The heat transfer performance and service life of the cooling stave, to a large extent, determines the efficiency and economy of the iron making process. In this paper, a 3-D heat transfer model of cooling stave was founded by ANSYS based on FEM theory, and the heat transfer characteristics of cooling staves which were made of different materials (cast iron, cast steel and copper) were analyzed. Then, several suggestions on choosing material of cooling staves were given according to the calculation results and the actual working conditions of the cooling staves.

Keywords Blast furnace · Heat transfer · Cooling stave · Numerical simulation · FEM theory

Introduction

According to the investigation result on blast furnace (BF) service life at home and abroad, the design performance of the cooling stave becomes an important factor influencing the BF life along with the popularization of intensified smelting and maximization of BF structure [1]. The BF cooling staves are usually made of gray cast iron, nodular cast iron, cast steel and copper. Among these materials, the heat conductivity of cast iron is around 25 W/(m K), and the gray cast iron and nodular cast iron are different in heat conductivity and mechanical properties. The heat

F. Li · R. Guo
School of Materials Science and Engineering, Hubei University of Automotive Technology, Shiyan 442002, Hubei, China
e-mail: lemon2006@126.com

J. Zhang (✉)
School of Automotive Engineering, Hubei University of Automotive Technology, Shiyan 442002, Hubei, China
e-mail: fionanzhang@163.com

J.-Y. Hwang et al. (eds.), *8th International Symposium on High-Temperature Metallurgical Processing*, The Minerals, Metals & Materials Series,
DOI 10.1007/978-3-319-51340-9_77

conductivity of cast steel is higher than cast iron, but it is merely 50 W/(m K), the mechanical properties are much excellent than cast iron. Copper has an outstanding heat conductivity of 380 W/(m K), but the cost of copper is extremely high [2–5]. Because of the different demands of costs and heat transfer performances, different blast furnaces, or the different areas of the same BF, should employ specific cooling staves [6]. Therefore, the heat transfer characteristics of different types of cooling staves should be analyzed and find out its most suitable application area. In this paper, a 3-D model cooling staves was built and the heat transfer characteristics of cooling staves made of different materials were calculated.

Establishment of the Model

Modeling with UG

In this paper, the 3-D solid model of cooling stave was built with Unigraphics NX. The cooling staves made of different materials had the same dimensions of L1970 mm × W907.4 mm × H115 mm (see Fig. 1). To reduce computational complexity, the model was simplified compared with the actual cooling staves, and the tiny parts like bolts were ignored. Once the solid model was built, it was imported into ANSYS to be meshed, and the meshing results were shown in Fig. 2.

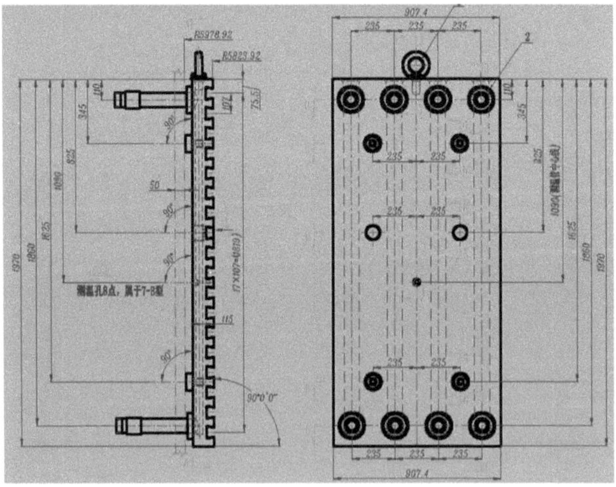

Fig. 1 Structure of the cooling stave

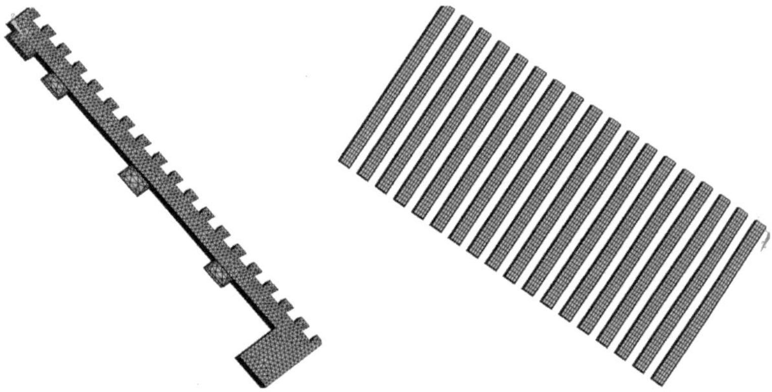

Fig. 2 Meshing of stave body and the insert bricks

Table 1 Properties of the materials

Materials	Density ρ/kg m^{-3}	Heat conductivity λ/W m^{-1} K^{-1}	Specific heat C_p/J kg^{-1} K^{-1}
Cast iron	7200	33	544
Cast steel	7800	50	500
Copper	8900	380	386
Insert bricks (SiC)	2640	14	1000
Furnace skull (steel)	7840	50	465

Material Properties

In this paper, the different materials of cast iron, cast steel and copper were considered, and their heat transfer performance were focused, so the single parameter of heat conductivity was needed. Since the many kinds of cast iron can be used to produce cooling stave, the heat conductivity of cast iron changes with the cast iron grade. Typical heat conductivity values of each material are used as the model parameters, shown in Table 1 [7, 8]. In Table 1, ρ is the density; λ is the heat conductivity; C_p is the mass specific heat of each material.

Boundary Conditions

Combined Heat Transfer Coefficient Between Stave Body and Cooling Water

The temperature of the hot face of the stave body mainly depends on the cooling power of the cooling stave. The cooling power can be characterized by the combined heat transfer coefficient between stave body and cooling water [9]. As shown in Eq. (1), the combined heat transfer coefficient is related to five thermal resistances [10]:

$$h_w = \frac{1}{R} = R_1 + R_2 + R_3 + R_4 + R_5 \tag{1}$$

In Eq. (1), h_w is the combined heat transfer coefficient between stave body and cooling water, W/(m^2 °C); R is the total thermal resistance, (m^2 °C)/W; R_1, R_2, R_3, R_4, R_5 are the thermal resistance between the interface of the cooling water pipe and the cooling water, the conductive resistance of the pipe scale, the conductive resistance of the pipe wall, the conductive resistance of the coating layer on the outer surface of the pipe, and the conductive resistance of the air gap between the water pipe and the stave body [10], (m^2 °C)/W, respectively.

(1) The thermal resistance between the interface of the cooling water pipe and the cooling water R_1, it was calculated with the following equation:

$$R_1 = \frac{1}{\alpha}\frac{d_o}{d_i} \tag{2}$$

$$\alpha = 0.023\frac{\lambda}{d_i^{0.2}}\frac{v^{0.8}}{\eta^{0.8}}Pr^{0.4} \tag{3}$$

In Eqs. (2) and (3): α is the forced convective heat transfer coefficient of water, W/(m^2 °C); d_0 is the outer diameter of water pipe, m; d_i is the inner diameter of water pipe, m; v is the water velocity, m/s; λ is the heat conductivity of cooling water, W/(m^2 °C); η is the dynamic viscosity of cooling water, Pa s; Pr is the Prandtl number of cooling water.

(2) The pipe scale was ignored in this model, so the conductive resistance of the pipe scale $R_2 = 0$.
(3) The conductive resistance of the pipe wall R_3, it is calculated with Eq. (4):

$$R_3 = \frac{d_o}{2\lambda_w} \ln \frac{d_o}{d_i} \tag{4}$$

In Eq. (4): λ_w is the heat conductivity of pipe wall, W/(m^2 °C).

(4) The conductive resistance of the coating layer on the outer surface of the pipe R_4, it is calculated with Eq. (5).

$$R_4 = \frac{\delta_c}{\lambda_c} \tag{5}$$

In Eq. (5): δ_c is the thickness of coating layer, m; λ_c is the heat conductivity of coating material, W/(m^2 °C).

(5) The conductive resistance of the air gap between the water pipe and the stave body R_5, it is calculated with Eq. (6):

$$R_5 = \frac{\delta_g}{\lambda_e} \tag{6}$$

In Eq. (6): δ_g is the thickness of air gap, m; λ_g is the heat conductivity of air, W/(m^2 °C).

(1) The calculation results of the combined heat transfer coefficient of cast iron cooling stave
All the parameters of the cast iron cooling stave were taken into Eq. (1)–(6) and the values of each thermal resistances was calculated: $R_1 = 0.16 \times 10^{-3}$(m^2 °C)/W, $R_2 = 0$,$R_3 = 0.13 \times 10^{-3}$(m^2 °C)/W, $R_4 = 0.11 \times 10^{-3}$(m^2 °C)/W, $R_5 = 3.90 \times 10^{-3}$(m^2 °C)/W. According to Eq. (1), $R = 4.3 \times 10^{-3}$(m^2 °C)/W, and $h_w = 232.6$ W/(m^2 °C).
(2) The calculation results of the combined heat transfer coefficient of cast steel cooling stave
When the cooling stave is made of cast steel, which share the same material with water pipe, the air gap between stave body and the water pipe won't form, and $R_5 = 0$. Take the parameters into Eqs. (1)–(6) and the values of each thermal resistances was calculated: $R_1 = 0.16 \times 10^{-3}$(m^2 °C)/W, $R_2 = 0$, $R_3 = 0.16 \times 10^{-3}$(m^2 °C)/W, $R_4 = 0.11 \times 10^{-3}$(m^2 °C)/W, $R_5 = 0$, $R = 0.4 \times 10^{-3}$(m^2 °C)/W, and $h_w = 2500$ W/(m^2 °C).
(3) The calculation results of the combined heat transfer coefficient of copper cooling stave
When the cooling stave is made of copper, the water channel is formed by drilling, so there is no water pipe wall nor coating layer, neither exist air gap. As a result, only the thermal resistance between the interface of the

cooling water pipe and the cooling water (R_1) should be taken into consideration., $R = R_1 = 0.16 \times 10^{-3}(m^2\ °C)/W$, and $h_w = 6250\ W/(m^2\ °C)$.

The Combined Heat Transfer Coefficient Between Stave Body and Gas

A serial of statistical results of the hot face conventional heat transfer coefficient under different conditions of changeable furnace temperature had been proposed by Пляшкевич et al. [11] according to an actual blast furnace. The fitted formula between the hot face conventional heat transfer coefficient and the furnace temperature is shown in Eq. (7):

$$h_g = -1.22 + 72.27e^{\frac{t_g - 507.26}{445.15}} \tag{7}$$

The possible furnace temperature of 800, 900, 1000, 1100 and 1200 °C were taken into Eq. (7) and the calculation results were shown in Table 2.

Other Boundary Conditions

The surfaces of the cooling stave are adiabatic except the hot face and the cold face, other boundary conditions was shown in Table 3.

Table 2 The conventional heat transfer coefficients under different temperature

Furnace temperature/°C	800	900	1000	1100	1200
The conventional heat transfer coefficients of hot face h_g/W m^{-2} $°C^{-2}$	138	173	218	272	341

Table 3 Parameters of the boundary conditions

Parameters	Values
The average water temperature, °C	35
Atmosphere temperature, °C	35
Cooling water velocity, m s^{-1}	2.5
Convectional heat transfer coefficient between cold face and air, W/(m^2 °C)	12
The heat conductivity of cast iron, W/(m °C)	33
The heat conductivity of cast steel, W/(m °C)	50
The heat conductivity of copper, W/(m °C)	380
The heat conductivity of insert bricks, W/(m °C)	12
Furnace temperature, °C	800, 900, 1000, 1100, 1200

Results and Discussion

Discussion on the Cast Iron Cooling Stave

When other boundary conditions kept invariant, the temperature fields of the cast iron stave body and the insert brick under different furnace temperature were shown in Fig. 3. In Fig. 3, the maximum temperature of the stave body were always found on the hot face of stave body near the top surface under different furnace temperature, and the temperature of the upper parts were higher than that of the lower parts (about 60 °C). Meanwhile, the temperature of stave body near the water pipe is much lower than the temperature of other area. The maximum temperature of the stave body is 989 °C when the furnace temperature is 1200 °C Query ID="Q2' Text="Please confirm the corresponding author is correctly identified and amend if necessary.' and the maximum temperature of the stave body is 527 °C when the furnace temperature is 800 °C. That is to say, when the cooling stave is made of cast iron, the maximum temperature of stave body will remain 527–989 °C when working in the blast furnace, which is very high.

Figure 4 is the maximum temperature-furnace temperature trend chart. Just as shown in Fig. 4, the temperature of stave body increases linearly with the furnace temperature, the temperature of stave body won't change sharply when the furnace temperature changes. That is to say, the temperature of stave body can be controlled easily by changing the furnace temperature. Since the phase transformation of cast iron will take place at the temperature of 760 °C, the temperature of the cast iron stave body should not be higher than 760 °C, that means the allowed furnace temperature when using cast iron cooling stave is about 1000 °C.

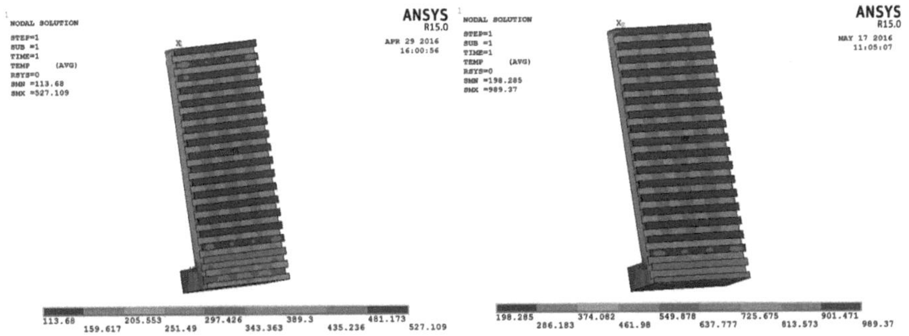

Furnace temperature 800°C **Furnace temperature 1200°C**

Fig. 3 Temperature fields of cast iron stave body and insert brick under different furnace temperature

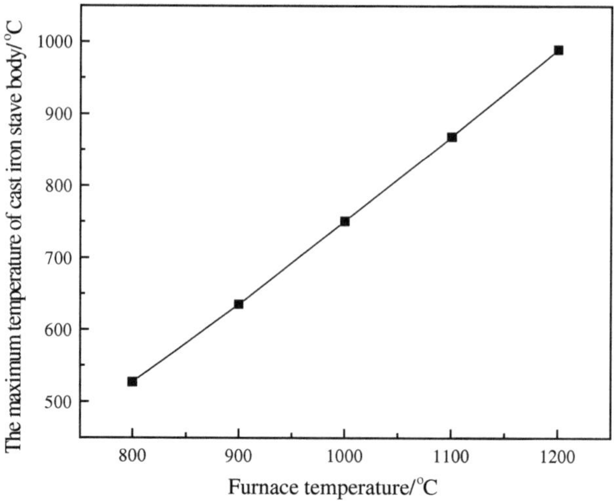

Fig. 4 The trend of maximum temperature of cast iron stave body along with the furnace temperature

Discussion on the Copper Cooling Stave

When other boundary conditions kept invariant, the temperature fields of the copper stave body and the insert brick under different furnace temperature were shown in Fig. 5. In Fig. 5, the maximum temperature of the copper stave body were also found on the hot face of stave body near the top surface under different furnace temperature, and the temperature of the upper parts were also higher than that of the lower parts. Meanwhile, the temperature of stave body near the water pipe is much lower than the temperature of other area. The maximum temperature of the stave body is 273.1 °C when the furnace temperature is 1200 °C) and the maximum

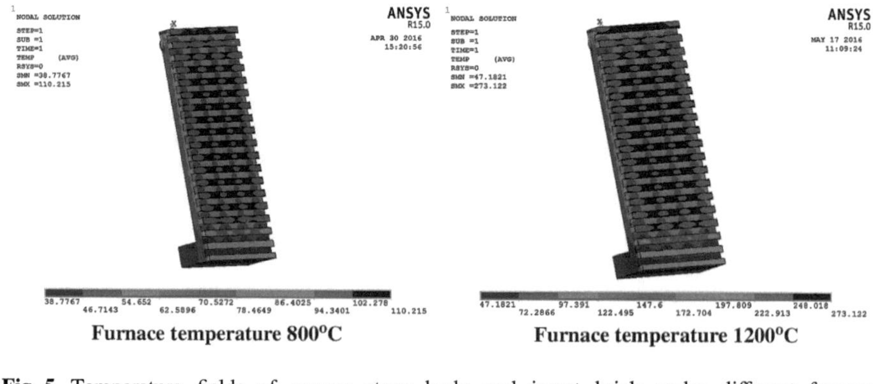

Fig. 5 Temperature fields of copper stave body and insert brick under different furnace temperature

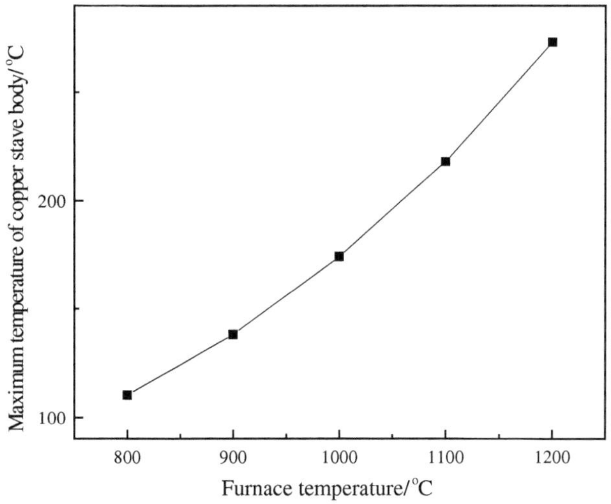

Fig. 6 The trend of maximum temperature of copper stave body along with the furnace temperature

temperature of the stave body is 110.2 °C when the furnace temperature is 800 °C. That is to say, when the cooling stave is made of copper, the maximum temperature of stave body will remain 110–273 °C when working in the blast furnace, which is very low.

Figure 6 is the maximum temperature-furnace temperature trend chart of copper stave. Just as shown in Fig. 6, the temperature of stave body increases nonlinearly with the furnace temperature, the temperature of stave body would change sharply when the furnace temperature increase, especially when the furnace temperature is high. That is to say, the temperature of stave body couldn't be controlled easily by changing the furnace temperature. Since the mechanical properties of copper will reduce rapidly when the temperature of stave body is higher than 250 °C, the temperature of the copper stave body should not be higher than 250 °C, that means the allowed furnace temperature when using copper cooling stave is about 1200 °C according to Fig. 6.

Discussion on the Cast Steel Cooling Stave

When other boundary conditions kept invariant, the temperature fields of the cast steel stave body and the insert brick under different furnace temperature were shown in Fig. 7. In Fig. 7, the maximum temperature of the copper stave body were also found on the hot face of stave body near the top surface under different furnace temperature, and the temperature of the upper parts were also higher than that of the lower parts. Meanwhile, the temperature of stave body near the water pipe is much lower than the temperature of other area. The maximum temperature of the stave

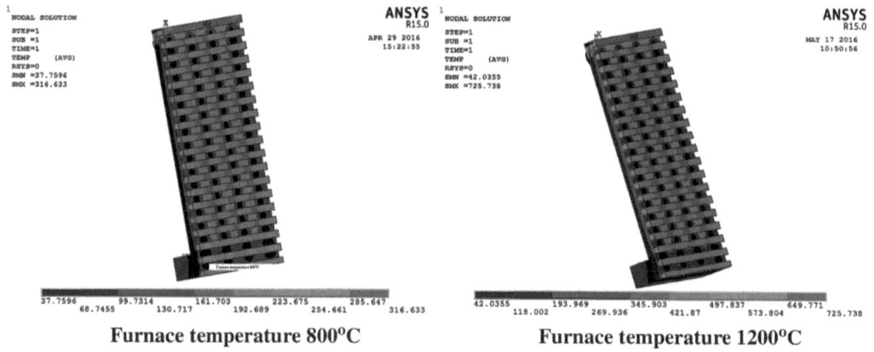

Fig. 7 The trend of maximum temperature of cast iron stave body along with the furnace temperature

body is 725.7 °C when the furnace temperature is 1200 °C) and the maximum temperature of the stave body is 316.6 °C when the furnace temperature is 800 °C. That is to say, when the cooling stave is made of copper, the maximum temperature of stave body will remain 317–726 °C when working in the blast furnace, which is moderate.

Figure 8 is the maximum temperature-furnace temperature trend chart of cast steel stave. Just as shown in Fig. 8, the temperature of stave body also increases nonlinearly with the furnace temperature, but the increase trend of the temperature of stave body is not so significant. The temperature of stave body can be controlled easily by changing the furnace temperature relatively. Since the mechanical properties of cast steel is excellent compared with cast iron and copper, the allowed range of the temperature of the cast steel stave body should be wider, that means the allowed furnace temperature when using copper cooling stave is higher than 1200 °C.

Comparison of the Different Types of Cooling Stave

Figure 9 shows the tendency of the maximum temperature of stave body made of different materials along with the changes of furnace temperature. According to Fig. 9, no matter what kind of material the cooling stave is made of, the temperature of stave body will increase when the furnace temperature increases. The temperature of the cast iron stave body increase most rapidly when the furnace temperature increases; the increase trend of cast steel stave body take second place; the increase tendency of copper-steel stave body is a little lower than that of cast steel stave body; the increase speed of the temperature of copper stave body is slowest. In another aspect, because of the difference of heat conductivity, the maximum temperature of stave bodies varies depend on its materials. The maximum temperature of cast iron stave body is highest and the maximum temperature of copper stave body is lowest, and the cast steel stave body and copper-steel stave body fall in

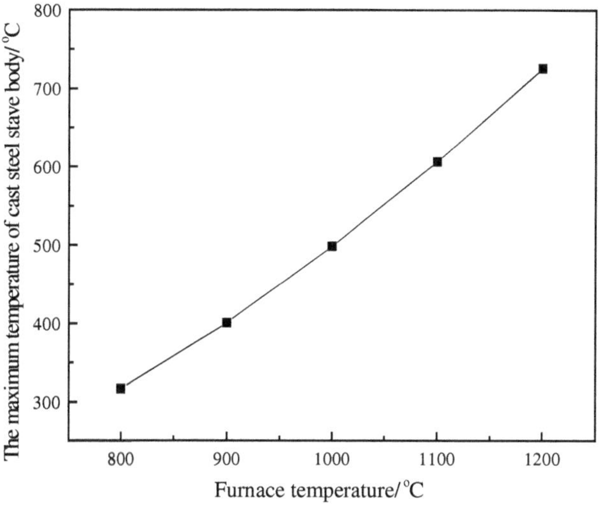

Fig. 8 The trend of maximum temperature of cast steel stave body along with the furnace temperature

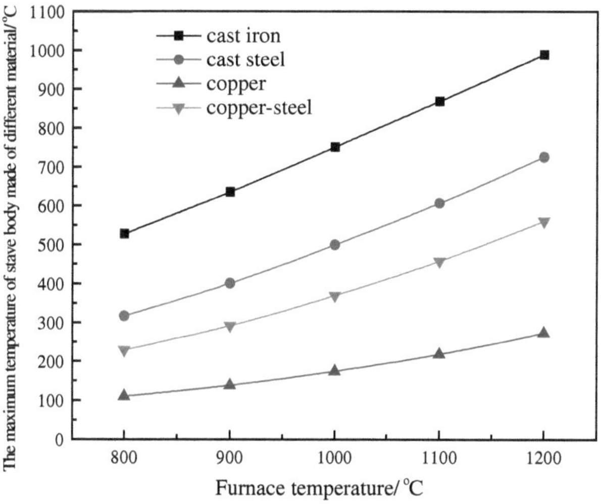

Fig. 9 The trend of maximum temperature of different kinds of stave body along with the furnace temperature

between them. That means the rank of the cooling staves' heat transfer performance is: copper stave > cast steel > cast iron. Since the mechanical properties of the materials are difference, so the choosing of types of cooling stave should give consideration to both heat transfer performance and mechanical properties. The heat transfer performance of each kind of cooling stave can get from Fig. 9, and it is an important reference foundation for type selection of cooling staves.

Conclusions

(1) The maximum temperature of cast iron stave body will remain 527–989 °C when working in the blast furnace, and the allowable furnace temperature when using cast iron cooling stave should not be higher than 1000 °C.
(2) The maximum temperature of copper stave body will remain 110–273 °C when working in the blast furnace, and the allowable furnace temperature when using copper cooling stave is about 1200 °C.
(3) The maximum temperature of cast steel stave body will remain 317–726 °C when working in the blast furnace, and the allowed furnace temperature when using cast steel cooling stave can be a little higher than 1200 °C.
(4) The rank of the cooling staves' heat transfer performance is: copper stave > cast steel > cast iron. The type selection of cooling stave should be performed considering both heat transfer performance and mechanical properties.

Acknowledgements This paper had been aided by the following foundations: (1) Natural Science Foundation of China (51604103): Study on induced cracking mechanism of BF copper cooling stave and constructing technology of hydrogen permeation suppression coating; (2) Natural Science Foundation of Hubei Province: Research on the cracking mechanism of BF cooling stave caused by hydrogen and the key technology of hydrogen permeation suppression; (3) The Doctoral Research Start-up Foundation of Hubei University of Automotive Technology (BK201607).

References

1. Z. Qian, H.E. Cheng, Heat transfer analysis of BF cast steel cooling stave based on ANSYS. Iron Steel Vanadium Titanium **26**(1), 55–59 (2005)
2. Z.Q. Guo, Research and analysis of modern BF lining cooling system. Shanxi Metall. **32**(5), 59–61 (2012)
3. J.H. Cui, Thermal test and temperature field simulation on blast furnace cooling stav. Master thesis, Zhenzhou University, 2006, pp. 30–32
4. Z.X. Liu, Coupled Thermo-mechanical analysis about blast furnace staves. Ph.D. thesis Northeastern University, 2009, pp. 52–55
5. N.C. Chen, Prediction of BF wall thickness using peripheral gas temperatures and stave heat flux intensity. Master thesis, University of Science and Technology Beijing, 2004, pp. 50–53
6. H.B. Zhao, Design and erosion prediction for BF long campaign. Master thesis, University of Science and Technology Beijing, 2006, pp. 20–27
7. J. Hong, Thermal test and simulation of BF cooling stave. Master thesis, University of Science and Technology Beijing, 2014, pp. 42–44
8. L. Jiang, Application and analysis of copper stave used in blast furnaces. Res. Iron Steel **31** (4), 1–5 (2003)
9. J.P. She, G. Chen, L.S. Xu et al., Several recognitions on copper cooling stave. Iron making **32**(4), 22–27 (2013)
10. S.S. Chen, Q.G. Xu, D.Q. Cang et al., Heat transfer analysis of blast furnace stave. Iron Steel **3**(5), 11–13 (1999)
11. А.С. Пляшкевич, К.К. Стрелов, А.С. фрейденберг. "Теплотехнический анализ оптималbного соотношения параметров футеровkи и системbl охлаждения шахтbl доменной печи," Сталb, **42**(3), 209–209 (1976)

Author Index

J.-Y. Hwang et al. (eds.), *8th International Symposium on High-Temperature Metallurgical Processing*, The Minerals, Metals & Materials Series,
DOI 10.17/978-3-319-51340-9

Subject Index

J.-Y. Hwang et al. (eds.), *8th International Symposium on High-Temperature Metallurgical Processing*, The Minerals, Metals & Materials Series,
DOI 10.17/978-3-319-51340-9

MIX
Papier aus verantwortungsvollen Quellen
Paper from responsible sources
FSC® C105338
FSC
www.fsc.org

If you have any concerns about our products,
you can contact us on
ProductSafety@springernature.com

In case Publisher is established outside the EU,
the EU authorized representative is:
Springer Nature Customer Service Center GmbH
Europaplatz 3, 69115 Heidelberg, Germany

Printed by Libri Plureos GmbH
in Hamburg, Germany